ELECTRONIC ENGINEERING AND INFORMATION SCIENCE

PROCEEDINGS OF THE 2015 INTERNATIONAL CONFERENCE ON ELECTRONIC ENGINEERING AND INFORMATION SCIENCE (ICEEIS2015), 17–18 JANUARY 2015, HAIKOU, P.R. CHINA

Electronic Engineering and Information Science

Editor

Dongxing Wang

Harbin University of Science and Technology, Harbin, China

CRC Press
Taylor & Francis Group
Boca Raton   London   New York   Leiden

CRC Press is an imprint of the
Taylor & Francis Group, an **informa** business

A BALKEMA BOOK

Published by: CRC Press/Balkema
P.O. Box 11320, 2301 EH Leiden, The Netherlands
e-mail: Pub.NL@taylorandfrancis.com
www.crcpress.com – www.taylorandfrancis.com

ISBN-13: 978-0-367-73339-1 (pbk)
ISBN-13: 978-0-8153-7868-6 (hbk)

Visit the Taylor & Francis Web site at
http://www.taylorandfrancis.com

and the CRC Press Web site at
http://www.crcpress.com

Typeset by diacriTech, Chennai, India

Table of contents

xi

Electronic Engineering and Information Science – Wang (Ed.)
© 2015 Taylor & Francis Group, London, ISBN: 978-1-138-02772-5

Preface

The 2015 International Conference on Electronic Engineering and Information Science (ICEEIS 2015) will be held on 17–18 January in Haikou, China. The ICEEIS 2015 is sponsored by the Harbin University of Science and Technology.

The main role of ICEEIS 2015 is to bring together innovators including engineering researchers, scientists, practitioners to provide a forum to discuss ideas, concepts, and experimental results related to all aspects of electronic engineering and information science. In order to meet the high standard of CRC Press, the organization committee has made their efforts to do the following things. Firstly, poor quality papers have been refused after a reviewing round by anonymous referee experts. Secondly, periodical review meetings have been held with the reviewers in six sessions to exchange review suggestions. Finally, the conference organization has had several preliminary sessions before the conference. Through efforts of different people and departments, the conference will be successful and fruitful.

In addition, the conference organizer will invite some keynote speakers to deliver their speech at the conference. All participants will have a chance to discuss with the speakers face to face, which is very helpful for the participants.

We hope that you will enjoy the conference and find the ICEEIS 2015 exciting. We are looking forward to seeing more friends at the next conference.

Dongxing Wang

Electronic Engineering and Information Science – Wang (Ed.)
© 2015 Taylor & Francis Group, London, ISBN: 978-1-138-02772-5

Organizing committee

Chairman of Scientific Committees

Prof. Dongxing Wang, *Harbin University of Science and Technology*

The members of Scientific Committees

Prof. Jiangping Liu, *Inner Mongolia Agricultural University*
Prof. Lili Ma, *Inner Mongolia Agricultural University*
Prof. Yongming Wang, *Kunming University of Science and Technology*
Prof. Shibo Zhang, *Xihua University*
Prof. Jianlin Mao, *Kunming University of Science and Technology*
Prof. Wei Xu, *Inner Mongolia University of Science & Technology*
Prof. Yanfei He, *Inner Mongolia Agricultural University*
Prof. Shuyan Jiang, *University of Electronic Science and Technology of China*
Prof. Jing Wu, *GanSu Agricultural University*
Prof. Yali Chen, *Southwest Petroleum University*
Prof. Yun Wei, *Lanzhou Jiaotong University*
Prof. Xun Liu, *Hebei University of Engineering*
Prof. Guangmiao Qu, *Northeast Petroleum University*
Prof. Yibin Huang, *Jiangxi Normal University*
Prof. Ruixue Zhou, *Guizhou Normal University*
Prof. Hongyu He, *Guilin Normal College*
Prof. Chenglin Sun, *Jilin University*
Prof. Feng Li, *Jilin University*
Prof. Xinlu Zhang, *Harbin Engineering University*
Prof. Yongkang Dong, *Harbin Institute of Technology*
Prof. Huimin Xue, *Hebei University of Engineering*
Prof. Yingzhi Wei, *Heilongjiang University of Science and Technology*
Prof. Moran Sun, *Zhengzhou University*
Prof. Kangle Ding, *Yangtze University*
Prof. Shigang Bai, *Northeast Agricultural University*
Prof. Wenbin Bu, *Northeast Normal University*
Prof. Hongchen Liu, *Harbin Institute of Technology*
Prof. Wenshun Li, *Heilongjiang Bayi Agricultural University*
Prof. Zhongheng Luan, *Haikou College of Economics*
Prof. Junfeng Pan, *Hainan University*
Prof. Bai Shan, *Agricultural University of Hebei Province*
Prof. Qingcheng Liang, *Changchun University of Science and Technology*
Prof. Jiquan Ma, *Heilongjiang University*

Electronic Engineering and Information Science – Wang (Ed.)
© 2015 Taylor & Francis Group, London, ISBN: 978-1-138-02772-5

Boundary control of Boost three-level converters

Q.K. Zhou, D. Zhao, H.C. Liu & E.H. Guan
School of Electrical Engineering and Automation, Harbin Institute of Technology, Harbin, China

ABSTRACT: In this paper, boundary control theory is applied to a Boost three-level converter. A perfect state-energy plan is used to instruct the switching action, so that a switching surface is derived with superior performance. This controller can easily control the Boost three-level converter in the stability region with very small output voltage ripple. And the proposed controller can control the converter and revert to the ideal steady state in 0.05s, when it is disturbed.

KEYWORDS: boundary control; multilevel converters; dc-dc conversion; Boost converters.

1 INTRODUCTION

The Boost converter is widely used in many fields, because of the advantages of simple topology structure, and easiness to control (Ogura et al. 2004). Compared to the traditional Boost two-level converter, Boost three-level converter has higher capacity, lower switching loss, smaller diode reverse recovery loss, and lower voltage stress of the switches etc. Therefore, the Boost three-level converters have wide application prospect in large power switching power. So an effective control method for Boost three-level converters is very necessary. Boundary control is a geometric-based control technique and it is very suitable for a strongly nonlinear system (Munzert et al. 1996, Leung et al. 2007, Sable, et al. 1991, Haiti et al. 1995). The Boost converter has been controlled by an improved boundary control using state-energy plane (Song & Chung 2008). This method achieved very good control effect. Now we apply this method to boost three-level converters.

In this paper, the working principle of Boost three-level converter circuit is learned. Then based on the state-energy plane method a switching surface is proposed. Finally, the MATLAB simulation results are given.

2 BOOST THREE-LEVEL CONVERTERS WORKING MODE ANALYSIS

The schematic diagram of the three-level Boost converter is shown in Figure 1.

Boost three-level converters have four kinds of different switching mode, as shown in Figure 2. The current flows from the solid line. There is no current in dashed lines.

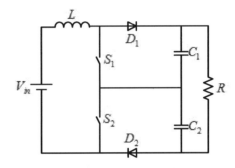

Figure 1. The schematic of the Boost three-level converter.

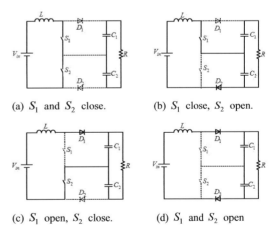

(a) S_1 and S_2 close. (b) S_1 close, S_2 open.

(c) S_1 open, S_2 close. (d) S_1 and S_2 open

Figure 2. Boost three-level converters switching mode.

When duty ratio D>0.5 and the output voltage V_o is larger than twice over V_{in}, the voltage on the each capacitor is greater than the input voltage V_{in}. At this time, the circuit has three kinds of working mode. The inductive current I_L increases if the two MOSFET are turned on. In other cases, the inductive current decreases. V_L is the voltage of the inductor (Figure 3a).

When D<0.5, likewise, if the two switches are turned off the inductive current will decrease. If there is a switch conduction, the inductive current will increase (Figure 3b).

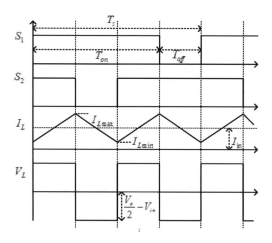

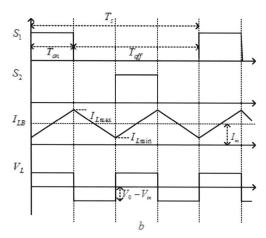

Figure 3. Boost three-level main parameter waveforms.

When the boost three-level converter is controlled, two switches need to have a certain Angle. This angle defined as α. From Figure 3, the inductive current ripple Δi_L is written as:

$$\Delta i_L = T(1 - D - \frac{\alpha}{2\pi}) \cdot \frac{V_o - V_{in}}{L} \qquad (1)$$

This control method can decrease the inductive current ripple, reduce the filter requirements, reduce the filter volume, and reduce costs, when choose the right phase shifted angle α. From the formula (1), when α equals to 180°, the inductor current ripple is minimized. So we choose α as 180°.

3 IMPLEMENTATION OF BOUNDARY CONTROL FOR THREE-LEVEL BOOST CONVERTERS

Now we have derived the mathematical equations of boundary control used state-energy plane for Boost three-level converters. In place of the state variables, the transient energy memory in E_{con} the converter to describe the converter trajectories. E_{con} is written as:

$$E_{con} = \frac{1}{2}C_1 V_{c1}^2 + \frac{1}{2}C_2 V_{c2}^2 + \frac{1}{2}L I_{in}^2 \qquad (2)$$

Where V_{c1} is the voltage of C_1, and V_{c2} is the voltage of C_2. When the circuit working at the operating point, that:

$$I_{ref} = \frac{P_{out}}{V_{in}} = \frac{V_{out}}{V_{in}} I_{out}$$

$$E_{con,ref} = \frac{1}{2}C_1 V_{c1,ref}^2 + \frac{1}{2}C_2 V_{c2,ref}^2 + \frac{1}{2}L I_{ref}^2$$

Figure 4 shows the time domain waveforms of inductive current I_L, the output voltage V_o, and the input power P_{in} and E_{con} in one switching cycle from t0 to t4.

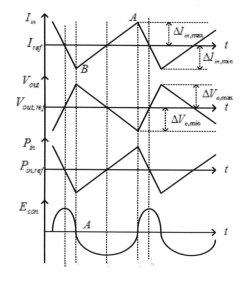

Figure 4. Waveforms and energy profile of the converter.

As shown is Figure 4, we know that power for the time integral equal energy, so we get the equation:

$$\int_{t_{on}}^{t} P_{in}(t) - P_{in,ref} \, dt = E_{con}(t) - E_{con}(t_{on}) \tag{3}$$

Among them:

$$P_{in} = V_{in} \cdot I_{in},$$

$$E_{con} = \frac{1}{2} C_1 V_{c1}^2 + \frac{1}{2} C_2 V_{c2}^2 + \frac{1}{2} L I_{in}^2$$

Where ton is the time instant at which S1 is on. When D>0.5 Two partial pressure capacitors charging and discharging in turn, equivalent to two basic boost converters in parallel. It can be assumed that the input inductive current increases or decreases linearly in the four topologies. So the switch opening conditions for:

$$\int_{t_{on}}^{t} P_{in}(t) - P_{in,ref} \, dt$$

$$= \frac{[P_{in}(t_{on}) + P_{in}(t)](t - t_{on})}{2} - P_{in,ref}(t - t_{on}) \tag{4}$$

Where

$$(t - t_{on}) = (P_{in}(t) - P_{in}(t_{on})) / (\dot{p}_{in,on})$$

$$\dot{p}_{in,on} = V_{in}(dI_{in}) / (dt)$$

By putting (3) into (4), the on-state trajectory can be written as:

$$\mathrm{Traj}|_{on} = E_{con}(t) - E_{con}(t_{on})$$

$$- \frac{L}{2} \left\{ \left[I_{in}(t) - I_{ref} \right]^2 - \left[I_{in}(t_{on}) - I_{ref} \right]^2 \right\} = 0 \tag{5}$$

Because of the ideal switch surface will be through the working point (Econ,ref,Iref) of the converter, when the state is above the load line and along the on-state trajectory. By taking

$$E_{con}(t) = E_{con,ref} \quad , \quad I_{in}(t) = I_{ref} \quad , \quad E_{con}(t_{on}) = E_{con} \quad ,$$

$I_{in}(t_{on}) = I_{in}$, we can get the new equation:

$$\mathrm{Traj}|_{on} = E_{con,ref} - E_{con}(t_{on})$$

$$- \frac{L}{2} \left\{ \left[I_{in}(t) - I_{ref} \right]^2 \right\} = 0 \tag{6}$$

By using the similar approach, we can get the off-state trajectory:

$$\mathrm{Traj}|_{off} = E_{con,ref} - E_{con}(t_{on})$$

$$- \frac{L V_{in}}{V_o / 2} \left\{ \left[I_{in}(t) - I_{ref} \right]^2 \right\} = 0 \tag{7}$$

At D < 0.5, the formula is the same as D > 0.5.

4 SIMULATED PERFORMANCE OF A BOUNDARY CONTROL THREE-LEVEL BOOST CONVERTERS

By the above equation, we can conclude that a set of control converter switch signal. This group of switch signal will control two switches, so we use the D flip-flop and two or door to divide the signal into two groups signals with a certain angle α (α=180°). Then we can get the minimum inductor current ripple. According to the above, we can find out the boundary control strategy of the Boost three-level converter using state-energy plane.

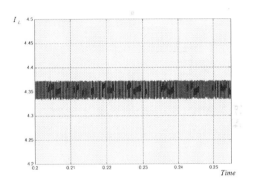

(a) Inductor current I_L

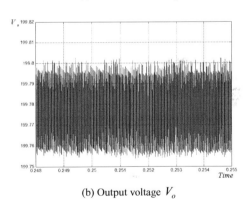

(b) Output voltage V_o

Figure 5. Boost three-level converter MATLAB simulation results.

Now we choose input voltage V_{in}=50V, reference voltage V_{ref}=200V, L=3ml, C_1=C_2=110mF. The MATLAB simulation results as shown in the Figure 5:

As we can see, from the result the inductor current ripple is about 0.04A, and output voltage ripple also very small, only about 0.2V.

Figure 6a and b shows the transient waveforms of V_o, when the input voltage V_{in} is changed from 48 to 40 and the load resistance from 200 to 100 respectively. The required time to the steady state is less than 0.005s. In Figure 6a, the output voltage and ripple deviation the average value by less than 4 V. In Figure 6b, the output voltage and ripple less than 2 V.

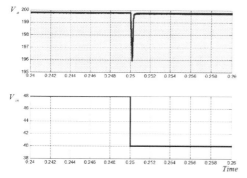

a) Input voltage from 48V to 40V at 0.25s

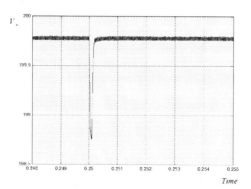

b) Load resistance from 200 to 100 at 0.25s

Figure 6. Dynamic response of the Boost three-level converter.

5 CONCLUSIONS

An improved boundary control theory using a state-energy plane is applied to a Boost three-level converter, and a second-order switching surface has been derived. Detailed mathematical analyses have been given. As the proposed switching surface, the ripples of output voltage and inductive current are very small, and the output voltage can be very quickly reverted to the ideally steady state during large-signal disturbances.

ACKNOWLEDGMENTS

This work is supported by the National Natural Science Foundation of China (no. 51107016), the National Key Basic Research Program of China (973 Program) under Grant (2013CB035605) and the Postdoctoral science-research developmental foundation of Heilongjiang province (no. LHB-Q12086). It is best to first retype the first words manually and then to paste the correct text behind. When the new file contains all the text, the old tags in the text should be replaced by the new Balkema tags (see section 3). Before doing this apply automatic formatting (AutoFormat in Format menu).

REFERENCES

Ogura K.& Nishida T, Hiraki E, et al. (2004). Time-sharing Boost chopper cascaded dual mode single-phase sinewave inverter for solar photovoltaic power generation system. *IEEE Power Electronics Specialists Conference. 6*, 4763–4767.

Munzert R. & Krein P. T. (1996). Issues in boundary control. *IEEE Power Electronics Specialists1*:810–816.

Leung K. & Chung H. (2007). A comparative study of boundary control with first- and second-order switching surfaces for buck converters operating in DCM. *IEEE Trans. Power Electron.22(4)*, 1196–1209.

Sable, D. M. Cho, B. H. & Ridley. R. B. (1991). Use of leading-edge modulation to transform boost and flyback converters into minimum-phase zero systems. *IEEE Trans. Power Electron.6(4)*, 704–711.

S. Hiti & Borojevic, D. (1995). Robust nonlinear control for boost converter. *IEEE Trans. Power Electron.10(6)*, 651–658.

Song T. T. & Chung H. S. H. (2008). Boundary Control of Boost Converters Using State-Energy Plane. *IEEE Transactions on Power Electronics.23(2)*, 551 563.

Electronic Engineering and Information Science – Wang (Ed.)
© 2015 Taylor & Francis Group, London, ISBN: 978-1-138-02772-5

Analysis of the three-paralleled-connected switched-inductor boost converter

E.H. Guan, D. Zhao, H.C. Liu & S. Yang
School of Electrical Engineering and Automation, Harbin Institute of Technology, Harbin, China

ABSTRACT: In this letter, the nonlinear dynamic characteristics of the three-paralleled-connected SI boost converter are analyzed. The discrete iterated mapping model under the continuous conduction mode of the circuit is derived. In order to capture the bifurcations which occurs in the system, we study the mechanism based on the loci of eigenvalues of the Jacobian matrix. The Power Electronics Simulator (PSIM) software is used to get the numerical results, which are provided to verify the diversity of the nonlinear behavior in the paralleled system.

KEYWORDS: Nonlinear; Boost Converter; Bifurcations; Paralleled System.

1 INTRODUCTION

The DC-DC converters with a high voltage transfer gain have been used for many applications such as fuel cell power system, high intensity discharge ballasts for automobile headlamps, backup systems and photovoltaic system battery for uninterruptible power applies. In order to achieve a high-voltage conversion ratio, the switched-inductor (SI) boost has been proposed (Alelrod & Berkovic 2008, Wang. Tang, He & Fu 2014). This kind of converter can reduce the current stress, the voltage stress of the switching devices and the loss of the circuit. What's more, in order to satisfy the applicable conditions of the large capacity situation, more switch converters are combined to form a parallel system in the practical engineering application. During the past two decades, the nonlinear phenomena of the conventional DC-DC converters have been investigated intensively (Wang, Zhou, Chen & Jiang 2008, Tse, Lai & to 2000), but there is little attention paid to the nonlinear behavior of the paralleled SI boost converter. In this letter, the difference of nonlinear dynamic characteristics between three-parallel-connected SI boost converter and the conventional SI boost converter is researched. Numerical simulation and PSIM simulation are provided to verify the influence of the parallel connected converters (Cheng & Sun 2008).

2 OPERATING PRINCIPLE AND MODELING OF THE CONVERTER

The schematic of the three-parallel-connected SI boost converter under current controlled is shown in Fig. 1, which is composed by three SI boost converters. When the three SI boost converters are connected in parallel, the current across the resistance would be triple of the one converter, while the voltage of the resistance and the capacitor is equal. Therefore, the three-parallel-connected SI boost converter can increase the current across the resistance with maintaining the voltage stress of the switch to satisfy the requirement of the large capacity situation.

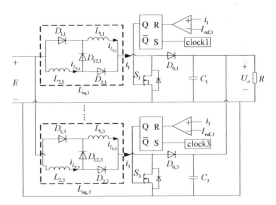

Figure 1. The current controlled three-parallel-connected SI boost converter.

In order to make sure that the three converters are all working in continuous mode, we vary the circuit parameters in a narrow range. Let $x = [i_1, i_2, \cdots, i_m, u_0]^T$ be the vector of state variables, assuming that the switch S_m (m=1, 2, 3) is on for [t_0, t_1] and is off for [t_1, t_2], then the state equation of the SI boost converter can be written as

$$\dot{x} = \begin{cases} A_{1,m}x + B_{1,m}E & t_0 \leq t \leq t_1 \\ A_{2,m}x + B_{2,m}E & t_1 \leq t \leq t_2 \end{cases} \qquad (1)$$

For $m = 3$, we can get the coefficient matrix

$$A_{1,3} = \begin{bmatrix} 0 & 0 & 0 & 0 \\ 0 & 0 & 0 & 0 \\ 0 & 0 & 0 & 0 \\ 0 & 0 & 0 & -\dfrac{1}{R(C_1 + C_2 + C_3)} \end{bmatrix}, \quad B_{1,3} = \begin{bmatrix} \dfrac{2}{L_1} \\ \dfrac{2}{L_2} \\ \dfrac{2}{L_3} \\ 0 \end{bmatrix}, \quad B_{2,3} = \begin{bmatrix} \dfrac{1}{2L_1} \\ \dfrac{1}{2L_2} \\ \dfrac{1}{2L_3} \\ 0 \end{bmatrix}^T$$

$$A_{2,3} = \begin{bmatrix} 0 & 0 & 0 & -\dfrac{1}{2L_1} \\ 0 & 0 & 0 & -\dfrac{1}{2L_2} \\ 0 & 0 & 0 & -\dfrac{1}{2L_3} \\ \dfrac{1}{C_1 + C_2 + C_3} & \dfrac{1}{C_1 + C_2 + C_3} & \dfrac{1}{C_1 + C_2 + C_3} & -\dfrac{1}{R(C_1 + C_2 + C_3)} \end{bmatrix} \qquad (2)$$

From the equation above, we can get

$$x(t_1) = f_1(x(t_0), t_1) = e^{A_{1,m}(t_1 - t_0)}[x(t_0) + \int_{t_0}^{t_1} e^{A_{1,m}(t_0 - \tau)} B_{1,m} E \, d\tau]$$

$$= \phi_1(t_1 - t_0)[x(t_0) + \int_{t_0}^{t_1} \phi_1(t_0 - \tau) B_{1,m} E \, d\tau] \qquad (3)$$

$$x(t_2) = f_2(x(t_1), t_2) = e^{A_{2,m}(t_2 - t_1)}[x(t_1) + \int_{t_1}^{t_2} e^{A_{2,m}(t_1 - \tau)} B_{2,m} E \, d\tau]$$

$$= \phi_2(t_2 - t_1)[x(t_1) + \int_{t_1}^{t_2} \phi_2(t_1 - \tau) B_{2,m} E \, d\tau] \qquad (4)$$

Substituting Eq. 2 into Eq. 3 and 4, we can obtain the discrete mode of the three-parallel-connected SI boost converter to analysis the influence of the variable parameters making to the nonlinear dynamic characteristics.

3 ANALYSIS FOR BIFURCATION

As is shown in the bifurcation diagram, the change of behavior can be seen clearly when a parameter is varied. In this letter, we vary the reference current and the output capacitor separately to analysis the bifurcation of the three-paralleled-connected SI boost converter.

3.1 Analysis for bifurcation with the variable reference current

Let the reference current I_{ref} vary from 0 to 50A with a step of 10A with the other parameters fixed at $E = 20$V, $R = 10\Omega$, $f = 10$kHz, all the capacitors are 10μF and all the inductances are 1mH. Fig.2 shows the bifurcation diagram of the three kinds of SI boost converters with I_{ref} as parameter.

6

From the simulation results, we can see that the three-paralleled-connected SI boost converter is much quicker to become period-doubling bifurcation and come into chaos.

3.2 *Analysis for bifurcation with the output capacitor*

With the other parameters fixed and equaled with the one above, choose the output capacitor to be variable. As shown in Fig. 3, the three-paralleled-connected converter has become chaos with the same range of the voltage and also has more complicated nonlinear dynamic behaviors.

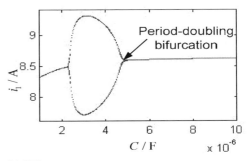

(a) SI boost converter

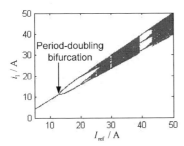

(a) SI boost converter

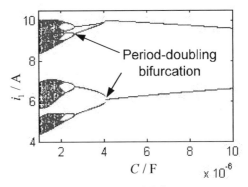

(b) two-paralleled-connected SI boost converter

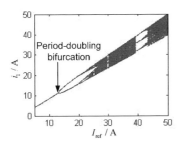

(b) two-paralleled-connected SI boost converter

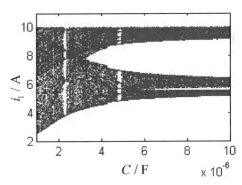

(c) three-paralleled-connected SI boost converter

Figure 3.　Bifurcation diagram with output capacitor as a parameter.

4 PSIM VERIFICATIONS

To verify the theoretical analysis and the numerical results, PSIM software has been used to study the three-paralleled-connected SI boost converter. Fig.4 shows the time-domain waveform and the phase

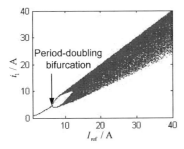

(c) three-paralleled-connected SI boost converter

Figure 2.　Bifurcation diagram with reference current as a parameter.

7

portrait of the three-paralleled-connected converter. Obviously, it is easy for the proposed converter to be unstable and work in the chaotic state.

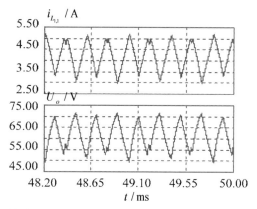

(a) time-domain waveform

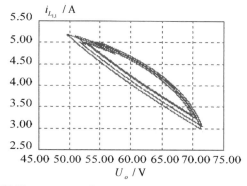

(b) Phase-space trajectory

Figure 4. The PSIM simulation of the three-paralleled-connected SI boost converter.

5 CONCLUSION

The bifurcation in a current controlled three-paralleled-connected SI boost converter is studied based on the discrete iterated mapping model. Comparing the nonlinear dynamic characters of the three SI boost converter, it finds that the three-paralleled-connected converter is much more easy to be chaos and the working state of the circuit is hard to predict. Consequently, it is very important to choose the parameters of the converters carefully to make sure the three-paralleled-connected SI boost converter working in stable state.

ACKNOWLEDGMENT

This work is supported by the National Natural Science Foundation of China (no. 51107016), the National Key Basic Research Program of China (973 Program) under Grant (2013CB035605) and the Postdoctoral science-research developmental foundation of Heilongjiang province (no. LHB-Q12086).

REFERENCES

Alelrod B, & Y. Berkovic (2008). Swithed capacitor switched inductor structures for getting transformless hybrid dc-dc PWM converters. *IEEE Trans. Circuit Syst*55(2): 687–696.

Wang T., Y. Tang, Y.H. He & D.J. Fu (2014). Multicell Switched-inductor Switched-capacitor Active-network Converter. *Proceeding of the CSEE*34(6): 832–838.

Wang S.B., Y.F. Zhou, J.N. Chen & X.D. Jiang (2008). Intermittency in High-order Switching Power Converters. *Proccedings of the CSEE*28(12): 26–31.

Tse C.K., Y.M. Lai & H.H.C. Tu (2000). Hopf Bifurcation and Chaos in a Free-Running Current-Controlled Cuk Switching Regulator. *IEEE Transactions on Circuits and Systems*47(4): 448–457.

Cheng Y. & Y.J. Sun (2008). Summary of power electronic converter modeling. *Electric Derive Automation*30(6): 4–9.

Electronic Engineering and Information Science – Wang (Ed.)
© 2015 Taylor & Francis Group, London, ISBN: 978-1-138-02772-5

Research on chaos characteristic of sound wave interactions in marine medium

C.F. Lan & D. L. Yu
School of Electrical and Electronic Engineering, Harbin University of Science and Technology, Harbin, China

C.F. Lan, M. Zhang & F.C. Li
Energy Science and engineering, Harbin Institute of Technology, Harbin, China

L.P. Zhang
SonoScape, Co., Ltd, Shenzhen, China

J.Y. Liu
Computer Science and Information Engineering, Harbin Normal University, Harbin, China

ABSTRACT: To investigate the dynamic and spectrum characteristics of sound wave interactions in the marine medium, classical Lorenz system bifurcation diagram, phase diagram, time-distance diagram and power spectrum diagram are analyzed, the parameter values leading to chaos and dynamic characteristics of system are given; The Lorenz equation describing atmospheric turbulence is regarded as the basic dynamic model in marine medium, under the effect of external excitation, bifurcation diagram, phase diagram, time-distance diagram and power spectrum diagram of generalized Lorenz system after interaction with sound waves are simulated. The results show that under the effect of external excitation, chaos phenomenon occurs more often compared with the original system and that if the frequency components are more and the amplitudes are higher, then chaos performances appears to be more obvious and line-spectrum expands wider.

KEYWORDS: Lorenz system; Chaotic anti-control; Band expansion.

1 INTRODUCTION

Chaos began in the sixties, American meteorologist E. N. Lorenz obtained a form of movement of fluid in the study of atmospheric convection experiment, and it is an important branch of nonlinear science (Chen and Lu, 2003). In the original concept dynamics, Lorenz system can be seen as velocity, density and temperature distribution function of time, the classical Lorenz dynamics model can be considered to be caused by the turbulence of the marine medium, and the turbulence notable feature is that the system exist the multiple frequencies of oscillation. When the fluid motion, the change of some parameters of the system makes the oscillation coupled, system generates a series of new motion coupling frequency, resulting in the occurrence of chaos (Lan et al., 2013). In recent years the study of underwater acoustic signals show that: when the ship sailing in the ocean medium, the noise generated by hydrodynamic, mechanical noise and propeller noise show a strong nonlinear characteristic, and there are some chaos characteristics(Lorenzo et al., 2002; Li et al., 2012). Therefore, if

the chaos theory is used in the spectral characteristics of the underwater acoustic signal analysis, this will be a significant thing (Vilenchik et al., 2008; Wang et al., 2013a).

In recent years, the nonlinear interaction between acoustic waves has made some achievements. The nonlinearity of two columns acoustic waves produce new frequency components after the interaction, and if it satisfies three columns waves can satisfy sonic momentum and energy conservation, resonance occurs between the acoustic waves, and the energy exchange is the largest(Xian and Xu, 2008; Wang et al., 2013b).

In this paper, the spread of the sound in the ocean which is disturbed by the ship disturbances spread is studied, first discussed the dynamics of the classical Lorenz system to provide a basis for comparative analysis of the parameters of generalized Lorenz system disturbances under the ship; ocean waves are added as the external excitation for the classical Lorenz system, the new Lorenz system will become a self-excited oscillator, when the waves with different frequency components work on the system, with the appropriate amplitude and frequency control, make it worked nonlinear interaction with sound waves of hydrodynamic systems, and generates a new frequency components, and the frequency band expansion.

2 PRINCIPLE LORENZ SYSTEM

The past two or three decades, in the chaotic motion of dynamic systems theory, drive – damping (driven-damped) three-wave interaction may play an important role in the plasma turbulence, which has attracted much attention. Discussion simple, the initial Lorenz non-autonomous system is change into autonomous systems, the system equation is

$$\dot{x} = \sigma(y - x) = M(x, y, z)$$
$$\dot{y} = rx - y - xz = N(x, y, z) \qquad \sigma, r, b > 0 \qquad (1)$$
$$\dot{z} = -bz + xy = P(x, y, z)$$

The autonomous Lorenz system can be considered similar as the dynamics of a simple model of the troposphere and closed-loop dynamics of convection. This equation can be used as an incompressible fluid model of two parallel horizontal boundary, and the lower boundary temperature is higher than the upper boundary temperature (Chen and Lu, 2003). Parameters forced fluid convection autonomous Lorenz system, equation (Wang et al., 2013b)

$$\dot{x} = \sigma(y - x)$$
$$\dot{y} = r(t)x - y - xz \qquad (2)$$
$$\dot{z} = -bz + xy$$

There, $r(t)$ is the cycle Rayleigh which is proportional to the temperature difference between the bottom and top boundaries. $r(t) = r_0 + r_1 \cos(\omega t)$, r_0, r_1, ω are parameters σ and b are proportional to the size of the region. x Is proportional to the intensity of convection, y and z are proportional to the temperature change of the horizontal and vertical directions.

Force model parameters can be understood as cycle heating at its bottom or as cycle cooling at its top to control fluid motion and the fluid of heat transformation in the convection. In other words, the fluid dynamic system attempts to control chaotic convection system boundaries, by applying sine signal or mediating the surface temperature.

3 STUDY FOR CYCLE MOTIVATION METHOD OF CHAOS CONTROL

3.1 Cycle motivation method principle analysis

Classical Lorenz system through a transformation, the system equation is

$$x'' + (\sigma + 1)x' - \sigma(r - 1)x + x^3/2 + xu/2 = 0 \qquad (3)$$

After adding more harmonic outside incentive, the system equations

$$x' = \sigma(y - x)$$
$$y' = rx - y - xz$$
$$z' = -bz + xy + \sum_{i=1}^{n} A_i \sin(\omega_i t \pm \varphi_i) \qquad (4)$$

There, A_i, ω_i, φ_i are the external excitation amplitude, frequency, and phase; u is the value of the variable changed with $y(t)$ and b values, so after variable substitution and after a given external excitation, Lorenz system is considered to be an essentially nonlinear dissipative non-autonomous self-excited oscillator.

A weak sinusoidal signal can be expressed as a Taylor series expansion

$$x(t) = b_1 y(t) + b_2 y^2(t) + b_3 y^3(t) + \dots \qquad (5)$$

The signal after non-linear element function, can be expressed as

$$x(t) = 1/2 \, b_2 \, B^2 + \left(b_1 B + 3/4 b_3 \, B^3\right) \sin(\omega_0 t + \varphi)$$
$$+ 1/2 b_2 \, B^2 \sin(2\omega_0 t + 2\varphi - \pi/2) \qquad (6)$$
$$+ 1/4 \sin(3\omega_0 t + 3\varphi + \pi)$$

Therefore, two or more columns sine signal through the nonlinear components will contain many frequency combination $\omega_n \pm \omega_m$, magnitude combination A_m and A_n, Thereby changing the energy of the original signal, the band expanding.

3.2 Study for numerical simulation

With or without external excitation, according to equation (4) dynamic characteristic simulation. Multi-frequency excitation frequency is a frequency axis in the frequency band selected, the maximum amplitude is 1. The following figure shows the output of dynamical systems on the plane, the x coordinate diagram, time diagram, phase diagram and power spectra. According to the principles given about Matlab/simulation, the model is shown in Figure 1.

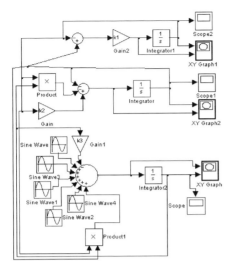

Figure 1. The structural diagram of system simulation.

Equation (4), single frequency signal as excitation, nonlinear interactions with sound hydrodynamic system, after that, the system output of the time-domain diagram, phase diagram and the power spectrum is shown in Figure 2; Equation (4), Take five single-frequency signals of different frequencies as the external excitation, signal frequency on the order of 10^{-5}, the maximum signal amplitude to 1, external excitation signal and hydrodynamic systems of sound waves occur nonlinear interaction, and the system output of the time-domain diagram, phase diagram and the power spectrum is shown in Figure 3.

The external excitation of the simulation compared with the Figures 2 and 3, harmonic amplitude increases 0.05, other parameters constant, get the system output dynamic diagram as shown in Figures 4 and 5. The external excitation of Figure 4 is a harmonic component, and the external excitation of Figure 5 is five harmonic components.

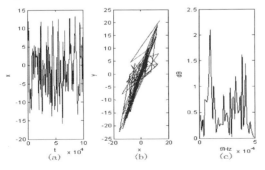

Figure 2. The dynamic rule of system output under with the amplitude 1.

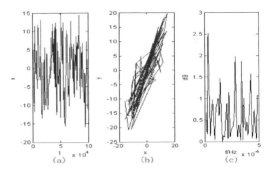

Figure 3. The dynamic rule of system output external stimuli under five different external stimuli with the amplitude 1.

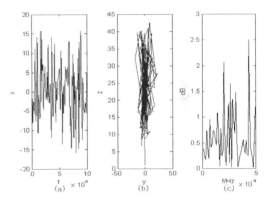

Figure 4. The dynamic rule of system output under single frequency signal with the amplitude 1.05.

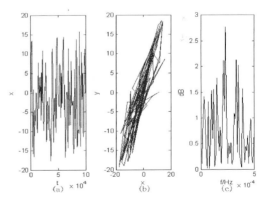

Figure 5. The dynamic rule of system output under five different external stimuli with the amplitude 1.05.

Learned from Figures 2–5, the external excitation applied to the system whether it is a single -frequency signal or multi-frequency signal, the system output chart is disorganized, no periodicity; Phase diagram contraction in a certain area and separate in local orbit, divergent, infinitely bending, folding; Power spectrum is similar to the broadband noise of the consecutive multi-peak spectrum line, show that system under the action of outer external excitation, kinetics of output in a state of chaos. From Figures 4 and 5 contrast can be seen, the external excitation, the more and more spectral components on the power spectrum, the easier the system of chaos; From Figures 2–5 contrast can be seen, when external excitation amplitude increases, chaos makes a certain frequency components distributed to the entire broadband segment. Spectral line to extend high frequency, low frequency energy decreases.

4 CONCLUSION

In this paper, the classical dynamics of the Lorenz system as a reference, and ocean internal waves in the medium as the classic Lorenz system outside incentive, then the system with acoustic wave happen nonlinear interactions in the ocean, and the paper study the dynamic characteristics. The results showed that: under the action of external excitation, the system is more prone to chaos phenomenon. The more acoustic frequency components, and the greater the amplitude, the more obvious the chaos, the wider the line spectrum to expand.

ACKNOWLEDGMENTS

This work was supported by the Scientific Research Fund of Heilongjiang Provincial Education Department [Nos.12541132,12541156,12541237].

REFERENCES

Chen, G. and J. Lu, (2003). Lorenz system dynamics analysis, *control and synchronization. Beijing : Science Press 9.*

Lan, C., D. Yang, & D. Lu, (2013). The theory and experiment of parametric amplification of three-wave nonlinear interaction in water. *Chinese Journal of Electronics. 22,* 308–312.

Li, A., L.Zhang, & J. Xiang, (2012). Influence of external forcing on the predictability of Lorenz model *Acta Physica Sinica 61,* 1311–1315.

Lorenzo, M., K. Holger, & H. Janusz, (2002). Optimizing of recurrence plots for noise reduction. *Physical Review E 65,* 021102.

Vilenchik, L.S., Ivanov, Y.V. & Trofimov, V.P. (2008). Digital Mathematical Modeling of Fractal Non-Linear Parametric Dynamics of Surface Sea Waves. *Mathematical Modeling 20,* 93–109.

Wang, W.,X. Zhang,& X.Wang(2013a). Based on independent component analysis and the empirical mode decomposition of chaotic signal noise reduction. *Journal of Physics 62,* 050201–1–8.

Wang, X., L.Zhang, & W. Chen, (2013b). Hopf bifurcation analysis and amplitude control the deformation of the Lorenz system. *Journal of Jiangsu University 34,* 121–124.

Xian, W. & X.Xu, (2008). A nonlinear noise reduction method for the underwater acoustic signal. *Marine Science Bulletin 27,* 17–21.

Electronic Engineering and Information Science – Wang (Ed.)
© 2015 Taylor & Francis Group, London, ISBN: 978-1-138-02772-5

Improvement and simulation of DV-HOP localization algorithm in wireless sensor networks

W.J. Zhou, J. Tang & C.T. Zhao

ABSTRACT: In the DV-HOP localization algorithm based on range-free for wireless sensor networks, in order to overcome the disvantage of the estimation of the average hop distance between beacon nodes and unknown nodes and insufficient three edge location selection in the process of positioning error, an improved DV-HOP localization algorithm is proposed, which refines the average hop distance and to locate the unknown node with the three edge location algorithm. The result of simulation demonstrates that the average hop distance is estimated more exactly in the improved DV-HOP algorithm, and the localization error of unknown nodes is reduced effectively.

KEYWORDS: wireless sensor networks; DV-HOP localization algorithm; improvement; the average hop distance; triangulation.

1 INTRODUCTION

Wireless sensor network WSN (Wireless Sensor Networks) is deployed in the monitoring region of the large number of cheap micro sensor nodes in the wireless communication way of self-organization and multi hop network system, in which the node localization technology is one of the key technologies. The node localization technology is mainly divided into two categories: range based localization technology and range free localization technology. Range based localization technology (Range-Based): the need to calculate the distance or angle between the different technology nodes, the calculation of the position of the node, and then use the corresponding positioning method. Range free localization technology (Range-Free): there is no distance or angle information between nodes, and the adjacent relationships between the nodes and the connectivity of positioning. The common positioning technology of centroid ranges free localization algorithm: DV-HOP algorithm, MDS-MAP algorithm, APIT algorithm (Sun et al 2005, He et al. 2003).

2 DV-HOP ALGORITHM DESCRIPTION

DV-HOP is a distance vector hop algorithm is very similar to the distance vector routing mechanism (Wang, Shi & Ren 2005) in traditional networks. Consists of three stages.

First stage calculate the unknown nodes and each node to node known, known packet broadcast neighbor node location information, including the hop count field (Romer K. & F.Mattern 2008). Connection records received information to each node is given the minimum hop count, packet ignores many hops from one node is given. Then hop value plus 1 and forwarded to the neighbor node. All the nodes through this method in network can record to the minimum hop every known node number

The second stage calculation of unknown nodes and known node distance of each hop in each node is given according to the first stages: recording other known location information of nodes and distance of hop, using type (1) the actual distance to estimate the average per hop.

$$Q^{size} = \frac{\sum_{i \neq j} \sqrt{\left(x_i - x_j\right)^2 + \left(y_i - y_j\right)^2}}{\sum_{i \neq j} \text{hop}_{ij}} = \qquad (1)$$

The third stage unknown nodes and unknown nodes by to hop all known node distance record in the second stage, calculates its own coordinates and the method to estimate the coordinate error by least squares using three-sided measurement.

As shown in Figure 1, the unknown node D and three known nodes A, B, C distance d_a, d_b, d_c respectively, three known nodes respectively (x_1, y_1), (x_2, y_2), (x_3, y_3), assuming that the D position (x, y) to create the following equations:

$$\begin{bmatrix} (x-x_1)^2 + (y-y_1)^2 \\ (x-x_2)^2 + (y-y_2)^2 \\ (x-x_3)^2 + (y-y_3)^2 \end{bmatrix} = \begin{bmatrix} d_1^2 \\ d_2^2 \\ d_3^2 \end{bmatrix} \qquad (2)$$

The equations above can be used to estimate the unknown node location (x, y).

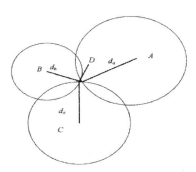

Figure 1. Schematic diagram of the three edge measurement of a typical figure.

3 IMPROCED DV-HOP ALGORITHM

The basis of the DV-HOP algorithm is to calculate the minimum number of hops and hop distance, in order to estimate the distance to a known node. There exists and the error between the true value error estimate value, will transfer to the three edge localization process, and in the three the process of locating, known to influence the position information of nodes the positioning accuracy greatly. This paper mainly aims at the distance of each hop on DV-HOP positioning algorithm is improved.

This paper proposes a new computational method to reduce the error introduced by the jump distance. In order to illustrate that the improved method in this paper, several parameters are defined first.

Definition 1 The actual between any two nodes i and j are known by the nodes to transmit information, defined by a number of hops, the hops between two known nodes denote by hop_{ij}.

Definition 2 Each hop distance between any two known nodes i and j, h_{ij}, e.g.

$$h_{ij} = \frac{d_{ij}}{hop_{ij}}. \qquad (3)$$

Among d_{ij} the two known node distance.

Definition 3 The average distance of each hop known node, denoted by distance, e.g.

$$distance = \frac{sum(h_{ij})}{BeconAmount} \qquad (4)$$

The $sum(h_{ij})$ for the sum of distance per hop known node, BeconAmount known as the number of nodes.

Definition 4 Estimation of the distance between the known node's value represented by the d_estimate, namely:

$$destimate=distance*hop_{ij} \qquad (5)$$

Definition 5 The actual distance is known among the nodes denote by d_true. The data obtained from the system simulation.

Set the several known nodes between the average distance per hop deviation factor (ROMER K,MATTERN F.), denoted as W, e.g.,

$$W=sum(\frac{d\_true\text{-}d\_estimate}{h_{ij}})/N \qquad (6)$$

From the above definitions and formulas, formulas of the actual distance per hop, denoted by d_rec, e.g.

$$d\_rec= d\_estimate +P*W \qquad (7)$$

4 THE EXPERIMENTAL STEPS OF IMPROVED DV-HOP ALGORITHM

Step one: Process as the initial algorithm.
Step two: To modify the distance, close to the real.
Step three: Process as the initial algorithm

5 THE SIMULATION EXPERIMENT AND RESULT ANALYSIS

The simulation network environments are as follows: the sensor nodes randomly distributed in a 100m* 100m square were studied in the different communication radius R, known under the condition of different number of nodes, the traditional algorithm and the improved location algorithm error.

In wireless sensor networks, the definition of average location error as the average estimate all unknown node value:

Consistency of style is very important. Note the spacing, punctuation and caps in all the examples below (Akyildiz et al. 2002).

$$e_{error} = \sum_{k} \sum_{i} \sqrt{(x_i' - x_i)^2 + (y_i' - y_i)^2} \qquad (8)$$

Type in (x_i, y_i) and (x_i', y_i') respectively to location estimation of unknown node and the actual position, K the number of simulation, i is the unknown node number.

Define the error coefficient of average location error and communication radius R (unit: m) ratio: distribution of known nodes and unknown nodes.

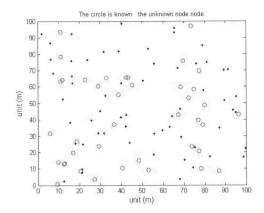

Figure 2. Node distribution system.

Figure 3 shows the 2 kinds of localization algorithms with increasing number of known nodes decreases, and tend to be stable, but the positioning error of the improved algorithm, but the positioning error of the improved algorithm than the original algorithm has greatly decreased than the original algorithm, about 10%~15%.

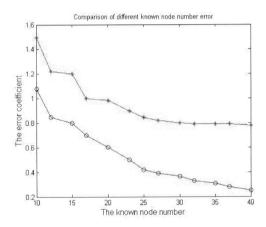

Figure 3. Mapping of quantitative error coefficient and known node relationship.

Figure 4 shows two kinds of localization algorithms are increased with the node communication radius increases.at the same radius, the positioning algorithm is superior to the traditional location algorithm.

6 SUMMARY

The simulation results show that the improved DV-Hop algorithm shows better performance without the need for additional hardware support, convenient and feasible.

REFERENCES

Sun L.M., J.Z. Li & Y. Chen et al. (2005).Wireless sensor net-works. Beijing: *Tsinghua University Press*.
He T., C.D. Huang & M. Brian et al. (2003). Range-free lo-calization schemes for large scale sensor networks. *MobiCom*, 81–95.
Wang F.B., L. Shi, & F.Y. Ren (2005). Self-localization systems and algorithms for w ireless sensor networks. *Journal of Soft-war*e 16(5): 857–868.
Romer K. & F.Mattern (2008). The design space of wireless Sotnetworks wireless communications. IEEE WireleSS Communications 11(6): 54–61.
Akyildiz L.F., W. Su & Y. Sankarasubramaniam (2002) Cayirci E.Wireless sensor network: Asurvey. *Computer Networ*ks 38(4): 393422.

Electronic Engineering and Information Science – Wang (Ed.)
© 2015 Taylor & Francis Group, London, ISBN: 978-1-138-02772-5

Eddy current loss analysis for half-speed large turbine generator

B.J. Ge, W. Guo & D.H. Zhang
Collage of Electrical and Electronic Engineering, Harbin University of Science and Technology, Harbin, China

ABSTRACT: The research presents comparatively deep analysis of the eddy current loss density generated by negative sequence current in the situation of using different rotor slot wedge materials. A variety of materials, such as aluminium and aluminum-bronze, which are in widespread use in slot wedge, are listed and compared in eddy current loss effects. Therefore, the different loss characteristic analysis processes for different material usage conditions are given. The calculation results show that the wedge material, actual location on the rotor surface, and asymmetry degree all should be taken into account in actual half-speed large generator design and electric machine industry.

KEYWORDS: Eddy current loss; wedge materials; aluminium; aluminum-bronze; half-speed large generator.

1 INTRODUCTION

Generally, turbine generators play an important role in network operation, but for a variety of reasons, generators carry three-phase unbalanced load sometimes. Generator stator windings will produce a negative sequence current in asymmetric operation (Li, Sun & Yang 2005, Bach et al. 2009). Since the rotary magnetic field generated by the negative sequence current is rotating at the same speed, inverting with the rotor, thus, the negative sequence field will induce a series of harmonics in the rotor winding (Ge, Wu & Zhang 2013). With the motor industry development, this situation has become increasingly prominent and urgent.

Due to the corresponding negative sequence component induced by each harmonic, stator and rotor will all contain multiple harmonics owing to this interaction. Meanwhile, because the negative sequence magnetic field speed is twice the synchronous speed frequency, eddy current on the rotor surface is induced, which can cause rotor heating and even damage generator system (Fujita et al. 2000, Hao & Sun 2010). Hence, some issues of eddy current loss for half-speed large turbine generators are analyzed and researched very closely in this paper.

eddy current loss density. The entire cross section of the generator is taken for electromagnetic field calculation, which is shown in Fig.1.

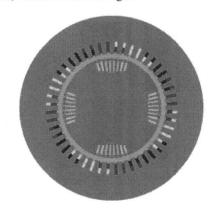

Figure 1. 2D physical model of solved region.

Rotor winding heating will not increase in the running process, which has little effect on the normal operation. Therefore, the eddy loss density calculations, in this paper, only contain the slot wedge and damping winding part.

2 EDDY CURRENT FIELD MODEL OF AP1000 GENERATOR

The typical half-speed, large generator, AP1000 nuclear generator, is adopted as an example in this paper, to carry out the analysis and calculation of

3 EDDY CURRENT LOSS CALCULATION OF SLOT WEDGES

From the loss density nephogram, as shown in Fig.2, by using the finite element method, the loss density values of each unit in the rotor are different.

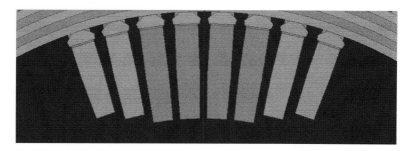

a. Slot wedges of one pole

  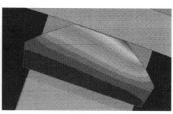

b. Wedge of the left part   c. Wedge of the middle part   d. Wedge of the right part

Figure 2.   The loss density distribution of one pole's wedges.

To bring a convenience and have the data done merely expediently, the same loss density value is given to deal with any part of the same rotor surface. Then, the loss density distribution of wedge's each part are getting in the case of using aluminum and aluminum-bronze as the wedge material, respectively. The results of 2-pole wedges are shown in Fig.3.

From Fig.2 and Fig.3 above, it can be seen clearly that, with the increase of depth from rotor surface, the eddy current generated by negative sequence magnetic field in each part of the rotor is decreased, and the loss of corresponding parts is also reduced. The eddy current values of wedge No.1, No.8, No.9 and No.16, which are near the big teeth, are much higher than other wedges. Furthermore, the eddy current loss density is reduced at different degree of the same depth according to different materials.

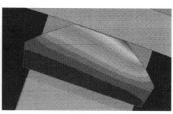

b. The loss density of the middle part

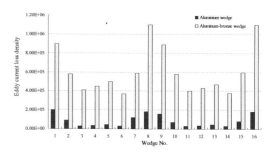

c. The loss density of the bottom

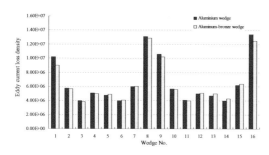

a. The loss density of the top part

Figure 3.   The loss density distribution of each portion of the wedge.

18

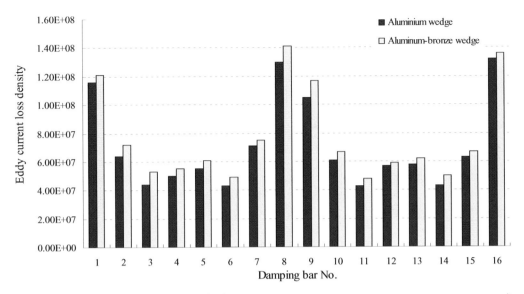

Figure 4. The loss density distribution of damping bars.

4 EDDY CURRENT LOSS CALCULATION OF DAMPING BARS

Additionally, the eddy current losses of damping bars are obtained, which are shown in Fig.4, by using the same method.

It can be seen from Fig.4 that under the same wedge material condition, the eddy current loss density of damping winding is much higher than the wedge's, that is also the reason why the damping bar is the most easily damaged part of half-speed large generator.

5 CONCLUSIONS

In order to get the eddy current loss density distribution of half-speed, large generator's rotor surface, the different loss density impacts on the rotor surface by using different slot wedge materials are deeply discussed. The finite element analysis method is used, and 2D mathematical model is established. Finally, the eddy current losses respectively, using aluminium and aluminium bronze are calculated and its variety law is obtained.

ACKNOWLEDGMENTS

This work was supported by the National Major Scientific and Technological Special Project of China (No. 2009ZX06004-013-04-01, Sub-project No. 2012BAF03B01-X).

The authors appreciate the financial support from the National Major Scientific and Technological Special Project of the Chinese Government. And Harbin Electric Machinery Company Limited, Harbin, China, is also acknowledged for the technical support for this work.

REFERENCES

Li, J.X., Y.T. Sun & G. J. Yang (2005). J. Calculation and analysis of 3D magnetic field for end region of large turbo generators. *Electrical Machines and Systems 3*, 2079–2082.

Bach T., D. Wohner & K. Takahashi (2009). J. Determining Negative Sequence Currents of Turbine Generator Rotors. *Electrical Machines and Systems*. 1–6.

Ge, B.J., G.Wu & D. H. Zhang (2013). J. Harmonic Analysis of AP1000 Large-capacity Turbo-generator Based on BP Neural Network. *International Journal of Control and Automation 6*, 163–175.

Fujita. M., Y. Tokumasu & H. Tsuda (2000). J. Magnetic field analysis of stator core end region of large turbogenerators. *Magnetics 36*, 1850–1853.

Hao L.L., Y. G. Sun & A. M. Qiu (2010). Analysis on the Negative Sequence Impedance Directional Protection for Stator Internal Fault of Turbo Generator. *Electrical Machines and Systems* 1421–1424.

Electronic Engineering and Information Science – Wang (Ed.)
© *2015 Taylor & Francis Group, London, ISBN: 978-1-138-02772-5*

The design and research of wafer-type tester based on microprocessor

J.M. Feng, Y.J. Cao & H. Luo
College of Applied Science, Harbin University of Science and Technology, Harbin, China

ABSTRACT: In this paper, a wafer conductivity type tester based on, the microprocessor is designed and produced, including parts of cold and hot probe temperature acquisition, thermal electromotive force acquisition, heating power control, microprocessor, input and output and so on. Temperature control is more stable and accurate by PID algorithm. Due to use of touch technology and liquid crystal display technology, it will make operation simpler, display more intuitive. The tester can not only measure wafer conductivity type, in the same test condition, can also compare the doping concentration of wafers in the same conductivity type.

KEYWORDS: Thermal electromotive force; Conductivity type; Doping concentration; PID algorithm.

1 INTRODUCTION

At present, the main methods of measuring the semiconductor conductivity type are single probe point contact rectifier, three probes, cold and hot probe and non contact type measuring with infrared light and so on (Dong et al. 2009). Commonly used method of cold and hot probe has not controllable hot probe, and detection and display method is relatively backward. This paper completes the design and manufacture of tester using temperature controlled hot probe, touch technology (Jain et al. 2013), liquid crystal display technology (LCD) (Liang et al. 2011) and C programming language. The tester can distinguish conductivity type of different wafers, can also compare the doping concentration of wafers with same conductivity type. The tester is especially suitable for experimental teaching in understanding cold and hot probe method to distinguish mechanism and methods of wafer conductivity type, and for observing experimental phenomena with the tester.

2 PRINCIPLE

As for p-type semiconductor, we contact p-type semiconductor using two probes. When one probe is heated, hole concentration of hot end is higher than that of cold end, because the intrinsic excitation degree of hot end substrate is higher than that of cold end. Holes diffuse to cold end in concentration gradient, and accumulate at the cold end, so the cold end potential is higher than hot end. Similarly for n-type semiconductor, cold end potential is lower than hot end.

Thermal electromotive force Θ_s of p-type semiconductor and n-type semiconductor is got by relevant theory. For p-type semiconductor:

$$\Theta_s = \frac{k_0}{q}(\frac{3}{2} - \ln\frac{p}{N_V})\Delta T \qquad (1)$$

For n-type semiconductor:

$$\Theta_s = -\frac{k_0}{q}(\frac{3}{2} - \ln\frac{n}{N_C})\Delta T \qquad (2)$$

k_0 is Boltzmann constant, q is the basic charge constant, ΔT is the temperature difference between hot and cold ends, p and n are majority carrier concentration of the p-type semiconductor and n-type semiconductor respectively, N_V and N_C are the effective density of states for the top of the valence band of the p-type semiconductor and for the bottom of the conduction band of n-type semiconductor respectively. For the specific substrate material, N_V and N_C are only related to temperature (Liu et al. 2009).

By the above formulas, direction of the thermal electromotive force of p-type semiconductor and n-type semiconductor can be derived opposite. To the same type of semiconductor materials doped with different doping concentrations, electric potential difference between hot and cold ends under the same test condition is related to the doping concentration. The greater doping concentration is, the greater the potential difference between hot and cold ends is.

3 DESIGN OF SYSTEM

Tester hardware includes thermal electromotive force acquisition module, Pt1000 temperature acquisition of the hot probe module, DS18B20 temperature acquisition of cold probe module, heating power control module, LCD display module, microprocessor module, the input module and power supply module and so on. The tester system structure is shown in figure 1.

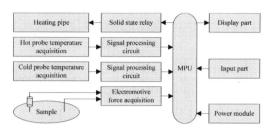

Figure 1. Diagram of system structure.

The tester uses a microprocessor as the core, with corresponding peripheral circuits to form an integral structure. The thermal probe temperature control and thermal electromotive force detection are key parts of the whole design.

Stable hot probe temperature is needed in order to accurately capture the thermal electromotive force, so hot probe temperature needs to be controlled after the tester is electrified. The hot probe temperature is getting through Pt1000 thermal resistance. Constant current source composed of OP07 drives Pt1000 to generate a voltage signal with temperature variation (Hahtela et al. 2014, Qin et al. 2013). A weak signal is amplified by amplifier INA118 with high precision instrumentation. The microprocessor will calculate the current temperature value based on the amplified signal. Then we compare current temperature to the set temperature value of user, and achieve temperature control through a solid state relay to control the heating power. The real-time temperature curve is getting on the LCD in the process of adjustment control.

The tester starts to get the thermal electromotive force of corresponding wafer when hot probe temperature reaches a relatively stable state. Direction of thermal electromotive force is associated with wafer conductivity type, and the collected signal is weak, so differential amplifier INA118 with high precision is needed to amplify the thermal electromotive force in order to ensure its signal strength. Then noninverting input summation operation circuit consisting of OP07 is used to ensure that thermal electromotive force signal in any case is positive. The processed signals are read by microprocessor, conductive type of wafer can be judged according to measured value of thermal electromotive force. On the other hand, by the formula (1) and the formula (2), under the same conditions, the size of doping concentration can be compared through distinguishing corresponding thermal electromotive force of same type wafers with different doping concentration. Real-time test results are getting on the LCD.

According to the actual measurement process, relevant programs of microprocessor are written (Schlich et al. 2011). The program flow chart is shown in figure 2.

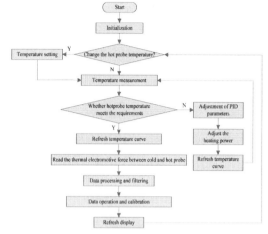

Figure 2. Program flow chart.

Initialization is processed after tester is electrified, this process includes configuration of some parts of the registers and displays. Then judge whether the temperature of the hot probe needs to be changed. If need be, enter into the temperature setting program, temperature measurement is processed after completing setup through the touch screen. If the hot probe temperature is not to be changed, enter into temperature measurement directly. PID parameters are adjusted according to the deviation between measurement value and set value. Temperature regulation is achieved through a PID algorithm to control heating power by using adjusted PID parameters (Liu et al. 2010). When the temperature reaches the set requirements, thermal electromotive force will be acquired, which can be converted into the digital signal through an internal analog-to-digital converter in the microprocessor. Then these signals can be operated and calibrated after filtering to obtain the final result of the calculation which will be printed on the display, and the next cycle will be processed.

4 DEBUGGING AND CALIBRATION

In order to compare the doping concentration of the same conductivity type wafers, test is needed in the same temperature conditions, which need high requirements for temperature control and need to calibrate DS18B20 temperature sensor and Pt1000 temperature sensor.

DS18B20 is a digital chip, we compare the output temperature value of sensor and measured value. At a temperature range of 10°C~70°C, take temperature pots to test every 6°C. Table of measured raw data is shown in Table 1.

Table 1.   Data of DS18B20 temperature calibration.

Actual value °C	Measured value °C	Deviation °C
12.3	12.8	0.5
18.5	19.1	0.6
24.1	24.6	0.5
30.7	31.2	0.5
36.1	36.5	0.4
42.6	43.1	0.5
48.2	48.6	0.4
54.3	54.8	0.5
60.4	60.9	0.5
67.0	67.5	0.5

From the table, we can see that deviation value of DS18B20 digital temperature sensor is about 0.5°C. This deviation needs to be corrected when writing programs.

Pt1000 of hot end is a resistance type analog output temperature sensor, calibration must be performed. In order to prevent the effect of air convection on the measurement results and make the data more accurate and reliable in the test, we will enclose test table in a certain space. Table of measured raw data is shown in Table 2.

Table 2.   Data of Pt1000 temperature calibration.

Temperature °C	AD value	Temperature °C	AD value
10.3	442	140.2	652
12.2	446	150.7	669
15.1	451	158.8	682
18.3	457	166.8	695
20.0	459	177.5	711
22.8	464	184.4	722
28.0	472	190.0	731
32.3	480	195.4	739
36.6	487	203.8	752
40.3	493	211.4	764
44.8	500	219.5	777
52.5	512	225.5	786
57.6	521	233.3	798
61.7	527	239.4	807
68.6	539	250.8	825
77.4	553	259.8	839
87.2	569	268.3	852
96.3	583	276.0	863
101.9	592	284.2	876
115.8	615	293.5	890
125.2	630	303.8	905
136.6	647	312.6	918

Especially, temperature in the table is contact temperature for the probe and sample, each group of data is read after the temperature is stable. The data above is processed by Origin software and curve fitting (Lv et al. 2013). The fitting results are shown in figure 3.

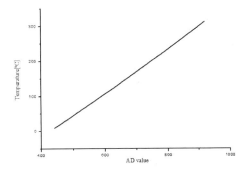

Figure 3.   Temperature as a function of AD value.

From the fitting curve, we can see that relationship between temperature and AD value are shown basically linear, the relation can be expressed by the following formula

$$T = (AD\ value - 442)/1.5746 + 10.3°C \qquad (3)$$

In addition, measurement circuit of thermal electromotive force is also needed for calibration. Corresponding thermal electromotive force can be obtained through adjustment of temperature difference between hot and cold probes, and can be converted into the AD value through the analog-to-digital converter, the raw data of the measurements is shown in table 3.

Table 3.   Relation between thermal electromotive force and AD value.

Voltage mV	AD value	Voltage mV	AD value	Voltage mV	AD value
−6.875	20	−1.800	328	2.350	628
−6.000	52	−1.575	344	2.500	640
−5.325	84	−1.450	356	2.650	652
−4.790	112	−1.300	364	2.850	660
−4.350	144	−1.150	376	3.050	676
−4.000	168	−1.100	380	3.275	692
−3.675	192	0.900	528	3.525	712
−3.400	208	1.000	532	3.850	736
−3.200	224	1.150	540	4.200	760
−3.000	240	1.250	552	4.650	792
−2.825	252	1.400	560	5.200	828
−2.675	264	1.600	576	5.850	880
−2.550	272	1.800	588	6.750	932
−2.250	296	2.050	608	7.900	976
−2.000	316	2.200	616	9.600	1016

The data above is processed by Origin software and curve fitting. The fitting results are shown in Figure 4.

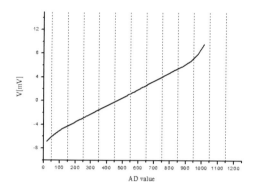

Figure 4. Thermal electromotive force as a function of AD value.

From the fitting curve, we can see that relationship between thermal electromotive force and AD value are shown relatively complex. We divide the curve into five sections for data processing. Each section uses a linear relationship as a mathematical model to conduct operations. Processed by microprocessor, we can get the corresponding AD value of thermal electromotive force. Under the same test conditions, we can distinguish relative doping concentration of samples simply by comparing the thermal electromotive force.

5 CONCLUSION

Wafer type tester made in this paper is developed in the embedded environment, the program is written in C language. The tester can not only measure wafer conductivity type, in the same test condition, can also compare the doping concentration of wafers in the same conductivity type. The tester adopts the interactive interface, operation is more convenient and flexible. A temperature control range of hot probe is room temperature to 170°C, control precision is ±0. 2°C. The detection range of thermal electromotive force is −7mV~10mV, minimum resolution is 0.017mV.

REFERENCES

Dong, S. & Cai, Y.K. 2009. Semiconductor materials to test the status of technology. *Solar & Renewable Energy Sources*, 3: 033.
Hahtela, O., Heinonen, M., Kajastie, H. et al. 2014. Calibration of Industrial Platinum Resistance Thermometers up to 700 °C. *International Journal of Thermophysics*, 1: 35.
Jain, A., Bhargava, D.B. & Rajput, A. 2013. TOUCH-SCREEN TECHNOLOGY. *International Journal of Advanced Research in Computer Science and Electronics Engineering (IJARCSEE)*, 2(1): 074–078.
Liang, M.L., Zhang, H.M. & Xu, B. 2011. ARM Microprocessor-based LCD Touch-screen Interface Design. *Computer Measurement & Control*, 3: 067.
Liu, E.K., Zhu, B.S. & Luo, J.S. 2009. Semiconductor physics. *National Defence Industry Press*: 311–319(In Chinese).
Liu, F., Wang, A.F., Sun, D.P. et al. 2010. Application of Origin Software in College Physics Experiment Data Processing. *Experiment Science & Technology*, 1: 006.
Lv, Y., Niu, C.H., Li, Y.Q. et al. 2013. Precision Temperature Control System Based on PID Algorithm. *Applied Mechanics and Materials*, 389: 483–488.
Qin, Y.Y., Yang, X.Q., Wu, B.Z. et al. 2013. High resolution temperature measurement technique for measuring marine heat flow. *Science China Technological Sciences*, 56(7): 1773–1778.
Schlich, B., Brauer, J. & Kowalewski, S. 2011. Application of static analyses for state-space reduction to the microcontroller binary code. *Science of Computer Programming*, 76(2): 100–118.

Electronic Engineering and Information Science – Wang (Ed.)
© *2015 Taylor & Francis Group, London, ISBN: 978-1-138-02772-5*

Design and implementation of SPI Flash controller based on Xilinx FPGA

Y. J. Gong & Q.G. Xiong
Wuhan University of Science and Technology, School of Information Science and Engineering, China

ABSTRACT: The SPI Flash controller is used for XILINX FPGA configuration of Flash programming, namely software access to the SPI Flash via the controller. The main purpose is to replace entirely the SPI Flash software program to improve reading and writing performance, and solve the problem of software upgrade the FPGA takes too long. The design has been used in many projects and achieved the desired performance.

KEYWORDS: field programmable gate array; SPI Flash; software upgrade; simulation.

1 INTRODUCTION

The traditional Flash programming is achieved through software. However, the low speed of reading or writing and the occupation of CPU resources make the online upgrade of the whole system to take a long time. And it is very difficult for direct operation on the chip owing to the Flash chip itself has too many function instructions. Based on the situation above, this paper presents an implementation scheme for the SPI Flash read and write based on Xilinx FPGA (Xu & Tian2012). The scheme uses the hardware to control the SPI Flash and can be very convenient to complete the Flash read and write operations. Also, the SPI Flash controller can be transplanted and reused easily (Zhang & Zhong 2010). SPI Flash controller is programmed with Verilog language, passed the functional simulation on Modelsim, and has been applied to some corresponding projects and achieved the goal expected.

2 DESIGN SCHEMES

2.1 *Design ideas*

We have had the detailed design of the whole system, and the hardware structure is shown in Fig.1 (Li et al. 2013). The system mainly consists of four parts, the host computer, microprocessor, FPGA and SPI Flash. The host computer provides a human-computer interaction, it can send an SPI Flash command to the microprocessor. Analysis of these instructions is mainly handled by the microprocessor, the microprocessor is capable of communicating with FPGA.

The FPGA part is the center of this design, which depending on the software configuration, generates SPI bus timing needed to achieve SPI read and write operation.

The software in the execution of an SPI command, should first query spi_busy. When the SPI bus isn't busy, the CPU can operate the SPI flash. At this point, according to the difference of the commands to be executed, the software need to write opcode_reg, address and data_wr register, finally, write to the spi_control registers to trigger a true SPI bus operation.

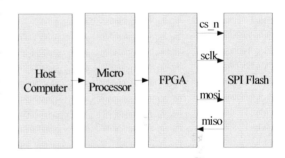

Figure 1. System architecture diagram.

2.2 *FPGA and SPI Flash*

The design was chosen FPGA Xilinx's Spartan-3A DSP family XC3SD3400A, the package uses FG676. Compared with competing FPGA products, Spartan-3A DSP 3400A LP has a 25% power efficiency advantage, the lowest cost device performance at 250Mhz clock speeds up to 4.06 GMACs per mW. This series of FPGA can finish signal processing function without spending extra logic resources, so designers can achieve performance and cost objectives in the case of higher efficacy. Used to construct a dedicated DSP circuitry Xtreme DSP DSP48A logic chip includes a dedicated 18 x 18 multipliers and 18 pre-adder and 48 post-adder / accumulator, at low cost DSP provides excellent performance. Spartan-3A DSP FPGA platform standby mode can also help prolong battery life in these applications.

Other applications that can benefit from the industry's lowest power, high-performance FPGA platform also includes military communications (MILCOM) portable and mobile tactical wireless devices and portable night vision equipment.

SPI Flash is used in Numonyx (Heng Yi) Company M25P16, the SPI Flash has more instructions, all instructions are 8 bits, when the chip select signal CS# is pulled low to select the device, and then enter the 8 bit operating instruction byte serial data pulled low, rising in the chip select signal CS# after the first clock the sampling, SPI Flash initiates the internal control logic, auto complete corresponding operation. Some operations need to enter the address byte and pseudo bytes in the input instructions, the last operation after the completion of the pull high chip select signal.

3 SPI CONTROLLER

It is relatively simple to realize the function of SPI controller in FPGA. According to the software configuration, the FPGA will generate SPI bus timing needed to achieve SPI read and write operations.

3.1 Interface description

Module interface signals about SPI controller are described in Table 1:

Table 1. Interface description table.

Name	Type	Function description
clk	Input	100M clock signal
reset	Input	Reset signal, active high
Signal with CPU interface module		
spi_req	Input	CPU request to operate on SPI Flash
opcode	Input	Operation code of CPU operated on spi_slave
address	Input	Address of CPU operated on spi_slave
data_wr	Input	Parallel data CPU write in
data_rd	Output	Parallel data CPU read out
spi_busy	Output	Signal of spi response for cpu operation
spi_control	Input	Analytic signal for opcodes
interface signals with spi_slave		
sclk	Output	Generated by the CLK
cs_n	Output	Chip select signal, active low
mosi	Output	Serial data input spi_slave
miso	Input	Output data from the serial spi_slave

3.2 Status description

According to the design, the state machine is divided into six states as follows:

IDLE: in this state, FPGA waits for CPU operating the spi_slave. IF the spi_en signal is "1", the next state will be the COMMAND state, otherwise, the same.

COMMAND: serial transmission 8 bit operation code state, if the operation code counter is less than 7, waiting for 8 bit operation code transmission is completed. And then through the spi_control to determine the next state. If spi_control [6] is "1", the next state is the ADDR state. Then FPGA will judge whether the spi_control [5:3] is "0", if it is not "0", the next state is the DUMMY state. And then it through spi_control [7] and spi_control [2:0] to determine the next state is WR_DATA or RD_DATA. If spi_control [2:0] is not "0" and spi_control [7] is "1", the next state is the WR_DATA state. If spi_control [2:0] is not "0" and spi_control [7] is "0", the next state is RD_DATA state, otherwise, is the IDLE state.

ADDR: serial transmit 24 bit operating address state, if the address counter is less than 23, waiting for 24 bit address code transmission is completed; Then a method of judging the COMMAND state description of the ADDR state to jump down.

DUMMY: any idle clock cycle state, this state of input and output data do not care about, waiting for the end of the state; and then the method of judging the COMMAND state description of the DUMMY state to jump down.

WR_DATA: write several bytes of data state, by comparing the size of the data counter and data bit reduction "1" to judge whether the state is over, the end of the state, the next state jump to IDLE sate.

RD_DATA: read several bytes of data state, by comparing the size of the data counter and data bits to determine whether the state of the end, the end of the state, the next state will be IDLE state.

4 FUNCTIONAL SIMULATION

SPI controller was implemented in a Xilinx ISE13.1 programming environment, using simulation software. Modelsim SE-64 10.1c to achieve functional simulation. Due to limited space, the function simulation part mainly tested the common read and write functions of the SPI controller module (Guan & Zhou 2012). Before the function simulation, the SPI Flash module and test module in the process were added, the simulation results were shown in Fig.2 and Fig.3.

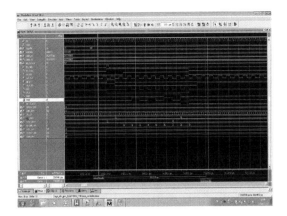

Figure 2. Write operation timing simulation diagram.

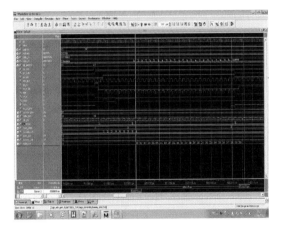

Figure 3. Read operation timing simulation diagram.

From Fig.2 and Fig.3, the simulation results have achieved the timing requirement of SPI Flash read and write instructions.

5 APPLICATION IN ENGINEERING AND ITS EFFECT

Before using this scheme, it took too long to upgrade the software of the entire system, which brought much inconvenience to the testing personnel. For example, in the8CPOS_B for the full voice project, not before adding the module, the realization of the software upgrade took around 30mins, adding the module after the upgrade, greatly shorten the time, only needed to spend about 7mins. Due to the module has a good portability and reusability (Wang & Zhou 2006), the module will be widely used in signaling data acquisition system based on Xilinx FPGA. In the future, it will be to bring the gospel for the upgrade of the series of products.

6 CONCLUSION

In this paper, an SPI Flash controller based on FPGA was achieved for practical application. This method has high portability, as well as simple and convenient user interface. This SPI controller implements a common framework, through software programming, accessing to a variety of SPI devices could be achieved through software programming, not limited to Flash. Therefore, it has high compatibility.

REFERENCES

Xu W.B. & Y. Tian (2012). Xilinx FPGA: development and practical tutorials. Beijing: *Tsinghua University Press*.

Zhang L.W. & H.M. Zhong (2010). Design and implementation of SPI Flash controller based on Xilinx FPGA. *MicroComputer Information 26 (6-2)*, 124–126.

Li M.B., M. Xie, G.M. Liu & X.C Liu (2013). A SPI FLASH-based FPGA Dynamic Reconfiguration Method. ICMTCE 2013–2013 IEEE International Conference on Microwave Technology and Computational Electromagnetics, *Proceedings*, 379–382.

Guan S.S. & J.M. Zhou (2012). Design and Verification of SPI Flash Controller Based on Xilinx FPGA. *Electronic devices*, 216–220.

Wang X.F. & J.P Zhou (2006). Method of On-line Configuring Flash Using FPGA. *Electronic devices 29(3)*, 902–904.

Electronic Engineering and Information Science – Wang (Ed.)
© 2015 Taylor & Francis Group, London, ISBN: 978-1-138-02772-5

The current-voltage characteristics of zinc phthalocyanine organic thin film transistor

Y.S. Zhang, D.X. Wang, X.C. Liu, Z.Y. Wang, Y.Y. Wang & J.H. Yin,
Key Laboratory of Engineering Dielectrics and Its Application, College of Applied Science, Harbin University of Science and Technology, Heilongjiang Harbin, China

H. Zhao
Harbin University of Science and Technology, Heilongjiang Harbin, China

ABSTRACT: In this paper, we have fabricated organic thin film transistors by vacuum evaporation and magnetron sputtering, the structure of ZnPc organic thin film transistor is ITO/ZnPc/Al/ZnPc/Cu. The active layer material is zinc phthalocyanine which has good photosensitive properties. As a collector electrode, ITO is the best choice which has high light transmittance. Detection this device and analysis its photoelectric characteristics, the device have a clear light amplification effect. The results show that the I-V characteristics of transistor are unsaturated. When $V_{ec}=1V$, current amplification factor changes in 7.35-18.18. We hope that this kind of organic transistor can be used as photoelectric detector.

KEYWORDS: zinc phthalocyanine; current-voltage characteristics; transfer characteristics; current amplification factor.

1 INTRODUCTION

For the past few years, the studies of organic thin film transistors (OTFTs) have grown dramatically due to their potential applications in integrated circuits, flat–panel displays. The organic electronic material is a new type of electronic material. The organic electronic material not only has the semiconductor, optical and electrical properties, but also keeps some good qualities of organic matter. We all know that the structure and material of the transistor had a great influence on the performance of the organic thin film transistor. Therefore, the choice of materials is very important. As everyone knows, the phthalocyanine organic semiconductor material is one of the most organic electronic material, it has received widespread attention. So we choose zinc phthalocyanine (ZnPc) as the active layer. The organic semiconductor materials ZnPc used in this work was with the purity of 99.9%.

As the anode materials of the transistor, the prerequisite are: (Florian E., K. Hagen & H. Marcus 2004) good conductivity; (Bialek, I.G. Kim & J. Lee 2003) good chemical and morphological stability; (Raoul S., M. Leszek & G. Martin 2004) its work function and the HOMO energy level of the hole injection material are matched. When the anode is used in transparent devices, another necessary condition is high transparency in the visible light area. The anode material has a transparent conductive oxide and metal two kinds big (Naoki H., O. Noboru & M. Nakamura 2008). Conductive oxide has high light transmittance, it is nearly transparent in the visible light region. Metal has high conductivity, but it is opaque. If you want to let the metal electrode pervious to light, its thickness should be thin enough.

Conductive oxide indium tin oxide (ITO) is a kind of N-type semiconductor material which has high conductivity, high transmission in the visible light region (close to 90%), high mechanical hardness and good chemical stability. ZnPc is the most commonly used material for transparent electrode. Therefore, we study the vertical structure of ITO / ZnPc/ Al / ZnPc/ Cu organic thin film transistor.

2 EXPERIMENT

The sample of this paper has a five-layer structure of ITO /ZnPc/Al/ZnPc/ Cu which fabricated by vacuum evaporation and magnetron sputtering. Shown in figure1 is the structure diagram of ZnPc organic thin film transistor. OLED multi-function multiple coating system was used to fabricate the device. The metal AI, Cu and ITO were fabricated by direct-current and radio- frequency magnetron sputtering respectively. To define the active layer, organic semiconductor material ZnPc was deposited by vacuum evaporation at the temperature of 35°C. This ZnPc organic thin film transistor is plated on the glass substrate. In this sample, Al is the base electrode, Cu is the emitter electrode and ITO is collector electrode.

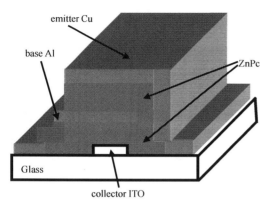

Figure 1. The structure diagram of ZnPc organic thin film transistor.

Keithley 4200 semiconductor parameter analyzer was used for analysis all electrical measurements. LP-130Xe type adjustable wavelength light source was the experimental source which can provide strong mono- chromatic beam and can be controlled by computer. All the experiments were carried out at room temperature. ZnPc materials have a good absorption coefficient in the wavelength range of 320~485nm and 550~750 nm. The absorption spectrum of ZnPc organic semiconductor materials has three absorption peak: 351nm, 618nm, 700nm, the absorption rate respectively is 0.595, 0.436 and 0.336. Therefore, in order to get better illumination characteristics, selection irradiation device under the 351 nm light.

3 DEVICE MEASUREMENT AND RESULT ANALYSIS

To measure the operating characteristics of ZnPc transistor, base-collector bias voltage (V_{bc}) was varied from 0 V to 1 V in 0.2 V steps, emitter-collector bias voltage (V_{ec}) was swept from 0 V to 3 V. The output characteristics measure results of ZnPc organic thin film transistor under 351nm light and darkness are shown in figure 2. I_{ec} is the abbreviation of operating current. The results indicate that the I-V characteristics of ZnPc organic thin film transistor are unsaturated. We can find that the operating current under 351nm light is bigger than the operating current in darkness. This is because electron-hole excltions are generated in ZnPc organic semiconductor layer by light, and then separated by built-in electric field,

electrons flow to base region, the holes flow to collector region, photocurrent is generated as a part of operating current, so operating current increase. For example, when V_{ec}=3V and V_{bc}=1V, operating current increases from17.74μA to 22.66μA.

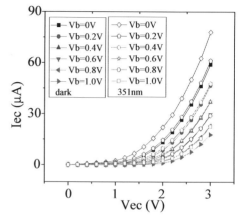

Figure 2. The output characteristics of ZnPc organic thin film transistor under 351nm light and darkness.

When emitter-collector bias voltage is 1V, the relationship between operating current and base-collector bias voltage is shown in figure 3. The curves show that with V_{bc} increasing, Iec gradually decreased and I_{ec} in darkness is much smaller than in 351nm light. The vertical structure thin film photoelectric dynatron can be seen as the two reverse p-n nodes are connected together through the Al gate electrode. When base voltage is 0V, the first p-n junction is in the forward bias state and the second p-n junction is in the reverse bias state. Because of Al film is very thin and the voltage on the second p-n junction is relatively bigger, the hole could be over ZnPc film reached collector to form operating current. With V_{bc} increasing, two p-n junctions are still in the original state, but the voltage on the first p-n junction is reduced. And then the hole number reached collector will be less, operating current will be smaller. Therefore, with the base-collector bias voltage is more and more big, the operating current is more and more small.

When emitter-collector bias voltage is 2V, the relationship between operating current and base-collector bias voltage is shown in figure 4. Comparing Figures 3 and 4 can be seen whether it is under the irradiation of 351nm light or without light, the operating current of Vec=2V is larger than the operating current of Vec=1V. Namely, operating current become larger with increasing emitter- collector bias voltage. This is because when the voltage between the collector and emitter increases, the Schottky barrier height of the first p-n junction has decreased. According to the carriers' movement, the more carrier overcomes the Schottky barrier, the more operating current will be obtained. But current amplification factor is getting smaller. The current amplification factor is defined as the radio of the currents which were obtained in351nm light and darkness, it is able to characterize the device's photoelectric conversion ability. The bigger V_{ec} is, the smaller current amplification factor. When V_{ec}=1V, current amplification factor changes in 7.35-18.18. When V_{ec}=2V, current amplification factor changes in 1.65-2.34.

The energy band diagram of the ZnPc organic thin film transistor is shown in figure 5. The organic materials ZnPc can be regarded as P-type organic semiconductor material, its HOMO level is -5.0eV and LUMO level is -3.4eV. According to the metal semiconductor contact theory, the Al should form good Schottky contact with both sides CuPc and CuPc should form good ohmic contact with Cu and ITO.

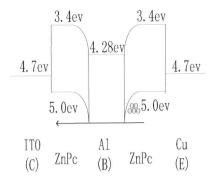

Figure 5. The energy band diagram of ZnPc organic thin film transistor.

4 SUMMARY

In a word, we have successfully prepared ITO/ ZnPc/ Al/ZnPc/Cu organic thin film transistor and measure its output characteristics under 351nm light and dark. The results exhibited excellent current characteristics. When V_{ec}=3V and V_{bc}=1V, operating current increases from17.74μA to 22.66μA. ZnPc material and ITO electrode can improve the performance of the organic thin film transistor. The ZnPc organic thin film transistor was irradiated by 351nm light, the current amplification factor is in the range of 7.35-18.18 when emitter-collector bias voltage is 1V, the current amplification factor is in the range of 1.65-2.34 when emitter-collector bias voltage is 2V. This device has the potential to become visible light detector because of operation current changes are very significant under light irradiation.

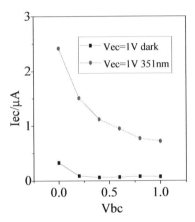

Figure 3. When Vec=1V, the transfer characteristics of ZnPc organic thin film transistor.

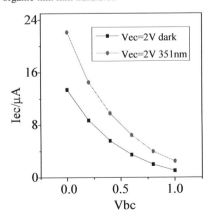

Figure 4. When Vec=2V, the transfer characteristics of ZnPc organic thin film transistor.

REFERENCES

Florian E., K. Hagen & H. Marcus (2004). Organic electronics on paper. *Appl. Phys. Lett. 84,* 2673–2676.
Bialek, B., I.G. Kim& J. Lee (2003). Electr- onic structure of copper phthalo-cyanine monolayer: a first-principles study. *Thin Solid Films 436,* 107–114.
Raoul S., A. Leszek & G. Martin (2004). Improving organic transistor performance with Schottky contacts. *Appl. Phys. Lett. 84,* 1004–1006.
Naoki H., O. Noboru & M. Nakamura (2008). Fabrication of Organic Vertical-Type Field-Effect Transistors Using Polystyrene Spheres as Evaporation Mask. *IPAP Conf. 6,* 158–160.

Electronic Engineering and Information Science – Wang (Ed.)
© 2015 Taylor & Francis Group, London, ISBN: 978-1-138-02772-5

Design and implementation of 8-bit MCU based on triple levels pipeline structure

W.M. Chen
School of Software, Harbin University of Science and Technology, Harbin, China

ABSTRACT: As the most widely used types of microcontrollers, 8-bit microcontroller (MCU) has high requirements on its area, operation speed, function, etc. Therefore, to design a high-performance MCU has been an important issue in the field of integrated circuit design. Based on the determination of instruction format, instruction system and addressing mode, a novel CPU structure is presented in this paper, which is divided into instruction fetch module, coding module and execution module according to the triple-levels pipeline. Then the peripheral module is designed and implemented. With the top-down method, the register transfer level model of MCU is built using Verilog, which is simulated by Quartus II. The results show that the designed MCU meets the requirement.

KEYWORDS: RISC; MCU; Pipeline; Quartus II.

1 INTRODUCTION

As we know, 8-bit MCU has widely applications in the field of integrated circuit design (Sebastian, M. J. 2003). But 8-bit MCU has highly requirements on engineering complexity, area, power dissipation, arithmetic speed, etc. (IEEE. 2000 & Wang, W. Y. 2004). Therefore, to design a high-performance MCU is always an important issue to be considered (Chandel, D. et al 2013). Based on the determination of instruction format, instruction system and addressing mode, the detail CPU structure is presented in this paper, which is divided into fetch module, coding module and execution module according to the triple-levels pipeline. In this paper, the research emphasis is to realize the pipeline parallelism. After accomplishing the design of entire CPU, the peripheral module is designed and implemented which include memory module, digital to analog converter (DAC) serial port control module, timer/counter module and watchdog module. With the top-down method, the register transfer level model of MCU is built using Verilog, which is simulated by Quartus II. Through the analysis of simulation waveform, the designed MCU realizes the corresponding function. The simulation results also show that the designed MCU can execute according to the expected instructions, and the timing sequence and pipeline operations are both correct, that meet requirements in general.

2 DESIGN AND IMPLEMENTATION OF CPU PIPELINE

Central processing unit (CPU) is the core component of MCU, which is in charge of controlling, commanding and scheduling the coordination work of the whole system. Figure 1 shows the integrated structure of the CPU.

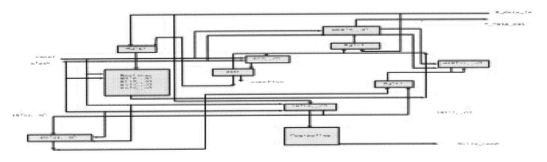

Figure 1. The integrated structure of the CPU.

The main function of the instruction fetch module is to fetch the instructions or operand according to the value in the program counter PC, and to store them in the instruction register IR. Figure 2 shows the circuit diagram of an instruction fetch.

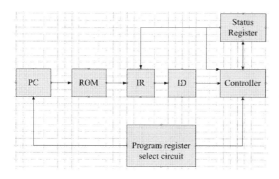

Figure 2.   Circuit diagram of instruction fetch.

The decoding module is decoded by hardwired logic. According to the requirements of decoding sequence, circuit is generated by a control signal which is composed by combinatorial logic unit and storage unit. Figure 3 shows the graphical diagram generated by this module.

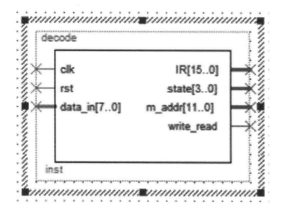

Figure 3.   Graphical diagram of decoding module.

ALU is the core component of microcontroller. As the data processing unit, all logical operations in microcontroller are completed in the ALU. In this paper, ALU includes 16 operations, and the operation code is 4-bit. ALU has 4 rhythms ranged from state [0] to state [3], and is controlled by state machines. Figure 4 shows the graphical diagram generated by this module.

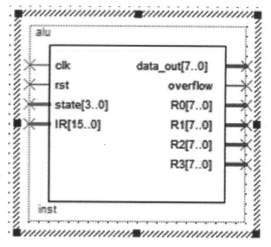

Figure 4.   Graphical diagram of ALU.

According to the flow described above, CPU designed in this paper includes instructions fetch module, decoding module, and AL module, which realize the CPU's triple-level pipeline structure.

3   DESIGN AND IMPLEMENTATION OF MCU PERIPHERAL MODULE

3.1   *Memory module*

The main function of memory module is to store instructions, instruction address and instruction data. The data storage unit in this paper applies Megawizard contained in Quartus II to implement the function. The memory space is 4KB, while the data bus width and address bus width are 8-bit and 12-bit, respectively. The definitions of memory module ports are described in Table 1.

Table 1.   Definition list of memory ports.

Name	Type	Function
WR_DATA	input port	write data signal
WRITE_EN	input port	write enable signal
WRJTE_ADDRS	input port	write address bus
READ_ADDRS	input port	read address bus
CPU_CLOCK	input port	write data effective clock
RD_CLK	input port	read data effective clock
RAM_DOUT	output port data	read port

3.2 Digital to analog converter serial port control module

The MCU designed in this paper extends a serial port control module included in the digital to analog converter DAC7512, which implements the logic control communication between DAC7512 and MCU. Figure 5 shows the interface connection diagram between DAC7512 and MCU.

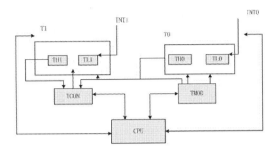

Figure 6.   Logic structure diagram of timer/counter.

3.4 Watchdog module

The watchdog is a timer circuit to prevent programs from endless loop. When the system starts, watchdog automatically counts. If the watchdog is not reset after a certain time, its counter will overflow and break off, and then cause the system to reset. RESET is reset signal and effective during low levels. SYSCLK is a clock signal. When RESET is "0", the counter starts to count; and when RESET is touched, system resets.

4   FUNCTIONAL VERIFICATION AND SIMULATION ANALYSIS

Figure.7 shows the arithmetic of ALU and oscillogram of CPU. From the figure, when CPU state = 0, the mission is instruction fetch. In this state, the instruction code (15) in the MEM output port M_q throws into the high 8-bit of instruction register IR, when the clock signal is jumping. And then the content of IR changes to 1500. When state = 1, it carries out the instruction in IR and throws 5 into R0 while the clock is rising edge; when state = 2, Write_Read = '0'.

Figure.8 shows the simulation results of watchdog. From the Figure.8, when the counter counts to 28, the reset signal of MCU will trigger and restart to count.

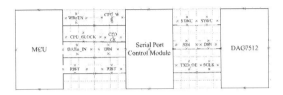

Figure 5.   Interface connection between DAC7512 and MCU.

3.3 Timer/counter module

Timer/counter is the important peripheral module in MCU, which mainly implements the function of timer and event counter. When working as a timer, the timing register adds "1" in each machine cycle. Otherwise, when working as event counter, the register does the addition counting to the external negative jump signal. The rule of sample a signal during each machine cycle is that when sampling to "1" in the previous cycle and to "0" in the afterword cycle, then counter adds "1". After detecting the jump signal, the new count value loads into counter register in the next state of the current cycle. Figure 6 shows the logic structure diagram of timer/counter.

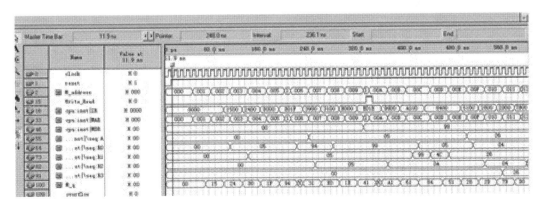

Figure 7.   Arithmetic of ALU and oscillogram of CPU result.

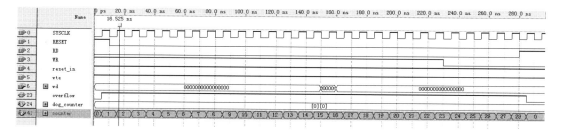

Figure 8. Simulation result of watchdog.

5 CONCLUSIONS

MCU designed in this paper adopts triple-level pipeline parallel structure, which includes instruction fetch module, decoding module and execution module. The peripheral module contains memory module, digital to analog converter serial port control module, timer/counter module and watchdog module. The main function of peripheral module is to monitor whether the MCU is under the normal working state and then to improve the MCU execution speed. In this paper, the designed MCU is simulated by Quartus II. The simulation waveform shows that the MCU realizes the corresponding function, and can execute according to the expected instructions while the timing sequence and pipeline operations are both correct.

REFERENCES

Sebastian, M. J. (2003). Application-Specific Integrated Circuit. *J. Electronic Industry Press.* 97–143.

IEEE. (2000). IEEE stardard hardware description language based on the verilog hardware description language. *IEEE computer society. NY. IEEE std.* 1364–2001.

Wang, W. Y. (2004). The Design of 8-bit MCU Based on Pipeline Structure. *J. North China University Press.* 3–8.

Chandel, D & Kumawat, G & Lahoty, P. (2013). Ease of Multiplication. *J. International Journal of Emerging Technology and Advanced Engineering.* 3, 326–329.

Electronic Engineering and Information Science – Wang (Ed.)
© *2015 Taylor & Francis Group, London, ISBN: 978-1-138-02772-5*

The operating characteristics measurements and analysis of CuPc thin film transistor

Z.Y. Wang, D.X. Wang, Y.S. Zhang & Y.Y. Wang
Key Laboratory of Engineering Dielectrics and Its Application, Ministry of Education, Department of Electronic Science and Technology, College of Applied Science, Harbin University of Science and Technology, Heilongjiang, Harbin, China

L. Wang, J.H. Yin & H. Zhao
Department of Electronic Science and Technology, College of Applied Science, Harbin University of Science and Technology, Heilongjiang, Harbin, China

ABSTRACT: Organic Thin Film Transistors (OTFTs) are fabricated through vacuum evaporation and DC magnetron sputtering technology. The OTFTs have a layered structure Au/CuPc/Al (semi-conductive)/CuPc/Au. The conductive channel length of the submicron order of magnitude of OTFTs is achieved, which depends on the thickness of CuPc active layer. The experimental results demonstrate that this device exhibits the non-saturating characteristics similar to the vacuum triode, has the advantages of a good frequency characteristic and large current density. In the lower operating voltage V_{DS}=3V, V_{GS}=0V, the current density is 0.365mA/cm^2. The drain/source current of OTFTs is controlled by the gate/source voltage, so this OTFT is suitable for forming high-speed switching devices.

KEYWORDS: organic semiconductor materials; CuPc; Schottky barrier; thin film transistor.

1 INTRODUCTION

The research and application of organic thin film transistors (OTFTs) made great progress (Zan & Tsai 2012), OTFTs are expected to become the core technology of a new generation flat-panel display. Compared with an inorganic thin film transistor, the OTFT has more advantages, such as more and newer preparation technology(Yun 2011, Adriyanto & Yang 2013); can be made into a large area display device; a good flexibility, easy to carry; the demand for conditions and purity of the gas is relatively low, can simplify the preparation process, lower costs etc. With the deepening of the research on OTFTs, we found that at present there are still many defects and problems, such as the mobility of most organic semiconductor material is very low (Yang & Cheng 2007), the existing semiconductor energy band theory is established on the basis of inorganic materials, we cannot find reasonable explanation for some phenomena in OTFT. In order to improve the performance of organic thin film transistor, the vertical structure organic thin film transistors (VOTFTs) with a vertical structure is investigated, the measurement results show that many of its performance is better than that OTFT device with a lateral structure. The VOTFT as switching element will be applied to large area active matrix flat display field, such as AMLCDs, AMLEDs, and memorizer, sensor array.

The organic thin film transistor with a vertical channel structure is charge injection voltage-controlled device. The effective channel length can be achieved submicron level, or even smaller (Watanabe et al. 2006). By changing the gate/source voltage, charge injection efficiency can be effectively controlled to achieve the purpose of modulation current between the drain and source electrode. We have reported the VOTFTs using organic semiconductor CuPc as the active layer, semi-conductive Al film as the gate electrode, the conductive channel length depends on the thickness of the active layer. The current-voltage relationship shows non-saturating characteristics. In this paper, we studied the static and dynamic operating characteristics of thin film transistor.

2 EXPERIMENTAL DETAILS

In the device structure Au/CuPc/Al/CuPc/Au of the VOTFT, back-to-back Schottky diodes are formed between Al and the two layer CuPc films. Good ohmic contact is formed between CuPc and Au film. The fabrication process was as follows: firstly, the bottom gold electrode was deposited on the glass substrate using DC magnetron sputtering. Second, the CuPc thin film was formed by vacuum evaporation. Third, semi-conductive Al gate thin film was fabricated. Fourth, the CuPc was deposited by evaporation. Finally, the gold electrode was prepared. During the fabrication, the temperature of glass substrate was kept at room temperature. The CuPc thin films were fabricated under a vacuum of 4×10^{-5}Torr, the source temperature of CuPc was 400°C, the evaporation rate was 0.5Å/s. The effective area of the VOTFT device was 0.025cm². Basic electrical measurements of the static and dynamic characteristics of the VOTFT were performed using a semiconductor characteristic analyzer (Keithley 4200), a current amplifier (Keithley 428), a function signal generator (33220A) and a digital storage oscilloscope (Tektronix 3021B). The structure of the VOTFT is shown in Figure 1. The average thickness of semi-conductive Al thin film, which evaporated on the CuPc was controlled about 10~20nm. The surface morphology of Al film is observed through AFM, AFM image of Al film gate was shown in Figure 2.

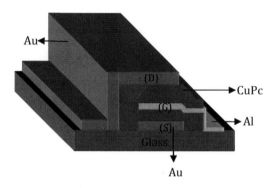

Au

(D)
(G)
(S)

→ CuPc
→ Al

Glass

Au

Figure 1.   Schematic diagram of VOTFT.

Figure 2.   AFM image of the semi-conductive Al thin film on CuPc film.

3   RESULTS AND DISCUSSION

The static characteristics of the VOTFT using CuPc as the active layer is shown in Figure 3. When V_{GS} is a constant, the drain/source voltage V_{DS} changes from -3V to 0V, the absolute value of drain/source current I_{DS} decreases. When the range of V_{DS} is 0 V to about 1V, the I-V characteristics follow an exponential behavior; under the V_{DS} of 1V~3V, I_{DS} increase rapidly in the form of square law. The increasing of drain/source voltage will lead to the depletion layer near the gate electrode becoming narrow, Schottky barrier reduces, more carriers from the source inject into the organic semiconductor layer, tunnel CuPc/Al/CuPc double Schottky barrier gate region, finally arrive the drain electrode, so operation current is formed. The current increases in the form of square law are mainly decided by the space charge limited current mechanism, it is associated with the nature of the CuPc. When the V_{DS} is a constant, the slope of the I_{DS}-V_{DS} curves decreases with increasing of V_{GS}. VOTFT has a conductive channel of the submicron order of magnitude, in the operating voltage of V_{DS}=3V, V_{GS}=0V, can get the larger current density 0.365mA/cm².

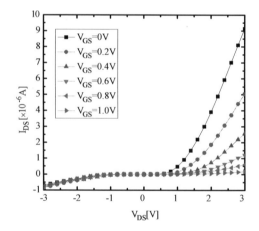

Figure 3.   Measuring results of static characteristics of VOTFT.

The current-voltage relations of the device exhibit the non-saturating characteristics. Whereas in the conventional field effect transistor, the series channel resistance is responsible for negative feedback effects resulting in saturation of DC characteristics. Expression of static characteristic parameters for the VOTFT:

$$g_m = \frac{dI_{DS}}{dV_{GS}}\Big|_{V_{DS}=Const} \qquad (1)$$

$$r_D = \frac{dV_{DS}}{dI_{DS}}\Big|_{V_{GS}=Const} \qquad (2)$$

$$\mu = -\frac{dV_{DS}}{dV_{GS}}\Big|_{I_{DS}=Const} \qquad (3)$$

According to the measurement results, characteristics parameters of VOTFT are as follows: when the gate voltage is 0V, the drain/source voltage is 3V, transconductance $g_m=2.045\times10^{-5}$S, output resistance $r_D=1.94\times10^{5}\Omega$, amplification factor $\mu=3.97$. Figure 4 shows the drain/source current I_{DS} characteristics curves as a function of the gate voltage of V_{GS} when the drain/source voltage V_{DS} is maintained at +3V. The drain/source current I_{DS} decreases with an increase of the forward gate voltage. The main reason is that V_{GS} made the space charge region become wide, the conductive channel narrow, the resistance of carrier transport increases, the current becomes small. Transconductance g_m can be calculated through the I_{DS}-V_{GS}. The current-voltage characteristics between the gate electrode and the source or drain electrodes are shown in Figure 5. In this case, the gate/source current (I_{GS}) increases with the increase of V_{GS}, the relationship between V_{GD} and I_{GD} is also like that. This result is same with that CuPc thin films exhibit P type semiconductor properties. Double Schottky barriers are formed between semi-conductive Al gate and the two side CuPc films, show excellent rectifying effect.

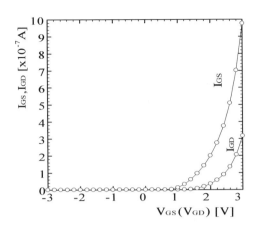

Figure 5.   I-V rectification characteristics of Al gate with both CuPc films.

source electrode, the signal frequency, amplitude and offset of V_{GS} are 1000Hz, 0.1 V and 0.1 V, respectively. When the frequency f is greater than 2.3 kHz, I_{DS} begin to decrease. The phase shift of I_{DS} also varies dramatically from 0 to 90°. The cutoff frequency of VOTFT is 2.3 kHz. Due to the conductive channel length with submicron, even if under the condition of high power, the VOTFT has advantages of fast transient response, small distortion, can make up for the insufficient of a low carrier mobility and high resistivity of organic semiconductor materials, which improve the operation speed of the device.

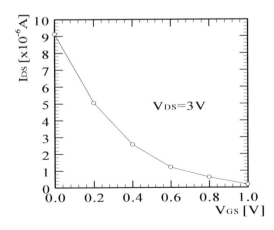

Figure 4.   Transfer characteristic curve of the CuPc OTFT.

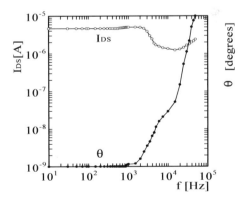

Figure 6.   Characteristic curves of I_{DS} as a of the VOTFT function of frequency f of gate/source voltage V_{GS}.

Measured I_{DS} as a function of frequency f of gate voltage V_{GS} is plotted in Figure 6. An alternating signal of the sine wave is applied to the gate/

The VOTFT is a kind of semiconductor device which uses gate/source voltage to control the majority-carrier injection. CuPc thin film of stable chemical property is used as the active layer, the active layer is not too thin, otherwise the surface state will increase the gap state concentrations of organic material to reduce the carrier mobility. Semi-conductive Al gate is embedded in organic semiconductor materials, it is sandwiched between the drain/source electrode. The vertical channel structure of VOTFT makes the carrier transport channel become short and wide. Therefore, there is no limit of current and voltage of common field effect transistors, VOTFT can withstand the high current, is suitable for manufacturing high power tube. By optimizing the device structure, VOTFT can obtain an excellent performance. The VOTFT can be applied to large scale integrated circuits that require a flexible substrate, such as smart cards, inventory tags, wireless electronic identification tag, commodity anti-theft tags and bar code etc.

4   CONCLUSION

The VOTFT is a kind of semiconductor device which uses gate/source voltage to control the majority-carrier injection. CuPc thin film of stable chemical property is used as the active layer, the active layer is not too thin, otherwise the surface state will increase the gap state concentrations of organic material to reduce the carrier mobility. Semi-conductive Al gate is embedded in organic semiconductor materials, it is sandwiched between the drain/source electrode. The vertical channel structure of VOTFT makes the carrier transport channel become short and wide. Therefore, there is no limit of current and voltage of common field effect transistors, VOTFT can withstand

the high current, is suitable for manufacturing high power tube. By optimizing the device structure, VOTFT can obtain an excellent performance. The VOTFT can be applied to large scale integrated circuits that require a flexible substrate, such as smart cards, inventory tags, wireless electronic identification tag, commodity anti-theft tags and bar code etc.

ACKNOWLEDGMENT

This work was supported by the science and technology research project of the Heilongjiang Province Education Department (11551100).

REFERENCES

Zan H.W. & W.W. Tsai (2012). Pentacene-Based Organic Thin Film Transistors for Ammonia Sensing. *Sensors Journal 12*, 594–601.
Yun C.G. (2011). High-Performance Pentacene Thin-Film Transistors Fabricated by Printing Technology. *Electron Device Letters 32*, 1454–1456.
Adriyanto F. & C.K. Yang (2013). Solution-Processed Barium Zirconate Titanate for Pentacene-Based Thin-Film Transistor and Memory. *Electron Device Letters 34*, 1241–1243.
Yang C.Y. & S.S. Cheng (2007). Pentacene-Based Planar- and Vertical-Type Organic Thin-Film Transistor. *Electron Devices 54*, 1633–1636.
Watanabe Y.K. & I. Hiroyuki (2006). Electrical characteristics of flexible organic static induction transistors under bending conditions. *Applied Physics Letters 89*, 233509- 233509–3.

Electronic Engineering and Information Science – Wang (Ed.)
© 2015 Taylor & Francis Group, London, ISBN: 978-1-138-02772-5

Research on the technology of static type induction heating

H.M. Wang, X.J. Li, H.Sun & P. Xue
Harbin University of Science and Technology

Q. Zhang
Shenzhen Power Supply CO.LTD

ABSTRACT: This paper introduces the dead zone problem of induction heating temperature, we describe this problem from two aspects of the theory and infrared imaging. In the theoretical aspect, we analyzed the distribution of magnetic induction intensity. On the other hand, we researched the infrared direction of the coil in different ways, coil solid ova, coil round and so on. And further put forward the solution. How to avoid it which is the dead zone.

KEYWORDS: Temperature dead; Infrared imaging; Induction heating; Magnetic induction intensity.

1 INTRODUCTION

Induction heating is vaas an emerging technology. Electromagnetic induction heating is mainly relied on alternating current induced magnetic field to realize the purpose of heating component. This article mainly is by an infrared thermal imaging method for induction heating theoretical analysis and practical study. Based on this, we propose a method to solve the dead zone.

2 THEORETICAL DEDUCTION AND ANALYSIS

IH (Induction Heating) technology is referred to as electromagnetic induction heating technology (Anderson et al. 2000). Its theoretic based on Faraday's law in physics. Faraday's law of electromagnetic induction is a specific form of application in real life. Alternating magnetic field around the coil will be formed. If the conductor is in which, this will appear that phenomenon of generate heat in the conductor surface by eddy currents. The distribution is shown by the temperature field distribution and eddy current field coil topology: If the high-frequency induction coil is passing alternating current. Then the flux will change in accordance with the sine law, the electromotive force and flux.

$$e = -\frac{d\Phi}{dt} \cdot (1) \quad \Phi = \Phi_m \sin 2\pi f t \qquad (1)$$

(Φ_m: the amplitude value of Φ; f: the magnetic field frequency) The resistance of the coil ring: (l the loop length, ρ the conductor resistivity, S the loop section area, h the thickness)

$$R = \rho \frac{l}{S} = \rho \frac{2\pi a_r}{d_r h} \qquad (2)$$

$r_2 > r_0$, when the ring is on the outside of C_2, it is the size of the induced eddy currents

$$I = \frac{e}{R} = -\frac{f d_r h}{\rho} \cdot \frac{\Phi_m}{a_r} \cos(2\pi f t) \qquad (3)$$

When the high frequency alternating current frequency is certain, the conductor thickness and resistivity are constants, K_1 is constants, $K1 = -f d r h / r$. If the a_r of the ring radius increases,Then the projection of external magnetic flux direction and internal scheduled flux reference are in the opposite direction (Zhou 2005). The flux by per area will be reduced. Therefore, the fact of that eddy currents will be reduced. According to the above analysis. In the flat conductor coil projection of external, if the center distance is smaller, the eddy current concentration is increasing. $r_2 < r_0$, when the ring is on the inside of C_2, B_{aM} represents amplitude values of the sinusoidal magnetic field.

$$e = -\frac{d}{dt} B_{aM} \pi r_2^2 \sin(2\pi f t) \, d_r \langle\langle a_r, a_r \approx r_2 \quad (4)$$

$$I = \frac{e}{R} = -\frac{\pi f d_r h}{\rho} B_{aM} a_r \cos(2\pi f t) \quad (5)$$

The high-frequency alternating current frequency is determined, conducting plate thickness and resistivity are constant, $K_2 = -\pi f d_r h/\rho$ If $B_{aM} a_r$ and a_r are setting the proportional relationship: If C_3 and C_4 are of two concentric circles, their projection is the internal projection of C_2, r_3 is the radius of C_3 S_3, and S_4 are the projection area of C_3 and C_4, $S_5 = S_3 - S_4$, B_{aM3}, B_{aM4}, B_{aM5} are the induction amplitude. The projection of the induction is the size of the radius, which is a monotone increasing function, $B_{aM4} < B_{aM5}$, So S_4 and S_5 flux on the projected area equal to the flux of S_3 on the projected area.

$$B_{aM3} \pi r_3^2 = B_{aM4} \pi r_4^2 + B_{aM5} \pi (r_3^2 - r_4^2) \quad (6)$$

$$(B_{aM3} - B_{aM4}) r_3 = (B_{aM5} - B_{aM4})\left(\frac{r_3^2 - r_4^2}{r_3}\right) \quad (7)$$

$B_{aM} a_r$ follows a_r as larger and larger. In the induction coil, the farther the distance between the center of projection radius, the eddy current density will be more concentrated. We know that the edge portion of the plate due to the boundary conditions of interference. Eddy current density is relatively weak, we do not avoid that intermediate induction magnetic field intensity is zero, In the case that the initial heating temperature of the dead zone occurs.

3 ANALYSIS OF INFRARED IMAGING TESTING

Infrared thermal imaging refers to the acceptance of infrared radiation being emitted by the test object, signal processing technology will be converted to thermal images of the visible (Shi & Hu 2013). By Stephen - Boltzmann law, for the object, the total radiation energy which is able to know.

$$W = \sigma T^4 \quad (8)$$

W is the total radiation energy of an object; T represents the temperature of the object; σ is a constant $(5.67 \times 10^{-8} \text{W/m}^2\text{K}^4)$. If the gray scale is determined, the equation can be calculated from the temperature of each point. The gray distribution of thermal imaging is the thermal field distribution. Several commonly used coil carries out an infrared image acquisition.

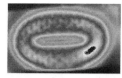

(a) Coil solid ova

(b) Temperaturedistribution

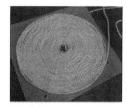
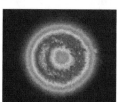

(c) coil round

(d) temperature distribution

(e) square coil

(f) temperature distribution

(g) elliptical hollow

(h) temperature distribution

Figure 1. Coil topology and the corresponding temperature distribution.

Figure 1, In the induction heating process, Edge temperature distribution in plate conductor is relatively weak. We can think about it from the viewpoint of structure. If the entire coil is completely flat under the component induction heating, the magnetic field distribution of the coil is also unable to satisfy the induction heating process for the heating balance requirements. The superposition effect of the magnetic field leads to the center of the coil magnetic induction intensity which is zero, we usually call the 'dead zone' refers to the area of magnetic induction intensity is zero.

4 THREE-DIMENSIONAL SURFACES

We know it from theory verification, conclusion of model simplification is put forward under the condition of negligible boundary conditions. In the plate coupling technology, the boundary conditions can't be ignored. In the flat-type induction heating process, the projection and the distribution of the eddy current are approximated. The temperature of rectangular coil is uniformly distributed, there is not the problem of the center of the hollow. To avoid the center of the coil temperature dead zone, we can't consider uncertain other confounding factors. By the principle of coil, the heating of straight coil can remove the deadzone. Therefore, the three-dimensional surface, the projection area is less than the area of the coil, that it can solve the problem of temperature dead.

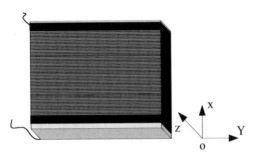

Figure 2. Cube induction coil topology.

In order to determine its magnetic field distribution is uniform or not (Figure 2). The number of turns increases to 2–3 times. The excitation current is 2.51A, operating voltage of the integrated Hall devices is 5.03V, 0.5mm of diameter, 2.50059V initial conditions for the next output voltage. 6mm: vertical height from top to bottom by equidistant intervals along the axis of X, direction of the border around the axis of X (1cm) to measure horizontal component value.

Table 1. The relationship of hall output voltage and X axis spacing.

X [cm]	Uo[V] IReverse	Uo[V] IClockwise	X [cm]	Uo [V] IReverse	Uo [V] IClockwise
−3	2.50045	2.50395	7	2.50648	2.49858
−2	2.50079	2.50359	8	2.50648	2.49865
−1	2.50162	2.50279	9	2.50646	2.49864
0	2.50278	2.50154	10	2.50648	2.49866
1	2.50423	2.50032	11	2.50649	2.49862
2	2.50516	2.49950	12	2.50645	2.49872
3	2.50585	2.49898	13	2.50638	2.49881
4	2.50611	2.49877	14	2.50595	2.49934
5	2.50632	2.49864	15	2.50533	2.50013
6	2.50646	2.49858	16	2.50426	2.50118

A cube surface current direction is not consistent with the axis of Y, it is the forward current. Table 1 envelope curve of the static magnetic field region is the value of an effective induction coil on the conductor plane projection. And it can be effectively simulated the dynamic state of the alternating magnetic field and resolve issues, such as equal distribution, this is the problem about effective distribution of the eddy current field. When the error is very small, To effectively meet the flat conductor are constant heating problems. And also to avoid the other coil topology of the dead zone in the center of the component.

5 CONCLUSION

Induction heating technology has a great development around home and abroad in recent years. In the process of induction heating, the dead zone of the induction heating can't be avoided. In this paper, induction heating technology has been proven temperature dead zone by theoretical analysis and infrared imaging. The test puts forward the pancake coil, which can solve the temperature problem of the dead zone and provide a theoretical basis for further studying on the static plate type of heating technology.

ACKNOWLEDGMENT

In the process of writing, thanks for teachers' guidance and help. I also owe my sincere gratitude to my friends and my classmates who gave me their help and time in listening to me and helping me work out my problems.

REFERENCES

Anderson J.O., T. Helander & L. Hoglund et al. (2000). Thermo-calc and DICTRA, computational tools for materials science. *Journey of Phase Equilibria 21(3)*, 269–280.

Zhou K.S. (2005). Test method for radiation of induction cooker. Conference proceedings of seventh international conference on electronic measurement &*indtruments, 1(1)*, 174–176.

Shi Q. & Y. Hu (2013). Application of infrared thermal imaging technology in oil refining industry. 30(1), 56–58. In Chinese.

Analysis of static operating characteristics and potential distribution of the CuPc-OSIT using finite element method

Y. Yuan, D.X. Wang, Y. Zhang & X.Y. Cui

Key Laboratory of Engineering Dielectrics and Its Application, Department of Electronic Science and Technology, College of Applied Science, Harbin University of Science and Technology

ABSTRACT: Organic Static Induction Transistors (OSITs) sample can be carried out by using vacuum coating method. According to the actual sample, choose appropriate structural parameters to fabricate a physical model. When a gate bias voltage changed, selecting the finite element method to achieve the corresponding potential in the conductive channel of OSIT. Then to analyze and solve the gate bias voltage of CuPc-OSIT affects current-voltage and the static operating characteristics.

KEYWORDS: organic static induction transistor; finite element method; potential distribution.

1 INTRODUCTION

In recent years, OSITs has developed rapidly as its properties different from Inorganic Thin Film Transistors (Jia, Wang & Zhang 2013, Wang et al. 2012, Wu & Yang 2013). Compared with the existing transistors and tubes, OSIT is a vertical-channel device and its structures are same to the junction field effect transistor. It adopts the pentacene as the active layer shows the best performance (Wan et al. 2011, Rahimi, Ronak & Korakakis 2010, Lohani et al. 2008). However, theoretical articles related to OSIT research so far are little. In fact, at the procession of its practical application, there are still a lot of work needs to be done.

There is a potential barrier in the channel of OSIT, so current voltage characteristics perform unsaturated. It works via gate voltage changes the barrier height to modulate channel current. Based on the CuPc-OSIT sample has been achieved, according to the semiconductor physics and electromagnetic theory and by the finite element method, this letter presents the gate bias voltage has a great influence on the current characteristics of OSIT.

2 THE STRUCTURE OF OSIT AND SAMPLE MAKING

Fabricating a thin film on the glass substrate by adopting a vacuum deposition, the structure of the device is composed of Au / CuPc / Al / CuPc / Au five layers, as shown in Figure 1. The P type copper phthalocyanine (CuPc) as the active layer of organic semiconductor

materials. The source and the drain electrode are made of Au, and the gate electrode is Al. CuPc and Al will form a Schottky barrier, but source electrode and drain electrode will form Ohmic contact.

Sample production procession: (1) Clean glass substrate, (2) Vacuum deposition. To avoid the influence of light exposure, the sample is placed in a sealed metal box. During the process of preparing, CuPc evaporation temperature at 420°C at a base pressure of about 10^{-5}Pa, and the substrate temperature is room temperature. CuPc film thickness between upper and lower layers is about 200nm, the evaporation rate is 0.05 nm/s, and effective area of sample is 0.025 cm².

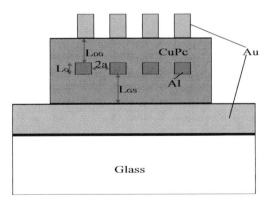

Figure 1. Structure sketch map of OSIT L_G is the length of the gate, L_{DG} is the distance between the drain and the source, L_{GS} is the distance between the gate and the source, 2a is the width of conduct channel.

3 THE EXPERIMENTAL TEST RESULTS AND ANALYSIS

By testing I-V characteristics for Al gate electrode and the Au electrode on both sides, make sure to form a Schottky contact between Al electrode and CuPc evaporated film interface. Then making gate bias voltage (V_{GS}) from 0V to 0.8 V and drain bias voltage (V_{DS}) is from −3V to 0V, all of them with the stride for 0.2V. Relations between V_{DS} and current between the drain and the source (I_{DS}) has been tested and shown in Figure 2, it performs unsaturated and same to SIT characteristics.

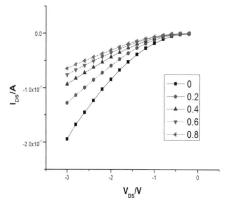

Figure 2. The actual measured static I-V characteristic curve of CuPc-OSIT at the voltage of the gate is from 0 to 0.8V, 0.2 in step and bias V_{DS} from 0 to 3V.

OSIT can still adopt an inorganic SIT theory to describe the expression referred to Equation (1). In the expression $V^*_G =\eta(V_{GS} +V_{DS}/\mu)$, where V^*_G represents the effectiveness of the gate voltage, I_0 stands for a constant, K is Boltzmann's constant, T is the thermodynamic temperature, W_G is the distance between the drain and the source, W is the distance between the gate and the drain.

$$I_{DS} = I_0 \exp(qV^*_G / KT) \qquad (1)$$

The following numerical solving and analysis about the potential distribution in the conductive channel of OSIT by using the finite element method, according to the structure of CuPc-OSIT achieved (Figure 1). OSIT with the structure of vertical conductive channel and multi-gate is composed of many repeating single channels. Here only select one channel as the research object and establish a three-dimensional physical model, as shown in Figure 3. To further simplify the problem, assume that the device has the same electrical characteristics in Z direction.

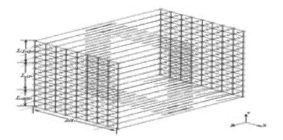

Figure 3. Three dimension model of OSIT.

Potential function inside OSIT can be described by the boundary conditions (2) and (3) and Equation (4) of Poisson equations.

$$\Gamma_1 : \varphi = \phi(x, y) \qquad (2)$$

$$\Gamma_2 : \frac{\partial \varphi}{\partial n} = \upsilon(x, y) \qquad (3)$$

$$\nabla^2 \varphi = \frac{\partial^2 \varphi}{\partial x^2} + \frac{\partial^2 \varphi}{\partial y^2} = -\frac{\rho}{\varepsilon} \ (x, y) \epsilon \Omega \qquad (4)$$

φ as the potential function in the internal model Ω, Γ_1 as the first boundary condition, which is the source grounded, $V_{DS}= -3V$ and $V_{GS}=1V$. Γ_2 as the second boundary condition, it removes the rest of the first kind of boundary edge in module boundary, it is the derivative along the outer normal direction, which fixed as derivative along the direction of the outward normal is 0. Further, the entire area of the model is built to meet the electrical neutrality condition, and the total charge density ρ is 0. Model structure parameter is set to L_{DG}=200nm, L_G = 300nm, L_{GS} = 200nm, 2a=400nm, ε=4. L_{DS} is the distance between the drain and the source, L_G is the length of the gate, L_{GS} is the distance between the gate and the source, 2a is the width of conduct channel, and ε is permittivity of OSIT. A three-dimensional potential distribution contours can be obtained by simulation and analysis, as shown in Figure 4. In order to facilitate, analyze, selecting its X-Y-sectional potential contours as described object (Figure 5). It can be seen from Figure 5, CuPc-OSIT potential distribution channel shows a saddle shape, which is similar to the inorganic SIT.

As can be seen from Equation (1), the gate bias is a major important factor affecting the current between the drain and the source. So using the finite element method to analyze and we can get different potential contours when V_{GS} as 0.2V, 0.4V, 0.6V, and 0.8V, as shown in Figure 6.

From Figure 6a–d, when drain bias voltage keeps unchanged, as gate bias voltage increasing, the

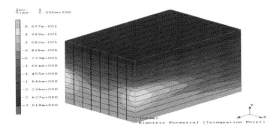

Figure 4.   Channel potential distribution of OSIT at the bias voltage of $V_{DS} = -3V$ and $V_{GS} = 1V$.

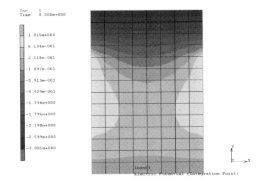

Figure 5.   Channel potential distribution of OSIT X-Y Section. Structure parameter of OSIT is L_{DG}=200nm, L_G=300nm, L_{GS}=200nm, V_{DS}=−3V, V_{GS}=1V, $\varepsilon = 4$.

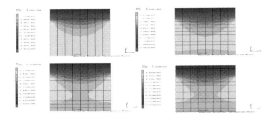

Figure 6.   Structure parameter of OSIT is L_{DG}=200nm, L_G=300nm, L_{GS}=200nm, 2a=400nm, V_{DS}= −3V ε=4. Channel potential distribution of OSIT X-Y Section when V_{GS} is changing at (a) 0.2(b) 0.4(c) 0.6 and (d) 0.8V.

channel potential barrier and the saddle point of the barrier is also increased. The channel potential barrier increasing leads to the number of carries crossed the barrier from the source to the drain decrease, that is to say, the current of channel decreased. The results meet to the actual measured C-V characteristic curve of CuPc-OSIT in Figure 2. In addition, when the gate voltage increasing the potential contour distribution of the channel can be observed becomes denser, it is also the voltage gradient becomes large. This shows

that the Schottky barrier built-in field increased with the gate bias voltage increasing.

Therefore, the operating mechanism of the device is the number of carrier flowing from the source to the drain is decided by the change of gaps of the gate electrode depletion layer region. It is same to the channel barrier height affect the number of carriers flowing from the source to the drain. With the high gate bias, the channel potential barrier is the harder for carriers (hole) to cross and smaller the current between the source and drain is. By changing gate bias voltage, the height of the channel potential can be changed, and the current between the source and the drain can be controlled. So CuPc-OSIT as a device can be controlled by voltage.

4   CONCLUSIONS

The finite element method has been adopted to establish CuPc-OSIT sample in three dimensional physical model. When the voltage of the gate changed, by resolving and analyzing the conductive channel potential distribution of CuPc-OSIT, we can get conclusions meet to actual measured characteristics. When drain bias voltage unchanged, the current between the source and the drain decrease while the voltage of the gate increase. The gate bias voltage control the source-drain current, which is also the principle of carries worked in the conduction channel.

REFERENCES

Jia P.F., D.X. Wang & H. Zhang (2013). The fabrication and operating mechanism analysis of CuPc organic thin film transistor. *ICMIC*.1, 173–176.
Wang D.X., X.L. Wang & C.H. Wang (2012). Fabrication and characteristics of sub-micrometer vertical type organic semiconductor CuPc thin film transistor. *ICPADM*. 85, 1–4.
Wu B.M., D.X. Wang & Y. Yang (2013). Device operation of organic semiconductor CuPc thin film transistor. *ICMIC*.1, 206–208.
Wang D.X., Y. Wang, B.B. Jiao & H. Zhao (2011). The fabrication and characteristics of vertical organic thin-film transistor. *CSQRWC*. 1.256–258.
Rahimi, Ronak & D. Korakakis (2010). Organic thin film transistors based on pentacene and PTCDI as the active layer and LiF as the insulating layer. *Materials Research Society Symposium Proceedings*. 1197. 59–64.
Lohani, Saho & Kumar (2008). Electrical characteristics of top contact pentacene organic thin film transistors with SiO$_2$ and poly as gate dielectrics. *Journal of Physics* 71, 579–589.

Electronic Engineering and Information Science – Wang (Ed.)
© 2015 Taylor & Francis Group, London, ISBN: 978-1-138-02772-5

Design on virtual experiment system for communication principle based on LabVIEW

H.Y. Wang, M.Z. Liu, Z.M. Chen & J.J. Qiao
The Higher Educational Key Laboratory for Measuring & Control Technology and Instrumentations of Heilongjiang Province, School of Measurement and Control Technology and Communications Engineering, Harbin University of Science and Technology, Heilongjiang, China

ABSTRACT: With the development of software engineering, the virtual instrument shows its advantages on the intelligent processing and operations. In this paper, the virtual experiment system is designed and established, based on the LabVIEW. This system makes good use of the graphic language and rich functions in the LabVIEW to design the experiments of communication principle and relevant virtual instrument. The framework, design method and operate of this system are discussed in this paper. This system has proven to simulate the process of those experiments, by running the experiment in it. The results of the experiments are correct and the operation is easy. So the experiments can be done without the hardware. The software is used to simulate the experiment in the designing of this system, which shows the practical applicability and broad prospects of virtual instrument technology.

KEYWORDS: LabVIEW; Communication principle; Virtual experiment; Virtual instrument.

1 INTRODUCTION

Communication principle (CP) course is an important professional basic course for the students majored in communication engineering. How well does the students' mastery this course will directly influence the following course that they are going to learn in the future. The communication principle course includes two parts of theory and practice, in which experiment curriculum has played an important role in developing student practical, application and integration abilities. Traditional experiment teaching of CP is confined by space and time. The virtual experiment teaching system (VETS) based on LabVIEW has broken through these limitations, and the students can do experiments expediently with the VETS. That can help students to complete verification, applicability and comprehensive engineering experiment, and it can also help students to build an integrity knowledge hierarchy of CP. At the same time, comprehensive abilities of students will be improved.

2 THE FRAMEWORK OF THE VETS

This VETS of CP is composed of three parts, which includes virtual instruments, simulation experiment of CP and assistant network module. Virtual instruments are composed of several modules which simulates the working principle of related instruments. These modules have been constructed by programmed and linked by LabVIEW. Simulation experiments of CP are programmed by using the function library and graphical programming language provided by LabVIEW. Those experiments have the same function as the experiment box. The figure 1 is the framework of VETS of CP.

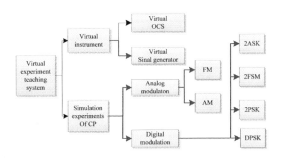

Figure 1. The framework of the VETS.

In this paper, the design of the VETS uses virtual experimental instruments, the conventional communication principle experiments and LabVIEW as the development platform. This teaching system provided a kind of convenient way not only for teachers to do experimental demonstration, but also for students to carry on the communication principle between related experiments. The design of this platform mainly includes three aspects, named respectively: 1) virtual instrument; 2) virtual experiment; 3) packaging and networking. In this paper, the design methods and processes of the first two parts will be introduced in detail as follows.

3 THE DESIGN AND REALIZATION OF THE VETS

3.1 The design idea of the teaching system

By analyzing the experiments of communication principle and the related instrument, the proposed teaching system mainly adopts LabVIEW as a development tool, and each module is connected by the LabVIEW graphical programming language, and then the modification of the front panel has also been beautified by it. Finally, a complete experiment platform will be structured by packaging and debugging.

3.2 The design of the virtual signal generator

3.2.1 Functional requirements of virtual signal generator

1 Can produce the basic function of different frequency ranges, such as sine wave, triangular wave, square wave, sawtooth wave and so on.

2 Can use the knobs on the front panel or input concrete numerical data by data input controls to adjust the parameters such as signal's frequency, amplitude, offset and phase position.

3 Sampling frequency and sampling points could be controlled through the data input controls.

3.2.2 Design on the front panel of the virtual signal generator

The front panel is a graphical user interface with switches, numerical inputs and output control buttons, which is used to simulate the real instrument panel. A method of design is to use the program block diagram code to control the front panel. The working state of the signal generator can be controlled by controlling shift knob on the front panel. The output waveform can be controlled by waveform selection button. Signal frequency, phase position, amplitude and DC offset of the signal output waveform can be adjusted by the knobs on the panel. At the same time, horizontal sliding bar can be used to fine tuning variate so the correct output waveform will be getting. The front panel layout diagram of virtual signal generator has been shown in figure 2.

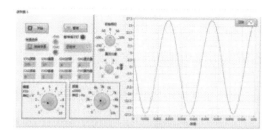

Figure 2.　The front panel of virtual signal generator.

3.2.3 Design on function program block diagram of the virtual signal generator

Program function block diagram is a kind of source program that can realize front panel device functions. The program chart includes input and output connection terminals on the front panel and it also contains functions, structure and connection. The input and output of the front panel button function can be controlled by editing of the source program. When writing a program, icons are usually used to express functions and the interconnecting links between input and output controls and display controls are used to express the direction of the program, which is the data flow. When using LabVIEW to write programs, creating a program block diagram and connecting the control process is the process of writing code. The block diagram of the application function of the virtual signal generator has been shown in figure 3.

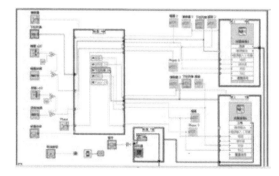

Figure 3.　The function program diagram of signal generator.

The total program block diagram of the virtual teaching system is included in a *while* cycle, at the same time, there are a number of *case* structure to select and connect the data stream within the program. The twin channel signal generator has been provided by a set of input controls which has been controlled in the *while* loop, and then using *case* selection structure to make single control output convert to double ones. And using a Boolean variable to control the output signal direction. The generating simulation signals have been controlled by a *case* condition structure. Each signal has been controlled by a condition variable which was provided by the previous data selection. The output of waveform data needs a real-time pause, so in the final output stage of the waveform adds pause and continue branch structure.

4 DESIGN INSTANCE FOR CP VIRTUAL EXPERIMENT

After the construction of virtual instrument, the virtual experiment VI should be programmed in order to achieve the functions of the VETS. Different experiments of CP have different characteristics, so every experiment needs a customized VI to process signals in different ways. Finally, the VIs of different experiments will be packaged to a project, and interface will be designed and built. Executable files will be generated if all the VI items run well. In this paper, digital modulation and demodulation VI will be taken as an example to illustrate the development process for the virtual experiment specifically.

The virtual digital modulation and demodulation experiment includes 2ASK, 2FSK, 2PSK, DPSK modulation and demodulation and so on. In these virtual experiments, modulation and demodulation process will be simulated clearly, and at the same time, its waveform and frequency spectrum will be displayed clearly.

The input form of digital modulation experiment VI is 8-bit binary number, and set the parameters such as carrier frequency, amplitude and initial phase to be constant. Firstly, Input variables can be composed by the Boolean input control. Secondly, Boolean variables are converted into integer variables with the convert function provided by LabVIEW. And then eight integer variables are bound to an array of the built-in function of LabVIEW. Finally, the converted data were inputted into the modulation module which contains several *for cycle* structure. The cores of those *for* structure have been programmed by formula nodes, which were provided by LabVIEW.

One of those *for* cycle program panel has been shown in figure 4. In which, three kinds of modulations have been achieved by these formula node because of the 2-ASK, 2-FSK, 2-PSK using the same data source.

After the modulation, demodulation operation should also be needed. The process of demodulation includes band-pass filter, multiplier, and low-pass filter. By displaying controls, the demodulation signal waveform for 2ASK after has clearly been shown in figure 5. By using of formula node, it can save a lot of data operation, compared with programming with the graphic language. It also can make the program panel simple, decrease the comments marked and improve the portability.

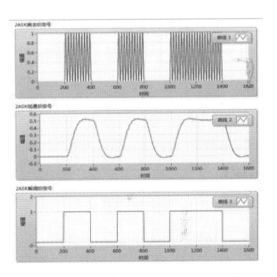

Figure 5. The waveform in the processing of ASK demodulation.

5 CONCLUSIONS

The development and use of virtual experiment teaching system has become an important way for the experiment teaching in colleges. Its utility, flexibility and convenience make the virtual experiment play an important role in the professional foundation course, such as CP signal and system and so on. In this paper, LabVIEW is considered as the development platform to explore the methods to build virtual experiment teaching system. The use of the virtual teaching system, especially combined with the network will help the students' autonomous learning and constantly improve the innovation ability a lot.

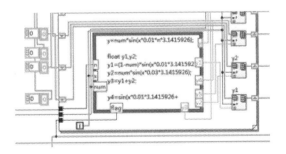

Figure 4. Program panel of digital modulation.

ACKNOWLEDGMENT

This paper is supported by the College Students' innovative venture fund of Harbin University of science and technology and Education teaching research project fund of Harbin University of science and technology (Z201300004).

REFERENCES

John, G. P. and Masoud, S. 2004. *Fundamentals of Communication Systems*. London: Prentice Hall.
L.P. Cheng. 2012. The Development of Virtual Experiment Teaching System Based on LabVIEW. *11-2739/N* 20:84–85.
G.S. Chen & T. Zhang et al. 2012. *Proficient in the Labview*. Beijing: Publishing House of Electronics Industry.
J. Ren & H.Y. Zhang. 2014. Reforming experimental teaching for Communication Principle course by using virtual simulation experiments. *Experimental Technology and Management* 31(1): 95–97, 104.
Y. Wu & X.X. Rong. 2013. Design of Virtual Signal Generator Based on LabVIEW9.0. *Computer Technology and Development* 23(1): 181–184.
F.L. Fan & J.X. Han. 2006. The design of virtual signal generator based on LabVIEW. *Journal of Zhongyuan Institute of Technology* 17(4):7–28.
M.G. Zeng & W.C. Lai et al. 2013. Design of virtual digital oscilloscope based on LabVIEW and sound card. *Electronic Design Engineering* 21 (3): 122–125.
Z.L. Yang & X.G. Hou. 2013. Research of "Digital Signal Processing" experiment based on LabVIEW. In *ICETMS* 2013, Nanjing, China, 551–554.
L. Yan. 2011. LabVIEW Experiment in the Teaching of Circuit Analysis Application, In *ICITIS*2011, Hangzhou, China, 434–438.
J. Yan & R. Liu et al. 2009. LabVIEW-based auto-timing counts virtual instrument system with ORTEC 974 Counter/Timer. *Nuclear Science and Techniques*. 20:307–311.

Electronic Engineering and Information Science – Wang (Ed.)
© 2015 Taylor & Francis Group, London, ISBN: 978-1-138-02772-5

Analysis of the working mechanism of organic static induction transistor

Y.H. Wu, D.X. Wang, X.C. Liu, X. Li, M.M. Li & X.X. Dong
Department of Electronic Science and Technology, College of Applied Science, Harbin University of Science and Technology, Heilongjiang Haribin, China

ABSTRACT: Based on OSIT samples prepared successfully, the potential distribution in the conduction channel of OSIT is simulated by establishing the physical model and using finite element method and the mathematical software MATLAB. In this paper, the calculation results show that, when there are diffusion and drift currents at the same time in the conduction channel and the bias between source and drain (V_{DS}) is small, 1D vertical and horizontal potential distributions meet the relationship of the second power near potential barrier in the conduction channel; and then it proved that the leakage current meets exponential distribution. The OSIT voltage amplification rate approximately meets the linear decreasing relationship with channel width, and a linear increasing relationship with the distance between the drain and gate.

KEYWORDS: electrostatic induction transistor, simulation, MATLAB, organic semiconductor.

1 INTRODUCTION

The static induction transistor is a kind of vertical structure channel device similar to junction type field effect transistor, with the advantages of high speed, high voltage, high power, low distortion and low loss, etc. In addition, SIT has the following advantages: first, SIT has the excellent exponential function characteristics which are better than all devices of the vacuum tube to semiconductor in the same era. It has low noise and it can zoom at very high frequencies. Second, another feature of SIT is further reducing waveform distortion (Li et al. 2008, Lv 2009). Therefore, SIT has been used for the final output part of the radio broadcast. SIT is widely used in various fields, such as high-power electric equipment used in high frequency induction heating, the radio transmitter in high frequency applications, etc.

At present, the static induction devices are mostly made of silicon materials, however, integration level of inorganic semiconductor device is increasing to physical processing limit, so in recent years, organic semiconductor device is widely studied. Organic semiconductor components are expected to make organic electronic components with excellent characteristics beyond the inorganic electronic components in some areas for its advantages of large area, low cost, ample material selectivity, etc. It laid the foundation of the development of new electronic devices and practical application. Organic static induction transistor (OSIT) has been prepared successfully. The OSIT samples using the P type organic semiconductor

materials-copper phthalocyanine (CuPc) is a Schottky gate structure. (Gao et al. 2005, Wang 2008)The conductive channel of the sample is shortened compared with the organic MOSTET in general. In this paper, based on OSIT samples successfully prepared and the existing theory of SIT, the physical model is established using the finite element method and the mathematical software MATLAB, the potential distribution in the conduction channel of OSIT is analyzed and simulation calculated, and the relationship between the OSIT amplification rate and its structure parameters μ is analyzed.

2 ANALYSIS OF POTENTIAL DISTRIBUTION IN THE CHANNEL OF OSIT

To take SIT on channel P tube as an example, when a small electric field exists near a barrier, namely V_{DS} is small. Supposing the leakage current in the channel is composed of the drift current and diffusion current, then the leakage current density can be expressed approximately as Eq.(1)and Eq.(2):

$$J_p = qp(x)\mu_p\frac{d\varphi(x)}{dx} - qD_p\frac{dp(x)}{dx} \tag{1}$$

After finishing

$$J_p = \frac{qD_pN_s}{\int_{x_1}^{x_2}\exp(-\frac{\varphi(x)}{V_T})dx} \tag{2}$$

Where $D_T=\mu_p V_T$ and $V_T=KT/q$, Ns is the source carrier concentration.

If it is assumed that the potential in the SIT channel is distributed as parabolic type and it approximately meets the second power relationship; The channel's equation of potential distribution is shown as Eq. (3) and Eq. (4):

$$\varphi(y) = \phi[1-(2\frac{y}{L}-1)^2] \tag{3}$$

$$\varphi(x) = -\phi[1-(2\frac{x}{W}-1)^2] \tag{4}$$

The equations of leakage current can be further derived as Eq.(5):

$$I_D = qD_p N_s Z\frac{W}{L}\exp(-\frac{\phi}{V_T}) \tag{5}$$

Where ϕ is the barrier height (relative to the potential of the source), Ns is the concentration of charge carrier (holes) in the source area, W/L describes the shape of saddle shaped barrier, Z is the length of the source.

Based on the above theoretical analysis, in order to simplify the resolution, the electrical characteristics in Z direction of the devices are assumed under the same conditions, the two-dimensional transverse profile field is as the research object. In Figure 1 of 2D model established, L_{DG} is the distance from drain to gate, L_G is the gate length, L_{GS} is the distance from gate to source, 2a is the width of conduction channel.

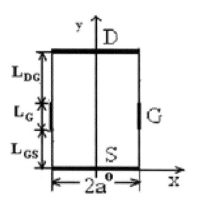

Figure 1. A planar graph of two-dimensional model of OSIT.

The structural parameters are set that, L_{DG}=500nm, L_G=100nm, L_{GS}=100nm, and 2a=400nm. The first boundary condition is set that the source electrode is connected to the ground, V_{DS}= −3V, V_{GS}=2V.

Using MATLAB simulation, the 2D potential conduction channel of OSIT distribution figure, is shown in Figure 2.

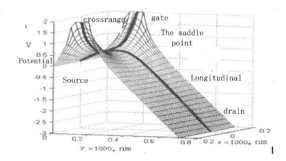

Figure 2. The two-dimensional potential distribution in the OSIT conduction channel.

From Figure 2, the potential near the conduction channel longitudinal center line that is greater than zero points are extracted and then draw into a line data1, as shown in Figure 3. The quadratic curve is gotten from the data1 for the quadratic fitting. And quadratic function is shown as Eq.(6):

$$\varphi(y) = -23y^2 + 6.6y - 0.029 \tag{6}$$

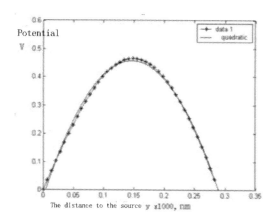

Figure 3. The potential distribution diagram of channel barrier near the longitudinal center line.

From Figure 3, L=287.1nm, the potential (barrier height ϕ) on the saddle point V_m=0.4542V.

Reorganizing the Eq.(6) we can get

$$\varphi(y) = \phi[1-(\frac{2.04}{L}y-1)^2]-0.029 \tag{7}$$

From Figure 3, extracting the points pass the saddle point in the conduction channel, we can draw into a line data1, as shown in Figure 4. The quadratic curve is gotten from the data1 for the quadratic fitting. And quadratic function is shown as Eq.(8):

$$\varphi(x) = 30x^2 - 0.0031x + 0.45 \tag{8}$$

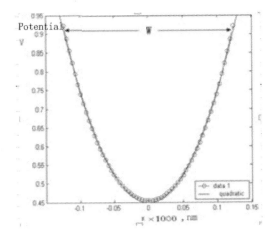

Figure 4. The transverse potential diagram near channel barrier.

From Figure 4, W=243.2nm, the electric potential (ϕ) on the saddle point V_m=0.4542V.

The Eq. (9) can be gotten from transforming the horizontal and vertical coordinates.

$$\varphi(x) = -\phi[1-(\frac{1.98}{W}x-1)^2]-0.0148 \qquad (9)$$

From the above figures and the equations we can explain that under the condition that V_{DS} is relatively small, one dimensional potential distribution of SIT and OSIT meets the second power relationship in the conduction channel of vertical and horizontal directions. And under the condition that V_{DS} is small, the leakage current meets exponential relationship.

3 ANALYSIS OF THE RELATIONSHIP BETWEEN THE VOLTAGE AMPLIFICATION AND STRUCTURAL PARAMETERS OF OSIT

Through Eq.(7) we can know the voltage amplification of SIT meets Eq.(10) and Eq.(11).

$$\mu = \frac{\partial V_{DS}}{\partial V_{GS}} \approx \eta\frac{L_{DG}}{L_{GS}} \qquad (10)$$

$$\eta = \frac{\partial V_m}{\partial V_{GS}} \qquad (11)$$

Where the voltage is Vm in the saddle point. Based on the following key points we analyze the relationship between η and 2a and L_{DG}.

3.1 Analysis of the relationship between η and channel width 2a

We set that L_{GS}=100nm, L_{DG}=500nm, L_G=700nm, V_{DS}=−3V, 2a =700nm to 1000nm, step is 100nm; when

the 2a is set to a certain value, V_{GS}=0.6V to1.4V,step is 0.2V,by simulation calculation we can obtain the height of barrier Vm, then the linear functional relationship between Vm and V_{GS} can be obtained. Then Eq.(11)can be obtained. The relationship between η and 2a, is shown in the Figure 5.

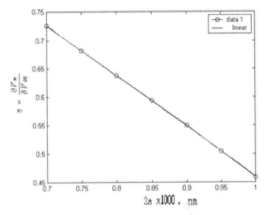

Figure 5. The relationship between η and 2a.

The curve data1 in the Figure 6 is connected from the actual points simulation calculated. The line linear is fitted out from linear curve data1. The two curves fit with each other. The following function fitting is as Eq.(12):

$$\eta = -0.89x + 1.3 \qquad x \in [0.7, 1.0] \qquad (12)$$

Where x is the channel width 2a,(unit: μm).From Figure 5 and Eq.(10), the linear decreasing relationship between η and the width of the channel 2a can be obtained. Then it can be inferred that, when the other structural parameters remain unchanged, the relationship between OSIT voltage amplification μ and channel width 2a is linear decreasing.

3.2 Analysis of the relationship between η and the distance from the drain to the gate L_{DG}

We set that L_{GS}=100nm, 2a=1000nm, L_G=500nm, V_{DS}=−3V, L_{DG}=500nm to 800nm, step is 50nm; when the L_{DG} is set to a certain value, V_{GS}=1.0V to1.8V, step is 0.2V, The relationship between η and L_{DG}, is shown in the Figure 6.

The curve data1 in the Figure 6 is connected from the actual points simulation calculated. The line linear is fitted out from linear curve data. The two curves fit with each other. The following function fitting is as Eq. (13):

$$\eta = 0.36x + 0.14 \qquad x \in [0.5, 0.8] \qquad (13)$$

55

Where x is L_{DG}(unit: μm). From Figure 6 and Eq.11, the linear increasing relationship between η and L_{DG} can be obtained. When the other structural parameters remain unchanged, the relationship between OSIT voltage amplification rate μ and L_{DG} is linearly increasing.

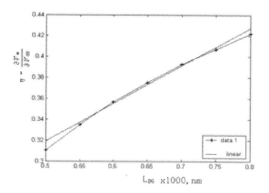

Figure 6.   The relationship between η and L_{DG}.

4   SUMMARY

In the conditions that the V_{DS} is small, the leakage current distribution of SIT is an exponential function. It has confirmed in the experiment. Based on OSIT successful samples, the physical model was established.

Using the finite element method and the mathematical software MATLAB, the potential distribution in the conduction channel of OSIT is simulated.

The calculation results show that, when diffusion and drift currents are all in the conduction channel, and V_{DS} is small, vertical and horizontal one-dimensional potential distribution near potential barrier meets the second power relationship in the conduction channel; and it was also suggested that the distribution of leakage current is exponential relationship. The OSIT voltage amplification and channel width approximately meet the decreasing linear relationship, and the L_{DG} approximately meets a linear increasing relationship. At the same time, it can be seen easily that the OSIT characteristics get a lot to do with the structure parameters.

REFERENCES

Li H.S & Y.N. Li (2008). Performance improvement for solution-processed high-mobility ZnO thin-film transistors. *J. Appl. Phys 41*, 125102–125105.
Gao H,H. & D.Y. Li (2005). ZnO Schottky barrier UV detector. *Chinese Journal of luminescence 26(1)*, 135–138.
Wang J.Z, E. Elangovan& P.B. Nuno (2008). Co-doping of aluminum and gallium with nitrogen in ZnO films deposited by RF magnetron sputtering. *J. Phys.: Condens. Matter 20*, 075220–075224.
Lv K. (2009). Study on ZnO film preparation method. *Cnet, 21*, 87–92.

Electronic Engineering and Information Science – Wang (Ed.)
© 2015 Taylor & Francis Group, London, ISBN: 978-1-138-02772-5

The photoelectric characteristics of vertical organic thin-film transistor

X.C. Liu, M. Zhu, Z.H. Jing, D.N. Yuan & H.R. Lv
Key Laboratory of Engineering Dielectrics and Its Application, Ministry of Education, Department of Electronic Science and Technology, College of Applied Science, Harbin University of Science and Technology, Heilongjiang, Harbin, China

ABSTRACT: In this paper, with vacuum evaporation and sputtering process, we prepared a photoelectric transistor with the vertical structure of Cu/CuPc/Al/CuPc/ITO. The material of CuPc semiconductor has good photosensitive properties. Exceptions will be generated after the optical signal irradiation in a semiconductor material, then transform into photocurrent under the built-in electric field by the Schottky contact, as the organic transistor drive current, makes the output current multiplication. The results show that the I-V characteristics of transistor are unsaturated. When the device was irradiated by a full band (white), 625nm, 700nm light, its working current significantly increased. In white light, when $V_{cc} = 3V$, the ratio of light and no light current was ranged from 2.9-6.4 times. In dark, $\beta = 16.5$. In 700nm, 625nm and white light, the photocurrent of the device was 0.0344μA, 0.0714μA, 0.122μA respectively, and the sensitivity of the device was 0.1632 A/W, 0.2023A/W, 0.0206A/W respectively.

KEYWORDS: CuPc; organic optoelectronic transistor; organic semiconductor.

1 INTRODUCTION

Since the advent of semiconductor devices, microelectronic technology to occupy a very important position in the ranks of high-tech. People all the time feeling the tremendous changes brought microelectronics to human society. It has become an important pillar of the field of social science and technology today (Michael et al. 2004). However, organic electronic materials are a new electronic materials, it is a new research field of combining semiconductor discipline and organic chemistry. Organic electronic materials have not only the optical and electronic properties of semiconductors, and retained some excellent properties of organic compounds, such as species-rich, easy processing and so on. Organic semiconductors have been extensively used in photoelectronics because of its advantages of lightweight, low-cost and compatibility with flexible substrate (Schmidt et al. 2001). People use organic electronic materials can produce an organic photoelectric sensor, organic solar cells, organic field-effect transistor, etc..

This paper studies the organic phototransistor. Through the design of the device structure and organic photosensitive functional layer, Excitons will be generated after the optical signal irradiation in a semiconductor material, then transform into photocurrent under the built-in electric field by the Schottky contact, as the organic transistor drive current, makes the output current multiplication.

2 PREPARATION OF ORGANIC PHOTOTRANSISTOR

The experimental apparatus of the device preparation is the multi OLED coating system in this project. The experiment of raw material of CuPc semiconductor material purity is 99.9%, evaporation temperature control at 350°C, Organic vacuum evaporation coating prepared by organic material CuPc, gate electrode(Al) and the collector (Cu) by DC magnetron sputtering preparation, emitter(ITO) by radiofrequency magnetron sputtered preparation. The production sequence is as follows. First, we should make gold film evaporation source electrode on the glass substrate. Second, protect the first layer of CuPc film evaporation, and the fabrication of thin film aluminum base and a two layer of CuPc thin films. Finally, making gold evaporated the drain electrode (Wang et al. 2010). The following sample preparation conditions, the evaporation temperature of CuPc is 400°C, the temperature of the substrate at room temperature of 20°C, CuPc film under two layer film thickness control by the evaporation time and the evaporation speed is about 3nm /min. CuPc thin film on the lower two layers of film thickness was about

70nm, 130nm, the base(Al) of the semiconductive thin film thickness is about 20nm. The device structure as shown in figure 1.

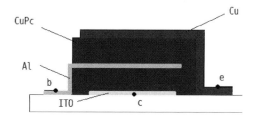

Figure 1.   The structure diagram of device.

3   CHARACTERISTIC ANALYSIS OF MEASUREMENT AND THE EXPERIMENTAL RESULTS OF ORGANIC PHOTOELECTRIC TRANSISTOR

Through the test of Schottky I-V rectifying properties which the base(Al) of the semi conductive thin film with the CuPc film of on both sides. The test results can explain both sides of the CuPc thin film with Al electrodes formed good Schottky contact, form good ohmic contact with Cu and ITO. This result with the phthalocyanine thin films exhibit the conductive properties of P type semiconductor, and Al electrodes formed resulting in the study of the Schottky barrier agreement (Ling et al. 2010). Determination of working characteristics of DC VOTFT, the bias voltage V_{bc} between the base and collector which they step is 0.25V, from 0V to 1V, the bias voltage Vce between the collector and the emitter which increased from 0V to 3V. When the device was irradiated by the absence of light, full band (white), 625nm, 700nm light. Determination of the changes relationship which emitter-collector voltage V_{cc} with the emitter-collector current I_{cc}. This paper lists a part of the measured results.

Figure 2 is the device of the output characteristic curve in the white light and in the dark. It can be concluded that in the same condition from the experimental data, the light makes the device of I_{cc} increased, but the light does not affect the change that when V_b increases, it will suppress the current I_{cc}. When the whole band of white light, the device current I_{cc} increased the most, followed by light is 625nm, 700nm; the devices current I_{cc} in the absence of illumination is minimum. This is because the light through the ITO film in the CuPc film. It will lead to the generation of an exciton, the exciton diffusion and subsequent dissociation into free charge carriers, thereby enhancing the carrier density within the CuPc film, then in light and without light, current I_{cc} size for the two different curves in the case of the same V_b.

Because the light exists, the current of the device Iec is greater than the curerent Ie when there is no light. When $V_{cc} = 2V$, current Iec in the light of 625nm with no light ratio range is 1.7–2.8 times, and white light with no light ratio range is 1.8–4.6 times. And in the base voltage of V_b is 0V, the current amplification factor is maximum. When $V_{cc} = 3V$, the current Iec in the light of 625nm with no light ratio range is 2.2–3.7 times, and white light with no light ratio range is 2.9–6.4 times, and the base voltage hour large amplification, the base voltage large magnification small. Through the experiment, we can know that the base voltage is more smaller, the current Iec of magnification is greater in the light; the voltage between the collector and emitter of V_{cc} is more larger, the current Iec of magnification is greater in the light [5]. In order to study the current multiplication photoelectric devices and photoelectric conversion efficiency, we set the Vb are open and 0V so that determine the relationship of I_b, I_c and the bias of Vec between the devices in different lighting conditions. The determination results as shown in figure 3 and figure 4.

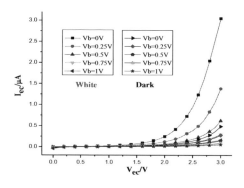

Figure 2.   The output characteristics of the device in the white light and in the dark.

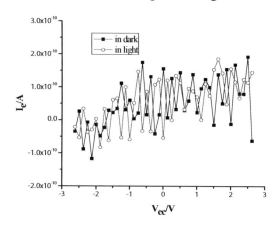

Figure 3.   When V_b = open, the characteristic curve of V_{cc}-I_c.

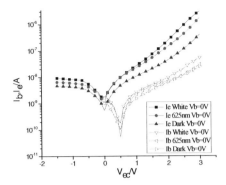

Figure 4. When $V_b = 0V$, the characteristic curve of V_{ec}-I_e.

The relationship between the movement of carriers and the current distribution by the transistor, and the relationship of the p-n-p triode base current Ib with the collector current Ie when no light:

$$I_{ec} = \beta I_b^{dark} \qquad (1)$$

However, in the light, because the light so that the electron-hole excitons were generated in the semiconductor layer of CuPc(Zaki.T., S.Scheinert. &I. Horselmann.: 2014), then they separated in the internal electric field, the flow of electrons to the base area, hole to flow in the collector region,and it can generate photocurrent IL and become a part of the device working current. So in the light irradiation, the current relationship is:

$$I_{ec} = \beta I_b^{dark} + I_L(1 + \beta) \qquad (2)$$

According to Fig. 4 , it can be obtained the value of the devices' base current I_{ec} with the working current when in the $V_b = 0V$, $V_{ec} = 3V$ as shown in table 1.

Table 1. Fund current in different light and the light generated current value.

$Vb = 0V$ $Vec = 3V$	$Iec(\mu A)$	$Ib(\mu A)$	β	$I_L\ (\mu A)$
No light	0.40	0.0247	16.5	–
700nm light	1.01	0.0272	–	0.0344
625nm light	1.74	0.0297	–	0.0761
White light	3.04	0.0548	–	0.15

In Table 1 the value of I_b with I_{ec} for testing income (Feng, Xu and Guo 2012), and the value of β can be calculated by Formula 1 (Table 1). It is substituted into the Formula 2, then the value of I_L can be calculated. When the device in the 625 nm light, the photo-induced current I_L is 0.0714 A, which is 2.56 times than the base current of I_b. In the white light generated current I_L is 0.122 A, which is 2.73 times than the base current.

The use of optical brightness measuring instrument to measure the 700nm light, 625nm light, full band (white) corresponding to the illumination light brightness. Their value in turn is 0.7915 cd/cm², 1.4145 cd/cm², 27.46 cd/cm². The device of effective area is 0.2 × 0.2 cm, where we can calculate the device sensitivity. Through the experimental data, we can know the devices of collector current and photocurrent in white light than in monochromatic light much, but its sensitivity is lower than that of monochromatic light. This is because in the white light its brightness value is high, and its white incident power to the light incident power is far greater than the monochromatic light. So the white light sensitivity is relatively low, only is 0.0206A/W. The main peak (625 nm) in the absorption spectrum of CuPc is more sensitive to the device, which is 0.2023A/W. Its sensitivity I about 10 times as much as the white light. The sensitivity is second in the absorption spectrum peaks (700 nm) of CuPc, which is 0.1632A/W. Its sensitivity is about 8 times as much as the white light.

4   CONCLUSIONS

In this paper, using vacuum evaporation and sputtering process, we prepared a photoelectric transistor with the vertical structure of Cu/CuPc/Al /CuPc/ITO. The results show that the I-V characteristics[8] of transistor are unsaturated. When the device was irradiated by full band (white), 625 nm, 700 nm light, its working current significantly increased. In white light, when V_{ec}=3V, the ratio of light and no light current were ranged between 2.9 and 6.4 times. In dark, β=16.5. In 700 nm, 625 nm and white light, the photocurrent of the device was 0.0344 μA, 0.0714 μA, 0.122 μA, respectively, and the sensitivity of the device was 0.1632 A/W, 0.2023 A/W, 0.0206 A/W, respectively. The current collector and the photocurrent of the device in white light was much larger than in monochromatic light, but the sensitivity of the device in white light was lower because of the incident power of white light which was much larger than in its monochromatic light.

REFERENCES

Michael, C., S. Martin & K. Jerzy (2004), Thin-Film Organic Polymer Phototransistors, silicon staggered-electrode thin-film transistor. *IEEE Trans. Electron Devices*, 6, 877.

Schmidt, M.L., A. Fechtenkotter & K. Mullen et al. (2001), Self-Organized Discotic Liquid Crystals for High-Efficiency Organic Photovoltaics. *Science*, 293(5532), 1119–1122.

Wang H.B., Z.T. Liu & F.L. Ming (2010). Organic-inorganic heterojunction field-effect transistor. J. *Appl. Phys.* 107, 024510.

Ling L., D. Maarten & G. Jan (2010). A compact model for polycrystalline pentacene thin-film transistor. J. *Appl. Phys.* 107, 024519.

Electronic Engineering and Information Science – Wang (Ed.)
© 2015 Taylor & Francis Group, London, ISBN: 978-1-138-02772-5

Design of ZigBee-based athletes pulse detection system

S.W. Dong
Department of Physical Education, Harbin University of Science and Technology, Harbin, China

Y.Z. Wang
School of Measurement and Control Technology and Communication Engineering, Harbin University of Science and Technology, Harbin, China

K.J. Dong
School of Automation, Harbin University of Science and Technology, Harbin, China

ABSTRACT: Zigbee-based technology introduces a new type of wrist-style pulse wave detection system, we can use a pressure sensor to extract the pulse wave signal, after pretreatment, send the signal to the CC2530 module for conversing analog to digital, to diagnose and deal with the data, and then using wireless sensor network-which is based on ZigBee technology- to send the packet data to the PC machine. After that, what needs to be done is processing, analyzing, and displaying data. This system can not only work with motion detection, but also the mobile monitoring in hospitals, and home portable monitoring.

KEYWORDS: ZigBee; CC2530; wristband style; pulse wave detection.

1   INTRODUCTION

1.1   *Type area*

Pulse signals (Johnson 1984)are important physiological signals. Based on non-invasive measurement and analysis of pulse wave, we can clearly see the athletes' physical condition, allowing coaches to adjust athletes' workouts and training methods. Through wrist-detection technology, we can overcome the limitations of conventional cable detection, and the personal life information can be detected in a routine training state, meanwhile, it can work with the existing personal communication module (such as mobile phone, PDA or laptop, etc.) achieving detecting in real-time. ZigBee (Pauca&O'Rourke&Kon 2001)wireless communications technology, which follows the IEEE802.15.14 standard (Texas Instrument 2009), can be very valuable with its low power consumption, low cost, low latency and long transmission distance and other characteristics. It is widely used in sports, medical and other industries. In this paper, we deeply looked into the ZigBee-based wrist-pulse wave detection systems, achieving real-time detection of pulse wave and wireless transmission with satisfactory results.

2   THE OVERALL DESIGN OF THE SYSTEM

This system consists of five modules: pulse signal acquisition module, signal processing module, CC2530 (Geer 2005)controlling module, ZigBee wireless module and the monitoring module. The system block diagram is shown in Figure 1.

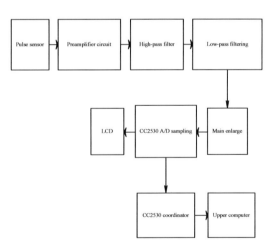

Figure 1.   System block diagram.

When a pulse sensor collects from physiological signals and converts them into electrical signals, through pretreatment like amplifying and filtering, the signal will be sent to the A/D sampling module in the CC2530 to achieve signal acquisition and processing. The results will be displayed on LCD, meanwhile, using the ZigBee wireless communications functions we can pass the data to the intensive care unit. These designs use the CC2530 ZigBee-compliant system-on-chip chips, built the ZigBee-based wireless sensor network, and then transmit the data to a base station, for coaches to view and analyze.

2.1 System hardware design

2.1.1 Pulse signal acquisition module
Pulse sensor(Perkins 2001) is used to sense the mixed-signal which includes the pulse wave and pressure, then translate it into a voltage signal, as the pressure-pulse output from the sensor is weak, whose amplitude is at 0.05~1 mV, we should first zoom in, and then proceed to the next step. The system pulse sensor uses SC0073B dynamic micro-pressure sensor, which uses piezoelectric film as a transducer material, and inside of the sensor, the amplifier circuits convert the signal to a voltage output. Its expanded working voltage is from 1.5V to 6 V.

2.1.2 Signal preprocessing modules
Bioelectrical signals have a high impedance, low signal strength, and low frequency characteristics, and usually are in the middle of significant background noise. Based on these particularities, the analog circuit should be qualified with high input impedance and high common-mode rejection ratio, and enough high-gain. Therefore, this design of pulse wave signal detection consists of by sensor circuits, pre-amplifier circuit, filter circuit, amplifier circuit, and through the A/D conversion (Carr&Brown 2008), it eventually converts the pulse wave signal to high- or low-level signal output. Signal preprocessing circuit is shown in Figure 2.

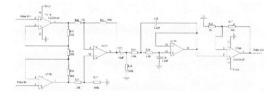

Figure 2. Signal preprocessing circuit.

2.1.3 ZigBee module
Zigbee (http://www.5lian.cn/html/2012/wulianpuji_0213/29975.html) (Figure 3) is the core of the whole pulse wave module, aiming at data interaction with external networks (ZigBee-based wireless sensor networks) usually contains an enhanced MCU. This design uses the CC2530 chip made by TI company to build the ZigBee-based star wireless network.

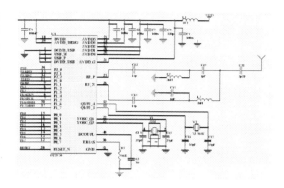

Figure 3. ZigBee module.

2.2 System software design

The software design is developed on the basis of Z-Stack1.4.0 protocol stack provided by the TI company. System software consists of two parts: the coordinator node and terminal nodes. The main role of the coordinator node is forming a network, accepting data sent by the terminal nodes and transmitting the data to the PC through the serial port. Pulse tester that works as a terminal node is responsible for collecting the pulse signals and processing them. Pulse tester can be used independently, they can also join a wireless monitoring network, and send the message to the coordinator in real time. Figure 4 is a coordinator node software flow chart, Figure 5 is a terminal node software flow chart.

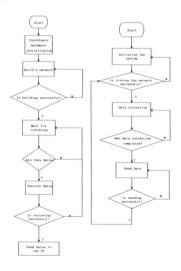

Figure 4. Coordinator node. Figure 5. Terminal node.

ADC has a selectable sampling frequency like 500Hz, 250Hz, 125Hz, 62.5Hz, subroutine is as follows:

    if (channal==HAL_ADC_DEC_064)

```c
{ADCCON3=0X00;
 ADCCON2=0X00;
}
else if (channal==HAL_ADC_DEC_128)
{ADCCON3=0X10;
 ADCCON2=0X10;
}
else if (channal==HAL_ADC_DEC_256)
{ADCCON3=0X20;
 ADCCON2=0X20;
}
else
{ADCCON3=0X30;
 ADCCON2=0X30;
}
ADCCON1=0X33;
ADCCON1=0XB3;//Starting A/D
while(!(ADCCON1&0X80));//Waiting for ADC
conversion completed
ADC_temp = ADCL;// Read the values of ADC
and converse the data
ADC_temp=ADC temp>>3;
ADC_16_temp=(unsigned int)ADC_temp;
ADC_temp=ADCH;
voltage_temp=(unsigned int)ADC_temp;
voltage_temp=voltage_temp<<5;
voltage_temp=voltage_temp+ADC_16_temp;
voltage_temp=voltage_temp* 0.0281;
voltage_value = (uint8)voltage_temp;
Calculate the peak value among sampling results
within 1s :
void mian()
{Init_All();
while(1)
{if(Set)
{ if(channal==ADC_DEC_500Hz)
  {T1CC0H = 0x01;
   T1CC0L = 0xf4;
  }
  else if((channal==ADC_DEC_250Hz)
  {T1CC0H = 0x03;
   T1CC0L = 0xE8;
  }
  else
  {T1CC0H = 0x0F;
   T1CC0L = 0xA0;
  }
 }
 else
 {if(channal==ADC_DEC_500Hz)
  {if(Counter >= 500)
   {printf(data_max);
    Counter == 0;
   }
  }
  else if((channal==ADC_DEC_250Hz)
  {if(Counter >= 250)
   {printf(data_max);
    Counter == 0;
   }
  }
  else
  {if(Counter >= 62)
   {printf(data_max);
    Counter == 0;
   }
  }
 }
}
}
#pragma vector = T1_VECTOR
 __interrupt void T1_ISR(void)
 {IRCON = 0x00;
//Clear interrupt flag, also can be done automati-
cally by hardware
  data_t0 = ADC_result();
  data_t1 = 0;
  if(data_t0 >= data_t1)
   data_max = data_t0;
  else
   data_max = data_t1;
  data_t1 = data_t0;
  Counter++;
 }
```

3 CONCLUSIONS

This paper describes the design process of Zigbee-based pulse wave detection system, including the pulse wave extraction and pre-processing circuits, and analyzes the workflow of ZigBee wireless sensor networks. The pulse wave detection achieves a real-time detection of the radial artery pulsation waveform via a wireless network is of great practical value.

REFERENCES

Carr,Joseph.J.&John.M.Brown(2008).*Introduction to Bio-medical Equipment Technology.* London: John Wiley & Sons Johnson,D.E.(1984).*Handbook of active filters.* U.S.: Electronic Industry.
Geer David.(2005).Users Make Beeline for ZigBee, *Sensor Technology.*Volume 38:12–16.
Information on http://www.5lian.cn/html2012/w-lianpuji_2013/29975.html.
Pauca,Alfredo.L.&Michael.F.O'Rourke&Neal.D. Kon.(2001). *Prospective Evaluation of a Method for Estimating Ascending Aortic Pressure from the Radial Artery Pressure Waveform.* Hypertension.
Perkins,Charles.E.(2001).*AdHoeNetworking.*London: Addison-Wesley.
Texas Instrument. (2009).*CC2530 Datasheet.*

Electronic Engineering and Information Science – Wang (Ed.)
© 2015 Taylor & Francis Group, London, ISBN: 978-1-138-02772-5

The design and implementation based on the Android platform of business process management system

L. Liu, D. Yang, S.M. Gao, C.L. Tang & P.D. Wang
School of Computer Science and Technology, Harbin University of Science and Technology, China

ABSTRACT: In order to achieve the mobility and momentariness of the business process management within the enterprise, a business process management system based on the Android platform is designed. Built on system requirements analysis and utilizing the Eclipse tools, the system function of business process management system based on the Android platform can be completed; meanwhile, it can achieve the client-side communicates with the server through socket programming. The system has many main functions as follows: to open all the BPM forms, to submit all sorts of check and approve and push personal task lists and to test system function and performance on the device simulator. The test result shows that the system with the strong practicability can not only accurately provide users with real-time information within the enterprise and complete personal operation, but also improve the work efficiency of the internal enterprise.

KEYWORDS: Business process management; Android; Socket programming; The server.

1 SUMMARY

With the booming development of mobile communication technology, smart phones began to be widely used, especially the smart phones with the Android operating system. Android, developed by Google company and released on November 5, 2007, is open source mobile phone software development platform based on the Linux kernel. It consists of the operating system, application software, user interface and middleware. Android gives the developers opportunity to develop novel mobile device application (Banggui Xia. 2011.). The Android operating system does not have any exclusive rights to hamper innovation, and has strong expansibility and inter connectedness so that it is popular among the manufacturers and operators. Because of its openness, flexibility and open Internet concept, it has many advantages in the mobile phone operating system(Gavalas D & Economou D . 2011).

With the development of the Internet, information communication is happening with increasing frequency. Traditional computer office cannot satisfy the users' requirements at any time and any place, and it raises the higher requirements for the communication method, so does the safety of the communication. In view of this, this paper introduces the enterprise process management system based on Android mobile phone. The system selects TCP/IP communications from many communication protocols, and adopts a new way of communication, which can realize mobile terminal for random access to internal network, and ensure the safety of the user communication.

2 SYSTEM STRUCTURE

The main function of this system is to achieve mobile office. The system is composed of the server, mobile phones (Android) and communication mode.

Servers have two main functions that are terminal equipment management and data transmission. It can respond to and handle requests, manage terminal state, and transfer application information. After receiving the request, it establishes a communication channel which can transfer data bi-directionally and realize data exchange between the terminal and server. Main function modules of the server including the following points: network transmission, account management, database management, message processing, terminal management, etc.

Mobile phone client software is developed on Android smart-phone, and the client is designed by the Android plug-in in Eclipse. Programming design is realized by using the JAVA language code and calling the Android API function library. Its main functions are to send requests to the server, receive the server data and complete the data conversion, to complete the necessary operations of users and to alert message for users, etc (Shiddiqi A & Pratama H & Ciptaningtyas HT . 2010).

For the choice of communication mode, Android API provides several supports for us about network programming(Yannan Zhang & Lu Yang . 2013).

A Network communication based on TCP/IP protocol. TCP/IP communication protocol is a reliable network protocol. It has respectively established a Socket at both sides of the communication to form the virtual links so that the virtual links can maintain the communication of the programs. Socket and Server Socket are underlying network programming modes, and other high-level protocols (for example: HTTP) are established on the basis of them. Socket programming is cross-platform programming, and it can communicate between heterogeneous languages. In the Socket programming mode, Socket class is to establish the client program, and Server Socket class is used to establish the server program.

B Network communication based on UDP protocol. UDP communication does not exist the concept of "connection, and it needs to establish Datagram Socket which enables to receive and send data through Datagram Packet. The data packet itself contains the IP address, port number, and data content. The UDP data will not be reset package on the opposite end, so the length of data sent at a time must be limited.

C Access to network resources by using the URL (Uniform Resource Locator) :Firstly, to call URL's open Connection() to create URL Connection which represents the communication connection between application and URL. Secondly, to set the parameters and common request attributes of the URL Connection. As for a GET request, it just need to call the connection method to set up the actual remote connection; If it is a POST request, the Output Stream is needed. Then to transfer request parameters to the network. Finally, the remote resource becomes available, and programming can access the header fields of remote resource or read remote resource data from the input stream.

D Access to network by using the HTTP. There are two methods: One is using Http URL Connection to make request or response; the other is using Apache HTTP client. The Android SDK has integrated Apache's Http Client module. Apache HTTP client calls Http Get or Http Post according to the corresponding request methods, and the response object is Http Response, and it uses Default Http Client to execute the request and get the response.

E Access to network by using the Web Service. Web Service is the mechanism that achieves the heterogeneous process to transfer application method. The API has nothing to do with any operation

platform and any programming language. XML language as the service description language, it adopts the SOAP (Simple Object Access Protocol) as the basic communication Protocol. For the small devices like mobile phone, its computing resources and storage resources are very limited. Therefore, the Android is unlikely to provide the external Web Service, just acts as a Web Service client, and invokes the remote Web Service. Android does not provide the operation of the library which can be direct transferred by Web Service, so it must rely on the third party libraries (soapy class library is used more commonly) to the Web Service application.

Comparing these methods and considering the security of the system and other factors, this paper selects the TCP/IP network communication. Adopting method that self-independent build mobile access the inner network server could complete the connection between mobile client and server. Before the Android mobile phone access to the server, the server will decide whether authorize access to the phone or not. Secondly, it should verify whether the network environment safety or not. If it could use the outer network to visit, VPN authentication is available to access. Finally, the data to be sent in the Intranet access point needs encrypting processing. The safety communication protocol and communication can guarantee the safety of system.

3 THE CLIENT DESIGN AND IMPLEMENTATION

3.1 *The client function design and implementation*

The system is used to implement the mobile process management. Login and register should be first taken consideration. The user name can confirm the user identity. After the success of the user authentication, the system could get user information including personal information, permission information, access module information and system into the navigation page. Secondly, it should analyze the system as a whole, such as the real-time, reliability, large number of data access, and the security analysis, etc. As shown in Figure 1.:

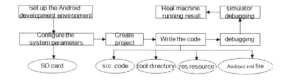

Figure 1. Based on the Android platform application develop software process.

Finally, this research is mainly in order to achieve the function as follows:

Open all Metastorm BPM forms. Different user permissions have different display and operational forms correspondingly. This part is mainly to achieve the operation that the different people can see overview the different forms. Everyone can also know it own personal tasks, etc.

Submit all kinds of examination and approve. This part is mainly for leaders at all levels, and each level of leadership has access to examination and approval authority for the individual tasks and the personal application. According to the level, it can request for examination and approval step by step as the main control part of the whole process, which will affect the whole task.

Push personal task list. For this part, the implementation of the function is to submit application to the superior according to individual circumstance, such as leave, reading material, to apply for the necessary requirements and so on. It can also include the work task application.

This design is mainly based on Java as a development language. It is necessary to study the overall framework and complete the writing and testing.

Throughout the project, under the SRC directory of the Java main program is in the driver's seat, by the layout file statement object will in root directory R.JAVA to automatically generate the corresponding number ID, Android applications get through R class to achieve the exhibition of resources. In Java main program, therefore, it need through the findViewById method to use the controls of layout, then triggers the corresponding event. For example let a Button control initialization and name, in "on Create" response function by statements MBT = (Button) the findViewById (R.I db utton01) to gain control, by setOnClickListener (new an OnClickListener () method to trigger the Button events. Using the socket to connect the, also to read and write, then using the Print Writer and Buffered Reader to send and receive data and message, the received message of String will be shown in the Text View, also can pass that information to the instruction by the connection between the Server and the Client (Dongjiu Geng & Yue Suo & Yu Chen & Jun Wen . 2011).

Android itself can receive two parameters that are the local network and remote network information, each TCP connection will create the new object of TcpClient class , TcpClient class object encapsulates the TCP connection, whenever after receiving the message, TcpClient will transfer the netDataListener DataRecieved method, also to response the message.

3.2 The design of user interface

Firstly, it should utilize the user language and feedback information and tip service. Finally, it should pay attention to the smallest information volume principle (Yi Du & Feng Tian & Cuixia Ma & Guozhong Dai & Hongan Wang. 2013).

According to the functional design to analyze the result, it mainly aimed at the user's interest degree and important degree to analyze the page layout. For example, we can adopt the following solutions on the landing page:

Landing page contains the enterprise image, the login information, copyright information and help information. Through the analysis of the user attention, information importance and content, the results can be drawn as the following Table 1:

Table 1. Margin settings for A4 size paper and letter size paper.

Information classification	User attention	The importance of information	Contains the content
The enterprise image	High	High	Corporate Logo; Corporate name; Background
The login information	High	High	The username; Password
Help information	Medium	Medium	Hint

According to the above form, the user's login information and attention and importance of the corporate image are both as the highest, so it is distributed on the top; the relative tip information lies at medium level, so it is put behind the logout button. Show the result, as shown in Figure 2. The design method is

Figure 2. The system login page.

simple and relaxed, the user friendly, matching the principle of minimum information, which the user can spend the shortest time in finding the content that they focus on. At the same time, by this way, it can achieve the sense and simplicity.

4 THE SERVER DESIGN AND IMPLEMENTATION

Server-side main function is to wait for the client's connection request and transmit information at the real-time. It needs to consider how to solve three problems of the mobile security and management before the connection. Namely, the problems are identification, date privacy and equipment management . In view of the identification, adopt dedicated access gateway and can unite the client and access control server to set up protection, to ensure the consistent strategy in different environments (company LAN, WLAN or remote access). In view of the data privacy, in the aspect of data transmission, SVN SSL VPN gateway can provide VPN high-strength encryption transmission to ensure the safety of data secret sex and prevent data malicious sniffing and manipulation. In view of equipment management, it can manage from the access to information, system operation, flexible deployment file, account recovery, as well as the release of the information, etc. In order to reduce loss of data and improve overall system performance in information engineering, on the server side use the double buffering mechanism.

The server to receive data service established by the socket to receive data, at the same time establish a socket send data services to the client to send data. TCP server system is created through the ServerSocket, The ServerSocket accept () function is constantly invoked to receive all the request from the client. Through the Socket InputStream, get-InputStream ()function returns the input stream of receiving Socket, then let the program through the data input stream rushed out of the Socket, after that to save the received data in the receive buffer. At the same time in the send buffer, using OutputStream, getOutputStream () returns send Socket object corresponding to the output stream, let the program send buffer data through the output flow Socket (Meihua Xiao & Jinli Yu & Pan Xiao . 2012).The system operation process, as shown in Figure 3:

5 THE EXPERIMENTAL RESULTS

The development of this program uses the Android 4.0 version of the SDK, the client and server get through Socket to proceed network communication. Mobile client run on XIAOMI 1 phone, the phone uses the

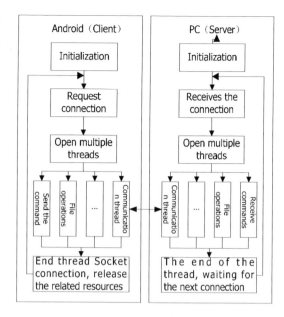

Figure 3. The system operation process.

2 g / 3 g network to set up connection with the server . Mobile phones connect to the server time is less than 3s and data delay is less than 5s, within the scope of the user can accept. In addition, the system adopts the modular design so as to maintain and extend the system function in the future. Test system and run successfully, and then we get the main interface, to-do tasks, the detailed to-do tasks , as shown in Figure 4:

Figure 4. The main interface and to-do tasks.

REFERENCES

Banggui Xia. 2011. The design and implementation of The library service system based on Android platform. New Technology of Library and Information Service.

Dongjiu Geng & Yue Suo & Yu Chen & Jun Wen . 2011. Design and implementation of Android phone based access and control in smart space. *Journal of Computer Applications.*

Gavalas D & Economou D . 2011 . Development Platforms for Mobile Applications : Status and Trends. *IEEE Software.*

Meihua Xiao & Jinli Yu & Pan Xiao. 2012. Model Extraction and Reliability Verification on SOCKET Communication Program. *Computer Science.*

Shiddiqi A & Pratama H & Ciptaningtyas HT. 2010. A Video Streaming Application Using Mobile Media Application Programming Interface. *Telkomnika.*

Yannan Zhang & Lu Yang . 2013 . Remote video surveillance system based on Android mobile phone. *Journal of Computer Applications.*

Yi Du & Feng Tian & Cuixia Ma & Guozhong Dai & Hongan Wang. 2013. A Mobile User Interface Generation Framework Based On Multi-scale Description. *Chinese Journal of Computers. 10.3724/sp.j. 1016.2013.0217.*

Electronic Engineering and Information Science – Wang (Ed.)
© 2015 Taylor & Francis Group, London, ISBN: 978-1-138-02772-5

The report based on OFDM power line communications system simulation platform

H. Ai, L.T. Wang & Z.X. Liu
School of Automation, Harbin University of Science and Technology, Harbin, China

L.H. Wei
Yichun Vocational College, Yichun, China.

ABSTRACT: The intrinsic noise interference, frequency selective attenuation and multi-path transmission characteristics of power line channel affect the performance of communication seriously. The paper established a simulation platform of the power line communication system based on OFDM. Some key technologies such as convolution code, interleaving encoding, error correction coding, carrier wave differential modulation, synchronization and reuse etc. had been studied and improved. MATLAB simulation results show that can effectively reduce the system BER, enhance the error correction ability and the spectrum utilization efficiency and realize high speed signal transmission. Significantly improve the overall performance of power-line communication system.

KEYWORDS: Power-line Communication; OFDM; channel; bit error rate.

1 INTRODUCTION

The modern communication services tend to show the characteristics of high-capacity, high-speed real-time data transmission. Power Line Communication (PLC) based on power line transmission medium refers to the data transmission and information exchange based on power line transmission medium (Ren 2013). By the power-line transmission to realize the automatic information in the field of voice, data, audio and video, for data exchange, processing (Wu 2010), control and monitoring, data communications network can remove the bother of special laying communication lines not only, but also facilitate the setup and hauling of the mobile communication terminal. Because of the power-line network, the coverage area is larger than any other networks (Zhang 2012). Thus, low voltage power-line communication has received great attention to the information appliances, building automation and broadband access, etc. However, the load on the power-line degeneration is very strong, sudden interference influence is very big (Zhang 2014). That makes it difficult to build an accurate channel model. Inherently unpredictable power-line channel noise attenuation and frequency selected multi-path interference, various propagation affect the communication performance seriously. At present PLC service quality is not stable. It is lack of an effective modulation method according to the characteristics of the PLC.

2 PLC SYSTEM PRINCIPLES

From the angle of the frequency bandwidth, power line communication can be divided into narrow-band PLC (NB - PLC) and broadband PLC (BB - PLC). The carrier frequency range of NB-PLC varies in different countries and different regions. It is 50 ~ 450 kHz in the U.S and 3 ~ 148.5kHZ in Europe. The frequency range of BB-PLC is 2 MHz ~ 30MHz. The system simulation is on the basis of European 3 ~ 148.5 kHz. The PLC simulation platform includes four parts. They are transmitter, receiver, channel and noise sources. The block diagram is as shown in Figure 1.

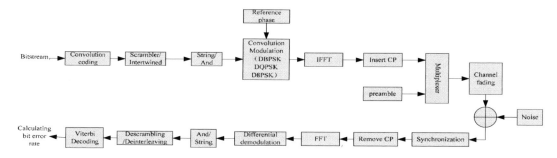

Figure 1. Block diagram of PLC communication system.

3 PLC PHYSICAL LAYER PARAMETERS

The working frequency band of this simulation platform was selected as 41992.1875Hz – 88867.1875Hz and the modulation method is OFDM. Its carrier number is 87-183,331-427. The OFDM sub-carrier allocation is shown in Figure 2:

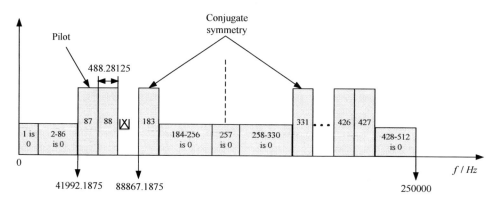

Figure 2. OFDM sub-carrier allocation.

The base band clock is 250 kHz, the sampling frequency is 250 kHz, the sampling period is 4us. The FFT points is 512 (FFT interval is 512 * 4 = 2048us, or 1 / 488.28125 = 2048us), the entire channel is divided into 512 sub-channels. The width of each channel is 250000/512 = 488.28125Hz.The cyclic prefix is 48 points, the cyclic prefix time of 48 * 4 = 192us.Symbol samples = FFT samples + Cyclic. Prefix samples = 512 + 48 = 560.Symbol interval 560*4 = 2240us.

4 EMITTING PART

4.1 Convolution coding

In order to enhance the capability of the system error correction the convolution codes are obtained from the binary bit stream. The generating polynomial of convolution code are $1 + D + D^2 + D^3 + D^6$ and $1 + D^2 + D^3 + D^5 + D^6$. Figure 3 is the convolution coding principle.

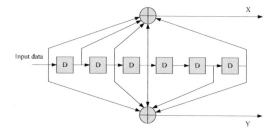

Figure 3. 1/2 convolution coding.

When decoding the Viterbi decoding is used. Taking DQPSK modulation as an example, when there is no Viterbi decoding and there is Viterbi decoding, the system error rate is as shown in figure 4:

Can be seen from the chart, after use the convolution code, the system BER is reduced greatly.

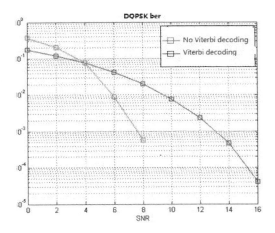

Figure 4. DQPSK signal decoding performance comparison.

4.2 Intertwined

Low-voltage power line channel belongs to reference channel, the channel characteristics vary with time and location, and because there are a lot of channel impulse noises, it is easy to generate traffic bursting error. To overcome this problem, interlaced encoding is used. The larger pieces of interleaving, the better the performance of the transmission system will be, but the greater the propagation delay will be. When interleaving pieces is too small it will not overcome the effect of bursting errors. After considering the system performance and transmission delay the interlaced piece is selected as the size of $16 * 8$.

4.3 Scrambler

The scrambler is obtained by a pseudo random code sequence multiplied by the bit stream Add the scrambler can avoid long even long 0 or 1 and easy to extract the clock signal. The data randomization is conducive to smooth the spectrum and reduce the influence of nonlinear. This system choice generation polynomial for pseudo-random sequence as $1 + x^{14} + x^{15}$, Pseudo-random code is as shown in Figure 5:

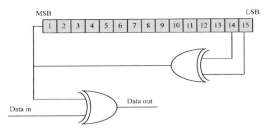

Figure 5. Pseudo-random code.

4.4 Differential modulation

Launch part can use DBPSK, DQPSK or D8PSK modulation, the reason for using differential modulation is differential demodulation is simple, and does not require the channel estimation to reduce the chip cost. Comparison of several modulations in AWGN channel bit error rate is as shown in Figure 6:

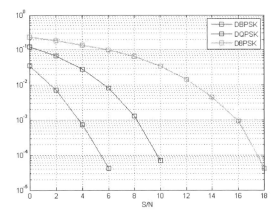

Figure 6. DBPSK, DQPSK, D8PSK BER comparison in the Gaussian white noise channel.

It can be seen that in the same noise environment the system BER gradually increased with the improvement of modulation efficiency.

4.5 Channel

The main task of the low voltage power line is to implement the allocation of 50 / 60Hz power within a short distance. Comparing with a dedicated communication line the channel environment is extremely bad. To build a mode of the low voltage power line communication channel the factors such as the time variation and location, etc., need to be considered. PLC communication is divided into narrow band PLC (NB-PLC) and broadband PLC (BB-PLC). This simulation platform is designed for narrow-band PLC. PLC narrow-band channel model can be expressed as a function of distance and frequency. It is shown in the following formula.

$$A = 12.6 + 0.055(d - 100) + 0.25(f - 60) \quad (dB) \quad (1)$$

Where f is the range of 60kHz–90kHz, the range of d is the 100m–550m.The simulation process designed filter, take d = 100 m. Filter frequency response shown in Figure 7.

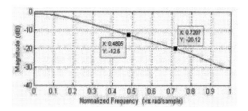
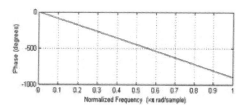

Figure 7.   PLC channel frequency responses.

5   THE RECEIVING PART

Due to the effect of attenuation, the signal envelope remains constant no longer when receiving. So the distribution of the signal constellation diagram appears uneven phenomenon after the OFDM be solved the difference. The phenomenon is shown in Figure 8.

The reason for this phenomenon is due to inter-symbol interference (ISI) effects. The solution is to advance the regular time lightly. The effect is shown in Figure 9.

The error rate of the PLC communication system after DQPSK modulation is as shown in Figure 10.

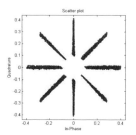

Figure 8a.   DQPSK signal (noiseless).

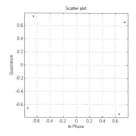

Figure 8b.   Demodulated QPSK signal.

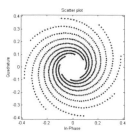

Figure 9a.   DQPSK signal (noiseless time advance 5).

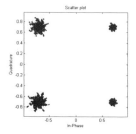

Figure 9b.   Demodulated QPSK signal.

REFERENCES

Ren Z. S. (2013). Simulation study of the physical layer power line carrier communication system, *Science and Technology Information 7*, 334–337.

Wu J. G. (2010). *OFDM* technology based on power line communication system, *Silicon Valley 17*, 23–27.

Zhang J. H. (2012). Ultra short wave communication system physical layer simulation modeling method accurately, *Telecommunications Science*, 223–226.

Zhang T. F. (2014). High speed narrow band OFDM power line communication in real-time dynamic spectrum allocation, *Chinese Society for Electrical Engineering*, 65–69.

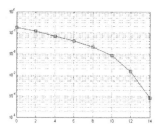

Figure 10.   DQPSK modulation mode error rate.

Electronic Engineering and Information Science – Wang (Ed.)
© 2015 Taylor & Francis Group, London, ISBN: 978-1-138-02772-5

Numerical analysis of the electric potential distribution of two dimension in the conductive channel of organic static induction transistor

X.Y. Cui, D.X. Wang, Y. Zhang & Y. Yuan
Key Laboratory of Engineering Dielectrics and Its Application, Ministry of Education, Department of Electronic Science and Technology, College of Applied Science, Harbin University of Science and Technology

ABSTRACT: Through the establishment of Organic Static Induction Transistor (OSIT) reasonable physical model and selecting the appropriate structure parameters, we explore the simulation calculation of the OSIT under different bias in its internal potential distribution by finite element method. According to the simulation calculation results, Getting internal potential distribution patterns of conductive channel of OSIT, then the functional characteristics of OSIT and the functional characteristics of OSIT with bias and structure parameters of dependencies will be analyzed.

KEYWORDS: finite element method; simulation; organic static induction transistor; Classification number; Document code.

1 INTRODUCTION

In this paper, we establish the physical simulation model of organic static induction transistor, take the finite element method and base on the given boundary conditions (Wang 2012), the simulation model will be gained by using a discrete numerical method by the computer, get the potential distribution patterns of model (Zhang et al. 2012). According to the functional characteristics of these pattern analysis OSIT and bias, and the relationship between structure parameters. Verifies the correctness of the result of the experiment, For OSIT practical theoretical preparation.

2 ESTABLISH A SIMULATION MODEL OF ORGANIC STATIC INDUCTION TRANSISTOR (OSIT)

The basic functional principle of OSIT, OSIT operates by changing the gate voltage (V_{GS}) and voltage (V_{DS}) to control the conducting channel leakage within the barrier height in order to achieve the modulation channel current. The structural parameters and operating characteristics of OSIT itself is conductive channel width, the length of the gate, the gate source spacing has a direct relationship.

Simulation model, OSIT structure as shown in Figure 1 (a), as shown in Figure 1 (b), Many gate vertical structure of OSIT is made up of many repeat single channel, Here are just a channel as the research object, and set up as shown in Figure 1 (c) of the flat structure (Guo et al. 2010). In the model gate (G) is made from aluminum coating and organic P type semiconductor material phthalic cyanide Copper Phthalocyanine constitute the Schottky gate (Zhao et al. 2012), The source (S) and drain (D), cyanogen and Copper Phthalocyanine evaporation film into ohm sexual contact gold evaporation film (Tang 2012). lsg as the distance from the source to the gate, lgd gate to drain away, 2a for conducting channel width, lg is the length of the gate.

OSIT model internal potential function can be made of poisson's equation (1) combined with boundary conditions (2) (3) to describe,

$$\nabla^2 \varphi = \frac{\partial^2 \varphi}{\partial x^2} + \frac{\partial^2 \varphi}{\partial y^2} = -\frac{\rho}{\varepsilon} \tag{1}$$

$$\Gamma_1 : \varphi = \phi(x, y) \tag{2}$$

$$\Gamma_2 : \frac{\partial \varphi}{\partial n} = \upsilon(x, y) \tag{3}$$

Φ as the potential function in the internal model, ρ is the charge density, ε is the dielectric constant, Γ_1 as the first kind boundary condition. Γ_2 for the second class boundary conditions. By the finite element method to solve the Poisson equation solution is the desires of the potential value of the internal.

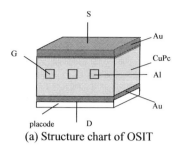

(a) Structure chart of OSIT

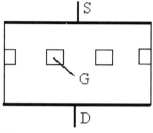

(b) Simulation model of the planar structure section

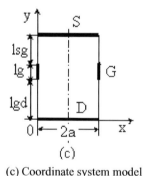

(c)

(c) Coordinate system model

Figure 1. Structure chart of OSIT.

3 THE FINITE ELEMENT METHOD EQUATION AND THE ANALYSIS OF THE RESULTS

The finite element method equation, Required by finite element method, first of all, using the variation principle the solution of the boundary value problem is transformed into the corresponding variation problem. Using subdivision interpolation the variation problem for general multivariate function extreme value problem,to get the results. boundary value problem(1), (2), (3) is equivalent to the variation problem following:

$$I[\varphi] = \frac{1}{2}$$ (4)

$$\iint_{\Omega} \varepsilon [\nabla \varphi]^T [\nabla \varphi] dx dy - \iint_{\Omega} \rho \varphi \, dx \, dy - \int_{\Gamma 2} \varepsilon \upsilon \varphi \, ds = \min$$

$$(x, y) \in \Omega \quad (5)$$

$$\varphi |_{\Gamma 1} = \phi(x, y)$$ (6)

The model structure parameter is: 2a=1000nm, lg=500nm, lsg=300nm, lgd=700nm. Milton Model, Drain −3 v voltage, The source ground, Gate and the voltage from 0.5 V to 0.5 V, stride length is 0.25 V, Internal potential distribution model of union line is used to describe, Respectively, as shown in Figue 2 (a) 0.5V, (b) 0.25V, (c) 0V, (d) −0.25V, (e)−0.5V

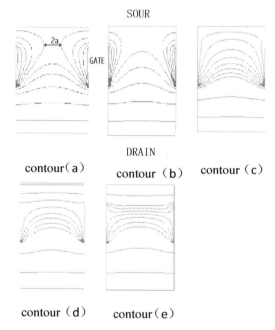

Figure 2. OSIT potential internal conduction channel distribution contours of different gate voltage.

According to the first set of data, lg, will only gate length from 500 nm to 500 nm, The length of the gate to drain lgd from 700 nm to 400 nm, Union line, respectively, as shown in Figure 3 (a) 0.5V, (b) 0.25V, (c) 0V, (d) −0.25V, (e) −0.5V.

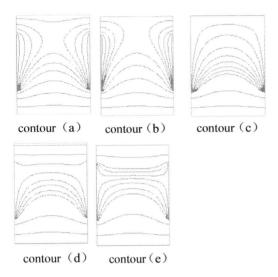

contour （a）   contour （b）   contour （c）

contour （d）   contour （e）

Figure 3. OSIT potential internal conduction channel distribution contours of different gate voltage.

According to the first set of data, Only the length of the gate to drain lgd change from 700 nm to 700 nm, The union line, respectively, as shown in Figure 4 (a) 0.5V, (b) 0.25V, (c) 0V, (d) −0.25V, (e) −0.5V.

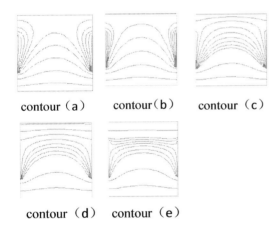

contour （a）   contour（b）   contour （c）

contour （d）   contour （e）

Figure 4. OSIT potential internal conduction channel distribution contours of different gate voltage.

According to the first set of data, will only the length of the gate to drain lgd from 1000 nm to 1000 nm, under the same voltage, the union line,

respectively, as shown in Figure 5 (a) 0.5V, (b) 0.25V, (c) 0V, (d) −0.25V, (e) −0.5V.

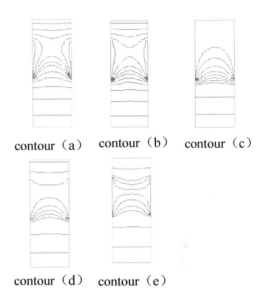

contour （a）   contour （b）   contour （c）

contour （d）   contour （e）

Figure 5. OSIT potential internal conduction channel distribution contours of different gate voltage.

Potential stereogram, according to the first set of data, Application MATLAB software commands of pdetool (Huang, 2012), Type, boundary conditions and coordinate mapping function models of the internal potential stereogram 6 (a) (b) (c), respectively corresponding to plan 2 (a) (c) (e), Describes the model of image's internal potential. X and Y coordinate unit to microns, the coordinates of the Z axis unit for volts.

4 CONCLUSIONS

Though these figures are compared with each other, organic static induction transistor operating characteristics and structural parameters of the device itself, gate bias and leakage source bias are closely linked, By changing the gate and the drain bias, the length of the gate, the gate leak distance and channel width can change the channel in the barrier height, so as to achieve the aim of the current size of a modulation channel. The simulation results with the experimental results is consistent, which proves that OSIT model is correct, and can be used as the basis of subsequent simulation Functional, applied in the simulation experiment.

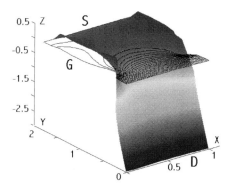

potential stereogram   (a) 0.5V

REFERENCES

Wang D.X. (2002). The development of organic semiconductor thin film transistor. *Semiconductor Journal 6*,
1–4.
Zhang Q,K., X.Z. Liu & J. Yang (2012). Generalized
extended finite element method and its application
in the analysis of crack propagation. *Computational.
Mechanics Journal 3*, 1–3.
Guo S.Y., D.X.Wang, C.J.Wang & G.Y.Wang (2010).
Organic static induction transistor ac small signal
detection circuit design. *Harbin University of Science
and Technology Journal 2*, 1–2.
Zhao S.Q., K.Sheng & Y. Liu (2012). New power electronic
devices- Invited Commentary editor. *Power Electronics
12*, 1–5.
Tang S.X. (2012). Mathematics test graphical user interface
system development based on Matlab. *Research and
Exploration in Laboratory 9*, 1–3.

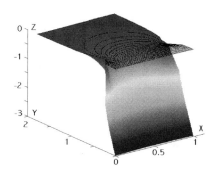

potential stereogram   (b)0V

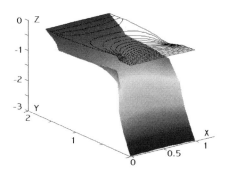

potential stereogram   (c)-0.5V

Figure 6.   Potential internal OSIT conductive channel of
distribution of the gate voltage is (a)0.5V,(b)0V,(c)−0.5V.

Electronic Engineering and Information Science – Wang (Ed.)
© 2015 Taylor & Francis Group, London, ISBN: 978-1-138-02772-5

The optimization of CAN communication network in automobiles

L.L. Wu, W. Wu, R.J. Pei & M.L. Zhou
College of Electrical & Electronic Engineering, Harbin University of Science and Technology, Harbin, Heilongjiang, China

ABSTRACT: Aiming at the status that automotive electronic control units increase brought the circuit complexity and wiring harness increase, a communication system scheme based on TMS320F2812's CAN module was put forward to simplify wiring, improve the communication speed between each unit and reduce the failure rate. According to the communication principle of CAN bus, the hardware circuits of network nodes were designed. Using TMS320F2812 chip built the hardware circuit of an automobile communication system, and conducted simulation experiments for the communication effect between electronic control units in the running process of automobile. Experimental results show that the design of communication between electronic control units features high speed, low failure rate and achieves high-speed CAN network communication.

KEYWORDS: automobile; controller area network bus; network optimization; digital signal processing; communication.

1 INTRODUCTION

With the rapid development of electronic technology and its wide application in automobile, automotive electronics play an important role in improving automobile power, economy performance, safety and driving stability. Many electric comfort separation modules are used, not only improve automobile comfort, but also cause cost increasing, failure rate rising, wiring complicating. Controller area network, which is called CAN bus for short has very high reliability, supports the distributed and real-time control of the serial communication network to adapt the automobile electronic equipment rapid increase (Han et al. 2011).

A design of DSP-embedded CAN controller node minimum system is given, which include just the example program of software design (Jiang et al. 2011, Qian & Huang 2014) rather than the actual design. At present, how to determine the communication priority is a hot topic in the field of CAN communication. According to the response time, which the parameters allowed, a good solution is made to solve competition priority. At the same time, the communication design is concreted and the communication results are given.

2 THE OVERALL DESIGN OF AUTOMOBILE

ECU connected to each CAN network receive the latest messages from bus to manipulate actuator as needed. For the receiving ECU, what they recently received are current messages. Since the message transmission rate between each ECU is changing, for a perfect automobile electronic control system, the transmission frequency must be synchronized with lots of dynamic messages. The CAN field bus is designed for meeting the real-time requirement of each subsystem.

If all control units simultaneously send messages to the bus, message conflict will occur. And the CAN bus protocol is proposed with identifier identification to arbitrate the priority of the message of the bus. Considering the complexity of automobile structures, automobile is divided into electronic fuel injection (EFI) system, electronic control system, anti-lock braking system (ABS), anti-skid regulation (ASR) system, the airbag system, the exhaust gas recirculation system, air conditioning system module. The overall scheme of vehicle network is shown in Figure 1.

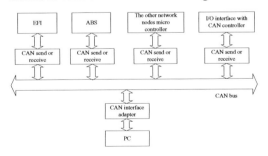

Figure 1.   The overall scheme of system.

3 THE HARDWARE DESIGN OF NODES

The node is the station of receiving and sending messages on a network. CAN bus system contains two types of node, including intelligent node and non-intelligent node. The so-called intelligent node is composed of the CAN control chip and a microprocessor programmable component, but the non-intelligent node is not. The design uses TMS320F2812 chip, which integrates CAN module and has a complete 32 peripherals eCAN controller, and fully support the CAN2.0B protocol (Li et al 2008). Taking the transmission as an example, the design of the subnodes hardware block diagram is shown in Figure 2.

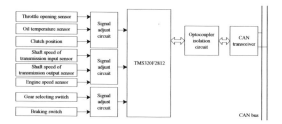

Figure 2. The hardware block diagram of subnodes.

It must be provided with external expansion interfaces to achieve CAN control, and constitute a complete CAN communication system. The PCA82C250 is selected as the interface of CAN controller and the physical bus, and provide differential transmitting and receiving ability to bus. The transceiver has a receiver that process messages from the CANH and CANL lines, and send processed messages to the CAN receiving area of control unit. The transceiver can send a task, and weak signal will control CAN controller units with the amplification. Besides, the signal level can be equal to the signal level and the input control unit on the CAN bus terminal. PCA82C250 is completely compatible with the ISO11898 standard, and could resist instantaneous interference, reduce radio frequency interference, protect power bus, prevent short circuit. The connected circuit of TMS320F2812 and PCA82C250 is shown in Figure 3.

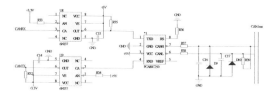

Figure 3. The interface circuit between TMS320F2812 and PCA82C250.

In order to enhance communication anti-interference and reliability, R39 is connected in parallel at both ends of the CAN bus network to matching bus. Meanwhile, 82C250 is connected with two isolation optocouplers to protect the chip and improve anti-interference capability. Considering the transmission rate of CAN network is high, the high speed optocoupler (6N137) is selected as isolation device.

The interface of 82C250 and CAN bus adopts a certain safety and anti-interference measures. After a resistor, 82C250's CANH and CANL pins are respectively connected with CAN bus to limit and protect 82C250 against overcurrent. Between the CANH and CANL, two small capacitors are connected with them in parallel to filter high frequency interference on the bus and prevent electromagnetic radiation. In addition, between the two CAN bus access terminal, a protection diode is connected to prevent overvoltage. In order to achieve power supply isolated completely, the small power supply isolation module or 5V isolation output switching power supply module can be used, moreover, these parts improve node stability and integrity while increasing the node complexity.

4 THE SOFTWARE DESIGN OF NODES

The CAN module in TMS320F2812 is a 32 bit peripherals full eCAN controller and contain mailbox module which has 32 mailboxes. Before the CAN frames are transmitted, the mailbox RAM saves these frames. However, the eCAN controller Module of TMS320F2812 must be initialized to work effectively. Setting the main control register MCR's CCR for 1, CAN module configuration mode can be set. Only register GSR state bit CCE is set 1, and the initialization request is confirmed, CAN could initialize. Then, the bit configuration register BCRn writes and confirms the baud rate of the configuration of the CAN communication, and sets the identifier of each mailbox, width of synchronization jump, sampling number and synchronization mode. Meanwhile, the CCR bit is set 0 by programming, and the CAN module normal working mode is activated again. After setting hardware reset, distribution mode goes into effect.

The other part of software design is the process and implementation of nodes. The system has a plurality of control units, when the CAN bus sends all control commands and messages, other nodes on the bus access to their corresponding messages by pre-programming (Dai & Zhang 2013). The flowchart of the main program of the CAN communication module is shown in Figure 4.

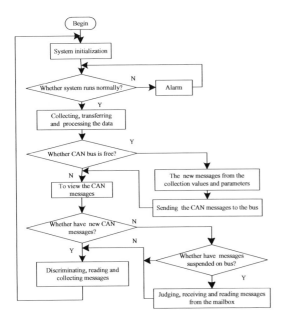

Figure 4. The flowchart of main program.

5 THE RESULTS OF CAN COMMUNICATION TEST

CAN communication simulations have accomplished and successfully achieved the CAN network communication. Node 1 successfully sends messages to the CAN bus, which is shown in Figure 5 (this figure shows messages from the position sensor of throttle). The communication result between Node 1 (throttle position) and node 2 (the input of transmission shaft speed) is shown in Figure 6.

In the Figure 5 and Figure 6, the abscissa represents time, the ordinate represents voltage. Figure 5 shows that the potential of CANH and CANL is always opposite, and the sum of the voltage is a constant. Figure 6 illustrates that messages which node 1 sent and node 2 received is consistent. In addition, the signal interference, which is shown in Figure 5 and Figure 6, mainly is pulse interference generated by high voltage electric spark ignition system of automobile engine and conduction interference from the lines. Moreover, automotive working in a bad environment will produce some interference signals. Anti-interference is a very complex and practical problem, which a kind of interference may be caused by several factors. Therefore, we should not only take anti-interference measures in advance, also should be a timely analysis encountered phenomena causes in the debugging process, continuously improve the circuit principle, system

specific wiring, shielding, and more, ameliorate the types of protection to improve system reliability and stability.

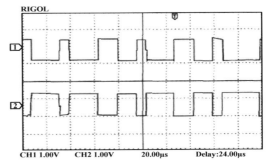

Figure 5. The information sent from single node to CAN.

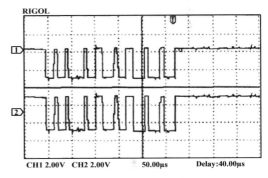

Figure 6. The communication between two nodes.

6 CONCLUSIONS

This paper presents the overall scheme of a vehicle network system and adopts the idea of modularization design. It is given that the specific hardware circuit of a node. According to the CAN communication protocol and node specific requirements, the software flow chart is shown. Experimentally verified that the scheme design can meet the requirements of the control instruction and data transmission nodes in automobile running process.

ACKNOWLEDGMENTS

This work was supported by the Heilongjiang Natural Science Foundation (E201302) and Heilongjiang Training Programs of Innovation and Entrepreneurship for undergraduates (No:201410214015).

REFERENCES

Han J. H. et al. (2011). A Study on the Application of the Controller Area Network Communication Protocol to Hybrid Electric Vehicle. *Automotive Engineering. 33(12)*, 1062–1066.

Jiang S.X. et al. (2011). In-vehicle Network Optimum Design Based on CAN /LIN Bus and CAN Communication. *Instrument Technique and Sensor. 6*, 48–51.

Qian S.P. & T.S. Huang (2014). CAN Network System Design and Applications on Hybrid Electric Vehicles. *Chinese Journal of Automotive Engineering.4(2)*, 109–115.

Li F. et al. (2008). Controllor Area Network Protocol for Powertrain System of Electric Vehicles. *Chinese Journal of Mechanical Engineering. 44(5)*, 102–107.

Dai P.J. & D.J. Zhang (2013). CAN Bus Communication Node Interface Design. *Popular Science and Technology 15(6)*, 1–3.

Electronic Engineering and Information Science – Wang (Ed.)
© 2015 Taylor & Francis Group, London, ISBN: 978-1-138-02772-5

Design and implementation of 16-bit high speed fixed-point multiplier based on FPGA

D.D. Han, H.G. Dong & M.Z. Lai
School of Software, Harbin University of Science and Technology, Harbin, China

ABSTRACT: High-performance multiplier is always considered as a significant part of the DSP. What`s more, it rules an indispensable function for the real-time DSP and image process. A 16-bit fixed-point multiplier is designed based on the optimization method in this paper, the multiplier is divided into partial product generator, partial products compressor and carry look-ahead adder, the topology structure of each module is given. Partial product compressor uses the Wallace tree structure with a higher degree of parallelism, and uses three methods to optimize the area. High carry-result of carry look-ahead adder is no longer dependent on low carry, so as to improve the speed of the whole circuit. Functional correctness is verified by Modelsim, further downloaded to the FPGA for debugging, the observation results showed that multiplier is correct. Due to compressor and adder were optimized, thus the performance of the multiplier is improved.

KEYWORDS: Fixed-point Multiplier; CLA; FPGA.

1 INTRODUCTION

High-performance multiplier rules an indispensable function for the real-time DSP and image process (Chandel, D. et. al 2013). A 16-bit fixed-point multiplier is designed based on the optimization method in this paper. Through the study of algorithms and optimization method, according to the top-down design method, multiplier is programmed using Verilog. Modelsim is utilized for function simulation, through the waveform diagram can be seen that the design of 16 bits fixed-point multiplier basic comply with the design requirements, meet the needs of the function (Dimitrakopoulos, G. & Nikolos, D. 2005). Finally download the designed multiplier to the FPGA for its implementation, through the observation on FPGA shows that as a result, verify the designed multiplier is correct. Designed multiplier algorithm has been conducted for compressors and adder were optimized, thus the performance of the multiplier is improved (Lyons, R. G. 2006).

2 HIERARCHY AND SIGNAL PORTS

The multiplier consists of three components: partial product generator (pp_generator), partial product compressor (compressor), look_ahead adder array (CLA_array). The main component of partial product generator is a plurality of Booth's encoder, the number of the encoder depends on bits of the multiplier (Uwe meyer-Baese. 2002).

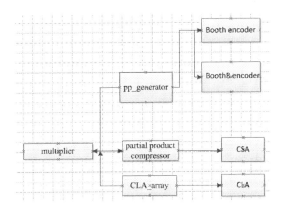

Figure 1. The hierarchy structure of multiplier.

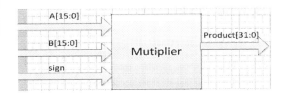

Figure 2. Port signals of multiplier.

When the input is a signed number, an encoder needs to be increased, that is, Booth8 in Fig.1. Partial product compressor is composed of majority full adders and a few half adders. CLA are grouped and carried out in two stages, each group is 4 bit. the hierarchy structure of multiplier is shown in Figure 1. Port signals of multiplier are shown in Figure 2.

3 DESIGN AND OPTIMIZATION OF MULTIPLIER

The topology of multiplier are given, optimization methods of partial product compressor and look-ahead adder are also given.

3.1 *Partial product generator*

By computing input signal A's consecutive three bits and the sign signal to give the corresponding partial product P0 ~ P7 in Booth sub-module, wherein A [−1] is 0. When the sign is 1, the output of P8 is 0, if sign = 0, the output of Booth8 is P8 or the product of A maximum bit and multiplicand B in Booth8 sub-module, the topology of the partial product generator is shown in Figure 3.

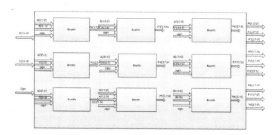

Figure 3. The topology of the partial product generator.

3.2 *Partial product compressor*

The compressor is constructed from the tree structure, the number of full adder minimized in order to achieve the purpose of saving the area. The final tree structure is shown in Figure 4.

The array time delay reach a high balance in Fig4, no extra time delay is substantially generated. Each CSA module in the figure is an array of adder, the number of adder is not the same. In the subsequent part of the area optimized, to reduce the number of adders in each CAS module in order to achieve the purpose of reducing the design area.

The area of partial product compressor is optimized from three aspects in this paper, namely, to reduce the sign bit extension, reducing the mantissa 0 filled and reasonable arrangements for "plus one" operation.

Addend of three highest weight is for sign-extended to 18 bit in this paper, the other addend is extended to this bit, adder function is implemented correctly. Without need to expand that the first place product is aligned with the addend's bit. Such a result can effectively reduce filling of the sign bit, so the design area is also reduced.

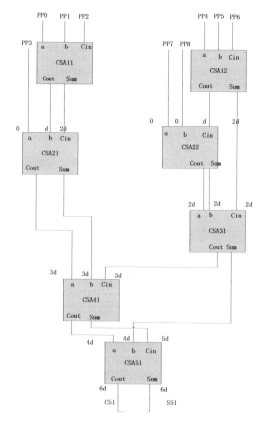

Figure 4. The tree structure of the partial product compressor.

Less than three addends addition is as possible postponed to the subsequent level adder for calculating, which can reduce bits of the intermediate results, the number of adders are fundamentally reduced, thus, the area has significantly improved.

Eight possible "add 1" operation is used as part of some addend in partial product compressor, addition operation is completed in dislocation gap of each summand, thereby avoiding the waste of hardware. Based on the above three area optimization methods, specific embodiments can be obtained in Figure 5.

		14	13	12	11	10	9	8	7	6	5	4	3	2	1	0
	PP0	14	13	12	11	10	9	8	7	6	5	4	3	2	1	0
CSA11	PP1			12	11	10	9	8	7	6	5	4	3	2	1	0
	PP2					10	9	8	7	6	5	4	3	2	1	0
	PP3							8	7	6	5	4	3	2	1	0
CSA21	S11					10	9	8	7	6	5	4	3	2	1	0
	C11						9	8	7	6	5	4	3	2	1	0
CSA12	PP4									6	5	4	3	2	1	0
	PP5											4	3	2	1	0
	PP6													2	1	0
	S12													2	1	0
CSA22	C12														1	0
	PP7															0
	S22															0
CSA31	C22															
	PP8			C7		C6		C5								
	S21							8	7	6	5	4	3	2	1	0
CSA41	C21								7	6	5	4	3	2	1	0
	C31										5	4	3	2	1	0
	S31									6	5	4	3	2	1	0
CSA51	S41									6	5	4	3	2	1	0
	C41										5	4	3	2	1	0
S51		14	13	12	11	10	9	8	7	6	5	4	3	2	1	0
C51			13	12	11	10	9	8	7	6	5	4	3	2	1	0

Figure 5. Optimized partial product compressor.

3.3 Look-ahead adder

Considered the area factor, the adder is divided into two segments. Then each segment as a new target and continue to practice CLA, so there are two CLA sub-modules in the array, carry item that is generated by p_q_generator sub-module and the module input item S[31:0]{C[30:0],0} perform XOR on two steps to obtain the final product [31:0]. The topology of look-ahead adder is shown in Figure 6.

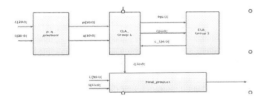

Figure 6. The topology of look-ahead adder.

4 FUNCTIONAL SIMULATION AND FPGA IMPLEMENTATIOT

The two inputs of the multiplier are added two random numbers, so the cycle 10000, if there is no error, the multiplier function is considered to be correct. Partial emulation result is shown in Figure7. Due to number of input switch less than 32,therefore, using a breadboard to expand the number of switches, expansion switch is access to GPIO1 area.

The number of led lights that represent 32 results are insufficient, so digital pipe is used to expand.

Input multiplier and multiplicand are complement form. -5 and 3 were randomized input, the results achieved in the FPGA is shown in Figure 8.

Figure 7. Partial emulation result.

Figure 8. The results achieved in the FPGA.

The Booth algorithm based 4 is used in this paper, at the same time, the calculation amount is reduced, and the subsequent consumption of the hardware circuit is reduced. The design of partial product compressor tree structure is optimized from three sides, so the operating speed of circuit is enhanced and the carry bit performance of multiplier is greatly improved.

5 CONCLUSIONS

A 16-bit fixed-point multiplier is designed in this paper. The multiplier is divided into partial product generator, partial products compressor and carry look-ahead adder. Partial product compressor uses the Wallace tree structure with a higher degree of parallelism, and uses three methods to optimize the area. High carry-result of carry look-ahead adder is no longer dependent on low carry, so as to improve the speed of the whole circuit. Functional correctness is verified by Modelsim, further downloaded to the FPGA for debugging, the observation results showed that multiplier is correct and improved performance.

ACKNOWLEDGMENTS

This paper is sponsored by Natural Science Foundation of Heilongjiang Province(Grant No. F201306)

REFERENCES

Chandel, D. & Kumawat, G. &Lahoty, P. (2013). Booth Multiplier: Ease of Multiplication. *J. International Journal of Emerging Technology and Advanced Engineering.* 3. 326–329.

Dimitrakopoulos, G. & Nikolos, D. (2005). High-speed parallel-prefix VLSI ling adders. *J. IEEE Trans Compute,* 54(2). 225–231.

Lyons, R. G. (2006). Understanding Digital signal Processing. *China Machine Press.* 121–144.

Uwe meyer-Baese. (2002). Digital Signal processing with Field Programmable gate Arrays. *Tsinghua University Press.* 21–77.

Electronic Engineering and Information Science – Wang (Ed.)
© 2015 Taylor & Francis Group, London, ISBN: 978-1-138-02772-5

FPGA implementation of the parallel CRC module in vehicular ad hoc networks

H.J. Yang
School of Software, Harbin University of Science and Technology, Harbin, Heilongjing, China

ABSTRACT: A novel parallel Cyclic Redundancy Check (CRC) method base on recursive formula was proposed in this paper. It meets that the data was transferred correctly in RFID system of vehicular ad hoc networks. The data error detecting module use CRC method to keep the data fields away from transmission errors. CRC is a very cheap and valid error detection method that often used in a wide variety of data storage devices, data communications and computer networks. The CRC module has been designed using Verilog HDL, and implemented by Field Programmable Gate Array (FPGA) of Cyclone IV EP4C30F23C8. It was being verified correctly and quickly in RFID communication system.

KEYWORDS: FPGA; Error Detecting; CRC.

1 INTRODUCTION

The sensors of Vehicular Ad Hoc Networks need a bridge for information transmission and the wireless information transmission module will play this role. The design of the wireless information transmission module is based on RFID technology, uses low-power MCU and high frequency RF chip as the core of the Wireless Information Transmission Module. To enhance the communication security, the security strategies have been designed. CRC has been used in the process of RFID communication adopts to ensure that the data were transferred correctly.

CRC is an error-checking block code that has been used for error detection only, in which the received word has to be divided by a predetermined number called the generator number (Y.PAN et al. 2007). If the remainder is zero, this means that there is no error detected, for nonzero reminder, this means that there is an error detected (S. Shukla, & N. W. Bergmann. 2004).

The traditional RFID system adopts serial CRC calculation Circuit method, this method is simple and reliability of communication (K. V. GANESH et al. 2011). But now the serial calculation method cannot meet the requirements of the information under the condition of high transmission speed, so it is necessary A faster method to calculate the complete data validation (G. Campobello et al. 2003). In this paper, we introduce the Parallel CRC algorithm to guarantee the reliability of the communication at the same time to meet the demands of the higher speed in data processing.

The Verilog HDL source code has been edited and synthesized using Quartus II 12.1, and then simulated and tested using modelsim 6.5. The design has been downloaded into Cyclone IV EP4C30F23C8 FPGA chip. The design has been tested in a hardware environment for different data inputs.

2 THE BASIC PRINCIPLE OF CRC CHECK METHOD

CRC is a common error-detection code adopted in storage devices and digital networks to detect transmission errors. The algorithm of the CRC will attach a short check bits to the data going through these systems, and to perform the remainder of a polynomial division on their contents. In retrieval the consideration is iterated to check for the correctness of the data.

There are two main types of CRC, the non-standard CRC defined by the user of the generation of CRC polynomial, and the standard CRC set by the international organization for standardization to generate polynomial. The second CRC is widely used, the international organization for standardization of several major common CRC as shown in Table 1.

CRC coding is a coding method based on the principles of the CRC, the basic idea is: using linear coding theory, according to the transmission of K-bit binary sequence to produce certain rules r-bit checksum used in the transmit direction of supervision do (CRC code), together with the information after a bit, to form a new binary code sequence of $n = k + r$ bits. On the receiving side, according to the rules of information between the code and the CRC checksum to determine whether transmission errors.

Table 1. The major common CRC of international organization standardization.

Name	polynomial	Briefing	Application
CRC-14	x4+x+1	0x13	ITU G.704
CRC-16	x16+x15+x2+1	0x18005	BM SDLC
CRC-CCITT	x16+x15+x5+1	0x11021	ISO HDLC, ITU X.25
CRC-32	X32+x26+x23+x22+x16+x12+x10+x8+x7+x5+x4+x2+1	0x104c11db7	ZIP, RAR, IEEE802 LAN/FFD I, PPP-FCS

Encodes a data, it is the original data through some algorithm, get a new data. The new data have fixed inner link with the original data. Through the original data and new data grouped together to form a new data, so the data has the ability of self calibration. To the original data is expressed as P (x), it is an order polynomial representation as Equation 1.

$$P(x) = a_{n-1}x^{n-1} + a_{n-2}x^{n-2} + ... + a_1x + a_0 \qquad (1)$$

In the expression, a_i is the data bit; x is a Dummy variables; x^i indicates the position of the data bit. When to encode the data, generated CRC polynomial G (x), and the coding of binary polynomial P (x) added 0 to the end, the number of 0 is r. So the corresponding binary polynomial ascending power for $x^r P(x)$. Then the new polynomial divided by generating polynomial G (x), and get the remainder that is r - 1 order the binary polynomial r (x). The CRC check code polynomial is R (x). Finally to mold method of minus 2 R (x), the corresponding binary sequence is sent string contains CRC check code, which is encoded data.

We can use a linear feedback shift register to implement the algorithm in hardware. This method is called LFSR. The common implementation in hardware was shown in the Figure 1. A clock drives the shifter register, and the input data were moving next to the register at each clock period besides to sending out the data. The shift register already contains the CRC bits, until all the input data have been handled. Then the CRC bits were moved out of the data line.

Presume that the check data are put in a register that is called the CRC register, the procedure of the algorithm implementation is as follows:

1　Initialize the CRC with 0.
2　Judging whether the CRC leftmost bit is equal to 1, if it is true, then move to the next message bit, and

XOR the generator polynomial with the CRC register; if it is false, just move to the next message bit.

3　Iterate step 2 when all bits of the expanded message have been moving in.

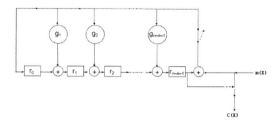

Figure 1. CRC generation using a linear feedback shift register.

3　THE DESIGN AND IMPLEMENTATION OF PARALLEL CRC METHOD

The second generation of RFID informative, high communication rate and recognition speed, high speed data processing circuit is required. CRC circuit, in particular, in a communication to decoding and encoding two CRC calculations, so to have higher data processing speed. The following is a kind of algorithm, the circuit is very simple, encoding and decoding speed high parallel CRC circuit design

We assume that the generator polynomial is G; the input is D; the output is X; the degree of polynomial generator is denoted by m; the length of the data to be handled is denoted by k; the number of bits to be handled in parallel is w. The sequence s is obtained in the final circuit, then the sequence S additional the zeros are transmitted to the circuit in every block of w bits. Waiting $\dfrac{m+k}{w}$ clock periods, the desired FCS are given throughout the FFs output.

$$\text{To construct a matrix } F = \begin{bmatrix} g_{m-1} & 1 & 0 & ... & 0 \\ g_{m-2} & 0 & 1 & ... & 0 \\ ... & ... & ... & ... & 0 \\ g_1 & 0 & 0 & ... & 1 \\ g_0 & 0 & 0 & ... & 0 \end{bmatrix}, \text{ the}$$

XOR operation is denoted by the symbol $\oplus$; the XOR and AND operators are denoted by the symbol $\otimes$; and X(0) to denote the initial state of the FFs, X'(0) to denote the second state. We can have the Equation 2.

$$X'(0) = F \otimes X(0) \oplus D \qquad (2)$$

Then we can recursive the Equation 3.

$$X' = F^m \otimes X \oplus D \qquad (3)$$

Where $F^m = \left[F^{m-1} \otimes g' | ... F \otimes g' | g' \right]$, $g' = (g_{m-1}...g_0g_1)^T$. The matrix F^m is the enable control symbols. Because the generator polynomial is fixed, the matrix F^m is also fixed.

We can obtain the parallel implementation of the CRC from the above depiction. Once again, it is composed of a particular register. In this construction, the specific sum of outputs and inputs of the FF are formed of the inputs of the FFs. The RTL architecture of the generic algorithm implementation is shown in Figure 2.

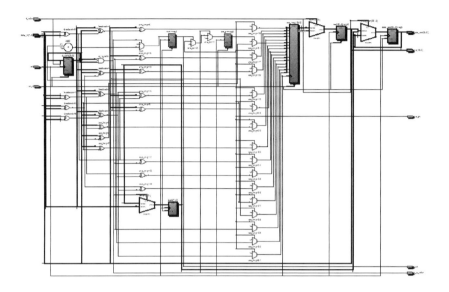

Figure 2. The RTL architecture of parallel CRC module.

4 SIMULATION AND RESULT ANALYSIS

In this thesis, we implement the novel parallel CRC module and The common LFSR CRC module with the same generator polynomial. We simulated the two modules with modelsim 6.5 In order to display the performance. The simulation diagram of the linear feedback shift register CRC module is shown in Figure 3, and Figure 4 is shown the simulation diagram of the parallel CRC module.

Figure 3. Simulation waveforms of the serial CRC module.

Figure 4. Simulation waveforms of the parallel CRC module.

5 CONCLUSIONS

In this paper the method of CRC calculation was researched, and a novel parallel CRC method was implemented using Verilog HDL base on the FPGA device. It is found that the experimental results are consistent with the design, which shows that the parallel CRC module circuit handles 1 byte data in a clock cycle and satisfies the requirements of RFIT system communication rate. It was used to help data delivery in vehicular ad hoc networks.

ACKNOWLEDGMENTS

In this paper, the research was supported by the Heilongjiang Science and Technology Research Project of the Education Department (Project NO. 12541162) and the Heilongjiang Natural Science Foundation (Project NO. F201314) and the Heilongjiang Natural Science Foundation (Project NO. F201232).

REFERENCES

Y.PAN, & N.GE, & Z.W.DONG. (2007). CRC Look-up Table Optimization for Single-Bit Error Correction. *J. Tsinghua science and technology. 12,*620–623.

S.Shukla, & N.W.Bergmann. (2004). Single bit error correction implementation in CRC-16 on FPGA. IEEE International Conference on Field-programmable Technology, 319–322.

K. V. GANESH, & D. SRI HARI, & M. HEMA. (2011). Design and Synthesis of a Field Programmable CRC Circuit Architecture. *J. International Journal of Engineering Research and Applications.* 1, 224–228.

G. Campobello. &. M. Russoand. &. G. Patanè. (2003). Parallel CRC realization. *J. IEEE Trans. Comput.* 52, 1312–1319.

Electronic Engineering and Information Science – Wang (Ed.)
© 2015 Taylor & Francis Group, London, ISBN: 978-1-138-02772-5

Optimization design of SDRAM controller based on FPGA in network-processor system

N. Du, C.Y. Wang & Z.L. Liu
School of Software, Harbin University of Science and Technology, Harbin, Heilongjiang, China

ABSTRACT: In network-processor system, massive data needs to be processed in real time. However, nowadays slow accessing speed of memory has become a bottleneck in enhancing the performance of the whole system. To resolve this problem, SDRAM is widely used instead of traditional SRAM to realize data storage. Due to the complexity of the control mechanism of SDRAM, it is necessary to design an independent controller to simplify the access from the system to the SDRAM. This paper puts forward a design of a high-performance SDRAM controller. Meanwhile, designs of sub-modules of are proposed. We also optimize the delay of read and write operation by adding the pre-instruction module. Finally, we describe the design with the Verilog language in code on modelsim and the simulation results prove the correctness and effectiveness of the design.

KEYWORDS: SDRAM; data storage; FPGA.

1 INTRODUCTION

Nowadays, with the development of high speed and large capacity data acquisition system, data storage technology is faced with a great challenge. In order to meet the data access requirement, most of the embedded systems need external memory to extend the storage space, such as network-processor system. Meanwhile, with the improvement of chip speed, slow accessing speed of external storage has become a bottleneck in enhancing the performance of the whole system (ZHUO L 2010). SDRAM is one of the most commonly used external storages (Jian Qituo 2013). So how to improve the performance of the SDRAM is crucial. SDRAM has many advantages, such as low price, large capacity and high processing speed (JEDEC 2010). However, the SDRAM also has more complex control logic and more stringent timing. So a dedicated controller is needed to simplify the access to the SDRAM. The stand or fall of the SDRAM controller design directly affects the efficiency of the SDRAM and external equipment.

Many researches have been undertaken in this field. A variety of different effective method was proposed to design the SDRAM controller. Seiji Miura and Satoru Akiyama propose two control modes to reduce the latent period of SDRAM cache, which are address queue control and virtual cache control

(Benny Akesson 2011).Benny Akesson and Kees Goossens propose three kinds of technologies to realize the predictability and combination of resources (Whitty 2008). Whitty and Ernst have proposed a controller, which was applied to MORPHEUS platform and interconnect bandwidth optimization SDRAM (Bonatto 2011).

In this paper, on the basis of the functions of conventional SDRAM controller, we put forward a design of a high-performance SDRAM controller. This design achieves the goal to control SDRAM basically and gives an optimization of the pre-instruction module to the memory devices. Besides, it describes the design with the Verilog language in the code. Finally, the simulation result proves not only the basic function of the controller and timing correctness of the logic, but also the effect of the controller after optimization by functional simulation.

2 DESIGN OF SDRAM CONTROLLER

2.1 *Framework of SDRAM controller*

The data storage system compose of six primary modules, including control interface module, CAS delay module, burst length, address generator, instructions queuing module and prefetching instructions module. The whole structure of the system is shown in Figure 1.

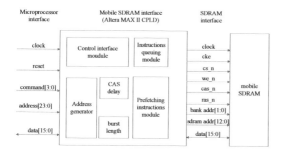

Figure 1. Structure of data storage system.

2.2 *Design of modules*

The CAS delay module is used to watching the instruction delay. The control interface module is used to control operation instruction. The address generator is used to address transition and convert the controller instruction into storage instruction. Instructions queuing module and prefetching instructions module is the optimization of the system, so we will discuss in detail later on. After all, the control module is the most important one.

The interior of the control module is FSM(finite-state machine) which sends instructions and address correspond to the timing acquiring from the microprocessor to the SDARM device. Then the SDARM device enters into corresponding state and executes the instruction. The control module decodes and registers the instruction from the host. Then it sends the decoded instructions which can be idle writable, readable, refresh or charging to instruction module.

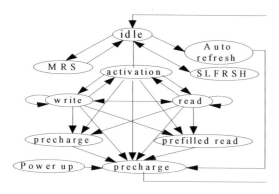

Figure 2. Procedure of the FSM.

The FSM will account for the timing requirement of the storage comprehensively and outputs various operation instructions in correct sequence. The state transition procedure of the FSM is shown in Figure 2. Before doing this apply automatic formatting (AutoFormat in Format menu).

3 OPTIMIZATION STRATEGY FOR PREFETCHING INSTRUCTIONS MODULE

In this paper, to improve the hit ratio of BANK, we adopt Ping-Pong operation to each BANK in the prefetching instructions. Pipeline structure is used for prefetching instruction, so the controller of the SDRAM has predictability. Thus, the opening BANK can hit effectively by the read and write instruction. The optimization strategy decreases the delay of charge in advance and line activation when the BANK is not hit.

3.1 *Optimization strategy of quick hit*

Nowadays, the design of the SDRAM controller always divides storage into multi-BANK to avoid from addressing conflict. Meanwhile, the Ping-Pong strategy is adopted to decrease the delay of shut down the BANK each time. When the SDRAM controller sends a read and write requirement to a BANK, another opening BANK executes pre-charge operation. This method confirms that the BANK hit correctly when next read and write request arrives.

To execute read and write instruction, we assume as Eq. 1:

$$T1 = tRP + tRCD \qquad (1)$$

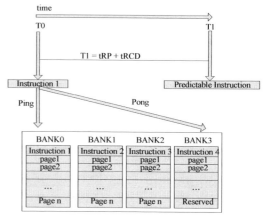

Figure 3. Analysis datagram of predictable quick hit.

T1 stands for delay time that one page of BANK opens and shuts down. If we can forecast the content of the instruction in the delay time which is longer than T1, we can know which line should be shut down. Then this instruction will be hit quickly when operating. As shown in Figure 3, at T0, BANK0 and BANK3 execute Ping-Pong operation. At this time, BANK3 not only execute charging operation, but also execute line activation operation. How can we know the activating line is the next instruction? We should add pipeline mechanism to improve the controller, the method we realize will be instructed as follows.

3.2 Design of the pipeline module

After study, we found that if the instructions which the SDRAM controller sends hit the opening page of the BANK line quickly, the condition N_bank≥N_ ppmust be met. This condition is decided by the storage device. Through inferencing, we need design three levels pipeline. The structure of the three level pipeline module is shown as Figure 4.

Each state and address-line from the initialization state machine and instruction state machine of the SDRAM controller are received by the three level pipeline module. Meanwhile, they will be sent into the instruction parsing module to determine whether to read and write. If it is a read instruction, it will be assigned to corresponding BANK to execute charging operation and line activation operation. The instructions will be sent into SDRAM after decoding in three level pipeline. The main function of the three level pipeline is to generate enough delay for reading and writing instruction. Meanwhile, three level structure makes the instruction to be predicated and will not delay each instruction execution.

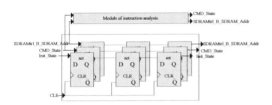

Figure 4.   Structure diagram of three level pipeline.

3.3 Simulation and verification

To verify the design scheme, we use modelsim to simulate the system. We select Verilog as programming language and convert it to block diagram for attachment and pin assignment. We verify all the operation, including initialization process, refresh operation, single write and read operation. At last, we finish the system level simulation. SDRAM can realize burst read and burst write operation. The simulation result of burst writing is shown as in Figure 5.

The four time points respectively represent four burst write instruction. The length of BL is four, which is equivalent to write four data at one time. In Figure 5, line write instruction is actually executed only once when continuous four times write operations are executed in inner of the SDRAM. The four writing operation is continuous and quick hits. The

result show burst write instruction efficiency is very high. In the same situation, as in Figure 6, burst read instruction efficiency is also very high. The simulation results show that the design scheme achieves the expected requirements and meet the goal of the SDRAM controller.

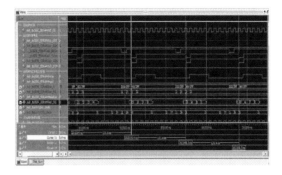

Figure 5.   Waveform figure of SDRAM burst write.

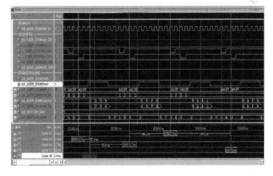

Figure 6.   Waveform figure of SDRAM burst read.

4   CONCLUSIONS

This paper adopts SDRAM technology to design a data storage system for network-processor instead of traditional SRAM. This paper designs a whole data storage system after analyzing the basic structure and knowledge of an inherent system. We primarily divide the system into six modules, including control module, CAS delay module, burst length module, address generator module instructions queuing module and prefetching instruction module. We predict the design and operation of the control module manually. Meanwhile, we give an optimization scheme for the SDRAM controller. At last, we implement software simulation to verify the scheme.

ACKNOWLEDGMENTS

In this paper, the research was supported by the Heilongjiang Science and Technology Research

Project of the Education Department (Project NO.12541162).

REFERENCES

ZHUO L & DUG M & ZHANG D Letal (2010). Design and Implementation of DDR2 Wrapper for Cluster Based MP- SoC. *International Conference on AntiCounterfeiting, Security and Identification*, 60–62.

Jian Qituo & Liu Liansheng & Peng Yu, Liu Datong(2013). Optimized FPGA-based DDR2 SDRAM Controller. *The 11th IEEE International Conference on Electronic Measurement & Instruments*, 786–790.

JEDEC(2010).DDR3 SDRAM Specification, *JESD*.79, 25–30.

Benny Akesson & Kees Goossens(2011). *SDRAM CONTRO-LLERS FOR MIXED TIME-CRITICALITY SYSTEMS*. EDAA, 145–166.

Whitty,S & Ernst(2008).R.A band width optimized SDRAM controller for the MORPHEUS reconfigurable archi-tecture. *Parallel and Distributed Processing. IP DPS*, 112–115.

Bonatto & A.C. & Soares & A.B & Susin(2011). Multichannel SDRAM controller design for H. 264/AVC video decodes r. *Programmable Logic (SPL)*, 898–903.

Electronic Engineering and Information Science – Wang (Ed.)
© 2015 Taylor & Francis Group, London, ISBN: 978-1-138-02772-5

The numerical analyze the change of gate length for the influence of organic static induction transistor potential distribution inside the conducting channel

Y. Zhang, D.X. Wang, Y. Yuan & X.Y. Cui
Key Laboratory of Engineering Dielectrics and Its Application, Department of Electronic Science and Technology, College of Applied Science, Harbin University of Science and Technology, Heilongjiang, Harbin, China

ABSTRACT: According to the actual Organic Static Induction Transistor (OSIT) establishing OSIT's physical model and selecting appropriate structure parameter, calculate the electric potential numerical value of OSIT by adopting the finite element method. According to electric potential distribution in OSIT electric channel, analyze the change of gate length for the influence of OSIT operating characteristics.

KEYWORDS: organic static induction transistor; finite element method; number resolution.

1 INTRODUCTION

In the modern society, a variety of electronic devices have been widely applied to various fields. As a kind of special structure, electronic devices, static induction transistor has many advantages, such as high speed, high pressure, high power, resistant to radioactivity, low distortion and low loss, compared with other types of semiconductor transistor (Shim 2010, Wang &Wang 2012, Wang 2001). The static induction transistor is widely used in high-power medium wave wireless radio stations, the power converters of space satellite and other fields of high reliability requirements. Apart from this, the static induction transistor is used in the devices of DC-AC conversion, audio power amplifiers and measuring instruments and makes high speed and high power integrated circuit chip, etc.

With the research of organic materials and organic semiconductor electronics, the organic semiconductor material used in the production of organic electronic devices, because of low manufacturing cost, variety of preparation methods, the variety of material selectivity and the particular conductive phenomenon discovered in single molecule (Jaehoon et al. 2013). The ranges of the organic semiconductor applications include: the flat panel display, the electronic merchandise tags and a large area of the sensor array, etc.

At present, the organic static induction transistor (OSIT) has been made the sample as a new power electronic device. It needs to do a lot of research work to reach the practical and commercialization level. On the one hand is looking for organic materials with good properties and the new production methods, on the other hand is analyzing the operating characteristics of the OSIT and designing the optimal device structure (Nurul 2011).

According to the actual organic static induction transistor (OSIT) establishing OSIT's physical model, calculate the electric potential numerical value of OSIT by adopting the finite element method. According to electric potential distribution in OSIT electric channel, analyze the change of gate length for the influence of OSIT operating characteristics; explore the OSIT operating characteristics and select appropriate structure parameter.

2 PHYSICAL MODEL

In this paper, establish the physical model according to the actual organic static induction transistor (OSIT). The actual OSIT structure is shown as in the Figure 1 (a), adopt P-type organic semiconductor material copper phthalocyanine, and fabricate the thin film by the vacuum vapor deposition method. The actual OSIT structure is using Au/CuPc/Al/ CuPc/Au five layers. At the gate of OSIT, the depletion is formed by a Schottky barrier with the Al / CuPc interface.

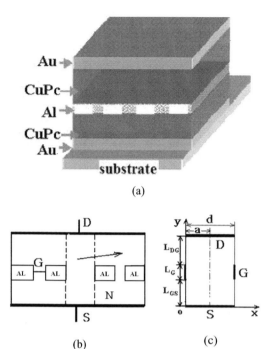

(a)

(b)   (c)

Figure 1. (a) The actual OSIT structure, Fig.1 (b) The sectional view of the structure of the two-dimensional model Fig.1 (c) Establishes two-dimensional models, and the coordinate system, L_{DG} is the distance between drain and gate, L_G is the length of gate, LDG is the distance between gate and source, d (d=2a) is the width of conducting channel.

In the physical model, under the assumption that the electrical characteristics of the device in the Z directions is same, get the two-dimensional transverse cross-section as the research object. The sectional view of the structure of the two-dimensional model is shown as in the Figure 1 (b). This paper discusses the example consists of one single channel, establishes two-dimensional models, and the coordinate system shown as Figure 1 (c). In this model, conducting channel consists of comb aluminum evaporation film gate, the source and the drain consists of gold evaporation films, form an ohmic contact with copper phthalein cyanide evaporation films. In the Figure 1 (c), LDG is the distance between drain and gate, LG is the length of the gate, LDG is the distance between gate and source, d (d=2a) is the width of conducting channel.

The potential distribution inside the model describes by Poisson equation (1) and the boundary conditions (2) (3):

$$\nabla^2 \varphi = \frac{\partial^2 \phi}{\partial x^2} + \frac{\partial^2 \phi}{\partial y^2} = \frac{q}{\varepsilon \varepsilon_0}[(n-p)-(N_d - N_a)] \quad (1)$$

$$\Gamma_1 : \phi = \phi(x, y) \qquad (x, y) \in \Omega \qquad (2)$$

$$\Gamma_2 : \frac{\partial \varphi}{\partial n} = \upsilon(x, y) = 0 \qquad (3)$$

In the equations, φ is the potential function of internal models, Ω is the copper phthalein permittivity, ε_0 is the vacuum permittivity, n is the free electron concentration, p is the hole concentration, N_d is the donor concentration, N_a is the dopant concentration. Γ_1 is the Dirichlet boundary condition, the voltage that drain, source and gate applied of the model is $\varphi(x, y)$, Γ_2 is the Neumann boundary condition, $\frac{\partial \varphi}{\partial n}$ is the derivative of φ outside the normal direction, the default is 0. In this model, the total charge density throughout the region is zero, that is $(n-p)-(N_d - N_a)=0$.

3 POTENTIAL VALUE CALCULATION AND RESULT ANALYSIS

In order to analyze the operating characteristics of the device, according to the parameters, calculate internal potential value, draw a potential distribution diagram, and analyze the relation of operating characteristic between the gate bias voltage and the drain bias voltage. The parameter of the model structure is set to: L_{DG}=800nm, L_{GS}=300nm, 2a=800nm.

1 When the LG change from 500nm to 900nm, the step is 200nm, V_{GS}=0. 5V, V_{DS} change from 0V to -3V, the step is -1V, the potential distribution diagram of any point in the channel centerline, which is parallel to gate, shown as in the Figure 2 (a) (b).

When the same bias voltage applied to the gate and drain of the device, the longer the gate length, the larger the potential value of the saddle point, but when the same amount of changing in the drain bias voltage, the longer the gate, the smaller the amount of changing on the potential value of the saddle point, this shows that, the effect is less on the device with longer gate than the device with shorter gate, under the condition that only change the drain bias voltage.

2 When L_G=300nm and L_G=900nm, the V_{GS}=0.5V, the V_{DS} change from 0V to -3V, the step is -1V, calculate the potential value of point in the perpendicular bisector between the two gates, draw a potential distribution diagram as shown in the Figure 3(a) (b).

It can be concluded from Figure 3 (a) (b), the channel potential decreases with the increasing of the drain bias voltage when the gate bias voltage is constant. From Figure (a) (b), it can be seen, the potential difference of the lowest point (Δ u1>Δ u2), when the gate bias voltage is constant, the effect is less on the device with longer only changes device with shorter gate, when only change the drain bias voltage. The result is the same as from Figure 2.

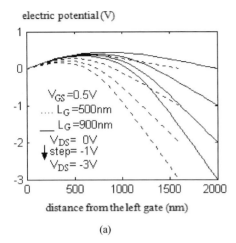

electric potential (V)

V_GS =0.5V
.... L_G =500nm
___ L_G =900nm
V_DS= 0V
↓ step= -1V
V_DS= -3V

distance from the left gate (nm)

(a)

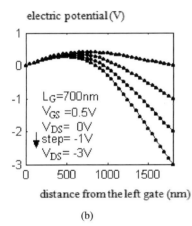

electric potential (V)

L_G=700nm
V_GS =0.5V
V_DS= 0V
↓ step= -1V
V_DS= -3V

distance from the left gate (nm)

(b)

Figure 2. Potential distribution diagram in OSIT channel centerline from source to drain, (a) is L_G=500nm, (b) is L_G=700nm.

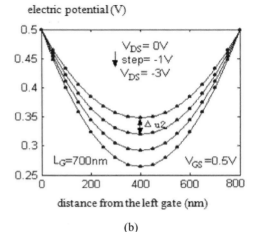

electric potential (V)

V_DS= 0V
↓ step= -1V
V_DS= -3V

Δ u2

L_G=700nm   V_GS =0.5V

distance from the left gate (nm)

(b)

Figure 3. Potential distribution diagram between the OSIT two gates, (a) is LG=300nm, (b) is LG=700nm.

3   When L_G=300nm and L_G=700nm, the V_{DS}=0V, the V_{GS} change from 0.1V to 0.5V, the step is 0.2V, calculate the potential value of point in the perpendicular bisector between the two gates, draw a potential distribution diagram as shown in Figure4(a) (b).

It can be concluded from Figure 5 (a), the channel potential decreases with the increasing of the gate bias voltage when keep the drain bias voltage constant. From Figure (b), it can be seen, the longer the gate, the larger the gate bias voltage. The potential difference of the lowest point (Δ u2> Δ u1), when the drain bias voltage is constant, the effect is more large on the device with longer gate than the device with shorter gate; under the condition that only change the gate bias voltage.

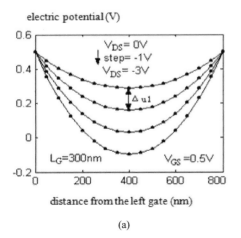

electric potential (V)

V_DS= 0V
↓ step= -1V
V_DS= -3V

Δ u1

L_G=300nm   V_GS =0.5V

distance from the left gate (nm)

(a)

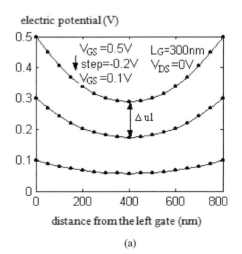

electric potential (V)

V_GS =0.5V   L_G=300nm
↓ step=-0.2V   V_DS=0V
V_GS =0.1V

Δ u1

distance from the left gate (nm)

(a)

97

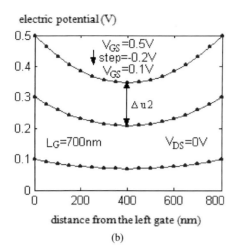

electric potential (V)

distance from the left gate (nm)

(b)

Figure 4. Potential distribution diagram between the OSIT two gates, (a) is L_G=300nm, (b) is L_G=700nm.

4 CONCLUSIONS

In this paper, according to the actual organic static induction transistor (OSIT) establishing OSIT's physical model and selecting appropriate structure parameter, calculate the electric potential numerical value of OSIT by adopting the finite element method. According to electric potential distribution in OSIT electric channel, analyze the change of gate length for the influence of OSIT operating characteristics. The result shows that, when the gate bias voltage is constant, the effect is less on the device with longer gate than the device with shorter gate, under the condition that only change the drain bias voltage; when the drain bias voltage is constant, the effect is more large on the device with longer gate than the device with shorter gate, under the condition that only change the gate bias voltage.

REFERENCES

Shim C.H. (2010). Structural Analysis on Organic Thin-Film Transistor With Device Simulation. *IEEE Transactions on Electron Devices*, 57–60.

Wang X.L. & D.X. Wang (2012). Fabrication and characteristics of sub-micrometer vertical type organic semiconductor copper phthalocyanine thin film transistor. *IEEE Properties and Applications of Dielectric Materials*, 3–7.

Wang D.X. (2001). Study and Fabricating of Organic Semiconductor Film Transistor. *Journal of Dalian railway institute 1*, 64 – 68.

Jaehoon P. & Keum C.M. et al. (2013). Photo-assisted molecular engineering in solution-processed organic thin-film transistors with a blended semiconductor for high mobility anisotropy. *Appl. Phys. Lett 102*, 3306–3310.

Nurul I.M. (2011). Impact of film thickness of organic semiconductor on off-state current of organic thin film transistors. *Journal of Applied Physics. 110*, 4906–4910.

Electronic Engineering and Information Science – Wang (Ed.)
© 2015 Taylor & Francis Group, London, ISBN: 978-1-138-02772-5

Motion characteristics of water droplet under AC and DC electric field

W. Liang, Q.G. Chen, C.H. Song, T.Y. Zheng & X.L. Wei
Key Laboratory of Engineering Dielectrics and Its Application, MOE, Harbin University of Science and Technology, Harbin, China

ABSTRACT: High voltage electric field dehydration is one of the most common technologies of crude oil dehydration. According to the types of applied electric field, dehydration can be divided into AC and DC electric field dehydration, and dehydration characteristics show different regulars upon electric field types. In order to investigate the effect of emulsion dehydration under different electric field types, research on water droplet motion characteristics under AC and DC electric field were carried out respectively. Characteristics of emulsion dehydration under AC and DC electric field were analyzed by observing the motion of water droplets. The results show that tensile deformation and coalescence of water droplets both occur in AC and DC electric field, and oscillation coalescence only occurs in AC electric field, which results in the faster dehydration speed under AC electric field. Electrophoresis coalescence occurs in DC electric field, which leads to the higher final dehydration rate.

KEYWORDS: emulsion; electric dehydration; oscillation; coalescence; electrophoresis.

1 INTRODUCTION

Electric dehydration is one of the most common technologies of crude oil dehydration process, and is widely used in oilfield with its high efficiency and economy (Feng & Guo 2006). In the mid-late period of crude oil exploration, the quality of crude oil becomes more and more heavy and poor. High contents of organic impurities, water and salts caused many problems during dehydration process, such as instability of electric field, large dehydration current and poor separation effect (Jia et al. 2010). These issues affect the efficiency and stability of dehydration operation in the oilfield.

According to the types of applied electric field, electric dehydration can be divided into DC electric dehydration and AC electric dehydration. The selection of suitable electric field type can avoid the instability of electric field and improve the dehydration rate. Therefore, it is necessary to study the dehydration characteristics under different electric field types. Many works on dehydration mechanism and motion characteristics of water droplets under different electric field types have been performed over the years (Eow & Ghadiri 2003, Alinezhad et al. 2010, Supeene et al. 2008, Yang et al. 2011, Zhang & He 2013, Sun et al. 2012). In this paper, the characteristics of crude oil emulsion dehydration under AC and DC electric field were analyzed by observing the motion of water droplets. The dehydration mechanism and processing capabilities under different electric field types are revealed in theory, which can provide theoretical guidance for improving the efficiency of electric dehydration.

2 EXPERIMENTAL SYSTEM AND METHODS

The experiment system for observing droplets motion characteristics are shown in Figure 1, which consists of experimental tank, high-voltage power supply, high-speed camera and computer.

The experimental tank is the glassware, and the flat electrodes are hung inside the tank with 1cm space. The electrodes are fixed by PTFE insulation column. The upper electrode is connected to high voltage, while the lower electrode is connected to ground, so that the high electric field can be established between the two electrodes. Dehydration power is provided by high-voltage power supply device with the DC voltage from -7 kV to +7 kV and AC voltage from 0 to 11 kV. The motion of water droplets is recorded by high-speed camera, which is connected to computer. During the experiment, the experimental tank is filled with white oil, and water droplet with a certain radius is injected into the tank between the two electrodes by micro syringe.

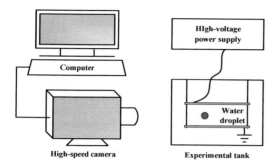

Figure 1.   Experimental system.

3 EXPERIMENTAL RESULTS AND ANALYSIS

3.1 *Motion characteristics of water droplets under AC electric field*

Deformation of water droplet with radius of 0.5mm is shown in Figure 2. It can be seen that the tensile deformation of water droplet occurs in the electric field. Deformation degree of water droplet increases with the increasing electric field strength. Once the electric field strength exceeding a certain range, the water droplet would be ruptured.

As a polar material, water droplet would be polarized in the electric field. There are same quantity bound charges on the opposite ends of polarized water droplet. Water droplet would be stretched by the electric force, as shown in Figure 3.

Deformation degree of water droplet is affected not only by electric field strength, but also by oil-water interfacial tension. Higher oil/water interfacial tension could lead to higher rupture electric field strength and larger maximum deformation degree of water droplet. The relation between electric field strength and deformation degree of water droplet can be expressed as

$$E = \sqrt{6\left(1 - \frac{b}{r}\right)\cdot \frac{b}{r^2}\cdot \frac{kTV_m \sigma}{\mu_0^2 N_A}}, \qquad (1)$$

where b is the minor axis of water droplet, r is water droplet radius, k is Boltzmann constant, T is absolute temperature, V_m is Molar volume, σ is oil-water interfacial tension, μ_0 is intrinsic dipole, N_A is Avogadro constant.

The coalescence of two water droplets with 0.5mm in radius is shown in Figure 4. It can be seen that two small water droplets gradually close to each other, and finally coalesce to a large water droplet. This coalescence of water droplets belongs to dipole coalescence.

There are many influence factors on dipole coalescence rate of water droplets. The dipole force between two water droplets can be expressed as

$$F_e = -\frac{8}{27}\cdot\left(\frac{\mu_0^2 E N_A}{kTV_m}\right)^2 \frac{\pi r_1^3 r_2^3 \cos^2\theta}{\varepsilon_w \varepsilon_0 d^4}, \qquad (2)$$

where E is applied electric field strength, r_1 and r_2 are water droplets radii, θ is the angle between center line and electric field direction, ε_w is relative dielectric constant of water, ε_0 is relative dielectric constant of oil, d is two water droplets spacing.

From equation (2), it can be seen that the attraction force between two water droplets is related to applied electric field, water droplet radii, spacing and the position of two water droplets. Higher applied electric field strength, larger water droplet radius or shorter spacing of water droplets could lead to bigger attraction force, faster speed of water droplets closing to each other and higher probability of collision coalescence.

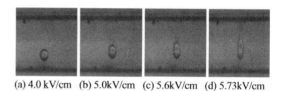

(a) 4.0 kV/cm (b) 5.0kV/cm (c) 5.6kV/cm (d) 5.73kV/cm

Figure 2.   Deformation of water droplet under AC electric field.

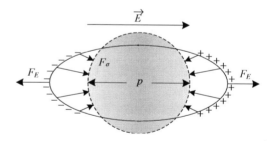

Figure 3.   Deformation of water droplet.

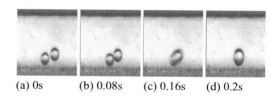

(a) 0s   (b) 0.08s   (c) 0.16s   (d) 0.2s

Figure 4.   Dipole coalescence of water droplets under AC electric field.

Figure 5. Oscillation of water droplet under AC electric field.

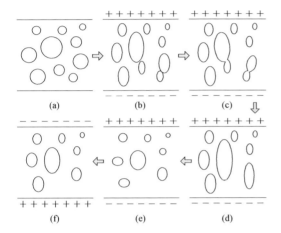

Figure 6. Oscillation coalescence of water droplets under AC electric field.

Oscillation of single water droplet under the electric field strength of 3kV/cm is shown in Figure 5. It can be seen that water droplet is stretched in the electric field and oscillated vertically near the original position. The quantity and polarity of polarization charge on the ends of the water droplet will change periodically with the change of electric field strength and direction under AC electric field. Therefore, the electric field force on the ends of the water droplet will change as well, which makes the water droplet stretched elastically and oscillated along the direction of electric field near the original position.

Oscillation coalescence diagram of water droplets is shown in Figure 6. At initial state, water droplets keep in sphere without electric field applied. After applying the electric field, water droplets would be oscillated along the direction of electric field. Large water droplets coalesce with neighboring small water droplets during the oscillation process, which makes their volume increase. With the change of electric field strength and direction, water droplets continue to oscillate and coalesce. Large water droplets would settle out of the emulsion until the electric field force is insufficient to overcome the gravity. This oscillation coalescence increases the probability of water droplet coalescence and makes the dehydration speed faster.

3.2 Motion characteristics of water droplets under DC electric field

Tensile deformation and dipole coalescence of water droplets will also occur under DC electric field, the process is the same as that under AC electric field. However, the oscillation of water droplets will not happen under DC electric field, because there are no changes of electric field strength and direction. Under this circumstance, the moving mechanism of water droplets under DC electric field can be contributed to electrophoresis.

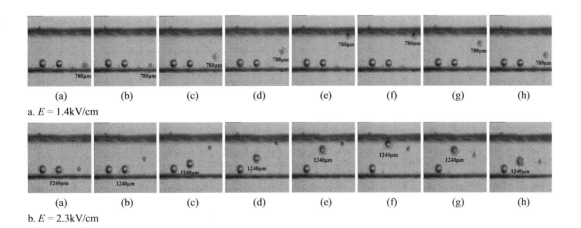

a. $E = 1.4$kV/cm

b. $E = 2.3$kV/cm

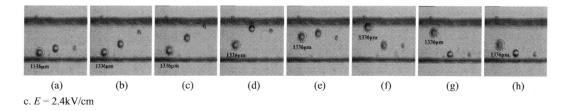

(a)   (b)   (c)   (d)   (e)   (f)   (g)   (h)

c. $E = 2.4\text{kV/cm}$

Figure 7.   The electrophoresis of water droplets with different radii under DC electric field.

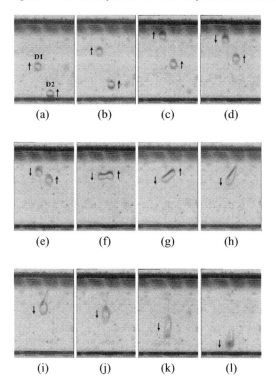

(a)   (b)   (c)   (d)

(e)   (f)   (g)   (h)

(i)   (j)   (k)   (l)

Figure 8.   The electrophoresis coalescence of water droplets under DC electric field.

$$v(t) = \frac{2r^2 g(\rho_o - \rho_w) + \pi^2 r \varepsilon_w \varepsilon_0 E^2}{9\delta}\left(1 - e^{-\frac{9\delta}{(\rho_o + 2\rho_w)r^2}t}\right), \quad (3)$$

where g is acceleration of gravity, ρ_o is density of oil, ρ_w is density of water, δ is viscosity of emulsion.

From equation (3), it can be seen that the speed of water droplet is related to water droplets radii, applied electric field strength and emulsion viscosity. When some factors are constants, the speed of water droplet will increase with time. When emulsion property parameters are constants, the speed of water droplet will increase with the increasing electric field strength. Therefore, electrophoresis speed of water droplet can be faster and coalescence efficient can be higher by improving the electric field strength under DC electric field.

Coalescence of water droplets during the process of electrophoresis is shown in Figure 8. In the figure, the left water droplet is named as D_1, and the right one is named as D_2. D_1 and D_2 move upwards respectively after charged negative charge on the lower plate, which is shown in Figure 8 (a)-(b). D_2 arrives at upper plate and move downward after charged the positive charge, at this time D_1 continues to move upward, which is shown in Figure 8 (c)-(d). When the spacing between two water droplets becomes smaller, two water droplets with opposite polarity charge would attract each other and coalesce into one droplet, as shown in Figure 8 (e)-(f). Because of the inertia of two small water droplets, the opposite ends of large water droplet still have the trend to move upward and downward respectively, which is shown in Figure 8 (g)-(i). With the neutralizing of charges in the large water droplet, electric field force on the water droplet cannot be sufficient to overcome the gravity, so that the water droplet settles out of the oil, which is shown in Figure 8 (j)-(l).

Through the above analysis, dehydration speed under AC electric field is faster than that under DC electric field. The reason is that oscillation of water droplets will occur under AC electric field, which improves the coalescence probability of water droplets and dehydration speed. Electrophoresis of water

The motion of water droplets with radii of 1336μm, 1240μm and 780μm are shown in Figure 7. Water droplet will be charged negative charges when contacting with the lower plate, then it will move upward under the electrostatic force by overcoming gravity. When water droplet reaches the upper plate, its negative charges are neutralized, and it will be charged positive charges and move downward, so that the water droplet reciprocates between upper plate and lower plate.

The starting motion electric field strengths of water droplets with different radii are different. The speeds of water droplets with different radii under the same electric field strength are different as well. Water droplet speed can be expressed as droplets will occur under DC electric field, and the moving water droplets are easily linked to chains when emulsion with higher water content, which can lead to short circuit between dehydration electrodes. Therefore, the dehydration electric field is instable and coalescence efficiency is lower under DC electric field. It's not easily to form a chain between upper plate and lower plate because of the oscillation of water droplets near original position under AC electric field, so that the dehydration electric field is more stable than that under DC electric field.

From the above experimental and discussion results, it can be seen that choosing a suitable electric field type and strength is very important to realize higher dehydration rate and prevent the breakdown happening during dehydration process. The use of AC electric field can improve dehydration speed when emulsion with higher water content, and DC electric field can improve finial dehydration rate when emulsion with lower water content.

4 CONCLUSION

The research on water droplet motion characteristics under AC and DC electric field were carried out in this paper. The results can be summarized as follows:

a. Tensile deformation and dipole coalescence of water droplets will both occur under AC and DC electric field. The dipole attraction force of water droplets is related to applied electric field, water droplets radii, water droplets spacing and position between two water droplets.
b. Oscillation coalescence of water droplets only occurs in AC electric field, which results in faster dehydration speed. Electrophoresis coalescence of water droplets only occurs in DC electric field, which leads to higher final dehydration rate.

c. The electrophoresis velocity of water droplet is related to water droplets radius, applied electric field strength and emulsion viscosity.
d. The use of AC electric field can improve dehydration speed when emulsion with higher water content, and DC electric field can improve finial dehydration rate when emulsion with lower water content.

REFERENCES

Alinezhad, K., Hosseini, M., Movagarnejad, K. & Salehi, M. 2010. Experimental and modeling approach to study separation of water in crude oil emulsion under non-uniform electrical field. *Korean Journal of Chemical Engineering* 27(1): 198–205.

Eow, J. S. & Ghadiri, M. 2003. Deformation and break-up of aqueous droplets in oils under high electric field strengths. *Chemical Engineering and Processing* 42(4): 259–272.

Feng, S. C. & Guo, K. C. 2006. *Oil & gas gathering transportation and processing in mines*. Dongying: China University of Petroleum Press.

Jia, P. L., Lou, S. S. & Chu, X. L. 2010. *Technology of crude oil electric desalting and dewatering*. Beijing: China Petrochemical Press

Sun, Z. Q., Jin, Y. H., Wang, L. & Wang, Z. B. 2012. Impact of high-frequency pulse electric field parameters on polarization and deformation of water droplet. *Journal of Chemical Industry and Engineer (China)* 63(10): 3112–3118.

Supeene, G., Koch, C. R. & Bhattacharjee, S. 2008. Deformation of a droplet in an electric field: nonlinear transient response in perfect and leaky dielectric media. *Journal of Colloid and Interface Science* 318(2): 463–476.

Yang, D. H., He, L. M., Ye, T. J. & Luo, X. M. 2011. Factors influencing single drop deformation in high-voltage AC electric field. *Journal of Chemical Industry and Engineer (China)* 62(5): 1358–1364.

Zhang, J. & He, H. Z. 2013. Dynamics of dispersed droplets in a demulsification process using high electrical voltage method. *Journal of Chemical Industry and Engineer (China)* 64(6): 2050–2057.

Electronic Engineering and Information Science – Wang (Ed.)
© 2015 Taylor & Francis Group, London, ISBN: 978-1-138-02772-5

A greenhouse seedling control system based on PLC and HMI

B.Q. Zhang, S.Q. Tian & Q. Wei
Jiamusi University, Heilongjiang, Jiamusi, China

ABSTRACT: In this article, through the greenhouse seedling to various environmental factors such as control index, the sensor technology, programmable control technology and configuration monitoring technologies applied to the greenhouse control system, and developed intelligent seedling greenhouse control system based on the touch screen and PLC. This system has high control, precision, improve the survival rate of seedling, and ensures that the seedling quality.

KEYWORDS: PLCl; real-time control; system.

1 INTRODUCTION

Vast of land and resources in our country, the development direction in the crops have a larger space, thus cultivating seedlings, become the main task of the development of agricultural product. The intelligent greenhouse system is gradually developed in recent years, a kind of resource saving and efficient facilities, agriculture, technology, it is on the basis of ordinary sunlight greenhouse, combined with modern computer automatic control technology and intelligent sensor technology developed high tech means. With the deepening of the research of greenhouse control technology, is widely used in industry of PLC (programmable controller) has begun to be applied in greenhouse seedling in the system. This design is to touch screen as human-computer interaction, PC and PLC in artificial intelligent control technology to realize the whole process of greenhouse seedling (Cheng & Wu et al. 2011). According to the requirements of the corresponding seedling and seedling growth environment, real-time control in a greenhouse environment, make it undisturbed, enhance the survival rate of seedlings, foster a better seedling quality.

2 SYSTEM COMPOSITION AND THE OVERALL DESIGN

For agricultural greenhouse seedling, the Siemens S7200 system as on-site control stations, through extension module and all kinds of sensor data collection and processing, completed the environmental control of greenhouses. Through the communication between PLC and touch screen, can directly reflect the current greenhouses, such as temperature, humidity, CO2 concentration in the data, draw curves and real time record. This system will be a complex seedling process simplification (Xian & Wang 2014), involved in the system of data conversion, data storage and data calls, etc. to ensure that the storage address within the PLC and touch screen memory address is consistent, otherwise unable to show the accurate data. The system flexible, efficient, wide range of applications, for different seedling can be set by the touch screen is suitable for all kinds of environment needed for the seedlings, and in strict accordance with the growth of the seedlings need real-time control of the environment (Li, Zhang & Li et al. 2012). At the same time, according to the actual situation, change set by the touch screen sensor range, strong flexibility.

3 SYSTEM HARDWARE DESIGN

In order to achieve the greenhouse temperature, humidity, our fleet online measuring the parameters such as concentration and light intensity, the greenhouse environment parameter computer monitoring and control system is designed (Wu & Shi et al. 2013). PLC and touch screen through the RS-485 communication, realize the online measurement. According to the requirements for temperature and humidity in the greenhouse, through screen give information to the PLC about the water pump, filling light fluorescent lamps, carbon dioxide air compensating valve, irrigation, electromagnetic valve actuator control. This system PC (HMI) adopts WEINVIEW MT6100i touch screen as a human computer interaction interface. Host workstation adopts Siemens PLC, CPU-226; Digital quantity expansion module on the 16 EM222; 4 input analog expands module USES EM231; Analog expands module adopts two EM235 (4 input/output) 1. The system hardware structure diagram is shown in Figure 1.

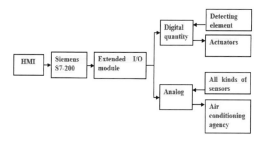

Figure 1. The system hardware structure.

4 SOFTWARE DESIGN OF THE SYSTEM

4.1 PLC program design

According to the control requirements of greenhouse seedling, respectively, within the greenhouse on light, temperature, PH, nutrient solution concentration and CO_2 concentration for data acquisition and processing, and through the air conditioner, fan, water pump, lighting lamp and CO_2 of executive components such as solenoid valve control and operation, the seedling in seedling stage primary respectively, the growth of seedling growth and transition in seedling stage optimal environment.

In greenhouse seedlings need these critical temperature, humidity, light, moisture conditions, the system through the installation of sensors inside the greenhouse feedback signal to the PLC, and then PLC compare the readings with the given value of touch screen make the corresponding judgment after to control the action of actuators, for example, the temperature control system use four temperature sensors placed within the greenhouse, four temperature sensors detected in different positions different, so the system will be concluded after the sum of the average temperature the average temperature as the actual detection, even if the individual sensor drops will not affect the average temperature, considering such holes on program design, so the test data is more accurate, indoor temperature adjusted by the air conditioning, if indoor temperature higher than the set value, the system through the detection of temperature and setting temperature contrast to make changes in temperature the reversing of the judgment, for indoor temperature control air conditioning refrigeration in order to reduce indoor temperature. This system has two kinds of patterns in the temperature control, can be chosen by the user, as the temperature needed for seedling request not high so can fuzzy temperature control, temperature too go refrigeration, heating temperature is too low, so temperature keep in an interval; Another mode is the PID control of temperature in the greenhouse can be relatively stable in

a temperature fluctuation, the control of this control mode is suitable for some of the seedlings of high temperature.

Also includes alarm link in this system, users can set according to his requirement what circumstances need to call the police, the police have a sound, two kinds of light signals. Acoustic signal is the bell, the light signal is light flashing, can remind the staff in the greenhouse of the system can't solve the problem. System of the main program flow chart as shown in Figure 2. Temperature control subroutine flow chart shown in Figure 3; Light subroutine flow chart shown in Figure 4.

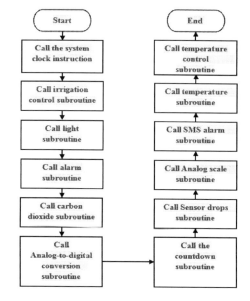

Figure 2. Main program flow chart.

4.2 Touch screen program design

According to design requirements to select touch screen for human computer interaction as the upper machine, in order to realize the monitoring of a machine. WEINVIEW EB8000 matching product development software as a platform, developed greenhouses seedling man machine interface window as shown in Figure 5. At the same time, this software can be flexibly set properties for each element, this is very useful it can be set in the type of data store and extract the data register location. At the side of touch screen program design, main grasp the two points. The first point is in an effort to expand the PLC module, data were collected a reflection of real and effective on a touch screen, so in the design to do touch screen in the text, attributes of components, components of collected data register address, and PLC of this relationship is one-to-one correspondence between a few to reflect the real scene environment conditions, touch

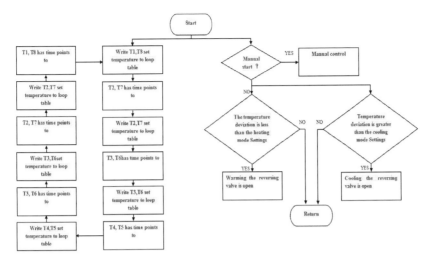

Figure 3. Temperature control program flow chart.

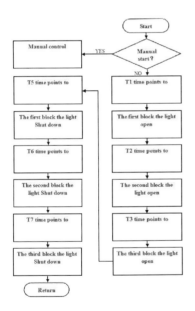

Figure 4. Light subroutine flow chart

Generally include categories within the touch screen picture (Xian & Wang 2014). The first is the total appearance, contains all the control screen, convenient user fast accurate find the need to set the parameters of the position; The second is to set up the picture, including temperature, humidity, illumination and CO_2 setting of various parameters such as the picture; , real-time trend the third kind is the trend picture according to the environment parameters form curve, convenient management personnel to observe the change of environmental parameters trend; The fourth class is alarm screen, responsible for monitoring site condition, the abnormal situation when the field operation of equipment or environment parameter data display and alarm when abnormal change box, at the same time of pro-viding audible and visual alarm display t-he location of the fault, and make history.

screen has nothing to do with data processing problem here, only involves the conversion of a PLC in after the completion of the data in the touch screen, and map the real-time curve; The second point is on the basis of the guarantee all functions as far as possible do beautiful, can download on the net some pictures about seedling appropriate to join in the touch screen, so that a more unique style, another touch screen due to the internal storage space is limited, so to optimize the file editor by touch screen, the advantage is that the touch screen reflects more quickly.

Figure 5. EB8000 software interface design window

5 CONCLUSION

This system based on PLC and touch screen principle, build a more solid platform for the automatic control, make the greenhouse seedling is in a relatively good cultivation mode. PLC is mainly responsible for all kinds of signal acquisition, data processing and control all types of actuators, the touch screen is primarily responsible for making these internal data visualization, and people can according to what you want to cultivate seedlings flexible change parameter, touch screen and data real-time monitoring work, when the parameters need to refer to history, staff are free to download the history file were analyzed, and the experience for the next cultivating seedlings. This system can improve the automation control of greenhouse seedling raising and management level, give full play to the greenhouse agricultural efficiency, all parameters of greenhouse production of decentralized control and centralized management. This system has a high degree of automation, the system is easy to expand, the characteristics of higher liability and economic efficiency, especially, can adjust control parameters, suitable for different crop seedling environment, have a broad market.

ACKNOWLEDGMENT

The paper was co-supported by Jiamusi University research project (ljz2012-23);The scientific research project of Educ-ation Department of Heilongjiang Province(12511556); Jiamusi University Science and Technol-ogy Innovation Team Building Project(Cx-td-2013-01).

REFERENCES

Cheng R. & Z.Q. Wu et al. (2011).The application of PLC to a greenhouse control system. *J. Agricultural mechanization research* (2), 167–169.

Xian Y.F. & C. Wang (2014). Wireless network sewage pumping station automatic control system based on the GPRS. *J. Techniques of Automation & Applications 33(7)*, 56–58.

Li L., L.X. Zhang & D.L. Li et al. (2012). Indicators selecting model for applicability evaluation of greenhouse intelligent control system. *Transactions of the CSAE 28(3)*, 143–153.

Wu X.W. & Z.Z. Shi (2013). Research Progress of On-line Control System in Domestic Greenhouse Environment. *Agricultural mechanization research (4)*, 1–7.

Electronic Engineering and Information Science – Wang (Ed.)
© 2015 Taylor & Francis Group, London, ISBN: 978-1-138-02772-5

Design of support vector machine recognizer based on single chip microcomputer

P. Jiang, Q. Liu, L. Zhao, Y. Tao, J. Xu & Z. Qi
Department of Electronic Science and Technology, School of Applied Science, Harbin university of Science and Technology, Harbin, China

ABSTRACT: This paper introduces a design of support vector machine recognizer who is suitable for linear separable case based on single chip microcomputer. It can classify the new input data into two opposite classes. Its input data are controlled by a digital key through single chip microcomputer. The data classification phase is accomplished by single chip microcomputer based on support vector machine to recognize the process. A set of LED displays the input data and the classification result. The single chip microcomputer used in this paper is STC89C52. It is tested by a set of data, and the results show that the recognizer can work very well.

KEYWORDS: Support vector machine; Single chip microcomputer; Recognizer; STC89C52.

1 INTRODUCTION

Support Vector Machine (SVM) is proposed by Corian cortes and Vapink in 1995. It shows many advantages in solving the small sample pattern recognition, and can introduce into other machine learning problems. Support vector machine is widely used in various fields. In the medical field, it can be used in genetic research and genetic map recognition; in the signal processing field, it can be used in tone recognition, the underwater target recognition; in the image processing field, it can be used in image classification, license plate recognition (Vapnik 2000, Shen 2012). In recent years, support vector machine has been successfully implemented in the related research area. Until now, most of the support vector machine is implemented based on the computer software system. There is a few of support vector machines specially used for image recognition is implemented on FPGA (RuizLlata et al. 2010). This paper implements a linear separable support vector machine recognizer based on the STC89C52 single chip microcomputer. The recognizer inherits the advantages of the single chip microcomputer, which makes this product more convenient and efficient to use (Ma 2007, Guo 2009).

2 SYSTEM DESIGN OF SUPPORT VECTOR MACHINE RECOGNIZER

The classification system is designed as shown in figure 1. The total system has four parts: keyboard controller, input data, support vector machine programs and digital tube display. The keyboard controller controls the single chip microcomputer to read input data since it detects the read signal from the keyboard. When the read control keyboard is pressed, the data which need to be processed are sent to the single chip microcomputer. The support vector machine program and digital tube display part are accomplished by a single chip microcomputer. The single chip microcomputer used in this design is STC89C52, which processing the input data according to the algorithm of support vector machine and send the values of input data to the digital tube display part. The classification result of support vector machine will also send to the digital tube display part (Stephen 2004, Cristianini & Shawe 2011). The first four digital tube displays the input data of support vector machine, the fifth digital tube displays the classification result. Assume all the input data belong to two classes: class A and class B.

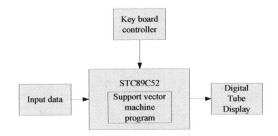

Figure 1. The design schematic diagram.

The main task of this design is implementing the support vector machine classification process on a single chip microcomputer. Therefore, a set of data should be chosen to evaluate this product.

3 HARDWARE DESIGN OF SUPPORT VECTOR MACHINE RECOGNIZER

In single chip microcomputer application system, a button is chosen to control data transmission, which is connected to the single chip microcomputer through the interface circuit. Each keyboard of single chip microcomputer has a fixed IO address. The single chip microcomputer can know whether it has an input signal of keyboard according to the way of interrupt input. If a control signal is detected by the single chip microcomputer through keyboard matrix, then the input data will send to the support vector machine program.

In this paper, a loading board with six digital tubes is used to display the input and output data. The control program of digital tube is written by software, it can control every digital tube to display the right value of the input data and results. The digital tube is lighted by dynamic common cathode method. Through the time-sharing control each COM port of digital tube, so that each digit tube displays the value according to the order of the single chip microcomputer.

4 SOFTWARE DESIGN OF SUPPORT VECTOR MACHINE RECOGNIZER

In order to implement the function of support vector machine recognizer which is suit for linear separable case, the software system design adopted module design idea. The main program flow chart of the recognizer is shown in figure 2. After the circuit is ready to work, when an instruction from a button of keyboard matrix is sent to the single chip microcomputer, the input data will transmit to the single chip microcomputer. And then, the support vector machine algorithm will compute the classification result, according to the input data. And finally, the input data and classification result will display on digital tubes. Support vector machine calculation flow chart is shown in figure 3. When the data are transmitted to support vector machine program, the predicted value Prey is calculated first, then the classification results is computed according to Prey. If Prey is larger than zero, then the input data belongs to class A, otherwise the input data belongs to class B. Class A will show 1 on digital tubes; class B will show 0 on digital tubes.

Debugging is a key process in the single chip microcomputer application system development. The software system program is written in C language (Zeng & Ma 2010), with Keil uVision3 compile software to compile the program and build the hex file for single chip microcomputer. software is STC_ISP_V483, which can download a hex file into the single chip microcomputer. By comparing the classification results from single chip microcomputer

with the results from software running, it is found that the result of this design is completely correct; it also means that the support vector machine recognizer is successfully implemented on the single chip microcomputer.

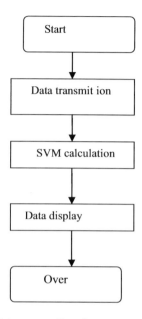

Figure 2. Main program flow chart.

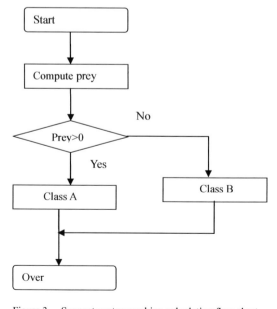

Figure 3. Support vector machine calculation flow chart.

110

5 CONCLUSIONS

This paper introduces the design of a linear separable support vector machine recognizer based on STC89C52 single-chip microcomputer. Its data transition is controlled by a keyboard. The input data and classification result are displayed on digital tubes. The main functions of this design are data analysis, data processing and data display. When using a set of data for support vector machine as a test set on the single chip microcomputer testing, all classification results are completely correct.

ACKNOWLEDGMENTS

This research is supported by college students' innovation experiment projects of Harbin University of Science and Technology in 2013. The number of this project is 30. Thanks all the team members and the teachers that gave their help to these projects.

REFERENCES

Vapnik, V.N. (2000). Essence of statistic learning theory. *M. Beijing, Tsinghua press,* 225–263.

Shen P. (2012). study of support vector machine algorithm *.M. NanJing university,* 5–7.

RuizLlata, M., Guarnizo, G. & Yébenes M. (2010). F-PGA implementation of a support vector machine for classification and regression, *the 2010 international conference on neural networks* (IJCNN), 1–5.

Guo T. (2009). C language of 51 single chip microcomputer. *M. Beijing, Publishing house of electronics industry,* 23–27.

Ma C. (2007). AVR single chip microcomputer theory and application. *M. Beijing, National defense industry press,* 55–65.

Stephen P. (2004). Primer Plus. *M. Beijing, Posts & Telecom press.*66–80.

Cristianini, N. & J. Shawe-Taylor (2011). Theory of support vector machine. *M. Beijing, China machine press.*

Zeng, W. & J. Ma (2010). Algorithm of support vector machine learning. *M. Beijing, Publishing house of electronics industry,* 242–268.

Electronic Engineering and Information Science – Wang (Ed.)
© 2015 Taylor & Francis Group, London, ISBN: 978-1-138-02772-5

Call identification and interception system based on STC MCU

Q. Li, Y.J. Cao, W. Liu, J.M. Feng & H. Luo
College of Applied Science, Harbin University of Science and Technology, Harbin, China

ABSTRACT: In this paper, a fixed telephone identification and interception system was designed and produced, combined with frequency-shift keying decoding chip HT9032D. The system uses STC12C5A60S2 MCU as the information processing and control unit, adopting modularized design idea, to realize the recognition and treatment of fixed telephone incoming call information. In blacklist mode, set to intercept blacklisted number. In white list mode, only connect setting important calls. This system also has functions of displaying the time and number of incoming calls. MCU program is written using C language in the Keil software platform.

KEYWORDS: MCU; frequency-shift keying; identify; intercept.

1 INTRODUCTION

The fixed telephone network has been widely used in China, but the telephone terminal that most users used only had the general functions of call remind and display (Zhao et al 2012), the users were easily vulnerable to be harassed by useless calls, such as marketing, fraud, etc. In order to effectively intercept useless calls, this paper designs and produces a call identification and interception system using frequency-shift keying mode. The system connects to the user terminal, telephone network, by identification and judgment of incoming information, the call is selected to switch on or intercept. This system uses simple concatenated way, has no need for refitting the existing telephone lines and terminal equipment; and has the advantages of high reliability, wide application range, is suitable for families and ordinary office.

2 OVERALL DESIGN

The system adopts the idea of modular design, mainly includes information processing and control module, interception module, FSK decoding module, keyboard module, display module and telephone condition examination module. Each module coordinates work in control of the information processing and control module, to realize the function of the system. The system structure diagram is shown in Figure 1.

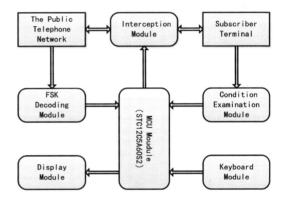

Figure 1. System structure diagram.

The system consists of two work modes, respectively blacklist mode and whitelist mode. The black list and white list can be input through a shortcut key when the call comes or directly using keyboard input. In blacklist mode, when the call is blacklist number, information processing and control module will control interception module to isolate telephone lines, which has the effect of intercepting the calls. When the call is not blacklist number, interception module will connect telephone network with user telephone, the telephone accesses to phone network to realize the normal call. In whitelist mode, when the call is white list number, information processing and control module will control interception module to connect telephone lines, so that the telephone accesses to the telephone network, and normal call is realized. When the call is not a white list number, interception module will isolate telephone network from user telephone. When the telephone condition examination module detects the telephone in the non call and on hook state, interception module will isolate telephone from the telephone network. System running status and incoming information can be shown on the display module in real time.

When the phone line of user has call signal, the FSK decoding module decrypts the call information (Bali et al 2014). Through the form of serial communication, FSK decosv vvvding module sends the decrypted call information (including the calling time and calling number information) to information processing and control module. Information processing and control module converts the received information into the actual phone number, and compares it with list data stored in EEPROM of MCU. According to the comparison results and working mode to choose whether to connect calls.

3 DESIGN OF SYSTEM HARDWARE

This system uses a STC12C5A60S2 microcontroller as the core of information processing and control module. The minimum system circuit includes reset circuit and oscillator circuit. In order to guarantee the accuracy of information transmitted and normal work of each module, we conduct reasonable functional partition for each pin according to the different functions of each pin of the microcontroller and control mode of each module (Xu et al 2011). The system is built after functional partition, in which the core modules are FSK decoding module, interception module and telephone condition examination module.

The FSK signal decoding circuit is completed mainly by the HT9032D chip (Fu et al 2010). The circuit diagram is shown in Figure 2, the FSK signal in the telephone network goes through a capacitor and a resistor coupled to the TIP and RING pins of the chip, and the FSK signal is decoded in the chip, after decoding is completed, HT9032D chip through DOUT pin sends decoded data to the serial communication port of the MCU in asynchronous serial communication, by which the decoding results is read (Zhao 2013).

The interception module circuit uses an electromagnetic relay as functional device. When the control signal is low, the electromagnetic relay is on-state, thus telephone accesses to the telephone network, telephone is switched on. On the contrary, when the control signal is high, electromagnetic relay disconnects, telephone is isolated from the telephone network. Telephone interception is realized by this way.

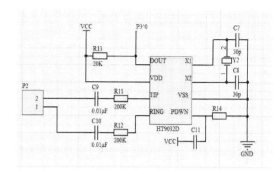

Figure 2. FSK decoding circuit.

Transmission line of the telephone network in China is two-wire analog line, through the DC loop signal, DC feeder, ringing signal, voice data and double audio frequency data are provided with analog phone. The corresponding standard is: when the line has no ringing, switchboard provides DC voltage of about 50V to telephone line. When a user has a call, telephone switchboard sends ringing signal. The ringing signal is a 25Hz AC signal, the intensity is about 75 ± 15V, with period 5s (1s sent, 4s off). When telephone is connected, voltage is low to about 8V (YD/T 1277.1-2003 2003). According to the above standard, the voltage between telephone lines can be divided through resistance, then detection of telephone on and off state will be realized by voltage comparator circuit build up by LM393. When the call is on hook, voltage comparator outputs a high level voltage and when the call is connected, voltage comparator output, low level voltage, when the telephone is ringing, it is a regular rectangular wave signal.

In addition, Display circuit uses LCD12864 of integrated Chinese font as the display, PNP type transistor is used in the circuit as switch device which controls module, the base electrode of the triode is connected with the control signal, when the control signal is low, the triode is on, LCD module works, correlation data information displays after receiving and executing instruction microcontroller sends. When the control signal is high level, the transistor is cut off, LCD module stops working.

4 DESIGN OF SYSTEM SOFTWARE

The system program is written using C language in Keil μVision3 software platform. The software design adopts the idea of modular design, according to the system function, the program is divided into the main program, FSK decoding routine, lookup subroutine, read and store subroutines, control subroutine, etc. The main program is the core of the software system, responsible for normal operation of the system and real-time scheduling of each subroutine. The flow chart of the main program is shown in Figure 3.

When the system is electrified, the whole system is initialized and the necessary configuration is for each module. After initialization, the system enters the standby mode, and real-time monitors the telephone network. When the system detects an incoming signal, it will make the corresponding action according to the working mode of the system. If working in the black list mode, decoding routine will decode and store new information, and then calls the lookup subroutine to compare the call number with numbers in the black list and returns search results. If the call number is in blacklist, intercept it, if not in the blacklist, then connect to the call and display call information, if the call is missed or on a hook, the system returns to standby mode and monitors the telephone network. If working in the white list mode, decoding routine will decode and store new information, and then calls the lookup subroutine to compare the call number with numbers in white list and returns search results. If the call number is in the white list, connect to the call and display call information, if not in the white list, intercept it, if the call is missed or on a hook, the system returns to standby mode and monitors the telephone network.

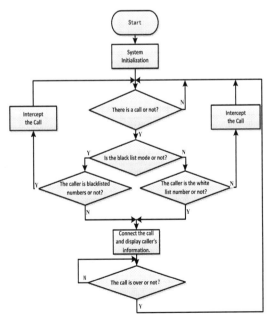

Figure 3.   Flow chart of main program.

5   ANALYSIS AND TESTING

The related standarwds of telephone network of China provide communication protocol of the ringing signal, timing diagram of information transmission is shown in Figure 4.

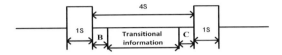

Figure 4.   Information transmission timing diagram.

The related standards provide that ringing signal is a discontinuous signal with 1 second on, 4 seconds off, before and after the transmission of information, time interval B and C are reserved, B and C are chosen according to specific conditions, but not less than 0.5 seconds. In order to efficiently complete intercept function, we hope that the response time of the system as quickly as possible, the best to complete the interception or connect operation within the time interval C. On this we conducted related tests.

The processor of this system adopts STC12C5A6-0SA MCU, the system uses external 12M crystal oscillator to provide the clock, a clock cycle is 83 ns. The lookup subroutine is used in one by one to search. We perform simulation for search time using built-in tools of compiling software. Simulation condition for the assumption stores 500 list numbers, test for the project is looking up a specific telephone number from them. From the simulation results we can see the best case time is about 0.0003 seconds, the worst case time is about 0.1428 seconds. Therefore, within the time interval C, completion of the interception or connect operation can be ensured to save the waiting time of user furthest. But due to the restrictions on transfer format of the related standard data, the first ringing is used to judge incoming information and must be isolated out. When the call needs to be switched on, the telephone will ring when the second ringing signal arrives.

In addition, when the first waiting tone of caller ends, telephone of the called party normally rings. We perform theoretical analysis of an output waveform of the telephone condition examination circuit and interception control terminal, and perform related test, the timing diagrams when the call is switched on and intercepted are shown in Figures 5 and 6, respectively.

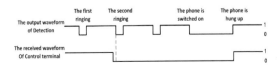

Figure 5.   Timing diagram when the call is switched on.

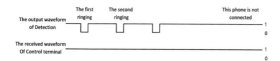

Figure 6. Timing diagram when the call is intercepted.

Examination circuit outputs high level in the standby mode. When a call is received, regular waveform is output. Low level is output with ringing, and high level is output without ringing. Low level is output when the phone is switched on, until the phone is hung up and high level is output. If the phone is not answered or intercepted, high level is recovered to ring after the ringing ends. Interception control terminal outputs a high level in the standby mode, when a call is received, low level is output if the system determines to be answered, high level continues to maintain if intercepted.

6 CONCLUSION

This system uses STC12C5A60S2 as the information processing and control core, through the frequency-shift keying decoding chip HT9032D and isolation circuit, recognition and interception of incoming information are achieved. When the system is on standby, MCU enters in hibernate mode, display module is at the outage, the interception module does not work, only FSK decoding module and condition examination circuit operates in low power state, thus effectively reduce the energy consumption. In addition, looking up a specific number in the worst case takes only 0.1428 seconds from stored 500 list numbers, it sufficiently reduces waiting time of user. Using this system, the functions of recognition of fixed telephone call information and selective connect of incoming call are achieved. Under the premise of ensuring not to miss important calls, the system can effectively prevent telephone harassment from affecting the normal life. The system also has functions of displaying time and call number.

REFERENCES

Bali, M.C. & Rebai, C. 2014. Optimum receiver of coded M-FSK modulation for power line communications. communications Computers and Communication (ISCC), 2014 IEEE Symposium on. IEEE: 1–6.

Fu, G.C., Kong, X.H., Zhang, W. et al. 2010. The design of the black and white lists about the telephone. 2010 International Conference of Information Science and Management Engineering, 1: 57–59.

Xu, T., Zhang, X.D. & Lu, Y.J. 2011. Onboard controlling system design of unmanned airship. Electronic and Mechanical Engineering and Information Technology (EMEIT), International Conference on. IEEE, 6: 3028–3031.

YD/T 1277.1-2003. 2003. Technical specification and testing method of caller identity delivery based on PSTN part 1: technical specification.

Zhao, P.Y. 2013. Design of Serial Communication in Special Circumstances. International Conference on Advanced Information Engineering and Education Science (ICAIEES 2013). Atlantis Press.

Zhao, W., Gu, X.P. & Wang, Y.H. 2012. DESIGN AND IMPLEMENTATION OF FSK FORMAT-BASED SMART CID MODULE. Computer Applications and Software, 29(2): 51–53.

Electronic Engineering and Information Science – Wang (Ed.)
© 2015 Taylor & Francis Group, London, ISBN: 978-1-138-02772-5

Design of fuzzy PID controller based on FPGA

J.H. Han & Z.P. Chen

Harbin University of Science and Technology, College of Computer Science and Technology, Harbin, China

ABSTRACT: Since the PID controller has many advantages, such as stability, relatively, simple structure and reliable work, it has been used more and more widely. But for complex systems, the performance of PID control often can't meet the requirements. In order to solve such problems, I combined the fuzzy algorithm with the PID controller. And owing to the original software method is not ideal, so I implement the controller by means of the hardware. It can solve the problem which is produced by the software method. In my paper, I regard FPGA as the core, and use the method lookup table to complete the fuzzy PID controller. According to the theoretical and experimental analysis, it has the following advantages: short development cycle, design flexibility, reliability, timeliness and good features.

KEYWORDS: Fuzzy control, PID algorithms, FPGA.

1 INTRODUCTION

The PID controller has so many advantages so it has been broadly applied to the field of industrial control. But in the actual process, the control (Li. 2007) led object often have the following characteristics such as time variation, nonlinear and uncertainty, so the conventional PID control is difficult to meet the requirements. The fuzzy controller can simulate the brain's way of thinking. It can deal with nonlinear and accurate information, and has a strong anti-interference. Most of the traditional, fuzzy control algorithm is realized by software, can't be good for real-time processing, and can't be completely avoided by the program fleet destructive impact on the whole control system.

The FPGA has the characteristic of field programmable (Zhao and Chou.2006), the hardware of the system can be programmed and reconfiguration. It has become one of the preferred scheme of hardware design because the development process of less investment, short cycle and can be repeatedly modified. We use FPGA to realize the fuzzy PID controller. It can easily modify the input and output variables with fuzzy rules. And its internal data processing is realized in parallel with hardware circuit, this way can not only solve the problem of the processing speed of the data, but also can solve the problem of susceptible which processed by using software to realize, this way has important application prospect

2 THE DESIGN OF FUZZY PID CONTROLLER

The system of fuzzy PID controller was devised according to the following method, the fuzzy PID controller use deviation and the change of deviation as the input, and then its output the parameters of PID controller. This controller can satisfy with the different time requirements for PID parameters self-tuning. According to the regulation of fuzzy control to revise the PID argument, we realize fuzzy adaptive PID controller. The design diagram of fuzzy regulator as shown in Fig 1. During the operation through constant tests and using the algorithm of fuzzy control to carry out the modification of argument online.

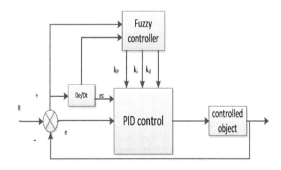

Figure 1. The design of fuzzy PID controller.

2.1 The control principle of PID

Since the PID controller has adjustable parameters, and the structure is relatively clear. It is Suitable with various kinds of control objects, the PID algorithm is relatively simple and more efficient, so it has been broadly applied to the dynamic of industrial control. The PID control expression as shown in formula (1):

$$u(t)=k_p[e(t)+\frac{1}{k_i}\int_0^t e(t)dt+k_d\frac{de(t)}{dt}] \qquad (1)$$

The e(t), u(t), kd, kp, ki in the formula, respectively, as the deviation, the control amount, the differential time constant, proportional gain and integral time constant. We use a summation way to achieve differential, so obtain discrete analog PID control expression is shown in equation (2)

$$u(k)=k_pe(k)+k_i\sum_{j=0}^k e(j)+k_d[e(k)-e(k-1)] \qquad (2)$$

Since the number of memory e(k) can be occupied by a relatively large so that the u (k) and u (k − 1) can be obtained by subtracting the control and organize the incremental-formula. It is shown in equation(3)

$$u(k)=u(k-1)+k_0e(k)+k_1e(k-1)+k_2e(k-2) \qquad (3)$$

In the formula(3),

$$k_0=k_p+k_i+k_d \qquad (4)$$

$$k_1=-k_p-2k_d \qquad (5)$$

$$k_2=2k_d \qquad (6)$$

We use a parallel structure to achieve the PID controller. As this approach has some shortcomings, we make some improvements. Improved methods are as follows:

Because the structural characteristics of FPGA are not suit with the arithmetic of floating-point (Gao & Xu 2012), so we need to convert floating-point conversion, there will be a sign bit coefficient increases, while negative by complement representation, with the addition instead of subtraction.

I extended the bit of adder to prevent the data overflow phenomenon of the adder. Since this system only allows the operation between integers, so we should convert real number to an integer. The data will be converted to an integer arithmetic by determining the coefficients of expansion of the multiples, until the end of the operation, then the output result of the determined multiple of the narrow. In this way can improve the calculation accuracy. The improved way to achieve parallel structure is shown in the following Fig. 2.

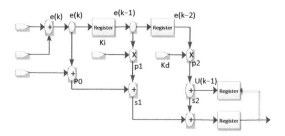

Figure 2. Diagram of improved parallel structure.

Compared with the previous, we reduce the use of logic devices and the cable, but it still can achieve the original effect.

2.2 The theory of fuzzy PID control

This control method is a digital control method which based on fuzzy set theory (Zhu & Su 2005), It uses the rules of fuzzy control to change PID parameters online. It can make the system more stable when the input is different. This controlled object which uses fuzzy control will be statically and dynamically.

We can use the module of Mega Wizard Plug-In Manager in Quartus II. We develop the fuzzy rules table through the measurement which Summary by expert's experience and the theory of PID parameter adjustment. And we realized it by Matlab simulation.

The fuzzy rule table is shown in Table.1. Fuzzy control rule table can be written in the ROM within the FPGA. The output of this controller is the fuzzy inference.

Since the deviation signal is 8 bit (He 2012),we can assume that the basic domain is [−255,255], the domain of integer is set to [3,3], so the basic domain is divided into seven sections, The basic domain can be divided according to 2n, the PID parameters expanded with 2n and then reduced the result of an output with 2n fold. It can be divided into the area of

[−255, −128], [−128, −64], [−64, −8], [−8,8], [8,64], [64, 128], [128, 255]. we can obtain the fuzzy result of e by the same way. Fuzzy control rules table can be written in the ROM table for the fuzzy reasoning process.

To choose the maximum and minimum operating as an operation of synthesis, Fuzzy reasoning uses the max-min synthesis reasoning method in Mandani, and then calculate the fuzzy output value.It also named the method of center of gravity.

3 THE IMPLEMENTATION OF FUZZY PID CONTROLLER WITH FPGA

The whole system is generated by the module of error, the module of address decoder, the module of output of the fuzzy control the module of parameter setting and the module of PID calculation, we designed the system by the method of pipeline processing, realized well of the function of each module.

3.1 The module of error and the variation of error

This module compares the value of output with the value of the set, then obtain the error and the variation of error. The input of the module is the value of set and the U of converted by A/D. The output is the deviation and the rate of deviation.

3.2 The module of address decode

The address decoding module generates an address coding according to the output of the module of the generated error, and use this address to find fuzzy control rule table.

3.3 The output of fuzzy controller

This module query the parameter of output increment based on the output of the module of address decode. The rule table saves the increment value which is calculated by the MATLAB, this paper used in the 49 control rules, it takes up the storage units of 49, the module uses a total of 64,the rules which did not use output 00000000.

3.4 The module of defuzzification

In the field of fuzzy control, the final part is module by removing the fuzzy. As the input of this module is the value of Fuzzy, so we should convert the value of fuzzy into the determinant value. This process of transformation is the solution of fuzzy out of

this module is the value of Fuzzy. In this article, the method of gravity is used to implement the remove fuzzy.

3.5 The module of PID control

In order to realize the PID control needs through a series of multiplier and adder. Because the FPGA's hardware resource is rich, can be directly call Mega Wizard Plug-In Manager which provided by Altera corporation to realize PID control.

3.6 The design of top module

The system consists of this module which introduced above, the module is based on the model Block design, generated all the modules by Verilog-HDL, and then phase interconnect components. The top of the module is shown in Fig. 3.

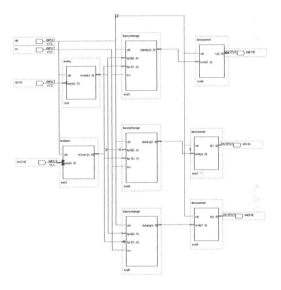

Figure 3. Top of the module.

4 SIMULATION

The simulation analysis was carried out by the ModerSim-Altera, the simulation result of PID module is shown in Fig.4,The sampling time is 40ns, The input of kp,ki and kd was respectively as 2,2 and 1,Assuming the set value r (k) is 100, the feedback value y(k) is 80,we can see from the simulation results. When it experienced about three sampling periods, the output of the system reaches a stable state. It accords with the basic principle of the algorithm of PID control.

119

Figure 4. The simulation of PID module.

The simulation of fuzzy PID module is shown in Fig.5. On the basis of the PID module we process the PID parameters of control faintly. The results of simulation consistent with the basic rules of the fuzzy controller, the input e change from 0 to 3 and then change from 0 to -3,ec to do the same change with e. From the output kp, ki and kd can be seen, the simulation results accord with the fuzzy control basic rules.

Table 1. The fuzzy rules table of kp.

		The deviation signal e						
Δk_p		NB	NM	NS	ZO	PS	PM	PB
The rate of change of the deviation signal e_c	NB	PB	PB	PM	PM	PS	ZO	NS
	NM	PB	PB	PM	PS	PS	ZO	NS
	NS	PM	PM	PM	PS	ZO	NS	NS
	ZO	PM	PM	PS	ZO	NS	NM	NM
	PS	PS	PS	ZO	NS	NS	NM	NM
	PM	PS	ZO	NS	NM	NM	NM	NB
	PB	ZO	Z0	ZM	NM	NM	NB	NB

Figure 5. The simulation of fuzzy PID module.

5 CONCLUSIONS

In this paper, we make full use of the advantage of FPGA to realized digital fuzzy PID controller. The part of PID adopts incremental algorithm, while the fuzzy control part adopts the offline calculation and look-up table online. Under the premise of without increasing the hardware resource consumption, compared with the previous method of completion the of PID control by software, This controller can solve the problem of the real time on industrial control, It can greatly improve the effect of the ordinary PID controller.

REFERENCES

Li C.J. (2007). Research and implementation of PID controller based on FPGA. *Dalian University of Technology*, 54–64.

Zhao G.S. & X.Q. Chou (2006). Development of the adaptive PID. *Chemical instrument and automation 33(5)*, 1–5.

Gao J.S. & X.G. Xu (2012). A study on the control methods based on 3-DOF helicopter model. *Journal of Computers*, 224–228.

Zhu J. & L.F. Su (2005). The principle and application of fuzzy control. *BeiJing: Machinery Industry Press*.

He J.Z. (2012). A study of fuzzy controller based on FPGA. DaLian: *Dalian University of Technology.*.

Electronic Engineering and Information Science – Wang (Ed.)
© 2015 Taylor & Francis Group, London, ISBN: 978-1-138-02772-5

The affection of different rotor structures on starting and operating performance in U-type single phase permanent magnet synchronous motor

M. Fu, Y. Chen, L. Shen, Y.M. Lin & K.K Gai
College of Electrical & Electronic Engineering, Harbin University of Science & Technology, Harbin, China

ABSTRACT: Permanent magnet synchronous motor is increasingly widely used in household appliances field. U type single phase permanent magnet synchronous motor is one kind of non conventional structure of the motor. In order to promote its scope of application, this paper proposes to change the structure of the rotor, and compares the results of air gap magnetic flux density, cogging torque and the speed during starting with different structure rotor based on the method of finite element analysis. It has an important significance for further optimization design and performs research on the motor.

KEYWORDS: Air gap flux density; Cogging torque; Single phase permanent magnet synchronous motor; Starting characteristic.

1 INTRODUCTION

The development of high performance permanent magnetic materials greatly promoted the exploitation and application of permanent magnet synchronous motor. Permanent magnet replace the electric excitation magnetic pole in the traditional synchronous motor, which simplifies the structure, carries out brushless structure and reduces the size of rotor by eliminating the slip ring and brush rotor. This motor is widely used in fans, small water pumps and other small power of household appliances for its cheap price and durability. H. Schemmann had put forward a kind of U-type single phase permanent magnet synchronous motor (SPPMSM) in 1970s, and expounded the motor advantages and disadvantages as well as application limitation. The structure of U-type single phase permanent magnet synchronous motor is simple. The stator consists of asymmetric U-type lamination-stacking silicon-steel sheet. The rotor adopts a cylindrical permanent magnet with two poles. Winding is centralized and connected with a single-phase power. Because the motor armature reaction magnetic contains lots of harmonic-component, the alternating torque generation makes the motor run unsteadily. The fluctuation of speed which is in a region near synchronous speed causes the vibration and noise of the motor (Wang et al. 2009, Fu et al. 2010)

In recent years, most scholars have studied the motor mainly focus on the analysis of the starting process of the motor, especially the research on the influence of stator asymmetric air gap structure for this type of motor starting and running. Such as literature, which presented the effect of the different air gap structure of starting performance of the motor and verified with the experiments. The literature deeply analyzed of the influence of air gap structure on cogging torque and rotor initial position angle, then selected the optimal air gap parameters to shorten the starting time, and optimized the starting performance. This paper analyses the influence on the motor starting and operation from the view of the changing motor rotor structure, which is significative for further improvement and optimization of this motor.

The material of rotor is a permanent magnet, and the circuits and magnetic circuits are asymmetric. Thus, it is difficult to build a mathematical model and do the analytical calculation by using an asymmetric operation analysis method (Fu et al. 2003). So we could simulate and analyze through the finite element method by using Ansoft software. This paper analyzes the influence of different rotor structures on the motor air gap magnetic flux density, cogging torque and starting and running situation of the U-type SPPMSM based on two-dimensional finite element method.

2 THE STRUCTURE AND PRINCIPLE OF THE MOTOR STARTING

2.1 The structure of U-type SPPMSM

The stator of U-type SPPMSM analyzed in this paper adopts the incremental step air gap structure of non symmetrical, stator winding adopts the centralized winding with 480 Ω, the coil inductance value is 0.22 H, and the rated voltage is 220V. The motor structure is shown in Fig.1.

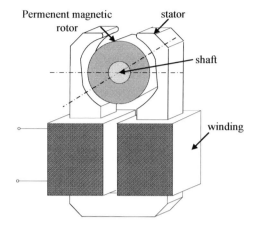

Figure 1. The structure of U-type single-phase permanent magnet synchronous motor.

2.2 The starting principle of U-type SPPMSM

When the stator winding of the U-type SPPSMS analyzed in this paper is not energized, permanent magnet magnetic field distributes in the air gap asymmetrically for its asymmetrical air gap. According to the principle of minimum reluctance, that carries out an angle between rotor pole axis A-B and stator magnetic field pole axis C-C when motor is not running, and the angle is called the starting angle as shown in Fig.2. So at the moment of power on, the electromagnetic force is proportional to the product of the stator and rotor magnetic field and the angle between them, for $T \propto B_s B_r \sin \theta_0$. So create the starting torque by the interaction of the magnetic field of the stator and rotor (Tang 1997).

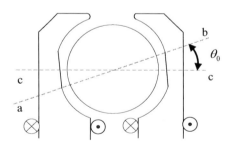

Figure 2. The diagram of starting angle.

2.3 The U-type SPPMSM with different rotor structures

The motor shaft effects of rotary inertia and the distribution of the magnetic field of the permanent magnet rotor is very weak for its very small radius, for about 1mm. So it can be ignored and the permanent magnet

rotor can be looked as solid cylinder. In order to analyze the influence of rotor structure on the operational effect of the motor starting performance, this paper respectively analyzes the four kinds of permanent magnet rotor structures on the electromagnetic torque and speed influence as follows.

The structure of I—a solid, permanent magnet structure;

The structure of II—the rotor with copper sleeve structure, copper cover thickness of 0.3mm;

The structure of III—the situation of permanent magnet rotor using hole structure;

The structure of IV—considers steel core, when the rotor diameter is large, with the diameter of 6mm as an example, we need to consider the influence of rotor shaft on structure parameters and magnetic field distribution. The rotor structures of II~IV are shown in Fig.3.

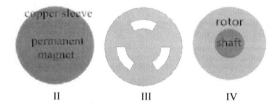

Figure 3. The rotor structure diagram.

3 THE STEADY MAGNETIC FIELD ANALYSIS OF U-TYPE SPPMSM

3.1 The calculation of air gap flux density in U-type SPPMSM

Due to the very little change in the axial direction of motor's electromagnetic force, the electromagnetic field is analyzed in the plane to simplify the calculation. In two-dimensional coordinate, U type motor satisfies the magnetic field equation expressed by the magnetic vector potential A_z, the equation as follows:

$$\begin{cases} \Omega : \dfrac{\partial}{\partial x}(\dfrac{1}{\mu}\dfrac{\partial A_z}{\partial x}) + \dfrac{\partial}{\partial y}(\dfrac{1}{\mu}\dfrac{\partial A_z}{\partial y}) = -J_z + \sigma \dfrac{\partial A_z}{\partial t} \\ \Gamma_1 : A_z = 0 \\ \Gamma_2 : \dfrac{1}{\mu_1}\dfrac{\partial A_z}{\partial n} - \dfrac{1}{\mu_2}\dfrac{\partial A_z}{\partial n} = J_s \end{cases} \quad (1)$$

In the above equations: μ is magnetic permeability. σ is conductivity. A_z is magnetic vector potential component of Z axis. J_z is current density component of the Z axis. Γ_1 is the first kind boundary condition. Γ_2 is permanent magnet equivalent surface current boundary. J_s permanent magnet boundary equivalent surface current density. Applying the first kind of boundary condition in solving regional peripheral air boundary.

This paper analyzes the motor starting angle through the steady field, under the condition of stator winding loading no excitation, $J_Z = 0$. Then makes the simulation of air gap magnetic flux density in condition of four kinds of rotor structures according to the formula (1), which only consider the effect of permanent magnets. Then the curves of air gap magnetic flux density changing following the position of the rotor could be gotten in condition of different rotor structures, as Fig.3 shows. We can know that curves of the rotor with copper sleeve and the rotor only for permanent magnet are coincident, which indicates that the copper sleeve structure do not change the value and distribution of magnetic field in the steady state. However, rotor using hole structure and the structure with large size shaft not only change the value of air gap magnetic flux density but also the distribution of magnetic field.

3.2 The calculation of cogging torque of U-type SPPMSM

In permanent magnet motors, there exist electromagnetic torque even without excitation on stator winding. Its value is equal to the value of torque that generated to make the rotor back from the other position to the rotor position of starting placed. The cogging torque can be deduced from the calculation of the magnetic field energy storage $W(\theta, i)$ at a different rotor position angle by using the finite element method, in the case of stator winding open-circuit. Finally get the cogging torque by using the virtual displacement method. The following formula (Wen et al. 2010).

$$T_c = \frac{\partial W(\theta, i)}{\partial \theta}\bigg|_{i=0} = \frac{\partial}{\partial \theta}\left[\int_v \left(\int_0^H B\, dH\right) dV\right]_{i=0} \quad (2)$$

Assume that the stator winding loads no excitation, only consider the rotor affection itself, do the calculation of every rotation. According to the above principle, take the simulation of cogging torque of four kinds of different rotor structures as they are in different position angles of the Ansoft software in transient field. The result curves are shown as Fig.4 and Fig.5.

It can be known from the Fig.4 that the starting angle of four kinds of rotor structure is all the same of 10.4 degrees, which indicates that the starting angle does not change with the transformation of the rotor's structure. The balance position of cogging torque is still at the axis of zero value. However, it can be known through the curve in the figure, that the electromagnetic properties and structural parameters of the rotor is changed resulting from the change of the rotor structure. And the cogging torque in different rotor

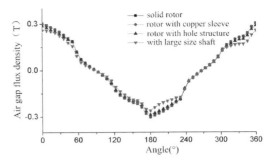

Figure 4. The distribution of magnetic field with different rotor structure.

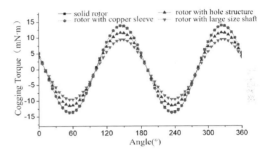

Figure 5. The cogging torque of different rotor structure.

structures is different either. And the structure of large diameter steel core shaft and the permanent magnet rotor with holes do not affect the equilibrium position of the corresponding angle. But the value of cogging torque decreases obviously for the reduced dosage of permanent magnet, especially the structure with holes.

4 THE TRANSIENT ANALYSIS OF U-TYPE SPPMSM

Assume that the stator winding load 220V voltage excitation, the rotor position in the initial angle is 10.4 degrees and the initial velocity is zero. The motion equations of the rotor for:

$$\begin{cases} J\dfrac{d\Omega}{dt} = T_e - R_\Omega\Omega - T_L \\ \theta = \theta_0 + \Omega t \end{cases} \quad (3)$$

In the above equations: J is rotor moment of inertia. T_e is electromagnetic torque. T_L is load torque. R_Ω is the coefficient of friction. Ω is Motor speed. θ is the angle of rotation.

Based on the above theory, speed results and starting data of four kinds U-type SPPMSM can be achieved through the time stepping finite element analysis in an electromagnetic field. The results are shown as Fig.6 and Table 1.

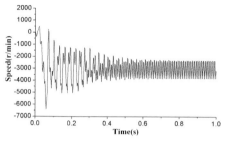

(a)The speed curve with solid rotor structure

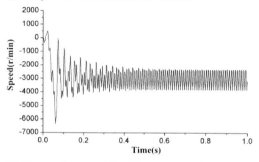

(b) The speed curve with solid rotor covered with copper sleeve

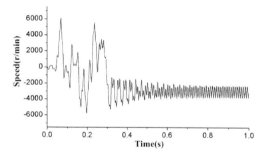

(c) The speed curve with perforated rotor

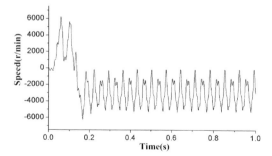

(d)The speed curve with large diameter steel shaft

Figure 6. The motor speed curve.

Table 1. Data of the motor starting.

Structure	Starting time [s]	The range of stability speed [r/min]	The MAX. speed fluctuation in starting [r/min]
I	0.45	2250–3750	6500
II	0.3	2300–3700	6500
III	0.6	2200–3800	6000
IV	0.2	200–5300	6000

It can be known that the speed fluctuation of U-type SPPMSM is volatile through the curves of four rotor structures. The speed jitters in start for a time, then starts to stabilize at around 0.6s with a speed about 3000r/min, and the range of speed fluctuation is 2250–3750r/min. In addition of copper sleeve structure, the speed fluctuation is suppressed in 0.1–0.3s after starting, and stabilized at about 0.4s. But the speed in steady state is almost no difference between solid rotor structure and the rotor with copper sleeve structure. This is because conductive property of copper sleeve is very good. The magnetic field generated vortex in the copper sleeve, thus forming a torque assist motor start. Saving dosage of permanent magnet and reduce rotor inertia through perforating on the rotor will cause strong jitter in the process of starting. That is due to the rotor hole so that the distribution of the magnetic field of the permanent magnet changed and reduced the energy product. But the speed fluctuation of the motor and the original structure is almost the same after 0.6s,which indicates that in the condition of reducing the dosage of permanent magnet, under the interaction of the moment of inertia and the magnetic energy product, may not change the stable speed. We can know from the figure (d) that in the condition of the large diameter motor shaft with no change of the rotor diameter, the rotor moment of inertia is increased and the product of energy and produce decreases. The motor running speed fluctuates a lot in the starting process and stable operation under the dual role. Thus, when the shaft diameter is large, it need consider the use of better magnetic performance permanent magnets, such as rare earth permanent magnet to improve the operation of the motor.

5 CONCLUSIONS

This paper simulated the air gap magnetic flux density, cogging torque and starting and running situation of the U-type SPPMSM with four kinds of rotor structures based on two-dimensional finite element method, and get the following conclusions:

1 The change of rotor structure could affect the moment of inertia and rotor air gap magnetic flux density.

2 The change of rotor structure does not change the initial position of the starting angle which is only related to the stator structure.
3 The change of rotor structure could cause great fluctuation of speed. This makes the motor produces a strong vibration, and seriously affects the motor service life. Thus, it needs to be fixed up by some other measures. But the copper sleeve structure has an inhibitory affection on speed fluctuation. What's more, the copper sleeve structure can also reinforce the permanent magnet. When the rotor structure changes greatly, the structure and excitation parameters may vary, in which condition all these changes may cause reversal, stall or unable to start.
4 The coordination among the rotary inertia of the rotor, rotating damping, load torque and excitation have great influence on starting and operating performance. Because the four rotor structures in this paper do not occupy the occurrence of the above situations, further studies are required to validate.

REFERENCES

Wang F.X., X.P. Wang & Y. Gao (2009). Starting performance of U-type single-phase permanent magnet synchronous motor. *Journal of Shenyang University of Technology 31(1),* 1–6+39.

Fu M., C.S. Yu, & H.Z. BAI (2010). Influence of step-air gap on line-start performance of U-shape single-phase permanent magnet motors. *Electric Machines and Control 06,* 39–44.

Fu D.J., Y.B. Yang & X.H. Wang (2003). The overview of Single phase line start permanent magnet synchronous motor. *Electrical Machinery Technology 01,* 34–37.

Tang R.Y. (1997). Modern Permanent Magnet Machines. Beijing, *Machinery Industry Press,* 4–6.

Wen J.B., H. Li & F.L. Zheng (2010). Calculation on Electromagnetic Torque of PM Synchronous Motors. *Explosion-Proof Electric Machine 01,* 1–3.

Electronic Engineering and Information Science – Wang (Ed.)
© 2015 Taylor & Francis Group, London, ISBN: 978-1-138-02772-5

Research on the personal area networking of wireless intelligent terminal equipment based on STM32

J.B. Xie, Y. Ning, S. Miao & H.Q. Gao
School of Electrical and Electronic Engineering, Harbin University of Science and Technology, Harbin, China

ABSTRACT: To realize the smart home service platform using Internet of Things technology, this paper constructs WIFI intelligent control system based on STM32. Firstly, the importance of WIFI signals to personal area networking of wireless intelligent terminal equipment is analyzed. Then whole system framework of WIFI networking is designed. Finally, the hardware circuit of WIFI intelligent control system is realized through the microcontroller STM32 and other circuit modules. We use android phone app developed by Eclipse and make it connected to the hardware system via WIFI to control LED on and off, verifying the feasibility and validity of whole WIFI intelligent control system that can achieve personal area near-field and far-field control and interconnection. This system has great practical significance.

KEYWORDS: STM32; Wireless Module; Intelligent Terminal; Android; TCP/IP.

1 INTRODUCTION

Smart home can also be referred to as smart home service platform. It effectively combines the home life of home intelligent control, information exchange and consumer service, and creates the highly efficient, comfortable, safe, and convenient personalized home life with the comprehensive utilization of computer, network communications, and appliance control technology (IEEE Std 802.11b. ed. 1999).

With the gradual development of information technology, increasingly sophisticated network technology, increasingly rich applicable network carrier, and the gradual promotion of large bandwidth indoor network home strategy, intelligent information service into home becomes possible. Residents can interact through the TV remote control (Gast, 2007), mobile phone and other terminals, and enjoy the smart, comfortable, efficient and secure home life quickly and easily.

As the high-tech industry, the smart home industry is supported by national industry (Leonid Batyuk, et al, 2009). National Twelfth Five-Year Plan has made it clear that wireless smart home industry and new energy, cultural and creative industries are tied for the strategic emerging industries (Zhang,Y.,J. Y. Ma, & X. Q. Yang, 2012). Smart home service platform system belongs to intelligent home, which will have broad market prospects in the future.

Juniper Research forecasts smart home service revenue in 2012-2017: smart home service revenue will be more than 58 billion US dollars in 2017, substantially increasing over the 25 billion US dollars this year, with the annual compound growth rate of 18%. Where, the entertainment market revenue will reach 48 billion US dollars, making up 82% of total revenue.

Most American families are single-family houses with a set of independent air-conditioning and boiler systems for heating and cooling. The thermostat device can be a good choice to satisfy the needs of these users, and can provide users with the constant suitable living environment.

In Japan, there exists more developed smart home. In addition to achieving indoor appliances automation networking, the automatic door is also achieved through biometric identification system. When you stand in front of a camera mounted on the entrance, it takes about 1 second to confirm your identity. If you the apartment dwellers, the door will open.

In Spain, when there is abundant natural light indoors, the fluorescent lamp with sensing function will automatically turn off, reducing energy consumption; Weather sensors placed on the roof can always get the weather and temperature data, when it rains, it will automatically turn off the lawn sprinkler and close pool; And when the sunlight is strong, it will automatically open the awning on room and yard.

Korea Telecom use 4A to describe the characteristics of their digital home system that allows the owner at any time and any place to operate any appliance at home, to get any service.

2 PERSONAL AREA NETWORKING OF INTELLIGENT TERMINAL EQUIPMENT

In China, smart home industry has entered a new round of development peak due to the implementation of Three-Network Convergence, Internet of Things national development strategies. At the Sixth China (Hefei) International Household Appliances Fair, as China's "two" fusion theory founder, Zhou Hongren made a speech on "smart home". He said the family information is a general trend. With the development of information technology, especially broadband and mobile, conditions for promoting the family information are gradually met, but family information will also become an important trend in the development of the information revolution and information technology. It improves people's lives quality and makes

technology are wireless access, high-speed transmission and long distance transmission. 802.11n can improve the transfer rate of WLAN from the current 802.11a and 802.11g 54Mbps offer up to 300Mbps and even up to 600Mbps. Some present microcontrollers have sufficient resources to manage 802.11 protocol and TCP/IP protocol, so that some embedded devices can use radio resources.

Based on the STM32 ARM microcontroller as core processor, the hardware part of wireless control terminal is developed and realized by building the peripheral circuits. We set up Personal Area Network via WIFI signals or connect to the Internet via WIFI hotspots. We can control another smart device of the networking, and research the mutual access and control capabilities of intelligent terminals in networking. The networking via WIFI is shown in Fig. 1.

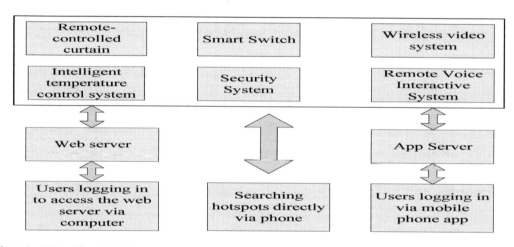

Figure 1. Networking via WIFI.

important contribution to further changing the way people work, learn and live.

There are not too many electronic control products based on WIFI signals. In China and abroad, they are still in development and exploring stage and are still far from using widely like infrared and Bluetooth signals. At present, some control devices based on infrared signals abound in our life. But the weakness of the infrared signal is small transmission range, especially when there are obstacles. WIFI signals are much larger than the infrared signals in transmission range. Therefore, the electronic control products based on WIFI signals have good prospects for development.

With the development of WIFI signals, they have not only been used as a network signal, they've also been emerging as a control signal for research and development. The main advantages of WIFI

3 HARDWARE IMPLEMENTATION OF WIFI INTELLIGENT CONTROL SYSTEM CIRCUIT

The hardware circuit of WIFI intelligent control system consists of four major parts. That is, control unit based on STM32F130 microcontroller, data acquisition system, WIFI module, and power section. WIFI module is mainly composed of WIFI chip WM-G-MR-09 and its auxiliary circuit. TCP/IP protocol, ICMP protocol and others protocols are embedded in WIFI module. IEEE802.11 b/g/n is supported and data can be transmitted transparently. Control unit STM32F130 microcontroller is the control core of system. Data acquisition system and the control unit complete processing and transmitting alarm signals. Power section provides power supply for all parts of the system.

Hardware production: we use software Altium Designer Winter 09 to draw schematic and component packaging, and obtain the system PCB and final printed circuit board, as shown in Fig.2. And we made the corresponding electronic devices welded on the circuit board. The electronic devices must be welded without any lack of fusion. And we must pay attention to whether there exist some unnecessary connections between the chip pins, which can be checked by a multimeter.

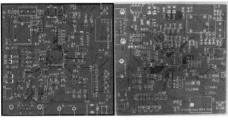

a. system PCB	b. printed circuit board

Figure 2.　System PCB and the printed circuit board.

WIFI module and STM32 are connected with USART1, thus when STM32 is initialized, we need to open USART1. The baud rate and interrupt type can be set in function USART_Configuration (void). The baud rate set in this project is 115200bps, and the interrupt occurs both in sending and receiving to ensure that data can be sent and received. The configuration process is shown in Fig. 3. And the format for each parameter is uniformed through data structure NetParaBuffer, whose specific structure is:

Typedef struct NetParaBuffer
  {
  unsigned char M_id; // Parameter name
  unsigned char cLength; // Parameter length
  unsigned char cInfo[64]; // Specific parameter content
  }

After configuring the WIFI parameters, the function WIFI_Para_set_auto () is called to make WIFI in automatic mode and networking, so that it can be connected to the network via the AP, later the data calling function SendDataToWIFI directly can be sent to the WIFI module through USART1. And the data is sent out via the AP in terms of the server IP address set previously.

1 When the connection is established, we can use two methods, one is the program initialization, using "static" or constructor; the other is user interaction, clicking a button to trigger events.
2 In the process of sending data, we can use function BufferReader to send large audio files quickly.
3 If the connection is established using the program initialization, we use function OutDestroy () to close the program; if using the user interaction, we just click a button. But both of the two processes are the same, closing DataInPutStream byte stream first and then closing the socket.

4　TEST

Eclipse is open source and it is a Java-based extensible development platform. On its own, it is only a framework and a set of services for building development environment by plug-in component. Eclipse comes with a standard set of plug-ins, including Java development tools (Java Development Kit, JDK). We use android phone app developed by Eclipse and make it connected to welded hardware systems via WIFI to test LED on and off, as shown in Fig. 4.

5　SUMMARY

This paper takes full advantage of Internet of Things technology, WIFI technology, and Android system, realizing Wireless Networking through WIFI

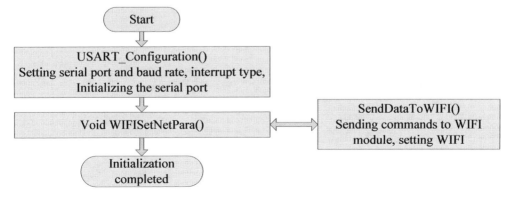

Figure 3.　Configuration process.

Figure 4.   Test for turning on LED.

intelligent control system based on STM32 for smart home devices. It makes intelligent terminal equipment in personal area into networking, achieving near-field and far-field control by cell phone or other smart terminal. Therefore, controlling home electrical equipment can be achieved anytime, anywhere, and really achieving the Internet of Things system.

ACKNOWLEDGMENT

In this paper, the research was sponsored by Overseas Scholars Fund Project of Heilongjiang Provincial Education Department (Grant No. 1253HQ019).

REFERENCES

IEEE Std 802.11b. ed. (1999). *Wireless LAN medium access control (MAC) and physical layer (PHY) specifications:higher-speed physical layer extension in the 2.4 GHz band.* New York: Institute of Electrical and Electronic Engineers.

Gast M.S ed.. (2007): *802.11 Wireless Networks Definitive Guide 2nd ed.* Tsinghua University Press.

Leonid Batyuk, Aubrey-Derrick Schmidt, Hans-Gunther Schmidt. (2009), Developing and Benchmarking Native Linux Applications on Android. *The 2nd International Conference on Mobile Wireless Middleware.*

Zhang,Y.,J.Y. Ma, & X. Q. Yang(2012). *J. Video Engineering* 36: 56.

Electronic Engineering and Information Science – Wang (Ed.)
© *2015 Taylor & Francis Group, London, ISBN: 978-1-138-02772-5*

Hardware implementation of communication and synchronization between tasks based on message-mailbox

Y. Li, G.H. Zhu, H.X. Cui, Z.G. Zhou & B. Wang
Harbin University of Science and Technology, Harbin, China

ABSTRACT: To improve the flexibility of information transfer between tasks in the μC / OS-II, in addition to using semaphore management hardware, the hardware realization and application of the message mailbox will make it quicker and agile for communication and synchronization between tasks. The paper proposed a way that the message mailbox hardware realization can obtain more flexible communication and synchronization between tasks. The whole design is using VHDL to describe the hardware realization modules, then simulating and verifying on ISE8.2, and at last making use of Xilinx's Virtex-II Pro FPGA to achieve it.

KEYWORDS: μC/OS-II; the Message Mailbox; Communication and Synchronization between Tasks; VHDL.

1 INTRODUCTION

μC/OS-II is an Embedded Real-Time Operating System (ERTOS), which has better real-time, deterministic, reliability and many other advantages. So it has been used in many fields, such as aerospace, industrial control, automotive electronics and nuclear power plant construction, etc. (Li, Cui & Li. 2010). But using an improved scheduling algorithm simply to improve the real-time of μC/OS-II can not obtain obvious effects (Hu & Chen 2007). To improve its scheduling efficiency, reliability and timeliness largely (Yin et al. 2008, Hou 2007), the RTOS μC/OS-II should be hardware converted, which means using hardware logic to realize its task management, semaphore management and interrupt management (Cui 2010).

The paper analyzed the message-mailbox of μC/OS-II in detail to grasp how they work at first, then hardware convert the communication mechanism. All parts of the message mailbox management modules are described by VDHL and then obtain timing simulation and verification on the ISE8.2, finally achieve implementation on Xilinx's Virtex-II Pro FPGA. Thus, it will be more flexibility of communication and synchronization between tasks.

2 THE MESSAGE MAILBOX OF MC/OS-II

The message mailbox which can send a variable pointer to a task or an ISR is a way of communication mechanisms in μC/OS-II (Jean & Labrosse 2003). And the relationship between the message mailbox, tasks and ISR can be achieved by the message mailbox management system calls. Tasks and ISR can apply and release the message.

The function of the message mailbox management system calls in μC/OS-II is shown in table 1, which includes creating a mailbox, deleting a mailbox, waiting for messages in mailboxes, sending a message to the mailbox, receiving a message from the mailbox without waiting and querying a mailbox status.

Table 1.   The message mailbox management system calls.

Functions	The Function
OSMboxCreate()	creating a mailbox
OSMboxDel()	deleting a mailbox
OSMboxPend()	waiting for messages in mailboxes
OSMboxPost()	sending a message to the mailbox
OSMboxAccept()	receiving a message from the mailbox without waiting
OSMboxQuery()	querying a mailbox status

3 THE HARDWARE DESIGN OF THE MESSAGE MAILBOX

As a communication mechanism of μC/OS-II, the message mailbox has made great contributions for communication and synchronization between tasks. The message mailbox management can be hardware converted to increase the speed of communication and synchronization between tasks. Namely, the series of system call functions like creating, deleting, pending, posting and querying a mailbox are described through the hardware description language

VHDL, and implement on the hardware logic. As it is shown in Figure 1, the working principle schematic of the message mailbox management includes CPU, the message mailbox management and the event control block ECB management.

The three lines in the above figure are control bus line CB, data bus line DB and address bus line AB. The message mailbox management and the ECB management above are independent of the CPU logical structure, that is, the message mailbox management and the ECB management can obtain data messages from DB directly and then process them. Meanwhile, in order to accelerate the communication between them without increasing the burden on the bus, we can establish a data channel between the storage module of the message mailbox management and the ECB management. It can reduce the workload of CPU through the hardware logical works independently, thus improving the speed of communication and synchronization between tasks, and making the system responsiveness improved further.

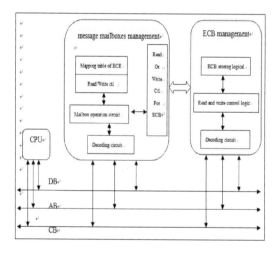

Figure 1. The working principle schematic of the message mailbox management.

4 SIMULATION AND EXPERIMENTAL RESULTS

The paper is using VHDL to describe every module of the message mailbox management. Through timing simulation for all functions in the environment of ISE8.2 software, it can validate the correctness and efficiency of the hardware design for system call functions of the message mailbox management. The functional simulation of creating, deleting, pending and posting a mailbox in the message mailbox

management hardware is shown in Figure 2 and Figure 3:

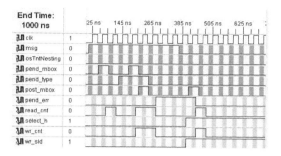

Figure 2. Creating/deleting a mailbox.

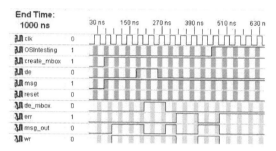

Figure 3. Pending/posting the mailbox.

5 SUMMARY

To improve the efficiency of communication and synchronization between tasks, in addition to the hardware semaphore management, this paper proposes a way that the message mailbox (another communication mechanism) is hardware converting to achieve the goal. In the paper, the hardware design of the message mailbox via VHDL to describe and the simulation results show that the hardware implementation of the mailbox message is correct, meanwhile reduce the system call execution time and the processor overhead, and then increase the flexibility of the communication and synchronization between tasks.

ACKNOWLEDGMENT

This study was supported by the National Natural Science Foundation of China (No. 61103149), the Education Department Foundation of China (No. 12521100), the Technology Innovation Talent Research Foundation of Harbin (No. 2013RFXXJ034) and Innovative and experimental program for college students of china.

REFERENCES

Li Y., X.Y. Cui & X.Y. Li (2010). Hardware Implementation of Task Management of uC/OS-II Based on FPGA. *Electronic Technology Applications*, 36 (2), 25–29.

Hu S.H. & J. Chen (2007). Analysis and Comparison of Several Embedded Real-Time Operating System. *Micro-Controllers & Embedded Systems* (5), 5–8.

Yin Z.Y.,H. Zhao, J.Y. Wang, J.Q. Xu & K. Lin. (2008). HOS Design on Embedded Processor. *Computer Engineering. 2008* (3), 268–270.

Hou M. (2007). *The Research Based on Hardware Real-Time Task Manager*. Shanghai: Shanghai Jiao Tong University.

Cui. X.Y. (2010). *Hardware Implementation of RTOS Based on FPGA*. Harbin: Harbin University of Science and Technology. 18–40.

Jean J. Labrosse. (2003). *MicroC/OS-II the Real-Time Kernel Second Edition*. Beijing: Beijing University of Aeronautics and Astronautics Press. 229–269.

Electronic Engineering and Information Science – Wang (Ed.)
© 2015 Taylor & Francis Group, London, ISBN: 978-1-138-02772-5

Research and hardware design of interrupt manager based on FPGA

Y. Li, T.X. Huang, G.W. Zhang & Z.K. Jin
Harbin University of Science and Technology, Harbin, China

ABSTRACT: In order to fulfill the hard real-time operating system, interrupt management method based on FPGA should be researched and give the interrupt management module structure model which used the VHDL hardware description language to describe. According to the different characteristics of the interrupt request and response mode, interrupt management is divided into two types named system interrupt and user interrupt. In addition, interrupt source and interrupt nesting and clock tick interrupt management were designed. The simulated data in the experiments had verified that the interrupt management module is separated reliably and designed correctly, both of which can improve system efficiency and meet the requirements of real-time operating system.

KEYWORDS: Hardware Operating System; Interrupt Manager; Clock Tick Interrupt; FPGA.

1 INTRODUCTION

With the development of aerospace, industrial control and the construction of nuclear power plants in the field of embedded operating system require higher and higher real-time, the embedded operating system is realized by hardware become possible when the speed and integration of FPGA improve. The traditional interrupt process includes the interrupt request, interrupt response and protection of the site, the interrupt service program execution, site recovery and return from interrupt.

This paper puts forward a kind of interrupt management method based on the PowerPC architecture, and implements interrupt management module in FPGA system.

2 INTERRUPT MANAGEMENT STRUCTURE MODEL RESEARCH

2.1 Interrupt management structure model

Interrupt management model of the overall structure is shown in Figure 1. Except in charge of notification system interrupt, interrupt system also send interrupt enable signal to the hardcore to identify the interrupt source (Jean J. Labrosse. & B.B. Shao 2003). In the hardcore, the interrupt service program called "interrupt task" can be scheduled the same way as common tasks by hardcore. When the external interrupt arrives, triggering the dispatching mechanism of hardcore, the interruption task is implemented in real time; and at the same time, the highest priority interrupt task in the processor task of the hardware

real-time operating system has been preempted the lower one (Pu, Liu & Ling 2003).

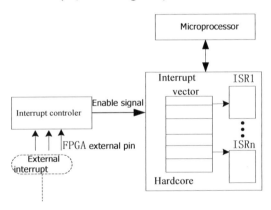

Figure 1. Interrupt management structure model.

2.2 Interrupt processing

According to different characteristics of the interrupt request and response mode, the interrupt management of system structure based on PowerPC includes the system interrupt management and the user interrupt management (Lei, Sang & Xiong 2004).

In the PowerPC architecture, in addition to external pin interrupt and the shielding chip module interrupt, the rest is called the system interrupt, and the external pin interrupt is triggered by the FPGA pins. Each pin corresponds to an external pin interrupt, then a plurality of external pins interrupt corresponding to an interrupt management task. The shielding chip module interrupts can be triggered by an on-chip module (Zhao & Jiang 2006). When a system interruption occurs, it will entry the unified system interrupt service program entrance before the execution, saving only a small amount of necessary register data. The handling process of system interrupt is shown in Figure 2.

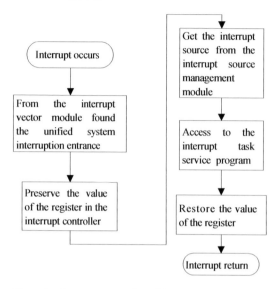

Figure 2. Interrupt processing of the system.

The external pin interrupt and shielding chip module interrupt are called user interrupt (Tian & Xu 2008). User interrupt disposal process is shown in Figure 3 :

First, interrupt occurs, interrupt task starts to execution from the unified system interrupt entry address which was found according to the interrupt vector register by hardware (Hu, Chen & Xie et al. 2007);

Second, saving the values of interrupt register and CPU register, and writing the values of registers to the corresponding stack space manager. If it breaks into from the task, the task stack needs switch to interrupt nesting stack, but it is not necessary to switch if it is a nested interrupt. At the same time shielding lower priority interrupt than the one;

Third, opening interrupt, and can obtain the interrupt source, then according to the data value, outputting to this kind of interrupt source entrance address from the interrupt vector model, finding the interrupt service program and executing;

Fourth, switching the stack pointer when the interrupt processing ends. The first thing is to determine the exit of the interruption, it needs to switch interrupt nesting stack to task stack if it is task interruption; but it is not necessary to switch if it is a nested interrupt;

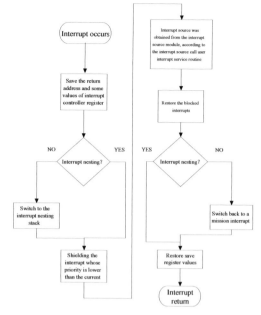

Figure 3. Interrupt processing of the user.

Fifth, recovery the register value saving in section Second;

Sixth, exit interrupt, then the interrupt program will be continued to execute.

3 THE INTERNAL STRUCTURE OF THE INTERRUPT MANAGEMENT MODULE

Interrupt management module internal structure is shown in Figure 4, interrupt management module is consist of the interrupt source management logic, interrupt vector management logic, interrupt nesting logic and beat the clock interrupt management logic. The interrupt source management logic generates the interrupt source Id, and the interrupt vector management logic provides a unified entrance address, and interrupt nesting logic judges user interrupt presences interrupt nesting or not, and the clock beat interrupt management logic belongs to interrupt system contains a clock counter register. When the clock beat arrives, interruption task executes.

4 THE HARDWARE DESIGN OF INTERRUPT MANAGEMENT MODULE

4.1 The interrupt source management logic

There are five types of interrupt sources, external device request interrupt ,fault forced interrupt, beat

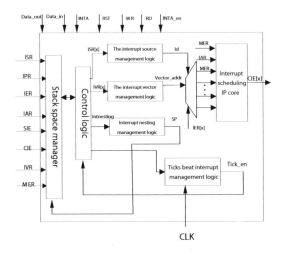

Figure 4. The internal structure of the interrupt management module.

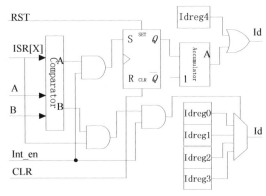

Figure 5. The figure of the interrupt source management.

Table 1. Interrupt type number to register table.

Register	Register's values
Idreg0	00000001
Idreg1	00000010
Idreg2	00000100
Idreg3	00001000
Idreg4	00010000

the clock request interrupt, data tunnel interrupt and program trap interrupt in the interrupt management logic. Among them, the external equipment request interrupt is defined as user interrupt, and fault forced interrupt, beat the clock request interrupt, data tunnel interrupt and program trap interrupt are defined as the system interrupt. An FPGA pin corresponds to a class of interrupt, and interrupt is likely to have different types of interruption, so it is designed like this can save hardware resources very well. The interrupt controller receives a kind of interrupt, then sends an interrupt enable, and interrupt source management logic, which is shown in Figure 5 can receive the interrupt enable signal to identify the kind of the interrupt source. If it is the same kind of interrupt source request interrupt, generating the trigger signal, and starting the logic, then assigning an appropriate interrupt source Id; if not, then assigning interrupt source Id directly.

It has been defined five kinds of interrupt sources, respectively ISR0, ISR1, ISR2, ISR3, ISR4 before designed. Among them, ISR0~ISR3 is the system interrupt, which is defined as the B signal; ISR4 is used for user interrupt, which is defined as the A signal. The type number of interrupt sources is defined in the Idreg0~Idreg4 register, and the definition of interrupt type number is shown in Table 1.

The size of interrupt type numbers is related to interrupt task priority, the smaller interrupt signal is, the higher interrupt task priority is. The working process of this logic is as follows: judging the kind of the interrupt source belongs firstly, it is the system interrupt or user interrupt, if the output A signal is effective, then the interrupt source is the user interrupt, or if the output B signal is effective, then the interrupt source is the system interrupt. If it's the user interrupt, a trigger generates a trigger signal when the interrupt controller gives the interrupt enable effective cases, then the accumulator receives the signal, and pluses 1, then creates a new Id from the accumulator or Idreg4, outputs the Id. The user interrupt supports up to 15 external interrupt. if it's the system interrupt, selector can choose the Id from the interrupt type number register and output it directly.

5 THE SIMULATION RESULTS AND ANALYSIS OF INTERRUPT MANAGEMENT

The simulation Figure 6 shows that, in the interrupt source management module, five kinds of interrupt sources are defined, and each type of interrupt source is corresponding to the corresponding bit in the ISR register. In other words, if ISR0 is effective, all of the values of the ISR register are 0; when ISR1 is effective, the first bit of the ISR registers is 1, and so on. Other values are invalid. The decimal mode is chosen in the simulation diagram. When the interruption is effective, the output's Id number is 1 if ISR0 is effective; the output's Id number is 2 if ISR1 is effective; the output's Id number is 4 if ISR2 is effective; the output's Id number is 8 if ISR3 is effective. ISR4 which is special, is a user interrupt.

It allots ID, responses three user interrupts and its output's Id number is respectively 16 to 18 by a trigger mechanism. The experimental data show that the same user interrupt may acquire different Id. These experimental results demonstrate that the design is correct and feasible.

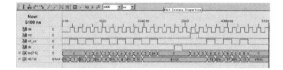

Figure 6. Timing simulation waveform figure of interrupt source management.

6 CONCLUSIONS

This article puts forward a kind of interrupt management method based on the PowerPC architecture, and the experimental data show the feasibility and stability of the interrupt management. The hardware implementation of interrupt management reduces the frequency of CPU scanning clock tick interrupt service program effectively, and improves the real-time of the system and improve the utilization rate of CPU effectiveness.

ACKNOWLEDGMENT

This study was supported by the National Natural Science Foundation of China (No. 61103149), the Education Department Foundation of China (No. 12521100), the Technology Innovation Talent Research Foundation of Harbin (No. 2013RFXXJ034) and Innovative and experimental program for college students of china.

REFERENCES

Jean J.Labrosse. & B.B. Shao translation. (2nd Edition). 2003. *Embedded real-time operating system UC/OS-II.* Beijing: Beijing University of Aeronautics and Astronautics Press.
H.L. Pu, H. Liu & M. Ling. 2003. Interrupt management implementation of small embedded operating system. *Electronic devices*,26(2):218–221.
H.W. Lei, N. Sang & G.Z. Xiong. 2004. Research on interrupt management technology of embedded real time system. *Microcontroller and Embedded Systems*:16–19.
M.D. Zhao & X.F. Jiang. 2006. Interrupt management method of embedded operating system based on PowerPC architecture. China, 200510060734.X.
Y. Tian & W.B. Xu. 2008. *Xilinx The utility of FPGA development tutorial[M].* Beijing: Tsinghua University Press.
W. Hu, T.Z. Chen & B. Xie, etc. 2007. A Component-based embedded operating system interrupt implementation methods. China, 200610154795.7.

Electronic Engineering and Information Science – Wang (Ed.)
© 2015 Taylor & Francis Group, London, ISBN: 978-1-138-02772-5

How to confirm IBeacon direction?

G.Z. Yan, N. Che, H. Liu & Y.Y. Tang
Harbin University of Science and Technology, Harbin, Heilongjiang, China

ABSTRACT: IBeacon is Apple's latest software and hardware technique by using a low-power Bluetooth Low Energy (BLE) technology which greatly improve indoor location accuracy to within one meter. IBeacon node takes the advantages of low power, low cost and easy-using, so it is more and more popular. However, recently most of them are used to measure distance to IBeacon node, direction judgment is not provided, thus the IBeacon node position can not to be computed out. This paper utilize a RSSI strength variation caused by user's rotation to determine the direction of IBeacon nodes, RSSI data are collected and analyzed to calculate the direction of the radiation source, thus we realize the IBeacon indoor positioning through the distance and direction of nodes.

KEYWORDS: IBeacon; Indoor positioning; RSSI; Direction.

1 INTRODUCTION

A currently positioning system commonly used in GPS, is not suitable for indoor positioning (Apple. 2014). IBeacon by Apple's latest an indoor positioning technology, through the use of low-power BLEva technology, software and hardware combination, not only greatly reduce the power consumption, achieve a battery can be maintained for several years, and also greatly improve the accuracy of indoor positioning accuracy of a few hundred meters from the original, tens of meters, improving to within one meter (Apple. 2014, Clancy 2014). However, most people only realize the calculation of the distance, and can not really achieve positioning. This article describes a method of utilizing IBeacon technology to confirm direction, so that genuine use IBeacon indoor positioning becomes possible.

IBeacon technology hardware is based on the BLE technology, July 7, 2010, the Bluetooth SIG announced the formal adoption of Bluetooth 4.0 (that is, Bluetooth Low Energy) Core Specification (Bluetooth Core Specification Version 4.0 (Grobart 2013). Compared with the previous Bluetooth, Bluetooth low energy advantage is that ultra-low power consumption, longer link distances and faster connection speeds.

IBeacon is one of the most important new features in IOS 7, March 11, 2014, Apple iOS 7.1 release also brings new capabilities for IBeacon agreement. IBeacon technology is by using a low-power Bluetooth technology, IBeacon area of the base station automatically creates a signal when the device enters the region IOS, the corresponding application would be prompted whether the user needs access to the signaling network, via a plurality of neighboring transceivers IBeacon, a precise positioning of the range of a few millimeters to 50 meters, and to locate the user in real time and data transfer in the background. Compared with NFC technology, obviously with precise positioning, more distant (NFC connection distance of only 4-20 cm), lower power consumption advantages (Hadwiger et al. 2005).

IBeacon technology was originally used in shopping malls commodity propaganda, the customer pre-installed a mall's App, and the mall to install a different IBeacon base stations in different locations, close to the corresponding base station when the customer will receive a presentation corresponding goods, in order to achieve the indoor positioning to customers and targeted advertising. But now achieved merely positioning function is to locate a certain range within the region, and cannot accurately determine the precise location, so it is important to determine the direction.

2 POSITIONING IBEACON LOCATION

2.1 *Distance measurement principle*

Received Signal Strength Indication (RSSI) indicating received signal strength, an optional part of the radio transmission layer, used to determine the quality of the link, received signal strength. The relationship between the reception signal and the signal strength of the wireless signal transmission distance can be represented by the equation (1), RSSI is the received signal strength, d is the distance between the send and receive nodes, *and* is a signal propagation factor.

$$RSSI = -(tx + 10 \cdot n \lg d) \tag{1}$$

Represented by the equation (1) can be seen, the value of the constant *tx*, and *n* determines the relationship between received signal strength and signal transmission distance. RF parameters *tx* and *n* is used to describe the network operating environment. RF parameters *tx* is defined as the absolute value of 1m from the transmitter and receiver when the average energy of the signal expressed in dBm. If the average received energy is -40 dBm, then the parameter *tx* is defined as 40. Parameter *n* indicates the radio frequency signal energy to the transceiver as the distance increases the rate of decay, the value of which depends on the size of the radio signal propagation environment. Attenuation of the received signal strength is proportional to dn, in an ideal environment, the signal long-term decline lognormal distribution.

When *tx* constant, *n* changes, when the *n* value is smaller, the smaller the signal attenuation in the communication process, the signal can travel longer distances. The value of the propagation factor depends mainly on the wireless signal attenuation in the air, reflection, multipath interference, etc., in the indoor environment is also affected by the reflective floor and walls and other obstacles, refraction and other interference. Interference smaller spread value factor n is smaller, the greater the distance the signal propagation, wireless signal propagation curve closer to the theoretical curve, the smaller the RSSI value at the same location.

This paper refers to "I Am the Antenna: Accurate Outdoor AP Location using Smartphones" which focuses on WIFI AP source direction ideological orientation. However, that article is based on the IEEE 802.11 standard for wireless networks, this paper is based on the BLE 4.0 of the IBeacon communication protocol stack, in principle, is completely different.

2.2   IBeacon direction confirm

Because the strength of the RSSI value at the same location by the impact of the spread factor, so after extensive testing found, in the same place, holding IBeacon node, when being on the emission source (such as Figure 1), the received RSSI value is always much larger than the received RSSI value facing away from the emission source (such as Figure 2). Therefore, when the handheld IBeacon node in place revolution, a lot of rotation in the process of collecting location data and RSSI value data can be calculated direction.

Figure 1.   Facing IBeacon.

Figure 2.   Back facing IBeacon.

After a lot of statistics, when the person holding BLE 4.0 equipment rotating in place (such as Figure 3), the received signal strength as the mean of every *Δ* degrees and the mean of the range for the remainder of the range of *360-Δ* to make difference comparator, and find the degree which the largest absolute value of the difference between the angle of 360 degrees. By the equation (2), *j* is calculated from the current angle, the size of 0 to 360 degree increments. *T* for each value of the angle corresponding to the RSSI strength. The constant is calculated from the exact value of the size range. *M* is the number of the range of the intensity of the collected RSSI values. *N* is outside the range of the intensity of the collected number of RSSI values. When *Δ* is 90, the range of 90 degrees corresponds exactly with the emission direction of the source node is located IBeacon contrast (see Figure 4), can be rotated many times to improve accuracy.

$$d = \underset{j\in[0,360]}{Max}\left(\left|\frac{\sum\limits_{i\in[j,j+\Delta]}t_i}{m_j} - \frac{\sum\limits_{i\in[0,j]\cup[j+\Delta,360]}t_i}{n_j}\right|\right) \qquad (2)$$

Figure 3.   The user rotating in place.

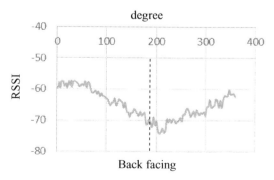

Figure 4.   Sampling results.

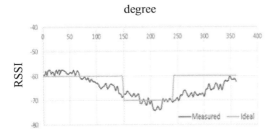

Figure 5.   Predicting confidence value.

2.3 Confidence limits

In order to verify the accuracy of the measured direction of the use of the equation (3) to calculate the degree of confidence. T is set according to the change of the signal intensity is determined by the angle of the projected (in the range of 0 to 90 degrees, within the remaining range of 270), r is the actual measurement of signal strength at each point in Figure 5, can be calculated the confidence p.

$$p = \frac{1}{m} \cdot \frac{\Sigma_{i \in [0, m-1]}(t_i - \bar{t})(r_i - \bar{r})}{\sigma_t \cdot \sigma_r} \tag{3}$$

3 POSITIONING EXPERIMENTS ON ANDROID

IBeacon is based on a technology BLE 4.0, so long as the use of modules with BLE 4.0 features and functionality of a having BLE 4.0 Android system equipment can be IBeacon communication protocol stack, through continuous scanning, so real-time access to be BLE module data on the measurement node, complete the orientation direction judgment. At present, many available on the market of portable IBeacon node, or even only a small button battery size, but they can maintain power over a year, and the price is only about tens of dollars (see Figure 6).

Figure 6.   IBeacon node.   Figure 7.   RFduino.

In this experiment, BLE module used is RFduino (see Figure 7), using of open source software Arduino programming, writing IBeacon configuration module. After the BLE module is powered on, open the Android test app, search for and connect current BLE module, handheld Android devices rotate in place many times, and calculate the direction of the record, according to equation (2), equation (3) calculate the direction and confidence in Figure 8. The experimental results are Figures 9 and 10.

Figure 8.   Android APP interface.

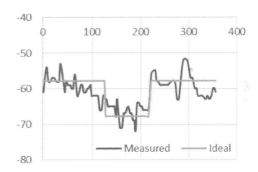

Figure 9.   Experiment 1, confidence: 0.72.

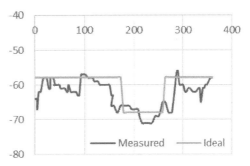

Figure 10.   Experiment 2, confidence: 0.65.

4 CONCLUSIONS

IBeacon is the latest one by using a low-power BLE technology, can greatly improve the accuracy of indoor positioning. Its low cost, low power, easy to use, so use IBeacon indoor positioning is a very important technology, but now most people can only measure the distance and direction can not be positioned, and this article is based on the characteristics of RSSI when the interference by environmental factors RSSI value will change in the same position, so that when people in the handheld IBeacon node rotation in place, through the acquisition RSSI value changes resulting analysis can be calculated starting radiation source direction, thus the perfect solution IBeacon orientation issues.

ACKNOWLEDGMENTS

Research is sponsored by the Science and Technology Research Project of Education Department in Heilongjiang Province (Grant No. 12531103, 12541162)

REFERENCES

Apple. (2014). Insight: Road Test - Apple's iBeacon grants retailers showroom control. *Campaign Asia-Pacific,* 50–51.

Apple. (2014). Watch out for iBeacon–because it's watching you. *Popular Science,* 23.

Clancy, H. (2014). Apple's iBeacon signals turning point for mobile engagement. *Fortune.com ,* 1–2.

Grobart, S. (2013). Apple's Location-Tracking iBeacon Is Poised for Use in Retail Sales. *BusinessWeek.com:* 39.

Hadwiger, M., C. Sigg, & H. Scharsach (2005). Real-time ray-casting and advanced shading of discrete isosurfaces. *Proceedings of Euro-graphics,* 23–24.

Electronic Engineering and Information Science – Wang (Ed.)
© 2015 Taylor & Francis Group, London, ISBN: 978-1-138-02772-5

Voltage sensitivity of band gap photonic crystal fiber of terahertz wave

Y.Y.J. Chen, G.J. Ren, P. Wu, Y.Y. Song & X.Y. Gao
School of Electronics Information Engineering, Tianjin Key Lab of Film Electronic & Communication Devices, Tianjin University of Technology

R. Ji
School of Management, Tianjin University of Technology

ABSTRACT: This paper analyzes the photonic crystal fiber core area filled with the nematic liquid crystal 5CB and studies the voltage sensitivity of band gap photonic crystal fiber of terahertz wave. The simulation results show that when liquid crystal refractive index increases, cutoff wavelength, core area and confinement loss all become larger. When the incident wavelength is longer, the core area and confinement loss are smaller while the voltage sensitivity becomes worse. It is more suitable to use voltage sensitivity in the design of switch filters for specific wavelengths.

KEYWORDS: Liquid crystal; Voltage sensitivity; Terahertz wave; Photonic crystal fiber.

1 INTRODUCTION

In 1974, Fleming first proposed terahertz (THZ), which is the electromagnetic wave whose frequencies range from 0.1 THz to 10 THz and wavelength is between 0.03 mm and 3mm (Yang, Dai, YiZhan, Wang, Wu, Xu, & Lin (2014). It connects electronics and shimmer macroscopic photon science and has many excellent characteristics. With the development of society and advanced technology, the research on terahertz has received worldwide attention.

In 1992, ST. J. Russle *et al.* first proposed the photonic crystal fiber (PCF) (Wang, Feng, Kong, Wang, and Ma 2014). PCF uses a linear defect as its core, which destructs periodic structures in the center of two-dimensional photonic crystal (Wang, Feng, Kong, Wang and Ma 2014; Feng, 2013). PCF can transmit different lights through the structure of the core and cladding. The transmission mechanism of the core and cladding depends on the relative refractive indices. It has many advantages like endlessly single mode of transmission, low loss, controllable dispersion, and controllable nonlinear. (Jian, 2013). In addition, it can transmit ultra-intense laser, liquid, gas and so on. Thus, it has already become the focus in the field of optics and photonics research since its inception.

In THz technology, photonic crystal is mainly used to produce some functional devices, such as THz photonic crystal fiber, waveguide, filter, wave switch, modulators and so on. (Zhou, Li, Yu, Jiang 2014). Some techniques and methods of photonic crystals used in the microwave and millimeter wavelength bands can also be used in the THz wave.

Liquid crystal photonic crystal of THZ device has broad application prospects (Zhou, Li, Yu, Jiang 2014; Zhou 2013)). The magnitude of PCF structural parameters used in THz waveguide is in the order of mm. Moreover, the flexibility of the plastic material is very good and the processing temperature is much lower than that of quartz material. What's more, the preparation of THz photonic crystal fiber is easier than that of visible light or infrared wave, but there are still many physical and technical problems remain for further study.

In this paper, we use the finite element method and the anti-resonance model to study the sensitivity of the band gap of photonic crystal fiber of terahertz wave. We study on PCF core area filled with nematic liquid crystal 5CB. On the one hand, its high refractive index can change the transmission mechanism of photonic crystal fibers. On the other hand, it can change the refractive index of the liquid crystal by regulating voltage for the adjustable output of PCF or the research of its sensitivity, etc.

2 LIQUID CRYSTAL REFRACTIVE INDEX VERSUS VOLTAGES

In this paper, the center hole of PCF is filled with nematic liquid crystal 5CB. Long axis of liquid crystal molecule is along the direction of fiber axis. The model takes the optical axis direction as Z and fiber cross section as X-Y plane. The ordinary and extraordinary refractive index are denoted by n0 and ne, n0=nx=ny, ne=nz.

When voltage is not applied to the liquid crystal, optic axis of liquid crystal and long axis of its molecule are consistent and along the Z axis of fiber. The transmitted light is commonly light, whose liquid crystal refractive index is common refractive index n.

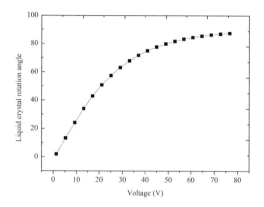

Figure 1.   Liquid crystal rotation angle versus voltage.

Seen from Fig.1, the liquid crystal rotation angle increases with voltage increases. If voltage is less than the threshold voltage, liquid crystal will not rotate. If voltage is larger than the threshold voltage, liquid crystal will rotate in the direction of applied electric field.

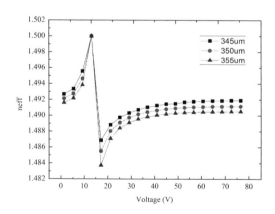

Figure 2.   Effective refractive index versus voltage.

Voltage sensitivity of different wavelengths (355 um, 350 um and 345 um) is simulated and the results of effective refractive index of PCF filled with liquid crystal versus voltage are shown in Fig.2.When voltage is the same, the longer wavelength is, the larger effective refractive index is.

When voltage increases, the effective refractive index becomes larger. There is a mutation around 13 V, which is corresponding to the maximum loss. When voltage increases to 65 V, the tendency of the effective refractive index becomes flat.

3   CONFINEMENT LOSS VERSUS VOLTAGES

Confinement loss represents the transmission properties. It can be calculated by using Comsol and given as:

$$L_1 = \frac{2 \times \pi \times 1000 \times 8.686 \times \text{Im}(n_{\textit{eff}})}{\lambda} (dB / km) \quad (1)$$

Where Im(neff) represents the imaginary part of the effective refractive index. Because confinement loss is very large when it is less than 20 V, we choose to calculate from 21 V. Confinement loss versus voltage of 355 um, 350 um and 345 um are shown in Fig.3.

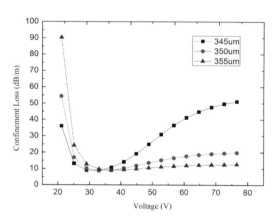

Figure 3.   Confinement loss versus voltage.

When voltage is more than 35 V, the shorter wavelength is, the greater confinement loss versus voltage changes. If confinement loss around 13 V is large, the switch is open. Otherwise, the switch is closed.

4   EFFECTIVE CORE AREA VERSUS VOLTAGE

The effective core area is defined as:

$$A_{\textit{eff}} = \frac{\left[\iint |E(x,y)|^2 \, dxdy \right]^2}{\iint |E(x,y)|^4 \, dxdy} \quad (2)$$

144

The effective core area of PCF is related to its non-linear coefficient and the relationship between them is:

$$\gamma = \frac{2\pi \cdot n_2}{\lambda A_{eff}} \tag{3}$$

Where n2 is the nonlinear coefficient of PCF. According to simulation for the wavelength of 355 um, 350 um and 345 um using Comsol, the results of effective core area of the PCF versus voltage are shown in Fig.4.

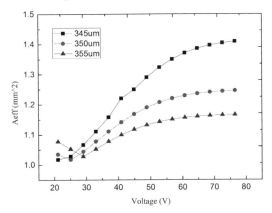

Figure 4. Effective core area versus voltage.

If liquid crystal refractive index increases, the wavelength will become longer. The effective core area becomes larger when liquid crystal refractive index increases. However, the increasing tendency is more and more flat. The overall tendency is also positively correlated to the liquid refractive index, except the mutation around the cutoff wavelength.

5 CONCLUSIONS

This paper analyzes voltage sensitivity of band gap photonic crystal fiber of terahertz wave. When the liquid crystal refractive index increases, cutoff wavelength increases, effective core area and confinement loss become bigger. Confinement loss and the effective core area become smaller when wavelength increases. Voltage confinement losses mutational curve provides reference for the design of switch filter.

ACKNOWLEDGMENT

This work is a national project supported by the College student innovation program of Ministry of Education (No.201210060037).

REFERENCES

P.L.Yang, S.X. Dai, C.S. Yi, P.Q. Zhang, X.S. Wang, Y.H. Wu, Y.S. Xu & C.G. Lin(2014) : *Journal of physics*, Vol. 1 , p. 201–208.

Wang ,R., C. Feng, L.P. Kong , Y.M. Wang , &D. Ma (2014): *Technology and enterprise*, Vol. 12 , p. 380.

Feng , X. G. (2013);Research of sensing characteristics based on the band gap of photonic crystal fiber (In Chinese). *Nanjing University of Posts and Telecommunications*.

Jian, D. (2013);Analysis and optimal design based on the characteristics of photonic crystal fiber (In Chinese). Chongqing University.

Zhou, L.D., X. F. Li , H. H.Yu , D.S. Jiang(2014) : *Functional materials.12 ,* 12006–12009+12018.

Zhou L.Y. (2013):Long-period gratings in photonic crystal fibers theory (In Chinese). *Journal of Yanshan University*.

Electronic Engineering and Information Science – Wang (Ed.)
© 2015 Taylor & Francis Group, London, ISBN: 978-1-138-02772-5

Analysis of working characteristics and modeling-simulation based on ZnO thin film transistors

Y.H. Wu, Z.Y. Wang, D.X. Wang, X. Li, M.M. Li, D.N. Yuan, X.X. Dong, H.R. Lv & X.C. Liu
Department of Electronic Science and Technology, College of Applied Science, Harbin University of Science and Technology, Heilongjiang, Harbin, China

ABSTRACT: According to the working characteristics of the ZnO TFT, the relationship between output I-V of the ZnO TFT has been studied; the transistor equivalent circuit model has been established. The transistor static characteristic parameters have been analyzed and simulated. The formulas and coefficients of the model required have been derived by the actual measurement data. The relationship between output current and bias voltage, $I_{ds}=0.465 \times 10^{-5} \exp[q(V_{ds}-V_{gs})/38.46KT]$, has been obtained. Using the formulas and coefficients, the model has been improved. The relationship between the output I-V of ZnO TFT which have been obtained by simulation and analysis is in good agreement with the actual measurement data.

KEYWORDS: TFT, working characteristics; simulation, analysis.

1 INTRODUCTION

As the important part of the display system of man-machine interface, the demand of display devices with special functions such as HD, transparent, thin, flexible, etc. has been higher and higher. In the current relatively competitive flat panel display, LCD flat panel displays, especially TFT-LCD, is the only superior to the traditional CRT display device on the comprehensive performances such as brightness, contrast, power and weight, etc. OLED display is expected to become the most competitive technology compared with TFT-LCD display. However, whether TFT-LCD or OLED display, all cannot works without the switch controlling pixel and driving unit - thin film transistor (TFT).

At present, the AMLCD and AMOLED TFT technologies are mainly the polycrystalline silicon(poly-Si) TFT and amorphous silicon(a-Si) TFT. The preparation process of a-Si TFT is simple, and its technology is mature. It has the absolute advantages in cost and large area display. However, its carrier mobility is low, in addition, it exists some problems such as photosentive reaction in the visible wavelength range, being opaque to the visible light. Polycrystalline silicon thin film transistor has high carrier mobility, quick response speed, strong anti-interference ability and good device performance. However, its preparation process is more complex, it needs to excimer laser rectystallization treatment, thereby the manufacturing cost is increased (Yu 2006, Fujimoto et al. 2005).

While ZnO carrier mobility is between polysilicon and amorphous silicon, the current meets the lighting requirements of OLED pixel. Therefore, the TFT of the sandwich structure-Al (emitter)/ZnO/ Ni (base)/ZnO/ Al (collector) is mainly studied in this paper. Through modeling and simulation using Multisim 12.0, the basic electrical characteristics of TFT are studied and the effect factors of the device characteristics are explored.

2 THE MAIN PRINCIPLE

The schematic diagram of the vertical structure TFT studied is shown in Fig.1.The coating sequence is followed by Al, ZnO, Ni, ZnO, Al. The gate Ni, the drain Al and the source Al are respectively connected with wires using indium. ZnO TFT is analyzed using Keithley-4200 semiconductor analyzer (Yu 2009).

According to the Schottky-Mott theory, the difference of between the work function of metal and that of the semiconductor is equal to the height of the barrier between metal and semiconductor contact in the ideal case, that also means $\Phi=\Phi m-\Phi s$. In this paper, the ideal energy band diagram of vertical structure ZnO TFT is shown in Fig.2.

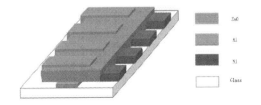

Figure 1. The schematic diagram of the glass/Al/ZnO/Ni / ZnO/Al vertical structure TFT.

The values of ZnO electron affinity χ and ionization energy I_E are 4.35eV and 7.72eV respectively, the band gap width is 3.37eV, the work function of Al is 4.2eV. The barrier height of hole injection is much larger than that of the electron injection, so the electrons are as carriers forming working current in the device.

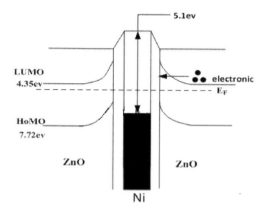

Figure 2. The ideal energy band diagram of vertical structure ZnO TFT.

According to the working characteristic data of actual measurement, the ZnO TFT micro variation equivalent circuit is established, as shown in Fig.3.

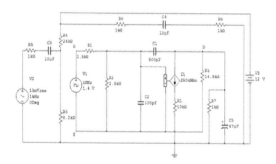

Figure 3. The micro variation equivalent circuit of ZnO TFT.

3 SIMULATION AND ANALYSIS ON CHARACTERISTICS OF ZNO TFT

Through analyzing the static output characteristics of ZnO TFT prepared, the equivalent microcircuit analyzing model of the TFT static parameters is established. The simulation circuit is established using Multisim 12.0, as shown in Fig.4.

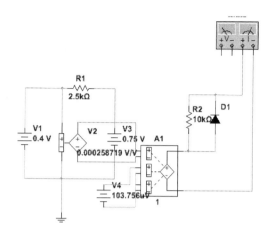

Figure 4. The model established by studying the transistor parameters.

The output characteristic data of ZnO TFT is analyzed using Origin 8.0. According to the actual effect gate voltage formula (1) combined with the formula (2), the coefficients a and b are derived. When $V_{gs}=0$V, Eq. (3) is obtained.

$$V_g^*=a(V_{gs}+V_{ds}/b) \tag{1}$$

$$I_{ds} = I_0 \exp(\frac{qV_g^*}{kT}) \tag{2}$$

$$I_{ds} = 0.465 \times 10^{-5} \exp\left(\frac{q(V_{ds}-V_{gs})}{38.46kT}\right) \tag{3}$$

Where $a = \dfrac{\partial V_{gs}(0)}{\partial V_{gs}}$, $V_{gs}(0) = \dfrac{W_{gs}}{W}|\Delta V_{ds}|$,

W_{gs} is the distance between the source and the gate, W is the distance between the gate and the drain, I_0 is constant, k is Boltzmann constant, T is thermodynamic temperature (Wang 2006).

When $V_{ds}=0$V, V_{gs} increases from 0V to 3 V in the step of 0.1V. The formula (3) is divided into the multiplication of two functions about V_{gs} and V_{ds}, as shown in the formula(4) and formula(5).

$$I_{ds_1} = 0.465 \times 10^{-5} \exp\left(\frac{qV_{gs}}{38.46kT}\right) \tag{4}$$

$$I_{ds_2} = 0.465 \times 10^{-5} \exp\left(\frac{-qV_{ds}}{38.46kT}\right) \tag{5}$$

When V_{ds}=0V, and in the case of V_{gs}=0V, the change of the formula(4) is needed to consider.

In order to verify the correctness of the Fig.3, when V_{ds} from 0V to 3V and V_{gs} in 0V, 0.25V, 0.5V, 0.75V, 1V, with the same method as above, a set of data-I_{ds} are obtained, the I_{ds}-V_{ds} curve is drawn, as show in Fig.5 (b).

Compared Fig.5 (a) with Fig.5 (b), it can be seen from Eq.(3) the calculated data and the actual measurement data are quite close, as shown in Fig.5 (b). It illustrates that the Eq.(3) can correctly describe the static output characteristics of the ZnO TFT prepared.

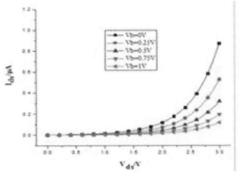

a) Output characteristics curve of actual measurement of ZnO TFT

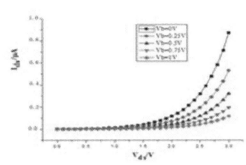

b) Output characteristics curve of simulation results

Figure 5.   Comparison between actual and simulation measurement.

It can be seen from Fig.5 that the theoretical value and the actual measurement value are roughly the same. Correctness of Eq.(3) has been verified. Further research on the Eq.(3) has been done. When V_{gs} is 0, 0.25V, 0.5V, 0.75V, 1V, the actual value of the model is compared with theoretical values. For convenience, two data sets are selected randomly.

When V_{ds}=0V, V_{gs}=0.1V, the value of I_{ds} =2.64100nA, and model theoretical value of I_{ds} is 2.64098nA. The difference is 0.00002nA; when V_{ds}=0.2V, I_{ds}=3.226nA, and the model theoretical value of I_{ds}=3.2257nA. The difference is 0.0003nA; when V_{ds}=0.4V, V_{gs}=0.75V, I_{ds}=10.842nA, and the model theoretical value of I_{ds}= 10.7374nA. The difference is 0.1046nA. When V_{ds}=0.5V, V_{gs}=2.2V, I_{ds}=64.796nA, and the model theoretical value of I_{ds}=64.789nA. The difference is 0.007nA. It proved that the equivalent circuit model can realize the characteristics of the device actually prepared.

4   CONCLUSION

In this paper, based on the actual measurement data and the corresponding analysis of the formulas, the relationship between current and voltage of transistor has been deduced. When V_{gs}=0V, I_{ds}=0.465× 10^{-5}exp[q(V_{ds}-V_{gs})/38.46KT]. And based on the

oxide TFT amplifier circuit, the static parameters analyzing model of the TFT has been established. The modeling and simulation of the ZnO TFT have been accomplished using the Multisim 12.0. Basic electrical characteristics of the TFT have been studied; the simulation results and the measurement data are very approximate. The actual measurement value is affected by many external factors, therefore, the data of model and the actual measurement data may not be completely consistent. So it can be regarded as the successful modeling. In further experiments, the experiment would be guided by changing the circuit parameters in model.

REFERENCES

Yu G., C. Zhang & A.J. Heeger (1994). Dual-function semiconducting Polymer devices: Light-emitting and Photodetecting diodes, *Appl Phys Lett*, 64, 1540–1542.
Fujimoto S., K. Nakayama & M. Yokoyama (2005). Fabrication of a vertical-type organic transistor with a planar metal base. *Appl Phys Lett*, 87, 133503-1335-6.
Yu G. & Z. Lei (2009). Field dependent and high light sensitive organic phototransistors based on linear asymmetric organic semiconductor. *Appl Phys Lett*, 25–30.
Wang D.X. (2006). Development of organic semiconductor thin film transistor. *Chinese Journal of Semiconductors*, 55–60.

Electronic Engineering and Information Science – Wang (Ed.)
© 2015 Taylor & Francis Group, London, ISBN: 978-1-138-02772-5

The design and implementation of intelligent electronic timer

Y.L. He, Y. Wang, L.Y. Zhang, Q. Liu & X. Wang
Harbin University of Science and Technology, Harbin, China

ABSTRACT: In this paper, an intelligent electronic timer is designed by using the voice control module, infrared module, high precision clock module and STC89C52 SCM which is regarded as the core. The switch display of the time, date and temperature can be achieved through a digital tape. In particular, the sound of a starter's gun shot can be used as a signal to start the stopwatch timing, and ending the timer with the signal of infrared beam is blocked when the athletes arrive at the end. The traditional timing usually uses a stopwatch manually. Because of the human have a response delay and different speed, manual timing will produce large error. The intelligent timer designed in the paper has been greatly improved in accuracy and fairness than the traditional timing method.

KEYWORDS: intelligent stopwatch; SCM; sound control; infrared light; temperature sensor.

1 INTRODUCTION

The single-chip microcomputer is the system that adopts technology of large scale integrated circuit, it puts the central processor with data processing capability, random access memory, RAM, Read-only memory ROM, a variety of I/O port, interrupt system, and the timer integrated on a silicon wafer (Baccherini & Merlini 2008). Various functions can be achieved with the combination of MCU and programming languages quickly and efficiently. A precise timer based on STC89C52 single-chip microcomputer is designed and implemented by this paper. Electronic timer is widely used in many sports events. It brings great convenience to people's study and work (Gu 2008). With the development of the society and the improvement of sports level, the time accuracy has never been requirements so highly (Auhmani et al. 2002). It can't be distinguished when players hit the line. Therefore, the research on intelligent electronic timer has a large realistic significance. The main characteristic of intelligent electronic timer is intelligent and precise. The function is very powerful and the operation is very simple, and the cost is low. The accuracy of the stopwatch is increased to 0.3s. The beginning and ending of the time can be measured automatically and accurately.

1.1 The system hardware design

The STC89C52 chip is used as the core controller of the hardware circuit, including the clock chip, LED digital tube, an infrared sensor module, a temperature sensor module and a voice control module (Wei & Zuo 2007). It can display the date, week, minutes and seconds through the eight LED digital tubes and can be individually set and modified. It can record 8 groups of stopwatch data. The DS1302 chip is used in order to achieve the exact timing of the operation. In addition to using the button control the stopwatch of start and suspension, it can also operate the stopwatch with the voice control module in record time with the sound, operate the stopwatch with the infrared induction module to record time with the body. Accuracy of the intelligent electronic timer is greatly improved compared with ordinary stopwatch. The function is very powerful and the operation is very simple. The hardware circuit diagram of the system is shown in Figure 1.

1.2 Acoustic control module

The voice module starts timing when the voice module detects the starting gun. Its working principle is: 5V voltage regulated by the LED (and the power indicator). After the capacitor filtering the output is about 1.8 V DC. Microphone receives the signal of the sound and converts it into an electrical signal. Voltage is coupled to the base electrode of the triode through a capacitor, making the triple tube conduction. And then output a low voltage signal through a pull-down resistor. The signal can control the circuit through the I/O port of the SCM. The circuit diagram of the acoustic control module is shown in Figure 2.

Its sensitivity is adjustable. It can detect the intensity of the sound environment, use note: this sensor

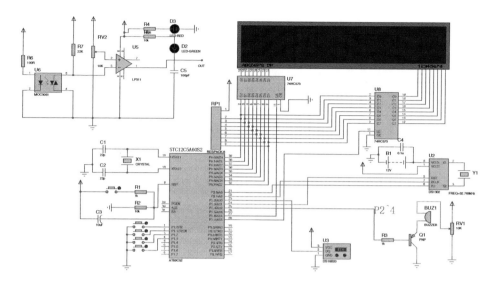

Figure 1.   Hardware circuit diagram of the system.

can only identify the presence of (according to the principle of vibration) of sound does not recognize the sound or the size of the particular frequency of sound. The beginning and ending of the time can be measured automatically and accurately and the cost is low.

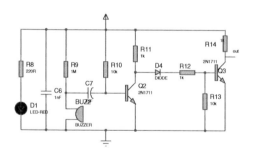

Figure 2.   Circuit diagram of the acoustic control module.

1.3   *Infrared induction module*

The infrared light detection module has a pair of infrared transmitting and receiving tube. When meeting with obstacles (reflecting surface) in detecting direction, the reflected infrared received by receiving tube. After a comparator circuit processing, LED indicator will light up indicate that the signal is received. The

circuit diagram of the infrared induction module is shown in Figure 3.

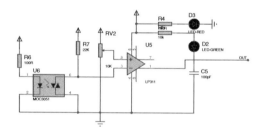

Figure 3.   Circuit diagram of the infrared induction module.

1.4   *Infrared induction module*

The system mainly consists of the following parts: The control system composed of reset circuit, a crystal oscillator and MCU. The clock system composed of DS1302 chip clock. The temperature detection system consists of DS18B20 temperature sensor. Infrared tube and voltage comparator composed of the infrared detection system. A system composed of electrets microphone and photosensitive resistance acoustic control system. A system consists of four separate button keyboard. The overall framework of the system is shown in Figure 4.

152

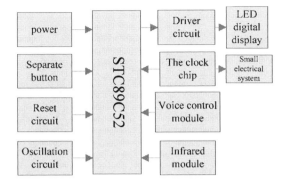

Figure 4. The overall framework of the system.

2 THE DESIGN OF MAIN PROGRAM

The program is written in C language. According to the requirement of design the stopwatch system the whole system is divided into several modules to design. System program mainly achieves the following functions: Display and set of the time, Stopwatch, detection and display of the temperature.

Chip and register are initialized when the system is powered on, then digital tube displays the current time and enter a state of key scanning and infrared detection. It can enter the corresponding functions by pressing the appropriate button. We use four 10 binary counters, to respectively in one percent seconds, 1/10 seconds, seconds and counting; We used two 6 binary counter: used for 10 seconds and very count. The design should counter with asynchronous reset enable end and end with the overall consideration. It can start and return to zero, and they stop and start. So consider the input method to realize these two kind of different modulus counters with a text. The flow chart of the main program is shown in Figure 5.

3 DATA MEASUREMENT AND RESULT ANALYSIS

Through the debugging of hardware and software, electronic timer can successfully complete all the functions of stopwatch and time. The Figure 6 shows that the system displays the current time when the power on or reset.

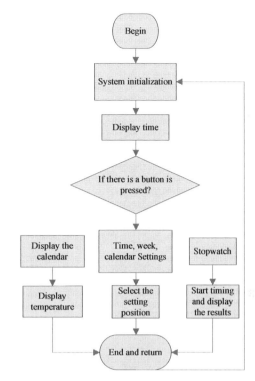

Figure 5. Flow chart of the program.

Figure 6. Display of the current time.

In the condition of the stopwatch, the voice control module can detect the sound to start timing. And then triggering the infrared module (analog athlete across the finish) can realize recording a group of several series. A stopwatch data is shown in Figure 7.

Figure 7. Display of the stopwatch.

153

The reaction time of normal people is 0.15s- 0.4s, therefore the manual timing error is very big. It is easy to cause inaccurate results in the 100 meter race. While the reaction time of the design of the intelligent timer is $t = \dfrac{1}{f} = \dfrac{1}{11.0592} \approx 1 \times 10^{-7} s$. It greatly reduced the error caused by human and equipment factors.

As shown in the reaction time of the test between intelligent induction time and manual timing, it can be seen that automatic timing begins almost immediately when a signal is detected in 0.1s, while the manual timing in 0.4s to timing due to the delay time of 0.3s. The two data of comparison of reaction time are shown in Figure 8.

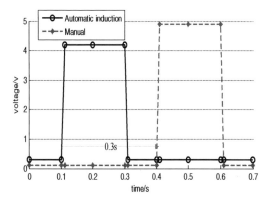

Figure 8.   Comparison of reaction time.

4   SUMMARY

In conclusion, Targets are implemented and the main characteristic of intelligent electronic timer is intelligent and precise. The function is very powerful and the operation is very simple, and the cost is low. The accuracy of the stopwatch is increased to 0.3s. The beginning and ending of the time can be measured automatically and accurately. It can be used in everyday life and sports events, and has a broad market application prospect.

ACKNOWLEDGMENTS

The work supported by the National Natural Science Foundation of China (Grant No. 61201075), and the Science and Technology Research Foundation of Heilongjiang Education Bureau of China (12521110), and by the China Postdoctoral Science Foundation (2012M511507), and by Science Funds for the Young Innovative Talents of HUST (No.201302), and by Frontier Research Foundation of State Key Laboratory Breeding Base of Dielectrics Engineering (DE2012B05).

REFERENCES

Baccherini D. & D. Merlini (2008). Discrete Mathematics. *Appl. Phys. Lett 184*, 4165–4169.
Gu M.Z. (2008). Electrical measurement and instumentation. *Nano Letters 30,* 522–527.
Auhmani, K., M. Benhayoun, M. Ankrim, & K. Quotb (2002). International symposium on acoustic remote sensing *8,* 248–252.
Wei J. & S. Zuo (2007). Electronic Design Engineering. *Thin Solid Films 238*, 1452–1457.

Electronic Engineering and Information Science – Wang (Ed.)
© 2015 Taylor & Francis Group, London, ISBN: 978-1-138-02772-5

Designing and implementation of barometric altimeter in navigation receivers

X. Wang & L. Hou
National Time Service Center, Chinese Academy of Sciences, Xi'an, China

J.L. Liu
University of Chinese Academy of Sciences, Beijing, China

ABSTRACT: Aiming at the problem that it is different to realize three-dimensional positioning in satellite navigation system, putting forward a aided-navigation method which is taking barometric altimeter. Using digital sensors to obtain real-time air pressure value and convert it to the digital signal, reading the signal from the micro-processor chip, designing an algorithm to research the relationship between air pressure, temperature and height, obtaining the height value finally, achieving the objective of the height measurement and add-navigation. Finally, the paper gives an experimental verification and data analysis of barometric altimeter.

KEYWORDS: STM32; air pressure sensors; height measurement.

1 INTRODUCTION

With the further development ofva GNSS (Global Navigation Satellite System) system, owing to provide all weather and continuous precise position, velocity and time information, the GNSS system become the significant symbol of the navigation technology (Xie 2011). However, there is some limitation in satellite navigation technology, such as the environment and weather conditions, these factors can affect position precision easily. For instance, in GPS system, only knowing at least 4 satellite's location information, the receiver can obtain a user's position. If the receiver affected by the non-artificial factors such as extreme environment or bad weather, the satellites which can be acquired will be only 3, even less, the receiver cannot determine the user's location. So, using the barometric altimeter to obtain the height, which is receiver located, the objective of navigation can be achieved.

Choosing a benchmark station Z_0 (Liu et al. 2004), and getting the station's height, the barometer is determined by measuring the relative height between the station and receiver, rather than measured the receiver's height directly, and realized positioning (Zhou 2003).

The height measurement is not only applying in the field of aviation, surveying and mapping, but is widely used in everyday activities such as mountain climbing and exploration. Nowadays, height measurement has been developed rapidly, for instance, laser ranging, satellite altimetry, radio range, and air pressure measurement were applied in many high technology fields. Among these fields, air pressure measurement is a common method. It has the advantages of fast speed and easy to operate, but the precision is limited, so it usually used in small scale of the regional gravity survey. With the deep development of sensor (Gao et al. 2012), because of its simple structure and high precision, the air pressure sensors apply in the field of flight height measurement.

2 THEORETICAL FOUNDATION

2.1 Pseudo-range positioning equations

The formula (1) gives the principle of GPS timing and positioning, which means to solve the four nonlinear equations:

$$\begin{cases} \sqrt{(x^{(1)} - x)^2 + (y^{(1)} - y)^2 + (z^{(1)} - z)^2} + \delta t_u = \rho_c^{(1)} \\ \sqrt{(x^{(2)} - x)^2 + (y^{(2)} - y)^2 + (z^{(2)} - z)^2} + \delta t_u = \rho_c^{(2)} \\ \sqrt{(x^{(3)} - x)^2 + (y^{(3)} - y)^2 + (z^{(3)} - z)^2} + \delta t_u = \rho_c^{(3)} \\ \sqrt{(x^{(4)} - x)^2 + (y^{(4)} - y)^2 + (z^{(4)} - z)^2} + \delta t_u = \rho_c^{(4)} \end{cases} \quad (1)$$

Each of the satellite's coordinate values can calculate from ephemeris which is broadcasted by its own satellites in equations. ρc (n) is measured by the receiver. So, there are receiver's coordinate values(x,y,z) and clock correction δtu leave in the equations. If the receiver obtains at least 4 visible satellites' pseudo-range, the equations can realize the position.

2.2 Definition of height

As the Figure.1 shows, H is expressed as height above sea level. h is also called elevation, which is the distance between datum ellipsoid and real landform in geodetic coordinate system. Nh is called geoid surface, its value can obtain by checking the relevant information (geoid surface's value is different in all over the world). The formula (2) gives the relationship between h, Nh, H:

$$h \approx H + Nh \qquad (2)$$

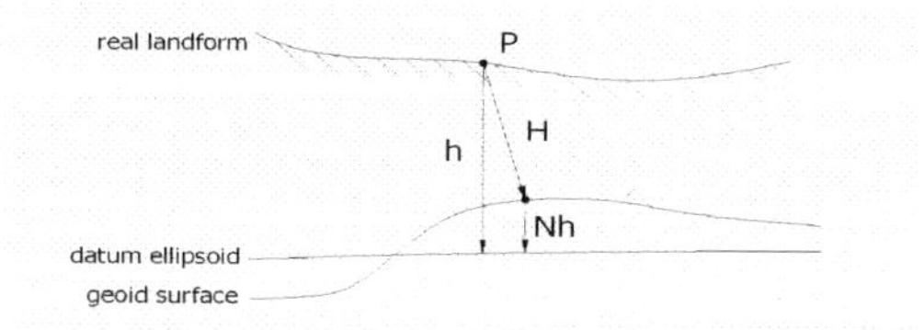

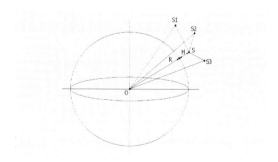

Figure 1. The relationship bewteen h,H and Nh.

2.3 The principle of aided-navigation by height measurement

As the formula (1) expressed, when the receiver encountered the bad environment such as multipath interference, bad weather or located in a city or forest, the signal strength of satellite will be attenuated seriously, the receiver may acquire three or fewer visible satellite signal, at this time, the receiver can only do two-dimensional positioning. And it does not meet the requirement of modern navigation technology.

Hypothesis the earth can express as a spheroid, as the Figure.2 shows, O is the earth's center,S1,S1 and S3 are visible satellites, S is the position where the receiver located, R is the earth radius, h is the receiver's height.

Figure 2. The principle of height measurement of air pressure in navigation.

If the height can be obtained, the formula (1) will transform into:

$$\begin{cases} \sqrt{(x^{(1)}-x)^2+(y^{(1)}-y)^2+(z^{(1)}-z)^2}+\delta t_u = \rho_c^{(1)} \\ \sqrt{(x^{(2)}-x)^2+(y^{(2)}-y)^2+(z^{(2)}-z)^2}+\delta t_u = \rho_c^{(2)} \\ \sqrt{(x^{(3)}-x)^2+(y^{(3)}-y)^2+(z^{(3)}-z)^2}+\delta t_u = \rho_c^{(3)} \\ \dfrac{x^2+y^2}{(a+H)^2}+\dfrac{z^2}{(b+H)^2}=1 \end{cases} \qquad (3)$$

As the formula (3) expressed, as a parameter, the height, which measured by a barometer is taking part into equations. When the visible satellites numbers is less than four, the numbers of equations remains four. The user's coordinate (x,y,z) will be obtained though calculate equations.

3 THEORETICAL FOUNDATION

$$H = H_0 + 67.4 \times (273.15 + t_m) \cdot \lg \frac{P_0}{P} \qquad (4)$$

The formula (4) gives Laplace Air Pressure Height Equations. P express the pressure value where the receiver is located. H_0 is the height of the datum plane (benchmark station), and P_0 is the pressure value of datum plane. t_m is an average temperature of the virtual atmosphere. Here are each of the parameters' measurement way:

1 Measurement way of P
Using the digital pressure sensor of receiver for real-time measurement.

2 Measurement way of H_0
Putting the receiver's location as the origin, establishing the coordinate system, searching the station, which pressure value is the most closest with P, making this station as datum plane H_0.

3 Measurement way of P_0
P_0 is the value of H_0.

4 Measurement way of t_m:

$$t_m = \frac{1}{2}(t_0+t_1) = \frac{1}{2}(t+t_{12}) + \frac{H_0}{400} \qquad (5)$$

t can express as the station's temperature at this time, t_{12} is the temperature 12 hours before, H is the height above sea level of the station.
Here gives the process of height measurement:

1 The barometer in receiver measured the pressure value P, and put P into formula (6)

$$H' = 44331 \times \left[1 - (\frac{P}{1013.25})^{0.19} \right] \qquad (6)$$

156

Solving the pseudo-altitude H', putting H' into formula (3) and obtaining coordinate (x',y',z').

2 Making the (x',y',z') as origin, searching the station around the (x',y',z'), solving the correspondent pressure value P' and temperature value T' though interpolation. The hypothesis that d1,d2,d3 and d4 are the distance from stations to origin. The formula of interpolation is:

$$P' = \frac{1}{2}\left(\frac{p_1 d_3 + p_3 d_1}{d_1 + d_3} + \frac{p_4 d_2 + p_2 d_4}{d_2 + d_4}\right) \quad (7)$$

$$T' = \frac{1}{2}\left(\frac{t_1 d_3 + t_3 d_1}{t_1 + t_3} + \frac{t_4 d_2 + t_2 d_4}{t_2 + t_4}\right) \quad (8)$$

3 Putting P' and T' into formula(4) and obtaining H, because the height of station is known, then putting sum into formula(3) to solve (x,y,z) again, the result of (x,y,z) is the receiver's real coordinate

4 HARDWARE DESIGNING

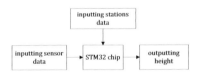

Figure 3. System principle block diagram.

The Figure 3 shows the system principle block diagram. Sending benchmark stations data and sensor data to STM32 chip, calculating height and outputting with serial ports. The process of height measurement is complete.

The MS5534B digital pressure sensor of the Switzerland Intersema Company has 6 pins, and communication with STM32 by SPI bus protocol.SPI bus protocol is a kind of high-speed duplex synchronous communication bus. The figure.4 shows the block diagram of MS5534B:

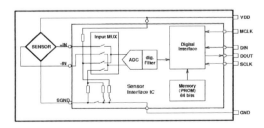

Figure 4. Block diagram MS5534B.

Table 1. Pin configuration of ms5534b.

Pin name	Pin	Type	Function
GND	1	G	Ground
SCLK	2	I	Series data clock
DOUT	3	O	Data output
DIN	4	I	Data input
MCLK	5	I	Master clock
VDD	6	P	Positive supply voltage
PEN(1)	7	I	Programming enable
PV(1)	8	N	Negative Programming voltage

SPI bus has four wires:

1 MISO: the master device input ,the slave device output;
2 MOSI: the master device output ,the slave device input;
3 SCK: clock signal, it is generated by the master, SCK=32. 768KHz;
4 CS: the chip chosen signal. The function of this wire is making the master communicating with the slave individually, and avoiding confliction on data wire

Figure 6 is the barometer altimeter PCB board, as the picture shows, part 1 is STM32 chip, part 2 is MS5534B pressure sensor, part 3 is external crystal which is providing 32.768K clock signal from the sensor. Here is the working principle of the barometer: the pressure sensor sends the real-time measurement pressure value and temperature value to STM32 chip by SPI bus, then the chip obtains the height value, though formula 2.3-2.5, and outputting the data to receiver with serial ports J1 or J2.

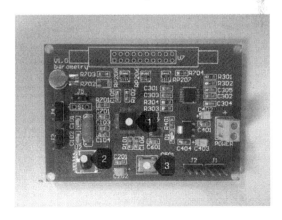

Figure 5. The barometer altimeter PCB board.

5 SOFTWARE DESIGNING

The MS5534B sensor converts the real-time uncompensated signal to 16 bits digital pressure signal D1

157

and D2. D1 is pressure value and D2 is temperature value. Here is the process of pressure and temperature measurement:

1 Reading the WORD1-WORD4 through MS5534B. Before the reading operation, the chip will reset one time.
2 Extracting compensated coefficients C1-C6 though the bit operation and bit manipulation.

Table 2. Coefficient C1 to C6.

C1	Pressure sensitivity	15 BITS	SENS
C2	Pressure offset	12 BITS	OFF
C3	Temperature coefficient of Pressure sensitivity	10 BITS	TCS
C4	Temperature coefficient of Pressure offset	10 BITS	TCO
C5	Reference temperature	11 BITS	Tref
C6	Temperature coefficient of temperature	6 BITS	TEMP

3 Reading the 16 bits pressure signal D1 and temperature signal D2.
4 Calculate calibration temperature:

$$UT1 = 8 * C5 + 20224 \qquad (9)$$

5 Calculate actual temperature:

$$dT1 = D2 - UT1 \qquad (10)$$

$$TEMP = 200 + dT * (C6 + 50) / 1024 \qquad (11)$$

6 Calculate temperature compensated pressure

$$OFF = C2 * 4 + [(C4 - 512) * dT] / 2^{12} \qquad (12)$$

$$SENS = C1 + (C3 * dT) / 1024 + 24576 \qquad (13)$$

$$X = [SENS * (D1 - 7168)] / 2^{14} - OFF \qquad (14)$$

$$P = X * 10 / 32 + 2500 \qquad (15)$$

Putting P and T into formula 3.1-3.5 and getting height value.

The Figure.6 shows the flow chart of software of the system:

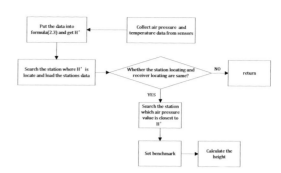

Figure 6. The flow chart of software.

6 TEST SCHEME

In order to test barometric altimeter's capability and performance, designing two experiments:

6.1 *Experiment to verify pressure and temperature's accuracy*

Comparing the pressure and temperature from sensor with the pressure and temperature from standard barometer and thermometer, analysis the deviation value(Zhou ,)of two groups.

Table 3. The test conditions about data collect.

Measurement Equipment		Time Interval	Conditions
NO.1	The reference device	2014.10.21– 2014.10.22	room temperature
NO.2	The device need to test		

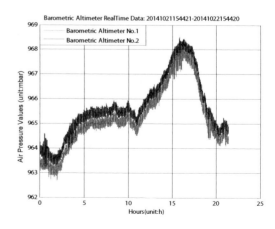

Figure 7. Air pressure data of two groups during 24 hours.

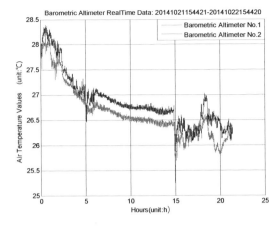

Figure 8. Temperature data of two groups during 24 hours.

The TABLE 4 shows the maximum and minimum deviation value:

Table 4. The altimeter's measurement data analysi.

Testing Parameter	Testing Content	Value
Air Pressure	Maximum deviation value	1.08mbar
	Minimum deviation value	−0.3mbar
Temperature	Maximum deviation value	0.51mbar
	Minimum deviation value	−0.31mbar

The TABLE 5 which is extracted from MS5534B datasheet shows the range of measurement error about air pressure and temperature:

Table 5. MS5534B allowed error range.

Parameter	Conditions	Min	Max
Pressure Accuracy	$P=750\ldots1100mbar$, $T_a=25°C$	−1.5 mbar	1.5mbar
Temperature Accuracy	$T=20°C$	−0.8°C	0.8°C

Comparing with two tables, the barometric altimeter's accuracy can meet the requirements.

6.2 Experiment to verify the height measurement accuracy

Putting the barometric altimeter in a place where the height is known and setting a station, choosing another place to set the barometric altimeter also, comparing the measurement height and its real height, analysis the deviation value of data.

Table 6. The height measurement test conditions.

Measurement Equipment		Time Interval	Conditions
NO.1	The reference device		
NO.2	The device needs to test	2014.10.28	Room Temperature

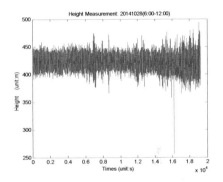

Figure 9. The height measurement data during 6 hours.

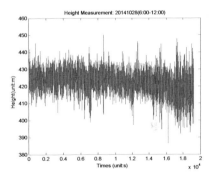

Figure 10. The data after using average moving filtering.

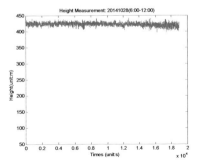

Figure 11. The data after using average filtering.

159

The TABLE 7 and 8 shows the maximum and minimum deviation value:

Table 7. The height measurement data analysis after average moving filtering.

Testing Parameter	Testing Content	Value
Height	Standard deviation	6.1255m
	Mean value	422.5920m

Table 8. The height measurement data analysis after average filtering.

Testing Parameter	Testing Content	Value
Height	Standard deviation	6.8047m
	Mean value	422.5682m

The result of the test shows the measurement precision can satisfy engineering demand.

7 CONCLUSION

The mainly conclusions are as follows:

1 Due to air pressure value is large changing in one day, and the changing is no obvious rules, so the measurement of height will be affected.
2 The result of data analysis shows that the standard deviations of height measurement can achieve $\sigma \approx 6m$ and satisfy the demand of engineer.

REFERENCES

Xie, G. (2011) Principles of GPS and receiver design. *Beijing: Publishing House Of Electronics.*

Liu G.Q, W.G. Liu & J.C. Lu. (2004) Measurement of UAV Barometric Height Calibrated by GPS Height and compensated by Atmosphere Temperature. *Journal of Electronic Measurement and Instrument (2),* 1161–1165.

Zhou Y.L. (2003). Measurement of Barometric Height apply in the regional gravity survey.

Gao, S.Q. & X.L. Wei (2012). Error analysis and correction for the atmospheric pressure measurement of altitude. *Electromic Measurement Technology,* 225–228.

Electronic Engineering and Information Science – Wang (Ed.)
© 2015 Taylor & Francis Group, London, ISBN: 978-1-138-02772-5

Fuzzy model of circulating fluidized bed boiler combustion system

N. Lv & L. Wang

Department of Automation, Harbin University of Science and Technology, Harbin, Heilongjiang, China

ABSTRACT: This paper presents a complete method of identifying Circulating Fluidized Bed Boiler (CFBB) combustion system TS fuzzy model, first the combustion system is decoupled into two Single Input and Single Output (SISO) subsystems, the input variables in the model is determined by the method of gray correlation, then the premise structure and parameters are also got by the method of Fuzzy C-means Clustering Method (FCM),at last the consequent parameters of rules are calculated by the Recursion of Least Square Estimation (RLSE), combustion system TS fuzzy model is obtained. In this paper, in order to verify the validity of the method and stability, the actual operation data of the boiler is used to be simulated.

KEYWORDS: CFBB; TS fuzzy model; Decoupling; FCM; RLSE.

1 INTRODUCTION

The CFBB combustion system is a complex object, it possesses the characteristics of strong nonlinear, strong coupling, therefore its control needs high precision mathematical model. The traditional method of establishing CFBB's model (Liu et al. 2010) cannot achieve the purpose of accurate control (Jiang, Liu & Yang, 2012). Since L. A. Zadeh has created a fuzzy theory, the theory has been making control enormous changes. This paper uses the fuzzy method to establish the T-S fuzzy model of CFBB combustion portion (Gao et al.2012).

2 THE T-S FUZZY MODEL

T-S model is established based on the fuzzy rules, the premise parameters of the rules are fuzzy variables, the consequent parameters are accurate. T-S fuzzy model is defined in the form:

$$R_i: \quad IF \quad (x_1 \text{ is } A_{i1}) \text{ and} \cdots \text{ and} (x_n \text{ is } A_{in})$$
$$THEN \quad y_i = P_{i1}x_1 + P_{i2}x_2 + \cdots + P_{in}x_n; \quad (1)$$
$$i = 1, 2, \cdots, c$$

Where R_i represents the $i-th$ fuzzy rule, c is the rule number, $x_1, x_2, \ldots, x_n$ are input variables, y_i is the output, P_{ij} is the consequent parameter, $A_{ik}(x_i) = \exp\left\{-(x_i - z_{ik})^2 / \sigma_k^2\right\}$, and $\sum_{i=1}^{c} A_{ik} = 1$, z_{ik} is the center of the $k-th$ fuzzy clustering, σ_k represents radius of the $k-th$ cluster.

3 FCM

On the basis of ensuring the input variables, using FCM, one can get the premise structure and parameters. FCM algorithm is given below: Given a sample $X = \{x_1, x_2, \ldots, x_r\}$ setting fuzzy clustering number c, the purpose of this algorithm is to find membership degree A_{ik}, clustering center z_{ik} and the clustering radius σ_k, this paper defines the objective function:

$$J(U,c) = \sum_{i=1}^{c} \sum_{k=1}^{r} (A_{ik})^m d_{ik}^2, \quad \text{where} \quad d_{ik}^2 = \|x_k - z_i\|^2,$$

i=1,2,…,c; k=1,2,…,r, m is the index weight for membership function value, $1 < m < \infty$, generally takes 2, d_{ik} is the distance from sample x_k to the center z_i. According to the literature documents, it presents a way to ensure the above objective function is minimized, where $U = [A_{ik}]$ (Hou & Gou 2013).

4 RLSE

According to the input variables and the premise parameters (Li, Chiang 2011), this paper uses RLSE to get the consequent parameters. Performance index is defined as:

$$J = \{\sum_{i=1}^{N} (y_i - \hat{y}_i)^2\} \Big/ N \quad (2)$$

Where y_i is the actual output, $\hat{y}_i$ is the desired output (i = 1,2,…,N).Setting:

$$X =$$

$$\begin{bmatrix} \lambda_{11},\cdots,\lambda_{R1},x_{11}\cdot\lambda_{11},\cdots,x_{11}\cdot\lambda_{R1}, & \cdots & , & x_{n1}\cdot\lambda_{11},\cdots,x_{n1}\cdot\lambda_{R1} \\ \vdots & & & \vdots \\ \lambda_{1N},\cdots,\lambda_{RN},x_{1N}\cdot\lambda_{1N},\cdots,x_{1N}\cdot\lambda_{RN},\cdots,x_{nN}\cdot\lambda_{1N},\cdots,x_{nN}\cdot\lambda_{RN} \end{bmatrix} \quad (3)$$

$$Y = \begin{bmatrix} y_1,\cdots,y_N \end{bmatrix}' \tag{4}$$

$$P = \begin{bmatrix} P_0^1,\cdots,P_0^R,P_1^1,\cdots,P_1^R,P_n^1,\cdots,P_n^R \end{bmatrix} \tag{5}$$

then $Y = XP$ (6)

Marking x_k is the k-th row vector of X, y_k is the k-th component of Y, the recursive algorithm is:

$$P_{k+1} = P_k + K_k(y_{k+1} - x_{k+1}\cdot P_k) \tag{7}$$

$$K_k = S_k \cdot x_{k+1}^T \cdot (1 + x_{k+1}\cdot S_k \cdot x_{k+1}^T)^{-1} \tag{8}$$

$$S_{k+1} = S_k - K_k \cdot x_{k+1}\cdot S_k \tag{9}$$

$$k = 0,1,\cdots,N-1 \tag{10}$$

5 THE SIMULATION EXPERIMENT RESEARCH

In this paper, the simulation object is CFBB combustion system, parameter variables collected is the amount of coal (B), the primary air volume (Q), bed temperature (Tb), the main steam pressure (PO). We use sample data to establish the T-S model. Considering the system has strong coupling, the system requires to be decoupled, patchwork method is adopted. We adopt Pade approximation methods to reduce the order. The system needs to be decoupled first. The research object is shown in the figure below:

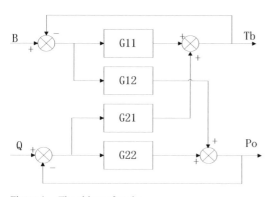

Figure 1. The object of study.

In data processing section, this paper adopts $zf = \mathrm{idfilt}(z,\mathrm{ord},\mathrm{Wn})$ and $zd = \mathrm{dtrend}(z)$ in Matlab to do means and filtering processing respectively. The four transfer function identifier is expressed as:

$$G_{11} = \frac{21.746e^{-30s}}{(1+0.001s)(1+281.53s)(1+285.13s)} \tag{11}$$

$$G_{12} = \frac{0.012(1+1.56s)e^{-27.6s}}{(1+17.7s)(1+0.4s)} \tag{12}$$

$$G_{21} = \frac{-0.008(1+15.4s)e^{-2.46s}}{(1+20.2s)(1+2.2s)} \tag{13}$$

$$G_{22} = \frac{0.000034e^{-0.0003s}}{(1+53.9s)(1+0.001s)(1+53.8s)} \tag{14}$$

In this paper, gray correlation method is used to determine the extent of the associated input output pairs, the required correlation table is shown in Table 1. It shows that B is greater than the correlation for Tb of Q, Q is larger than B's correlation for Po. Therefore, this article system is decoupled into two systems: subsystem one: B for input, Tb for output, subsystem two: Q for input, Po for output.

Table 1. The degree of correlation between input and output.

output \ input	B	Q
Tb	0.8780	0.7516
Po	0.7399	0.7493

The system cascade structure is shown in figure 2, decoupling structure is shown in figure 3:

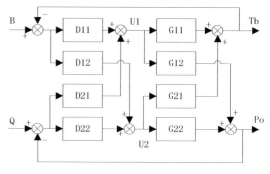

Figure 2. Cascade structure.

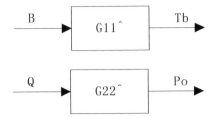

Figure 3. Two subsystems.

Patchwork method is adopted. Calculated as follows :

$$A = 0.0536s^8 + 107.3s^7 + 53740s^6 + 71640s^5$$
$$+ 25440s^4 + 1031s^3 + 14.38s^2 + 0.09497s$$
$$+ 0.0008354$$

$$B = 73240s^{10} + 1.467 \times 10^8 s^9 + 7.369 \times 10^{10} s^8$$
$$+ 2.276 \times 10^{11} s^7 + 1.163 \times 10^{11} s^6 + 1.427 \times 10^{10} s^5$$
$$+ 7.157 \times 10^8 s^4 + 1.643 \times 10^7 s^3 + 1.719 \times 10^5 s^2$$
$$+ 714.8s + 1$$

$$G_{11}^{\wedge}(s) = G_{22}^{\wedge}(s) = \frac{A}{B} e^{-30s} \quad (15)$$

In order reduction part, the reduced-order transfer function $G_{11}^{\wedge}(s)$ is

$$G^{\wedge} = \frac{2.26 \times 10^{-6} s^2 + 4.194 \times 10^{-9} s + 1.782 \times 10^{-10}}{s^3 + 0.02414s^2 + 0.0001332s + 2.133 \times 10^{-7}} \quad (16)$$

The figure below shows that step response of comparison of the original transfer function and the reduced order one.

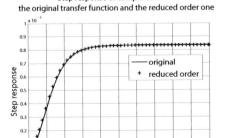

Figure 4. Verify the order reduction effectiveness.

It can be seen from Figure 4 that reduced order is valid, and the system stability characteristic doesn't change.

Input variables should be selected from all input variables in consideration. From the reduced order model in this article, you can see that the order is 3, so all input variables in consideration are B(t), B(t-1), B(t-2), B(t-3), Tb(t), Tb(t-1), Tb(t-2), Tb(t-3). The model input variables are determined by the method of grey correlation, the correlation of each determined input variable and output Tb(t) is shown as follows:

Table 2. The correlation between established input variables and output variables.

output \ input	B(t)	B(t-1)	B(t-2)	B(t-3)
Tb(t)	0.8780	0.8852	0.9573	0.9604

output \ input	Tb(t)	Tb(t-1)	Tb(t-2)	Tb(t-3)
Tb(t)	1	0.9843	0.9011	0.8544

From the table above, all the input variables which have the more correlation to the output variable Tb(t), such as {B(t-2), B(t-3), Tb(t-1)}, we call it the input variables. After many experiments, the 288 groups of data are divided into three fuzzy clusters, as follows:

Table 3. Clustering centres and radius.

The fuzzy clustering	The input variable B(t-2)	The input variable B(t-3)	The input variable Tb(t-1)	Radius σ_k
1	0.3246	0.3910	10.1169	0.2403
2	−0.1185	−0.1020	1.3552	0.5609
3	−0.1558	−0.1798	−7.7608	0.2321

Gaussian membership functions of different input variables are shown below:

$$A_{11} = \exp(-\left|\frac{B(t-2) - 0.3246}{0.2403}\right|^2) \quad (17)$$

$$A_{21} = \exp(-\left|\frac{B(t-3) - 0.3910}{0.2403}\right|^2) \quad (18)$$

$$A_{31} = \exp(-\left|\frac{Tb(t-1) - 10.1169}{0.2403}\right|^2) \quad (19)$$

$$A_{12} = \exp(-\left|\frac{B(t-2) + 0.1185}{0.5609}\right|^2) \quad (20)$$

$$A_{22} = \exp\left(-\left|\frac{B(t-3)+0.1020}{0.5609}\right|^2\right) \tag{21}$$

$$A_{32} = \exp\left(-\left|\frac{Tb(t-1)-1.3552}{0.5609}\right|^2\right) \tag{22}$$

$$A_{13} = \exp\left(-\left|\frac{B(t-2)+0.1558}{0.2321}\right|^2\right) \tag{23}$$

$$A_{23} = \exp\left(-\left|\frac{B(t-3)+0.1798}{0.2321}\right|^2\right) \tag{24}$$

$$A_{33} = \exp\left(-\left|\frac{Tb(t-1)+7.7647}{0.2321}\right|^2\right) \tag{25}$$

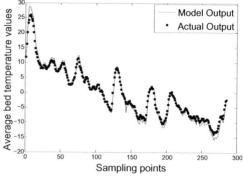

Figure 5. The comparison of model output and actual output

Using the recursive least squares, this article identifies the consequent parameters, identification result is shown as follows:

$$P^1 = [0.1631 \ 0.0488 \ 0.9611] \tag{26}$$

$$P^2 = [0.3992 \ -0.3220 \ 0.9226] \tag{27}$$

$$P^3 = [0.1441 \ -0.0372 \ 0.9916] \tag{28}$$

From the above calculation, we have established in this paper the amount of coal - bed temperature subsystem T-S fuzzy model of CFBB combustion system is:

$$R^1: \ IF \quad B(t-2) \ is \ A_{11} \ and \ B(t-3) \ is \ A_{12}$$
$$and \quad Tb(t-1) \ is \ A_{13}, \tag{29}$$
$$THEN \quad y^1 = 0.1631B(t-2)+0.0488B(t-3)$$
$$+0.9611Tb(t-1)$$

$$R^2: \ IF \quad B(t-2) \ is \ A_{21} \ and \ B(t-3) \ is \ A_{22}$$
$$and \quad Tb(t-1) \ is \ A_{23}, \tag{30}$$
$$THEN \quad y^1 = 0.3992B(t-2)-0.3220B(t-3)$$
$$+0.9226Tb(t-1)$$

$$R^3: \ IF \quad B(t-2) \ is \ A_{31} \ and \ B(t-3) \ is \ A_{32}$$
$$and \quad Tb(t-1) \ is \ A_{33}, \tag{31}$$
$$THEN \quad y^1 = 0.1441B(t-2)-0.0372B(t-3)$$
$$+0.9916Tb(t-1)$$

Figure 5 shows an output curve of comparison of the amount of coal-bed temperature subsystem and T-S fuzzy model established in this paper. We can clearly see that the model output and the actual output are consistent, which verifies the validity of

the model, at the same time, the model output stably tracks system actual output, it indicates the model is good for stability, meets the actual needs.

6 CONCLUSIONS

Based on the complexity of the CFBB combustion system, this paper adopts the fuzzy c-means clustering algorithm and the recursive least squares method, establishes the amount of coal - bed temperature subsystem T-S fuzzy model, it's output basically consistent with the actual output, the accuracy of the model is very good. At the same time, the stability of this model is also very good. The model is effective, it can be used in practical application.

REFERENCES

Liu, C.Y., J. Wang, Q. Li, X.L. Song, & Z.Y. Song (2010).The study of the control of the bed temperature in the circulating fluidized bed boiler based on the fuzzy control system.*2010-2010 CCTAE International Conference on Computer and Communication Technologies in Agriculture Engineering 1*, 285–288.
Jiang, Y.P., S. Liu & Y.H. Yang (2012).Application of advanced process control in Circulating Fluidized Bed Boiler. *Applied Mechanics and Materials, Advances in Science and Engineering II. 135*,305–308.
Gao Q., X.J. Zeng, G. Feng, Y. Wang & J.B Qiu (2012). T-S-fuzzy-model-based approximation and controller design for general nonlinear systems. *IEEE Transactions on Systems, Man, and Cybernetics, Part B: Cybernetics 42(4)*, 1143–1154.
Hou F. & J. Gou (2013). Fuzzy Rule Generation Based On CoWM And FCM Algorithms. *International Journal of Applied Mathematics and Statistics 45(15)*, 20–27.
Li C.S. & T.W. Chiang (2011). Complex fuzzy model with PSO-RLSE hybrid learning approach to function approximation. *International Journal of Intelligent Information and Database Systems. 5(4)*, 409–430.

Electronic Engineering and Information Science – Wang (Ed.)
© 2015 Taylor & Francis Group, London, ISBN: 978-1-138-02772-5

The characteristics analysis of ZnO thin films transistor

X.Y. Wang, Y.S. Zhang & D.X. Wang
Key Laboratory of Engineering Dielectrics and Its Application, College of Applied Science, Harbin University of Science and Technology, Heilongjiang, Harbin, China

ABSTRACT: At room temperature, we make a vertical type ZnO thin film transistor with the structure of Al/ZnO/Ni/ZnO/Al. The ZnO thin film transistor has the advantages of fast reaction speed, low operating voltage and high carrier mobility. Volume of Transistor is smaller than that of MOSFET and its operating current (IDS) can reach milliampere and the gate electrode can control of operating current. In this device, ZnO (active layers) and Al (source/drain electrode) to form ohmic contact, ZnO and Ni(gate electrode) to form Schottky contact. When gate-source voltage is 0.2 V and drain-source voltage is 3 V, operating current can reach 9.95×10-3A. Its electrical conductivity is 6.4288×10-8S/cm and carrier mobility is $32.93 \mathrm{cm}^2 \mathrm{V}^{-1} \mathrm{s}^{-1}$. The Schottky contact depletion layer space charge density is $1.22 \times 10^{10} \mathrm{cm}^{-3}$.

KEYWORDS: vertical structure; driving voltage; carrier mobility; ZnO.

1 INTRODUCTION

The transparent oxide thin film transistors used widely have the advantage of transparency, electrical properties and can be made in low temperature. It may take over non-transistor used in active matrix liquid crystal display and light-emitting diode. Jiahong Wu (2012) makes the MIZO thin film by solution-gel method in 2012. That lead to that its carrier mobility reaches $0.064 \mathrm{cm}^2 \mathrm{V}^{-1} \mathrm{s}^{-1}$ and subthreshold swing reach 2.93V/dec and threshold voltage is 5.17V and on-off ratio about 8.80×10^3. Cong Wang (2013), Yurong Liu, Xinghuo Li, Jing Su and Ruche Yao find that the minimum value of leakage current is 0.58 and the maximum of relative variation is 36.29 by study on the light's influence of leakage current and relative variation. American Institute finds that a-IGZO carrier density can be changed by making over the dielectric surface of organic material in 2014. Xianfeng Xiong (2014) finds that thin film to have better continuity, uniformity, if it is made by nozzle printing. And organic thin-film transistors have higher mobility and better environmental stability. We know that structural morphology will influence operating performance in transistor by reading literature. The structure of the traditional ZnO transistor is planar structure, which can be made by lithography. These devices have big volume, high operating voltage and lower operating current. So its operating current is hard to reach the demand of organic light emitting diode. So we can improve the performance by changing the structure. Therefore, we adopt a vertical structure to prepare ZnO thin film transistor, its structure is Al/ZnO/Ni/ZnO/Al. ZnO is the active layer and Ni is

gate electrode. Conduction channel is short, so it has low threshold voltage, high operating current, small volume and fast reaction (Serkan & Gao 2005, Chao & Meng 2006). It is proved that this structure can make operating current achieve milliamp grade.

2 EXPERIMENT

As shown in Figure 1, the structure diagram of ZnO thin film transistor. Firstly, we clean the quartz glass. Secondly, Al was deposited on a quartz glass substrate by DC sputtering, the sputtering time is 40s. The purity of Al is 99.99%. Thirdly, put argon and oxygen with 2:1. ZnO was deposited on an Al substrate by RF sputtering, the sputtering time is 25s. Then Ni was deposited in 25s. Second ZnO is made like first ZnO and sedimentary Al as a drain. The thickness of ZnO is 200nm.

According to theory of metal semiconductor contact, ZnO forms Schottky contact with Ni and ohmic contact with Al. When we give forward voltage in drain and make source grounding, the electron of source cathode come to ZnO. It has two Schottky contacts, one is in positive bias state, and the other come reverse bias, when we give forward voltage in drain. When electronic comes to Ni from ZnO, this Schottky contact is positive skew, so electronic pass to ZnO. But when it comes to ZnO from Ni, that Schottky contact is reverse bias. Because ZnO is thin and voltage of the gate is high, electronic tunneling ZnO to drain. From the result, when we give voltage to drain, the current between drain and source will be controlled by the voltage of the gate.

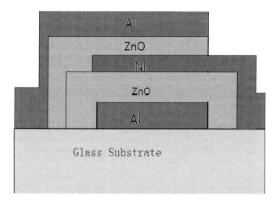

Figure 1. the structure of ZnO thin film transistors.

3 MEASUREMENT AND RESULT ANALYSIS

We give the electric capacity-voltage between Ni/ZnO/Al by picture 3. We change the voltage of Al from −2 to 2 and the step is 0.05V for getting this picture. For making junction capacitance have Q value and it meets noise-signal ratio, we make the frequency of the signal is 1MHz. Let V<KT/q, so we choose 15mV.

According to ideal metal semiconductor materials Schottky barrier theory, though Poisson equation, we get the formula (1), (2) and (3).

$$C = A\sqrt{\frac{q\varepsilon\varepsilon_o N_n}{2(\Phi_B + V)}} = \frac{A\varepsilon\varepsilon_o}{W} \qquad (1)$$

$$\frac{1}{C^2} = \frac{2(\Phi_B + V)}{A^2 q\varepsilon\varepsilon_o Nn} \qquad (2)$$

$$\frac{d(\frac{1}{C^2})}{dV} = \frac{2}{A^2 q\varepsilon\varepsilon_o Nn} \qquad (3)$$

We know that the depletion layer thickness becomes digger with voltage comes bigger. When drain-source voltage (VDS) is 0V, W is 13.9nm. When VDS is 2V, W is applied voltage 32nm. According to formula (3), we get a picture of 1/c2 and voltage V. According to fighter 4, we can get the curve slope. According to formula (2), we know the curve slope is 8.32962×1019.

$$\text{So } \frac{2}{A^2 q\varepsilon\varepsilon_o Nn} = 8.32962 \times 10^{19} \qquad (4)$$

A=0.04cm2, q=1.6×10-19C, ε=8.656F/m, εo= 8.8542×10-12F/m. So we know that the space charge density of ZnO the depletion layer of Schottky contact is n=1.22×1010cm-3.

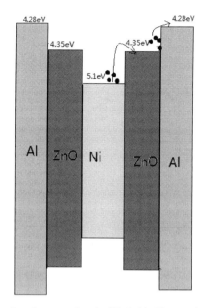

Figure 2. the energy band of ZnO thin film transistors.

We use Keithley 4200-SCS/F semiconductor instrument to get output characteristics. We test the device at room temperature. To measure the operating characteristics of ZnO transistor, gate-source voltage (V_{GS}) was varied from 0 V to 1 V in 0.2 V steps, V_{DS} was swept from 0 V to 3 V.

Figure5 is the I-V characteristics of ZnO thin film transistor. Grid voltage plays a role in control drain current, when it smaller than 1V. We can find that when voltage is low, I-V characteristics show I∝V; When V_{DS} is high, I-V characteristics show I∝V². Space charge limited current, when voltage is bigger than threshold voltage V_{th}, so I-V shows I∝V², we know V_{th}=1.2V though count. I_{DS} come bigger with V_{DS} increase, when V_{GS} be kept. Between ZnO and Ni will be lower with V_{DS} increase, when V_{th} be kept. So I_{DS} becomes bigger. The voltage between ZnO and Ni will be smaller when V_{DS} with V_{th} reduce. So I_{DS} becomes smaller.

It is 9.95×10⁻³A, when V_{GS} is 0.2V and V_{DS} is 3V. I_{DS} are 9.95×10⁻³ A, so it can be used as the drive of light emitting diode.

We get figure 6 when we take the logarithm about I_{DS} and V_{DS} in figure 5, formula (5), (6). Though formula (5), (6), we can get oxidation zinc carrier mobility. A is ZnO film size, d is the thickness of oxidation zinc films, σ is conductivity, μ is carrier mobility, q is electron charge, n is carrier concentration. We can get these though experimental. A=0.04cm², d=200nm, q=1.6×10⁻¹⁹C, n=1.22×10¹⁰cm⁻³. We use these in formula (5),(6). We can know that σ is 6.4288×10⁻⁸S/cm, μ=32.93cm2V⁻¹s⁻¹. So the carrier mobility of ZnO film is higher than organic material.

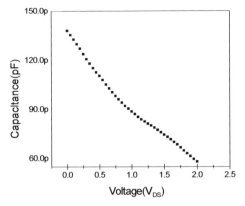

Figure 3. The electric capacity—voltage between Ni/ZnO/Al.

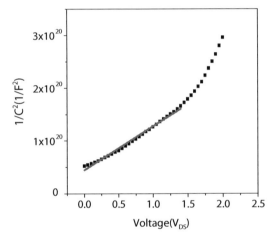

Figure 4. The V-1/C² between Ni/ZnO/Al.

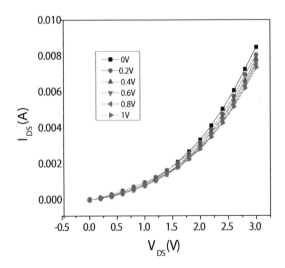

Figure 5. Output characteristic curve of transistor.

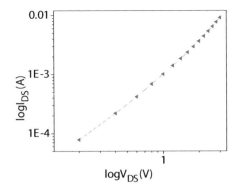

Figure 6. Curve of log (I_{DS}) -log (V_{DS}).

4 SUMMARY

In a word, we have successfully prepared a vertical structure ZmO thin film phototransistor and measure its light-electrical characteristics. We use Al as source and drain and ZnO as active layer in this passage. Its carrier mobility is $32.93 cm^2 V^{-1} s^{-1}$. When $V_{GS} < 1$, V_{GS} can control I_{DS}. When V_{GS} is 0.2V and V_{th} is about 1.2V. We prove that device work in low voltage, and put out high current. So it can be used as active luminescence display driving.

REFERENCES

Wu J.H. (2012). Research of MgInZno transparrent oxide thin transistor. *Research & Progress of SSE.* 32, 511–516.

Wang C., Y.R. Liu, X.H. Liu, J. Su & R.H. Yao (2013). Light induced instability of oxide thin transistor, Light induced instability of oxide thin transistor. *Natural Science Edition 41,* 12–18.

Xiong X.F. (2014). Modification of the substrate surface insulation in ink jet printing organic thin film transistor. *Chiness Journal of Luminescence 35,* 106–110.

Serkan Zorba & Y.L. Gao (2005), Feasibility of static induction transistor with organic semiconductors, Appl. Phys. Lett. 86, 193508–193511.

Chao Y.C. & H.F. Meng (2006). Polymer space-charge-limited transistor. *Appl. Phys. Lett. 88,* 223510–223515.

Electronic Engineering and Information Science – Wang (Ed.)
© 2015 Taylor & Francis Group, London, ISBN: 978-1-138-02772-5

Study about the legal protection of personal data and its effective transmission in Electronic Commerce

Y.J. Ye

The School of Marxism in Wuhan University of Technology, Wuhan, Hubei, China
The Social Management Department of Huanghuai University, Zhumadian, Henan, China

ABSTRACT: With the rapid development of the Electronic Commerce, the contradiction between the legal protection of personal data and its effective transmission is getting more and more serious. In order to solve the problem about the personal data circulation and its legal protection, firstly master the meaning of the meaning of the personal data in Electronic Commerce. Then learn the methods of ensuring the security of account passwords and other personal information. At last take such measures as strengthening self-discipline of the merchants from their own perspectives, strengthening the self-protection of consumers from their own perspectives and so on. Through these efforts, try to promote the harmonious development of the Electronic Commerce.

KEYWORDS: Personal Data; Circulation; Legal Protection; Electronic Commerce.

1 INTRODUCTION

Along with the fast development of the information technology and international internet, the consumers have to deal with the misuse of their personal data in the process of Electronic Commerce transaction, in addition to facing up to the ecological environment, food, disease risk, which brings a great deal of psychological distress to them. According to the survey report of Beijing Youth Daily, in Beijing, Shanghai, Guangzhou, the proportion of personal data infringement respectively is 60.4%, 60.4% and 39.1% (Zhou, H.H. 2006). It's getting more and more serious of the infringement of personal data. The Electronic Signature Law has been promulgated to solve the authentication problem of the development of Electronic Commerce in China, but it has only solved the authentication problem of the development of Electronic Commerce and there are also many other problems. So the consumers' anxiety and perplexity have increased about his personal information leakage and illegal use of personal data (Qi, A.M.2012). The personal data circulation conflicts bitterly with the privacy protection, so it's very important to strive for balance between data circulation and personal privacy protection.

2 THE MEANING OF THE PERSONAL DATA IN ELECTRONIC COMMERCE

The personal data in Electronic Commerce means any data concerning the personal or material circumstance of an identified or identifiable individual (the data subject) in Electronic Commerce. Sometimes the personal data are called personal information or personal privacy. Personal data in Electronic Commerce has the characteristic of rights and interests. Firstly, it is the embodiment of the right of personality. The elements of personal data and other components of individual identity information, such as name, reputation, privacy, portrait personality, carrying the ethical value of equality, freedom, dignity, and so on, are the important parts of the man in the society, so the personal information has strong personality character in Electronic Commerce. In addition, it is also the object of personality right (Ma, J.J.2006). So when the intruders gain unauthorized access to personally identifiable information on Electronic Commerce, he also violates the right to personal dignity. Secondly, personal data has the property. Nowadays, in Electronic Commerce, because of the property of the personal information, it's very common to violate the right to information to make a profit, which has caused a great deal of trouble for people in Electronic Commerce.

3 THE METHODS OF ENSURING THE SECURITY OF ACCOUNT PASSWORDS AND OTHER PERSONAL DATA

The development of the internet affects our daily life and our future. There is no safety rules or safety organizations since the techniques in many companies are different, leaving the Internet everywhere "buried unsafe mines". CCTV "315" shows big bugs in mobile phone app and cookies in ads. The Internet security issues affect not only individuals, but also Internet companies. So, it is urgent for us to have a set of security rules to follow.

3.1 Security issue

When a person opens an account with a website, he may be attacked by DNS hijack, proxy server hijack, xss hijack, web database security etc. The website access structure diagram is shown in Fig.1.

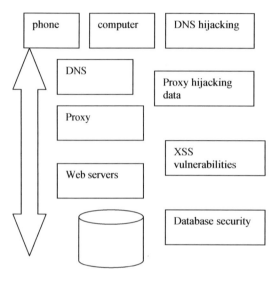

Figure 1. Website access structure diagram.

DNS serves a base in internet and plays an important rule in the site operation and maintenance. Fig.2 shows the domain name resolution process.

It is easy to find in Figure 2 that if DNS server returns a fake IP address, the user will visit a fake

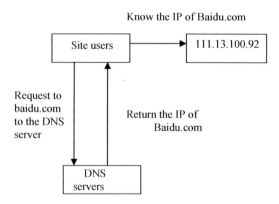

Figure 2. Domain name resolution process.

website and this is DNS hijack. DNS hijacking is the practice of subverting the resolution of the Domain Name System (DNS) queries. This can be achieved to point at a rogue DNS server under the control of an attacker, or through modifying the behavior of a trusted DNS server so that it does not comply with internet standards (Xu T. 2007).

3.2 Storage security password

The user's password and account information are saved in the database server. DBA as well as system developer get access to these data easily. The hackers can also access if there is a site security loophole. The risk is pretty high to store users' passwords in plain text. For example, 6000000 users' passwords are let out in plain text in a famous csdn.net in 2011.

It is a basic rule to encrypt the user's password. An irreversible hash algorithm md5 is used in most websites, for example, md5("1234567") =fcea920f7412b-5da7be0cf42b8c93759. Though md5 is an irreversible hash, some md5 hash password can be cracked with the development of CPU and storage devices, as is shown in Table 1(Zhang, Y.X. , Y. Zhao, 2008).

Table 1. The original password and the corresponding MD5 table.

Original Password	MD5
1234567	FCEA920F7412B5DA7BE0CF42B8C93759
abc	900150983CD24FB0D6963F7D28E17F72
abcd	E2FC714C4727EE9395F324CD2E7F331F
1213abcd	9C98DF872D24244696C393A1D26AB749

It is easy to get md5 password with the help of www.cmd5.com.cn, md5.com.cn, etc. For example, input "fcea920f7412b5da7be0cf42b8c93759"in Table 1 and get the original password "1234567".To increase the security and the difficulty, the salt is added in md5 encryption. As an example of md5(password+salt),m-d5("1234567pN9PEczWrczKk")= de1e6ed-28b5a706bb2637ba6907bcbb3, cannot be cracked by "de1e6ed28b5a706bb2637ba6907bcbb3". Is it safe to save encrypted password? The answer is no. It is found that high frequency used password has great coincidence, e.x. "123456789", listed top 10. This is statistically analyzed though csdn.net and weibo.com. The same way can be used to statically analyzed md5 password and get high-frequently used original password.

The requirement to store a secure password. NOT to store the original data; The same original data are saved in different ways even in the same database; Increase the difficulty of dictionary crack.

3.3 *The password transportation security problem*

As is shown in the following.

<form action="http://hostname/login" method="post"> <label>account</label> <input id=" username"type=" text"name="username"value=""/> <label>password</label> <input type ="password" name=" password" value="" /> <input value="login" type="submit" /> </form> This is a basic login form in internet. When the user typed username and password and clicked login button, account information and password will be sent to http://hostname/login with http protocol in plain text username=username&password=password. The account information and password will be easily got by attacker with some tools through transportation. In some website, the password interaction of the system and SSO is shown in Fig.3.

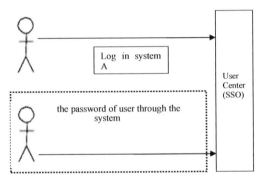

Figure 3. The interaction of password between system and SSO.

The SSO, Single Sign On, is provided by the user' center. SSO means the user just logs in once and can visit many trusted applications. It is one popular solution in internet with multi service systems. When a use longs in A system, the input account and password are directly transferred to SSO. SSO will send its own password with cookie to all trusted systems and authorize the user to access them and after it verifies the legal user. The same way is for B system. That is to say, when a use longs in B system, the input account and password are directly transferred to SSO. SSO will send its own password with cookie to all trusted systems and authorize the user to access them and after it verifies the legal user. In the transformation system, the account and password may be easily let out if the tcpdump is used. If B system can write a cookie to the user's browser, B system can also write the fake account cookie, then to log in A system(Wang, X.L. 2011).

We should follow the below to secure data transportation. The website should be encrypted when users log in, to make sure that the account and password in plain text cannot be found with some network tools; The account and password provided by user are verified by the original verification system or website, NOT through any transit server.

4 THE LEGAL PROTECTION OF PERSONAL DATA AND ITS EFFECTIVE TRANSMISSION

There are many other measures in the Electronic Commerce, which can be taken to protect the personal data as well to promote the circulation of data, such as strengthening self-discipline of the merchants from their own perspectives, strengthening the self-protection of consumers from their own perspectives and so on.

4.1 *Strengthen self-discipline of the merchants from their own perspectives*

Enhancing the self-discipline of the merchants is very important to solve the contradiction between the free movement and the legal protection of personal data, to promote the harmonious development of the Electronic Commerce. Enhancing the self-discipline of the merchants mainly refers to standardizing the storage and management of personal data of consumers. Standardizing the storage and management of personal data of the consumer means the business website takes appropriate measures to store the personal data collected by them. The main content is just as the following:

First of all, the period of the storage. The period of the storage of personal information should be consistent with the existence period of the collection of the personal data. Once the purpose of collecting loses its reason for existence, the commerce website will not correspondingly have the necessity of keeping the personal data of the consumer. So the website shall accordingly take measures to erase or destruct the personal data. Secondly, data replacement. Personal information which is dated will affect the purpose of the collection of business websites. So once consumers change their personal information of contact, business site shall update the personal data and make the new ones stored in the computer. Finally, data management. The aim of management is to ensure the safety of the personal information management, as well as to prevent such activities from occurring as alignment or combination, blocking, erasure or destruction of the information or data. For example, give different access privileges to different people having different levels of use, in order to prevent unauthorized persons involving accessing unauthorized data, as well as from entering the database of the consumer's personal information without authorization, unauthorizedly reading, copying and modifying.

In addition, in order to protect the personal data against infringement, the merchants shall make corresponding rules and systems, which require that the dealer shall implement appropriate technical and organizational measures to protect personal data against accidental or unlawful destruction or accidental loss, alteration, unauthorized disclosure or access, in particular where the processing involves the transmission of data over a network, and against all other unlawful forms of processing. In addition, the followings shall be contained in the rules and systems: It is prohibited to process personal data that reveals race, ethnic origin, religious belief, health or sex life, etc, but such sensitive personal data as the racial origin, health or sex life may be collected, processed and assigned only when, for reasons of general interest, this is so provided for by law or the consumer has given his explicit consent; Personal data shall be processed only for a specified purpose, in exercise of a right or in compliance with an obligation. In the course of the entire processing this purpose shall be complied; Data processing based on the compulsory supply of information shall be ordered in favor of public interest; No personal data shall be processed other than those indispensably required for satisfying the purpose of processing and only in a way compatible with that purpose.

4.2 Strengthen the self-protection of consumers from their own perspectives

From the customers' perspectives, internet has reduced space limitation they are facing, but their personal data are easy to be impaired seriously in the Electronic Commerce. The onus should mainly be on consumers to protect themselves, not only on merchants. The consumers shall protect them by themselves, so it's important to strengthen the self-protection of consumers from their own perspectives. The self protection mode of network consumers should be a comprehensive system , including Self - Control, Self Selection, etc.

Self Control means that the consumers in networks make use of the technical means to strengthen the control of the relevant privacy of personal information, including deleting or disabled the Cookies, using anonymous registration and browsing, and application of the technology of software and so on. Self selection means that the consumers should try their best to know the regulation of the Electronic Commerce enterprise about the protection personal data, such as the content and categories of the personal data collected by the Electronic Commerce enterprise, the manner and purpose of collecting personal information and so on.

5 SUMMARY

Along with the fast development of the information technology and international internet, the Electronic Commerce is getting faster and faster, but the protection of the persona data is becoming worse and worse. So the contradiction between the consumer's personal information communication and effective legal protection has become increasingly prominent. In order to solve the problems, China shall make laws about the personal data protection through learning from foreign countries. At present, the measures should be taken from such aspects as strengthening self-discipline of the merchants from their own perspectives, strengthening the self-protection of consumers from their own perspectives and so on. Only in this way, can the Electronic Commerce be guaranteed healthy and ordered development.

ACKNOWLEDGMENTS

The financial support of the flowing fund projects is gratefully acknowledged: the project about the theory and practice of the China Disabled Persons Federation 2014-2015 annual Chinese disabled person enterprise. (item number:2014&ZZ0004); 2014 annual bidding project of the People's Government of Henan Province (item number:2014181); 2014 annual project about humanities and social sciences of the Education Department of Henan Province(item number: 2014-qn-009). This paper is also one part of the projects supported by the Young Backbone Teachers of the Colleges and Universities in Henan Province .

REFERENCES

H.H. Zhou,(2006). Study on the frontier issues of personal information protection. *Law Research.* 16,43–46.
A.M. Qi(2012).The information from the perspective of private Law. *Law Science.*7,58–60.
J.J. Ma(2006). The definition of the object of legal relations about the personality right.*Hebei Law.*10,43–44.
T. Xu(2007). The principle of DNS attacking and prevention based on IPv4.*Chines Technology Information* .15, 125–126.
Y.X. Zhang, Y. Zhao(2008). Research on MD5 algorithm .*Computer Science.* 7, 295–296.
X.L. Wang(2011). A prevention method of XSS client based on the behavior.*Graduate University of Chinese Academy of Sciences.*5,668–669.

Electronic Engineering and Information Science – Wang (Ed.)
© 2015 Taylor & Francis Group, London, ISBN: 978-1-138-02772-5

Genetic algorithm nested with interval-indexed formulation heuristic procedure for a single machine scheduling problem

Y.M. Wang
Faculty of Management and Economics, Kunming University of Science & Technology, Kunming, China

H.L. Yin
School of Computer Science and Information Technology, Yunnan Normal University, Kunming, China

ABSTRACT: Many optimization problems in manufacturing systems are very complex and difficult to solve by common optimization methods. Because the complexity of the models or algorithms, formal techniques are limited to solving them. In this paper, interval-indexed formulation based heuristic is adopted to improve the population of a genetic algorithm. The interval-indexed, based heuristic is applied to each iteration of the genetic algorithm, and the population is updated based on the obtained solutions. We validated the presented algorithm with a single machine scheduling problem with total weighted tardiness. Experiment results show that the proposed approach has promising for application to some traditional hard problems, especially to single machine weighted tardiness problem. The algorithms discussed in this paper could be generalized to solve many other combinatorial scheduling problems, with small modifications in the algorithms.

KEYWORDS: Interval-indexed formulation; Big scheduling problem; Genetic algorithm; Heuristic algorithm.

1 INTRODUCTION

Scheduling is a kind of combinatorial problems; its optimization goal may be stated as, minimizing flow time, minimizing makespan, minimizing the number of tardy jobs, etc. (Topaloglu 2010 Wang 2013). Single machine scheduling also called sequencing determines the order of jobs. In the modern competitive world, a company has to meet the demand come from customers in time. A job is said to be tardy if it is not ready on its due date (Akrout 2012). Often there are contract penalties applied to the supplier for tardy jobs. The penalties are generally greater for strategic jobs, and they are usually linear.

Abdul-Razaq et al. gives a comprehensive review of several branches and bound, and dynamic programming algorithms for this problem (Abdul-Razaq 1990). Because of the computational burden, heuristics are more commonly used for 1||TWT. Commonly used scheduling heuristics are Weighted Shortest Processing Time (WSPT), Earliest Due Date (EDD), Minimum Slack (MS), and Weighted Processing Time and Due Date Index (WPD). Meta-heuristics are also widely used to solve 1||TWT. Local search is a meta-heuristic which starts from a candidate solution and moves to a neighbor solution.

In this paper, interval-indexed formulation is used to formulate a single machine total weighted tardiness problem. If a solution can be obtained through a well-defined modification of another solution, we call two solutions are neighbors. Crauwels et al. The method combines the merits of interval-indexed formulation. Compare several meta-heuristics, including Genetic Algorithms (GA), Simulated Annealing (SA), and Tabu Search (TS), in which TS dominates other methods (Crauwels 1998). Maheswaran et al. propose a GA and local search based meta-heuristic namely, memetic algorithm (Altunc 2008). Potts and Van Wassenhove (Potts 1985) have solved exactly problem instances with up to 40 jobs. However, for big problems, such as problem with 50 or more jobs, the needed computational time is unaffordable.

2 PROBLEM STATEMENT AND MODELING

The n-job single machine weighted tardiness problem is the problem with sequence jobs on one machine while minimizing the weighted sum of the tardiness. Given that the jobs are available on time t=0, and given for each job i (i=1, 2, n) processing times pi , due dates di and weights wi the problem is to find a sequence such that $\sum w_i T_i$ is minimal, where Ti is the tardiness of job i. It is a classical machine sequencing problem with NP-hard computational complexity.

To reduce the size of the time-indexed formulation, Hall et al. introduced an interval-indexed formulation. The core idea is to divide the time horizon into a polynomial number of intervals, and the decision variables indicate the intervals in which the jobs are completed. The interval set suggested in the paper is for some small positive constant.

Let τl be the end point of the nth interval, and the nth interval is defined by (τl-1,τl]. The interval-indexed formulation introduced by Hall et al. (1997) can be modified for the 1‖TWT problem as follows:

$$Min \sum_{j=1}^{n} \sum_{l=1}^{L} w_j c_{jl} x_{jl} \qquad (1)$$

$$S.t \sum_{l=1}^{L} x_{jl} = 1 \, , \; j = \{1, \cdots, n\} \, , \quad x_{jl} = 0 \, ,$$

$$if \; \tau_l < p_j \qquad (2)$$

$$\sum_{j=1}^{n} \sum_{l=1}^{L} p_j x_{jk} \le \tau_l \, , \qquad l = \{1, \cdots, L\} \qquad (3)$$

$$x_{jl} \in \{0,1\} \, , \quad j = \{1, \cdots, n\} \, , \; l = \{1, \cdots, L\} \qquad (4)$$

Where x_{jl} is a binary variable which is equal to 1 if job j is completed in interval l, i.e., $C_j \in (\tau_{l-1}, \tau_l]$, c_{jl} is the tardiness of job j, if it is finished in interval l, and L is the number of intervals. First constraints ensure that each job finishes exactly in one interval. Second constraints are to ensure that a job j cannot complete before its earliest completion time (earliest completion time of a job j is its processing time). Third constraints state that the sum of the processing times of the jobs that are completed by time τ must be less than or equal to τ.

3 INTERVAL-INDEXED FORMULATION BASED HEURISTIC PROCEDURE

Figure 1 gives a formal framework description of our heuristic approach. The following section will discuss the details of these steps.

Step 1: Determine the intervals.

Step 2: Solve the interval-indexed formulation and obtain solution.

Step 3: Apply post-processing to the schedule obtained in the corresponding Step 2.

Figure 1. Interval-indexed formulation based heuristic procedure.

3.1 Formulation of the intervals

Each of these solutions will give us n time points that the jobs in the given solution finish. Let S be the union of the time points given by these feasible solutions. Note that S will contain at most $m \times n$ time points. We define the end points of the intervals with the time points in S. We call the intervals found by this method as construction heuristics intervals.

3.2 Solving the interval-indexed for formulation

When we solve the LP-relaxation of the interval-indexed formulation, we are not guaranteed to obtain the solutions that are as good as the initial solutions at least and therefore we used both over and under estimation models. Following example illustrates the formation of the intervals and finding the α-point schedule.

3.3 Post-processing

In our solving method, there can be more than one job assigned to an interval in the solution given by the interval-indexed formulation and we use EDD rule to break the ties. In addition to this tie- breaking rule, we also apply different improvement heuristics to this solution to search the neighborhood of the obtained solution. We can apply this approach in two different ways: forward and backward.

4 GA NESTED WITH INTERVAL-INDEXED FORMULATION HEURISTIC PROCEDURE DESIGNING

The genetic algorithm is based on meta-heuristic and starts with a set of solutions, called chromosomes. Interval-indexed formulation can be used to improve a set of solutions obtained by other heuristics. We will be using a basic genetic algorithm to test this approach and implementing interval-indexed formulation based heuristic as a local search tool in every iteration of a basic GA and test our approach on the single machine total weighted tardiness problem.

Each chromosome of the proposed GA represents a feasible solution. Interval-indexed formulation is generated using the whole population, and then the population is updated based on the solution of the interval-indexed formulation or a -point solution obtained by the LP-relaxation.

Interval-indexed formulation heuristics based genetic algorithm combines interval-index based heuristics with a genetic algorithm. At each iteration, interval- indexed formulation is generated and solved using the whole population or a set of chromosomes, Chromosomes are represented by random key encoding. Each position gives a random number, and jobs are assigned to the positions in nondecreasing order of these random numbers. Crossover and mutation are performed using the random keys, and the jobs are reassigned to the positions using the new random keys. Figure 2 presented a good description of the interval-indexed formulation heuristics based genetic algorithm.

Step 1: Form an initial population using feasible
   solutions and apply post-processing.
Step 2: Apply crossover, mutation and post-
   processing.
Step 3: Form clusters, generate intervals for each
   cluster using the chromosomes in them and solve
   the LP-relaxation for each cluster.
Step 4: Update clusters by replacing the worst
   chromosome in each cluster with the
   schedule obtained by the solution of the
   LP-relaxation solved over the intervals
   obtained by that cluster.
Step 5: Generate the top cluster using the best
   chromosome of the population and the
   solutions of the LP-relaxation solved for
   each cluster.
Step 6: Solve the interval-indexed formulation
   over the top cluster.
Step 7: Update the population if the criteria for
   update is met.
Step 8: Stop if the stopping criteria is met, else go

Figure 2. Interval-indexed formulation heuristics based genetic algorithm.

5   EXPERIMENTS AND RESULTS

Since our algorithm is based on both linear programming(LP), and integer programming(IP), population size should be selected such that it allows IP to be solved in an affordable number of times. We performed preliminary experiments to determine the parameters of the GA.

We obtain an initial population of 20 chromosomes. Stepped optimization is applied as post-processing. To benefit most from the stepped optimization, we apply backward and forward stepped optimization. Random keys of the parents are swapped to obtain two child chromosomes.

We did extensive computational experiments to test the effect of interval-indexed formulation heuristics based genetic algorithm (IIFHGA). We compared our approach with two cases. We call the first case as a Pure GA and it is a GA without solving the LP-relaxation We call the second case as IIGA and it is a GA without solving the LP-relaxation but with an interval- indexed formulation solved. The numbers given in the cells are cumulative. The first column gives the percent deviation. Following three columns give the results of pure GA, IIGA and IIFHGA, respectively.

As seen from Table 1, we are able to find optimal solutions for 2 and 10 instances by pure GA and IIGA, respectively, and the maximum deviation is 14.966% for pure GA and 3.733% for IIGA. As expected, IIGA is better than the pure GA. Using IIFHGA improves the results significantly. We are able to solve 20 instances to optimality while 95 instances out of 96 instances have a percent deviation less than 0.7%. Maximum deviation is 2.265% for IIFHGA.

Table 1.   Results for pure GA, IIGA and IIFHGA for big single machine scheduling problems.

% Deviation	Pure GA	IIGA	IIFHGA
≤ 0	3	11	21
≤ 0.1	5	46	63
≤ 0.2	9	63	83
≤ 0.3	10	78	89
≤ 0.4	11	87	96
≤ 0.5	12	89	100
≤ 0.6	13	92	101
≤ 0.7	15	95	103
≤ 1.0	19	101	104
≤ 2.0	41	104	105
≤ 3.0	52	106	107
≤ 10	106	108	108
≤ 15	109	109	109
> 24	0	0	0

6 CONCLUSIONS

We investigated methods for improving a population of a genetic algorithm by using the interval-indexed formulation based heuristics as a local improvement tool which is applied at each iteration of a genetic algorithm. We introduced a basic interval-indexed formulation heuristics based genetic algorithm and tested our approach on single machine total weighted tardiness problem from OR-Library.

ACKNOWLEDGMENT

The work is supported by Natural Science Foundation of China (NO. 71262029, NO. 71362030, NO. 71362025, NO. 71362024), Natural Science Foundation of Yunnan (NO. 2013FB029, NO. 2013FZ048). Foundation for Food Safety of Yunnan Province (NO. 2013SY05).

I would express special thanks and gratitude to reviewer and editor for their efforts in reviewing and correcting this paper manuscript.

REFERENCES

Abdul-Razaq TS. Potts CN. & Van Wassenhove LN(1990). A Survey of Algorithms for the Single Machine Total Weighted Tardiness Scheduling Problem. *Discrete Applied Mathematics.26*, 235–253.

Akrout H. Jarboui B. & Siarry P(2012). A GRASP based on DE to solve single machine scheduling problem with SDST. *Computational Optimization and Applications. 51*, 411–435.

Altunc ABC(2008). Hybrid algorithms for combinatorial optimization problems. Doctoral Dissertation, *Arizona State University, USA*.

Crauwels HAJ., Potts CN. & Van Wassenhove LN(1998). Local Search Heuristics for the Single Machine Total Weighted Tardiness Scheduling Problem. *INFORMS Journal on Computing*. 10, 341–350.

Hall LA. Schulz,A S. Shmoys DB. & Wein J(1997). Scheduling to Minimize Average Completion Time: Off-line and On-line Approximation Algorithms. *Mathematics of Operations Research,, 22*, 513–544.

Topaloglu S. & Selim H(2010). Nurse scheduling using fuzzy modeling approach. *Fuzzy Sets and Systems, 161(11)*, 1543–1563.

Potts CN. & van Wassenhove LN(1985). A branch and bound algorithm for the total weighted tardiness problem. *Operations Research, 33(3)*, 63–77.

Wang YM. & Yin HL(2013). A Novel Genetic Algorithm for Flexible Job Shop Scheduling Problems with Machine Disruptions. *International Journal of Advanced Manufacturing Technology, Springer London, 68(5–8)*, 1317–1326.

Electronic Engineering and Information Science – Wang (Ed.)
© 2015 Taylor & Francis Group, London, ISBN: 978-1-138-02772-5

Design of low power CMOS instrumentation amplifier

M.Y. Ren, B. Yu & Z.P. Jiang
School of software, Harbin University of Science and Technology, Harbin, China

ABSTRACT: A low power CMOS instrumentation amplifier used for biomedical applications is presented in this paper. It consists of a low power operational amplifier with CMFC structure. By analysis and optimization of the parasitic effects and parameters, instrumentation amplifier has better performance in every aspect. This simulation result shows that, in the case of guaranteed bandwidth, the amplifier can suppress flicker noise interference effectively. The instrumentation amplifier designed in 0.5μm CMOS technology with 3.3V power supply shows CMMR of 92dB.

1 INTRODUCTION

Instrumentation amplifiers are very important circuits in many sensor readout systems where there is a need to amplify small differential signals in the presence of large common-mode interference. Application examples include automotive transducers (Bernard & Lawrence, 1981), industrial process control (V. Schaffer, 2009), linear position sensing (Rahal, 2009), and biopotential acquisition systems(Yazicioglu, 2007). We have a particular interest in the design of integrated instrumentation for piezo-resistive sensors using accelerometer measurements(Michel, 1987). With the continuing downsizing of submicron CMOS technology and reduction of power supply voltage, a single-chip sensor would require all circuitry to operate under low power supply voltage.

Therefore, there is a need for a low power interface circuit, which serves as a bridge between the on-chip sensor and the backend digital processor. The aim of this work is to realize a low power interface circuit for a CMOS piezo-resistive accelerometer sensor. The sensor resistance is changed when an acceleration variation is applied. The resistance change is traditionally measured by using the Wheatstone bridge circuit, the sensitivity of which depends on the excitation voltage or current. For high bridge sensitivity, high excitation voltage or current is needed, which prevents low-voltage and low power operation.

In this paper, a new low power Instrumentation amplifier is proposed. The paper is organized as follows. In section II, principle of instrumentation amplifier is described. Sections III describe the architecture and the simulation of the proposed Instrumentation amplifier. Conclusions are given in sections IV.

2 PRINCIPLE OF INSTRUMENTATION AMPLIFIER

An instrumentation amplifier is one that is used in electronic measuring instruments or in electronic systems in which the highest precision is required. A popular instrumentation amplifier is shown in Fig. 1. It can be built from three op amp integrated circuits but more often the three amplifiers are manufactured already connected on the same chip. As a 'black box' the instrumentation amplifier, or 'in amp', has the same input and output terminals as a single op amp, but its performance is superior.

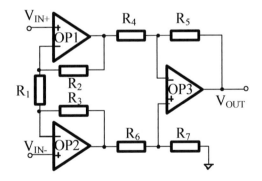

Figure 1. A standard type of instrumentation amplifier comprises three op amps.

When $R_4 = R_5 = R_6 = R_7$, the signal transfer function can be simplified to

$$A_D = \frac{V_{OUT}}{V_{IN+} - V_{IN-}} = (1 + \frac{R_2 + R_3}{R_1}) \qquad (1)$$

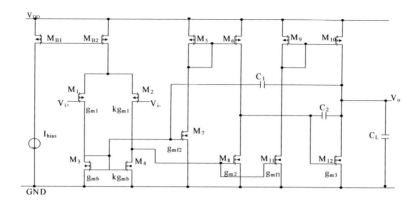

Figure 2.   Three-stage operational amplifier with capacitor-multiplier frequency compensation.

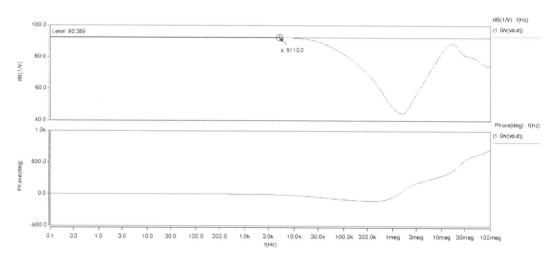

Figure 3.   Simulation of CMMR.

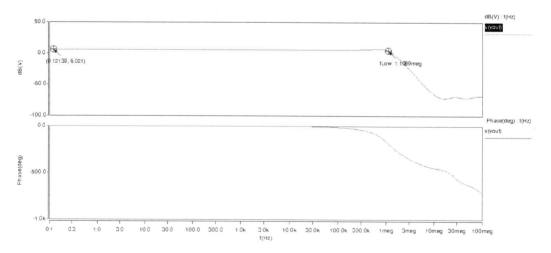

Figure 4.   Simulation of AC gain.

where R_2 and R_3 is generally set to the same value. Ideally, by simply adjusting the value of R1, the entire gain of the circuit can be adjusted to any value.

To calculate the common mode gain, we assume the input common-mode voltage is V_{CM}.

$$V_{IN+} = V_{IN-} = V_{CM} \qquad (2)$$

Therefore, no voltage drop across R_1, OP1 and OP2 output voltage is equal to V_{CM}. If we assume OP1, OP2 and OP3 is the ideal op amp, the amp's common-mode rejection ratio is

$$CMRR(dB) = 20 \log \left(\frac{A_D}{Ac} \right)$$

$$= 20 \log \left(\frac{(3)Gain}{\dfrac{R_5 + R_4}{R_4} \dfrac{R_7}{R_6 + R_7} - \dfrac{R_5}{R_4}} \right) \qquad (3)$$

where amp's common-mode rejection ratio mainly is decided by matching of several resistance. The resistance of the relative accuracy can very high through the layout of symmetry techniques in CMOS process, so the kinds of structures is suitable for CMOS technology.

3 ARCHITECTURE AND SIMULATION

The operational amplifier plays an important role in performance of system. Because the amplification magnitude of the sensor with the electric bridge structure is quite big, good noise restriction and low offset voltage are desirable for an amplifier. In order to guarantee the linearity of system, the operational amplifier must have a gain high enough. Besides, the total power consumption of the integrated system is also needed to be considered. Three-stage operational amplifier with capacitor-multiplier frequency compensation meets all these requirements. It could effectively avoid the low DC gain and high power consumption, which are draw-backs of typical two-stage operational amplifier. The diagram of the three-stage operational amplifier is shown in figure 3. The performance of the actual three stage op-amp is shown in table 1.

The simulation of CMRR with the instrumentation amplifier is 92dB from 0 to 1kHz, it is shown in Figure 3. The simulation of AC gain with the instrumentation amplifier is 6 dB and 3dB frequency is 1.1 MHz, it is shown in Figure 4.

Table 1. The performance of three-stage operational amplifier.

Unity-gain-bandwidth	1.699 MHz
Open-loop gain	120 dB
Phase margin	64°
Slew rate	1.6 V/µs
CMRR	93 dB
Power	0.33 mW

4 CONCLUSION

A low power CMOS instrumentation amplifier used for measuring applications is presented in this paper. It consists of a low power operational amplifier with CMFC structure. By analysis and optimization of the parasitic effects and parameters, instrumentation amplifier has better performance in every aspect. The design of instrumentation amplifier is realized with low power and low noise, so it meets the expectant design requirements.

ACKNOWLEDGMENT

This work is supported by Science and Technology Research Funds of Education Department in Heilongjiang Province under Grant Nos. 12541174.

REFERENCES

Bernard, DM and Lawrence, RS (1981) Instrumentationamplifier IC designed for oxygen sensor interface requirements, *IEEE Journal of Solid-State Circuits*, 16(6), 677–681.
V.Schaffer, M. F. Snoeij, M. V. Ivanov and D. T. Trifonov (2009) *IEEE J Solid-State Circuits*, 44, 2036.
M. Rahal and A. Demosthenous(2009), *IEEE Transactions on Instrumentation and Measurement*, 58, 3693.
R. F. Yazicioglu, P. Merken, R. Puers and C. V. Hoof(2007), *IEEE J Solid-State Circuits*, 42, 1100.
Michel S J. Steyaert, Willy M C, CHANG Z Y(1987). A Micropower Low-Noise Monolithic Instrumentation Amplifier For Medical Purposes. *IEEE Journal of Solid-State Circuits*, 22(6):1163–1168.

Electronic Engineering and Information Science – Wang (Ed.)
© 2015 Taylor & Francis Group, London, ISBN: 978-1-138-02772-5

The design of CMOS rail-to-rail operational amplifier

M.Y. Ren, B. Yu & Z.P. Jiang
School of Software, Harbin University of Science and Technology, Harbin, China

ABSTRACT: The rail-to-rail operational amplifier was designed. The circuit design is realized in 0.5μm CMOS technology and HSPICE simulation results. Simulation results show that the design of operational amplifier working in 5v power supply can get 20.502MHz band-width, phase margin is 62.473°, low-frequency open loop voltage gain of 113.16 dB. Operational amplifier input common-mode voltage range and output voltage swing basically reached the rail-to-rail.

1 INTRODUCTION

Operational amplifier is the most important unit circuit in the analog circuit. It is widely used in all kinds of analog and mixed-mode circuit. Recently, with the popularity of the portable electronic products and the decrease of the threshold voltage by the progress in the CMOS fabrication technology analog circuits become more and more popular. Therefore analog circuits design techniques are also a hot spot in the research of analog circuit. With the ongoing reduction of the supply voltage and the dynamic range, the rail-to-rail input/output is necessary in the operational amplifier(Botma, 1993, Carrillo, 2003, Satoshi, 1996, Chih, 2004, Stockstad, 2002, Giuseppe, 1997).

2 STRUCTURE OF RAIL-TO-RAIL AMPLIFIER

The input stage is consist of complementary differential input pair, constant g_m circuit, and a bias circuit. M_1, M_2 and M_3, M_4 respectively composed of NMOS, PMOS input pair, and $M_1 \sim M_4$ form complementary input differential pair. M_5, M_6 are current switch, M_9, M_{10} and M_{11}, M_{12} are respectively composed of two 1:3 Current mirror. When the current mirror M_9, M_{10} action, $I_{10} = 3I_9$. Similarly, when the current mirror M_{11}, M_{12} action, $I_{11} = 3I_{12}$. The width to length ratio of NMOS and PMOS differential input transistor

satisfied $\dfrac{(W/L)_P}{(W/L)_N} = \dfrac{\mu_N}{\mu_P}$ and $I_7 = I_8 = I_{ref}$.

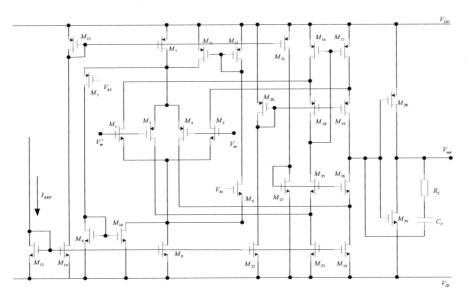

Figure 1. The whole circuit of rail-to-rail amplifier.

Current mirror transistors M_{15}, M_{16}, M_{21}, and M_{22} as a PMOS differential pair and NMOS differential pair of active load, such a structure not only to achieve the function of the circuit, but also saves the layout area of the circuit and reduce the power consumption. From the input stage differential current output of the post into the intermediate gain stage, two current mirrors by the M_{15}, M_{16}, M_{21}, and M_{22} to achieve a double-end consisting of single-ended conversion, outputs to the next-stage circuit.

Common source amplifier is a complementary push-pull common source output structure, the output voltage range can be achieved substantially full swing.

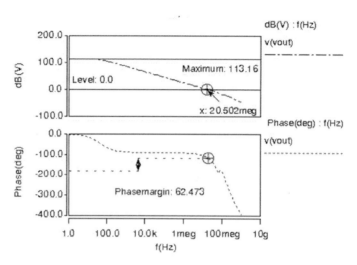

Figure 2. The curve of open-loop gain and phase margin.

3 SIMULATION

The simulation results is shown in Figure 2, it can get a unit gain bandwidth of 20.502MHz, phase margin of 62.473 degrees, low frequency voltage gain of 113.16dB.

It can be seen by figure 3, the output range of the rail to rail amplifier is from 0.012256V to 4.9728V.

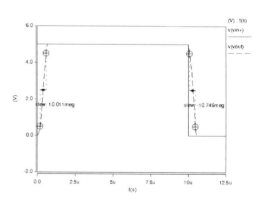

Figure 4. The slew rate of rail to rail amplifier.

From the simulation results by figure 4 can be seen, the positive slew rate of rail to rail amplifier is 10.011V/ s, the negative slew rate of it is 10.746V/ s.

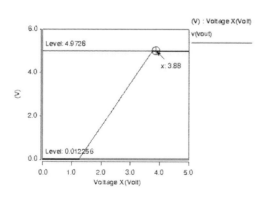

Figure 3. The curve of range of input and output.

4 CONCLUSION

A low power CMOS instrumentation amplifier used for measuring applications is presented in this paper. It consists of a low power operational amplifier with CMFC structure. By analysis and optimization of the parasitic effects and parameters, instrumentation amplifier has better performance in every aspect. The design of instrumentation amplifier is realized with low power and low noise, so it meets the expectant design requirements.

ACKNOWLEDGMENT

This work is supported by Science and Technology Research Funds of Education Department in Heilongjiang Province under Grant Nos. 12541174.

REFERENCES

Botma J.H. Wassenaar R.F, Wiegerink R. J(1993), A Low-Voltage CMOS Operational Amplifier with a Rail-to-Rail Constant-gm Input Stage and a Class AB Rail-to-Rail Output Stage. *IEEE Transactions on Systems*, 22(3): 1314–1317.

Carrillo, J.M.; Duque-Carrillo, J.F.; Torelli, G.; Ausin, J.L. (2003); Constant-gm constant-slew-rate high-bandwidth low-voltage rail-to-rail CMOS input stage for VLSI cell libraries; *IEEE Journal of Solid-State circuits*, 38(8), 1364–1372.

Satoshi Sakurai, Mohammed Ismail(1996); Robust Design of Rail-to-Rail CMOS Operational Amplifier for a Low Power Supply Voltage; *IEEE Journal of Solid-State circuits*, 31(2),146–156.

Chih, Wen Lu(2004), High-Speed Driving Scheme and Compact High-Speed Low-Power Rail-to-Rail Class-B Buffer Amplifier for LCD Applications, 39(11),1938–1947.

Stockstad, Troy; Yoshizawa, H(2002), A 0.9-V 0.5-uA rail-to-rail CMOS operational amplifier, *IEEE Journal of Solid-State circuits*,37(3), 467–470.

Giuseppe Ferri; Willy Sansen(1997), A Rail-to-Rail Constant-gm Low-Voltage CMOS Operational Transconductance Amplifier. *IEEE Journal of Solid-State circuits*,32(10), 1563–1567.

Electronic Engineering and Information Science – Wang (Ed.)
© 2015 Taylor & Francis Group, London, ISBN: 978-1-138-02772-5

The design and implementation of a low temperature coefficient bandgap reference voltage source

M.Y. Ren, B. Yu & Z.P. Jiang
School of software, Harbin University of Science and Technology, Harbin, China

ABSTRACT: In this paper designed a kind of low temperature coefficient of bandgap reference source, first on the background of the bandgap reference voltage source and introduces the current situation of the development. Then this paper introduces the basic structure of bandgap reference source and the analysis the basic principle. To improve the power supply rejection ratio of bandgap reference as well as the temperature compensation principle of bandgap reference voltage source is studied. The bandgap consist of the bias current source circuit, operational amplifier circuit and the core bandgap reference circuit, design index as the temperature changes -40°C~80°C, within the scope of the temperature coefficient of less than 10ppm/°C. The PSRR is higher than 80dB, then using HSPICE simulation tool for the design of bandgap voltage reference source circuit simulation and analysis results are compared with those of the indicators.

1 INTRODUCTION

With the development of CMOS process, CMOS circuit tends to be low cost, low power consumption and high speed. CMOS circuit design progress as a bandgap reference source circuit, has always been the hot spot of the research. The Bandgap reference source is affected by power, process, the temperature is small, its impact on the accuracy and stability of system, it is widely used in the A/D, D/A converter, filters and phase lock loop circuits (Widlar, 1971, Robert, 1990, Leung, 2003, Najafizadef, 2004).

2 STRUCTURE

The architecture of the bandgap reference circuit is shown in Figure 1, which is used by the compensation method with temperature coefficient of resistance. In order to ensure the bias current of the two transistors having same temperature characteristic, we use the PMOS current mirror instead of resistance to transistor bias. The voltage of between base area and emitter area of bipolar has a negative temperature coefficient.

$$\Delta V_{BE} = V_{BE1} - V_{BE2}$$

$$= V_T \ln \frac{nI_0}{I_{S1}} - V_T \ln \frac{I_0}{I_{S2}} = \frac{K \ln n}{q} T \qquad (1)$$

$$V_{BE}(T) = V_{BG} - (V_{BG} - V_{BE0})\frac{T}{T_0} - (\eta - \alpha)V_T \frac{T}{T_0} \qquad (2)$$

η is related to the transistor structure, whose value is about 4. α is related to a current flowing through the transistor, whose value is 1 when the PTAT current flows through the transistor, whose value is 0 when the current without associated temperature flow through transistor; T_0 is a reference temperature; V_{BG} is a bandgap voltage of silicon. It can be seen by formula (2) that V_{BG} is having a negative temperature coefficient.

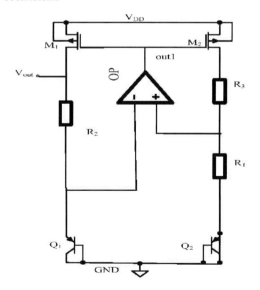

Figure 1. The core circuit of bandgap reference voltage source.

Proportional to the absolute temperature difference between the emitter voltage of - transistors operate at different current densities, their base. If two similar transistors ($I_{S1} = I_{S2}$) bias collector current respectively nI_0 and I_0, and their base currents are ignored, then equation (1): ΔV_{BE} exhibits a positive temperature coefficient, and the temperature coefficient is temperature-independent constant. The weighted positive voltage, the negative temperature coefficient of the sum, can get an approximate temperature-independent reference voltage. Here direct output reference voltage. By bipolar transistors Q_1, Q_2 and R_1 form a current proportional to absolute temperature:

$$I = \frac{V_T \ln(N)}{R_1} \tag{3}$$

Among them, $V_T = KT/q$, the current flowing through the resistor R2 form the output voltage of a bandgap reference voltage source, that is:

$$V_{REF} = V_{RE1} + \left(\frac{R_2}{R_1}\right) \cdot \ln(N) \cdot V_T \tag{4}$$

It is obtained by analyzed the ratio of resistor R_2 and R_1.

$$V_{REF} = V_{RE1} + b_1 T + b_2 T^2 + b_3 T^3 \tag{5}$$

According to $(1 + R_3 / R_1) \ln n \approx 17.2$, if n = 8, $R_1 = 1.5k, R_3 = R_2 = 13k$.

The archeture of is shown in Figure 2. The open-loop gain of this operational amplifier is 104dB, phase margin of it is 63 degrees, the power supply rejection ratio is 130 dB. The simulation results of amplifier meet the actual needs.

3 SIMULATION

Figure 3 shows the temperature coefficient curve in VDD=3.3V. It can be seen from the graph, the changes of the output voltage only are 0.001V with temperature changes from -40°C to 80°C. The calculation of temperature coefficient is

$$T_c = \left[\frac{V_{max} - V_{min}}{V_{ref} \times (T_{max} - T_{min})}\right] \times 10^6$$

$$= \left[\frac{0.001}{1.198 \times 120}\right] \times 10^6 = 6.9 \text{ppm/°C} \tag{6}$$

Figure 4 shows the power source rejection ratio (PSRR) of the bandgap, which reflects the ability of the circuit to suppress supply noise at a certain frequency. It is can be seen from the Figure 4, PSRR is 89dB and it meets the design specifications.

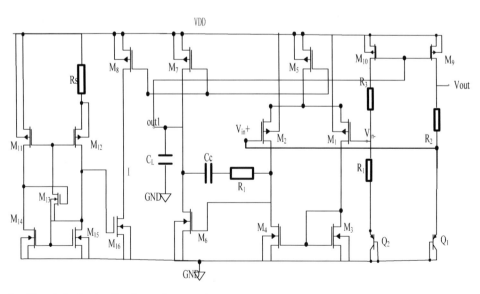

Figure 2. The whole circuit of bandgap reference voltage source.

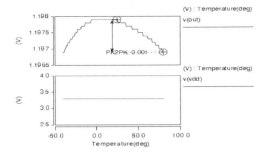

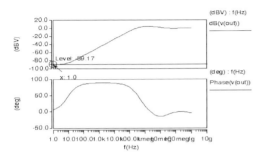

Figure 3. The curve of temperature at VDD=3.3V.

Figure 4. The PSRR of bandgap.

The simulation results of bandgap reference circuit meet the design specifications. The temperature range of bandgap reference is from -40°C to 80°C, the changes of output voltage changes are only 0.001V, the temperature coefficient is 6.9ppm / °C, PSRR is 89dB, the bandgap reference voltage is 1.2V.

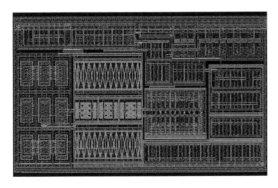

Figure 5. Layout of bandgap reference voltage source.

4 CONCLUSION

A kind of low temperature coefficient of bandgap reference source is presented in this paper. By analysis and optimization of the parasitic effects and parameters, a bandgap reference source has better performance. The design of bandgap reference source is realized with low temperature coefficient and high PSRR, so it meets the expectant design requirements.

ACKNOWLEDGMENT

This work is supported by Science and Technology Research Funds of Education Department in Heilongjiang Province under Grant Nos. 12541174.

REFERENCES

Widlar R J.(1971) New developments in IC voltage regulators. *IEEE Journal of Solid-State Circuits*, 6(1): 2–7.
Robert A. Pease(1990). The Design of Band-Gap Reference Circuits: Trials and Tribulations,*IEEE. Bipolar Circuits and Technology Meeting*, 214–218.
Ka Nang Leung, Philip K T, Chi Yat Leung(2003). A 2-V 23-uA 5. 3-ppm/°C Bandgap Voltage Reference. *IEEE Journal of Circuits*, 38(3): 561–564.
Najafizadef L, Filanovsky L(2004). Towards a sub-1V CMOS voltage reference. *In:Proceedings of the 2004 Internatinal Symposium on Circuits and Symposium on Circuits and Systems. Montreal: IEEE*, 2004. 53–56.

Electronic Engineering and Information Science – Wang (Ed.)
© 2015 Taylor & Francis Group, London, ISBN: 978-1-138-02772-5

A complete electrical equivalent model for micromechanical gyroscope

C.C. Dong
School of Software, Harbin University of Science and Technology, Harbin, China

L. Yin & Q. Fu
MEMS Center, Harbin Institute Technology, Harbin, China

ABSTRACT: Analysis of the four non-ideal factors micromachined vibratory gyroscope performance improvement and its related theory analysis, including four non-ideal factors sensor structure modeling, build a complete electrical equivalent model of the micromechanical gyroscope, and based on its Hspice simulation software, so that not only simulate the performance and operating characteristics of structural analysis, but also with specific transistor-level circuit co-simulation, verification nonidealities ASIC interface circuit to eliminate the correctness of the circuit module.

KEYWORDS: Electrical equivalent model; Hspice simulation; Gyroscope.

1 MICROMECHANICAL GYROSCOPE STRUCTURAL IMBALANCE ERROR ANALYSISN

When considering the detection output offset equivalent capacitance deviation caused by the detection of the equivalent capacitance of the capacitor values are not equal, the effect of driving the equivalent capacitance of the capacitor value is not the same bias (Sangkyung 2009). Figure 1 for the connection between the structure and micromechanical gyroscope sensor interface circuit, this time to detect the equivalent capacitance value of C1, C2 are not equal (Sadik 2007).

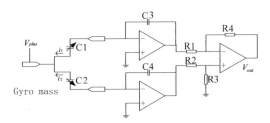

Figure 1. Sketch of connection between micromachined gyroscope sensor and ASIC.

The capacitor C1 in Figure 1 currents i1 and the current i2 in the capacitor C2 can be expressed as

$$i_1 = V_{plus} \cdot j\omega C_1 \qquad (1)$$

$$i_2 = -V_{plus} \cdot j\omega C_2 \qquad (2)$$

Where V_{plus} is high frequency square wave modulation signal, and the currents for the

$$i = i_1 + i_2 = V_{plus} \cdot j\omega(C_1 - C_2) = V_{plus} \cdot j\omega\Delta C_s \qquad (3)$$

Current i through the integrator voltage output of the integrator is generated

$$V_{err-s} = V_{plus} \cdot j\omega\Delta C_S \cdot \frac{1}{j\omega C_S} = V_{plus} \cdot \frac{\Delta C_S}{C_S} \qquad (4)$$

Formula, the value of capacitor C3 is equal to the capacitance C4, by the formula (4) can be seen, when the detection of the equivalent capacitance of C1, C2 is the capacitance value of the deviation, the output of the integrator to produce a size of the coupled signal, which is the integral capacitor(Lota 2007). Coupling high frequency square wave signal with a frequency modulated signal is the same, this high-pass filter coupled signal powerless. Formula (4) coupled signal through the first stage of phase-sensitive demodulation circuit switch, the reference signal K is a square wave signal, to expand the frequency of the sine function and odd harmonics, multiplied with the signal, results is

$$V_{err-s} \cdot K = -\frac{16}{\pi^2} \frac{\Delta C_s}{C_3} V_{amp} (\sin \omega_1 t + \frac{1}{3}\sin 3\omega_1 t + \frac{1}{5}\sin 5\omega_1 t + ...)^2 \qquad (5)$$

Formula (5) contains many frequency components, it may be a high-frequency component through a low pass filter to filter out high frequency components, leaving only a DC signal(Aaltonen 2010). Thus, when the equivalent capacitance of the capacitance detecting deviation too, in the detection circuit of the first stage low-pass filter offset errors generated at the output signal of a size, which is the integrating capacitor, the high-frequency amplitude modulation of the square wave(Tsai 2010).

Offset error and then after the second stage switch phase sensitive demodulation circuit, the second stage phase-sensitive demodulation circuit switch reference signal is usually formed by a square wave from the comparator, which multiplied the result can be expressed as

$$V_{err-dc} \cdot K' = \frac{4}{\pi} \frac{\Delta C_s}{C_3} V_{amp} (\cos \omega t + \frac{1}{3} \cos 3\omega t + \frac{1}{5} \cos 5\omega t + ...) \quad (6)$$

2 MICROMECHANICAL GYROSCOPE NONLINEAR STRUCTURAL ANALYSIS

In the design of the micro gyroscope structure, in order to improve detection sensitivity, usually a method to improve the driving displacement, in addition, to reduce the driving of the detected quadrature coupled, usually using thin elastic beam, this approach is equivalent to the structure introduction of nonlinear factors, because the long thin elastic beam response is not linear change, resulting in the displacement of the driving force and driving is not a first-order linear relationship (Raman 2009).

Driving nonlinear modal analysis is usually performed in the field of nonlinear vibration classical Duffing equation

$$F = k_1 x + k_3 x^3 \quad (7)$$

Wherein the elastic force F is freely movable end support beam suffered, x its displacement, k1 and k3 are linear and cubic elasticity(An 1999).

The formula (3-1) into the driving direction gyro kinetic equation

$$M_d \frac{d^2 x}{dt^2} + \lambda_d \frac{dx}{dt} + k_d x + k_d' x^3 = F_o \sin \omega_d t \quad (8)$$

Theoretical analysis shows that the offset error, the offset error in the original is equal to the vibration displacement, plus a small amount of change in vibration displacement, and therefore, Rs and Cs and the change amount of the inductance Ld another constitutes a simulation using the model in the electrical changes in the amount of electrical vibration displacement model, the voltage across

the inductor changes its form of voltage-controlled voltage source is applied to the detection of the original electrical model, equivalent to the amount of vibration displacement caused by the offset error (Chong 1994).

Theoretical analysis of the specific structure of the nonlinear analysis in 2.4.4, mainly on account of structural deformation nonlinear elastic beam, this time driven modal kinetic equation becomes

$$C_d \frac{d^2 V_1}{dt^2} + \frac{1}{R_d} \frac{dV_1}{dt} + \frac{1}{L_d} V_1 + \frac{1}{L_d} (V_1)^3 = \frac{dI_1}{dt} \quad (9)$$

3 ELECTRICAL EQUIVALENT MODEL FOR MICROMECHANICAL

So consider the structure of the nonlinear deformation of the electrical model shown in Figure 2. In drive mode equivalent electrical model, using three voltage-controlled current source is introduced term displacement(Guo 2010).

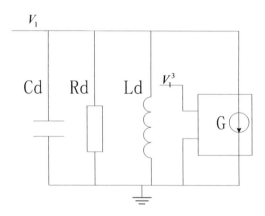

Figure 2. Total equivalent circuit model including nonlinearity.

4 SIMULATION RESULTS

Spectrum for electrical simulation model-driven direction of the curve, to verify its narrowband spectral characteristics of the filter, the same rate of increase in the electrical model sweep signal with its corresponding software research at different frequencies speed waveform amplitude, frequency sweep range set at 1kHz to 12kHz, the results shown in Figure 3.

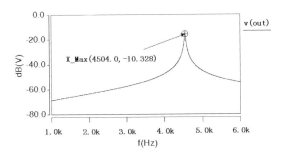

Figure 3. Simulated result of the equivalent circuit model of micromachined gyroscope driving.

The figure can also get the drive mode resonant frequency of the electrical model of 4504Hz, the resonance frequency corresponding to the driving structure of the MEMS gyroscope. Figure amplitude-frequency curve is a narrow bandpass filter response curve, the Q value is large, the biggest gain narrow band center frequency, the corresponding maximum passband signal strength, pass-band signal is rapidly attenuated. Therefore, the drive signal frequency micromechanical gyroscope should be within the narrow band for maximum drive capability.

When considering the non-ideal factor of four, the simulation results shown in Figure 4, it can be seen from Figure 4 quadrature error as long as the signal is present, cannot be seen from the simulation results in the envelope signal, and that the fact is consistent.

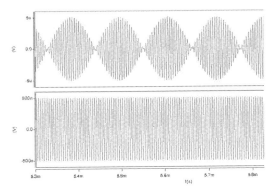

Figure 4. Comparison of the simulation results of Vs signal when frequency matching.

5 MEASURED RESULTS

Applications built gyroscope complete electrical equivalent model, using this electrical model Hspice simulation software, through communication and electrical transient simulation model to verify the correctness of one hand, on the other hand to analyze the performance and operating characteristics of micromechanical gyroscope.

ACKNOWLEDGMENT

This work is supported by Science and Technology Research Funds of Education Department in Heilongjiang Province under Grant Nos. 12541115.

REFERENCES

Sangkyung S, Woon T S, Changjoo K, et al(2009). On the Mode-Matched Control of MEMS Vibratory Gyroscope via Phase-Domain Analysis and Design. *IEEEASME Transactions on Mechatronics*, 14(4): 446–455.

Sadik A Z, Hussain Z M, Yu X(2007). An Approach for Stability Analysis of a Single-bit High-Order Digital Sigma-Delta Modulator. *Digital Signal Processing*, 17(6): 1040–1054.

Lota J, Janabi M A, Kale I(2007). Stability Analysis of Higher-Order Delta-Sigma Modulators for Sinusoidal Inputs. *Instrumentation and Measurement Technology Conference. Warsaw, Poland*, 1–5.

Aaltonen L, Halonen K(2008). Integrated High-Voltage PID Controller. *Baltic Electronics Conference. Tallinn, Estonia*, 125–126.

Tsai N C, Sue C Y(2010). Experimental analysis and characterization of electrostatic-drive tri-axis micro-gyroscope. *Sensors and Actuators A*, 10(11): 231–239.

Raman J, Cretu E, Rombouts P. A(2009) Closed-Loop Digitally Controlled MEMS Gyroscope With Unconstrained Sigma-Delta Force-Feedback. *IEEE Sensors Journal*, 9(3): 297–305.

An S, Oh Y S, Lee B L(1999). Dual Axis Microgyroscope with Close Loop Detection. *Sensors and Actuators A*, 73(1-2): 636–645.

Chong G Y, Randall L G(1994). An Automatic Offset Compensation Scheme with Ping-Pong Control for CMOS Operational Amplifiers. *IEEE Journal of Solid-State circuits*, 29(5): 601–610.

Guo Z Y, Yang Z C, Lin L T, et al(2010). Decoupled Comb Capacitors for Microelectromechanical Tuning-Fork Gyroscopes. *IEEE Electron Device Letters*, 3(13): 26–28.

A sensitivity study of fluxgate excitation fed by a square wave voltage

C.C. Dong
School of Software, Harbin University of Science and Technology, Harbin, China

L. Yin & Q. Fu
MEMS Center, Harbin Institute Technology, Harbin, China

ABSTRACT: This paper studies the impact of incentives square wave fluxgate sensitivity. A dynamic transfer function of its analysis, the relationship between the amplitude of the excitation voltage and sensitivity model, combined with experimental results verify the theoretical analysis, while the other test results corresponding excitation waveform analysis.

KEYWORDS: Sensitivity; Square wave excitation; Fluxgate.

1 INTRODUCTION

Studies have shown that the sensitivity of the probe with the specific fluxgate excitation waveform and a current source or voltage source drive and change, and therefore, as well as incentives for different excitation parameters cause sensitivity study has been one of the major changes in research direction fluxgate, among them, Spain's Lucas P, et al. (Perez 2006), the domestic Zhang Fu-based sinusoidal current incentives calculate the sensitivity of the probe (Zhang 1995), the United States Gordan DI, et al for the sensitivity of the triangular wave current and voltage sources incentives were calculated (Gordon 1965), sensitivity American Burge JR Research Triangle wave excitation ring probe (Burger 1972), generally speaking, theoretical research output fluxgate before focused on incentives sine and triangular wave, square wave excitation basically no way involved. The square wave excitation mode due to its easy to achieve, especially existing fluxgate sensor integrated fluxgate sensor-driven approach is widely used, and thus to study its impact excitation parameters of sensitivity has certain practical significance(Liu 2010).

Based on the above considerations, this paper square-wave voltage excitation mode, the theoretical analysis of the square-wave voltage source excitation parameters for fluxgate probe the sensitive influence on this basis, the use of cobalt-based amorphous melt pump drawing produced a fluxgate sensor, then by experimental tests have proved the correctness of the theory.

2 EXCITATION PARAMETERS AFFECT THE PROBE SENSITIVITY ANALYSIS

Square-wave voltage source fluxgate driver is currently the most widely used and popular drivers, this approach has several advantages: First, centralized incentives, the square wave generating circuit is the easiest to design, especially now in the research hot integrated fluxgate, the square wave oscillator circuit is easy to use CMOS technology, and does not require external components; Second, if using sine wave excitation mode, the second harmonic circuit selection method requires a sinusoidal signal multiplier, multiplier in general, multipliers, because the accuracy of the multiplier will be biased, thus affecting the accuracy of phase-sensitive demodulation, and square wave excitation mode can be used very sophisticated and accurate CMOS technology divide, much higher than the phase sinusoidal excitation mode Min demodulation accuracy, but generally not high bandwidth multiplier is not suitable for high frequency applications, this restriction does not exist a square wave. Therefore, the square-wave voltage source-driven approach fluxgate sensor now become the main incentives. Based on this, the square-wave voltage source excitation fluxgate probe sensitivity study has some practical significance.

Because the input dynamic transfer function of the probe in the magnetic field strength H, and therefore needs to derive waveform fluxgate magnetic field strength H probe, according to the excitation voltage waveform, and then analyze the sensitivity of the dynamic transfer function. As shown in Figure 1, because the excitation voltage is a square wave, its core internal excitation waveform can be obtained by integrating the curve through the thrust reverser, get inside the core magnetic field intensity versus time curve, the specific process As shown in Figure 1.

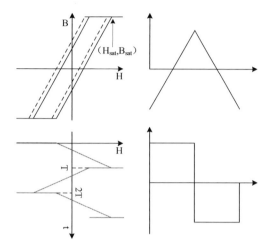

Figure 1.   Variation of H with time induced by drive electromotive force.

According to the magnetic field intensity curve shown in Figure 1, the use of the dynamic transfer characteristic curve, the final output of the probe can be positive and negative voltage pulse, shown in Figure 2.

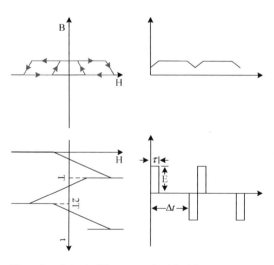

Figure 2.   Output of fluxgate probe drived by square wave.

When the current through the cyclical dynamic transfer function is converted to B-T function, its derivative as in Figure 2. Positive and negative pulses,

make time for each other intervals, by calculating the positive pulse Fourier fundamental component with the Fourier fundamental wave component of the negative pulse manner in accordance with the vector sum, to obtain the second harmonic component fluxgate sensor output. By Fourier transform, the fundamental component positive and negative pulses is:

$$V^{\pm} = \frac{2E}{\pi} \sin \frac{\pi \tau}{T} \tag{1}$$

where E is the amplitude of the positive pulse or negative pulse, τ is positive pulse or negative pulse width, T is cycle square wave excitation signal.

Figure 2 shows that the positive and negative pulse time difference Δt between pulses is:

$$\Delta t = T - \tau \tag{2}$$

The positive difference between the pulse and the negative pulse:

$$\varphi = \frac{\Delta t}{T} 2\pi \tag{3}$$

Therefore, the final output of the probe for the second harmonic amplitude V_{2f} is:

$$V_{2f} = 2V^{+} \sin\left[(T - \tau)\frac{2\pi}{T}\right] = 2V^{+} \sin\frac{\pi\tau}{T} \tag{4}$$

Equation (1) into the equation (3), may have:

$$V_{2f} = \frac{4E}{\pi}\left(\sin\frac{\pi\tau}{T}\right)^{2} \tag{5}$$

By the formula (5) shows that want to calculate the square wave excitation of the second harmonic component of the probe output, you need to first calculate the amplitude and pulse width to calculate the magnitude of the magnetic field inside the core probe needs to derive strength with time curve changes.

Figure 2 shows that either a positive pulse or negative pulse, the area are equal and constant, can be expressed as:

$$E\tau = N \times A \times B_{\Delta} \tag{6}$$

where N is probe excitation coil turns, A is probe core cross-sectional area, B_{Δ} is magnetization caused by the applied magnetic field strength variation.

$$\mu_{\Delta} = \frac{B_{sat}}{H_{sat} - H_c} \tag{7}$$

then,

$$B_\Delta = 2\mu_\Delta H_0 \tag{8}$$

The relation (8) into the equation (6), may have a pulse area and the number of turns, core cross-sectional area and the equivalent permeability of:

$$E\tau = N \times A \times 2\mu_\Delta H_0 \tag{9}$$

To calculate the pulse width, the driving voltage waveform function within the first cycle can be expressed as:

$$e = E_p \quad 0 \le t \le T/2 \tag{10}$$

Where E_p is square wave drive voltage amplitude of the first half cycle.

Therefore, the amount of change in magnetization can also be expressed as the driving voltage waveforms of the first integration cycle, specifically:

$$\Delta B = \int \frac{e}{NA} dt = \frac{E_p}{NA} t \tag{11}$$

so

$$\Delta H = \frac{\Delta B}{\mu_\Delta} = \frac{E_p}{NA\mu_\Delta} t \tag{12}$$

As can be seen from Figure 1, the variation in the magnetic field strength corresponding to the width of the positive pulse, there may be:

$$H_0 = \frac{E_p}{NA\mu_\Delta} \tau \tag{13}$$

It can be solved pulse width, and its expression is:

$$\tau = \frac{NA\mu_\Delta H_0}{E_p} \tag{14}$$

In order to solve the formula (8), United vertical (9) and (13), may have:

$$E = 2E_p \tag{15}$$

The second harmonic voltage equation (15) and (14) into (15), can be well-wave excitation fluxgate sensor outputs is:

$$V_{2f} = \frac{8E_p}{\pi} \left(\sin \frac{\pi NA\mu_\Delta H_0}{TE_p} \right)^2 \tag{16}$$

Excitation frequency of the probe are a few kilohertz, so less than 1, for square wave excitation amplitude, in order to ensure the excitation coil into saturation region. The excitation amplitude is also greater than at least one volt, therefore, less than 1, while the maximum of 10^{-4}T, the core area A is calculated according to 100 μm radius, only about 10^{-8}, except for the constant, the number of turns N is greater than 1, the maximum is thousands of turns, the magnitude of 10^3, the coil magnetic permeability is not relative permeability, and therefore is not large, it was found from the above analysis, the equation (15) sinusoidal part is very small, close to zero, it can be approximated as:

$$V_{2f} = \frac{8E_p}{\pi} \left(\frac{\pi NA\mu_\Delta H_0}{TE_p} \right)^2 = \frac{32\pi N^2 f^2 A^2 \mu_\Delta^2 H_0^2}{E_p} \tag{17}$$

The sensitivity of the probe for:

$$S = \frac{V_{2f}}{H_0} = \frac{64\pi N^2 f^2 A^2 \mu_\Delta^2 H_0}{E_p} \tag{18}$$

In equation (17), the number of turns, area, and performance ratios can be regarded as constant, if the change is ignored, can be found, the sensitivity is inversely proportional to the excitation voltage amplitude of the second harmonic output of the probe . That is, the relationship between sensitivity and excitation voltage amplitude for:

$$S \propto \frac{1}{E_p} \tag{19}$$

where

$$E_p \frac{T}{2} = N \times A \times 2B_{sat} \tag{20}$$

B_{sat} is Magnetic saturation of the core material., so:

$$S = \frac{V_{2f}}{H_0} = \frac{8\pi NfA\mu_\Delta H_0}{H_{sat}} \tag{21}$$

3   MEASURED RESULTS

Figure 3 is a square wave excitation probe sensitivity change with the excitation voltage amplitude. As can be seen, the excitation frequencies of 2kHz, 4kHz, and when 6kHz, the sensitivity of both the square wave excitation voltage amplitude with increasing reduced.

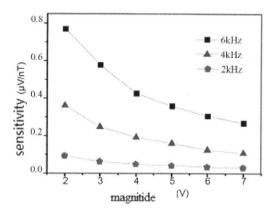

Figure 3. Variation of sensitivity with magnitude on excited by square wave.

4 CONCLUSION

The establishment of a fluxgate sensitivity model square-wave voltage excitation. This model uses the square-wave voltage inverse magnetic field strength in the core, based on the dynamic transfer function, sensitivity to build relationships with the core material parameters. For square wave excitation mode, the probe sensitivity is inversely proportional to the amplitude of the excitation voltage relationship, and experimental results verify the proposed model for the correctness of the square wave sensitivity.

ACKNOWLEDGMENT

This work is supported by Science and Technology Research Funds of Education Department in Heilongjiang Province under Grant Nos. 12541115.

REFERENCES

Perez L, Lucas I, Aroca C, et al (2006). Analytical model for the sensitivity of a toroidal fluxgate sensor. *Sensors and Actuators A*, 130–131:142–146.

X.F. Zhang, Y.L. Lu (1995). Fluxgate technology. *Beijing: Defense Industry Press*, 1995: 34–41.

Gordon D I, Lundsten R H, Chiarodo R A(1965). Factors Affecting the Sensitivity of Gamma-Level Ring-Core Magnetometers. *IEEE Transactions on Magnetics, MAG-1*(4):330–337.

BURGER J R(1972). The Theoretical Output of a Ring Core Fluxgate Sensor. *IEEE Transactions on Magnetics, MAG-8*(9):791–796.

Liu S, Cao D P, Jiang C Z (2010). A Solution of Fluxgate Excitation Fed by Squarewave Voltage. *Sensors and Actuators A*, 2010, 163: 118–121.

Electronic Engineering and Information Science – Wang (Ed.)
© 2015 Taylor & Francis Group, London, ISBN: 978-1-138-02772-5

Nonlinear analysis on gyroscope phase-sensitive circuit

C.C. Dong
School of Software, Harbin University of Science and Technology, Harbin, China

L. Yin & Q. Fu
MEMS Center, Harbin Institute Technology, Harbin, China

ABSTRACT: In the drive circuit and the detection circuit with high accuracy gyroscope sensors, nonlinear problems is a key factor restricting its performance. This chapter nonlinear conversion process micro gyroscope interface circuit drive mode and test mode in the presence of in-depth theoretical analysis, through theoretical analysis proposed reducing the interface circuit nonlinear optimization methods, and specific circuit design scheme, and the corresponding simulation.

KEYWORDS: Nonlinear; Phase-sensitive demodulation; Gyroscope.

1 INTRODUCTION

In the drive circuit and the detection circuit with high accuracy gyroscope sensors, nonlinear problems are a key factor restricting their performance (Mikko 2008). For a relatively high sensitivity gyroscope, the drive signal and angular velocity signal is present in many non-linear signal conversion process micro gyro system, these will drive the nonlinear modal vibration stability, detection modal noise ratio, and impact on the scale factor of the sensor and so on. In the design of the interface circuit, especially for sensor performance plays a key role in the detection circuit, and the lack of solutions to the nonlinear circuit optimization design related issues, and these nonlinear indirectly restrict the performance of the gyro system, affecting the sensor performance upgrade (Ji 2006).

Because the gyro sense mode usually includes two demodulation circuits, so the need to focus on the analysis of the nonlinear demodulation circuit Dienger 2007). Phase-sensitive switch circuit in two ways to signal through the introduction of non-linear components, on the one hand, the linear transfer characteristics of the switching phase-sensitive demodulation circuit reference signals may affect the system, on the other hand switch the sign of the phase-sensitive demodulation circuit transfer characteristic asymmetry will affect the system linearity(Raman 2006).

2 NONLINEAR ANALYSIS OF THE PHASE-SENSITIVE DEMODULATION CIRCUIT

Phase-sensitive demodulation full-wave rectification circuit shown in Figure 1, the transfer characteristic at

this time by the reference signal determining circuit, with a positive or negative reference signal input conversion value, the circuit outputs the corresponding transfer characteristics of the positive and negative transmission characteristic, Since the positive and negative transfer characteristics are not equal, and thus would be treated introduced into the non-linear component of the demodulated signal (Petkov 2005).

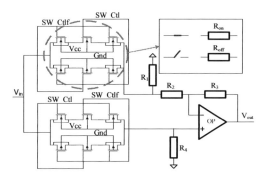

Figure 1. Switch phase-sensitive demodulation circuit.

The circuit shown in Figure 1, to select the circuit by CMOS switch in the forward amplification state or negative amplification status, under ideal conditions, the switch on-resistance is 0, shutdown infinite resistance, then the phase-sensitive demodulation circuit Forward the same gain and reverse gain, without introducing non-linear elements(Choop 2009).

In actual CMOS circuits, on-resistance CMOS switches is not close to 0, but not greatly off resistance approaches infinity, the equivalent circuit shown in Figures 2 and 3, can have its forward transfer characteristics are:

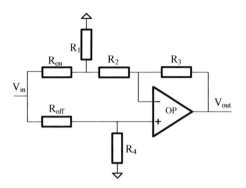

Figure 2. Equivalent circuit of phase-sensitive demodulation circuit.

$$V_{out}^{+} = \frac{1 - \dfrac{R_3}{R_{off}}[1 + \dfrac{R_{on}}{R_4}] + [\dfrac{1}{R_{off}} + \dfrac{1}{R_1}](R_2 + R_3)}{[1 + \dfrac{R_{on}}{R_4}] \times [(\dfrac{R_2}{R_{off}} + \dfrac{R_3}{R_1}) + 1]} \qquad (1)$$

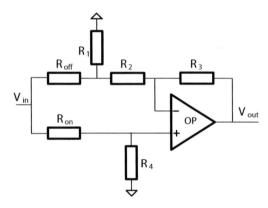

Figure 3. Equivalent circuit of phase-sensitive demodulation circuit.

$$V_{out}^{-} = \frac{-1 + \dfrac{1 + \dfrac{R_{on}}{R_1}}{1 + \dfrac{R_{off}}{R_4}}[1 + \dfrac{R_2}{R_3}] + \dfrac{R_{on}}{R_3} \times \dfrac{1}{1 + \dfrac{R_{off}}{R_4}}}{[1 + \dfrac{R_{on}}{R_1}] \times \dfrac{R_2}{R_3} + \dfrac{R_{on}}{R_3}} \qquad (2)$$

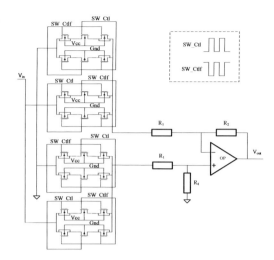

Figure 4. Improved phase-sensitive demodulation circuit.

In this case, the forward transfer characteristics of the circuit can be calculated as:

$$V_{out}^{+} = \frac{1 - \dfrac{R_3}{R_{off}}[1 + \dfrac{R_{on}}{R_4}] + [\dfrac{1}{R_{off}} + \dfrac{1}{R_1}](R_2 + R_3)}{[1 + \dfrac{R_{on}}{R_4}] \times [(\dfrac{R_2}{R_{off}} + \dfrac{R_3}{R_1}) + 1]} \qquad (3)$$

The circuit reverse transfer characteristics are:

$$V_{out}^{-} = \frac{-1 + \dfrac{1 + \dfrac{R_{on}}{R_1}}{1 + \dfrac{R_{off}}{R_4}}[1 + \dfrac{R_2}{R_3}] + \dfrac{R_{on}}{R_3} \times \dfrac{1}{1 + \dfrac{R_{off}}{R_4}}}{[1 + \dfrac{R_{on}}{R_1}] \times \dfrac{R_2}{R_3} + \dfrac{R_{on}}{R_3}} \qquad (4)$$

From equations (3) and (4) can be seen, the circuit is equal to the positive and the negative transfer characteristic to transfer characteristics, that is $|V_{out}^{+}| = |V_{out}^{-}|$, to suppress the switching phase-sensitive demodulation circuit in the secondary demodulation process changes in non-ideal characteristics of nonlinear switch resistance introduced.

3 MEASURED RESULTS

Figure 5 is the value of input amplitude of 0.1 volts to 1 volt sinusoidal signal through phase-sensitive demodulation circuit, the output of the low pass filter curve, in order to verify the non-linear output curve, where the signal is relatively stable portion of the sampled signal, taking 0.38 seconds time voltage values shown in Figure 6, fitting a straight line as shown in Figure 7, may have linearity before switching phase-sensitive demodulation is 0.124%.

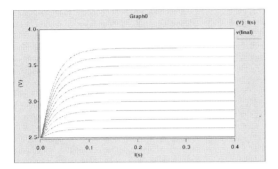

Figure 5. The nonlinearity test output of phase-sensitive demodulation circuit.

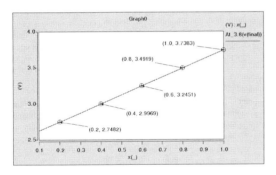

Figure 6. The nonlinearity test output of phase-sensitive demodulation circuit in 0.38s.

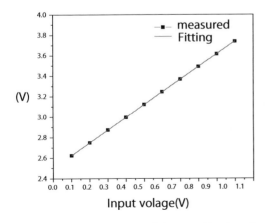

Figure 7. The input and output of phase-sensitive demodulation circuit before improved.

4 CONCLUSION

This article analyzes the nonlinear detection mode, the main inconsistency caused by the negative transfer characteristic demodulation circuit for phase-sensitive demodulation circuit to improve and optimize the structure parameters. Simulation results show that for the improved nonlinear reached 0.010 percent, compared with 0.124% improvement before the nonlinear increase by nearly an order of magnitude.

ACKNOWLEDGMENT

This work is supported by Science and Technology Research Funds of Education Department in Heilongjiang Province under Grant Nos. 12541115.

REFERENCES

Mikko S, Lasse A, Halonen K(2008). Effects of Synchronous Demodulation in Vibratory MEMS Gyroscopes:A Theoretical Study. *IEEE Sensors Journal*, 8(10): 1722–1733.

Ji X S, Wang S R, Xu Y S, et al(2006). Application of the Digital Signal Procession in the MEMS Gyroscope De-drift. *Proceedings of the 1st IEEE International Conference on Nano/Micro Engineered and Molecular Systems. Zhuhai, China*, 218–221.

Dienger M, Buhmann A, Northemann T, et al(2007). Low-power, Continuous-time Sigma-delta Interface for Micromachined Gyroscopes Employing a Sub-nyquist-sampling Technique. *The 14th International Conference on Solid-State Sensors, Actuators and Microsystems. Lyon, France*, 1179–1182.

Raman J, Cretu E, Rombouts P, et al. (2006) A Digitally Controlled MEMS Gyroscope with Unconstrained Σ-Λ Force-feedback Architecture. *Micro Electro Mechanical Systems 2006. Istanbul, Turkey*, 10–713.

Petkov V P, Boser B E(2005). A Fourth-Order Σ-Λ Interface for Micromachined Inertial Sensors. *IEEE Journal of Solid-state Circuit*, 40(8): 1602–1609.

Choop P M, Hamoui A A(2009). Analysis of Clock-Jitter Effects in Continuous-Time ΔΣ Modulators Using Discrete-Time Models. *IEEE Trans. on Circuits and Systems*, 56(6): 124–128.

Electronic Engineering and Information Science – Wang (Ed.)
© 2015 Taylor & Francis Group, London, ISBN: 978-1-138-02772-5

An effective English word segmentation method based on interconnected domain search

F. Yin, L. Zheng & J.H. Han
College of Computer Science and Technology, Harbin University of Science and Technology, Harbin, China

F. Yin & X.Y. Yu
Instrument Science and Technology Postdoctoral Workstation, Harbin University of Science and Technology, Harbin, China

F.H. Jin
College of Computer Science and Technology, Harbin University of Science and Technology, Harbin, China

ABSTRACT: The effect of word segmentation directly affects the recognition accuracy in English text recognition. This paper presents a word segmentation method based on interconnected domain search. Firstly all the interconnected domains are found and recorded through comprehensive image search. Then the interconnected domains are merged according to the position, size and other judgment conditions of them. Interconnected domain pieces and noise are filtered, and English word areas are divided accurately. Experimental results show that the method can solve the problem of word segmentation of characters overlapping which vertical projection method can not deal with.

KEYWORDS: Text Recognition; Word Segmentation; Interconnected Domain Search.

1 INTRODUCTION

Word segmentation is the preprocessing for word recognition. Word recognition based on characters need that the word is segmented into completely separate characters to be recognized.so word segmentation section must ensure the integrity of the characters to ensure the acuracy of recognition (Chen Tao & Yang Chenhui & Qing Bo 2009). Currently text segmentation methods often used by researchers include vertical projection method and improved one (Jiao Pengpeng & Guo Yizheng 2013, Liu Yangxing 2001, LI Zuo & Wang Shuhua & Cai ShiJie 2001), curve segmenting path method (Liu Yu, & Zhang Yanduo & Lu Tongwei 2011), integrated segmentation method (Meng Qingyuang & Bai Yanping & Hu Hongping 2011), clustering method (WANG J & JEAN J 1994, Wu Rui & Yin Fang & Tang Xianglong, et al 2010) and recognition feedback method (Wang Jiangqing, & Cao Wei 2011). While all these algorithms either can not well process touching and kerned samples or not applicable for printed text recognition because of high complexity, large computational cost and low efficiency and accuracy.

Aacording to the statistics, the wrong of English OCR mainly for error segmentation of kerned characters(Yin Fang & Wang Weibing, Chen Deyun 2008).

The so-called kerned character(Yang Wuyi & Zhang Shuwu 2010) refers to two independent characters overlapping. There is no doubt an incorrect recognition result for a wrong segmentation one such as the "fe" in Figure 1 and Figure 2. Therefore, the correctsegmentation directly affects the recognition accuracy in recognition system.

feasible

a) Original image of feasible

b) Effect image of feasible segmented

Figure 1.   Segmentation effect diagram of vertical projection method.

Interconnected domain search method is widely applied in image segmentation and achieves good effects (Yin Fang & Chen Deyun & Wu Rui 2011,

ff tr tu

leftwards from

opposite physical

Figure 2. Possible overlapping patterns of English characters.

Yu Ming & Guo Qian & Wang Dongzhuang 2013). Based on the traditional vertical projection histogram segmentation method, this paper presents segmentation method based on interconnected domain search. Firstly all the interconnected domains are found and recorded through comprehensive image search. Then according to the location, size and other judgment conditions of them, the interconnected domains are merged. The method can solve the problem of word segmentation of characters overlapping that vertical projection can not do.

2 INTERCONNECTED DOMAIN

For a domain named D, if any two points can be connected using broken line in D, D can be called as interconnected domain. Interconnected domain is divided into simply interconnected domain and multiply interconnected domain. If any closed curve not through points outside of D can continuous shrink to any point of D, D is simply interconnected domain, otherwise D is multiply interconnected domain that also is called as complex interconnected domain, has been shown in Figure 3.

Processed data are discrete digital lattice for Digital Image Processing, which is different from Math processing continuous number field. So the concept of interconnected domain in Digital Image Processing differs in Math. Interconnected domain can be divided into 4-interconnected domain and 8-interconnected domain in Digital Image Processing. In Figure 4 let P be any pixel in the image, and the eight pixels next to P called the 8-neighborhood. P0, P2, P4 and P6 called 4-neighboring point of P and P0 ~ P7 called 8 -neighboring points of P.

For an image region, if we can reach any point of the image starting from any point of it and going through the up, down, left and right direction of it, the image region can be called as 4-interconnected domain. Similarly if we can reach any point of the image starting from any point of it and going through

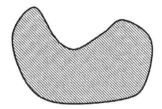

a) Simply interconnected domain

b) Multiply interconnected domain

Figure 3. Interconnected domain.

the up, down, left, right, upper left, upper right, lower left and lower right direction of it , the image region can be called as 8-interconnected domain.

P3	P2	P1
P4	P	P0
P5	P6	P7

Figure 4. 8-neighborhood.

3 THE ENGLISH WORD SEGMENTATION ALGORITHM BASED ON INTERCONNECTED DOMAIN SEARCH

This paper uses the growth algorithm to seek interconnected domain by the black pixels growing to the adjacent pixels until finding all black pixels belonging to the same interconnected domain.

Algorithm implementation

For an $m * n$ image, let an array named area $[m * n]$ corresponding to its all the points. And initial value of area $[m * n]$ is set to 0. Let an int type variable *areanum* = 0.

If there are some points that are not processed in addition, make the following operations.

Loop:

Getting a point from the image of the i-th row and the j-th column and checking the value of area [i * n + j] corresponding the point, if the value is not 0, it proves this point is already processed. Skipping. If the value is 0 and pixel value is 0, the point is the background. Skipping . If the value is 0 and pixel value is not 0, the point is not processed in interconnected domain. And i*n+j is pressed into stack, areanum + 1, area[i*n+j] = areanum. Continuous. If the stack is not empty, the operation is as follows.

Sub-loop

A position popping from the stack and seeking 8 adjacent points up, down, left, right, upper left, upper right, lower left and lower right of the point corresponding the position, if the point of pixel value is 0, it is background. Skipping. If the value is 1, it proves the point and the points of the stack are in the same interconnected domain, pressing the position into stack, *area[i*n+j] = areanum*, processing like above.

Sub-loop end;

Loop end.

At the same time, let four arrays in the program. Two of them respectively deposit x0 and y0 that are the starting value of a interconnected domain. And the other two respectively deposit the width and height of the interconnected domain. This is the important basis for the interconnected domain restructured again.

In the kerned sample, a character fracturing may be segmented into more than one interconnected domains. And two or more interconnected domains belong to the same interconnected domain for the touching sample. Due to the above, the interconnected domains should be restructured after all interconnected domains are found at the first time.

For the problem that a character may be segmented into more than one interconnected domains, the interconnected domains are restructured according to the following mainly.

a. For the all interconnected domains, if the height of a interconnected domain less than 1/3 of the image height, the interconnected domain must not be a complete image. Then finding the other interconnected domain most close to the interconnected domain, and the distance overlapping between the two interconnected domains is worked out. Distance overlapping refers to the common field between *(x11, x12)* and *(x21, x22)* that are the defining field of two interconnected domains in the horizontal direction. As defined as follows:
  if *(x21 > x11)* and *(x22>x12) Dis=abs(x12-x21);*
  if *(x21 > x11)* and *(x22<x12) Dis=abs(x22-x21);*
  if *(x21 < x11)* and *(x22<x12) Dis=abs(x22-x12);*
  if *(x21< x11)* and *(x22>x12) Dis=abs(x12-x11).*

  The greater *Dis* expressing the distance overlapping is, the more likely they belong to the same character. If *Dis* is greater than a fixed value, two interconnected domains will be merged and amending the value of area and the interconnected

domain properties including the value of *Width*, *Height*, x0 and y0. If *Dis* is less than the fixed value, the interconnected domain is thrown out as noise.

b. There are also some images of the characters fracturing vertically. For instance *m* is divided in *l* and *n*. The height of images fracturing is enough. But its aspect ratio is very small generally. The character fragments are similar with character image l, character image I and character image i. So it tries to distinguish the characters above with analyzing positions or the ratio of the fragments in the image. For example, character image l, character image I and character image i are all occupying the position between top line and base line. But these characters fragments are most between the top and the bottom base line. It is sure that whether it should be merged with other interconnected domains by the above comprehensive judgment.

c. As the image properties of character fragments as same as that of complete characters and both aspect ratio and the size of character fragments are right, distance overlapping is worked out. As the distance overlapping is enough great. For example, if it is more than 1/2 of the image of character fragments, the two interconnected domains belong to the same character image. And if the character image of A is segmented into two interconnected domains, left oblique line and horizontal line in the middle of it belonging to a interconnected domain, right oblique line belonging to another interconnected domain, the two interconnected domains should be merged with this judgment that the region overlapping between the two interconnected domains is very large.

Algorithm process logic is described in Figure 6.

The Figure 4 is got by original image of feasible being segmented in Figure 5. The 'fe' overlapping above and below can be segmented apart with the segmentation method based on interconnected domain, which is more effective than the vertical projection segmentation method.

feasible

a) Original image of feasible

b) Effect image of feasible segmented

Figure 5.   Interconnected domain segmentation.

4 EXPERIMENTAL RESULTS ANALYSIS

Comparing the proposed method with the vertical projection segmentation method using the same recognition platform and samples of 100 words, by which words are preprocessed. The final result is shown in table 1.

Table 1. Segmentation result comparing.

Total number of words	Projection method	The proposed method
100	18	97

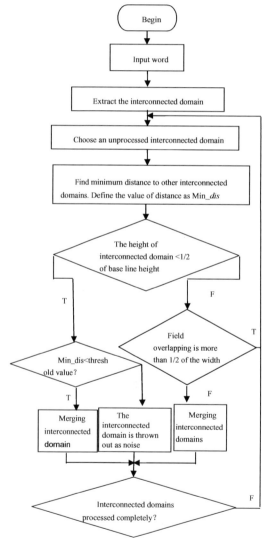

Figure 6. Algorithm process logic diagram.

5 CONCLUSIONS

The English word segmentation method based on interconnected domain presented in this paper solves the problem of word overlapping above and below that the vertical projection method can not do. Firstly all the interconnected domains are found and recorded through comprehensive image search. Then according to the position, size and other judgment conditions of them, the interconnected domains are merged. Interconnected domain pieces and noise are filtered. English words are segmented accurately. Experimental results show that the method is a practical word segmentation method and provides a good performance for text recognition.

ACKNOWLEDGMENT

This paper is supported by the research project of science and technology of Heilongjiang provincial education department (No. 12541119).

REFERENCES

Chen Tao & Yang Chenhui & Qing Bo (2009). Characters segmentation of license plate based on combina-tion of projection and intrinsic characteristics. *Computer Technology and Development*, 19(5):45–47.

Jiao Pengpeng & Guo Yizheng (2013). An algorithm for license plate location and character segmentation based on mathematical morphology. *Techniques of Automation and Applications*, 2013, 32(6): 57–59.

Liu Yangxing (2001). Non-linear partitioning path based approach for touching and kerned charac-ter segmentation. *Application Research of Computers*, 28(10): 3998–4000.

LI Zuo & Wang Shuhua & Cai ShiJie (2001). An Algo-rithm for segmentation of merged characters based on fore-part prediction and recognition. *Journal of Computer Research and Development*, (11): 1337–1344.

Liu Yu, & Zhang Yanduo & Lu Tongwei (2011). Method of connected domain denoise based on least absolute deviation in OCR. *Journal of Wuhan Institute of Technology*, 33(1): 84–87.

Meng Qingyuan & Bai Yanping & Hu Hongping (2011). Vehicle template character recognition technol-ogy based on connected domain characteristic. *Journal of Test and Measurement Technology*, 25(1): 87–92.

WANG J & JEAN J (1994). Segmentation of merged characters by neural networks and shortest path. *Pattern Recognition*, 27(5): 649–658.

Wu Rui & Yin Fang & Tang Xianglong, et al (2010). Spectral clustering based text image segmentation using fuzzy logic. *Journal of Harbin Institute of Technology*, 42(2): 268–276.

Wang Jiangqing, & Cao Wei (2011). Vertical projection characters segmentation based on minimum threshold and curve-fitting. *Journal of South-Central University for Nationalities(Natural Sci-ence Edition)*, 30(4): 83–85.

Yin Fang & Wang Weibing, Chen Deyun (2008). Architecture and implementation of the printed English documenta-tion recognition system. Jour-nal of Harbin University of Science and Technology, 13(6): 9–12.

Yang Wuyi & Zhang Shuwu (2010). An integrated seg-men-tation and recognition algorithm for text in vide. *Acta Automatica Sinica*, 36(10): 1468–1476.

Yin Fang & Chen Deyun & Wu Rui (2011). Improved method of image segmentation using spectral clustering. *Computer Engineering and Appli-cations*, 47(21): 185–187.

Yu Ming & Guo Qian & Wang Dongzhuang (2013), et al. Improved connectivity-based layout segmenta-tion method. *Computer Engineering and Ap-plications*, 49(17): 195–198.

Zhang Zhenhui & Liu Sai (2011). Design and realization on character segmentation Algorithm for women writings. *China Science and Technology In-formation*, 47(21): 185–187.

Electronic Engineering and Information Science – Wang (Ed.)
© *2015 Taylor & Francis Group, London, ISBN: 978-1-138-02772-5*

Random forest algorithm for spam filtering based on machine learning

W.B. Wang, F. Yin, H. Sun & P. Li
College of Computer Science and Technology, Harbin University of Science and Technology, Harbin, China

ABSTRACT: A mass of spam greatly affects the use of email. The machine learning techniques achieve success in tackling spam. This paper employed a random forest algorithm to tackle spam. The algorithm is appropriate for this task as it runs fast on real-time setting. In order to speed up the performance of the automatic task, a method of novel feature selection is proposed to reduce the feature vector dimension. Evaluate these methods on three TREC spam datasets, the experimental results demonstrates that random forest is appropriate to spam filters, term frequency variance offers a far cheaper alternative with performance improvement.

KEYWORDS: Information Security, Spam Filtering, Learning-based Method, Random Forest Algorithm.

1 INTRODUCTION

1.1 Type area

Traditional content-based filtering is a natural solution for combating spam (Baojun. Su & Chongfu. Xu 2009). In spam classification tasks, algorithm based on machine learning can help the system to get precise prediction. Spam filtering using machine learning is a research hotspot. Good-man developed a very simple technique for performing logistic regression (D. Sculley 2007). Sculley proposed an efficient spam filtering using a relaxed on-line support vector machine (ROSVM), which greatly reduced computational cost and achieved strong classification performance (D. Sculley & G. M. Wachman 2007). Metsis discussed five different versions of Naive Bayes and compared them on six datasets (G. V. Cormack 2008). Su described a not so naive Bayesian (NSNB) that weakens the independent assumption and was no longer "naive" (Salton, C. Buckley 1997). Liu presented Lasso regression for building directly with high-dimensional and sparse email data depicted by the vector space model with Chinese emails (G. V. Cormack 2007). In this paper, we consider the automatic email filtering by taking random forest algorithm to filter spam. As a forest building it is a generalization error generated by an internal unbiased estimate. In addition to pro-duce a highly accurate classifier for large-scale data, a novel feature selection method is applied for improving the system performance.

2 RANDOM FOREST ALGORITHM

Random forest (RF) composed of many decision trees as an ensemble classifier that (Hastie. T & Tibshirani. R & Friedman, J 2009). In the classifiers each member is a decision tree, therefore the collection of these trees generate a "forest". Random forest tries one's best to improve bagging property by de-correlating equalization algorithm on the trees. With this method, it makes each decision tree in RF has the same expectation. It predicts the vote of the classifier output by individual trees (H. Drucker & V. Vapnik & D. Wu 1999). In order to create a set of decision trees, the method uses some controlled variations to compose bagging idea with the random selection of features. In view of the given original training dataset D, that d refers to the number of samples and m refer to the number of the available feature. The pseudo-code of the whole RF algorithm is summarized in Algorithm I. That is a process of generating k decision trees. At its iteration step (i=1, 2 ...k), from the original dataset they are randomly selected that d examples of every ensemble member with replacement. That is, every Di is the set of the bootstrapped instances. This makes some examples appear many times, while others may not be once in Di.

RF is a combination of two random precautions of decision trees constructing; at each node, the vectoring procedure duplicates each tree examples and samples a features random subset. At each tree split, to draw a random sampling of features, and to split only those f (f << m) features are considered .The best split is calculated only within this subset Di. For each tree grown on a vectoring procedure sampling, to monitor the error rate for observations can be left out of the bootstrap sample. This is called the out-of-bag (OOB) error rate. As no pruning step is performed opposite to the standard decision tree algorithm, so all the trees of the forest are maximal trees.

For large-scale dataset, RF can achieve a classifier with high accuracy; therefore it is robust for error and outlier. It is a good method for missing data estimation. And in the situation of that the data missing are a large proportion it can also maintain high accuracy. In the random forest, those effects lean on the strength of individual trees. The effect also leans on the correlation between every pair of trees. Ideally the optimization of RF aims to improve individual classifier's ability rather than their correlation. These factors are often also sensitive to the number of attributes f. typically f = log (m) +1, where m is the number of available features.

Algorithm I Random Forest

Input

Given dataset D;

d – the number of examples, m – the number of available features, f – the number of sampling features in RF, k – the number of decision trees;

Create Random Forest

For ith iteration (i=1,2,...k)

Bagging:

Generated d bootstrap samples;

Form D_i from bootstrapped training examples;
Random forest:

Build one tree per bootstrap samples;

Randomly pick f features to split at each node without pruning;

Prediction

New examples pass down each random tree until it reaches a leaf node. RF outputs the classification that is the majority voting by individual trees.

3 FEATURE SELECTION METHOD

3.1 *Feature mapping*

We choose character-level n-grams as feature space. Character-level n-grams has been proofed that it can achieve strong performance, which has a more valid and robust for a variety of spam (J. Goodman & W.-T. Yih 2006). Scullery used a simple binary 4-gram feature extraction drawn from the first 3,000 characters of each email (Baojun. Su & Chongfu. Xu 2009), but term frequency weights implicitly contain more information than the binary. For instance, the "free" appearing 5 times in an e-mail is a distinct revelation than that occurs once. Here, in each email only first 2500 features were extracted by the 4-gram feature extraction method, and the same features were added up the term frequency in each email.

3.2 *Term frequency variance*

Feature selection seeks to sift the informative features that have the aggressive ability for classification. It removes the noisy and useless features to reduce the computational cost. Koprinska proposed the method of term frequency variance (TFV) (Koprinska. I &

Poon. J & Clark. J, & Chan. J 2007), but here 4-gram features are defined as term instead of the word-based. For each term we compute the variance equation 1.

$$TEV(f) = \sum_{i=\{sm,hm\}} [tf(f,c_i) - mean\_tf(f)]^2 \quad (1)$$

The TF scoring method has been weighted some (predominantly more informative) words more heavily in information retrieval (LIU, Zunxiong & Xianlong ZHANG & Shujuan ZHENG 2012). TF emphasizes the contribution of the term to the current document. Initially, it isn't designed for categorization task. Going even further, TFV is considered informative term with high variance across categories. If a feature often occurs in spam category, that mean a higher distinction with high variance.

3.3 *Information gain*

Almost all of the feature selection, it is always to quantify the importance of the features and then choose. So as to test the reliability of TFV, we do a comparison with information gain. In IG process, the criterion is that how much contribution they can bring to the classification system. The more knowing, the more crucial these are. A feature f is calculated as equation 2 from the information gained.

$$\begin{aligned} IG(f) &= H(c) - H(c \mid f) \\ &= -\sum_{i=\{sm,hm\}} p(c_i) \log_2 p(c_i) \\ &\quad + p(f) \sum_{i=\{sm,hm\}} p(c_i \mid f) \log_2 p(c_i \mid f) \quad (2) \\ &\quad + p(\bar{f}) \sum_{i=\{sm,hm\}} p(c_i \mid \bar{f}) \log_2 p(c_i \mid \bar{f}) \end{aligned}$$

Each feature is estimated by the accessing to information. A feature f is considered well classified and should be remained when IG(f) is greater than a fixed threshold value.

4 EXPERIMENTS

4.1 *Data sets*

In order to verify the novel method, three benchmark datasets are used, which ware developed for the TREC spam filtering competitions: trec05p-1, trec06p, trec07p. The details of three large benchmark corpora are shown in Table 1. In addition, all of them is used canonical ordering.

Table 1.	Test data.		
Data Sets	Spam	Ham	Totality
TREC06p	24912	12910	37822
TREC05p-1	52790	39399	92189
TREC07p	50199	25220	75419

Table 2.	The performance with TFV and IG.		
Data set	Method	Lam%	1-ROCA%
Trec05p-1	RF	0.43	0.0121
	RF-IG	0.33	0.0094
	RF-TFV	0.26	0.0083
Trec06p	RF	0.51	0.0837
	RF-IG	0.44	0.0595
	RF-TFV	0.39	0.0395
Trec07p	RF	0.42	0.0251
	RF-IG	0.24	0.0148
	RF-TFV	0.10	0.0096

4.2 Evaluation measures

In order to evaluate this method the measures data set TREC 2005 is taken for spam tracking (D. Sculley 2007). Here, the Ham % (hm) is referred to the percentage of misclassification, means all part of the ham to the junk mail; the Sm% is referred to the percentage of spam misclassification (sm), means part of all the junk mail delivered to ham. Logistics as a percentage of average error classification to the harmony (lm %) is between hm% and sm%. The graphical representation shows a receiver operating characteristic (ROC) curve. Here, the area between (1-ROCA) % and above the regional ROC curve shows the probability of the random spam scored lower on message will receive random ham news (V. Metsis & I. Androutsopoulos & G. Paliouras 2006). hm% and sm %, and consistent with the lm % and fabricated (1 - ROCA) % will serve as a standard performance measures, 0 is the best place. For consistency with hm% and sm%, lm% and (1-ROCA %) will be reported as the standard performance measures, where 0 is optimal.

4.3 Performance and discussion

The experiments adopted 10-fold cross validation. The dataset is divided into ten non-overlapping parts of equal size. Each part is then tested on the data from the other segments which were used as training data. The error estimate is then averaged over the 10 classification process. For the RF, we set k = 12, f =24. For feature selection, all numeric values were normalized in the given dataset.

To summarize, the ROC curves are shown in Fig 1-3, the result reported in Table 2. They vary significantly across all the datasets. First, RF is reliable and can achieve high accuracy on all large benchmark corpora. Single decision tree of RF increases via additional randomization. Every tree is endowed with special preferences in certain aspects. The combination of multitudinous members produces a sufficiently faithful and accurate classifier. Second, IG and TFV obviously promote the system's accuracy.

This upgrade in performance is accompanied by the deploy of TFV and IG. Both can be matched

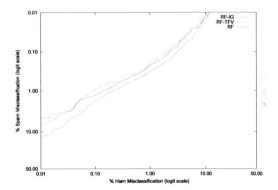

Figure 1. The ROC with trec05p-1 corpus.

and TFV outperformed the IG. Moreover, the TFV typically takes O(n) time, IG has quadratic time complexity. For content-based spam detection these experiments indicate that random forest is appropriate to spam filters. Term frequency variance offers a far cheaper alternative with performance improvement.

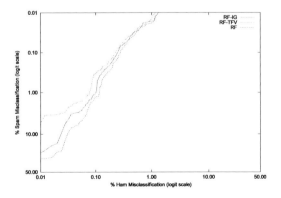

Figure 2. The ROC with trec06p corpus.

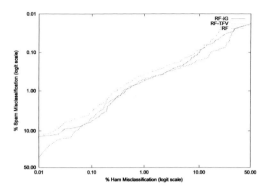

Figure 3.   The ROC with trec07p corpus.

5   CONCLUSIONS

In this paper, we have considered a learning-based algorithm for spam filtering and explored the feature selection approaches for improving the prediction performance. In order to speed up the performance of the automatic task, a novel feature selection method is proposed to reduce the feature vector di-mension. Evaluate these methods on three TREC spam datasets, the experimental results demonstrates that random forest is appropriate to spam filters, term frequency variance offers a far cheaper alternative with performance improvement. The random forest algorithm is an ensemble classification technique, which involve developing numerous decision trees from randomly sampling subspaces of input instances. RF is employed to filter spam. RF runs very efficiently on large-scale datasets with high number of features, which makes it very suitable for spam detection.

REFERENCES

Baojun. Su & Chongfu. Xu (2009), Not so naive online Bayes spam filters, *In Processings of the 21st Innovative Applications of Artificial Intelligence Conference.*

D. Sculley (2007), Online active learning methods for fast label-efficient spam filtering, *In Proceedings of the Fourth Conference on Email and Anti-Spam,* 2007.

D. Sculley & G. M. Wachman (2007), Relaxed online support vector machines for spam filtering, *in 30th ACM SIGIR Conference on Research and Development on Information Retrieval,* Amsterdam.

G. V. Cormack (2008). Email Spam Filtering: A Systematic Review, *Foundations and Trends in Information Retrieval,* 1(4):335–455.

G. Salton, C. Buckley (1997), Improving retrieval performance by relevance feedback, *in information retrieval,* pages 355–364.

G. V. Cormack (2007), TREC 2007 spam track overview, *In To appear in: The Fifteenth Text Retrieval Conference (TREC 2007) Proceedings.*

Hastie. T & Tibshirani. R & Friedman, J (2009), Random Forests, *In the Elements of Statistical Learning, Springer Series in Statistics,* pp. 1–18, Springer New York, 2009.

H. Drucker & V. Vapnik & D. Wu (1999), Support vector machines for spam categorization, *IEEE Transactions on Neural Networks.*

J. Goodman & W.-T. Yih (2006). Online discriminative spam filter training, *In the Third Conference on Email and Anti-Spam,* Mountain View, CA.

Koprinska. I & Poon. J & Clark. J, & Chan. J (2007), Learning to classify e-mail, *Information Sciences,* 177(10), 2167–2187.

LIU, Zunxiong & Xianlong ZHANG & Shujuan ZHENG (2012), Lasso-based Spam Filtering with Chinese Emails, *Journal of Computational Information Systems* 8.8 (2012): 3315–3322.

V. Metsis & I. Androutsopoulos & G. Paliouras (2006), Naive Bayes-Which Naive Bayes?, *in Proceedings of CEAS 2006 - Third Conference on Email and Anti-Spam,* Mountain View, CA.

Electronic Engineering and Information Science – Wang (Ed.)
© 2015 Taylor & Francis Group, London, ISBN: 978-1-138-02772-5

Research on linkage collaboration of snort and WFP based on Windows 7

L. Ding, J.Q. Zhai & H.Z. Qi
School of Computer Science and Technology, Harbin University of Science and Technology, Harbin, China

ABSTRACT: Most recent studies of the linkage of Snort IDS and Firewall on Windows are still based on Windows XP series platform. In this paper how to filter or block the malicious network traffic by configurating Windows Filtering Platform automatically when Snort is triggered by suspicious network data on Windows 7 series OS is presented. The WFP Filters created by Snort can setup the parameters source and destination IP address, source and destination port, data flow direction, blocking time respectively according to the specific intrusion actions. And the implement of linkage collaboration of WFP and Snort is introduced and analyzed in detail, then the rewritten IDS based on Snort is tested amply, the experimental results testify this novel linkage method can block the dangerous network packets successfully on Windows 7.

KEYWORDS: Snort; WFP; Collaboration; Filter.

1 INTRODUCTION

Snort is the most famous and free Network Intrusion Detection software based on misused detection. Snort can perform real-time network traffic analysis, suspicious network action logging, TCP/IP hierarchy network protocol analysis, special content matching. It also can be used to detect the buffer overflow, port scan, OS fingerprint probe, DOS/DDOS attacks etc. Snort can run on the Linux/Unix and Windows etc. Snort has good Cross-Platform and it is one of the most widely used open sources Intrusion Detection System in the world at present (Qing-Xiu Wu 2012). However, according to Common Intrusion Detection Framework (HJ Liao & CH Richard Lin & YC Lin 2013), the response unit of Snort IDS is relatively inefficient because of just having log alarming mechanism and simple active response ability by using Flexresp2.

And the Operating System of the majority of computer nodes in current organization network and enterprise network is Windows 7 series, especially after Microsoft Corp announced on April 8th 2014 that no longer providing technical support and updates for Windows XP, Windows 7 has become the most widely used personal computer terminal Operating System. So the new method is presented in this paper to filter or block the malicious network traffic by configurating WFP (Windows Filtering Platform) automatically when Snort is triggered by suspicious network data on Windows 7 series OS. The created filters can setup the parameters source and destination IP address, source and destination port, data flow direction, blocking time respectively according to the specific intrusion.

2 IMPLEMENTION OF LINKAGE OF SNORT AND WFP

Windows Filtering Platform (WFP) is the novel network filter driver framework starting in Windows Vista designed to replace the Windows XP and Windows Server 2003 network traffic filtering interfaces (Govind Kritika & Vivek Kumar & Selvakumar 2012). WFP is composed of a set of hooks into the network stack and a filtering engine that coordinates network stack interactions, which provides a packet filtering infrastructure that enable independent software vendors to plugin specialized filtering modules for data filtering, modification, and re-injection (Huang Junsheng & Fan Bingbing 2013).

For implementation of linkage collaboration of WFP and Snort on Windows 7 series, the linkage code should be appended to the OutPut Plugins (Brian Caswell & Jay Beale & Andrew Baker 2006) of Snort source code. The workflow of linkage collaboration module of Snort is described detailedly as follows .

2.1 *Trigger linkage*

At first the linkage collaboration module is triggered promptly when detection engine of Snort matches one or several network data packet with a linkage collaboration rule. The corresponding code is as follows.

```
if(CheckList(FilterQueueHead,str)==0)
WFPBlock(&testpacket);
```

2.2 *Create a timer*

Secondly, the linkage collaboration module call the API function timeSetEvent to create a multimedia timer to dominate the validity time of the WFP Filter according to linkage rule. Unlike the windows timer, multimedia timer executes its own thread instead of windows message, so the resolution can reach the millisecond level. The corresponding code is as follows.

```
nIDTimerEvent=
timeSetEvent( bd->duration* 1000, 0, TimeProc,
0, (UINT) TIME_ PERIODIC);.
```

2.3 *Configure the WFP filter*

Then configure the WFP filter according to the link-age rule, which is the most important and substantial function of the linkage collaboration module. The essential code is abbreviated as follows.

```
switch(bd->Mode&WFP_HOW)
{case WFP_HOW_THIS:
        if(bd->protocol==6 || bd->protocol==17){
snprintf(cmd1, sizeof(cmd1)-1," /k netsh adv-
firewall firewall add rule name=Filter-%d dir=in
action=block enable=yes localip=%s local-
port=%d remoteip=%s remoteport=%d proto-
col=%s interfacetype=any", bd->dstip, bd->d-
stport,bd->srcip,bd->srcport,bd->protocol== 6?
"TCP":"UDP");
snprintf(cmd2, sizeof(cmd2)-1," /k netsh adv-
firewall firewall add rule name=Filter-%d
dir=out action=block enable=yes localip=%s
localport=%d remoteip=%s remoteport=%d
protocol=%s        interfacetype=any",bd->d-
stip,bd->dstport,bd->srcip,bd->s-
rcport,bd->protocol== 6?"TCP":"UDP");}
        if (bd->protocol==1){
        snprintf(cmd1, sizeof(cmd1)-1," /k netsh
advfirewall firewall add rule name=Filter-%d
dir=in action=block enable=yes localip=%s
remoteip=%s protocol=icmpv4 interfacetype=
any", nIDTimer Event,bd->dstip,bd->srcip);
snprintf(cmd2, sizeof(cmd2)-1," /k netsh adv-
firewall firewall add rule name=Filter-%d
dir=out action=block enable=yes localip=%s
remoteip=%s protocol=icmpv4 interfacetype=
any", nIDTimer Event,bd->dstip,bd->srcip);}
        break;
case WFP_HOW_IN:
if((bd-Mode&WFP_WHO)==WFP_WHO_ SRC)
        snprintf(cmd1,sizeof(cmd1)-1," /k netsh adv-
firewall firewall add rule name=Filter-%d
dir=in action=block enable=yes localip=any
remoteip=%s protocol=any interfacetype=any",
nIDTimer Event,bd->srcip);
        else
```

```
snprintf(cmd1,sizeof(cmd1)-1," /k netsh adv-
firewall firewall add rule name=Filter-%d
dir=in action=block enable=yes localip=%s
remoteip=any protocol=any interfacetype=any
", nIDTimer Event,bd->dstip);
        break;
case WFP_HOW_OUT:
if((bd->Mode&WFP_WHO)==WFP_WHO_ SRC)
        snprintf(cmd1,sizeof(cmd1)-1," /k netsh adv-
firewall firewall add rule name=Filter-%d
dir=out action=block enable=yes localip=any
remoteip=%s protocol=any interfacetype=any",
nIDTimer Event,bd->srcip);
        else
snprintf(cmd1,sizeof(cmd1)-1," /k netsh adv-
firewall firewall add rule name=Filter-%d
dir=out action=block enable=yes localip=%s
remoteip=any protocol=any interfacetype=any",
nIDTimer Event,bd->dstip);
case WFP_HOW_INOUT:
        if((bd->Mode&WFP_WHO)==WFP_
WHO_
SRC) {
snprintf(cmd1,sizeof(cmd1)-1," /k netsh adv-
firewall firewall add rule name=Filter-%d dir=in
action=block enable=yes remoteip=%s proto-
col=any      interfacetype=any",nIDTimeEvent,
bd->srcip);
snprintf(cmd2,sizeof(cmd2)-1," /k netsh adv-
firewall firewall add rule name=Filter-%d
dir=out action=block enable=yes remoteip=%s
protocol=any interfacetype=any",IDTimerEv-
ent, bd->srcip);}
        else {
snprintf(cmd1,sizeof(cmd1)-1," /k netsh adv-
firewall firewall add rule name=Filter-%d
dir=in action=block enable=yes localip=%s
remoteip=any protocol=any interfacetype=any",
nIDTimerEvent,bd->dstip);
snprintf(cmd2,sizeof(cmd2)-1," /k netsh adv-
firewall firewall add rule name=Filter-%d
dir=out action=block enable=yes localip=%s
remoteip=any protocol=any interfacetype=any",
nIDTimerEvent,bd->dstip);}
        break;}
```

2.4 *Append new WFP Filter*

The fourth step is to append new WFP Filter to the Filter queue which stores all the active and valid WFP Filter. The corresponding code is as follows.

```
FilterDesc=(Data *)malloc(sizeof(Data));
FilterDesc->TimerId=nIDTimerEvent;
strcpy(FilterDesc->FilterName,cmd);
FilterQueueHead=InsertList(FilterQueueHead,
FilterDesc);
OutputList(FilterQueueHead);.
```

2.5 *Delete the WFP Filter*

The final step is to delete the corresponding WFP Filter after timeouts to release the block and filter of suspicious network traffic. The corresponding code is as follows.

```
void CALLBACK TimeProc(UINT uID, UINT
uMsg, DWORD dwUser, DWORD dw1,
DWORD dw2)
{
TimeKillEvent(uID);
snprintf(cmd,sizeof(cmd)-1,"Filter-%d",uID);
netsh    advfirewall    firewall    delete    rule
name=Filter-uID;    FilterQueueHead=DeleteList
(FilterQueueHead,uID);}
```

3 LINKAGE COLLABORATION RULES

It is necessary to add linkage configuration to the conventional rule files of Snort so as to implement the collaboration. The following is the example of linkage collaboration rules which is modified based on the original icmp-info rule file.

alert icmp $EXTERNAL_NET any -> $HOME_NET any (msg: "ICMP PING"; icode:0; itype:8; detection_filter: track by_src, count 10, seconds 60; classtype: misc-activity; sid:384; rev:5; WFPFilter: SRC[in], 1 minitues;)

WFPFilter is a linkage collaboration keyword defined by the users themselves, it implements register through the function RegisterRuleOption ("WFPFilter ", AlertFWsecOptionInit, NULL, OPT_ TYPE_ ACTION).

The above Linkage Collaboration rule means that when the frequency of ping operation from internet to intranet more than 10 times per 60 seconds, Snort would cooperate with WFP to block the data packet from the IP address of initiating ping connection for one minute. It is obvious that the similar rules can mitigate and avoid DOS/DDOS (Jose Nazario 2008) attack effectively.

4 LINKAGE COLLABORATION RULES

Our team have analyzed the new Snort Rules Library snortrules2.8 (Khamphakdee & Nattawat 2014) and rewritten nearly 2000 rules. And the rules and Snort that have been modified were tested on the Windows 7, Windows Server 2008, Windows Server 2008R2 and Windows Vista platform respectively. The experimental result shows that the linkage collaboration module operates stably, which could interdict the intrusion and filter dangerous data packet effectively. The following experiment was implemented on Windows 7 Ultimate.

Figure 1.   Linkage collaboration test of ping.

From Figure 1. we can see that the host computer with IP address 192.168.1.6 sends continuously ping connections to the host computer with 192.168.1.72. When the ping action's frequency is more than 10 times per minute, the linkage collaboration module of rewritten Snort automatically create two WFP Filter to block the data packets which come from 192.168.1.6, which is described clearly in Figure 2.

Figure 2.   The WFP Filter created by collaboration module.

And the linkage collaboration mudule would delete the WFP Filter one minute later to release the block of the computer with 192.168.1.6.

5 CONCLUSIONS

With exhaustive tests and analysis, the experimental results demonstrate that the collaboration of Snort and WFP has been implemented based on Windows 7 successfully. When rewritten IDS based on Snort is triggered by suspicious network data, it can automatically create and configure WFP Filter to block the corresponding dangerous traffic. This novel method can insulate and mitigate network intrusions and attacks efficaciously without using the third party firewall (Liao Guangzhong 2009) and modifying Windows System Kernel (Yongqiang Zhang & Hai Bi 2011). This study extends intrusion response function of snort and would contribute to improve the security of the whole Windows 7 series network platform.

ACKNOWLEDGMENT

This research is sponsored and supported by Science and Technology Research Project of Heilongjiang Educational Committee under grant No.12531121. Any opinions or findings of this work are the responsibility of the authors, and do not necessarily reflect the views of the sponsors or collaborators.

REFERENCES

Qing-Xiu Wu (2012). The Network Protocol Analysis Technique in Snort. *J. Physics Procedia*, 25, 1226–1230.

HJ Liao & CH Richard Lin & YC Lin (2013). Intrusion Detection System: A Comprehensive Review. *J. Journal of Network and Computer Applications*, 36(01), 16–24.

Govind Kritika & Vivek Kumar & Selvakumar S. 2012. Pattern Programmable Kernel Filter for Bot Detection. *J. Defence Science Journal*, 62(03),174–179.

Huang Junsheng & Fan Bingbing 2013. Research and Realization of WFP-based Montoring System for Terminal Information Leakage. *J. Computer Applications and Software (China)*, 30(03), 315–318.

Brian Caswell & Jay Beale & Andrew Baker 2006. Snort Intrusion Detection and Prevention Toolkit. M. Chapter 2, Introducing Snort 2.6, 31–67.

Jose Nazario 2008. DDoS attack evolution Original. *J. Network Security*, (07), 7–10.

Khamphakdee & Nattawat 2014. Improving intrusion detection system based on Snort rules for network probe attack detection. *Proceeding of the 2nd International Conference on Information and Communication Technology*, 9–74.

Liao Guangzhong 2009. Linux-based Network Intrusion Defense System Research Design. *J. Computer Technology and Development (China)*, 18, 134–136.

Yongqiang Zhang & Hai Bi 2011. Anti-rootkit Technology of Kernel Integrity Detection and Restoration. C. *Proceeding of International Conference on Network Computing and Information Security*, 01, 276–278

Electronic Engineering and Information Science – Wang (Ed.)
© *2015 Taylor & Francis Group, London, ISBN: 978-1-138-02772-5*

Research on building multi-campus adult education platform model based on VPN technology

Z.Y. Luo & Z.W. Qin
School of Computer Science and Technology, Harbin University of Science and Technology, Harbin, Heilongjiang, China

ABSTRACT: Building adult education platform network in a university with multi-campus becomes difficult and unsafe. The paper proposes a model to build it based on VPN technology. The networking model uses the IPSec VPN to solve the problems of the data secure communication and network connection between the various campuses, and uses SSL VPN to resolve the security problems for the external person using electronic resources in an adult education platform. The paper uses DynamipsGUI to simulate networking model. The simulating result shows that this model has a good maximum number of connections and network throughput, and it cans the data transporting safely.

KEYWORDS: VPN; IPSec VPN; SSL VPN; Adult Education.

1 INTRODUCTION

In the process of the formation of multi-campus adult education information platform, there have some problems such as software and hardware facilities inadequate. These problems do not make the security network connect. It also makes the multi-campus adult education resource not share securely. A VPN is a virtual technology. It can establish a secure data transmission channel in an open public network. It expands the campus private network greatly(Luo Z.Y. 2011). We use the VPN technology to build the multi-campus adult education information platform. It not only ensures the secure transmission of data, but also saves the cost of accessing remote trainees.

2 TECHNICAL ANALYSIS OF BUILDING A VPN NETWORK MODEL

2.1 *VPN technology and principle*

Currently, the realization of VPN network technology has mainly two measures (Hu W.N. 2013): IPSec VPN and SSL VPN.

1 IPSec VPN. IPSec VPN builds side to side network communications on the IP layer. The principle is similar to the packet filtering firewall. It handles the data packets security through using the encryption, authentication and integrity check technology. It ensures the security of data communications (Liu Z.L. 2013).

2 SSL VPN. SSL VPN uses SSL protocol to achieve point to side network communications. This technique uses the security control policies for mobile clients to build a secure access channel to visit the network resources which is from outside to inside (Luo H.Q. 2013).

2.2 *Key problems of building the multi-campus adult education VPN network*

According to the characteristics of multi-campus in the author's college, the following key issues of the establishment of multi-campus adult education VPN networks need to be solved (Luo Z.Y. 2012):

1 The problems of users visiting various types of server in the school inside the network. Because of the college has many variety servers in the inside network, how to control the assign permissions of inside and outside users visiting the electronic resources of the variety servers is a worth considering question.

2 The access problems of the mobile client. Because of teachers or staff traveling, how to let they visit the school network more safety and fast is a very worth considering problem.

3 The problems of the remote access easily. Complex configuration, program setup, and equipment operation will add the maintenance of the managers and the network cost. Teachers and students are away from the campus network, and their computer skills are not the same. If more complex and

requiring a lot of efforts to learn the entire access process, it will increase the burden on the teachers and students. It reduces the availability of the plan.

3 A MODEL OF THE MULTI-CAMPUS ADULT EDUCATION PLATFORM BASED ON VPN NETWORK

According to the characteristics of multi-campus in the author's college, the authors' design the network topology, which is shown in Figure 1.

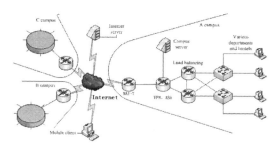

Figure 1. Topology of multi-campus adult education VPN Network.

4 SIMULATION OF MULTI-CAMPUS ADULT EDUCATION VPN NETWORK

4.1 *The topology of simulation*

According to the design requirements, we choose the DynamipsGUI simulator to simulate the communication process of the entire network which is based on Windows 7 system. In DynamipsGUI, a campus network consists of R1, R2,SW1, SW2, SW3, SW4, C1, C2, R8, R9, B campus network consists of R4, R5, and C campus network consists of R6, R7. The Internet uses the R3 to simulate. The topology simulation result is shown in Figure 2.

4.2 *Multi-campus adult education VPN network simulation*

In the main network addresses, the address of router R1 is 202.168.101.1 and the address of router R2 is 202.168.101.2, and the cores of the pseudo-code are as follows:

```
vpn(config)#int f0/0
vpn(config-if)#ip add 192.168.10.1 255.255.255.0
vpn(config)#int f1/0
vpn(config-if)#ip add 202.168.102.1 255.255.255.0
vpn(config)#int lo0
vpn(config-if)#ip add 1.1.1.1 255.255.255.0
vpn(config-if)#no sh
vpn(config-if)#exit
```

Router(vlan)#vtp domain lzy@com Changing VTP domain name from NULL to lzy@com

4.3 *IPSec VPN simulations*

In the ISAKMP / IKE Phase, we need to complete the following configuration:

1 Configure crypto ACL. We use the crypto ACL to match IPSec VPN network flow.
2 Configure crypto map. Its function is to organize all the information together to build IPSec sessions. Router interface has only one Crypto Map, but a router can achieve flow protection in multiple interfaces, then it may need more than one Crypto Map.

IPSec VPN core pseudocode between R1 and R4 are as follows:

```
Router(config)#crypto isakmp policy 1
Router(config-isakmp)#group 1
Router(config)#crypto isakmp key password
address 10.20.0.2
Router(config-crypto-map)#match address 110
```

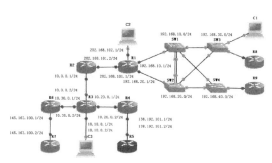

Figure 2. The topology simulation in DynamipsGUI.

4.4 *SSL VPN simulations*

SSL VPN is mainly used to complete that the function of the traveling users to access inside network resources. And its implementation steps are as follows:

1 We set up an inside network server, and its IP address is 202.168.102.2. Then we set up DNS servers and Web servers.
2 Configuration prerequisite: AAA, DNS and certificate. AAA is used to authenticate Web VPN users; DNS is used to resolve the name information of the URL and gets a SSL certificate for the router, the certificate is used to protect user data between the desktop and the router.
3 Create a URL and port forwarding entry for home page.
4 Maintenance, monitoring and fault diagnosis and troubleshooting of Web VPN connection.

5 SIMULATION RESULTS

After completion of the multi-campus adult education VPN network, the network research groups carry out the following tests.

1 Connections number tests: Authors used Loadrunner software testing tools. Authors test the maximum connections number of the multi-campus adult education VPN network servers in 1 second, and the test results are shown in Figure 3. Through the analysis, the average of the maximum connections number is 600 in the VPN, which can meet the application needs of the campus.

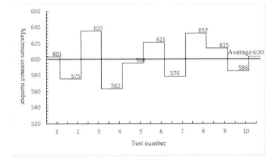

Figure 3. Connections number texts.

2 Throughput tests: The authors select the Win7 system and 10Mb/s network card on the host. On this host, the authors test the network throughput in the IPSec VPN mode and in the SSL VPN mode. We use data 110088018Byte to test, and simulation results are shown in Figure 4. Through the analysis, the average throughput of the SSL VPN is 397.36KB/s, and the average throughput of IPSec VPN is 615.03KB/s.

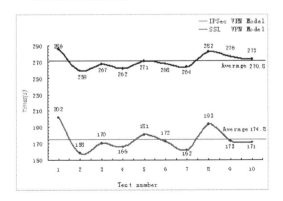

Figure 4. Throughput of model.

6 CONCLUSIONS

This paper presents a VPN network model for building multi-campus adult education, using this technology can better solve some multi-campus adult education issues. This network form uses IPSec VPN to guarantee teaching data communication safety between the different campuses of adult education, meanwhile using SSL VPN to ensure communications security for migrant workers or students to use teaching resources in an adult education platform, which provides needs of building the multi-campus adult educational.

Supported by Foundation of Heilongjiang Educational Committee (NO:12521108).

REFERENCES

Luo Z.Y, Z.H Duo& P.L Qiao (2011). Formal description of IPSec security policy in VPN networks. *Journal of Huazhong University of Science and Technology(Natural Science Edition).4*, 65–68.
Hu W.N(2013). Key technologies of VPN test in IPv6 environ ent. *J. Information Security and Technology 4*, 56–59.
Liu Z.L &W.D Ren (2013).Research and development of communication power monitoring system based on VPN technology. *Chinese Journal of Power Sources 37*, 141–143.
Luo H.Q&R.H Nie (2013). Research of SSL VPN secure access to campus network based on the IVE. *Computer Security.2*, 41–46.
Luo Z.Y, G.L Sun, J.H.Liu &W.B.Wang (2012). Application of attack graphs algorithms in invasion prevention system. *Journal of Yunnan University (Natural Sciences Edition), 34*,271–275.

Electronic Engineering and Information Science – Wang (Ed.)
© 2015 Taylor & Francis Group, London, ISBN: 978-1-138-02772-5

Extract Chinese term based on linguistic knowledge

D.S. Sun

School of Computer Science and Technology, Harbin University of Science and Technology, Harbin, China

ABSTRACT: Term extraction is an important task in natural language processing. It has been key factors which influence many systems' performance. In order to improve the extraction accuracy, a new method of term extraction is proposed in this paper, in which Chinese linguistic knowledge is used. At the same time, the architecture of term extraction is given. Part of speech sequences are adopted as patterns to extract candidate terms from the corpus. Compactness degree is constructed to extract terms. In experiments, the proposed method is used to extract terms from Chinese sentences. Experimental results show that the accuracy of the proposed method is improved.

KEYWORDS: Term Extraction; Natural Language Processing; Linguistic Knowledge; Part of Speech.

1 INTRODUCTION

Term extraction is important for natural language processing, and widely applies to text classification, information retrieval and lexicon construction. A hybrid system of term extraction is given, which integrates a linguistic filter on Ezafe construction with a statistical filter on C-value algorithm (Hossein 2013). A model-driven approach of term extraction is presented in which business vocabularies are extracted from databases of those existing software systems. At the same time, a transformation framework is given to obtain knowledge discovery metamodel (Kestutis 2014). A new algorithm is proposed for extracting terms based on the analysis of web resources and user behaviors. Body of page data, anchor texts and the information about user query data are adopted to get terms in specific domains (Yan 2013). A method is given to find relationships between two terms based on snippets which are extracted from Wikipedia. Connectors between the two terms are applied to find these snippets in which terms may not only occur in both articles, but also link both articles (Mathiak 2013). A tunable term extractor is developed in which linguistic-based rules and reuses of existing terminologies are combined (Aubin 2006). A new method is presented to extract glossary terms automatically from unconstrained natural language requirements in which linguistic techniques are introduced to identify process nouns, abstract nouns and auxiliary verbs (Dwarakanath 2013). Topics in corpus are extracted based on latent dirichlet allocation, in which variational inference and expectation-maximization algorithm are applied to estimate model parameters (Yu 2012). A verb-object phrase extraction algorithm is proposed for a specific domain (Liu 2011). Firstly, a context-based method is applied to extract separators. Secondly, separators and contextual terms are combined to acquire verb-object phrases, in which the improved NC-value term extraction algorithm is applied. A new Chinese term extraction method is given in which part-of-speech analysis and string frequency statistics are combined to find terms in documents (Yu 2010). A systematic framework is presented to recognize and extract Chinese terms based on hidden Markov model (HMM) from a specific domain. Training corpus is roughly segmented and every word is tagged with part-of-speech, from which HMM parameters are learned (Cen 2008). An opinion target network is given for modeling opinion patterns in two layer graphs and a unified collocation framework for extracting opinion collocations is presented (Xia 2009). A statistical term extraction method is proposed for specific domains in which information entropy is adopted (Liu 2007).

A new approach of term extraction based on linguistic knowledge is proposed in this paper. The frame of term extraction is described, in which word segmentation tool, part-of-speech tagging tool and linguistic resources are used. Term patterns are applied to extract candidate terms from the corpus. The evaluation function of terms is given.

2 ARCHITECTURE OF EXTRACTING CHINESE TERMS

The term is a word or a phrase that describes general concepts in a special domain. It is often used in any scientific domain, which describes concepts or relationships in this domain. For example, information retrieval.

A part of speech is a linguistic category of words and definitions by syntactic or morphological behavior of lexical items including noun, verb, adjective and etc. It is used to classify words better by function. Chinese term is often comprised of several isolated words. But, part of speech for every Chinese word is still identified. So, a term can be determined by several isolated units with its component word's part of speech. Part of speech labeling sequences can be applied to find terms from Chinese texts. For every Chinese word, its part of speech is located in lexicon which is a kind of linguistic resources. The frame of extracting terms based on linguistic knowledge is shown in figure 1.

Firstly, stopping words are deleted from Chinese corpus, in which stopping word table is used. Stopping words are characters whose frequency are high and whose contribution to term extraction is low. These words have little information. For example, Chinese word liao, shi, and etc. Secondly, illogical words are filtered from Chinese corpus, in which illogical word table is adopted. Linguists are invited to collect illogical words and compile an illogical word table in practice. Illogical words refer to characters which cannot be used when terms are extracted from the corpus. For example, digits, hyphens, punctuations and etc. Thirdly, each sentence in Chinese corpus is segmented into a sequence of words. In order to segment sentences, Chinese lexicon and statistics word table are used. But, new words appear with the development of human language. These new words cannot be found in Chinese lexicon. Statistical methods are

adopted to obtain new words from a text corpus, including newspapers, Internet and documents. When a new word's frequency is over a threshold, it is added into a statistics word table. With lexicon resources and statistics word table, more Chinese words are segmented and the precision of word segmentation is improved. Fourthly, part-of-speech tagging tool is applied to label words in Chinese sentences. Every word is labeled with its part of speech. Then, part of speech sequences are obtained.

Terms' components are often satisfied with some patterns. For example, adjective+noun, noun+noun, adjective+noun+noun, noun+noun+noun and etc.

For the following Chinese sentence, the process of extracting terms is shown as shown:

Chinese sentence: bai zuo zi shang fang le yi ge ping guo bi ji ben dian nao bao .

Word segmentation results: bai/ zuo zi/ shang/ fang/ le/ yi/ ge/ ping guo/ bi ji ben/ dian nao/ bao/ ./

Part of speech tagging results: bai/a zuo zi/n shang/f fang/v le/u yi/m ge/q ping guo/n bi ji ben/n dian nao/n bao/n ./w

Matched patterns: a+n, n+n, n+n+n, n+n+n+n

Candidate terms: bai/ zuo zi/, ping guo/ bi ji ben/, bi ji ben/ dian nao/, dian nao/ bao/, ping guo/ bi ji ben/ dian nao/, ping guo/ bi ji ben/ dian nao/ bao/

3 CHINESE TERM CLASSIFIER

An important characteristic of the term is that it occurs frequently in the corpus. If two or more Chinese words co-occur frequently, the probability that they constitute a term is higher. So, frequency can be used to measure the probability that a sequence of Chinese words is a term. For a sequence of Chinese words $S=w_1, w_2, ..., w_n$, its frequency is denoted as $freq(S)$. It can be estimated from a large corpus, which is computed as shown in formula (1). Here, $count(S)$ is the number that S appears in the corpus and n is the number of Chinese words in the corpus.

$$freq(S) = \frac{count(S)}{n} \qquad (1)$$

For Chinese word sequence S, the compactness degree between its component words is also an important indicator. If theses n words constitute a phrase compactly, the probability that S is a term is higher. The compactness degree of word sequence S is computed as shown in formula (2).

$$CDeg(S) = \log_2 \frac{freq(S)}{freq(S_l) freq(S_r)} \qquad (2)$$

Here, $S_l=w_2, w_3, ..., w_n$, $S_r=w_1, w_2, ..., w_{n-1}$. If relationships among $w_1, w_2, ..., w_n$ are very compact,

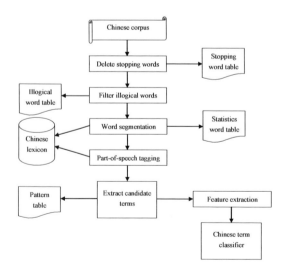

Figure 1. The frame of extracting terms based on linguistic knowledge.

there is little difference between $freq(S_l)$ and $freq(S)$. At the same time, the difference between $freq(S_r)$ and $freq(S)$ is little. So, the value of $CDeg(S)$ is larger. If the difference between $freq(S_l)$ and $freq(S)$ is large or the difference between $freq(S_r)$ and $freq(S)$ is large, the value of $CDeg(S)$ is less.

Here, $freq(S)$ is normalized and discriminative feature $freq_{norm}(S)$ is gotten. If the value of $freq_{norm}(S)$ is larger, the probability that S is a term is higher. $CDeg_{norm}(S)$ is normalized and discriminative feature $CDeg_{norm}(S)$ is gotten. If the value of $CDeg_{norm}(S)$ is larger, the probability that S is a term is higher.

4 EXPERIMENTS

100 Chinese sentences are collected and applied to acquire terms. Two human linguists are asked to label terms in these 100 Chinese sentences. Firstly, stopping words are deleted from theses 100 Chinese sentences. Secondly, illogical words are filtered from these sentences. Thirdly, the word segmentation tool applies to segment Chinese sentences and the corresponding word sequences are obtained. Fourthly, part-of-speech tagging tool is used to label every Chinese word. According to pattern table, candidate terms are extracted. For candidate term S, its frequency $freq(S)$ and compactness degree $CDeg_{norm}(S)$ are respectively computed. Then, they are normalized. Discriminative features $freq_{norm}(S)$ and $CDeg_{norm}(S)$ are gotten.

In order to evaluate the performance of the proposed method in this paper, two groups of experiments are conducted. In experiment 1, discriminative feature $freq_{norm}(S)$ is used to compute the evaluation score of candidate term S. In experiment 2, discriminative feature $CDeg_{norm}(S)$ is utilized to calculate the evaluation score of candidate term S. Then, the top 500 terms whose evaluation scores are highest are selected as terms. Automatically-labeled terms are compared with human-annotated terms. Then, accuracy is adopted to measure the classification performance. Accuracies in two groups of experiments are described respectively in Table 2.

Table 2. Performances in two experiments.

Experiment number	Accuracy(%)
1	50.0%
2	51.0%

From table 2, it can be seen that when discriminative feature $freq_{norm}(S)$ is used, its accuracy is 50.0%. When discriminative feature $CDeg_{norm}(S)$ is used, its accuracy is 51.0%. The accuracy in experiment 2 is higher than that in experiment 1. This shows that the discriminative ability of $CDeg_{norm}(S)$ is better than that of $freq_{norm}(S)$.

5 CONCLUSIONS

Term extraction is an important research topic in natural language processing. In this paper, linguistic knowledge is applied to acquire terms automatically from corpus, including Chinese lexicon, stopping word table, illogical word table, statistics word table, pattern table, word segmentation tool and part-of-speech tagging tool. At the same time, the architecture of term extraction is described. Frequency and compactness degree are utilized to evaluate the degree that a word sequence is term. Experimental results show that the accuracy of the proposed classifier achieves at 51.0%.

ACKNOWLEDGMENTS

This work is supported by Science and Technology Research Funds of Education Department in Heilongjiang Province under Grant Nos. 12531106.

REFERENCES

Hossein B. F. (2013). A reliable linguistic filter for farsi term extraction. In *Proc. 2013 5th Conference on Information and Knowledge Technology*, pp. 328–331.

Kestutis N. (2014). Extracting term units and fact units from existing databases using the knowledge discovery metamodel. *Journal of Information Science*. 40(4): 413–425.

Yan X. L., Liu Y. Q., Fang Q., Zhang M., Ma S. P. and Ru L. Y. (2013). Domain-specific terms extraction based on web resource and user behavior. *Journal of Software*. 24(9): 2089–2100.

Mathiak B. and Peña V. M. M. (2013). Extracting term relationships from Wikipedia. In *Proc. Web Information Systems and Technologies-8th International Conference*, pp. 267–280.

Aubin S. and Hamon T. (2006). Improving term extraction with terminological resources. In *Proc. of Advances in Natural Language Processing 5th International Conference on NLP*, pp. 380–387.

Dwarakanath A. (2013). Automatic extraction of glossary terms from natural language requirements. In *Proc. 2013 21st IEEE International Requirements Engineering Conference*, pp. 314–319.

Yu J., Wang J. L. and Zhao X. D. (2012). An ontology term extracting method based on latent dirichlet allocation. In *Proc. 2012 4th International Conference on Multimedia and Security*, pp. 366–369.

Liu L. (2011). A verb-object compound extraction algorithm based on separator and contextual term. *Journal of Computational Information Systems*. 7(10): 3415–3421.

Yu J. and Dang Y. Z. (2010). Chinese term extraction based on POS analysis & string frequency. *System Engineering Theory and Practice*. 30(1): 105–111.

Cen Y. H., Han Z. and Ji P. P. (2008). Chinese term recognition and extraction based on hidden markov model. In *Proc. 2008 Pacific-Asia Workshop on Computational Intelligence and Industrial Application*, pp. 219–224.

Xia Y. Q., Hao B. Y. and Dai L. L. (2009). Term extraction from web reviews with opinion heuristics. In *Proc. 2009 International Conference on Machine Learning and Cybernetics*, pp. 3516–3521.

Liu T., Liu B. Q., Xu Z. M. and Wang X. L. (2007). Automatic domain-specific term extraction and its application in text classification. *Acta Electronica Sinica*. 35(2): 328–332.

Electronic Engineering and Information Science – Wang (Ed.)
© 2015 Taylor & Francis Group, London, ISBN: 978-1-138-02772-5

On the development of a novel music player with RGB-LED

Z.Q. Hu, H. Zhang, G.Y. Lin, R.R. Zhao & H.B. Wang
School of Computer Science & Technology, Harbin University of Science & Technology, Harbin, China

ABSTRACT: A novel music player with Light Emitting Diode (LED) lights is introduced in this paper. The device possesses U disk and SD card to store data of audio files which are read under the microprocessor control and sent to an audio decoder VS1003 whose audio outputs drive two speakers. In order to solve the problem of non-correspondence between music and color while the colorful music shows, the paper investigates the relations between music chords and color on the basis of research on music and color, through the spectrum analysis and algorithms, to determine the corresponding color, by circuit to control light color, at least to achieve real-time accurate music and color matching. The Fast Fourier Transform (FFT) method is optimized to a suite for the application. File system compatible with FAT32 is implanted in the microprocessor system. Colors of LED lights correspond with played music.

KEYWORDS: Audio signal; Microprocessor; Frequency spectrum analysis; RGB.

1 INTRODUCTION

Sound and color have induction in common with each other, which is called synesthesia (Grossenbacher 2001, Schubert, 2007). Color music includes music, optics, electronics and other disciplines, and then develops a new musical visual form; the color music playback device has more application in the urban landscape lamp. Now some audio players can only play music, although some of them have light color display, but music and light color have no relevance. Music emotion and the color are difficult to be corresponded with, while some reports show that colors of lights flash randomly, and several others with fixed colors are also carried out by the color of the light according to pre-programmed procedures. But the one with the perfect combination of music and colors, has not been reported. In order to solve the problem of non-correspondence between music and color while the colorful music shows and achieve the effect of color that coincides with the playing music. This paper establishes the corresponding relations between music chords and colors on the basis of research on music and color, through the spectrum analysis, and research on music chords and playing music, then the similar chords with playing music and the corresponding color are entirely determined, by circuit to control light color, at least to achieve real-time accurate music and color matching, and good color rendering of the music.

2 SYSTEM RESEARCH METHOD

2.1 *Music chords spectrum analysis*

Alone to the spectrum of playing music, it is not sure of the corresponding color. Therefore the analysis of the standard of music chords is needed to carry out. According to the statistics, among all kinds of music, the total chord as many as thousands of species, but the amount of 90% of modern music chords is only 20 kinds, based on the study of music research, through the statistics, these common music chords on behalf of the color is determined. It is hard to evaluate audio waveforms in the time domain. But the frequency spectrum graph can reflect some distinctive characteristics, shown as in Figure 1.

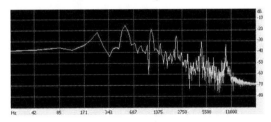

Figure 1. Frequency spectrum diagram.

2.2 *Spectrum comparison*

Method for the playing music, through A/D sampling, its audio signal does fast Fourier transform

(FFT) conversion, and then obtain its spectrum. The Decimation-in-time radix-2 FFT algorithm, which is one of the FFT efficient algorithm, is chosen to do spectrum analysis, FFT algorithm is faster than the Fourier transform DFT, FFT mainly use the symmetry and properties of twiddle factor, FFT make DFT algorithm be decomposed into less points DFT, thereby improve the algorithm and reduce the computational complexity (Zhang, 2013). The spectrum of playing audio signal compares with the spectrum of sampled music chords, that is, through the method of frequency comparison, it is determined that the spectrum of an audio signal is similar to or corresponding with what kind of music chords spectrum, and then determine the corresponding color for the playing music. Bayesian formula, with its deeper ideological features, is widely used in scientific research. The natural state θ has k kinds, $\theta_1, \theta_2, \theta_3 \ldots \ldots \theta_k$. $P(\theta_i)$ represents the natural state θ_i, the prior probability $P(x \mid \theta_i)$ represents the state condition for θ_i, the probability of events for x. $P(\theta_i / x)$ is the posterior probability of the occurrence of θ_i. Total probability P is the probability of x that may appear in various states, that is,

$$P(x) = \sum_{i=1}^{k} P(x \mid \theta_i) P(\theta_i) \tag{1}$$

Bayes formula is used to construct the probability model, according to the frequency spectrum of the audio signal. It is easy to confirm the corresponding music chords for playing the audio signal via statistical calculation to realize the spectrum comparison between sampling audio signal and standard music chords.

3 THE DESIGN OF EXPERIMENTAL SYSTEM PLATFORM

3.1 The overall design

System component includes an embedded microprocessor control system, in the process of MPU selection, from cost efficient, now the widely used MPU are the STC series, such as STC12C5A60S2. Implementation of audio file operation and reading, through the audio decoder to play audio files is the main function of the system. The key components of the system are processor MPU, memory, and audio decoder chip (H. Zhang, 2013).

Audio files require large capacity storage, so SD/TF cards are used, and MPU connects with SD/TF card interface via SPI synchronous serial bus (Y. J. Zhu, 2010). And the U disk is widely used by small volume, lightweight, hot-swappable, and rewriting. U disk data also plays a role of commonality of USB protocol, so that the U disk realizes the portable

features, which is more suitable for audio file storage. CH375 is a general USB bus interface chip, which has 8-bit data bus and read, write, chip select control lines and interrupt output. CH375 also provides serial communication via serial input, output, and interrupt output and MCU / DSP / MCU / MPU to connect (L. J. Wang(2007). MPU hooks up to SD / TF card through the SPI bus, and reads audio files from the U disk via CH375, then the audio files are sent to an audio decoder VS1003(Deng Xueming 2013). CH375 interface circuit is as shown in Figure 2.

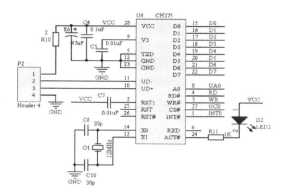

Figure 2. CH375 interface circuit.

3.2 RGB-LED driver design

RGB means red, green, blue, the three primary colors of light, each of them has a different wave length, according to the International Commission on Illumination CIE in 1931, and it is defined by the standard color matching function that the red, green and blue LED wavelength is respectively set to 700 nm, 546.1nm, 435.8 nm. Through experimental analysis, the geometric center of an isosceles right triangle is regarded as the observation point. Considering about the uniformity, firstly, symmetrical distribution

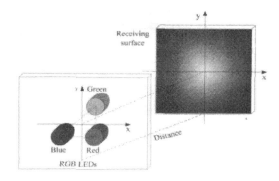

Figure 3. RGB-LED relative positions in an isosceles right triangle.

is necessary, and the RGB are located in the three vertices of the triangle respectively in chromaticity diagram, then the several triangles are simulated, and the mixed white light speckle has a large range and better uniformity with an isosceles right triangle, so the isosceles right triangle is chosen in the RGB array shape selection. Location is as shown in Figure 3.

The three kinds of LED luminous tubes with the color of red, green and blue are chosen as the light source, through the PWM output to drive three LED light-emitting tubes, then a variety of colors will be seen, which forms a mixed light effect.

3.3 File system transplantation

Since the audio data is stored in file format on an SD card and U disk, so it is convenient to download audio files from the Internet, the computer, so in the ARM system must be able to perform file system operations. The most basic ideas is to use the FAT file system to record the location of the data file by using the chain structure, this allows storage of file data cannot be continuous, thus the utilization rate of the disk is greatly increased. The FAT file system is divided into different versions, with the latter figure to distinguish, they respectively are FAT12, FAT16, FAT32, the different versions, supports different partition size, cluster size, etc. ZnFAT file system makes the function of reading and writing on the SD card in the bottom of the sector to abstract the interface between the FAT32 file system and the underlying storage device drive.

4 CONCLUSIONS

Although there are many researches on the relationship between music and color, the most of them still remain in the theoretical analysis, while the problem of mismatch between playing music and color has not been resolved. And this paper adopts the method of spectrum comparison to solve the problem of matching between playing music and color. Spectrum analysis is an effective tool to grasp the audio signal, but only by the spectrum of the playing audio signal, it is still difficult to determine the corresponding color. Through music chords spectrum and playing audio frequency spectrum, spectrum comparison method, identification of the similar chords with the playing audio signal to determine the corresponding color of the playing music. The wavelet transform is used to effectively eliminate the high frequency interference,

improve the spectrum analysis and the accuracy of spectrum identification. Through the experiment testing, it reaches the RGB-LED mixed light and playing music in real-time matching effect. Patterns of RGB-LED lights mixed matching music playing is shown in Figure 4.

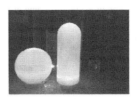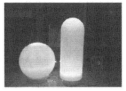

Figure 4. RGB - LED mixed light and playing music matching test.

ACKNOWLEDGMENT

This work is partially supported by Heilongjiang Province Undergraduate Training Programs for Innovation and Entrepreneurship (201410214036), and Project of Education Department of Heilongjiang Province (12541177).

REFERENCES

Grossenbacher P G, Lovelace C. T.(2001). Mechanisms of synaesthesia: Cognitive and physiological constraints. *Trends in cognitive sciences*, 36–41.

Schubert E.(2007). The influence of emotion, locus of emotion and familiarity upon preference in music. *Psychology of Music*, 499–515.

H. Zhang, C. W. Tian, G. Y. Lin(2013). Implementation of a wireless controlled device with RGB driving based on FFT. *International Journal of Multimedia and Ubiquitous Engineering*, 151–158.

H. Zhang, G. Y. Lin(2013). Design and implementation of mixed light effect based on the audio signal frequency spectrum analysis. *Proc. of 8th International Forum on Strategic Technology 2013, IFOST 2013*, 68–74.

Y. J. Zhu, Y. Luo(2010). The audio broadcast system based on SOPC. *Audio Engineering*, 34–36.

L. J. Wang(2007). The design of MP3 player based on U disk with CH375. *Modern Electronics Technique*, 58–60.

Deng Xueming, Liu Qiaowei, Yan Dongsong, Zhang Yifei, and Fu Kanjun(2013). A Design of Automatic Bus Station Reporter Based on VS1003 and SD card. *Proc. of 2013 International Conference on Structures and Building Materials*, 2847–2850.

Electronic Engineering and Information Science – Wang (Ed.)
© 2015 Taylor & Francis Group, London, ISBN: 978-1-138-02772-5

Research on classification and technology of honeypots

J.Q. Zhai, L. Ding, J.H. Liu & Y.J. He
School of Computer Science and Technology, Harbin University of Science and Technology, Harbin, China

ABSTRACT: The honeypot is a network deception system under strict surveillance based on the inveiglement theory, which could attract attacks by real or virtual network and services so as to analyze the blackhat's activities during honeypot being attacked by hackers, protect the genuine productive network. Nowadays many honeypot approaches have been proposed and applied in practice. However, there is no classification of these existing approaches to understand their common characteristics and limitations. In this paper, we first analyze the existing approaches and technology of honeypots, then based on these research we present a classification of the state-of-the-art honeypot approaches. The study will enable practitioners and researchers to differentiate among existing honeypot approaches and provide a guideline to develop new honeypot approaches.

KEYWORDS: Honeypot; Interaction; Deception; Classification.

1 INTRODUCTION

Unlike the conventional defensive means for attack detection such as Intrusion Detection Systems which are based on knowledge about specific attack patterns and thus are designed to detect known attacks by detecting signatures or anomalies, honeypots technology represents a different and complementary approach. Honeypot is a new proactive network security technology based on the inveiglement theory which could attract attacks by real or virtual network and services so as to analyze the blackhat's activities during honeypots being attacked by hackers, delay and distract attacks in the meantime (Zhuge Jian-Wei & Tang Yong & Han Xin-Hui 2013). The honeypots are valuable for researching and developing new IDS signatures and rules, analyzing new hacker's attack tools, detecting new methods of stealthy communications and Distributed Denial of Service (DDoS) tools (Mairh Abhishek & Barik Debabrat & Verma Kanchan 2011).

Currently many honeypot approaches have been proposed and applied in practice. However, there is no comprehensive classification of these approaches to understand their common characteristics and limitations. In this paper, we present a classification of the state of the art honeypot approaches employed for alluring and dispersing attacks, monitoring networks, capturing and analyzing of network traffic, extracting and generating attack signature. We outlined the differences between the basic honeypot types and the corresponding trade-offs. Based on these analyses we

created a taxonomy of honeypots which considers the metrics: role in architecture, level of interaction, deployment and targeted attack vector. Our classification distinguishes between LI server honeypots, HI server honeypots, LI client honeypots and HI client honeypots.

2 HONEYPOT TECHNOLOGY

Honeypots could offer a broad spectrum for using the findings of the observed attacks in order to take appropriate countermeasures. For instance, various honeypots could be used for long-term gathering of attack data, capture of zero-day exploits which are attacks using undisclosed attack vectors and vulnerabilities, the creation of signatures for the detected novel attacks for future detection, observing botnets, automated collection of malware binaries and enabling the identification of infected systems. Moreover honeypots produce much less information than conventional means for attack detection, such as IDS, while the gathered information has more valuable at the same time. Therefore the honeypots are increasingly accepted as a legitimate and valuable sensor for attack detection complementing traditional security measures, also in productive environments.

The honeypots are not a standardized solution dedicated to solve a specific problem but rather a highly flexible and effective means to detect and track abusive and malicious activities. Depending on the scope and the use case the honeypots can be deployed in many ways and act in various roles. Therefore a couple of technologies have been used and implemented in honeypots, which are often combined on demand. The most common ones are as follows.

2.1 Darknets

A darknet is commonly also referred to as network telescope, which is a portion of allocated and routed address space while no hosts or services reside in it (Simon Woodhead 2012). Thus all the traffic hitting the addresses within the darknet is observed and any packet can be treated as suspicious by its presence. Therefore the traffic may be forwarded to one or more honeypots hosts which handle and manage this traffic. It offers an insight into unwanted network activities which may result from misconfiguration or malicious actions such as DDoS backscatter which occurs as a side effect when spoofed IP addresses are used, port scans or malware propagation.

2.2 Vulnerability emulation

Instead of just passively monitoring the address space like darknet, the honeypot may pretend having a specific vulnerability. It does so by behaving like a given OS when it is probed remotely using TCP/IP fingerprinting (Medeiros Paulo & De Medeiros Brito Jr Agostinho & Motta Pires Paulo 2011) techniques announcing a given software and version. Therefore it is not necessary to emulate a full service or platform but sufficient to emulate the parts of computer system indicating the vulnerability which is necessary to trick the attacker into believing that he interacts with such a real vulnerable system. This approach goes one step further by provoking an interaction with an attacker or malware, thus it could gather more information than passively monitoring the network traffic merely.

2.3 Rootkit techniques

The central collection, capture and control of the data gathered by a honeypot are fundamental in order to ensure the data's integrity. Possible modifications to the honeypot system made by an attacker or a malware may include changes at low system levels. Therefore it is necessary that components for data collection and analysis utilized by honeypot operate at equally interior level in order to remain undetected and unmodified. Therefore several means are used such as a transparent bridge or the rootkits kernel module (Arati Baliga & Liviu Iftode & Xiaoxin Chen 2008) to log all actions and secretively send the data to another host.

2.4 Tarpits

A tarpit is a service intended to delay network connections as long as possible. Within the context of honeypots this technique may be utilized to slow down outgoing connections initiated by malware in order to mitigate further propagation or other malicious activities. Another usage is to slow down outgoing SMTP connections of a honeypot acting as open relay (so-called SMTP-tarpit). This makes Spam delivery unattractive while still being able to monitor abusive activities performed via the open relay.

2.5 Deception

In order to distract an attacker, the various deception techniques may be applied in honeypots to trick the attacker into believing that he is on a real system. The intention is to hold up this illusion as long as possible and thus to track preferably much of his activities. Therefore some data that seem to be true, such as fake accounts, files, database entries or log files, is placed on the honeypot. As there is no legitimate usage of this data this approach equals the honeypot concept. Accordingly this type of data is commonly referred to as honeytokens (Dominic Storey 2009).

2.6 Auxiliary software

In addition to the generic techniques described above several tools have been deployed adding novel and useful capabilities to honeypots, which are called auxiliary software. This refers to software that is not a dedicated honeypot or a software specifically designed to support honeypot operation but either a commodity software. For examples, the auxiliary software include QEMU (Chandra Shekar & Bhukya Wilson Naik 2011) which is a processor emulator commonly used for virtual machines and libemu (Georg Wicherski 2010) which provide x86 microprocessor architecture emulation and shellcode detection within honeypots.

3 CLASSIFICATION OF HONEYPOTS

Within the last years honeypots have received a lot of attention in the research community. Therefore various honeypots approaches have been introduced which are classified in this sections. The honeypots can be classified by several aspects:

- Whether client- based or server-based.
- Level of interaction with attacker.
- Whether deployed within a virtual or on a physical machine.
- Whether running on a productive system or not, respectively corresponding research and production honeypots.

Basically there are two types of honeypots with fundamental different approaches:

Server honeypots are the traditional honeypot variant providing a complete or partial vulnerable machine such as a given service or OS and wait passively for incoming attack attempts.

Client honeypots are referred to as honeyclients, which actively search malicious systems by visiting websites acting as a vulnerable system in order to discover whether or not they experience attack attempts.

Today predominant attack vector is client-based and therefore sensors that are capable of detecting this type of attack are required. Server honeypots are not able to detect client-side attacks, since they aim at luring attackers to their exposed vulnerabilities. Thus client honeypots play an increasingly important role in detecting state of the art attacks because their approach is at some points fundamentally different compared with server honeypots. Client honeypots simulate or run client-side applications and therefore they need actively probe possible malicious servers and interact with them. This implies two more major differences. First it is necessary to determine the servers which are suspected to be malicious in an efficient way since probing all servers on the Internet is not feasible. Second, a client honeypot has to determine on its own whether an examined server is malicious or not compared with a server honeypot in which every connection attempt can be regarded as malicious or at least suspicious.

Regarding the level of interaction (Lance Spitzner 2002), the honeypots could be classified into low-interaction, medium-interaction and high-interaction honeypots. Low-interaction (LI) and high-interaction (HI) honeypots thereby represent the two ends of a broad spectrum which include various trade-offs considering complexity, the cost for deployment and maintenance, the type and amount of the gathered information as well as risk posed by the operation of the corresponding honeypot type. Thus the level of interaction offers one metric for comparing and measuring different honeypots.

3.1 Low-interaction honeypot

A LI honeypot has thereby the lowest complexity as it is easiest to deploy and maintain due to its simple design. LI honeypots typically emulate services and vulnerabilities, so that the attacker's interaction is restricted to this emulated functionality. Therefore their scope is also limited to simple malicious connection attempts such as port scans,brute force of login and autonomous spreading malware.

The operation of LI honeypots poses the lowest risk, since there is little functionality offered which

is furthermore just emulated. Therefore it is unlikely that such a system gets successfully compromised and abused for attacks against other systems. LI honeypot implementations are mainly intended to act as deception systems and to provide protocol stack emulation.

3.2 High-interaction honeypot

HI honeypots require the highest effort to build and maintain, and pose also the highest risk, since they are usually real systems. That is, a HI honeypot exposes an entire vulnerable system to an attacker to interact with rather than merely emulating one functionality, system component, vulnerability or service. Therefore an attacker may be able to fully compromise the system and gain full control over it including the execution of arbitrary commands and usage of his own tools. Thus HI honeypot offer the highest level of interaction and need to be closely monitored in order to prevent harming other systems in case the honeypot got compromised. Furthermore HI honeypot produce a vast amount of information which is two-edged. On the one hand they allow deep insights into attacker's activities and on the other hand the these gathered information has to be handled and analyzed.

Since HI honeypots are generally real machines even production system and network, they have several drawbacks, such as disruption of personal data and dedicated host would certainly be deployed. So a feasible method is to setup a virtual machine dedicated to this task using commodity virtualization solutions such as VMware, Xen or KVM. Several solutions have been implemented in order to ease the deployment of HI honeypots and add extra functionality, specifically with regard to data capture and monitoring.

3.3 Medium-interaction honeypot

Medium-interaction honeypots aim at combining the advantages of LI and HI honeypots and merely are an emulated resource rather than a real system. A common usage of medium-interaction honeypots is malware collection since it is necessary to provide slightly more interaction than offered by a LI honeypot. However the boundaries between LI and MI honeypots are often fluid and this terms are not widely used.

3.4 Low interaction client honeypots

In contrary to server honeypots, which passively wait for an incoming attack, the basic idea behind client honeypots is to actively search malicious content.

Therefore client honeypots simulate the behavior of a user in order to determine whether it is exploited. Generally client honeypots aim at detecting attacks against client-side applications such as web browsers. Thus passive client honeypots would be thinkable, such as based on applications for Email and instant-messaging and P2P-networks. However most currently known implementations are active client honeypots and focus on web-based attacks since these attacks are currently considered the most popular ones. Thereby the concept of LI honeypots can be applied to client honeypots as well. Rather than operating the corresponding real application, a LI client honeypot just emulates the application or necessary parts of it.

3.5 High interaction client honeypots

HI client honeypots are commonly used to crawl the content of a website while using a real resource rather than an emulated one. Since the operation of HI honeypots pose more risk compared to LI honeypots, so the resources of HI client honeypots are typically virtualized, for instance running within a VM.

Our classification of honeypots is depicted in Figure 1.

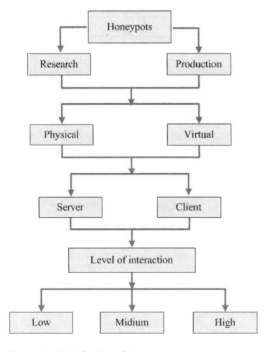

Figure 1.   Classification of honeypots.

4   CONCLUSIONS

In this paper we introduced the concept of honeypots and pointed out their value for intelligence on novel attacks and efficient malware collection. Next we classified the existing honeypot implementations in general by taking several aspects into account, such as the level of interaction (low, medium, high), the role within a multi-tier architecture (client, server), the deployment (physical or virtual) and the operating environment (production or research). The study enables practitioners and researchers to differentiate among existing honeypot approaches and provide a guideline to consider the desired characteristics while applying and researching the honeypot approaches.

ACKNOWLEDGMENT

This research is sponsored and supported by Science and Technology Research Project of Heilongjiang Educational Committee under grant No.12531121. Any opinions or findings of this work are the responsibility of the authors, and do not necessarily reflect the views of the sponsors or collaborators.

REFERENCES

Zhuge Jian-Wei & Tang Yong & Han Xin-Hui (2013). Honeypot technology research and application. *J. Ruan Jian Xue Bao/Journal of Software*, 24(04), 825–842.

Mairh Abhishek & Barik Debabrat & Verma Kanchan (2011). Honeypot in network security: A survey. C. *Proceedings of International Conference on Communication, Computing and Security, ICCCS 2011*, 600–605.

Simon Woodhead (2012). Monitoring bad traffic with darknets. *J. Network Security*, 2012(01), 10–14.

Medeiros Paulo S & De Medeiros Brito Jr Agostinho & Motta Pires Paulo S (2011). A qualitative survey of active TCP/IP fingerprinting tools and techniques for operating systems identification. C. *Lecture Notes in Computer Science*, 6694 LNCS, 68–75.

Arati Baliga & Liviu Iftode & Xiaoxin Chen (2008). Automated containment of rootkits attacks. *J. Computers & Security*, 27(7–8), 323–334.

Dominic Storey (2009). Catching flies with honey tokens. *J. Network Security*, 2009(11), 15–18.

Chandra Shekar & Bhukya Wilson Naik (2011). Forensic analysis on QEMU. C. *Proceedings of Communications in Computer and Information Science*, 250, 777–781.

Georg Wicherski (2010). Placing a low-interaction honeypot in-the-wild: A review of mwcollectd. *J. Network Security*, 2010(03) ,7–8.

Lance Spitzner (2002). Honeypots: Tracking Hackers. M. Chapter 1 The Sting: My Fascination with Honeypots, *Addison Wesley Professional*, 1–8.

Electronic Engineering and Information Science – Wang (Ed.)
© *2015 Taylor & Francis Group, London, ISBN: 978-1-138-02772-5*

Digital library network based on the Internet of Things

L. Ma

Harbin University of Science and Technology, Harbin, China

ABSTRACT: In order to solve the problems of single operation mode and low efficiency in traditional library, this paper studied a network of digital library based on the Internet of Things. The framework of digital library network was researched and designed emphatically using the technology of the Internet of things. By simulating this network model on the NS2 network simulation platform, the feasibility analysis and performance evaluation were performed. The results show that this kind of network based on the Internet of Things is feasible. It could have great significance for digital library network to apply the Internet of Things, and could be an effective method for improving the work of library.

KEYWORDS: Internet of Things; digital library network; RFID.

1 INTRODUCTION

With the development of information technology and the Internet, the technology of the Internet of Things has become up-and-coming worldwide (Atzori L. 2010). Many advanced countries have regarded the Internet of Things as a new industry, and taken strategic measures to support it. The necessity of accelerating the research is clearly stated, development and application of the Internet of Things in the Report on the Work of the Government in the third month, 2010, which means that the Internet of Things will be included in the focus of the revitalization of industry in China(Niu Y. 2010). At present, in more than 10 countries, including Singapore, Australia, India, the Netherlands and Malaysia, nearly 100 organizations have adopted the technology, -radio frequency identification (RFID)(Liu Q. 2010), a key technology based on the Internet of Things, in library automatic management system. Domestic researchers in this field have also undertaken a number of studies, which reflect from different aspects that the technology of the Internet of Things will have a tremendous impact on the development of future libraries, from different aspects.

Based on previous researchers' studies, this paper proposes a digital library network system based on the Internet of Things, and designs a system framework for the network. Then it builds a model of this network through NS2 network simulation platform, and conducts a preliminary simulation evaluation to its performance. The results show that the library network performs well, and will have good application prospect in the environment of the Internet of Things. This kind of library network will reflect the major changes of library management by applying the Internet of Things.

2 OVERVIEW OF THE INTERNET OF THINGS

Massachusetts Institute of Technology (MIT) put forward the concept of the Internet of Things first in 1999. In 2005, International Telecommunications Union (ITU) enriched the connotation of the conception in its annual report and pointed out that the development aim of information and communication technologies have changed from connecting anybody, at any time, in any place, to connecting any object, the Internet of Things thus coming into being(Kong N. 2010, Wang, X. 2010& Liu X. 2012). So, the Internet of Things is a network that connects various information sensing devices and systems, such as wireless sensor network (Wireless Sensor Network, WSN), RFID tags reading devices, bar code and two-dimensional code equipment, global positioning systems and other short-range wireless, self-organizing network based on Matter-matter (M2M) communication mode, which is connected to the Internet through a variety of access networks to form a huge intelligence network.

The Internet of Things achieves the connection between any object and any person at any time and place, through any information path, network or device. Therefore, the relevant attributes of the Internet of Things include concentration, content, collection, calculation, communication and scene connectivity, all the attributes embodying a seamless connection between people and objects or between objects and objects.

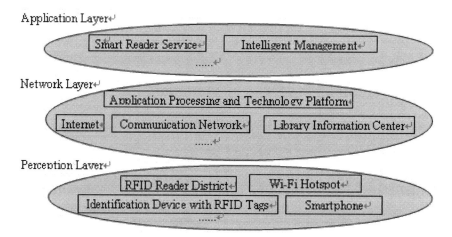

Figure 1. System framework of digital library under the Internet of Things environment.

The Internet of Things technology involves many areas, and these techniques often have different application requirements and technological forms in different industries, which can be summarized as four technique systems: perception and identification technology, network and communication technology, computing and service technology, and management and technical support technology. At present, the development of technologies of the Internet of Things is not balanced. In the area of digital library, techniques, like radio frequency identification tag, bar code and two-dimensional code technology and so on, have been very mature, and are widely used, while the WSN technology still has larger development space. The key technologies in WSN mainly include data acquisition, signal processing, protocol, management, security, network access, design verification, intelligent information processing and information fusion, etc.

3 DIGITAL LIBRARY NETWORK BASED ON THE INTERNET OF THINGS

In the network environment, the digital library should take the customers as the center, so that the users, at any time and any place, can obtain any resource and service they want in any way. The Internet of Things can further extend and expand library services of the Internet age to physical library resources and realistic environment to build an all-around, three-dimensional library service system which integrates various resources and services, and combines the Internet and reality.

Considering the present situations of libraries, we combine advanced the Internet of Things technology

with the digital library network, and design an intelligent library network system. This system has achieved the Intelligent of library management and service by means of locating readers, through wireless sensor network and transmitting the monitoring data, and by means of identifying and locating books, through RFID and processing the data by calculating, which effectively improves library management efficiency, and simplifies the management process. Therefore, the network can bring more convenience to readers. It can realize intelligent management. As is shown in Figure 1, the Internet of Things framework based on a digital library contains three layers.

3.1 The perception layer

In Perception Layer, the system perceives books and readers. During this process, the perception of books is through RFID tag sensing devices, while the perception of readers is through perceiving intelligent mobile phones they carry, through wireless positioning technologies, like Wi-Fi, and so on.

3.2 The network layer

System, at the network layer, transmits and processes information acquired by perception layer to provide appropriate services.

3.3 The application layer

This layer is used to achieve a variety of service functions of the future digital libraries, such as intelligent book inventory and pushing services of reader information based on locations.

Table 1. Specific simulation parameters.

Parameters	Values	Parameters	Values
MAC Protocol	IEEE802.11	Propagation Mode	TwoRayGround
Scene	$1000 \times 1000 m^2$	Business Class	CBR
Channel Type	Wireless Channel	Antenna Type	OmniAnten
Channel Capacity	2Mbps	Queue Type	PriQueue
Simulation Time	1000s	Queue Length	50
Transmit Rate and Maximum Moving Speed	2Mbps and 0m/s 384Kbps and 1.2m/s 144Kbps and 27m/s	Num of Subscriber Stations	10, 20, 30, 40, 50

4 THEORETICAL ANALYSIS AND COMPARISON OF RESULTS

Because the framework of future digital library network in the Internet of Things environment relies on mobile communication network, the mobile communication network based on WSN is the focus of this research.

In order to evaluate the feasibility of the digital library network, the following part will have a simulation test on the performance of this network. In Linux10.0 operating system, we use NS2.34 network simulation platform to establish the network model. In addition to that, this network has 1 gateway node which is connected to the library server. The number of other mobile users not based on WSN is 10-50, and the information is transmitted at the speed of the third generation of mobile phone (3G). And the commonly-used LEACH protocol will be used as router protocol as in WSN. The specific parameter settings are shown in table 1.

Each numerical point is simulated 10 times, and the average value of the numbers acquired is gained when the experiment is completed. Use programming language to program and conduct a statistical analysis to the simulation results, and we'll figure out the changes in the relationship, between the numbers of message delivery rate and average end-to-end delay and the number of mobile users. Then, we use Matlab to draw corresponding performance comparison parameter graph (shown in Figure 2 and Figure 3).

As is shown in Figure 2 and Figure 3, all the performance indexes of the network are better, so it can meet the requirements of application. However, with the increase of the number of mobile users of the network, the network message delivery rate drops, and the end-to-end delay increases. This is because the increasing network users results in the rise of overall business traffic, and thus causes more packet losses and longer average end-to-end delay in the same bandwidth condition, which surely leads to the decline of the network performance. On the other hand, when the motion rate of nodes goes up, message delivery rate will decline accordingly, and the average end-to-end delay increases as well. This is because the increase of node movement rate makes the network topology

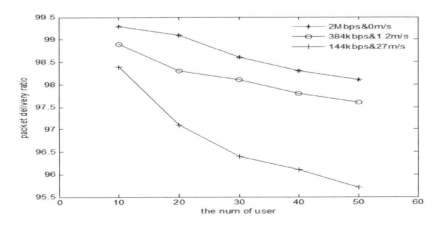

Figure 2. Relationship between network packet delivery ratio and the number of mobile users.

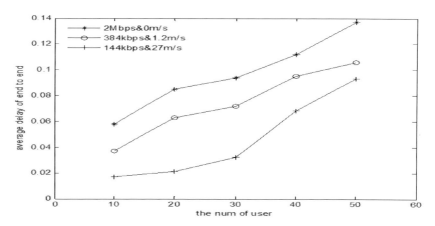

Figure 3.   Relationship between network average end-to-end delay and the number of mobile users.

change frequently, resulting in the decline of the network performance.

5   CONCLUSIONS

The appearing of the Internet of Things makes it possible to achieve efficient and automatic library management, using the Internet of things technologies. The digital network system, based on the Internet of things, proposed in this paper, provides a solution for the application of the Internet of Things in the library field.

Through simulating the system model and extracting performance data, after analyzing and comparing these data, we prove that the model is feasible and superior to other ones. This network can realize automatic and intelligent library management, enhance the core-competitiveness of the libraries, and improve the efficiency of library service. According to the analysis above, it will have good application prospect.

Therefore, this research provides a reference for the exploring the developing mode of the Internet of Things technologies in library field. In a word, the extensive application of the Internet of things technology in libraries, will also promote the development of library network system, so as to improve work efficiency and service level.

ACKNOWLEDGMENT

This work is supported by Project of Steering Committee for Academic Libraries of Colleges and Universities of Heilongjiang Province(No.2013-B-045); Supported by Science and Technology Research Project of the Department of Education of Heilongjiang Province(No.12541125).

REFERENCES

Atzori, L., A. Iera & G. Morabito (2010). The internet of things: A Survey. *Computer Networks. 54*, 2787–2805.

Niu, Y (2011). On the library of internet of things. *Theory and Practice of Library. 1*, 15–16.

Liu Q, L. Cui, & H. m. Chen (2010). Key technology and application of the internet of things. *Computer Science. 36*, 1–4.

Kong N., X. dong Li, & W. M. Luo (2010). Addressing model of the internet of things' resources. *Journal of Software. 7*, 1657–1666.

Wang, X., Z. H. Qian, & Z. C. Hu (2010). The research of RFID anti-collision algorithm based on binary tree. *Journal of communication. 6*, 49–57.

Liu X., Y (2012). Key technologies and applications of internet of things. *2012 Fifth International Conference on Intelligent Computation Technology and Automation. 01.*

Electronic Engineering and Information Science – Wang (Ed.)
© 2015 Taylor & Francis Group, London, ISBN: 978-1-138-02772-5

Research and application of cloud of virtual network data security

J. Lin

Sichuan TOP IT Vocational Institute, Chengdu, Sichuan, China

ABSTRACT: The virtual network security is an important component of cloud computing security. In order to control and ensure the safety of virtual network traffic, this paper proposes a security solution for virtual network encryption and compression technology based on Ethsec. At the same time, between the user and the server identity authentication and authentication solution of combining between the server and the server, improving the reliability of cloud services. Virtual safety switches to monitor and analyze all the virtual network traffic.

KEYWORDS: virtual network; identity authentication; network security; cloud computing.

1 INTRODUCTION

Generally speaking, cloud computing service architecture as shown in figure. 1, It includes 3 levels: first, the virtual machine (Virtualization) based on the multiInfrastructure as a service (IaaS); second, multi tenant model which based on the service platform. Service (PaaS); third, end user oriented application software as a service (SaaS).

Figure 1. Cloud computing service system.

Cloud computing security service system also to serve as the center, around the service, service security based on virtual security, data security, and rely on them. As shown in Fig.2.Bear the security is reflected and a cloud service provider. There is a close relationship between the cloud service provider role, the role rank more base, More security responsibility to undertake. The cloud computing service system must solve good resources dynamic reconfiguration, monitoring and automatic deployment, and these need to virtualization technology, high performance storage technology, processor technology, high-speed Internet technology as the foundation. The identity

trust cloud services, cloud computing is to establish identity trust service platform, computing, parallel processing, data storage technology, virtualization technology, cloud technology platform management technology to provide a service for the user password, authentication service and service through a distributed electronic seal.

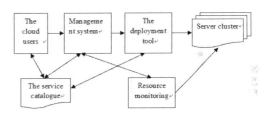

Figure 2. The identity trust cloud computing architecture.

In cloud computing, virtualization technology is one of the basic supporting technology of modern Cloud Computing Data Center Virtualization Technology, great scalability and easy management features for the data center, it is a great driving force of the evolution of the data center. On virtualization technology, people often pay attention to security issues in the virtual machine itself and its own, but ignore the virtual network is closely related to the. Virtual network with virtualization technology was born, running dozens or even hundreds of virtual machines on a single physical server, and by a few hundred virtualization servers that constitute the data center.

2 VIRTUAL NETWORK SECURITY SOLUTIONS

2.1 *Frames*

The framework of secure virtual network as shown in Fig. 3, including virtual security layer, Virtual safety switch, secure virtual network management platform and secure virtual network, Network key distribution system.

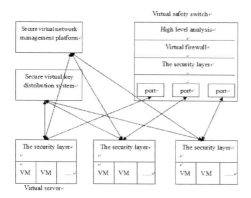

Figure 3. Secure virtual network framework.

2.2 *Virtualization security layer*

The virtual machine monitor virtualization security layer is positioned on the virtual server (Hypervisor) and virtual safety switch, its role is to process the virtual machine to send or receive Ethernet MAC frame. The security layer in Hypervisor Ethernet MAC frame compression and encryption, sent to the virtual safety switch, and Ethernet MAC frame received from the virtualization security switch, decrypt and decompress. The method of Ethernet MAC frame processing is called Ethsec.

2.3 *Virtualization security switch*

The structure of a virtual safety switch is shown in fig. 4. It includes physical port, port, virtual exchange logic, a security layer, encryption and decryption, security control strategy and other controls. Virtual safety switches according to the security policy setting, obtain the encrypted data frames from a virtual port, during the decryption and decompression, virtualization, security switches can be monitored, filtering and analysis of data, after data compression encryption transmission to the target virtual machine.

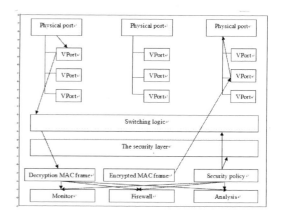

Figure 4. Virtualization security switch.

2.4 *Secure virtual network management platform*

Secure virtual network management platform is used to set the security strategies, monitor the status of network security, security risk management. Create a security policy of security, virtual network management platform, creating strategy, strategy is sent to the server virtualization and virtual safety switch. Management of data communication between the virtual network security management platform and server virtualization, virtualization security switch uses secure sockets layer (SSL, Secure Socket Layer) encryption. In this scheme, can be further used for classification of the virtual machine monitor network behavior measures. These measures include: monitoring the risk analysis for a period, classification of virtual machine monitor, the use of any network port and protocol for the virtual machine to focus on monitoring and take the mode of processing association.

2.5 *Key security, virtual network distribution system*

The key management system of virtual network security is the role of the security layer using Ethsec security channel release key. The system generates a pair of keys for each virtual server, and the public key to the virtual server through the SSL protocol to send, the private key is through the SSL protocol to send the virtual switch.

3 THE PRINCIPLE OF ETHSEC TECHNOLOGY

The realization process of this technology in the solution is as follows:

a. Virtualization security exchange calculation of a key factor, is encrypted with the private key by SM2 algorithm, is sent to the server virtualization.

b. In a time of quiet period, with virtualization security switch virtual server use the same algorithm to generate the symmetric encryption key.
c. Both sides shook hands, verify the symmetric key is the same, if the same will establish an encrypted channel.
d. Virtual server will be Ethernet MAC frame compression, and through the SM4 algorithm for the symmetric key encryption for Ethsec frame.
e. The Ethsec frame by the sending host physical port to virtualization security switch.
f. Virtualization security switch through SM4 algorithm for the symmetric key will decrypt and decompress Ethsec frame, as the Ethernet MAC frame.
g. The Ethernet MAC frame through virtual firewall filtering and high level analysis, and then transformed into Ethsec frame is passed to the target virtual machine.

The method of Ethernet MAC frame issued each virtual machine is encrypted with SM4 algorithm, to obtain valid data, it must be through virtualization security switch. The data frame sent force virtual machine must be through the switch, so cannot directly attack the local virtual machine, the local virtual machine before may attack could not effectively untie the attack data send a local virtual machine to package, it must acquire the data packet from the virtualization security switch. Therefore, both the packet attacking or ordinary data packets, must go through a virtual securities exchange, can be to determine whether exchange and sent according to the attack characteristics, in order to prevent the attack, if the new virtual machine and not open the built-in firewall, can block this virtual machine to receive any packets, until the built-in firewall open.

4 CONCLUSIONS

The scheme is a new solution, through the Ethsec technology, the original Ethernet MAC frame data according to the safe load.

Strategies for compression and encryption, and all data through virtualization security switch analysis and monitoring, so as to realize the visibility, virtual network traffic security, consistency. To solve the safety problem and the control problem of virtual network traffic.

REFERENCES

Yi T. (2012). Research on the technology of. *information and communication security virtualization security in cloud computing.5,*63–65.
Du Q.L. (2012). The realization of. communication technology. *design of enterprise information management of cloud platform, 45 (6),*110–112.
Huai J. P., H.M. Wang & C.M. Hu (2007). Virtual machine virtual computing environment research and design based on. *Journal of software.18 (8),*2016–2026.
IBM virtualization and cloud computing group.(2009) Virtualization and cloud computing. *Beijing: Publishing House of electronics industry.*
Zhang Y.Y, Q.J Chen& S.B. Pan (2010).Analysis of cloud computing. *Telecom Science and security key technology. 5,* 34–37.
Wang P. (2011). Cloud computing the key technique and its application. *Beijing: People's post Electric press,* 123–150.
Wu X.Z. & B. Wang (2012). The next generation data center design of. *communication technology, based on cloud computing.45 (6),*107–109112.
Yang Y. & Q. Wang (2012). The cloud service data isolation technology of. *information and communication security. 2,*57–59.
Ge Q.J. (2011) Virtualization: technology, applications and challenges of. *communication technology. 44 (10),*91–93.

Electronic Engineering and Information Science – Wang (Ed.)
© *2015 Taylor & Francis Group, London, ISBN: 978-1-138-02772-5*

Design of information management system for college sports meeting

B. Bai

Harbin University of Science and Technology, Harbin, Heilongjiang, China

ABSTRACT: In this paper, an information management system is designed based on the Microsoft database technology, using the actual management process of college sports meeting as research object. Such system is used to service college sports meeting and assist the managers to complete many works including registration, programs schedule as well as scores calculation and query. Results show the system could improve the operation efficiency of college sports meeting greatly.

KEYWORDS: Information management system; database; college sports meeting.

1 INTRODUCTION

Nowadays, the information technology develops rapidly and shows broad application prospect in the field of sports. It is well known that the college sports meeting has not only the student group and the staff groups, but also the athletics project and popular game, so the artificial arrangement of college sports meeting was very complex and time-consuming, and the efficient information management system must replace the traditional way. In this paper, the information management system applying for college sports meeting is investigated and designed with ASP. NET4.0, Visual Fox9.0 and SQL Server2008, which provided a simple, convenient, effective way for the sponsors to organize and manage the college sports meeting. Moreover, the sponsors can decrease operating costs while increase work efficiency by taking the advantages of computer and network.

2 ANALYSIS ON THE STRUCTURE OF SYSTEM

Through the analysis and comparison of the previous research results, the B/S and C/S combination model is adopted for the system. The structure of B/S model is used to meet geographical dispersal and apply for the diverse communications media (Brian J. 1996). The structure of C/S model can make the application process closer to the users; on the other hand, it can alleviate the amount of network data transmission, and reduce the load on the server, so that the whole system can gain better performance. The system supports geographically dispersed pattern and decreases the amount of client maintenance as the greatest extent as possible on the basis of the network structure. The C/S model is suitable for high security requirements, strong interactive, small scope and location fixed and has been used in the management subsystem such as the schedule, the event settings, printing and advanced operating systems because of its relative independence and security properties. The B/S model is suitable for the requirements of widely application, location flexibility, security and interactivity less demanding situations and mainly used for online registration, competition results release, information inquiry and so on. The system with the new structure model combines the respective advantages of the two models mentioned above, which has the properties of high security, strong interaction, massive data flow, wide application and flexible location.

The system consists of online registration, generating schedule, processing results and publishing scores of four subsystems. The online registration and publishing the results are adopted by B/S model which design with the technology of ASP.NET4.0. The generation of schedule and score processing subsystem are used by B/S model with the Visual FoxPro9.0. Each subsystem is relatively independent, but it closely links with the SQL Server2008 database.

3 COLLEGES AND UNIVERSITIES SPORTS MEETING SYSTEM DESIGN

3.1 *Systematic architecture design*

System architecture is the macro framework which describes the sports meeting management information system constitution, and it is the overall structure of the run system. It can be expressed in three layers of logical structure. In performance layer, website subsystem part using IE browser as various users of unified interface, at the time when it reaches unity, friendliness, ease of use, it also makes the future deployment of applications to upgrade more convenient, the backstage subsystem section uses the Windows Forms technology and with Windows Forms components to design interface, server-side Web server and application server show service information to the end users and collect business data, at the same time they deal with various requests which the users send through the campus network, and compete the work by calling backstage business processes or data processing components(Libkin L. 2002). The business layer is the executive part of practical business rules and data processing. It applies the formal process and business rules to the relevant data to make the business request which users issue through the presentation layer come true. Data Layer is the depository of business data. Use the SQL Server data base to store data, and centrally manage the data to achieve the Integrity, security, and disaster protection of business data. This system design is a three-layer database management system which is based on C/S and B/S. The C/S uses Windows forms technology which under the framework of .NET, and B/S uses Web forms technology which under the framework of .net. The application server and data server use ADO.NET by middle layer to achieve.

3.2 Online registration subsystem

Online registration system use b/s structure, apply ASP.NET4.0 to make the front page, and as SQL Server2008 the back-end database. The subsystem is mounted to the specified server before the online registration starting. As is shown in Fig.1 the administrator can log management page, set the contestant, sports and deadline. Participating units can be used to specify the user name and password login website for online registration. Add the information of the company, the administrator from the input interface to input participating units, companies abbreviation, players start numbering, contacts, contact telephone number and choose whether to professional team and other information. These information will sent to the system, then the system judges whether the added information conformance to specification, if do not conform to the input, if meet the specification is

calling a stored procedure, it will enter the information into the database. If it saves successfully, it will show the successful information, if not; it will display the error message.

3.3 Generate order booklet subsystems

Generate order booklet subsystem is developed using visual fox9 which is under the structure of C/S. According to competition rules and application situation of all participating teams to statistics about the number of participants and make a plan of competition grouping; Grouping based on the situation of athletes entered for the races; Then arrange for the game schedule on the basis of grouping plan, competition groupings, simultaneous entry condition and exciting level; setting the agenda and schedule; Edit and print various of competition information, forms, order booklet and so on. As is shown in Fig.2 the system puts the work of edit and print various of race information, forms, order booklet into the next module, pre-match layout module is consist of the number of participates statistics, simultaneous entry statistics, development of group plans, competition grouping and the group list, competition schedule arrangement assisting and the activity schedule of the Games.

The function of the number of participant's statistics module is to statistics for the personnel participating games and displays the results, this is prepared for competition grouping and the administrator could not modify (Ben-Asher Y. 2002). Simultaneous entry statistics is one of the basis work before arrange competition schedule. Due to players enter the events doesn't regularly, therefore, this factor must be consider when arranging the competition schedule, try to avoid the situation that there are several games for one athlete in a period of time, player must have enough time to have a rest after one race. Grouping plan is based on the contest regulations, statistics the athletes' number of each games according to the participating situation then determine the number of groups and promotion in each projects, do the basis work for the next race schedule arrangement. We use the Random function in Visual Fox9.0 to do series of

Figure 1.　Online registration.

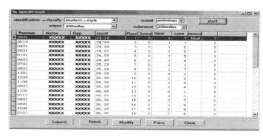

Figure 2.　Generate order booklet.

algorithm design and after repeated resting and verification to realize this feature eventually.

3.4 Result processing subsystem

The C/S model of the software Visual FoxPro9.0 is used to complete inputting results, ranking and aggregating scores in the stage of competition. As is shown in Fig.3 referees will input each individual athlete's results to database. After press the submit button by referee, the system sort by score for all the results in the database at first, and then add scores according tot the player's rank by order. The system will auto search for the top 8 what are they department and add scores to them according to their corresponding rank. In addition, we also do the corresponding design for athletes who break the record. The system can automatically determine whether athletes break records and automatically add a record-breaking score.

Figure 3.   Process result.

3.5 Result publishing subsystem

Publishing online achievement is undoubtedly the best way in order to ensure the competition results quickly and accurately release, so that all athletes and spectators can query the results conveniently and timely. But we separated the publishing results subsystem from the system to avoid the whole management system communication block as for visit too much. We apply it to the network. As is shown in Fig.4 the final list, competition scores and statistics and other information publish on the web in real time. Publishing the

game information and results make the game more open and transparent relied on network of data.

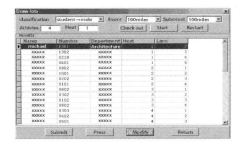

Figure 4.   Publish result.

4   CONCLUSIONS

To the problems appeared in the work of university sports meeting, this thesis put forwards systematic solutions and management styles, and makes the whole design project, which includes systematic architecture, choices of operating system, net structure, system functions, database design and so on. This thesis analyses the signing up of the athletes, the input of the competing score and introduces the concrete implement process of them.

Through analysis and design, this system can deal with the management of the sports meeting information. It is a college sports meeting information management system which really relies on computer itself and owns strong information dealing capacity.

REFERENCES

Brian J (1996). Enterprise Client/Server computing. *Information System management.2,57–62.*
Libkin L (2002). Expressive power of SQL. *Theoretical Computer Science.296(14),* 379–404.
Ben-Asher Y(2002).The parallel client-server paradigm. *Parallel Computing.28(3),*503–523.

Electronic Engineering and Information Science – Wang (Ed.)
© *2015 Taylor & Francis Group, London, ISBN: 978-1-138-02772-5*

A variety of auxiliary facilities of key techniques of VANET

C.Y. Wang & Y.Y. Tang
School of Software, Harbin University of Science and Technology, Harbin, Heilongjiang, China

ABSTRACT: This paper the research contents and key issues of the research is focused on a variety of types of VANET research on auxiliary facilities, the type structure, various types of network infrastructure between heterogeneous characteristics, the network QOS evaluation index to depict, resources (including communication resources and energy resources) limited analysis, dynamic network changes and carry?Relationship between forward routing way, network load change and the relationship between network performance, height of node mobility problem such as impact on the communication link has become the basic content in the research of this topic.

KEYWORDS: VANET; AP; Qos; Drive-thru Internet.

1 INTRODUCTION

In recent years, the explosive growth of the vehicle and the ubiquity of information needs combining communication network and vehicles. VANET in addition to the application of intelligent transportation aspect, it as a terminal to access the Internet, for car users to download data to meet the demand of its communication (Network on Wheels 2006, Internet ITS Consortium 2006). Drive through the Internet refers to being in the process of moving through the side of the road vehicular node of the wireless Access Point (Access Point, AP) connected to the Internet, download the required data and share advice another VANET application scenario. Vehicles in the process of moving between application of VANET technology can realize the vehicle and the communication between vehicles and roadside Internet access points. At present, the widely used onboard wi-fi broadband wireless access technology, for studying the Drive - thru Internet data transmission. The main reason is the use of 3 g, 4 g network is costly, relative to the high cost of its communication performance is not ideal (J.Ott, D. Kutscher. 2004). So on-board nodes by road network in the deployment of the Internet access point connected to the Internet, in order to satisfy the demands of the Drive - thru the Internet. Despite a massive deployment of VANET WLANs can give an area to provide high quality data service, But due to the number of AP required too much with very high cost, it is not feasible in practice (Z. Zheng, Z. Lu, S. Prasun, K. Santosh. 2010). In order to solve the problem of the high cost of the current AP deployment, literature adds a combination in VANET in Mesh and WIFI communication technology of the

auxiliary infrastructure to improve network performance, extend the coverage of VANET (B. Nilanjan, C. Mark, T. Don, L. Brian 2008).

2 THE KEY TECHNOLOGY AND SOLUTION

This topic is proposed to solve the key problem:

1 The QOS guarantee of AP and W2M infrastructure deployment cost minimization problem
2 To ensure that the network connectivity Relay node deployment cost minimization problem
3 The network load time and space changes periodically rechargeable power energy utility maximization problem
4 On the Internet cover empty vehicle data downloads a maximization problem
5 Special vehicle node auxiliary data transmission with QoS guarantees

The solution:

1 Based on the MoXiao use function and set covering model to guarantee QOS of AP and W2M infrastructure deployment cost minimization.

Vehicle along a path in a 2 d space, Internet coverage of this path will experience more infrastructure and vehicles can only when coverage to receive services, So this topic to be covered by the path path length of total length ratio to evaluate VANET for along the path of the vehicle service quality.

Thus, we define L for the set of all child line after the partition, For any one segment $l \in L$, $d(1) \in R^+$ represent the length of the line. For any kind of infrastructure deployment, consists of two parts: ① for the

AP deployment way, already in A full selection of A subset S as AP deployment location; ②for W2M equipment deployment, already in A full selection of A subset S' as W2M equipment deployment location. For L $L_c \subseteq L$ be covered in the set, the existing formula:

$L_c(S,S') = \{l \in L \mid l \in (\bigcup_{a \in S} C_a) \bigcup (\bigcup_{b \in S'} C_b')\}$. On two different equipment to make unified deployment way, D possible deployment of collection, $D = \{d_1, d_2, d_3 \ldots d_n, d_1', d_2', d_3' \ldots d_n'\}$, Element d1 means can deploy a AP in $a_j \in A$ position, d_j' means can decorate a W2M on $a_j \in A$ location equipment, This can be simple $|D| = 2|A|$, An arbitrary element is a ÎD coverage corresponds to a line segment set $C_a \subseteq L$ and $w_a \subseteq R^+$ deployment costs, So for any $M \subseteq D$, can be converted into $L_C(M) = \{l \in L \mid l \in \bigcup_{a \in M} C_a\}$.

The research of QOS guarantee infrastructure deployment cost minimization problem QMIDC can be defined as follows:

Definition 1 QMIDC: Known possible deployment set D, $a \in D$ corresponding to each element in the cover line segment set $C_a \subseteq L$ and $w_a \subseteq R^+$ deployment costs, Indicators of service quality lower Q(QOS), and $M \subseteq D$ meet deployment way:

Minimize: $w(M) = \sum_{a \in M} w_a$

s.t.: $\min_{p \in P\eta_p}(M) \geq Q$ (1)

2 Based on the theory of graph approximation algorithm to achieve the minimum guarantee network connectivity Relay node deployment

This paper intends to add minimum number in heterogeneous WSNs relay node makes any a node in a Mesh network and a communication path between an arbitrary AP make it up to each other. So this topic to solve the problem, namely the Mesh network connectivity problem CIP - Mesh formalization description is as follows:

Definition 2 CIP-Mesh: In the plane, undirected graph $G = \langle V, E \rangle$, The node set V by End node set S_E and HA-sensor nodes of S_{HA}, namely $V = S_E \bigcup S_{HA}$, And the location of the known V function for the loc: $V \to R \times R$, and edge set E is composed of two vertices of communication radius constraint edge, the edge set meet formula (2) is as follows:

$E = \{e(i,j) \mid d(i,j) \leq R_E$

且 $i \in S_E \parallel j \in S_E\}$ $\bigcup \{e(i,j) \mid d(i,j) \leq R_{HA}$ 且

$i \in S_{HA} \&\& j \in S_{HA}\}$ (2)

When join in G relay node set Srelay generate graph $G^+ = \langle V \bigcup S_{relay}, E^+ \rangle$, E^+ collection also meet

communication radius constraint, in joining S_{relay} to satisfy graph G^+ point set $V \bigcup S_{relay}$ can reach between any two points in the case, the S_{relay} Position function loc': $V \to R \times R$, making the formula (3) set up:

$Minimize(|S_{relay}|)$ (3)

3 By a single node periodically in spatial and temporal change of network load analysis, rechargeable power supply energy utility maximization realization for the whole network energy utility maximization.

This topic from a single node needle under the environment of network load time cyclical change a single node rechargeable power energy scheduling problem of utility maximization MPU - SN, through analyzing the characteristics of the network structure, topological characteristics, data transmission in order to find more than one node energy consumption under collaboration for the influence of the network transmission utility, the periodic change and solve the network load time and space environment network scheduling rechargeable power energy utility maximization problem MPU - WN. First of all for a single node energy use, we define a utility function $u(k)$, k for energy consumption in time t e and the ratio of energy you need to meet the network load b, $k = e(t)/b(t)$. The utility function $u(k)$ is not strictly decreasing and properties, and meet the requirements of $u(k) = 0$. And r(t) for the node in t time to collect the energy and the battery on the initial energy is 0, the MPU - SN can be described as follows:

Definition 3 MPU-SN:

For $\vec{r} = \{r(1), r(2), r(3), \ldots = \{\}$, corresponding network load change $\vec{b} = \{b(1), b(2), b(3), \ldots\}$, find energy $\vec{e} = \{e(1), e(2), e(3), \ldots\}$ U meet the overall utility:

Maximize:

$U = \sum_{t=1}^{\infty} u\left(\frac{e(t)}{b(t)}\right)$ (4)

Because of VANET has the characteristics of network load time cyclical change so $\vec{b}$ in MPU - SN problem is predictable, and $\vec{r}$ depends on a variety of external factors is not known, so it had to find a suitable online algorithm to solve the MPU - SN. For 1 to T define this time $\vec{e}^* = \{e^*(1), e^*(2), e^*(3), \ldots e^*(T)\}$ to solve the problem of the optimal solution of MPU - SN, which can be:

$U^*(T) = \sum_{t=1}^{T} u\left(\frac{e^*(t)}{b(t)}\right)$ (5)

4 Combined with resource allocation and transmission scheduling maximize the Internet cover empty vehicle data downloads.

A vehicular node in the process of moving through access Internet infrastructure to meet the communication needs and the on-board nodes and downlink communications infrastructure, can be summed up in different amount of data downloaded. To Drive - thru the Internet application scenarios, nodes within range of the communication infrastructure response of Drive - thru the Internet request, for resource scheduling. The question that the infrastructure resources, how to schedule the channel is the most efficient way of using a V2V data transmission, maximize the infrastructure service ability. Found in the actual research, car users Drive - thru internet request has a lot of repetitive data download, if the infrastructure of the precious resources to duplicate data download, cannot effectively use the scarce resources of infrastructure channel. This topic is put forward based on the optimization, scheduling method for prediction of links between nodes, through the optimization of channel resource scheduling, a more effective use of vehicular node removed V2V after communication coverage of AP self-organized communication, improve the AP to Drive - thru Internet communication service ability, achieve their research objectives.

When a set of nodes into an infrastructure, wireless communication area, the optimized selection of k relay nodes by AP connected to the Internet to download files. When removed from the group node communication coverage, the relay nodes by using Ad hoc links between nodes, will forward the downloaded content to other nodes. In order to maximize meet the demand of Drive - thru Internet users, the key to solve the channel resources assigned to which nodes. In this paper, the objective function of formalized as follows:

$$S_k^* \leftarrow \arg \max_{j=1}^{c_n^k} \{\sum_{i=1}^{k} |s_{(i)}|, n_{(i)} \in S_k^{(j)}\} \tag{6}$$

$$\text{subject to } p_{ij} \geq p_0$$

5 Based on the state space diagram and Markov decision method to realize special vehicle node auxiliary data transmission with QoS guarantees

This research use the bus routes features can be predicted and dense nodes and lines distribution features, the bus as a movement characteristics of infrastructure: Mobile gateway MG, this paper proposes a new data transmission method based on MG forwarding MGF, under the condition of satisfy user's QoS requirements, minimize data transmission delay. First of the state - space figure, road network

model as shown in Figure 1, using markov decision method is established based on MG forward V2I optimization model of data transmission, and then through the model V2I optimum forwarding decision of data transmission, the optimal forwarding decision means the corresponding optimal sequence of each state, final selection on purpose vehicle track meet transmission threshold, the success rate and minimize the V2V transmission delay of intersection nodes as packets and purpose vehicle optimal gathering node, the destination node.

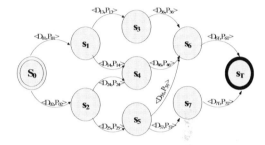

Figure 1.   Probability of state - space diagram example.

3   CONCLUSIONS

Core solution of this research is to use a variety of infrastructure auxiliary ways to reduce the cost of VANET network deployment supporting VANET, and embodies in two aspects:

1   The project is given in view of various infrastructure support of VANET, fully considering the deployment cost minimization problem, energy and infrastructure to the variety of the scheduling, communication resource scheduling, mobile gateway scenarios of information transmission, and then puts forward the corresponding solution, can effectively reduce the cost of VANET, improve the effectiveness of the entire network, reduce the amount of data transmission, improve the performance of VANET.
2   This Project research VANET can provide to car Drive - thru the Internet, download the required data and share advice of application scenarios. Vehicles in the process of moving between application of VANET technology can realize the vehicle and the communication between vehicles and roadside Internet access points.

ACKNOWLEDGMENTS

In this paper, the research was supported by the Heilongjiang Science and Technology Research

Project of the Education Department (Project NO. 12541162) and College Students' Innovative Entrepreneurial Training in Heilongjiang Province (Project NO.201410214011).

REFERENCES

Network on Wheels. (2006). *http://www.network-on-wheels. de*.

Internet ITS Consortium. (2006). *http://www.internetits.org*.

J.Ott, D. Kutscher. (2004). The "Drive-thru" Architecture: WLAN-based Internet Access on the Road. *Proceedings of Vehicular Technology Conference*. 2615–2622.

Z. Zheng, Z. Lu, S. Prasun, K. Santosh. (2010). Maximizing the Contact Opportunity for Vehicular Internet Access. *Proceedings of IEEE INFOCOM*.1032–1035.

B. Nilanjan, C. Mark, T. Don, L. Brian. (2008). Enhancing Mobile Networks with Infrastructure. *Proceedings of ACM MOBICOM*. 81–91.

Electronic Engineering and Information Science – Wang (Ed.)
© 2015 Taylor & Francis Group, London, ISBN: 978-1-138-02772-5

A universal framework design for the network test instrument

M. Ma, C.J. Dong & H. Yan
School of Automation Engineering, University of Electronic Science and Technology of China

Y.H. Guo
EMC Lab of Advanced Technology Generalization Institute of China Ordnance Industries

ABSTRACT: Network test instrument is a instrument, which introduces the computer network technology into test instrument and uses the computer communication network as a transmission medium to transfer data, achieve test, share resource, and coordinate work. The current domestic network test instruments lack of uniform standards and manufacturers who design those instruments are just based on their own understanding, thus making the instrument not compatible and standardized. In view of this kind of phenomenon, the authors raise the hardware and software constitution's framework of network test instrument which is based on years of related development experience, and it will lay the foundation of standardized design.

KEYWORDS: Network test instrument; Software and hardware framework; Normalization; Standardization.

1 INTRODUCTION

During these years, the development of network test instrument (Gutterman L.. 2005) in our country has gotten stronger support from the National Industrial Sector. Network test instrument not only has increased in variety and quantity, but also has improved substantially in quality. While compared with foreign mature network test instrument, the specification is not unified, and the design is a bit confusing. Most manufacturers design, network test instrument hardware and software for specific tasks and needs. This kind of measurement instruments is usually dedicated in software and hardware design. Therefore, it lacks standards and norms.

The irregular of network test instrument brings potential dangers when distributed automatic test system (Burch J. 2008) is being built.The lack of standardization in instrument hardware and software design has an effect on instrument 's network identification and program control; the confusion in the design of instrument network functions will hinder the effective communication and interaction among instruments, the nonstandard of the instrument network communication protocol will make it difficult to achieve accurate and real-time data transmission between the instrument and the remote test. If the instrument data formats are not unified, there is also a problem in realizing deeply information sharing between the instrument and data fusion. The non-synchronization among test instruments will make system test data meaningless.

Based on the above analysis, a universal design framework of network test instrument is given in this paper, the universal design framework will rule the design of network test instrument and promote standardization, thus providing convenience for the formation of the distributed test system.

2 THE DESIGN OF SOFTWARE FR-AMEWORK

The design of network test instrument software framework is shown in Figure 1. It's mainly divided into two parts: hierarchical design and modular design.

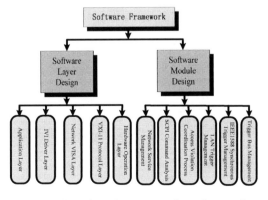

Figure 1. Network test instrument software framework.

2.1 Software -level design

Many manufacturers are unclear in the hierarchy of the network test instrument software design. The software-level architecture firstly divides software into different levels, then the different level software will be linked in a certain way, thereby completing entire software function.

The software of the network test instrument includes PC software and network test instrument software. PC software should provide users with operator interface to transmit control commands, and the own software of network test instrument is responsible for parsing the control commands sent by the PC, and providing users with the required instrument function. The application layer of PC has provided interactive interface and control interface. The users operate the application layer, and then the application layer will call the IVI driver (IEEE Std. 802.3af.2003) function which is based on VISA I / O resource (Sundaramurthy R. 2010) pool. The following will describe the various components of the software-level architecture.

2.2 The design of software module

After a good division of different levels, the overall function of the instrument is divided into different functional modules to play all levels' complete instrument functions, thus improving software development efficiency and reusability. Fig. 2 shows the software module composition of network test instrument.

The PC of network test instrument should be able to access the instrument via the Web, at the same time the PC can send SCPI commands to the instrument for programming control, and trigger the instrument through the LAN packets. 1) Network Service Management Module: Network test instrument offers network service management function to ensure the transmission of network data, and provides users with the Web access function. Network Service Manager includes LAN configuration, the realization of network discovery instrument protocols, like VXI-11, mDNS, DNS-SD, and other mechanisms. 2) Instrument Instruction Parse Module: Network test instrument should be able to parse the SCPI commands (Wubbena.H.2008). 3) Access Violation Coordinate Process Module: Network testing should be able to coordinately conflict situations. There are various ways to access the instrument simultaneously. Besides, network testing should also make certain restrictions for the number of users accessing the instrument at one time. 4) LAN Trigger Management Module: the module provides proposal on the format of LAN messages among inter-module communications. Network testing should provide LAN messaging trigger function. LAN trigger is equal to LAN event package message trigger. To be specific, it means that it triggers the instrument through LAN message which includes

trigger messages (Liu Z.Q. 2013). 5) IEEE1588 Synchronous Trigger Management Module: Internal of network test instrument should contain synchronous trigger management module to ensure accurate timing synchronization of the entire test system. 6) Hardware Bus Trigger Module: Network test instrument provides a hardware bus trigger to provide higher accuracy than synchronization timing trigger. Hardware bus trigger uses specialized hardware trigger bus, and provides eight physically separate trigger channels. Each hardware bus channel can work in one of two modes: drive mode and wired and mode.

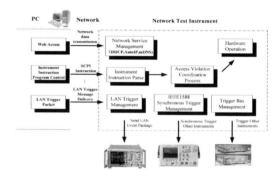

Figure 2. Network test instrument software modules composition.

3 THE BUILD OF HARDWARE FRAMEWORK

At the core of network test instrument design, there are no detailed requirements for hardware design in this field. We build the network test instruments' hardware design framework from two aspects, including hardware structure design and hardware module design shown in figure3.

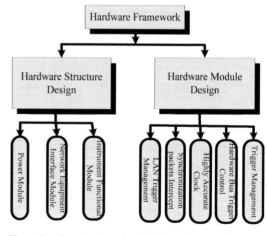

Figure 3. Construction of the hardware framework.

3.1 The design of hardware architecture

The hardware architecture of network test instrument should be composed of three parts: power module, network instrument interface module and instrument function module. The power part is mainly responsible for the supply of various parts, instrument power needs to conform to IEEE Std 802.3af standard Network instrument interface modules are used to process related interface functions, which includes LAN port, hardware trigger bus interface, second pulse interface and external trigger output port.The second pulse interface is mainly used to monitor multiple instruments' network synchronization performance. External trigger output can be utilized to test instrument hardware trigger synchronization performance. Instrument function modules are mainly responsible for unique functions of the instruments.

3.2 The design of hardware module

The hardware module of the network test instrument is divided into a central control module, synchronous trigger module and instrument function modules. The central control module is responsible for the dispatch of instrument internal functions. Synchronous trigger module completes the management of relevant synchronous trigger. Instrument function module completes the instrument itself functions, like data acquisition, waveform generation, etc.

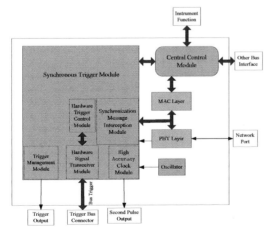

Figure 4. Hardware modules composition of the network test instrument.

It can be seen from Fig. 4 that synchronous trigger module is the core part of the network instruments' network section, and it is composed of LAN message trigger management module, synchronization message interception module, high accuracy clock module, hardware bus trigger control module and

trigger management module, etc.1) LAN Message Trigger Management Module: Network test instrument provides LAN message triggers management module to process the received message, and generate a LAN message. LAN message trigger management module should determine whether a LAN message frame meets the conditions, thus generating the trigger signal.2) Synchronization Message Interception Module: Network test instrument should be able to intercept event messages to get timestamp. 3) High Accuracy Clock Module: Network test instrument should provide a highly accurate time reference to other modules through high accuracy clock module. 4) Hardware Bus Trigger Control Module: Network test instrument should provide the management and control of hardware bus trigger to receive hardware bus trigger signal and generate hardware bus trigger signal. Hardware bus ought to possess M-LVDS (Multipoint low Voltage Differential Signaling) interface circuit electrical characteristics. The standard of M-LVDS can support a data rate as high as 500Mbps and a wide common-mode voltage range(± 2). Besides, it has strong ESD protection features which enable the hardware bus to support hot swap function.5) Trigger Management Module: Network test instrument should provide management modules for the management of hardware bus trigger, external trigger and a trigger associated with a particular instrument, meanwhile, it provides the selection of input and output.

4 SUMMARY

In the view of network test instrument in this field, this paper presents a more standardized network test instrument universal framework, and provides a more detailed description of the software and hardware. Moreover, it actually validates the feasibility and versatility of the framework. A standardized framework configuration of hardware and software will make the design and the test personnel convenient to introduce the computer network technology into testing instrument, thus achieving the formation of a network test instrument system truly unified, rapid and practical. Then it also provides a theoretical norm for the fast and real-time of a test.

ACKNOWLEDGMENT

The paper is supported by Sino-German joint research project of the Sino-German Center for Science(No. GZ817) the Fundamental Research Funds for the Central Universities (No. ZYGX2012J090). The National Natural Science Foundation of China (No. 61271035).

REFERENCES

Gutterman L(2005). Integrating VISA, IVI and ATEasy to migrate legacy test systems. *Aerospace and Electronic Systems Magazine, IEEE. 20(6)*, 36–38.

Burch J, A. Cataldo & J. Eidsone (2008). LAN-based LXI Instrument Systems- the Next Step in the Evolution of Measurement System Technology. *IEEE International Instrumentation and Measurement Technology Conference.5.*

IEEE Std. 802.3af (2003). IEEE Standard for Information technology – Telecommunications and information exchange between systems-Local andmetropolitan area networks-Specific requirements.

Sundaramurthy R. & P. Dananjayan(2010). Ser-vice Provider Architecture for Dynamically Reconfigurable Virtual Instruments in Networked Environments. *Control Automation Robotics & Vision (ICARCV) International Conference,* 1035–1039.

Wubbena.H. (2008). Emerging Synthetic Instruments And Ivi Driver Solutions. *Elevaluation Engineering 47(12)*, 14–19.

Liu Z.Q ,S.W. Pan&Y.G. Zhang(2013). A design and implementation of and triggering system for LXI instruments. *Measurement. 46(8)*, 2753–2764.

Zhengwei Z, H.H. Z&S.Lin (2010).The application of structure arrays and files in the SCPI parsing system. *IEEE Intelligent Computation Technology and Automation (ICICTA) International Conference*, 710–711.

Electronic Engineering and Information Science – Wang (Ed.)
© 2015 Taylor & Francis Group, London, ISBN: 978-1-138-02772-5

Low latency routing algorithm simulation and hardware verification of NoC

S.Y. Jiang, L. Chen, Z. Lu, J. Peng & Y.Y. Mao
School of Automation Engineering, University of Electronic Science and Technology of China, Chengdu, China

G. Luo
Chengdu Technological University, Chengdu, China

ABSTRACT: Network on Chip (NoC) avoids a series of troubles of System on Chip (SoC) based on bus structure, such as clock synchronization, limited address space, unable flexible expanding of system, etc. A low latency routing algorithm, *odd-even-XY* (*OE-XY*) routing algorithm is proposed in this paper. Hardware model and feasibility simulation of this algorithm indicate that the delay of data package transition needs more time when the Manhattan Distance between a source node and destination node becomes bigger, this demonstrates the rationality of the algorithm. According to the verification and theoretical analysis of the low latency algorithm which we proposed in the actual circumstance of networks on chip communication, considering the maximal network block to the effect of delay, we derived that this is a low delay routing algorithm of NoC, better than *XY* routing algorithm and *XY-YX* routing algorithm.

KEYWORDS: Network on Chip; *OE-XY* Routing Algorithm; Low Latency; Isim; Verilog HDL.

1 INTRODUCTION

SoC design method integrates multiple functional modules integrated on a single chip, provides efficient integration performance, and solves many challenging problems in the fields of electronics and communications (Jiang 2010). But the SoC system based on bus structure has exposed some inevitable problems, such as poor extensibility, single clock synchronization (Jiang 2013). In order to solve these problems, a new integrated circuit architecture – Network on Chip (NoC) was proposed, which introduces the concept of computer communication network to chip design, and uses the routing and packet switching technology to transfer the data (Benini 2005), providing a new interconnection scheme for on-chip communication as well as solving a series of problems on the bus from the architecture.

In this paper, we focus on studying a kind of low latency routing algorithm for NoC, simulating and verifying the algorithm on Isim. Finally, we summarize the design, and analyze the shortcomings in the design.

2 OE-XY ROUTING ALGORITHM

The proposed routing algorithm in this paper is inspired by *XY-YX* and *XY* algorithm. *XY* algorithm is deterministic routing algorithm (Jiang S. Y. 2013). The *OE-XY* algorithm chooses packet is along *X* or

Y direction transmission according to the location of the destination node and the current node, when the *Y* value of the destination node and the current node are not equal, transmission direction is also different. The algorithm routing direction selects *NE*, *ES*, *WS* and *NW* of Turn Model, that is, its routing process only permits the above four steering, so there is no ring for the algorithm.

The idea that based on odd or even column divided was first proposed in this model, that is, to perform different operations depends on which columns the reference points in. Thus, reduces averaged flow and congestion furthermore, but also reduces the latency.

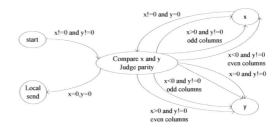

Figure 1. Simulation model of *OE-XY* algorithm.

OE-XY routing algorithm is described as fol-lows: (1) Wait for the data packet. (2) Receive the data packet. (3) Transmit the packet along the direction of the x-axis or y-axis. (4) Judge x values of the source node and the destination node. (5) If x value of destination node is grea-ter than that of the source node, and the source node in the even column, then using the y first x behind routing algorithm, otherwise using the x first y behind routing algorithm. (6) If x value of destination node is smaller than that of the source node, and the source node in the even column, then using the x first y behind routing algorithm, otherwise using the y first x behind routing algorithm. (7) Send the packet, back to Step 1.

3 SIMULATION AND VERIFICATION

In order to verify whether the algorithm of *OE-XY* presented in this paper is correct, we need to analyze the feasibility simulation of the algori-thm at first. The simulation tool we used here is ISE's own Isim, and the language is Verilog HDL.

The flow chart of the algorithm is mainly divided into the following categories: the judgment of ΔX and ΔY, the judgment of odd-even columns for-warding, local delivery, and the choice of XY routing successively. Therefore, the choice of XY routing suc-cessively can be summarized as two kinds of state: walk along X and walk along Y. In conclusion, the state machine has the following five kinds of state: State 1: start; State 2: compare Δx and Δy with 0, then determine the size of odd-even columns; State 3: walk along X; State 4: walk along Y; State 5: local delivery.

And the conversion between each state are as fol-lows: State 1 to state 2: Enter uncondi-tionally; State 2 to state 3: $\Delta x > 0$ and $\Delta y < 0$, $\Delta x > 0$ and $\Delta y \neq 0$, and Δy in odd columns; $\Delta x < 0$ and $\Delta y \neq 0$,and Δy in even columns; State 3 to state 2: Enter uncondition-ally; State 2 to state 4: $\Delta x = 0$ and $\Delta y \neq 0$, $\Delta x < 0$ and $\Delta y \neq 0$,and Δy in odd columns; $\Delta x > 0$ and $\Delta y \neq 0$, and Δy in even columns; State 4 to state 2: Enter unconditionally; State 2 to state 5: $\Delta x = 0$ and $\Delta y = 0$; State 5 to state 1: Enter unconditionally. From the above analysis, we can get the state machine as shown in Figure 1.

The purpose of this simulation is to get the trans-mission delay of data packets from source node to destination node under the algorithm (Feng C. C. 2010). So the end of a transmission process is when data packets are under the local delivery status. As odd-even-*XY* algorithm simulation model show in Figure 2, we set state1=s0, state 2 to 5 for s1 to s4. In turn, the operation of each states are: s0: Compute values of Δx and Δy, set flag=0, enter s1; s1: According to the values of ΔX and ΔY with 0, and the odd-even of the current source node in the column, then choosing the next state to enter. s2: If the cur-rent $\Delta x > 0$, then $\Delta x = \Delta x - 1$, and enter the s1, if the current $\Delta x < 0$, then $\Delta x = \Delta x + 1$, and enter the s1. s3:

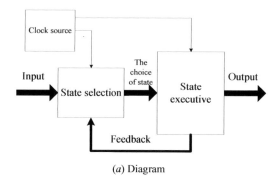

(*a*) Diagram

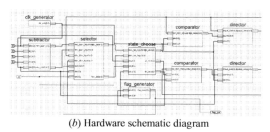

(*b*) Hardware schematic diagram

Figure 2.   Simulation model.

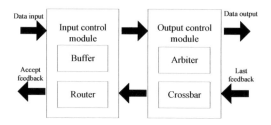

Figure 3.   Routing unit structure.

If the current $\Delta y > 0$, then $\Delta y = \Delta y - 1$, and enter the s1, if the current $\Delta y < 0$, then $\Delta y = \Delta y + 1$, and enter the s1. s4: Set flag=1, enter s0.

From the procedure of established simulation model, we can draw the conclusion that the model is constituted of clock source, state selection and state execution with the term of function. Therefore, we can have the princi-ple block diagram as shown in Figure 2a. In the diagram, taking the coordinates of the source node and destina-tion node as the input. The input is the selected state. According to the states, we can take some other opera-tions. These changes will be feedback to the state selec-tion. In a certain state, executing diagram can produce the output. Without the stable clock source, the system can't run normally. Schematic is shown in Figure 2b.

Compared with *XY* and *XY-YX* algorithm, the pro-posed *OE-XY* algorithm in this paper causes lower network congestion, thus reducing latency (Jiang S.Y.

2014). The above store-and-forward nodes are namely routing nodes in network topology, and routing unit structure is shown in Figure 3, cache and routing calculation are in the input control module, arbitration and switch are in the output control module.

And then, we analyze *OE-XY* routing algorithm based on 6×6 *2D* mesh by comparing to *XY* and *XY-YX* algorithm, concluding switch nodes have lower congestion probability (Shu Y.J. 2013).

Assuming that there are three pairs of com-munication nodes in Figur 4a, namely *S1 (1, 1)* and *D1 (5, 2)*, *S2 (2, 1)* and *D2 (5, 3)*, *S3 (3, 1)* and *D3 (5, 4)*. We use gray represents the switch nodes entered by two packets, black re-presents the switch nodes entered by three packets at the same time (Wang Y.M). Figure 4 gives routing results with the *XY* algorithm and *OE-XY* routing algorithm. Based on the analysis mentioned above, we can draw the conclusion: compared to *XY* and *XY-YX* algorithm, the probability of congestion of these nodes is the lowest. Therefore, under the non-ideal condition, *OE-XY* algorithm improves the *XY* and *XY-YX* algorithm improves the performance in the terms of latency.

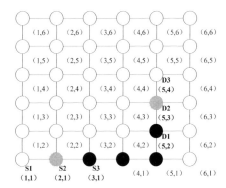

(*a*) *XY* algorithm

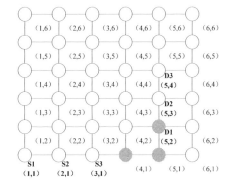

(*b*) *OE-XY* algorithm

Figure 4.   Communication results of three pair of nodes.

4   CONCLUSIONS

The main work was proposed and verified the *OE-XY* algorithm in this paper. This algorithm only focuses on delay of NoC, and no dead-locks. Based on the traditional *XY-YX* algorithm, the algorithm adds a key factor in determining the *XY* routing order, namely odd-even columns of the source node. So when packets are transferred into the network, the routing behavior is mainly determined by two factors at the same time, namely odd-even columns of the source node as well as the position relationship of the source node and destination node.

ACKNOWLEDGMENTS

This work was supported by the Program of National Nature Science Foundation of China under Grant No. 61471407, 60971036 and 60934002, and the Application Fundamental Research Funds of Department of Science and Technology of Sichuan Province under Grant No. 2013JY0192.

REFERENCES

Jiang S.Y (2010).Design and development of wiring harness test based on S3C44B0.*The International Journal for Computation and Mathematics in Electrical and Electronic Engineerin. 29(2)*, 362–369.
Jiang S.Y&G.Luo& S.Chen&W.H.Zhao&Q.Z.Zhou, (2013).Study of synchronization test methods of NoC at multi-clock domains.T*he International Journal For Computation And Mathematics In Electrical And Electronic Engineering.32(2)*, 504–515.
Benini L.&D. Bertozzi (2005).Network on chip architectures and design methods. *IEEE Proc.–Comput. DigitTech. 152*, 261–272.
Jiang S. Y.&Y. Liu&J. B. Luo&H. Cheng(2013).Study of fault-tolerant routing algorithm of NoC based on 2D-mesh topology.*IEEE International Conference on Applied Superconductivity and Electro-magnetic Devices*. 189–193.
Feng C. C.&Z. H. Lu&A. Jantsch(2010).A reconfigurable fault-tolerant deflection routing algorithm based on reinforcement learning for network-on-chip. *NoC Arc'10, Atlanta, Georgia, USA,* 1–6.
Jiang S.Y.&G.Luo&Y.Liu&S. S. Jiang&X.T. Li(2014). Fault-Tolerant routing algorithm si-mulation and hardware verification of NoC. *IEEE Transaction on Applied Superconductivity.24*, 9002805.
Shu Y.J.& G. Luo&Xiu Tang Li, Xing Di Sun, Yue Liu. (2013).Research of Online Detection of Routing Transmission Errors of NoC, *Proceedings of 2013 IEEE International conference on Electromagnetic Devices Beijing, China,* 194–197.
Wang Y.M, Yin HL. A Novel Genetic Algorithm for Flexible Job Shop Scheduling Problems with Machine Disruptions, *The International Journal of Advanced Manufacturing Technology, Springer London, 68(5-8),*1317–1326.

Electronic Engineering and Information Science – Wang (Ed.)
© 2015 Taylor & Francis Group, London, ISBN: 978-1-138-02772-5

Design of dual mode radix4 SRT divider

W.L. Guo, H.J. Xing, Y.X. Tang & Z.J. Wang
College of Computer Science and Technology, Harbin University of Science and Technology, Harbin, China

ABSTRACT: In this paper, a dual mode radix-4 SRT divider which could support IEEE754 floating point standard is designed and applied using Verilog. It could process one double precision floating point division at one time or two independent single precision floating point divisions at the same time. Finally, the simulation result from Cadence SimVision and the synthesis reports from Synopsis Design Compiler are demonstrated. The performance of this design related to timing, area and power is discussed.

KEYWORDS: Divider; Division Algorithm; SRT Divider; Floating Point Number.

1 INTRODUCTION

All the division algorithms are based on the recursion and mainly fall into two categories: data recursion and function recursion. The data recursion algorithm realizes the slow division which generates one bit or several bits at a time based on radix number. The non-restoring division, restoring division and SRT division are based on the data recursion which is always realized by several full additions, comparisons and shifts (Gallagher & Swartzlander 2000). The hardware is simple to realize, which means less power and area. As it only generates one bit (generally no more than two bits considering the hardware complexity), however, the latency is considerable.

As the high speed multipliers gets available in modern VLSI design, the function recursion algorithm, which realizes fast division using multiplication, are also getting more attention (Wey & Wang C.P,1999). The Newton/Raphson algorithm and Goldsmith algorithm are examples of the function recursion algorithms (Harris et al. 1997). Even with the enhancement in multiplication, however, the hardware is still more complex than the data recursion method, which means more power and area can be consumed though it can be faster (Upadhyay & Stine, 2007).

2 REVIEW OF ALGORITHM OF DIVISIONRESTORING DIVISION

The restoring division is based on the data recurrence division formula:

$$Q = N / D,$$

$$P_{j+1} = P_j R - q_{n-(j+1)} D \qquad (1)$$

Where, P_j is the partial remainder of the division, R is the radix of the division, $q_{n-(j+1)}$ is the quotient bit, D is the divider, and N is the dividend. Initially, $P_0 = N$.

For radix2 restoring division, the flow chart is shown Fig.1.

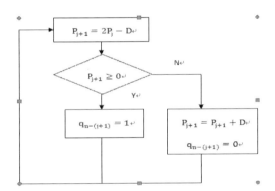

Figure 1. Flow chart of restoring division.

The quotient set is expands to +1, 0 and -1 compared to normal non-restoring divider. The quotient is selected based on the range of the partial reminder. New partial reminder is calculated by multiplying radix (realized by left shifting by one bit) and then subtracting the divider. As the comparison between value with a large amount of bits are time consuming, only reasonable sets of bits from partial reminder take into consideration in the selection of the quotient. The quotient estimation determines the accuracy and performance of the division (Burgess, 1991). The quotient converting and the final quotient correction is also needed to get the binary result. As quotient 0 is introduced, some iteration may save the addition or subtraction when the partial remainder is very close to zero.

3 DESIGN OF SRT RADIX-4 DIVIDER

In this paper, we implemented radix-4 SRT division. Our divider supports both IEEE754 standard single precision floating point operands and double precision floating point operands. For single precision operation, two 32-bits divisions are computed at the same time. For double precision operation, one 64-bits division is calculated at one time.

3.1 IEEE754 floating point number

Single precision floating point number has three parts, including the 1 sign bit, 8 exponent bits and 23 mantissa bits. Double precision floating point number has 1 sign bit, 11 exponent bits and 52 mantissa bits. The sign bit indicates whether the number is negative or positive. When the sign is 1, the operand is negative. When the sign is 0, the operand is positive. The 8 bits exponent indicates the exponent of the operand. The actual value of the exponent bits needs to be subtracted by 127 from 8bits exponent value for single precision FP number and subtracted by 1023 from 11bits exponent value for double precision FP number. The value greater than 127 (1023) indicates positive exponent and the value smaller than 127 (1023) indicates a negative exponent. The mantissa part indicates the value after the point. According to the IEEE standard, the value before the point is always one.

3.2 Quotient selection algorithm

In the SRT radix4 division, the partial remainder is shifted left by two in each round (two bits quotient generated). Then, the most significant four bits of the divider (most significant bit is always one, therefore only three are taken into consideration) are used to decide the quotient selection boundary and the most significant seven bits of the partial remainder are used as the inputs of the quotient selection from {−3, −2, −1, 0 and +1, +2, +3} (Harris et al. 1997). There are eight sets of quotient selection boundary according to the divider and all of them need to be calculated and stored in our system. The next partial remainder can be obtained according to the quotient. During the quotient calculation, on-the-fly conversion is also used to convert the quotient to the binary complementary form. The Taylor diagram for radix-4 SRT division with quotient set {0, +1, −1, +2, −2, +3, −3} when the divider is 1 is shown in Fig. 2.

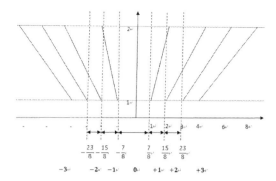

Figure 2. Taylor diagram for SRT radix-4 division.

3.3 On-the-fly conversion

The on-the-fly conversion is used to convert the quotient to the binary complementary number. It performs conversions during each round which is obviously more efficient than doing the conversion after all bits are generated. To realize this, two registers are used and updated together at each round: A register stores the current quotient and B register stores the quotient in the condition that the next quotient is negative. For the single precision processing, the size of A and B are 27 bits respectively. TABLE I shows the conversion table for SRT radix-4 division.

Table 1. On-the-fly conversion table.

Quotient	Register A	Register B
0	{A, 0, 0}	{B, 1, 1}
+1	{A, 0, 1}	{A, 0, 1}
−1	{B, 1, 1}	{B, 1, 0}
+2	{A, 1, 0}	{A, 0, 1}
−2	{B, 1, 0}	{B, 0, 1}
+3	{A, 1, 1}	{A, 1, 0}
−3	{B, 0, 1}	{B, 0, 0}

3.4 Hardware implementation

The block diagram of the SRT divider is shown in Fig. 3.

The remainder register holds the value of the partial remainder and the divisor holds the value of the divider. In each round, the most significant 7 bits from the remainder register and 4 bits from the divider register are sent as the input of the quotient selection logic. The quotient selection logic provides the two bits quotient and one bit indicating the sign of the quotient. The generated quotient is then sent

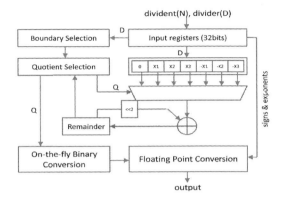

divident(N), divider(D)

Boundary Selection

Input registers (32bits)

Quotient Selection

| 0 | X1 | X2 | X3 | -X1 | -X2 | -X3 |

Q

<<2

Remainder

On-the-fly Binary Conversion

Floating Point Conversion

signs & exponents

output

Figure 3. Block Diagram of Radix4 SRT Divider.

to the on-the-fly conversion module which finally produces two binary complementary quotient bits at each round. The estimated quotient bits will also be sent to a MUX to select the next operand for the full adder. Another input of the adder is the value of the remainder register which is shifted left by 2. The result of the adder stored as a partial remainder for the next iteration. After completion, the remainder needs to be examined. Quotient correction is needed if the reminder is negative (Pham & Swartzlander, 2006).

For single precision processing, two identical 32bits processing blocks are used. For double precision processing, the adders and registers for 32bits processing are shared and used with some extra overhead.

4 PERFORMANCE RESULTS

The Radix4 SRT divider is then synthesized in Synopsys Design Compiler with TSMC 0.18um process. For timing, there is no timing violation at clock period 3 ns, resulting in 333.3 MHz maximum clock frequency. The area is about 99,400 um2 at 3ns clock period. Power consumption is about 13.5MW, which mainly due to dynamic power consumption. Since the minimum clock period is about 3ns, and one operation needs 15 cycles for 32bits processing and 28 cycles for 64bits processing, the latency is about 45ns for 32bits processing and 84ns for 64bits processing.

Even though the radix4 SRT can achieve half number of cycles to finish a division compared to the radix2 SRT, the quotient selection logic of radix4

SRT is more complex than the radix2 SRT, which leads to longer clock period needed. The complex quotient selection logic also causes larger area. The radix-4 SRT will be faster than the radix-2 with increased complexity, but not twice as fast as radix2 SRT because of the longer clock cycle.

5 CONCLUSION

A radix-4 SRT divider is realized supporting two parallel single precision divisions or one double precision division at one time. The quotient selection boundary is obtained from Taylor diagram, and on-the-fly quotient conversion is applied to obtain binary results. RTL level design is realized using Verilog and SimVision is used for simulation. Synthesis report is provided by Synopsys Design Compiler which indicates a satisfactory performance.

ACKNOWLEDGMENT

This work was supported by Science and Technology Project (NO:12521108) from the Education Department of Heilongjiang Province, China.

REFERENCES

Gallagher W.L & E.E. Swartzlander (2000). Fault-tolerant Newton-Raphson and Goldschmidt dividers using time shared TMR. *Computers, IEEE Transactions on.49, (6)*, 588–595.
Wey C.L. & C.P. Wang (1999). Design of a fast radix-4 SRT divider and its VLSI implementation. *Computers and Digital Techniques, IEE Proceedings .146(4)*, 205–210.
Harris D.L., S.F. Oberman & M.A. Horowitz (1997). SRT division architectures and implementations. Computer Arithmetic, 1997. Proceedings., *13th IEEE Symposium on*, 18–25.
Upadhyay S. & J.E. Stine (2007). Pipelining high-radix SRT division algorithms. *Circuits and Systems, 2007. MWSCAS 2007. 50th Midwest Symposium on*, 309–312.
Burgess N. (1991). Radix-2 SRT division algorithm with simple quotient digit selection. *Electronics Letters*.27(21), 1910–1911.
Pham T.N.& E.E. Swartzlander (2006). Design of Radix-4 SRT Dividers in 65 Nanometer CMOS Technology. *Application-specific Systems, Architectures and Processors.*

Electronic Engineering and Information Science – Wang (Ed.)
© 2015 Taylor & Francis Group, London, ISBN: 978-1-138-02772-5

A method based on word feature selection for relation extraction in EMRs

X.B. Lv, Y. Guan & J.W. Wu
School of Computer Science and Technology, Harbin Institution of Technology, Harbin, Heilongjiang, China

ABSTRACT: Relations between entities in electronic medical records play an important role in understanding the content of EMRs. A CRF-based machine learning method was applied in this paper. In order to select a better feature set, we first merged the jargon words and common words in the preprocessing step, then discriminate category matching and entropy-based feature ranking was used to optimize the features selected. Experiment results show that our method outperforms 0.023 than the baseline system.

KEYWORDS: clinical relation extraction, machine learning, feature selection.

1 INTRODUCTION

Texts in electronic medical records (EMRs) contain a lot of valuable information about the status of one person's health during a period. As a result, there is a rising need to automatically analyze and obtain certain information from the texts in EMRs. Among which, the task of extracting semantic relations between clinical entities is a key component to understand the content of EMRs and has been attracting more and more researchers in recent years.

Relation extraction is always seen as a classification problem, many machine learning methods are attempted to solve this task. Roberts et al. (2008) used SVM to extract relations in their information extraction system CLEF, they compared the performance of relations within a sentence and across a sentence, the results showed that relation extraction across sentence is more difficult. Wang et al. (2009) annotated corpus for entities and relations based on SNOMED-CT by themselves and combined CRF and SVM together to extract the relations. The 2010 Informatics for Integrating Biology and the Bedside (i2b2)/Veteran's Affairs (VA) challenge provided an opportunity for all participants to demonstrate their methods on relation extraction (Uzuner et al. 2011). In this challenge, DeBruijn et al. (2011) ranked high among all the submitted systems with an F-score of 0.7313, they found unlabeled data and syntactic dependency structures are useful to improve the performance. Grouin et al. (2010) imported linguistic patterns to improve their system and gained an F-score of 0.709.

The main object of this paper is to discuss the preprocessing and feature selection strategies to improve the performance of relation extraction. For the former, we merge the jargon words and common words respectively, while for the latter, we use discriminating category matching and entropy-based feature ranking to optimize the features that were selected.

2 TASK DEFINITION

Relation extraction in electronic medical records aims to automatically identify whether two clinical entities have some kinds of relationship and assign the relation type to them. Formally, given a clinical text snippet x written in natural language and two clinical entities E_1 and E_2 in x, the task aims to identify whether there is any relationship between E_1 and E_2, all the possible relationships are defined in set Y. The 2010 i2b2 challenge defined eight kinds of relations based on three marked entities (treatment, problem and test), that is, treatment applies, not applies, improves, worsens, and causes a problem (TrAP, TrNAP, TrIP, TrWP and TrCP); test reveals a problem and test is required to examine a problem(TeRP and TeCP); problem indicates another problem(PIP). We will follow the definition in the 2010 i2b2 challenge, as a result, Y={TrAP, TrNAP,TrIP, TrWP, TrCP, TeRP, TeCP, PIP} in our task.

To predict input (x,E_1,E_2) to one of the Y relations, statistical machine learning methods require N training instances in the form of (x^i,E_1^i,E_2^i,y^i), i=1...N, then a predict model will be constructed based on the training instances to annotate test data (x,E_1,E_2).

3 PREPROCESSING THE EMR

3.1 *Merge jargon words*

In EMRs text, there are a lot of jargon words written in different ways. For example, "Atrial fibrillation",

"Atrial fibrillations" and "Auricular fibrillation" are all referred to the same concept "Atrial fibrillation". However, these words will be treated as different words in the statistical procedure and thus may lead to bad results. As a result, we should merge these words with the same one. We utilize UMLS to convert all the jargon words with the same meaning to the same style.

3.2 Merge common words

There are a lot of common words in the electronic medical records, for these words, they may also be written in different ways while having the similar meanings. For example, "improve" and "improvement" play the same role in relation extraction. As a result, we should merge them to some extent. Specifically, we merged these words through the following three ways. First, we merged the words based on their word shapes. We consider the words that have the same word shape may express the same semantic meaning; as a result, they will play a similar role in the relation extraction task. Second, we merged the words that have a similar context because these words have a similar meaning as their contexts are analogous. We set a window size to construct a context vector for a specific word, and then we calculate similarity between two context vectors. Last, we use WorldNet to recognize the similar words and merge them.

4   FEATURE SELECTION

In this paper, we introduce two feature selection methods: 1.discriminative category matching; 2. Entropy-based feature ranking.

4.1 Discriminative category matching

Discriminative Category Matching (DCM) (Gabriel 2002) is a method that estimates a feature's importance for a specific category by the means of statistical method; it is always used in the document classification task. The fundamental idea of DCM is that if a feature appears frequently in every category, that is, its distribution among different categories is not significant. In this case, this feature is not effective enough for the task of classification. On the contrary, if a feature just appears frequently in several specific categories, that is, its distribution among different categories is obvious. In this case, this feature is a key feature that is important to distinguish between different categories. To estimate a feature's importance, there are two frequently used methods, one is to estimate the feature's importance in one category, and the other is to calculate the diversity of the feature's

distribution among other categories. We will introduce the two methods in the following.

We use Eq. 1 to estimate the degree of importance of a feature f_i in category k. The $pf_{i,k}$ is the number of instances that have feature f_i in category k. The N_k is the number of the whole instances in category k. To prevent the data sparsity problem, the "add one" smooth is imported in Eq. 1. The feature f_i will have a greater influence in category k if the $WC_{i,k}$ is larger.

$$WC_{i,k} = \frac{\log_2(pf_{i,k}+1)}{\log_2(N_k+1)} \tag{1}$$

$WC_{i,k}$ is used to evaluate the importance of a feature within one category, but it cannot estimate the distribution of features among the whole categories. However, it is important to realize the distribution from the perspective of all categories. For example, one feature that appears frequently in many categories is not more valuable than the one that appears only in a few categories. As a result, we use Eq. 2 to evaluate the distribution of a feature f_i in different categories. In this equation, N is the number of all categories; C_i is all the categories that have feature f_i; 1/logN is used for normalization to ensure CC_i belongs to [0, 1].

$$CC_i = \log\left(\frac{N\max_{k\in C_i}\{WC_{i,k}\}}{\sum_{k=1}^{N}WC_{i,k}}\right)\frac{1}{\log N} \tag{2}$$

Finally, we used Eq. 3 to combine the two equations described above. $W_{i,k}$ considers not only a feature's effectiveness in one category but also its distribution in the whole categories. The larger $W_{i,k}$ is, the more effective f_i is.

$$W_{i,k} = \frac{WC_{i,k}^2 \cdot CC_i^2}{\sqrt{WC_{i,k}^2 + CC_i^2}} \cdot \sqrt{2}, 0 \le W_{i,k} \le 1 \tag{3}$$

4.2 Rank feature based on entropy

The features described above are all from words that are restricted in a small window size, however, words that are far from the entities may also play an important role in relation extraction task. This section will describe our attempts using entropy to rank these features and add to the final feature set.

For an entity pair E_1 and E_2, let $P=\{p_1,p_2,...,p_n\}$ be the context vectors for all the occurrences of E_1 and E_2. $W=\{w_1,w_2,...,w_N\}$ is all the words occurred in P. p_i is the word vector that consists of all the words in the context and its dimension is N. Each element in p_i has a value 0 or 1 depending on whether the context includes the word w_i. The object of this section is to select the key features (represented by words) from W, as a result, we should rank all w_i in W.

260

The entropy based feature ranking method (Chen 2005) is involved in our task. The similarity between p_i and p_j is evaluated by the equation $S_{i,j}=\exp(-\alpha \cdot D_{i,j})$, in which $D_{i,j}$ is the Euclidean distance between the two vectors, the value α is a normalization parameter, $\alpha =-\ln 0.5/D$, D is the average value of all the context vectors' distance. As a result, the entropy of P is calculated by Eq. 4.

$$E=-\sum_{i=1}^{N}\sum_{j=1}^{N}(S_{i,j}\log S_{i,j}+(1-S_{i,j})\log(1-S_{i,j})) \quad (4)$$

To evaluate one feature's importance, first remove the feature from feature space. Then E is calculated according to Eq. 4 using the new feature space. If E changes a lot, that means the discarded feature is important; otherwise, if there is no significant change, the feature may be not an effective one. By this way, we can estimate all features and get the rankings of them.

5 EXPERIMENT

5.1 Data and evaluation

The data in our experiment are from the 2010 i2b2/VA challenge, which consists of three sets constructed by three institutions, named Beth, Partners and UPMC respectively. The statistical information of the three sets is described in table1.

Table 1. Statistical information of data sets.

Set Name	Documents	Lines	Tokens	Relations
Beth	73	8727	88722	1973
Partners	97	7515	60819	1147
UPMC	256	27507	267130	6292
Total	426	43749	516671	9412

Table 2. Results of our method on relation extraction.

	P(%)	R(%)	F(%)
TrAP	79.3	95.9	86.8
TeRP	90.1	97.8	93.8
TrIP	74.1	13.2	22.3
TrCP	61.8	50.3	55.5
TrWP	83.3	4.6	8.7
TrNAP	70.7	25.9	37.9
PIP	100	100	100
TeCP	71.6	34.3	46.4

We used part of this data set (177 documents and 3120 relations) for training data and the rest (259 documents and 6293 relations) for test data. The usually used metrics P(precision), R(recall) and F(F-score) were imported to evaluate the performance of our system.

5.2 Results and analysis

Table 2 shows the results of our system based on the methods described above. We choose CRF from our machine learning model. From this table, we can see TrAP, TeRP and PIP gained the top 3 results because relations in these three types frequently appear. In contrast, the performances of identifying relations TrWP, TrIP and TrNAP are low because these three types of relations are uncommon in the data set. This result shows that the number of instances will impact the performance of the final system, with more instances, we may have a better performance.

Figure 1 shows the comparison of our method and the method which just uses the basic feature set. From this figure we can see that our method outperforms in almost all types of relations, especially, TrNAP and TrCP improves best. The total F-score is improved by 2.3 percent. From the result, we believe the methods we introduced are effective.

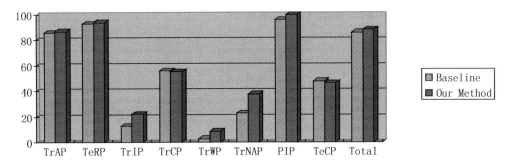

Figure 1. Comparison of our method and the baseline method.

ACKNOWLEDGMENTS

De-identified clinical records used in this research were provided by the i2b2 National Center for Biomedical Computing funded by U54LM008748 and were originally prepared for the Shared Task for Challenges in NLP for Clinical Data organized by Dr. Ozlem Uzuner, i2b2 and SUNY

REFERENCES

Chen, J., D. Ji, C.L. Tan, & Z. Niu (2005). *Unsupervised feature selection for relation extraction.* In Proceedings of IJCNLP 2005.

DeBruijn, B., C. Cherry, S. Kiritchenko, J. Martin, & X. Zhu (2011). Machine-learned solutions for three stages of clinical information extraction: the state of the art at i2b2 2010. *J Am Med Inform Assoc 18(5),* 557–562.

Grouin, C., A.B. Abacha, & D. Bernhard (2010). *CARAMBA: concept, assertion, and relation annotation using machine-learning based approaches.* Proceedings of the 2010 i2b2/VA Workshop on Challenges in Natural Language Processing for Clinical Data. Boston, MA, USA.

Gabriel, P., F. Cheong, & X.Y. Jeffrey (2002). *Discriminative category matching: efficient text classification for huge document collections.* ICDM2002. Japan.

Roberts, A.R., M. Hepple, & Y. Guo (2008). Extracting clinical relationships from patient narratives. *BMC Bioinformatics 9(11),* 3.

Uzuner, Ö, B.R. South, S. Shen, & S.L. DuVall (2011). 2010 i2b2/VA challenge on concepts, assertions, and relations in clinical text. *J Am Med Inform Assoc 18(5),* 552–556.

Wang, Y. (2009). *Annotating and recognizing named entities in clinical notes.* Proceedings of the ACL-IJCNLP Student Research Workshop. Singapore: Suntec.

Electronic Engineering and Information Science – Wang (Ed.)
© *2015 Taylor & Francis Group, London, ISBN: 978-1-138-02772-5*

Brain oscillations mechanism for cognitive control process

B. Yu, H.F. Li, L. Ma & X.D. Wang
School of Computer Science and Technology, Harbin Institute of Technology, Harbin, China

B. Yu
School of Software, Harbin University of Science and Technology, Harbin, China

ABSTRACT: Frequency bands (0.5-70 Hz) have been evidenced to play an important role in our brain cognition, but evidence that links a frequency band to a specific function during a cognitive control process is largely lacking. The present research investigated the general cognitive function of frequency bands and their role for a cognitive control process. The results in our study indicated that brain oscillations in different frequency band may all correlate with a complete cognitive control process and have different role in the process. The study first established a brain oscillations model for a complete cognitive control process. The model is very important for intelligent information processing, cognitive psychology resolving, etc.

KEYWORDS: Cognitive Control; Brain Oscillations; Oscillation Model; Frequency Bands.

1 INTRODUCTION

In the Information science Era, our brain are equipped with the ability of cognitive control allowing us interact with our complex environment in a goal-directed manner and ignore the irrelevant things. For example, we are listening to the lectures while ignoring the noise. Study on cognitive control has important implications for intelligent information processing, electronics engineering, information science, etc.

Many previous studies have evidenced that frequency bands 0.5~3.5 Hz (delta band), 3.6~7 Hz (theta band), 8~13 Hz (alpha band), 14~29 Hz (beta band), and 30~70 Hz (gamma band) are real brain oscillations reflecting specific cognitive process (Balconi & Pozzoli 2008, Lakatos et al. 2008, Klimesch et al. 1999, Engel & Fries 2010, Tallon-Baudry & Bertrand 1999, Putman et al. 2014). Therefore, decreases or increases of power in these brain oscillations might all be reflections of a complete cognitive control processes.

Few studies reported the power change of delta band during cognitive control task. The previous studies definitely suggest that cognitive control demands lead to an increase in medial frontal theta band power (Cavanagh et al. 2010, Cohen and Cavanagh 2011, Nigbur et al. 2011, Cavanagh, Zambrano-Vazquez & Allen 2012, Cavanagh & Frank 2014). Klimesch's study found alpha ERS during cognitive control task (Klimesch et al. 1999), but alpha ERD was found in the other studies (Compton et al. 2011, Tang and Chen 2013). Furthermore, a few studies found that cognitive control demands lead to beta ERD (Wang et al. 2014), whereas gamma ERS (Kieffaber and Cho 2010, Tang et al. 2011). In the above studies, the specific functions of the brain oscillations in a general cognitive control process remain unexplained. Therefore, the frequency domain mechanism of cognitive control has not been fully elucidated.

The aim of the current study is to give frequency domain biomarker of cognitive control processes. We presented the general cognitive functions of the brain oscillations, and discussed their function in cognitive control. Therefore, a brain oscillations model was set up to elucidate a complete cognitive control process.

2 BRAIN OSCILLATIONS (DELTA, THETA, ALPHA, BETA AND GAMMA) CORRELATES TO COGNITIVE CONTROL PROCESSING

2.1 *Delta oscillations (0.5~3.5 Hz)*

It's well known that delta band occurs in deep sleep. However, various cognitive processes can be assigned to delta band. The amplitude of delta oscillations is increased during visual or auditory oddball experimental paradigm, in which the delta activity is thought to reflect conscious stimulus evaluation and memory updating. Accordingly, it confirms that delta band is related to motivation, noticeable things detection and decision making. Moreover, Balconi and Pozzoli reported that delta band is associated with decision processes and updating functions (Balconi & Pozzoli 2008).

Yordanova et al. proposed that the performance monitoring system is specific for the detection of performance errors and operates in the delta oscillations reflecting error-specific processing in medial prefrontal cortex (mPFC) (Yordanova et al. 2004). Furthermore, Lakatos et al. proposed that delta oscillations are related to attention selection of stimuli, and the phase of delta also determine instantaneous power in higher-frequency oscillations (Lakatos et al. 2008).

2.2 *Theta oscillations (3.6~7 Hz).*

A variety of studies have evidenced that theta band is involved in many cognition functions such as working memory (Maurer et al. 2014), learning (Cavanagh et al. 2010), attention (Putman et al. 2014), and emotional regulation (Knyazev et al. 2009) , etc.

Theta power in the mPFC (FCz) was increased after an error trials during a Flanker task, and theta phase synchronization between FCz and F5/F6 electrodes was increased during error trials (Cavanagh et al. 2009). They also evidenced that theta bands of the medial frontal and lateral prefrontal synchronized in the conflict trial (Cohen & Cavanagh 2011). Time-frequency analysis methods indicated that theta bands in the Anterior Cingulate Cortex (ACC) increased for interference trials during Stroop task (400-500ms) (Hanslmayr et al. 2008). In addition, Nigbur observed increased theta activities with medial frontal cortex in different kinds of cognitive control paradigms (Simon paradigm, Flanker paradigm, Go-No-Go paradigm) (Nigbur et al. 2011). Recently, Mas-Herrero and Marco-Pallares proposed that theta band activity over midfrontal region increased for the unsigned prediction error (Mas-Herrero & Marco-Pallares 2014). Other researches also proposed that ERS of theta band over centro-frontal region, lateral prefrontal cortex and sensory-motor areas reflect the neural mechanism for monitoring conflict or errors (Cavanagh et al. 2010, Cohen & Cavanagh 2011, Cavanagh et al. 2012). Base on the previous studies, Cavanagh et al. provided evidence theta band activities over the medial frontal region as a compelling candidate for cognitive control (Cavanagh & Frank 2014).

2.3 *Alpha oscillations (8~13 Hz)*

ERS of alpha band plays an important role for the inhibition control and timing of cortex processing whereas ERD of alpha band represents the release of top-down inhibition control is related to the emergence of spreading activation activities (Klimesch et al. 1999). To increase the SNR (Signal to Noise Ratio) within the cerebral cortex, alpha band inhibits conflicting or interfering information to make the subject to his task at hand (Pfurtscheller & da Silva 1999). During auditory memory encoding, the power of alpha band (10-12 Hz) typically increases whereas alpha desynchronization is observed during memory retrieval (Krause et al. 2001).

Klimesch et al. proposed that ERS of alpha emerged when the subject control or withhold the execution of a response, and is generated from electrodes on human scalp that probably are exert, or under top-down control. Contrast analysis revealed that a large burst of power in the low alpha band after an error trial during a Flank task (Cavanagh et al. 2009). Recently, Wang et al. also found that both Stimulus-Stimulus and Stimulus-Response conflicts enhanced the power of theta and alpha band frequencies by the method of event-related spectral perturbation (ERSP) (Wang et al. 2014). But some studies found that cognitive conflicts suppressed the power of alpha band. For example, the power of alpha reduce after following Stroop trials with conflicting stimuli compared to trials with neutral ones, suggesting that alpha oscillation is involved in the process of neural rhythms modulation (Compton et al. 2011). Moreover, Tang et al. found that the significant alpha ERD (480~980 ms) were observed for the incongruent conditions in the Flanker task, whereas alpha-band ERS in these regions were observed for the congruent ones (Tang & Chen 2013).

2.4 *Beta oscillations (14-29 Hz)*

Like alpha band, ERD of beta can also be considered as a representation of aroused neural structures of activated cortex regions. Some studies have suggested that beta band may subserved important functions in cognitive processing, such as memory (Park et al. 2010), attention (Putman et al. 2014), etc. Engel et al. gave a unifying hypothesis that beta-band activities are in related to the maintenance of the current cognitive state (Engel & Fries 2010).

Instead of enhanced oscillation power as seen in the theta and alpha band, Stimulus-Response conflicts reduced spectral power in the beta band (Wang et al. 2014).

2.5 *Gamma oscillations (30-70 Hz)*

Gamma oscillations have been in relation with sensory integration and object representation (Tallon-Baudry & Bertrand 1999), working memory (Yamamoto et al. 2014), selective attention (Doesburg et al. 2012). The review concluded that Gamma oscillations possibly represented a universal code of central nervous system communication (Basar et al. 2000).

Cho et al. evidenced that higher desire of cognitive control were related to increases in gamma band activities over the prefrontal region for healthy subjects but that cognitive control-related adjustment

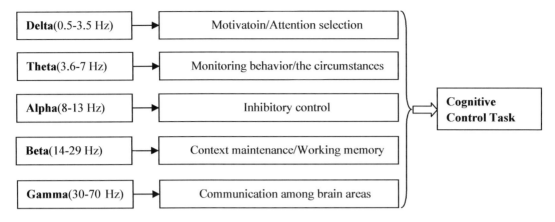

Delta(0.5-3.5 Hz)	→	Motivatoin/Attention selection	
Theta(3.6-7 Hz)	→	Monitoring behavior/the circumstances	
Alpha(8-13 Hz)	→	Inhibitory control	**Cognitive Control Task**
Beta(14-29 Hz)	→	Context maintenance/Working memory	
Gamma(30-70 Hz)	→	Communication among brain areas	

Figure 1. Brain oscillations model for a complete cognitive control process.

of prefrontal gamma band activities was absent for schizophrenia ones (Cho et al. 2006). Furthermore, Kieffaber et al. evidenced that induced cortical gamma-band activities would be increased during the anticipatory cognitive control when the demand for that cognitive control was only made implicitly available to subjects (Kieffaber & Cho 2010). Moreover, Tang et al. proposed that induced gamma activity represented cognitive control during detecting emotional expressions (Tang et al. 2011).

3 BRAIN OSCILLATIONS MODEL FOR A COMPLETE COGNITIVE CONTROL PROCESS

According to the mental operations needed by a complete cognitive control and the cognitive function of the brain oscillations, we proposed a brain oscillations model for a complete cognitive control process (See Figure 1).

Delta mainly contributes to the goal-directed process and attention selection. Theta monitoring our behavior and the circumstances is a important biomarker of cognitive control. Alpha plays a role of withholding or controlling the execution of a response. In addition, the cognitive function of beta is to maintain the context information of the current task and form the working memory. At last, gamma is used to communicate among brain areas before decision making. The model can account for our real cognitive control task such as Stroop task, Flank task, Simon task , etc.

4 CONCLUSIONS

In this paper, we have considered the specific role of different brain oscillations observed in neural activity during a general cognitive control process. We summarized the general cognitive function of brain oscillation and their power change during a cognitive control task. The main contribution of this work is to establish a brain oscillations model for a complete cognitive control task. The model is important for studying the mechanism of cognitive control, analyzing brain activity, processing EEG signal and resolving cognitive psychology process, etc. Although studies in this paper achieved some initial results, much in-depth research works were required to do in future.

ACKNOWLEDGMENT

The authors are grateful for the anonymous reviewers who made helpful comments. This work is supported by the National Natural Science Foundation of China (NO. 61171186 & NO. 61271345), the Natural Science Foundation of Heilongjiang Province (NO. F201313) and the Harbin Technological Foundation for Innovative Talents (NO. 2012RFQXG089).

REFERENCES

Balconi, M., & U. Pozzoli (2008). Event-related oscillations (ERO) and event-related potentials (ERP) in emotional face recognition. *International Journal of Neuroscience 118 (10)*,1412–1424.

Basar, E., C. Basar-Eroglu, S. Karakas, & M. Schurmann (2000). Brain oscillations in perception and memory. *International Journal of Psychophysiology 35 (23),* 95–124.

Cavanagh, J. F., M. X. Cohen, & J. J. Allen (2009). Prelude to and resolution of an error: EEG phase synchrony reveals cognitive control dynamics during action monitoring. *J Neurosci 29 (1),* 98–105.

Cavanagh, J. F., M. J. Frank, T. J. Klein, & J. J. Allen (2010). Frontal theta links prediction errors to behavioral adaptation in reinforcement learning. *Neuroimage 49 (4),* 3198–209.

Cavanagh, J. F., L. Zambrano-Vazquez, & J. J. B. Allen (2012). Theta lingua franca: A common mid-frontal substrate for action monitoring processes. *Psychophysiology 49 (2),* 220–238.

Cavanagh, James F., & Michael J. Frank (2014). Frontal theta as a mechanism for cognitive control. *Trends in Cognitive Sciences 18 (8),* 414–421.

Cho, R. Y., R. O. Konecky, & C. S. Carter (2006). Impairments in frontal cortical gamma synchrony and cognitive control in schizophrenia. *Proc Natl Acad Sci U S A 103 (52),* 19878–83.

Cohen, M. X., & J. F. Cavanagh (2011). Single-trial regression elucidates the role of prefrontal theta oscillations in response conflict. *Front Psychol 2,* 30.

Compton, R. J., D. Arnstein, G. Freedman, J. Dainer-Best, & A. Liss (2011). Cognitive control in the intertrial interval: evidence from EEG alpha power. *Psychophysiology 48 (5),* 583–90.

Doesburg, S. M., J. J. Green, J. J. McDonald, & L. M. Ward (2012). Theta modulation of inter-regional gamma synchronization during auditory attention control. *Brain Res 1431,* 77–85.

Engel, A. K., & P. Fries (2010). Beta-band oscillations--signalling the status quo? Curr Opin Neurobiol 20 (2), 156–65.

Hanslmayr, S., B. Pastotter, K. H. Bauml, S. Gruber, M. Wimber, & W. Klimesch (2008). The electrophysiological dynamics of interference during the Stroop task. *J Cogn Neurosci 20 (2),* 215–25.

Kieffaber, P. D., & R. Y. Cho (2010). Induced cortical gamma-band oscillations reflect cognitive control elicited by implicit probability cues in the preparing-to-overcome-prepotency (POP) task. *Cogn Affect Behav Neurosci 10 (4),* 431–40.

Klimesch, W., M. Doppelmayr, J. Schwaiger, P. Auinger, & T. Winkler (1999). 'Paradoxical' alpha synchronization in a memory task. *Brain Res Cogn Brain Res 7 (4),* 493–501.

Knyazev, G. G., J. Y. Slobodskoj-Plusnin, & A. V. Bocharov. 2009. Event-Related Delta and Theta Synchronization during Explicit and Implicit Emotion Processing. *Neuroscience 164* (4):1588–1600.

Krause, C. M., L. Sillanmaki, A. Haggqvist, & R. Heino. 2001. Test-retest consistency of the event-related desynchronization/event-related synchronization of the 4-6, 6-8, 8-10 and 10-12 Hz frequency bands during a memory task. *Clin Neurophysiol 112* (5):750–7.

Lakatos, P., G. Karmos, A. D. Mehta, I. Ulbert, & C. E. Schroeder. 2008. Entrainment of neuronal oscillations as a mechanism of attentional selection. *Science 320* (5872):110–113.

Mas-Herrero, E., & J. Marco-Pallares. 2014. Frontal theta oscillatory activity is a common mechanism for the computation of unexpected outcomes and learning rate. *J Cogn Neurosci 26* (3):447–58.

Maurer, U., S. Brem, M. Liechti, S. Maurizio, L. Michels, & D. Brandeis. 2014. Frontal Midline Theta Reflects Individual Task Performance in a Working Memory Task. *Brain Topogr.*

Nigbur, R., G. Ivanova, & B. Sturmer. 2011. Theta power as a marker for cognitive interference. *Clinical Neurophysiology 122* (11):2185–2194.

Park, H. D., B. K. Min, & K. M. Lee (2010). EEG oscillations reflect visual short-term memory processes for the change detection in human faces. *Neuroimage 53 (2),* 629–637.

Pfurtscheller, G., & F. H. L. da Silva (1999). Event-related EEG/MEG synchronization and desynchronization: basic principles. *Clinical Neurophysiology 110 (11),* 1842–1857.

Putman, P., B. Verkuil, E. Arias-Garcia, I. Pantazi, & C. van Schie (2014). EEG theta/beta ratio as a potential biomarker for attentional control and resilience against deleterious effects of stress on attention. *Cogn Affect Behav Neurosci 14 (2),* 782–91.

Tallon-Baudry, C., & O. Bertrand (1999). Oscillatory gamma activity in humans and its role in object representation. *Trends in Cognitive Sciences 3 (4),* 151–162.

Tang, D. D., & A. T. Chen (2013). Neural oscillation mechanisms of conflict adaptation. *SCIENTIA SINICA Vitae 43,* 992–1002.

Tang, Y., Y. Li, J. Wang, S. Tong, H. Li, & J. Yan (2011). Induced gamma activity in EEG represents cognitive control during detecting emotional expressions. *Conf Proc IEEE Eng Med Biol Soc 2011,* 1717–20.

Wang, K., Q. Li, Y. Zheng, H. Wang, & X. Liu (2014). Temporal and spectral profiles of stimulus-stimulus and stimulus-response conflict processing. *Neuroimage 89,* 280–8.

Yamamoto, J., J. Suh, D. Takeuchi, & S. Tonegawa (2014). Successful execution of working memory linked to synchronized high-frequency gamma oscillations. *Cell 157 (4),* 845–57.

Yordanova, J., M. Falkenstein, J. Hohnsbein, & V. Kolev (2004). Parallel systems of error processing in the brain. *Neuroimage 22 (2),* 590–602.

Electronic Engineering and Information Science – Wang (Ed.)
© 2015 Taylor & Francis Group, London, ISBN: 978-1-138-02772-5

Research on intelligent test paper generation based on knowledge point and paper template

S.M. Gao

School of Computer Science and Technology, Harbin University of Science and Technology, Harbin, China

ABSTRACT: Aiming at the intelligent algorithm of middle school examination, the key characteristic parameters for test paper and test question are analyzed and extracted. And an idea of using a test paper template to express the existing experience in test paper is put forward to improve the speed of auto generating paper. And knowledge point relationship model is constructed and applied to improve the quality of test paper. Finally, an intelligent algorithm of generating test paper is realized based on these.

KEYWORDS: middle school examination; intelligent test paper algorithm; template; knowledge point.

1 INTRODUCTION

Intelligent test paper, which has considerable related researches nowadays, is the core part in Computer-Aided Instructions' test field. At present intelligent test paper's realization methods, mainly are the following several kinds of types: randomized algorithm (Yuying Wang et al. 2003, Zelin Chen et al. 2010), backtracking algorithm (Bingyi Mao 2002, Peiguang Lin et al. 2006), genetic algorithm (Xiaodong Chen et al. 2005, Lian Xiao et al. 2008), particle swarm algorithm (Xinran Li et al. 2013) and fuzzy algorithm (Yun Qing 2013). Most of them are genetic algorithm.

When different algorithms are implemented, the control parameters of them come with a little difference. Within the process of development, in order to achieve better results, there are some researches which introduced a test question score as a control parameter (Yuying Wang et al. 2003, Zelin Chen et al. 2010). Meanwhile, test question time is introduced as a control parameter (Bingyi Mao 2002). Moreover, some researches adopt these two control parameters at the same time (Guixia Xiao et al. 2012, Li Zhou et al. 2013).

At present, the examination mainly includes the level examination, selection examination and diagnosis examination. Social examination (such as the CET-4 or CET-6) and the university courses' final examinations are level examination. While the senior high school entrance examination, the college entrance examination and the postgraduate examination, are all selection examinations. Diagnostic examinations are generally prepared for their corresponding examinations, and their forms are same as the corresponding test's forms, such as midterm examination, unit quiz, and so on. From the open

literature, most of these current intelligent algorithms are prepared for level examination.

According to the information gathered from the Internet, there are also some researches based on middle school examination's test paper. However, this kind of intelligent algorithms in the open literature is hardly recorded.

Attempting to directly apply those intelligent algorithms in the open literature to middle school's test paper, it is found that they are not appropriate enough. Therefore, intelligent algorithm of the middle school exam is analyzed and developed in detail in this paper. The best test paper standard and the key parameters affecting algorithm's performance will be demonstrated and cleaned out in following part as a start.

2 KEY PARAMETERS OF THE TEST PAPER AND TEST QUESTIONS

Examination paper is usually measured by the following factors: reliability, validity, difficulty, and discrimination. As a result, a paper's question type will be generated variety, distribution of difficulty should be equilibrium, covering the knowledge should be wide.

For selection examination, discrimination is more important. It is certain that the knowledge point coverage and important knowledge point reflection are also very important. If a test paper's discrimination is good, it is natural that the hardship and differentiation of entire test should be distinguished gradient. Theoretically speaking, a high level candidate should have high score; a low level candidate should have a low score in a same examination. In addition, good test paper can not have repeated questions, or several questions have the same knowledge points, or a lot of knowledge points concentrate to a small number of chapters etc..

Conventional test paper factors include total score, total time, difficulty coefficient, and chapter coverage. Question factors include question type, difficulty coefficient, chapter, and knowledge point.

In addition, from all kinds of test papers of many years, it is easy to see that question's score, difficulty and test time are not directly proportional to each other, whether they are the same type or not. For example, the same candidate at the same time does choice questions or comprehensive questions; the obtained score difference should be very distinct. Usually, the score obtained from choice question should be much higher than the score obtained from the comprehensive question per unit time. Even when the questions are same type and their difficulty is different, in most circumstances their scores are not directly in proportion to their times. For example, for many years in many Chinese provinces, score of choice questions are same in the college entrance examination papers, but their difficulty and estimated times are all not same. This situation also exists upon other types of questions. The reason is whether the question's difficulty and estimated time are proportional; it is not directly related to discrimination of test paper. Especially up to now there is no specific conclusion how to quantify these attributes such as difficulty. Factors affecting the difficulty of the questions include knowledge range, depth, complexity, etc. It is difficult to quantify these question's properties. Particularly for this kind of exam such as middle school mathematics, there are more differences. For example, the same math travel problems, their difficulty, time, and score in the first final exams are not same for that in the senior high school entrance examination. So directly using the fixed score and estimated time to restrict questions is not appropriate. However, the difficulty attributes of question can be defined under a more extensive unified framework.

It was found that adding the difficulty distribution control parameter of the same question type can improve test paper's quality. In addition, the relationship between question's score and difficulty in the same question type is added to control the scores of the same type questions of different difficulty are same or not.

According to the analysis and research above, aiming to the intelligence test paper composition of middle school examination, finally determined the test paper's parameters that directly affect the test paper generating algorithm are: subject, grade, semester, total score, total time, the overall coefficient of difficulty, difficulty distribution, chapters cover ratio, relationship between question's score and difficulty in the same question type. The question's control parameters are: question types, difficulty coefficient,

located chapter, used knowledge points. Score and time will not be considered as questionable control parameters in this article.

3 KNOWLEDGE POINT TREE AND TEST PAPER TEMPLATE

At present, knowledge point appears as a string in question attribute in current algorithms. It cannot play its full role. The knowledge point tree is constructed and shown in Figure. 1. The leaf node of the tree has chapter and importance attributes. And there can be more than one knowledge point ID in question attribute.

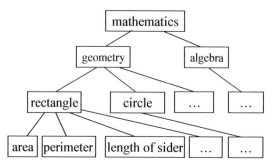

Figure 1.   Knowledge point tree.

In the absence of intelligent test paper generation algorithm for years, more reasonable examination paper generating modes are continually researched, including test time, test total score, question type, total question amount, test paper's difficulty, the difficulty of each question, and proportion of knowledge coverage. Naturally, a large number of related all kinds of examination questions are also accumulated. The test paper modes are constantly revised, verified and inspected for many years. So test paper template is put forward to refine and express these experience modes.

The major content of the test paper template is: total score, total time, the overall coefficient of difficulty, difficulty distribution proportion, question type, scores of all kinds of question types, total question amount, and the relationships between question scores and question difficulty.

Subject, grade and some other parameters are not adopted in the paper template, but need to be provided to the algorithm before generating test paper.

4 INTELLIGENT TEST PAPER ALGORITHM

Amount of preparation works needs to be done before generating test paper. The first is to construct, test and test question database. Many database tables also need to be created, including auxiliary subject, grade, question types, difficulty coefficient, knowledge points, chapters and sections, test paper template, and so on. Base data are also prepared. In addition, in order to facilitate the test paper, question's attributes are saved separately. The attributes closely related to the algorithm are placed in one table; others are placed in another table. The factors such as test paper's name, date, title and so on are not emphasized here. The algorithm is demonstrated as:

1 Set paper parameters and select a template;
2 Initialization.
   2.1. Construct question array. The array elements are objects of question class. The properties of the object include question ID, type, chapters, knowledge points, importance level and difficulty level. These question data are sorted by type, chapter, knowledge point, importance level and difficulty level when searching from database by SQL statement.
   2.2. According to known conditions, a two-dimensional control parameter array is calculated. One dimension is question type; the other is chapter. The elements of the array are objects which attributes are: distributive question score; minimum assigned scores proportion (0 or 1); optimal assigned scores proportion. In the algorithm these are adopted as control parameters.
   2.3. Confirm a judgment whether the test paper can be generated. If not (such as there are not any questions in any type), then tips or hints will be provided and proceed to (5).
3 Start the cycle treatment of the type which single unit score is the highest.
   3.1. Perform extraction processing algorithm. The input parameters of the algorithm are: question array, control table, current question type, questions test estimated difficulty. A specific question ID is the output. If the result is null, then proceed to (3.2); otherwise proceed to (3.3);
   3.2. Adjust the difficulty coefficient variable by increasing or decreasing a minimum difficulty unit), then proceed to (3.1).
   3.3. Modify the control parameters after a question ID is got.
   3.4. If there are other question types need to be processed, then get the next question type and turn back to (3.1); otherwise proceed to (4);
4 According to these specific question IDs, these questions' detail information can be searched from the database. Then the test paper can be generated and displayed.
5 End.

5 CONCLUSION

As researched, detailed above, in the current circumstances that kinds of question types exist, adding scores and time to question control parameters is not suitable. Thus, these two parameters will not be applied in the algorithm. At the same time, templates are adopted to express mature test paper experience, which makes the algorithm's speed substantially enhanced. In addition, because of the construction of knowledge points in the form of hierarchical model, the probability of question repetition is reduced. Therefore, the quality of test paper is improved.

REFERENCES

Yuying Wang, Shuang Hou, Maozu Guo (2003). Algorithm for automatic test paper generation. *Journal of Harbin Institute of Technology*. 35(3), 342–346.
Zelin Chen, Qingbiao Zhang (2010). Test—bank Design and Algorithm of Test System Based on JAVA. *Journal of Chongqing Institute of Tech-nology*. 24(3), 48–55.
Bingyi Mao (2002). An Intelligent Test Combination Algorithm of Test-sheets Based on Target Tree. *Computer Engineering and Applications*. 38(23), 245–246.
Peiguang Lin, Xiaozhong Fan (2006). Algorithm of On-line Generating Paper Based on Knapsack Problem. *Computer Engineering and Applica-tions*. 42(5), 165–166.
Xiaodong Chen, Hongyu Wang (2005). A exam-paper generating algorithm based on improved genetic algorithm. *Journal of Harbin Institute of Technology*. 37(9), 1174–1176.
Lian Xiao, Duwu Cui (2008). Design and imple-menttation of creation system for test paper based on genetic algorithm. *Journal of Computer Applications*. 28(5), 1362–1364.
Xinran Li, Yongsheng Fan (2013). Study on Intelligent Test Paper Generation Strategy through Improved Quantum-behaved Particle Swarm Optimization. *Computer Science*. 40(4), 236–239.
Yun Qing (2013). Research on fuzzy-Logic algorithm of intelligent generating test paper. *International Journal of Emerging Technologies in Learning*. 8(4), 47–50.
Guixia Xiao, Wuchu Zhao (2012). Iterant Problems Replacement Method Based on Genetic Algorithm with Test Paper Intelligent Generation. *Computer Engineering*. 38(11), 150–152.
Li Zhou, Jue Wang, Yong Zhou (2013). Application of Test Paper Generation Based on Genetic Algorithm. *Journal of Jiangxi Normal University (Natural Sciences Edition)*. 37(6), 579–583.

Electronic Engineering and Information Science – Wang (Ed.)
© *2015 Taylor & Francis Group, London, ISBN: 978-1-138-02772-5*

Determine weights of evaluation indices for E-Commerce websites ranking based on fuzzy AHP

H. Liu, H.W. Xuan & X. Cui
School of Software, Harbin University of Science and Technology, Harbin, China

V.V. Krasnoproshin
Faculty of Applied Mathematics and Computer Science, Belarusian State University, Minsk, Belarus

ABSTRACT: Some enterprises need E-commerce experts to rank their website alternatives. Recent related researches do not serve this problem. As a well-used multi-criteria decision making methods, a fuzzy variant of analytic hierarchy process has been used to calculate evaluation weights. A multi-criteria decision-making model has been build to rank alternatives based on opinions from experts. Method and case study have been clearly described. In conclusion, fuzzy analytic hierarchy process is presented to evaluate electronic commerce websites.

KEYWORDS: fuzzy logic, fuzzy inference, fuzzy AHP, electronic commerce website.

1   INTRODUCTION

Nowadays, economic commerce (E-Commerce) provides more convenient, faster and cheaper way of shopping, internet banking, employing etc. As the result, it becomes necessary for both companies and customers to evaluate the E-Commerce websites. However, evaluation of an E-Commerce website is not familiar with most of people. It includes quite a lot of technical and professional knowledge. As well as the evaluation method is not the tradition one which is using Boolean logic. However, it is a kind if logic related artificial fuzzy logic (Zadeh 1965). It attempts to model our sense of words, our decision-making and our common sense (Chen 2001). In the evaluation process, there are many criteria and sub-criteria. So it is natural to bring in analytic hierarchy process (AHP) (Saaty 1980), which is commonly used multiple criteria decision-making problem.

To achieve the previous aims, we decided to research an approach of evaluating E-Commerce website based on Fuzzy-AHP. Furthermore, a case study has been described to rank the evaluation weights of criteria and sub-criteria for e-commercial website evaluation.

2   BACKGROUND

The weights determine methods of E-commerce website evaluation include artificial intelligence (AI) techniques, such as expert systems, artificial neural networks (ANNs), and fuzzy sets theory used in inference will also be reviewed (Achabal 1982, Arnold 1983, Goodchild 1984).

Recent research has showed that the applications of ANN techniques in domain of knowledge based system are very promising. A fuzzy rule-based neural network instead of the traditional steepest descent technique was presented for multiple criteria decision making (Hanss 2005, Hung 2004). The ANN achieved good results in evaluating and ranking alternatives. In addition, the trainability and applicability of ANN techniques to addressing general multi-attribute utility methods problems were also confirmed. In order to evaluate the capability of ANN with error back-propagation learning algorithm in decision analysis, three types of multi-attribute functions (additive, quadratic and Chebyshev) were implemented and got excellent solutions for the presented problems.

Barnes (2002) designed a "WebQual" system to assess E-commerce quality with no interests on quantity considering. Miranda (2006) designed a quality evaluation method with criteria of functionality, usability, efficiency and reliability, but it can not combine experts opinions. Abd El-Aleem (2005) constructed a mathematical model to compare website traffic, but had not covered the synthesizing the weights of several experts. Chu (2006) presented the ranking of websites from best to worst using fuzzy logic, however could not know the absolute value of each website. Hung (2004) developed an evaluation instrument for E-commerce websites, which was only user's satisfaction from the first-time buyer's view and discussed less about quality analysis.

3 WEIGHTS DETERMING METHOD

The standard AHP method, and its extension Fuzzy-AHP (Laarhoven 1983), proved to be an efficient tool for approaching performance evaluation. AHP uses the pair-wise comparison matrix to evaluate the uncertainty in MCDM problems as in formula (1):

$$A = [a_{ij}] = \begin{array}{c} \\ C_1 \\ C_2 \\ \vdots \\ C_n \end{array} \begin{array}{cccc} C_1 & C_2 & \cdots & C_n \\ \begin{bmatrix} 1 & a_{12} & \cdots & a_{1n} \\ a_{21} & 1 & \cdots & a_{2n} \\ \vdots & \vdots & \vdots & \vdots \\ a_{n1} & a_{n2} & \cdots & 1 \end{bmatrix} \end{array} \quad (1)$$

where C_1, C_2, ..., C_n represent the set of elements, while a_{ij} denotes a enumerated decision on a pair of elements C_i, C_j. The relative importance of two elements is rated using a scale with the values 1, 3, 5, 7, and 9, where "1" represents that two elements are equally important, while the other extreme "9" represents that one element is important than the other.

There are more than a few ways to denote fuzzy numbers. One unusual way of fuzzy numbers is triangular fuzzy numbers, which is easy to model and is used well with most problems. Geometric mean method is used for triangular fuzzy numbers. In this research, triangular fuzzy numbers are used to denote personal pair-wise comparisons of evaluation process in order to capture the imprecision. A triangular fuzzy number is signified just as (L, M, U). The constraints L, M and U, correspondingly, represent the smallest value, the most probable value and the largest value describing a certain event (Laarhoven Van 1983).

The triangular fuzzy numbers $\tilde{u}_{ij}$ are recognized as follows:

$$\tilde{u} = (L_{ij}, M_{ij}, U_{ij}) \quad (2)$$

where $L_{ij} \leq M \leq U_{ij}$ and $L_{ij}, M_{ij}, U_{ij} \in [1/9, 1] \cup [1,9]$.

$$L_{ij} = \min(B_{ijk}) \quad (3)$$

$$M_{ij} = \sqrt[n]{\prod_{k=1}^{n} B_{ijk}} \quad (4)$$

$$U_{ij} = \max(B_{ijk}) \quad (5)$$

where B_{ijk} represents a decision from expert k for the relative importance of two elements C_i-C_j. The triangular fuzzy numbers, $\tilde{1}$-$\tilde{9}$, are applied to progress the predictable scaling scheme. The fuzzy AHP method includes the following steps.

- Step 1. Hierarchical modeling.

Defining the evaluative criteria used to evaluate the E-Commerce website, and establish a hierarchical framework.

- Step 2. Comparing pair-wise and constructing each fuzzy judgment matrix.

We took triangular fuzzy numbers to show the relative strength of each pair of elements on the same level. The relative strength can be the relative importance of criteria along with the relative importance of features related with each criterion. The fuzzy judgment matrix $\tilde{A}$ is constructed as below via expending triangular fuzzy numbers:

$$\tilde{A} = [\tilde{a}_{ij}] = \begin{array}{c} \\ C_1 \\ C_2 \\ \vdots \\ C_n \end{array} \begin{array}{cccc} C_1 & C_2 & \cdots & C_n \\ \begin{bmatrix} 1 & \tilde{a}_{12} & \cdots & \tilde{a}_{1n} \\ \tilde{a}_{21} & 1 & \cdots & \tilde{a}_{2n} \\ \vdots & \vdots & \vdots & \vdots \\ \tilde{a}_{n1} & \tilde{a}_{n2} & \cdots & 1 \end{bmatrix} \end{array} \quad (6)$$

where $\tilde{a}_{ij}$ represents a triangular fuzzy number for the relative strength of two elements C_i and C_j.

- Step 3. Defuzzification.

The defuzzification method was consequent from (Hus 1997, Liou 1992). As shown in Eq. (7), this method can clearly express fuzzy perception. Outstanding to the ability of this method to obviously show the preference (a) and risk tolerance (λ) of decision makers, they can more thoroughly understand the risks they face in different situations.

$$\left(a_{ij}^{\alpha}\right)^{\lambda} = \left[\lambda \cdot L_{ij}^{\alpha} + (1-\lambda) \cdot U_{ij}^{\alpha} \right], \quad (7)$$

$0 \leq \lambda \leq 1$, $0 \leq \alpha \leq 1$, $i < j$

where $L_{ij}^{\alpha} = (M_{ij} - L_{ij}) \cdot \alpha + L_{ij}$ denotes the left-end value of α-cut for a_{ij} and $U_{ij}^{\alpha} = U_{ij} - (U_{ij} - M_{ij}) \cdot \alpha$ denotes the right-end value of α-cut for a_{ij}.

$$\left(a_{ij}^{\alpha}\right)^{\lambda} = 1 \big/ \left(a_{ji}^{\alpha}\right)^{\lambda}, 0 \leq \lambda \leq 1, 0 \leq \alpha \leq 1, i > j \quad (8)$$

where α can be seen as a stable or fluctuating condition (Hus 2000, Saaty 1990). It reflects the uncertainty of the judgment and ranges from 0 to 1. Eleven values, 0,..., 1 are used to emulate. The range is the greatest when $\alpha = 0$. $\alpha = 0$ represents the upper-bound U_{ij} and lower-bound L_{ij} of triangular fuzzy numbers, and $\alpha = 1$ represents the geometric mean M_{ij} in triangular fuzzy numbers. λ can be viewed as the degree of pessimism of decision maker and ranges from 0 to 1. $\lambda = 0$ means that the decision maker is more optimistic and, thus, the expert accord is upper bound U_{ij}. In contrast, $\lambda = 1$ means that the decision maker is pessimistic (Ayag 2006, Wu 2006). The single pair-ware comparison matrix can be expressed as following:

$$[(A^\alpha)^\lambda] = [(a_{ij}^\alpha)^\lambda] = \begin{bmatrix} 1 & (a_{12}^\alpha)^\lambda & \cdots & (a_{1n}^\alpha)^\lambda \\ (a_{21}^\alpha)^\lambda & 1 & \cdots & (a_{2n}^\alpha)^\lambda \\ \vdots & \vdots & \vdots & \vdots \\ (a_{n1}^\alpha)^\lambda & (a_{n2}^\alpha)^\lambda & \cdots & 1 \end{bmatrix} \quad (9)$$

- Step 4. Calculation of the eigenvalue and eigenvector.

Remarkably, $\bar{\lambda}$ is assumed to be eigenvalue of the single pair-ware comparison matrix $(A^\alpha)^\lambda$ [10].

$$(A^\alpha)^\lambda \cdot W = \bar{\lambda} \cdot W \quad (10)$$

$$[(A^\alpha)^\lambda - \bar{\lambda}] \cdot W = 0 \quad (11)$$

where W signifies the eigenvector of $(A^\alpha)^\lambda$, $0 \le \lambda \le 1$, $0 \le \alpha \le 1$.

Using the comparison matrix, the eigenvectors were calculated according to Eqs. (10) and (11).

- Step 5. Total rating.

Calculate the total level weight to evaluate the E-commerce websites.

4 CASE STUDY

In this part, a detailed case study has been presented. The proposal of fuzzy AHP to determine the weights of criteria and subcriteria for E-Commerce website evaluation is given to prove its applicability (Torfia 2010, Luo 2011, Chamodrakas 2010, Kayaa 2011, Büyüközkana G. 2011, Wanga 2011).

The overall goal has been stated as determining the weights of criteria and sub-criteria for E-Commerce website evaluation. Finally, the best non-fuzzy performance values have been calculated, according to which the rank of E-Commerce websites alternatives has been presented. Based on the Matlab, a general consensus among experts can be reached to establish a three-leveled hierarchical structure (see Table 1).

The overall weights of criteria and sub-criteria for a certain application have been listed in Table 2.

Similarly, the BNP values for all possible alternatives with a view to their subsequent comparison were prepared. Results of calculations are shown in Table 3.

The best on weight is the alternative at number 5, apparently from results of an estimation of alternatives (in Table 3). Results of table 3 reflect the fact that changes of weights of criteria in the certain degree influences results of ranking. It is clear, that the order of ranking as a whole depends on weights of criteria, which are used for an estimation of alternatives.

Table 1. The hierarchical structure of criteria for E-Commerce website evaluation.

Level 1: Goal	Level 2: Criteria	Sub-criteria
Calculating the weights of all criteria and sub-criteria for E-Commerce website	Usability (C_1)	Accuracy (C_{11})
		Authority (C_{12})
		Current information (C_{13})
		Efficiency (C_{14})
	Reliability (C_2)	Security (C_{21})
		Functionality (C_{22})
		Integrity (C_{23})
		Navigation (C_{24})
	Design (C_3)	Aesthetic features (C_{31})
		Contents (C_{32})
		Layout (C_{33})
		Standard conformance (C_{34})

Table 2. Weights of the 3 criteria and 12 sub-criteria.

Criteria	Weights of Criteria	Sub-Criteria	Weights of Sub-Criteria	Total Weights
usability	0.315	C_{11}	0.274	0.086
		C_{12}	0.390	0.123
		C_{13}	0.172	0.054
		C_{14}	0.164	0.052
reliability	0.462	C_{21}	0.211	0.097
		C_{22}	0.115	0.053
		C_{23}	0.178	0.082
		C_{24}	0.495	0.229
design	0.223	C_{31}	0.341	0.076
		C_{32}	0.295	0.066
		C_{33}	0.198	0.044
		C_{34}	0.166	0.037

273

Table 3. Estimation and rank.

Alternatives	BNP_i	Rank
A-1	161.23	3
A-2	145.17	4
A-3	140.08	5
A-4	169.56	2
A-5	176.21	1

5 CONCLUSION AND ACKNOWLEDGMENT

Because ambiguity is an unnecessary and predictable step of evaluation process, the fuzzy AHP is used to allow people to account for the impression of ambiguity. The whole progress of Fuzzy-AHP approach has been discussed and used. The major lead of this research is that it can be used for both qualitative and quantitative criteria. The results show that the model has the capability to be flexible and apply in E-Commerce website evaluation. And also the Fuzzy-AHP method can be extended to various evaluations involving human subjectivity.

This research was supported by "Science and Technology Research Project of the Education Department of Heilongjiang Province", under the project "Intelligent multi-criteria decision-making method and applied research based on fuzzy AHP and neural network" with project No.12541150.

REFERENCES

Achabal D.D., W.L. Gorr, & V. Mahajan (1982). MULTILOC: a multiple store location decision model. *J. Retail. 58 (2)*, 5–25.

Arnold D.R., L.M. Capella, & G.D. Smith (1983). Strategic Retail Management. Addison-Wesley, *Reading, MA.* 295, 297.

Ayag Z. & R. G. Özdemir (2006). *J. Int. Manu.* 17, 179.

Barnes S. J. & R.T. Vidgen (2002). J. Electron. Commer. 3, 114.

Büyüközkana G., G. Çifçia, & S. Güleryüz (2011). *Exp. Syst. App.* 38, 9407.

Chamodrakas I., D. Batisa, & D. Martakosa (2010). *Exp. Sys. App.* 37, 490.

Chen S. (2001). *Fuzzy Sets and Systems.* 118, 75.

Chu F. & Y. Li (2006). In Proc. Int. Conf. "Management Science and Engineering", *Lille, France*, pp.111–115.

El-Aleem A., W. El-Wahed, N. Ismail, & F. Torkey (2005). *World Acad. Sci. Eng. Tech.* 4, 20.

Goodchild M.F. (1984). ILACS: a location-allocation model for retail site selection. *J. Retail.* 60(1), 84–100.

Hanss M. (2005). Applied Fuzzy Arithmetic: An Introduction with Engineering Applications. *Springer, Netherlands.*

Hung W. & R. J. McQueen (2004). Electron. *J. Inform. Syst. Eval.* 7, 31.

Hus T. H. & S. H. Nian (1997). *J. Trans. Taiwan.* 10, 79.

Hus T. H. & T. H. Yang (2000). *J. Man. Syst.* 7, 19.

Kayaa T. & C. Kahraman (2011). *Tech. Eco. Dev. Eco.* 17, 313.

Laarhoven Van, P. J. M., & W. Pedrcyz (1983). *Fuz. Sets Syst.* 11, 229.

Liou T. S. & M. J. Wang (1992). *Fuzz. Sets Syst.* 50, 247.

Luo Q. & Y. Wang (2011). *Adv. Mat. Reseach.* 181, 987.

Miranda F., R. Corte's, & C. Barriuso (2006). *Electron. J. Inform. Syst. Eval.* 9, 73.

Saaty T. L. (1980). The Analytic Hierarchy Process. *McGraw-Hill, New York,*

Saaty T. L. (1990). *Euro. J. Oper. Res.* 48, 9.

Torfia F., R. Z. Farahanib, & S. Rezapourd (2010). *App. Sof. Com.* 10, 520.

Wanga X., H. K. Chana, R. W.Y. Yeeb, & I. Diaz-Raineya (2011). *Int. J. Prod. Econ.*

Wu C. R., C. W. Chang, & H. L. Lin (2006). *J. Amer. Ac. Bus.* 9, 201.

Zadeh L. (1965). *Information and Control.* 8, 338.

Electronic Engineering and Information Science – Wang (Ed.)
© 2015 Taylor & Francis Group, London, ISBN: 978-1-138-02772-5

Building of numerical models and simulation research for magnetite core inductive sensor

L.J. Wang
College of Measure-Control Technology and Communication Engineering, Harbin University of Science and Technology, the Higher Educational Key Laboratory on Measuring & Control Technology and Instrumentations of Heilongjiang Province, Harbin, China

Y.B. Han
Harbin Normal University High School, Harbin, China

H.Q. Chen
College of Geoscience and Surveying Engineering, China University of Mining and Technology, Beijing, China

S.Y. Xiao, D.D. Hui, J.X. Li & L.K. Sun
College of Measure-Control Technology and Communication Engineering, Harbin University of Science and Technology, Harbin, China

ABSTRACT: From perspective of magnetite iron core inductive sensor, iron grade's quick detecting method for magnetite is given. Sensing mechanism of the sensor is analyzed. Magnetic field model is built, and numerical analytic expressions are given among magnetic induction intensity, inductance value and induced electromotive force of the iron core inductive sensor. By using the finite element energy method, numerical model of inductance of cylindrical solenoid iron core winding is solved. Boundary conditions are set according to geometric characteristics of the winding, and adopting ANSYS finite element analysis software, simulation models for windings of the iron grade sensor are built; influence regulations of changing of self-inductance coefficients of the winding on output inductance value of the sensor are gotten, furthermore, optimal combination of self-inductance coefficients of structure of the magnetite core inductance sensor is determined. The research lays foundation for ultimate realization of real-time monitoring of iron grade for the magnetite.

KEYWORDS: Inductive sensor; Magnetite core; Numerical models; Finite element analysis.

1 INTRODUCTION

Real-time monitoring to iron grade during iron ore dressing could follow up changes of iron grade of materials; accordingly, timely adjusting process parameters can not only reduce the loss of product in tails, but also improve and ensure the quality of iron ore products. Routine testing of the iron grade for the iron ore usually are arranged after completing dressing and processing, and generally are made by samples drawing. Testing methods of chemical analysis are commonly used in the routine testing above.

It is the accurate test results of the chemical analysis methods provided by the national standard methods, but these analysis methods usually require making in special laboratory, and completing measuring using special chemical reagents. So the chemical analysis methods are not adequate for real-time, on-line and quick measuring of the iron grade for iron

ores. In this paper, new method for detecting total iron grade of magnetite based self-inductive sensor is given. Analyzing from the perspective of sensing mechanism of the magnetite core inductance sensor, the inductance value solving models are built for inductive winding of the magnetite core iron grade inductive sensor. It can provide foundation and basis of the research work above for solving key problems such as the subsequent measuring and modeling, experimental designing, and the structure and performance optimization during the development of quick detecting apparatus for iron grade of iron ores.

2 DETECTING METHOD AND MECHANISM

Inductance is an inherent character of the inductive winding itself, it is a parameter related to number of turns, size, shape, and magnetic medium of the winding. Inductance is a magnitude of inertia of the inductive winding, and it is not relevant to impressed current. Adopting measured iron ore to be iron core of the cylindrical screw pipe inductive winding, according to experience formula that is shown in (1), when self-inductive coefficients including outer diameter, inner diameter, height, number of turns of the inductive winding remain unchanged, the inductance value of the winding is proportional to the relative permeability of the iron core.

$$L = \frac{kN^2 \mu_r \mu_0 A}{l} \qquad (1)$$

where L is inductance value; N is number of turns; μ_0 is permeability of vacuum of the iron core, and $\mu_0 = 4\pi \times 10^{-7}$; μ_r is relative permeability, and $\mu_r = 1$ when air core winding; A is cross-section area of the winding; l is height of the winding; value of k is decided by ratio of radius R and height l of the winding, and some values of the relations between $2R/l$ and k is shown in Table 1.

Table 1. Relations between $2R/l$ and k.

$2R/l$	0.1	0.2	0.3	0.4	0.6	0.8	1.0	1.5	2.0
k		0.96	0.92	0.88	0.85	0.79	0.74	0.69	0.60	0.52

According to (1), when the self inductive coefficients remain unchanged, the inductance value of the winding is related to existing or no existing of the magnetic iron core. The measuring iron ore samples are set in the air core winding to be core, inductance value will increase, and quality factor of the inductive winding will improve. When iron grade of the measured iron ore samples changing, relative permeability also change, the inductance value and resistance value of the winding are also changed too.

Adopting triangular wave AC excitation signal to act on the two ends of the winding to excite magnetic field, according to the law of electromagnetic induction:

$$\varepsilon = L \frac{di(t)}{dt} \qquad (2)$$

where ε is self induced electromotive force of the inductive winding; L is the inductance value of the winding; $i(t)$ is excitation current.

According to (2), when applying the triangle wave excitation current to the inductive winding, self induced electromotive force ε is proportional to L within half a cycle.

As have discussed above, changing of the self induced electromotive force of the winding reflects the big or small of the inductance value, and the inductance value reflects big or small of the iron grade value.

Integrating triangular wave AC excitation circuits, iron ore core, and the cylindrical screw pipe inductive winding, the iron ore core self inductive sensor is designed. Designing and assembling the inductive winding keel, the iron grade sensor, and external shield, information acquisition testing probe based the iron grade inductive sensor is developed. After iron grade information is acquired by using the probe, it is transferred to the sensor testing circuits. Then, corresponding voltage signal is processed by the signal conditioning circuit. Finally, corresponding digital signal is input into the single chip microcomputer control system (Zheng L.P. & J.J. Chen 2002), so the quick and real-time detecting of the iron ore grade during dressing, processing or trading of the iron ore products will be realized.

3 MEASURING SAMPLES OF MAGNETITE

Magnetite is most commonly used raw materials of smelting iron. Chemical composition of the magnetite is Fe_3O_4, and it can also be expressed as $FeO \cdot Fe_2O_3$. Iron elements in magnetite presents form of ferric iron, and form of ferrous is also exist. Magnetite belongs to paramagnetic ferromagnetism material, it has strong magnetism, and its magnetism is the strongest in minerals.

In our studies, raw ore of the selected magnetite is broken into 0~50mm, by using the chemical analysis method, content of total iron (TFe) of useful chemical composition of the ore is 21.0~41.2%, content of TFe is usually more than 30%, and contents of the harmful ingredients such as sulfur, phosphorus are very low. After processed such as ball grinder, magnetic separation, and dehydration, fine iron powder whose grade is 65.0~66.5% is obtained, and granularity achieves to 60%~70% under 200-mesh sieve. It is shown of multi-element chemical analysis results for fine iron powder sample of the magnetite that content of SiO2 is 4.2%, maximum content of Al_2O_3 is 0.21%, maximum content of S is 0.05%, maximum content of P is 0.07%, and content of Moisture is 8.0%~9.0% (Kong L.T. 2003, Luo L. F. et al. 2014).

4 MAGNETIC FIELD MODEL

When triangle wave AC excitation signal is applied to the inductance winding, alternating electromagnetic field will be come into being. Schematic diagram of the magnetic field in vertical eccentric line at the

XOY plane for circular winding is shown in Figure 1 (Yellishettya M. et al. 2010).

In Figure 1, radius of the circular winding is ρ, current-carrying of the winding is I. dH is magnetic field generated by electric current unit $d\alpha$ at space dot $P(l,0,h)$. Line segments b is in the XOY plane, and the line is parallel to the radius of $d\alpha$. Length of connecting line between intersection of b and tangent line of the circle at the unit current $d\alpha$ and the dot P is a. Therefore,

$$dH = \frac{I\rho d\alpha}{4\pi r^2}\sin\theta = \frac{I\rho d\alpha}{4\pi r^2}\frac{a}{r} = \frac{I\rho d\alpha}{4\pi}\frac{\sqrt{h^2+b^2}}{r^3}$$

$$= \frac{I\rho d\alpha}{4\pi}\frac{\sqrt{h^2+b^2}}{\sqrt{(h^2+c^2)^3}}$$

$$= \frac{I\rho d\alpha}{4\pi}\frac{\sqrt{h^2+b^2}}{\sqrt{(h^2+l^2+\rho^2-2\rho l\cos\alpha)^3}}$$

$$= \frac{I\rho d\alpha}{4\pi}\frac{\sqrt{h^2+\rho^2+l^2\cos^2\alpha-2\rho l\cos\alpha}}{\sqrt{(h^2+\rho^2+l^2-2\rho l\cos\alpha)^3}} \quad (3)$$

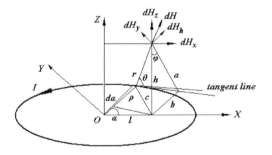

Figure 1. Schematic diagram of the magnetic field of circular winding at the XOY plane.

If making integration to (3), magnetic field intensity of the arbitrary spatial dot P of circular current-carrying winding would be obtained.

Furthermore, dH can be decomposed into dH_x, dH_y, and dH_z. It is shown in Figure 1 that dH_h is composition of dH_x and dH_y in horizontal direction, and dH is composition of dH_h and dH_z in vertical direction. Due to:

$$dH \perp a, \quad dH_h \perp h$$

So,

$$dH_h = dH\cos\varphi = dH\frac{h}{a}$$

$$= \frac{I}{4\pi}\frac{\rho h d\alpha}{\sqrt{(h^2+\rho^2+l^2-2\rho l\cos\alpha)^3}} \quad (4)$$

Supposing number of turns of the inductor winding is N, cross-sectional area perpendicular to the axial direction is S, relative magnetic permeability of magnetite is μ_r, and air relative permeability is 1, so radial micro component of magnetic induction intensity of arbitrary spatial dot P is gotten in (5).

$$dBr = \mu_r N dH_h = \frac{\mu_r NI}{4\pi}\frac{\rho h d\alpha}{\sqrt{(h^2+\rho^2+l^2-2\rho l\cos\alpha)^3}} \quad (5)$$

Making integration to (5), corresponding radial component of magnetic induction intensity is obtained according to (6).

$$B_r = \frac{\mu_r NI}{4\pi}\int_0^{2\pi}\frac{\rho h d\alpha}{\sqrt{(h^2+\rho^2+l^2-2\rho l\cos\alpha)^3}} \quad (6)$$

5 SIMULATION ANALYSIS AND EXPERIMENT

5.1 Numerical analysis for using finite element energy method to solve winding inductance value

Theoretical basis of the finite element energy method is the finite element method of structural matrix analysis. It is advocated of the approach that approximate function instead of each unit is continuous, and unit boundaries are also continuous. Therefore, through the integration of individual solutions, all solutions of the system can be gotten.

Taking air core winding for example, medium is considered linear, energy density ω of any point is

$$\omega = \frac{1}{2}BH = \frac{1}{2}B\frac{B}{\mu_0} = \frac{1}{2\mu_0}B^2 \quad (7)$$

where B and H are respectively magnetic induction intensity and magnetic field intensity.

Making integration to magnetic energy density of magnetic field in the whole volume v, then, magnetic energy of the inductor can be gotten.

277

$$W = \int_v \omega \, dv = \int_v \frac{1}{2\mu_0} B^2 \, dv \qquad (8)$$

During parsing to magnetic field distribution using finite element method, firstly, continuous space field is discrete into several units, and finite number of nodes is set in each unit; secondly, nodes at the boundaries are linked to each other as a whole. Therefore, total magnetic energy W within a region solved for the hollow inductive winding can be calculated.

$$W = \sum_{k=1}^{n} \frac{1}{2\mu_0} B^2(k)v(k) \qquad (9)$$

where n is total number of the subdivided units; $B(k)$ and $v(k)$ are respectively magnetic induction intensity and volume of the unit k.

According to the theory of electromagnetic field, magnetic energy is the result of work done by the external sources in the flowing process of loop current, and is distributed in the whole space domain of the existence of the magnetic field (Yang X.M. et al. 2006). Cylindrical inductance winding can be thought of as a loop or whole, then, total magnetic energy W stored in the magnetic field of the inductance is

$$W = \frac{LI^2}{2} \qquad (10)$$

where I is current flowing through the inductive wing; L is inductance value of the winding.

According to (10), numerical calculation formula of inductance of the cylindrical hollow inductive winding can be obtained.

$$L = \frac{2W}{I^2} = \frac{2}{I^2} \sum_{k=1}^{n} \frac{1}{2\mu_0} B^2(k)v(k) \qquad (11)$$

5.2 Analysis of the influence of change of self inductive coefficients and design of structure parameters of the sensor

Hollow inductance is linear, design process is simple, and inductance value is constant. Iron grade sensor is designed to set magnetite medium as core of the inductive winding; the permeability will present nonlinear variation regular in the process of testing (Xie X.L. & J.Q. Xu 2011). It is a process of first increasing and then reducing of the changing of magnetic permeability with changing of the magnetizing current. To get a smaller magnetic permeability, it can be when small magnetizing current; it can also be at the time of magnetization curve being close to saturation. So, the design of

the magnetite core inductive sensor becomes more flexible. Orthogonal test design scheme of the experimental design methods is adopted for structure designing of the sensor winding. In order to get stable inductance value, when testing, it is need to make the inductive sensor work in the region that the permeability is relatively stable.

Magnetic permeability of ferromagnetic material is high, usually, 100 to 10000. It is set to be 650 of magnetic permeability of magnetite core in simulation. Setting different combinations of the self inductive coefficients of winding, geometrical characteristics of magnetite core inductive winding and the corresponding boundary conditions are determined, and making use of the ANSYS finite element simulation software, different simulation models for the inductive windings of the iron grade sensor are set up. Changing curves of output inductance value of the windings following the respectively changing of structure parameters including inner diameter, outer diameter, height, and number of turns are achieved, and it is shown in Figure 2 to Figure 5.

According to changing regulations as shown in Figure 2 to Figure 5, linear variation area of the self inductive coefficients including inner diameter, outer diameter, height, and number of turns are determined respectively. For the above four factors, three testing levels are accordingly set for each factor. Orthogonal experimental design $L_9(3^4)$ is made to optimization design, experiments and test analysis for the structure parameters combinations of the sensor. Eventually, the structure size parameters configuration of the magnetite core inductive sensor is determined that the outer diameter is 66mm, the inner diameter is 16mm, the number of turns of the winding is 1000, and the height is 20mm.

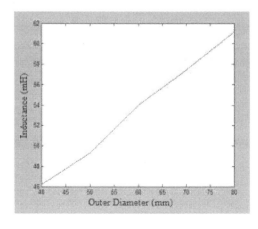

Figure 2. Curve of outer diameter and inductance value of the winding.

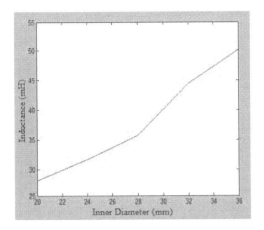

Figure 3. Curve of inner diameter and inductance value of the winding.

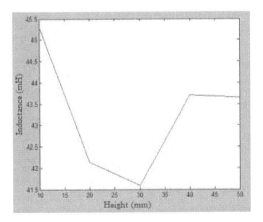

Figure 4. Curve of height and inductance value of the winding.

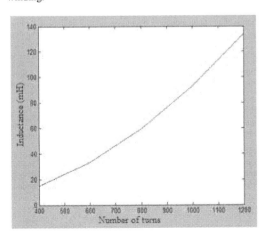

Figure 5. Curve of number of turns and inductance value of the winding.

6 CONCLUSIONS

Quick detecting method is researched for the iron grade of the iron ore during the iron ore processing and trading. By parsing the electromagnetic inductive mechanism of the iron grade sensor, based theory of electromagnetic induction type iron grade detecting method is explored, and it can provide the reference to rapid and real-time detection to magnetite and similar products. To meet to the new development request of sensor testing technology and iron ore processing testing technology (Ou Y.Y. & X.J. Hu 2006, Tang Y.H. & L.C. Qiu 2011, Chen Z.B. 2008), it is of great significance of exploring the iron grade detecting theory and method based on the self inductive sensor to improve the detecting level of China's iron ore processing testing. Overall management level, controlling and analysis level of the iron ore and steel manufacturing enterprises will be improved by means of realizing automation controlling to the iron ore processing and dressing testing. From the viewpoint of development, the rapid detection method research to iron grade of magnetite conforms to the trend and direction of development with high quality and finish machining of international iron ore dressing and processing industry.

ACKNOWLEDGMENTS

This work was financially supported by the project, Scientific Research Fund of Heilongjiang Provincial Education Department of China (12531142).

REFERENCES

Zheng L.P. & J.J. Chen (2002). A calculation method of the flux and inductance of dry-type air reactors. *High Voltage Engineering 28(10)*, 18–22.

Kong L.T. (2003). Progress and outlook of scientific and technological for technology of China's ironmaking raw material. *China Metallurgical 4*, 1–4.

Luo L. F. et al. (2014). Study of vibration modal of inductive filtering transformer. *Advanced Technology of Electrical Engineering and Energy 33(5)*, 27–31, 43.

Yellishettya M. et al. (2010). Iron ore and steel production trends and material flows in the world: Is this really sustainable. *Resources, Conservation and Recycling 54(12)*, 1084–1094.

Yang X.M. et al. (2006). Study on on-line analysis measurement system for grade of iron concentrate. *Microcomputer Information 22(11-1)*, 197–198, 292.

Xie X.L. & J.Q. Xu (2011). *Fundamental Theory of Electrical Engineering*. Beijing: Tsinghua University Press.

Ou Y.Y. & X.J. Hu (2006). Development review of the sensing technology. *Electrotechnics 11*, 1.

Tang Y.H. & L.C. Qiu (2011). Research on the measurement system for grade of magnetic iron powder. *Process Automation Instrumentation 32(4)*, 53–58.

Chen Z.B. (2008). Introduction to discussion for the national standard of iron ore chemical analysis method. *Metallurgical Standardization and Quality 38(6)*, 58–59.

Electronic Engineering and Information Science – Wang (Ed.)
© 2015 Taylor & Francis Group, London, ISBN: 978-1-138-02772-5

Determine sense of ambiguous word based on part of speech

C.X. Zhang
College of Information and Communication Engineering, Harbin Engineering University, Harbin, China
School of Software, Harbin University of Science and Technology, Harbin, China

L.L. Guo
College of Information and Communication Engineering, Harbin Engineering University, Harbin, China

X.Y. Gao
School of Computer Science and Technology, Harbin University of Science and Technology, Harbin, China

ABSTRACT: Word sense disambiguation is important in natural language processing fields. In this paper, a new word sense classifier is proposed, where part of speech in ambiguous word is extracted as discriminative features. At the same time, the frame of word sense disambiguation is given, where part of speech is integrated. Then, the construction method of word sense classifier is described and the structure of decision table is given. The method of organizing parameters is described. Experimental results show that the accuracy of the proposed word sense classifier is 42.86%.

KEYWORDS: Word Sense Disambiguation; Word Sense Classifier; Part of Speech; Discriminative Features; Decision Table.

1   INTRODUCTION

Word sense disambiguation is essential for natural language processing applications such as machine translation, answering question, and natural language interface. The purpose is that the performance of such applications depends on senses of lexicons. A new word sense disambiguation algorithm is proposed for simple semantic units based on dynamic programming, in which the semantic computation model based on semantic relevancy is given, the definition of simple semantic units is given, and characteristics of simple semantic units are analyzed (Liu 2014). A machine learning method is proposed and features are designed to recognize metaphors, whose purpose is to make a comparison with a word sense disambiguation task. Experimental results show that the method achieves much higher accuracy (Jia 2014). A composite kernel is presented, which is a linear combination of two types of kernels (Wang 2014). They are respectively bag of words kernel and sequence kernel for word sense disambiguation. Its purpose is to integrate heterogeneous sources of information in a simple and effective way. Experiments show that the composite kernel can consistently improve the performance of WSD. Main existing methods for word sense disambiguation problems are investigated and a genetic algorithm is proposed to solve it. At the

same time, it is applied to modern standard Arabic. Its performance is evaluated on a large corpus. The prediction of the genetic algorithm is improved (Menai 2014). A novel approach for word sense disambiguation in English to Thai is proposed. The approach generates a knowledge base in which information of local context is stored. Then, this information is applied to analyze probabilities of several meanings of a target word (Keandoungchun 2013). WSD methods based on machine learning techniques with lexical features can suffer from the problem of sparse data. A hybrid approach is given which copes separately with an error-prone data due to sparsity, in which a data is regarded as error-prone if its nearest neighbors are relatively distant and their senses are uniformly distributed (Han 2013). The connection between lexical chains and word sense disambiguation is investigated. A system that extracts words from unstructured text is given in which sets of lexical chains are provided. At the same time, the process of disambiguation is implemented based on WordNet's synsets (Dumitrescu 2011). The coreference resolution technique is incorporated into a word sense disambiguation system for improving disambiguation precision. With the help of coreference resolution, the related candidate dependency graphs at the candidate level and the related instance graph patterns at the instance level in instance knowledge network are connected together (Hu 2011). A novel WSD model based on distances between words is proposed, which is built on the basic of traditional graph WSD model (Yang 2012). The method can make full use of distance information. A distributional thesaurus which is computed from a large parsed corpus is used for lexical expansion of context and sense information. It bridges the lexical gap that is seen as the major obstacle for word overlap-based approaches (Miller 2012).

In this paper, a method of word sense disambiguation is given, in which part of speech is adopted as discriminative features for deciding correct senses of ambiguous words. A word sense classifier based on part of speech is constructed. At the same time, the architecture is described, in which part of speech labeling tool is integrated.

2 ARCHITECTURE OF WORD SENSE DISAMBIGUATION

In grammar, a part of speech is a linguistic category of words, which is generally defined by the syntactic or morphological behaviour of the lexical item. It is viewed as a word class, a lexical class, or a lexical category. Common linguistic categories include noun, verb, adjective, and etc. Part of speech labeling is an important step in natural language processing field. It can process the problem of word sense disambiguation in a higher level above lexicons, which is key to build effective WSD classifiers. For Chinese sentence CS which contains ambiguous word 'face', part of speech labeling results are shown as follows:

Chinese sentence CS: it/ is/ to/ see/ face/ ./
Word segmentation results: it/ is/ to/ see/ face/ ./
Part of speech labeling results: it/p is/v to/a see/v face/n ./w

Here, 'face' is an ambiguous word. Its sense is determined by its part of speech. For word 'face', its part of speech is n. The correct semantic category should be Bk02 in TongYiCi CiLin which is a Chinese semantic dictionary. So, part of speech of the ambiguous word can be applied to determine its sense. The frame of word sense disambiguation combined with part of speech labeling tool is shown in Figure 1.

In Figure 1, training corpus is processed by word segmentation tool. Then, every Chinese word in training corpus is tagged with part of speech. Word sense classifier is trained on training corpus tagged with part of speech and decision table is gotten. The optimized classifier is gotten. When the classifier is applied to determine the correct sense of an ambiguous word, the decision table is input into the classifier. When an ambiguous word is disambiguated, the test sentence including it is segmented by word segmentation tool. Then, the test sentence is processed by part-of-speech tagging tool. The optimized classifier gets parameters

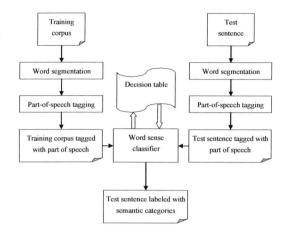

Figure 1. Frame of word sense disambiguation.

from decision table and determines the semantic category of the ambiguous word automatically.

3 WSD CLASSIFIER BASED ON PART OF SPEECH

For ambiguous word w, the process of determining its correct sense is described in formula (1).

$$S = \arg \max_{i=1,2,...,m} P(S_i \mid pos_w) \qquad (1)$$

Here, ambiguous word w has m meanings and their semantic categories include S_1, S_2, ..., S_m. They are respectively categories of w in TongYiCi CiLin and pos_w is its part of speech. $P(x)$ is the probability that x occurs. Parameter $P(S_i \mid pos_w)$ is estimated as shown in equation (2).

$$P(S_i \mid pos_w) = \frac{n_i}{n} \qquad (2)$$

In training corpus, Chinese sentences containing ambiguous word w whose part of speech is labled with pos_w are collected. The number of these Chinese sentences is n. Here, n_i is the number of sentences in which the category of ambiguous word w is S_i in these n Chinese sentences.

These parameters are stored in decision table. The decision table is organized according to ambiguous words. The structure of decision table is shown in Figure 2. When ambiguous word w is disambiguated, parameters of the word sense classifier are found according to w from decision table.

$\#w_1$	pos_{11}	$P(S_{11} \mid pos_{11})$ ...	$P(S_{n1} \mid pos_{11})$
	pos_{12}	$P(S_{11} \mid pos_{12})$ ...	$P(S_{n1} \mid pos_{12})$
	...		
	pos_{1m1}	$P(S_{11} \mid pos_{1m1})$ ...	$P(S_{n1} \mid pos_{1m1})$
......			
$\#w_i$	pos_{i1}	$P(S_{i1} \mid pos_{i1})$ ...	$P(S_{ni} \mid pos_{i1})$
	pos_{i2}	$P(S_{i1} \mid pos_{i2})$ ...	$P(S_{ni} \mid pos_{i2})$
	...		
	pos_{imi}	$P(S_{i1} \mid pos_{imi})$ ...	$P(S_{ni} \mid pos_{imi})$
......			

Figure 2.   Structure of decision table.

4   EXPERIMENTS

In order to evaluate the proposed method, 200 Chinese sentences including ambiguous words are collected. Ambiguous words in these 200 Chinese sentences are labeled with semantic categories according to its contexts. These 200 Chinese sentences are divided into two parts. One is training data and the other is test data. Then, the proposed method is used to construct word sense classifier and its parameters are computed on training data. The optimized classifier is applied to determine correct senses of ambiguous words in test corpus. We use accuracy to evaluate the performance of the classifier. Its performance is shown in Table 1.

Table 1.   The performance of classifier on test corpus.

Number of instances	Number of correct instances	Accuracy(%)
21	9	42.86%

From Table 1, we can find that the accuracy of word sense classifier achieves at 42.86%. It shows that part of speech has some discriminative ability in process of word sense disambiguation.

5   CONCLUSIONS

In this paper, a new WSD classifier is proposed, in which ambiguous words' part of speech is used as discriminative feature. The frame of integrating Chinese part of speech labeling tool into word sense disambiguation is given. Experimental results show that the performance of word sense classifier is 42.86%.

ACKNOWLEDGMENTS

This work is supported by China Postdoctoral Science Foundation Funded Project under Grant Nos. 2014M560249.

REFERENCES

Liu Y. T., Xiong J. and Cui J. L. (2014). A word sense disambiguation algorithm for the simple semantic units based on semantic relevancy. *Journal of Computational Information Systems*. 10(4), 1555–1563.

Jia Y. X., Chen Z. Y., Zan H. Y., Fan M. and Wang Z. M. (2014). A comparative study of metaphor recognition and word sense disambiguation. *ICIC Express Letters*. 5(5), 1391–1396.

Wang T. H., Zhu W. S., Zhang Q. and Xie H. H. (2014). A composite kernel for word sense disambiguation. In *Proc. 2014 International Conference on Sensors Instrument and Information Technology*, pp. 522–525.

Menai M. E. B. (2014). Word sense disambiguation using an evolutionary approach. *Informatica*, 38(2): 155–169.

Keandoungchun N. and Thammakoranonta N. (2013). A word sense disambiguation approach for English-Thai translation. In *Proc. 2nd International Conference on Information Technology and Management Innovation*, pp. 287–290.

Han Y. J., Lee S. J., Park S. Y. and Park S. B. (2013). Detection of error-prone cases for word sense disambiguation. In *Proc. 20th International Conference on Neural Information Processing*, pp. 98–105.

Dumitrescu S. D. and Gainaru A. (2011). A study on lexical chain identification and word sense disambiguation. *UPB Scientific Bulletin*. 73(4), 197–212.

Hu S. F. and Liu C. F. (2011). Incorporating coreference resolution into word sense disambiguation. In *Proc. 12th International Conference on Computational Linguistics and Intelligent Text Processing*, pp. 265–276.

Yang Z. Z. and Huang H. Y. (2012). Graph based word sense disambiguation method using distance between words. *Ruan Jian Xue Bao*. 23(4), 776–785.

Miller T., Biemann C., Zesch T. and Gurevych I. (2012). Using distributional similarity for lexical expansion in knowledge-based word sense disambiguation. In *Proc. 24th International Conference on Computational*, pp. 1781–1796.

Electronic Engineering and Information Science – Wang (Ed.)
© *2015 Taylor & Francis Group, London, ISBN: 978-1-138-02772-5*

Background threshold Gaussian estimation image segmentation algorithm

D. Lu, X. Lin, H.Y. Zhang, J. Yan, C.J. Wen & J.X. Liu
Electrical and Electronic Engineering College, Harbin University of Science and Technology, Harbin, China

ABSTRACT: Image segmentation algorithm is the basis of image processing and text recognition. However, current segmentation methods are all focused on high-quality document images. Aiming at the characteristics of degraded document images, a segmentation method based on local histogram and Gaussian estimation is proposed. First of all, the distribution characteristics of local histogram are used to calculate the local threshold, the original image can be divided into background and foreground. Then, recalculate the background and foreground surface using the pixels in a small range, and a more accurate segmentation threshold can be obtained. At last, re-divide original image into a more accurate background and foreground image. Through the simulation, the proposed method is verified more suitable for processing degraded document images.

KEYWORDS: background threshold; image segmentation; low-quality image.

1   INTRODUCTION

Image segmentation is an important aspect of image analysis and image recognition. Its purpose is to divide the image into a background image and a foreground image, to facilitate image analysis and image recognition. Therefore, the value results directly determine the accuracy of the entire system. In general, the image segmentation methods can be divided into two types: global threshold methods and local threshold methods.

The global threshold methods obtain threshold by analyzing and processing the entire image. Using the threshold as a split point, the image is divided into background and foreground. The largest feature of the algorithm is simple, and the calculation speed is fast. However, due to the use of single threshold, the processing result of the non-uniform illumination image is not desired. The commonly global threshold algorithm is Otsu algorithm (Otsu, 1979).

Local threshold methods search threshold for each pixel rather than one threshold for the entire image. Through the analysis of different image location, suitable local independent thresholds can be identified, and these thresholds can be used as the basis for image segmentation. In common, local threshold methods include Niblack algorithm (Niblack, 1986) and Sauvola algorithm (Sauvola & Pietikainen, 2000). In (Valizadeh et al. 2009), the authors proposed a new local threshold binarization method based on improved Rainfall algorithm. In (Zhou et al. 2009), the authors designed a local threshold binarization method based on Laplician-Gauss (LOG) filter. Normally, these methods are only able to handle

a relatively clear image, and can obtain the desired effect. However, for low-quality images obtained from an ordinary digital camera, the processing result is not satisfactory. This is mainly because that low quality image tends to have higher interference noise and lower resolution, which can result in undesirable background and foreground partition result of the commonly binarization method. In addition to the above-mentioned binarization methods, there are also binarization algorithm based on color classification (Tsai & Lee, 2002), morphology (Tang, 2010) and gray gradient (Shivakumara, et al. 2011).

In recent years, digital cameras have emerged, image acquisition method is gradually increasing. The digital images are widely used. But compared to specialized equipment acquired image, the image acquired by the conventional digital camcorder is clearly insufficient. That mainly reflected in the uneven illumination, low resolution, rasterization noisy and color cast. However, the existing image binarization methods are mainly used on high-quality images, when used in a low-quality image, the binarization results are far from ideal.

Through the analysis of the document images, a low-quality document images binarization method based on document image grayscale statistical properties is proposed. In order to analyze document images in theory, the gray model based on gray characteristics is established. The proposed image binarization method involves two main steps: first, the image local histogram features are used to obtain a rough estimate of the segmentation threshold, and get the rough estimated background of the image, then, through calculating each pixel and neighbor pixels, the foreground and background of the image are estimated, to obtain a precise segmentation threshold based on the rough estimated background.

2 GRAY MODEL

In order to analyze the histogram of the image, image gray value model is as follow:

$$I(x,y) = B(x,y) + F(x,y) + N(x,y) \qquad (1)$$

where x is the horizontal coordinate of the pixel, y is the vertical coordinate of the pixel; $B(x,y)$ represents the background grayscale of the image; $F(x,y)$ is the foreground grayscale of the image, that is, the gradation of characters in the document; $N(x,y)$ represents the noise grayscale. The background of the image is slowly changed within a small range, so it can be considered as a constant. For document images, the foreground $F(x,y)$ has a mutation characteristic, and there is a large difference between $F(x,y)$ and slowly changing background $B(x,y)$, i.e. background and foreground gradation difference is obvious. Noise $N(x,y)$ is mainly due to the slight differences of document background and noise in rasterization process. It follows normal distribution, namely

$$N(x,y) \sim \frac{1}{\sqrt{2\pi}\sigma} e^{-\frac{x^2}{\sigma^2}} \qquad (2)$$

where σ is the variance of noise. After establish of gray model, the image histogram can be analyzed according to the model. Since background and foreground have a difference of certain grayscale, it is easy to distinguish them visually from the histogram. With the influence of noise in the image, the background portion is no longer a constant in the histogram. The foreground portion of the image is a region connected together with the background in the histogram. Therefore, a suitable curve can be found which fitting parameters, and thus the image background and foreground can be estimated.

3 PROPOSED METHOD

The proposed binarization method is a local threshold method based on statistical features. The statistical characteristics of gray model are used for local analysis to get the rough estimates of image background. Using the rough estimates, the final binary image is obtained and the influence of uneven illumination to the binarization result can be reduced.

3.1 Rough estimate of the image background

The rough background extraction method used here is based on local histogram. According to the gray model established in Section 2, by traversing all gray level, the gray level with maximum sum of neighboring 2N+1 gray in the histogram can be identified, as the background gray of the image. Namely:

$$B_1 = \max(j, \sum_{i=-N}^{N} hist(j+i)) \qquad (3)$$

$hist(j)$ represents the gray distribution in the histogram; N is the estimated radius of the window; $j=0,1,2\ldots,255$, is the distribution of the center point. In order to find the range of the background, the complete background distribution curve should be estimated. The mean of background is as follow

$$\mu = \frac{\sum_{i=-N}^{N} \left[hist(B_1+i) \times i \right]}{\sum_{i=-N}^{N} hist(B_1+i)} \qquad (4)$$

the variance σ is:

$$\sigma^2 = \eta \frac{\sum_{i=-N}^{N} \left[hist(B_1+i) \times i^2 \right]}{\sum_{i=-N}^{N} hist(B_1+i) \sum_{i=-N_2}^{N_2} hist(B_1+i)} \qquad (5)$$

where η is the coefficient of variance estimation, normal value is 1.4; N_2 is the window radius of variance estimation, and the experimental value is 20, representing the approximate range has a large amplitude distribution value in the histogram. The estimated background distribution curve is obtained by as follow:

$$f(x) = \lambda \frac{1}{\sqrt{2\pi}\sigma} e^{-\frac{(x-\mu)^2}{\sigma^2}} \qquad (6)$$

and

$$\lambda = \sum_{i=-N_2}^{N_2} hist(B_1+i) \qquad (7)$$

thereby, the radius of the noise can be estimated as:

$$r_t = r_e \sqrt{-2\sigma^2 \times \ln\left(\sqrt{2\pi}\sigma r_h\right)} \qquad (8)$$

r_e is set to 1.4, r_h is set to 0.00001. Using the above estimated values, the background is:

$$B(x, y) = B_1(x, y) \qquad (9)$$

then

$$BW(x, y) = \begin{cases} 0 & |I(x, y) - \mu| < \gamma_r \\ 1 & \text{else} \end{cases} \qquad (10)$$

3.2 Accurate calculation of image segmentation threshold

Estimation of accurate image segmentation threshold is dependent on the background image obtained in the previous step. The threshold is calculated using the following formula:

$$t(x, y) = \delta \left(\frac{1}{p(x, y)} + kp(x, y) \right) d(x, y) \qquad (11)$$

$d(x,y)$ represents the average gradation difference between the background and foreground, and

$$d(x, y) = \frac{\sum\limits_{i=-w}^{w} \left[BW(x, y) \times |I(x, y) - B(x, y)| \right]}{\sum\limits_{i=-w}^{w} BW(x, y)} \qquad (12)$$

$p(x,y)$ represents the image proportion of foreground, which is calculated as follow:

$$p(x, y) = \frac{\sum\limits_{i=-w}^{w} BW(x, y)}{2w + 1} \qquad (13)$$

δ is a threshold factor, usually selected as 0.1; k is the amplitude coefficient, select 9. Then, the binary result can be calculated as follows:

$$Bina(x, y) = \begin{cases} 0 & |I(x, y) - B(x, y)| < t(x, y) \\ 1 & \text{else} \end{cases} \qquad (14)$$

4 SIMULATION RESULTS

In order to increase the difficulty of the experiment, the low-quality documents photos acquired under natural non-uniform irradiation light, by Android phones are selected for the experiment. The resolution is 2592×1456. A 70×70 window is used as a local window in Niblack method. And k is −0.2. The proposed method also uses a 70×70 window, the other parameters are mentioned in previous. Fig.1 shows the binarization results: a) is the original image; b) is the proposed method result; c) is the Otsu method result; d) is the Niblack method result. The simulation results show that the binary image obtained by the proposed method can clearly distinguish the characters. Therefore, the proposed method is suitable for the low-quality image binarization. Fig.2 shows histogram estimation of several different

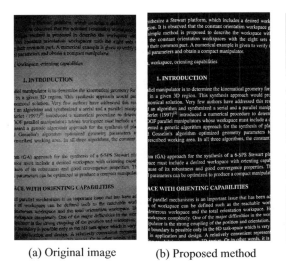
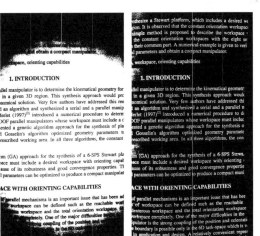

(a) Original image   (b) Proposed method   (c) Otsu method   (d) Niblack method

Figure 1. Binarization results comparison.

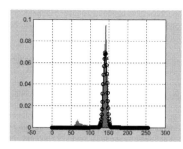 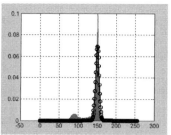 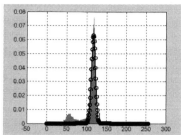

(a) Local histogram at coordinate (800,800)

(b) Local histogram at coordinate (1100,800)

(c) Local histogram at coordinate (500,1000)

Figure 2.   Histogram estimation of several different locations.

locations. The figure shows the local histogram of the local image, the curve with o represents the estimated curve of the proposed method. As can be seen from the figure, although there is still some difference between the estimated curve and the actual histogram, the estimated curve is still relatively satisfactory.

By using different type images for test, the difference in performance between these two binarization methods is: 1) For Otsu method, due to its global threshold can not handle the change of the local background, therefore, its uneven illumination image processing results are very poor, and cannot be applied to low-quality images acquired from digital devices. 2) Niblack method uses local threshold, so it can solve the uneven illumination problem. But this method is based on the image fluctuation variance, it will get unpredictable results in the image fusion and reduce the accuracy of final binarization result. Therefore, this method can not be applied to low-quality photos.

5   CONCLUSION

An image segmentation method focused on low-quality digital images is proposed. Uneven illumination influence to binarization result is reduced by using local threshold; the noise is also reduced by using the statistical characteristics of histogram; finally, the accuracy estimation reduces the isolated noise point of binarization result. Comparing the experimental results of Otsu algorithm, Niblack algorithm and the proposed method, the proposed method result is much better on processing low-quality images. Therefore, the method described herein is more suitable for low-quality document image binarization processing.

ACKNOWLEDGMENTS

This work has been supported by research of Harbin science and technology innovative talents (2014RFQXJ163) and College Students' innovative entrepreneurial project of Harbin University of Science and Technology.

REFERENCES

Niblack, W. (ed.). 1986. *An introduction to digital image processing:* 115–116. Englewood Cliffs: Prentice Hall.
Otsu, N. 1979. A threshold selection method from gray-level histograms. *J. IEEE Trans. Systems Man Cybernet* 9(1):62–66.
Sauvola, J. & Pietikainen, M. 2000. Adaptive document image binarization. *Pattern Recognition* 33:225–236.
Shivakumara, P. , Bhowmick, S. , Su, B. , Tan,C.L. & Pal, U. 2011. A new gradient based character segmentation method for video text recognition. *2011 International Conference on Document Analysis and Recognition (ICDAR)* :126–130.
Tang, J.2010. A color image segmentation algorithm based on region growing. *2010 2nd International Conference on Computer Engineering and Technology (ICCET)* 6: 634–637.
Tsai, C.M., Lee, H.J. 2002. Binarization of color document images via luminance and saturation color features. *J.IEEE Transactions on Image Processing:* 434–451.
Valizadeh, M. , Komeili, M., Armanfard, N.& Kabir, E. 2009. A contrast independent algorithm for adaptive binarization of degraded document images. *CSICC 2009, 14th International CSI*: 127–133.
Zhou, S.T. , Liu , C.P. , Cui, Z.M. & Gong, S. R .2009. An improved adaptive document image binarization method. *2nd International Congress on Image and Signal Processing* : 1–5.

Electronic Engineering and Information Science – Wang (Ed.)
© 2015 Taylor & Francis Group, London, ISBN: 978-1-138-02772-5

Dynamic gesture recognition algorithm based on skin color detection and depth information

W.L. Guo, Y.F. Cai, J.L. Zhang & Y.X. Tang
Harbin University of Science and Technology, Harbin, China

ABSTRACT: Dynamic hand gesture recognition technology based on computer vision is a key technique to a new generation of human-computer interaction technology and a hot to computer research field. Combining traditional skin color detection with image depth information in the process of gesture image segmentation can effectively reduce the disturbance under complicate and noisy background which gestures picture is split, tracking gestures. Camshift algorithm of gesture tracking further reduces the complex background of effects on the environment and improves the accuracy and efficiency.

KEYWORDS: Hand gesture recognition; Depth information; Skin color detection; HMM.

1 INTRODUCTION

In today's rapid development of science and technology, Von Neumann's computer already can't satisfy people's needs, human natural interactions become a development trend. Human-computer interaction technology developments in the direction from the people adapt to machine to machine adapt to people, (Chen H. et al.2014). Through gestures or multiple input channel users interact more explicit is the development trend of the human-machine interaction, (Zhu X.Y. et al.2009).

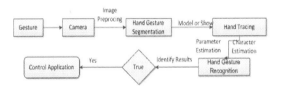

Figure 1. Flow chart of gesture recognition based on vision.

Traditional gesture recognition into equipment based on data gloves and computer vision based on two methods. The recognition system based on data glove has a good recognition effect, but due to require users to wear special sensor gloves that there are no natural interaction and user experience effect is not ideal, (Xu X. et al.2011). Gesture recognition systems based on computer vision is easy to operate, simple and intuitive, more in line with the user habits. This paper mainly studies, gesture recognition based on computer vision, put forward depth image information and skin color detection algorithm combination method under complex background, (Ann et al. 1998). The flow chart of computer vision based gesture recognitions shown in Fig.1.

2 GESTURE SEGMENTATION

2.1 Gesture segmentation based on skin color

Skin color detection to select the appropriate color space, commonly used color space with RGB color space, YCrCb and HSV color space, etc., (Zhi Z.Q. et al.2014). RGB is based on the human eye to identify the color of the definition of space, can represent most of the color. It makes the hue, brightness and saturation three quantities together, says it is difficult to separate, so skin color detection generally don't use RGB color space. YCrCb belongs to orthogonal color space, it divides into the color, brightness and tone, brightness is weighted by calculating the RGB of each color value obtained, the tone is passed from the blue and red value minus the 1 brightness component. H and S two components in HSV space to contain the image color information, and shows the depth of the color, synthesis of Chroma and V component said brightly, has nothing to do with the image color information. So abandon V component, only consider the influence of H and S components of light conditions. RGB color space according to the formula into HSV color space:

$$
\left.
\begin{aligned}
H &= \begin{cases} \theta, B \le G \\ 2\pi - \theta, B > G \end{cases} \\
\theta &= \arccos\theta\left(\frac{\frac{1}{2}[(R-G)+(R-B)]}{\left[(R-G)^2+(R-B)(-B)^{\frac{1}{2}}\right]}\right) \\
S &= 1 - \frac{\min(R,G,B)}{V} \\
V &= \max(R,G,B)
\end{aligned}
\right\} \quad (1)
$$

2.2 Based on the depth information of skin color segmentation

Solely using skin color detection results will be affected by a face and interference with the skin close objects, so choose the combination with the image depth information method of image segmentation. This paper, by using a binocular camera using the parallax method for image depth information, establishes the Gaussian mixture model consisting of two single Gaussian model and then using the depth image for the bimodal Gaussian mixture model. Now the probability density function:

$$P_d(x) = \alpha_h g(\chi, \mu_h, \sigma_h) + \alpha_f g(\chi, \mu_f, \sigma_f) \qquad (2)$$

Its parameters are α_h, α_f, μ_h, μ_f, σ_h, σ_f. μ_h in $p_d(x)$ is the depth of the gestures threshold, μ_f is the depth of body. And, it satisfies the condition that $\alpha_h + \alpha_f = 1$. According to the gesture in front torso, threshold setting depth $t_d = (\mu_f + \mu_h)/2$. Get rid of $dep(p) \geq t_d$ corresponding in the original image pixels, keep only points which the depth value meet $dep(p) \geq t_d$ he pixels in image deal with depth threshold input color model, then compute the probability of pixel points for skin and image binary with threshold method, complete the gesture segmentation.

3 GESTURE TRACKING

Mean shift based image segmentation algorithm is a kind of kernel density estimation based feature space analysis algorithm, and the nature of it is statistical optimization, the Mean Shift refers to the average vector migration result, (Munox-Salinas et al.2008). The iterative process of iterative algorithm can be summarized as the following three steps: firstly, step calculates a bit of Mean Shift, secondly step, according to the calculated Mean Shift moves the point and last is under the condition of certain constraints, repeat the above steps that point to the new starting point, (Kai Zhang et al. 2013). Mean shift algorithm because of its less iterative times, and the characteristics of small amounts of calculation are widely used in the real-time target tracking.

Improved Mean Shift algorithm-Camshift-adapted Mean Shift algorithm was cited. The basic idea of the method is: all frames do Mean Shift algorithm for image. And the results of the last frame as the initial value of the search window for next frame, so iterative. The key of Camshift algorithm is when the target size is changed the algorithm is based on an adaptive adjustment of target area and continue to track the target, but for occlusion and the background change situation cannot achieve accurate tracking, (Comaniciu et al, 2000).

First, choose the search window, and then calculate the size and position of the search window to make it contains the target. Calculation of window center of gravity: Set I(x,y) is (x,y) pixel coordinate value, the zero order moment of search windows and the first order moments are:

$$M_{00} = \sum_x \sum_y I(x, y), \quad M_{10} = \sum_x \sum_y xI(x, y),$$
$$M_{01} = \sum_x \sum_y yI(x, y) \qquad (3)$$

The barycentric coordinates:

$$x_c = \frac{M_{10}}{M_{00}}, \quad y_c = \frac{M_{01}}{M_{00}}, \qquad (4)$$

Resize the search window:

$$s = 2 * \sqrt{\frac{M_{00}}{256}}, \qquad (5)$$

Complete the search window center moved to the center of gravity. To switch the selected track gestures to HSV space, sampling on H channel, the color histogram H channel is obtained and then get the color probability distribution. Calculate tracking target direction and size, complete gesture tracking. Make:

$$a = \frac{\mu_{20}}{\mu_{00}} - \overline{x}^2, \quad b = 2\left(\frac{\mu_{11}}{\mu_{00}} - \overline{xy}\right), \quad c = \frac{\mu_{02}}{\mu_{00}} - \overline{y}^2 \qquad (6)$$

The target is moving direction and the horizontal direction angle:

$$\theta = \frac{1}{2}\arctan\left(\frac{b}{a-c}\right) \qquad (7)$$

The target area of long axis l and short axis w respectively for:

$$l = \sqrt{\frac{(a+c) + \sqrt{b^2 + (a-c)^2}}{2}} \qquad (8)$$

290

$$w = \sqrt{\frac{(a+c)+\sqrt{b^2+(a-c)^2}}{2}} \qquad (9)$$

4 DYNAMIC GESTURE RECOGNITION

Dynamic gesture recognition is the last step of the gesture recognition system, by the dynamic gesture recognition can be determined by the meaning of the gesture of dynamic sequence, and then according to the meaning of gestures to execute the corresponding operation. Dynamic gestures are a combination of static gestures posture in chronological order, so each state traversal of a dynamic gesture has, calculating the joint probability of them, classify for dynamic gesture. In this paper, we adopt Hidden Markov Model (HMM) based on the state space method to study dynamic gesture recognition.

5 CONCLUSION

This paper mainly researches the dynamic gesture recognition based on computer vision, and proposed a segmentation method for skin color detection and depth information based gestures, the interference problem in complex environment brings to gesture recognition.

ACKNOWLEDGMENT

This work was supported by Science and Technology Project (NO:12521108) from the Education Department of Heilongjiang Province, China.

REFERENCES

Chen.H.2014.A review of the gesture recognition based on depth image. *Journal of Inner Mongolia university.*
Zhu, X.Y.2009.Video target tracking research. Xi'an: *Xi'an university of electronic science and technology.*
Rafael, M.S & RMedina-Carnicer, F .J (eds) .2008. Depth Slihouettes for Gedture Recognition. *Pattern Recognition Letters,* 29(3):319–329.
Xu, X & Xu,X.M.2011. Hand Gesture Recognition based on Hidden Markov Module. *SIP, 2011.5, 45–60.*
Ann.A .1998. Valued Processes and Interacting Particle Systems Application to Nonlinear Filtering Problems. Appl. Probab.1998,p.438–495.
Zhi, Z.Q & Xu,C.2014. The gesture recognition based on computer vision. 2015.5.41–45.
Zhang, K.2013. Depth Data Fused Human Computer Interaction Hand Gesture Recognitions. *Educational Technology,* 9, 25–40.
Comanicuu,D & Ramesh.V(eds). 2000.Real-time tracking of non-rigid objects using Mean Shift [C]//Proc IEEE Conference on Computer Vision and Pattern Recognition, Hilton Head Island: SC, p.142–149.

Electronic Engineering and Information Science – Wang (Ed.)
© *2015 Taylor & Francis Group, London, ISBN: 978-1-138-02772-5*

Spare parts transportation in clone immune algorithm

H.Wei & P.L. Qiao

School of Computer Science and Technology, Harbin University of Science & Technology, Harbin, PR China

ABSTRACT: In this study, cost function and customers' satisfaction quantization are used to investigate spare parts transportation supply chain problem. Since the traditional genetic algorithm is limited by the large population scale, it evaluates slowly inevitably. Considering this limitation, this study comes up with an advanced algorithm, and it uses an adaptive selection method for merit-selection. The superior individuals are being cloned adaptively. Thus, it improves the operation efficiency and decreases the iteration times. As a result, the optimal solution could be got more quickly.

KEYWORDS: Clone Immune; Transportation Problem; Spare Parts; Satisfaction.

1 INTRODUCTION

Single-path vehicle arrangement belongs to single loop transportation. It is defined that in transportation route optimization, which is assumed to be a set 'A', it is possible to find a loop path to traverse all nodes, eventually return to the initial node, aiming to achieve the shortest driving distance. Traveling Salesman Problem (TSP) is one of the most typical representatives (Grahovae J. & Chakravarty A. 2001). Scholars who are researching on this problem divided the model into single-objective model and multi-objective model, early used branch-bound algorithm to figure them out. But, due to these accurate algorithms were seldom used to deal with practical application, recently they tend to use a heuristic algorithm (Chart L. M. A. et al. 2002) to solve such problems.

Considering the importance of customer satisfaction, this study improves the service spare parts logistics network transportation model using satisfaction quantization method which based on time. According to the model characteristics of spare parts, this study quotes the method of Clone Immune Genetic Algorithm (CIGA) (F. G. Yang 2011 & H. H. Zhang et al. 2008) that limits the number of clone antibody dynamically in order to promote the traditional CIGA. Therefore, the phenomenon of partly optimal which is caused by precocity is decreased significantly. On the other hand, the computational efficiency of getting the global optimal solution can be improved significantly. In this study, the algorithm is verified by numerical samples, and so are the models.

2 MODELING

Because of the importance of delivery time to customers' satisfaction, this study cites a time-satisfaction function (Y. F. Ma 2005). After quantization, this function is applied to the model for the algorithm solving. In this model, There are four time point (Te_p, TE_p, TL_p, Tl_p) in the model, and they divides the delivery time into five time intervals. The acceptable interval is [Te_p,Tl_p], and the satisfactory interval is [TE_p,TL_p]. It can assume that the customers are non-satisfactory, whether the delivery time factor locates out of the acceptable interval.

The penalty function is found in this study in order to change the aim of maximizing the satisfaction to minimizing the penalty cost. The functions are as follows:

$$S(t_p) = \begin{cases} W(Te_p - t_p), 0 \le t_p < Te_p \\ 1 - \left(\dfrac{t_p - Te_p}{TE_p - Te_p}\right)^{\alpha_k}, Te_p \le t_p < TE_p \\ 0, TE_p \le t_p < TL_p \\ 1 - \left(\dfrac{Tl_p - t_p}{Tl_p - TL_p}\right)^{\alpha_k}, TL_p \le t_p < Tl_p \\ M, t_p > Tl_p \end{cases} ; \quad (1)$$

Equation 1, M is a positive number of infinite. if delivery time is smaller than Te_p, the waiting cost is $W(Te_p - t_p)$, which the value range is [0,1]. t_p is the arrival-time of a customer. α_k is a sensitive factor with a positive integer.

Customer satisfaction and the cost are two objective factors of the model. The Objective function can be expressed as:

$$E = Min(\theta_1 (F/F_{max}) + \theta_2 S) \quad (2)$$

Including

$$F = \sum_{i \in I} \sum_{j \in I} \sum_{k \in K} \sum_{t \in T} C_{ij} H_t D_{ij} Z_{ijkt} + G_c \sum_{k \in K} E_k \qquad (3)$$

$$S = \sum_{i \in I} \sum_{j \in I} \sum_{k \in K} S(t_j) Z_{ijkt} \qquad (4)$$

$$t_j = \sum_{i \in I} \sum_{j \in I} (t_{ij} + T_i + S_i) Z_{ijk} \qquad (5)$$

Equation 2, θ1 and θ2 is the weight of the two functions, θ1≥0, θ2≥0, and θ1+θ2=1. The value of θ is positively correlated with the priority consideration.

Subject to

$$\sum_{i \in I} Q_{it} Z_{ikt} \le A_k, \forall k \in K, \forall t \in T; \qquad (6)$$

$$\sum_{i \in I} Z_{ikt} = 1, \forall k \in K, \forall t \in T; \qquad (7)$$

$$\sum_{i \in I} Z_{ijkt} = Z_{jkt}, \forall i, j \in I, \forall k \in K, \forall t \in T; \qquad (8)$$

$$Z_{ijkt} \le \frac{Z_{ikt} + Z_{jkt}}{2}, \forall i, j \in I, \forall k \in K, \forall t \in T; \qquad (9)$$

$$\sum_{i,j \in I} Z_{ijkt} \le N_k - 1, N_k \subset I \text{ and } \sum_{i \in S_k} Z_{ikt} < \sum_{i \in I} Z_{ikt}; \qquad (10)$$

$$S(t_j) \le \alpha, \forall j \in I; \qquad (11)$$

$$Z_{ikt}, Z_{jkt}, Z_{ijkt}, E_k \in \{0,1\}, \forall i, j \in I, \forall k \in K, \forall t \in T; \qquad (12)$$

Parameter Description:

I - Customers set;
i - Customer i;
j - Customer j;
α - The highest value limits of customer satisfaction penalty function;
C_{ij} - The unit distance cost from customer i to j;
D_{ij} - The shortest distance from customer i to j;
H_t - The unit cost of spare part-t;
G_c - Fixed transportation cost of a vehicle travels a loop path;

A_k - The maximum storage capacity, while a vehicle's k-th departure;
Q_{it} - The amount of spare parts-t for customer i needs;
N_k - The amount of chosen customers, while a vehicle's k-th departure;
t_{ij} - Travel time from customer i to j;
T_i - The time of a vehicle arrival customer i;
S_i - Service time for customer i;

Equation 6 represents the demand of customer-i cannot exceed this vehicle's maximum storage capacity in a loop trip; Eq. 7 indicate each customer should be serviced once only in a loop trip; Eq. 8 and Eq. 9 limited the visit times of each customer once only in a loop trip; Eq. 10 limited the start and finish points are the same distribution center; Eq. 11 represents the highest constraint of the satisfaction penalty function; Eq. 12 is the decision variables.

3 CLONE IMMUNE GENETIC ALGORITHM

According to this study, different cloning scales evolve from different fitness and different affinity antibodies. Thus, it guarantees the population diversity and avoids the algorithm premature convergence as well. Therefore, calculation accuracy was well increased in this method. The algorithm procedures are as follows:

Step 1, Coding. There are 4 decision variables as Z_{ikt}, Z_{jkt}, Z_{ijkt} and E_k composed a coding sequence 'C', assuming the number of customers as 'n' and the number of Types of spare parts as 'm'. The first bit represents the type of this spare part with the range from [1,m]; the second bit represents the number of times for the vehicle currently departure with the range from [0,n]; the variable Z_{ikt}, Z_{jkt} and Z_{ijkt} reflect situation of transportation, which are conducted by binary coding method, with encoding length is 'n+1', and each bit indicates the chosen customer in a vehicle loop trip, that '1' means selected and '0' means unselected. The coding sequence is shown in Table 1:

Table 1. Coding sequence.

1	2	3	...	3+n
t	k	Z_{0i} ...	Z_{ij} ...	Z_{j0}

Step 2, Vaccine Extraction. Using overall immune method, this study calculates the current population fitness. And the highest fitness antibody is picked out.

Step 3, Calculation of Fitness and Affinity. In this calculation, we treat the objective function 'E(C)' as fitness function, and change it into maximum function F(C):

$$F(C) = \begin{cases} M - E(C), & E(C) < M \\ 0, & E(C) \ge M \end{cases} \qquad (13)$$

The affinity μ_i of antibody-i is as follows:

$$\mu_i = \min(D_{ij}), i \neq j; i, j = 1, 2, ..., N \qquad (14)$$

include

$$D_{ij} = \sqrt{\sum_{\substack{1 \leq i \leq n \\ 1 \leq j \leq n}} (F(a_i) - F(a_j))^2} \qquad (15)$$

Where, $F(a_i)$ and $F(a_j)$ are the fitness of antibody a_i and antibody a_j. The minor value of D_{ij} is, the higher affinity between these antibodies is.

Step 4, Clonal Proliferation. There are 3 factors that influenced the clone number of antibodies: the clone scales M, the fitness of antibodies and antibody affinity. The function is as follows:

$$Q_i = \text{int}\left[M \cdot \frac{F(a_i)}{\sum_{i=1}^{M} F(a_i)} \cdot \mu_i \right], i = 1, 2, ..., N \qquad (16)$$

Step 5, Vaccination and Update. Select the antibody sample randomly and pick out the highest fitness antibody. Then, compare the vaccine's fitness with the highest fitness. Lastly, choose the antibody which with the highest fitness as the new vaccine.

4 EXPERIMENTS RESULTS AND ANALYSIS

An example is used to verify the model in this part. There are two kinds of spare parts, four customers and one distribution center compose a distribution logistic network. G_c is 100 and A_k is 100. Simulation data from Table 2 to Table 5 are showed below:

Table 2. The distance cost for transporting spare part-t from customer i to j.

$C_{ij}H_tD_{ij}$ (t_1,t_2)	j_1	j_2	j_3	j_4
0	(135,256)	(185,320)	(146,313)	(130,245)
i_1		(155,250)	(137,244)	(160,265)
i_2	(155,250)		(132,240)	(141,253)
i_3	(137,244)	(132,240)		(155,258)
i_4	(160,265)	(141,253)	(155,258)	

In this part, the initial conditions are specified in the algorithm. θ_1 and θ_2 in objective function are assigned to 0.6 and 0.4. Gen is the termination condition which is 300, and the popsize is 50. Moreover, the improved algorithm and the standard Genetic Algorithm are

Table 3. The demands of customer i.

$Q_{it}(t_1,t_2)$	i_1	i_2	i_3	i_4
Q_t	(15,40)	(15,55)	(30,50)	(45,20)

Table 4. The travel time from customer i to j.

t_{ij}	j_1	j_2	j_3	j_4
0	11	16	13	9
j_1		14	12	16
j_2			11	14
j_3				15

Table 5. The service time of customer-i.

Time	i_1	i_2	i_3	i_4
T_i	3	2	4	2

compared with the same conditions. After calculating, the optimal solution of the improved algorithm and traditional algorithm is 0.732 and 0.768 respectively. The result is shown as Table 6:

Convergence curvatures of CIGA and GA are shown in Figure 1:

Table 6 shows that the optimal solution which gets by the new algorithm is more accurate. And the running time of the new algorithm is apparently decreased and the convergent generation is reduced as well. Obviously, the new algorithm is properly improved. According to Figure 1, because of the improvement of the new algorithms, the antibody sample is much superior and multiple. Therefore, the improved algorithm is more optimized to get more accurate optimal solutions.

5 CONCLUSION

In this study, integrating customer satisfaction model is used to research spare parts supply chain transportation. The clone scale of superior antibodies is self-adjusted in the calculation. In conclusion, the improved algorithm averts the confusion of local optimal solution and avoids premature convergence. Moreover, the population's evolution efficiency is promoted impressively.

ACKNOWLEDGMENT

This study is supported by scientific and technological projects of Harbin (2011AA1CG063).

295

Table 6. Comparison of two algorithms results.

Algorithms	Optimal solution	Optimal path		Iterations /Gen	Running time/s
CIGA	0.732			125	2.231
		$t_1 \begin{cases} 0 \to 4 \to 2 \to 3 \to 1 \to 0 \\ 0 \to 1 \to 0 \end{cases}$	$t_2 \begin{cases} 0 \to 4 \to 2 \to 0 \\ 0 \to 1 \to 3 \to 0 \end{cases}$		
GA	0.768			151	3.455
		$t_1 \begin{cases} 0 \to 1 \to 3 \to 2 \to 4 \to 0 \\ 0 \to 4 \to 0 \end{cases}$	$t_2 \begin{cases} 0 \to 1 \to 3 \to 2 \to 0 \\ 0 \to 4 \to 2 \to 0 \end{cases}$		

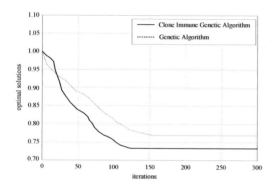

Figure 1. Curve diagram of two algorithms.

REFERENCES

Chart L. M. A. & Muriel A., et al. 2002. Effective zero-inventory-ordering policies for the single warehouse multi-retailer problem with piecewise linear cost structure. *Manage. Sci.* 48(11): 1446–1460.

F. G. Yang. 2011. Adaptive clone and suppression artificial immune algorithm. *Appl. Res. of Comput.* 28(2): 481–484.

Grahovae J. & Chakravarty A. 2001. Sharing and lateral transshipment of inventory in a supply chain with expensive low-demand items. *Manage. Sci.* 47(4): 579–594.

H. H. Zhang & P. J. Wang. 2008. Status & Outlook of immune-algorithm-based production scheduling. *Comput. Integr. Manuf. Syst.* 14(11): 2081–2091.

Y. F. Ma. 2005. Satisfaction based on time coverage in allocation problem. Huazhong Univ. of Sci. & Tech.: 46–48.

Electronic Engineering and Information Science – Wang (Ed.)
© 2015 Taylor & Francis Group, London, ISBN: 978-1-138-02772-5

Predicting message propagation based on logistic regression

T.G. Ren & H.F. Ding
Harbin University of Science and Technology, Harbin, Heilongjiang, China

H.L. Qi, B. An & H.H. Yu
Heilongjiang Institute of Technology, Harbin, Heilongjiang, China

ABSTRACT: The objective of the retweet predicting is to predict forwarding times of the target message in social networks. This study helps us to predict and control the scale of information dissemination. However, with large user bases and complex user relation, it's a very difficult task to construct the topology of user networks. Most research on retweet predicting with the building machine learning model were conducted from the perspective of classification, this paper used regression analysis to study the retweet predicting. Based on the Logistic Regression algorithm, build a regression analysis, prediction model towards the retweet prediction to infer the retweet scale. Experimental results based on the actual data set show the regression analysis predicting model has a good predicting accuracy in dealing with retweet predicting, the proposed method is effectiveness.

KEYWORDS: Social network; Retweet predicting; Logistic Regression.

1 INTRODUCTION

Retweet is one of the important propagation mechanisms for information diffusion in social networks, users can retweet the interesting message to their personal account, and the message can be seen by more people. The objective of the retweet predicting is to predict forwarding times of the target message in social networks, in order to achieve the purpose of predicting the scale of information dissemination. This study is very important for many fields, such as viral marketing, personalized recommendation and others.

Currently, the research on retweet prediction of a social network is divided into two directions. One is through studying the information propagation path topology to build prediction model, most research based on dynamic propagation and virus propagation theory. The other way to build a prediction model is based on a machine learning algorithm, consists mainly of Support Vector Machine, Bayesian networks and Logistic Regression.

In the real social network, with large user bases and complex user relation, it's a very difficult task to construct the topology of user networks. Wu has built an information dissemination network topology, and get a good prediction results, but this topology contains only 2000 users (Wu et al. 2013). Obviously, there is a lot of difference compared with the real situation in the number of users.

It is precisely because there are enormous difficulties to build prediction models by information propagation path topology, the other scholars choose to use machine learning algorithm to build the prediction model. Sasa used the PA algorithm in the study (Petrovic et al. 2011), zaman used Collaborative Filtering algorithm (Zaman et al. 2010), and Hong used Logistic Regression (Hong et al. 2011), they successfully build the prediction model and enhanced the user scale, the number of users was increased to 2.5 million by Hong. However, they only studied whether the message will be retweeted without taking into account the forwarding times of the message will be received.

In summary, these previous results are unsatisfactory, further studies are still necessary. This paper argues that retweet predicting should be regression analysis problem, not just the classification problem. Because we need through the forwarding times of the message to determine the propagation scale, rather than whether retweeted. This paper reports on building regression analysis model based on a Logistic Regression algorithm for predicting the scale of information dissemination. We selected some features that have an important impact on the retweet behavior and divided into four categories, including user features, text features, temporal feature and metadata feature. To note is that we take into account the effect of the text content of the retweet behavior and obtain the topic distributions for each tweet based on Latent Dirichlet Allocation.

2 RELATED WORK

With the sweeping of social network, the associated research achieves widespread concerns from academia to the business sector. How to analyze and study the scale of information dissemination has become an important research direction. Researchers made some research results in retweet predicting from a different perspective.

Different social networks have different propagation mechanisms, Boyd analyzed twitter background, conventions, and The syntax of a retweet, also revealed some salient motivations users have to retweet: amplify or spread tweets to new audiences, entertain or inform a specific audience, publicly agree with someone, mark the tweets for future personal access and so on (Boyd et al. 2010).

2.1 Based on propagation topology

To build prediction models by information propagation path topology, the topology should include all the users who have seen the tweet, then through analyzing the similarity between users and user history behavior to decide whether the user will retweet. Macskassy build four prediction models including recent communication model and homophily model, and got best prediction performance when using all models. Yantao also proposed a new algorithm to establish the information propagation path topology. The limitations of their research are that their model is only suitable for small scale users, however, the amount of users on social network is so huge that cannot build prediction model.

2.2 Based on machine learning

In order to prove building prediction model based on machine learning algorithm is feasible, sasa also conducted an experiment with human subjects, and confirmed that modeling based on machine learning algorithms to predict better than the human subjects (Petrovic et al. 2011). In addition, Chen used Bayesian network modeling process, Luo used Support Vector Machine and pointed out that there is a better prediction results when using all of the features. Hong mentioned the model based on the Logistic Regression algorithm is better than the other algorithms after testing different algorithms (Hong et al. 2011). However, these previous results are controversial because they formulated the research into a classification problem, but not a regression analysis problem.

Considering it is too difficult to obtain the accurate forwarding times of a tweet, Hong proposed a new definition for forwarding times. The tweets which have the same MD5 value will be considered to be the same, then sort all such same tweets by ascending time order, forming a chain of tweets, then the number of tweets following the target tweet is treated as the forwarding times the tweet received (Hong et al. 2011). Thus, the forwarding times of other tweet which are not in any chains will be considered as zero.

3 RETWEET PREDICTING BASED ON REGRESSION ANALYSIS

3.1 Why is regression

As previously mentioned, there are some problems in previous researches, we confirm that building prediction model based on the machine learning algorithm is worthy of recognition, however, we build the regression analysis model, but not a classification model, Then we can predict the exact forwarding times of a particular tweet. The difference in regression and classification is the output variable of regression takes continuous values, and the output variable of classification takes class labels. Obviously, it is regression problem, because the forwarding times of tweets are continuous.

3.2 Logistic regression model

In statistics, Logistic Regression is a statistical classification model, used to predict the results of classification based on one or more of the features. Logistic Regression measures the relationship between independent variables and the variable. We determine its category by the probability scores of the dependent variable which the model predicted. Thus, Logistic Regression not only be used for solving classification problems also be used to solve the regression problem, that is why we use Logistic Regression modeling. For Multinomial Logistic Regression, the logistic function can be written as:

$$P(Y = k \mid x) = \frac{\exp(w_k \bullet x)}{1 + \sum_{k=1}^{K-1} \exp(w_k \bullet x)}, k = 1, 2, ..., K-1 \quad (1)$$

$$P(Y = K \mid x) = \frac{1}{1 + \sum_{k=1}^{K-1} \exp(w_k \bullet x)} \quad (2)$$

Where, $x \in R^{n+1}$ is the input, $Y \in \{1, 2, ..., K\}$ is the output, $w_k \in R^{n+1}$ is the parameter, called weight, $w_k \bullet x$ is the dot product of w_k and x.

In the Logistic Regression model parameter estimation, we used the gradient descent method which can find the minimum of the likelihood function. Gradient descent starts with a set of initial weights, and decreases each weight in proportion to its partial derivative in each step.

3.3 Features

Many researchers have conducted studies tweet feature selection, and pointed out what features have a large impact on the retweet behavior. Suh noted that the number of followers, friends, URLs, and hashtags have a positive impact for retweet behavior, but whether mention others and the number of historical statuses have negative effects. In addition, is the user verified, number of times the user was listed and other features be widely used.

Finally, we use the following features and divide them into four distinct sets. User features contain a number of followers, number of friends, number of times the user was listed, is the user verified, the number of historical statuses. Text features contain whether to include hashtags, URLs, mention others, and the content filter. Temporal feature refers to the time period when the information release. Metadata feature refers to the user is verified or not.

We believe that the forwarding times of tweets which include different topics is not the same, for example, the forwarding times of the tweets about sport topic are different compared to the tweets about military topics, so we should analysis the tweet belongs which topic. Then, we obtain the topic distributions for each tweet based on Latent Dirichlet Allocation (LDA).

3.4 Latent Dirichlet Allocation topic model

Latent Dirichlet Allocation is a topic model and presented as a graphical model for topic discovery by David Blei, Andrew Ng, and Michael Jordan in 2003 (Blei et al. 2003). In Latent Dirichlet Allocation, each sample can be viewed as a mixture of different topics, and is assumed to be feature by a specific set of topics, the topic distribution is assumed to have a dirichlet prior.

As shown in Figure 1, the LDA model is represented as a probabilistic graphical model where, K represents the number of topics, M represents the number of documents and N_m represents the length of document m, $\vec{\varphi}_k$ represents the probability distribution of words in topic k, $\vec{\theta}_m$ represents the topic distribution of document m, $w_{m,n}$ is the word n of document m and $z_{m,n}$ is its topic. In addition, $\vec{\alpha}$ and $\vec{\beta}$ is the hyper-parameters of topic model, as is usually the case, they are fixed values and symmetrically

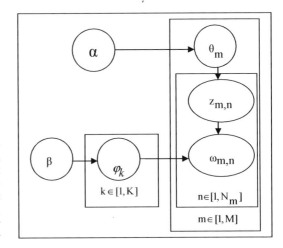

Figure 1.   The graphical model representation of LDA.

distributed. In the parameter deduction of LDA model, we use Gibbs Sampling algorithm, because it has better operation efficiency.

$$p(z_i = k \mid \vec{z}_{-i}, \vec{w}) \propto \frac{n_{m,-i}^{(k)} + \alpha_k}{\sum_{k=1}^{K}(n_{m,-i}^{(k)} + \alpha_k)} \cdot \frac{n_{k,-i}^{(t)} + \beta_t}{\sum_{t=1}^{V}(n_{k,-i}^{(t)} + \beta_t)} \quad (3)$$

Then, we the Gibbs Sampling formula of LDA model, in the formula 3, $\vec{z}_{-i}$ represents the topics except topic z_i, the right half of the formula, it represents $p(topic \mid doc) \bullet p(word \mid topic)$, and we can get the topic distribution of the target document from the matrix of doc-topic.

4   RETWEET PREDICTING BASED ON REGRESSION ANALYSIS

4.1 Datasets

We run our experiments with tweets collected in February and March 2013 which obtained from Twitter API by ourselves, and we retain only the English text tweets, because we need to build LDA model on the dataset. After some other preliminary treatments, the final data set contains approximately 136.8 million tweets and 24.56 million users.

By definition of forwarding times, the time span of training set and test set need to be consistent, furthermore, in order to reduce the influence of data set on the results, the data set is divided into four parts. Thus, each training set or test set contains a week of tweets, and to ensure that the maximum forwarding times of the training set is greater than the test set. As shown in table 1, we counted the number of tweets,

Table 1. Detailed information of each dataset.

	number of tweets	retweet rate	maximum number of forwarding times
Part one	30774166	9.062%	1551
Part two	32531931	7.465%	2353
Part three	31710733	7.545%	1761
Part four	31786975	7.770%	1743

retweet rate and the maximum number of forwarding times to each part. We can also find that each part contains about 31.7 million tweets and there are only 7.77% of the tweets has been reweeted in our datasets.

4.2 Experimental settings

We build a classification model and regression model by each part of a dataset, all models based on a Logistic Regression algorithm. In addition, we use MSE to evaluate the predicted results of different models, the smaller value of MSE, the better prediction results of model, and the final results should be the average value of four experiments.

When extracting the feature, we use the LDA topic model to determine the topic for each tweet then can get content feature, and other features can be obtained directly from the sample. We set the topic number at 90 in our experiments, because there is the best classification results when the number of topics equal to 90 in the LDA model.

For classification model, instead of directly predicting the forwarding times, we divide the messages into different classes by the forwarding times: a: zero, b: between 1 and 100, c: between 101 and 1000, d: more than 1000. When calculating the MSE value, we need to give a value for each category as their predicted forwarding times: a: 0, b: 50, c: 550 and d: 1000. For regression model, we take forwarding times of tweets as the dependent variable of the model, and calculate the mean square error of prediction value and the real value for tweets.

4.3 Experimental results and analysis

The detailed information of each data set is shown in Table 1, we can see that the max of forwarding time only 2353, this is not consistent with the real situation. Getting all tweets which user generated is an impossible task, because Twitter did not fully open the data interface, we can only crawl the data randomly in a time sequence from the Twitter API. Thus, the experimental data will be biased to the actual situation, but it will not affect the experimental results, to some extent, this will reduce the advantage of the regression model.

The data in Table 2 also show that the MSE of regression model based on Logistic Regression is

Table 2. The experimental results.

	Classification model	Regression model	Performance improvement
Part one	314.262	218.726	30.40%
Part two	385.072	298.673	22.44%
Part three	443.020	345.279	22.06%
Part four	428.825	340.068	20.70%
mean	392.795	300.687	23.45%

smaller than classification model, the average of the former reached 392.795, the latter only 300.687, the prediction performance of the model has improved about 25%. In addition, there are 92.04% tweets of the dataset have not been retweeted, that their forwarding times is 0, both regression model and classification model will predict the forwarding times is 0 if the prediction is correct, then will narrow the gap of MSE between regression model and classification model. Despite this, the predicted performance of regression model still better than the classification model.

5 CONCLUSION

Through building a regression model based on a Logistic Regression algorithm to predict the forwarding times of a tweet, which is the focus of this paper. This study can help us to predict and control the scale of information dissemination. We selected some features which have an important impact on the retweet behavior and build the regression analysis predicting model. Experimental results based on the actual data set show the regression analysis predicting model have a better predicting accuracy in dealing with retweet predicting than the classification model, our method is effective.

ACKNOWLEDGMENTS

This work was supported by the National Natural Science Foundation of China under Grant Number 61370170 and National Social Science Foundation of China under Grant Number 14CTQ032.

REFERENCES

Wu, K. & Ji, X.S. & Liu, C.X. (2013). Modeling information diffusion based on behavior predicting in microblog. *J. Application Research of Computer. 2013, 30(6)*, 1809–1812.

Petrovic, S. & Osborne, M. & Lavrenko, V. (2011). RT to Win! Predicting Message Propagation in Twitter. *C. The International AAAI Conference on Weblogs and Social Media(ICWSM)*.

Zamam, T.R. & Van, G. (2010) Predicting information spreading in twitter. *J.NIPS*.

Hong, L.G. & Dan , O. & Davison, B.D. (2011)Predicting Popular Messages in Twitter. *C. Proceedings of the 20th international conference companion on World Wide Web, ACM*.

Boyd, D. & Golder, S. & Lotan, G. (2010) Tweet, tweet, retweet: Conversational aspects of retweeting on twitter. *C. Proceedings of the 43rd Hawaii International Conference on System Sciences*.

Blei, D.M. & Ng, A.Y. & Jordon, M.I. (2003). *J. Latent Dirichlet Allocation. Journal of Machine Learning Research 3, 993–1022*.

Electronic Engineering and Information Science – Wang (Ed.)
© *2015 Taylor & Francis Group, London, ISBN: 978-1-138-02772-5*

The analysis of biological characteristics of real trajectories in the logistic map

J.H. Liu & Zhiyong Luo
College of Computer Science and Technology, Harbin University of Science and Technology, Harbin, China

J.H. Liu
*Research Center of Computer Network and Information Security Technology,
Harbin Institute of Technology, Harbin, China*

Dahua Song
Center of Educational Technology and Information, Mudanjiang Medical University, Mudanjiang, China

ABSTRACT: This paper investigates the biological characteristics of real trajectories in the logistic map, when the value of the lowest digit position is equal to five to the initial iterative value and the control parameter. Experimental results are shown that the property of the logistic map is like the tree ring which can calculate the age of a tree. The common gene positions are increasing by one digit position, while the root gene position can replicate itself.

KEYWORDS: Chaos; Logistic map; Numerical simulation; Gene position.

1 INTRODUCTION

Chaos and fractals are part of an attractive topic known as dynamics. Generally speaking, it is the subject that is concerned with change, with systems that evolve in time. Chaotic phenomena have been widely researched in different scientific fields (Ditto 1995).

This research area is still quite active. Many research reports are dedicated to analyzing properties of chaos (Bouza & Liu 2009, Shi 2008, Stein & Isambert 2011).

However, it is an interesting problem. Whether the real and true trajectories of nonlinear dynamic systems can be characterized by numerical simulative ones or not (Hammel et al. 1987). Plenty of puzzles need to be solved in the continuous real space of chaos and discrete simulative one.

Blanck (2005) used iterative maps as a test of the applicability of exact real arithmetic. Liu & Zhang (2011) reported a parallel computing method of chaotic, random sequence based on the logistic map in scalable precision. Besides, Liu et al. (2012) introduced the iterative method of lower positions which is used to observe the real trajectory of the nonlinear dynamical system. As mentioned above, it is the base to reveal the truth of chaos.

Moreover, in mathematical biology Schreiber (2001) presented chaos and population disappearances in simple ecological models. In recent years, Liu et al. (2014) reported a half-life characteristic in the nonlinear dynamical system. Besides, the natural structure was revealed in the real continuous trajectories.

Chaotic dynamical systems have the sensitive feature and statistical randomness. Deterministic system exhibits aperiodic behavior which makes long-term prediction impossible. In comparison with the existing and known properties, the natural structure of chaos has been little reported. It suggests that the biological characteristics of chaos need to be explored deeply.

To address this issue, we exploit the iterative method with lower positions of true numerical solutions to investigate the biological characteristics of the logistic map in real trajectories. The major contributions of this work are as follows.

i We explore the biological characteristics of the nonlinear dynamical system. The value of initial iterative conditions is specified to five for the lowest position of numerical solutions.
ii The experimental results are shown that the property is like the tree ring which can calculate the age of a tree.
iii We find that the common gene positions will increase by one digit position for each step of iterations.

The plan of the work is organized as follows. Section 2 observes real trajectories of the nonlinear dynamical system, and demonstrates the feature of the chaotic dynamical system. Section 3 gives the analysis of the biological feature of real trajectories. The last part summarizes this work.

2   REAL TRAJECTORY

In this section, we explain the real trajectory and observe the regularity of lower positions of the nonlinear dynamical system.

The logistic map known as a typical nonlinear dynamical system has the form as follows.

$$y_{n+1} = c\, y_n\, (1 - y_n), \tag{1}$$

where y_n represents a real number in the range of zero to one.

y_0 is specified to an initial iterative value for y_n.

n stands for the number of iterations, i.e. nth iteration.

c is the control parameter of the nonlinear dynamical system, and its value ranges from zero to four in real number.

The real chaotic trajectory of the logistic map is shown in Figure 1.

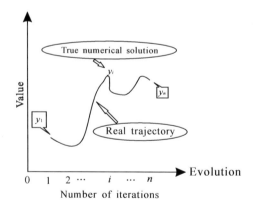

Figure 1.   A real chaotic trajectory in evolution is plotted for the nonlinear dynamical system.

The chaotic trajectory consists of the continuous real points which are computed in infinite precision. The iterative procedure is similar to the evolution. The number of iterations is continuous points in real numbers. Note that there is not any computer error. In other words, the true numerical solutions are computed without round-off errors.

2.1   Example 1

We provide examples to explain the real trajectory. Assume that $y_0 = 0.15$ and $c = 3.25$, the real trajectory of the logistic map is shown as follows.

$$y_1 = c * y_0 * (1 - y_0) = 0.414375.$$

$$y_2 = c * y_1 * (1 - y_1) = 0.78867216796875.$$

$$y_3 = c * y_2 * (1 - y_2) = 0.54167223318072080612\\1826171875.$$

$$y_4 = c * y_3 * (1 - y_3) = 0.80685613119062780362\\9587285233526472438825294375419616699921875.$$

$$y_5 = c * y_4 * (1 - y_4) = 0.50647777293984070979\\3679456992821531765290390282754904621886351\\7561071973201071715368556946934575080376816\\91348552703857421875.$$

$$y_6 = c * y_5 * (1 - y_5) = 0.81236362498765456920\\5922634034525098563561406115439774070189852\\3420652814113136915336203523830618534432385\\4756621727843002149817003572420661364562174\\5726250208766069838102086347076935311529196\\8370682760648317916312571851378265819221269\\3393230438232421875.$$

The regular feature in the chaotic trajectory of the nonlinear dynamical system is listed in Table 1.

From Table 1 the common digit positions of true numerical solution start from x_2 and end in x_{10}, which are located in the lower digit position. It is equal to "875". Besides, note that upper two digit positions sway like a pendulum, which range from "81" to "49".

Table 1.   The feature of lower positions is shown for $y_0 = 0.15$ and $c = 3.25$.

y_n	Value
y_1	0.414375
y_2	0.78…875
y_3	0.54…1875
y_4	0.80…21875
y_5	0.50…421875
y_6	0.81…2421875
y_7	0.49…82421875
y_8	0.81…482421875
y_9	0.49…9482421875
y_{10}	0.81…69482421875

2.2 *Example 2*

We provide other instance for observing the regular feature.

Assume that initial iterative value y_0 is changed, while the control parameter c is invariable. Assume $y_0 = 0.0015$ and $c = 3.25$, the real trajectory of the logistic map is shown in Table 2.

Table 2. The feature of lower positions is shown for $y_0 = 0.0015$ and $c = 3.25$.

y_n	Value
y_1	0.0048676875
y_2	0.0157429776348076171875
y_3	0.05035...4169921875
y_4	0.15542...8857421875
y_5	0.42662...8232421875
y_6	0.79500...6982421875
y_7	0.52966...4482421875
y_8	0.80963...9482421875
y_9	0.50090...9482421875
y_{10}	0.81249...9482421875

We use C programming to compute the true numerical solutions. The computational results are stored in the array with an integer type. We give the upper five digit positions. Besides, middle digit positions are omitted. The lower ten digit positions are provided.

As seen above, the regularity also maintains.

From Table 2, when the value of the lowest position is equal to 5 for y_0 and c, the common gene positions are increasing by one digit position in the iterations of the logistic map.

2.3 *Example 3*

We change the control parameter of c, and maintain a constant value for y_0 in comparison with Example 1. Assume that $y_0 = 0.15$ and $c = 3.35$, the real trajectory of the logistic map is shown in Table 3.

When the value of the lowest digit position of y_0 and c is equal to 5, the regularity of the logistic map is that each true numerical solution includes the root gene positions of "625". Moreover, the similar situation occurs again. Note that upper two digit positions sway regularly in the range of "83" to "46".

3 ANALYSIS OF BIOLOGICAL CHARACTERISTICS

In this section, we analyze the biological characteristics of the logistic map.

The biological characteristics for Example 3 are shown in Figure 2.

Table 3. The feature of lower positions is shown for $y_0 = 0.15$ and $c = 3.35$.

y_n	Value
y_1	0.427125
y_2	0.81970893515625
y_3	0.495083759217210054840087890625
y_4	0.83741...2431640625
y_5	0.45609...4619140625
y_6	0.83104...8994140625
y_7	0.47037...7744140625
y_8	0.83455...5244140625
y_9	0.46253...0244140625
y_{10}	0.83279...0244140625
y_{11}	0.46647...0244140625
y_{12}	0.83373...0244140625

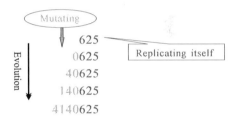

Figure 2. The evolution of the lower positions for example 3 is plotted, which includes the procedure of replicating and mutating.

The evolution represents the iteration of nonlinear dynamical system evolved in time. The common digit positions of "625" are passed on to the next generation. The gene positions are replicated by the next generation. The gene positions are mutating in evolution.

The root gene position exists in each true numerical solution of real trajectories. It is equal to "625". Note that the delay part of the real trajectory will be ignored.

A tree ring is shown in Figure 3.

We can calculate the age of a tree by the tree ring. It is a similar situation that we can deduce the number of iterations by lower positions of true numerical solutions. After one step is iterated the logistic map, the common gene position is converted into "0625", which is shown in Figure 4.

This feature of the logistic map maintains in the next iteration.

4 CONCLUSIONS

In this paper, we analyze the biological feature of the nonlinear dynamical system in the case of the value of the lowest digit position is equal to 5 for y_0 and c.

Figure 3. A tree ring is shown, which can calculate the approximate age of a tree.

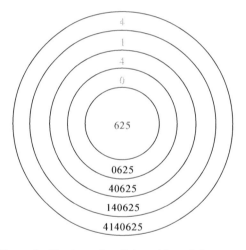

Figure 4. The increasing digit position of the common gene position as the tree ring is plotted.

The experiments imply that the property is like the tree ring, which can deduce the number of iterations in evolution for the logistic map. Besides, we find that the common gene positions will increase by one digit position at each step of iterations.

The biological feature is a new natural characteristic of the logistic map. It may be worthy to further reveal the feature of nonlinear chaotic systems.

ACKNOWLEDGMENTS

This paper is provided by the Natural Science Foundation of Heilongjiang Province with Grant No. F201304, China.

REFERENCES

Blanck, J. 2005. Efficient exact computation of iterated maps. *The Journal of Logic and Algebraic Programming* 64: 41–59.

Bouza, M.K. & Liu, J. 2009. Implementation of a logistic mapping in a parallel system. *Informatics* 22: 131–140. (in Russian).

Ditto, W. & Munakata, T. 1995. Principles and applications of chaotic systems. *Communications of the ACM* 38: 96–102.

Hammel, S.M., Yorke, J.A. & Grebogi, C. 1987. Do numerical orbits of chaotic dynamical processes represent true orbits. *Journal of Complexity* 3: 136–145.

Liu, J. & Zhang, H. 2011. A parallel computing method of chaotic random sequence based on logistic map with scalable precision. *Journal of University of Science and Technology of China* 41: 837–846. (in Chinese).

Liu, J., Zhang, H., Song, D., Buza, M.K., Yang, B. & Guo, C. 2012. A new property of logistic map with scalable precision. *Chaos-Fractals Theories and Applications; Proc. 5th International Workshop, Dalian, China, 18–21 October 2012*, 67–71.

Liu, J., Zhang, H. & Song, D. 2014. The property of chaotic orbits with lower positions of numerical solutions in the logistic map. *Entropy* 16: 5618–5632.

Schreiber, S.J. 2001. Chaos and population disappearances in simple ecological models. *J. Math. Biol.* 42: 239–260.

Shi, P. 2008. A relation on round-off error, attractor size and its dynamics in driven or coupled logistic map system. *Chaos* 18: 013122.

Stein, R.R. & Isambert, H. 2011. Logistic map analysis of biomolecular network evolution. *Physical Review E* 84: 051904.

Electronic Engineering and Information Science – Wang (Ed.)
© 2015 Taylor & Francis Group, London, ISBN: 978-1-138-02772-5

The improvement of virtual machine

S.L. Zhang & J.L. Sun
College of Electronic and Information Engineering, Hebei University, BaoDing, China

ABSTRACT: With cloud computing booming, cloud security arises. The distrust between cloud providers and cloud users greatly influenced the development of cloud computing. While trusted cloud platform has been put forward, the cloud security issues were preliminarily solved. On the basis of the basic model, adopted the combination of active authentication and two-way authentication to intermittent supervision have created a good node whether there is incredible behavior, so as to solve the disadvantages of the passive verification and one-way authentication. Taking this active validation method greatly improves the speed of the live migration, and can reduce the time to establish trust between the network and the network in a certain degree. Rather than the previous passive authentication, When need to be migrated to verify whether the node to be trusted. By way of a combination of active and two-way authentication can timely and effective to eliminate the threat to the platform of malicious nodes, so as to insure the safety of the platform.

KEYWORDS: Trusted platform; Cloud security; Active authentication; Two-way authentication.

1 INTRODUCTION

Cloud computing is one of the hot topics of the current IT field. Cloud computing not only significantly reduces the operating costs but also significantly improve resource management effects, which can quickly meet the business requirements and meet the requirements of contemporary social development. Cloud service providers will be a lot of computing resources, storage resources, software resources are linked together to form a huge scale shared virtual pool of IT resources, to provide users with unlimited IT services (Wu J.Y., L.D. Ping & X.Z. Pan 2012). With the development of cloud computing, cloud security is becoming a matter of growing concern. Users can not verify the safety and integrity of uploading to the cloud information, users lose direct control over the data, service providers do not disclose how the data is processed on the cloud, so users for the cloud service providers produce distrust, and cloud service providers also worry about the content of the user will be deployed in the cloud services produced damage. These issues of mistrust directly hindered the development of cloud computing.

Trusted Computing technology combination of virtual machine technology, the proposed TCCP trusted cloud computing platform (Santos N., K. P. Gummadi, & R. Rodrigue 2011), this is for infrastructure as a service model proposed trusted cloud computing platform. The trusted third coordinator authenticates the built-in trusted chip TPM and then manage the active list of trusted services and participate in a virtual machine (VM) migration of dynamic and virtual machine starting. The trusted chip can ensure the security of virtual machines running in a cloud computing environment. Secure virtualization environment (data center) for cloud computing can ensure user privacy not to be leaking, to ensure data integrity. This paper is based on the TCCP model to improve it, and adopted the combination of active and two-way authentication validation to ensure the security of the platform. Using the active method to replace the passive validation, the node transitions can reduce the time when establishing trust in a certain extent. Meanwhile, using two-way authentication instead of one-way authentication, the platform of the node and PCA have played an important role in the real-time supervision and increase the security of the platform.

2 THE IMPROVEMENT OF VIRTUAL MACHINES IN TCCP MODEL

2.1 *Improvements to TCCP model*

In Iaas, Peas and Saas these three models Iaas is at the bottom and is closer to the users. Only the VM in the Iass allows users to have control, so in the Iaas model it is easier for the user's data security and privacy protection. TCCP model framework (Towards trusted cloud computing). As is shown in Figure 1.

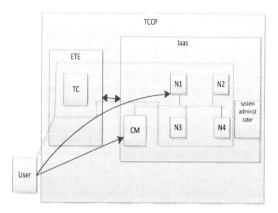

Figure 1.   TCCP model framework.

In this model, ETE is a trusted third party, TC is a trusted coordinator in the third party management and maintenance. TC is used to manage a range of trusted nodes TN list, which can safely run the VM user. CM is the untrusted cloud manager, because CM could be manipulated. If TN is to become a trusted node, it must be located in security domain and running on the trusted virtual machine monitor (TVMM). TVMM can control each node runs in the background and prevent prying eyes and tampering by privileged users. TVMM can protect their own safety and to comply with TCCP protocols, nodes are embedded TVMM which by TPM proven and through the secure boot process. Where TPM chip as a trusted base, It is solidified in hardware products. TPM contains a recognized engraved with a unique identification of the TPM private key, which is (EK) and some function that cannot be modified and encrypted. The manufacturer signs the corresponding public key, in order to ensure the correctness of the chip and the effectiveness of the secret key, the corresponding public key can allow a trusted third party to do the authentication. System from the root of trust, trust root is part of the BIOS is started gaining control of the first module, which can be completely trusted, its trustworthiness and integrity guaranteed by physical means. From a trusted root to start building a chain of trust (Zheng L. 2013), and gradually extended to the entire system. The establishment and maintenance of the trust chain by continually measuring, the next subject will receive the right to be manipulated value, this value can represent the current state of that subject, usually in binary image for Hash calculation, Will get after the Hash value through operation extension to the original TPM of PCR. So, start the process computer system having a TPM is a "first measure, and then run" process. Trust chain is in the main body of this control has to measure in the future to have the body of the control process. TPM maintains the credibility

of the entire system model by making each compute node in the TCCP model to become a trusted node.

In the original model of TCCP message transfer process from a dynamic VM migration [6] can be seen, in the process of migrating virtual machines, all certified behavior should interact with TC, TC participated in several calculations, TC is the bottleneck of the whole TCCP model. Once the trust list of TC is broken, the whole credibility of the interaction process cannot guarantee and this will lead to the collapse of the entire TCCP trust model.

The following will present a program to improve the TC, that is based on Privacy CA (Privacy Certification Authority) strategy. Limitations of the original TCCP platform, namely performance and security bottlenecks TCCP model for ETE internal TC. Improvement plan is a TC management function of TNs will be moved to the cloud internal TN, namely the cloud has an internal TC (Internal TC, ITC) assumed the management functions ETE internal TC, where ITC's role is clearly needed for election between TNs. Due to a single server platform performance and the limitation of resources coupled with an internal cloud resources Zone also has a massive number of servers, therefore a single TN is not up to the ITC role of work.This article will decentralize workload of ITC to multiple TNs, each TN is able to take appropriate workload, the paper treats this TN as Privacy CA. The privacy CA can manage part of the TNs list in the Zone and it will be able to accommodate the size of the TNs in this small area, and report to the ITC this information. The program makes a part of the TC's paralysis does not affect the entire system, that part of the TC cannot be trusted can not affect the safety of the entire system. According to the difference of Privacy CA performance and hardware configuration, the number of TNS, which are managed by privacy CA is inconsistent, and the available resources are dynamic when the platform is running, so the TNs which is managed by the Privacy CA is dynamic. In the improved scheme, Privacy CA assume the responsibility of the ITC, ITC is responsible for the contact each Privacy CA, collecting it will be able to accommodate the number of TNs in different management area, and the remaining resources can be assigned to the VM and report the information to the TC, the TC based on these nodes information to specify the VM getting goal area. Privacy CA and ITC replace the responsibilities TC manage TNs in ETE, but TC is responsible for the TN credibility certification function unchanged, because only TC can judge the TPM EK whether it is valid. TC is responsible for TN assigned Privacy CA roles, Privacy CA in the cloud interior elected ITC. When TN which is managed by Privacy CA reaches the maximum number, PCA will report the situation to the ITC and ITC report to the TC, and then TC specifies the other TN containing an effective TPM assume the Privacy CA role. Improved TC scheme shown in Figure 2.

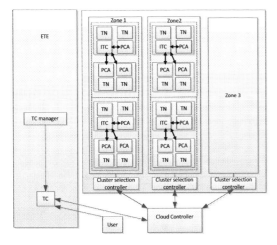

Figure 2. Improved virtual machine.

As can be seen from Figure 2 shown, the ITC in each Zone report to the TC through the cloud controller, TC by the ITC feedback information to assign tasks, decided which zone the ITC should be allocated. ITC manages Privacy CA, Privacy CA manages TN node. In this way, will be the responsibilities of the TC distributed to nodes in the cloud, when a node is attacked, Privacy CA becomes incredible, it cannot affect the safety of the entire cloud.

2.2 *Improvements on validate node credible way*

When users finish creating virtual machines (Yang S. 2013), in order to prevent the trusted node is equipped with a virtual machine is compromised, the following improvements can be made in the existing model: adopted the combination of active authentication (Sadeghi A.R. & C. Stüble 2014) and two-way authentication to intermittent supervision have created a good node whether there is incredible behavior, so as to solve the disadvantages of the passive verification (Wang H.Z. 2011) and one-way authentication. Assuming that each Privacy CA manages N nodes, the nodes after being created, Privacy CA uses intermittent send their management area of the node Verification information. Privacy CA uses the private key of the PCA to signature a random number and with the PCA public key encryption, then send to the node, specifically to send a message to: $\{\{n1\}\ pkPCA\}$ pkAIK, where the PCA public key encryption is used AIK strategically to prevent exposing the TPM. The privacy CA's public key is publicly available. After the node receives the message decrypts the message to obtain n1, the node also generates a random number n2, and n1, n2 encryption $\{\{n1, n2\}\ pkAIK\}$ pkPCA sent to Privacy CA, Privacy CA decrypted and checks if the received and sented out n1 are the same, If not consistent, PCA will kick out this node, if it is consistent, the node is trusted, then Privacy CA discard the used n1, n2 left while generating a random number n3, and sends the message $\{\{n2, n3\}\ pkPCA\}$ pkAIK to the node, Node decrypted and test whether n2 is the last time send out random number, If yes, then the Privacy CA is credible, when after verification PCA trusted, node pass the list of their management to PCA. If not, the Privacy CA has been compromised. When PCA untrusted ,nodes with the same method generate a random number pass to the ITC, If we confirmed the ITC trusted, would report the untrust of PCA to ITC, ITC re-elected PCA and kicked out the original PCA. If the ITC also not to be trusted, then continue to pass the random number to the PC. Like this again and again, Privacy CA can continuously check whether its manage node is credible, and the node can also continuously check whether the Privacy CA can be trusted. Taking this active validation method greatly improves the speed of the live migration, and can reduce the time to establish trust between the network and the network in a certain degree. Rather than the previous passive authentication, When need to be migrated to verify whether the node to be trusted. Meanwhile, With the method of the active validation and help to know resource usage in the region, PCA obtain node credible service list and report to the ITC, different regions of the ITC reports to the TC, eventually the TC with this table to arrange the nodes create, join and kick, achieve their management capabilities.

3 PERFORMANCE AND SECURITY ANALYSIS OF TCCP 'S IMPROVED MODEL AND IMPROVED THE WAY OF VALIDATION

TCCP improved model for the proposed theoretical model, turning it into the actual operation of the platform requires a lot of resources and work. First, due to the VMM must run on barel computer, model tests conducted by TCCP virtualized environment is not feasible, even if you have enough servers running VMM, vTPM internal code must be modified to comply with the requirements of the TCCP. Therefore, this article only gives qualitative analysis on the security and performance of TCCP model and improves the way of validation.

3.1 Performance analysis of TCCP improved model and improved the way of validation

Performance bottlenecks in the original TCCP model are over-reliance on TC, TC had to play a major role in the process of VM's creation and migration, if the TC paralysis, and the entire model with paralysis. The TCCP model using Privacy CA in this paper puts the functions of TC manages TNs inside the cloud, then the Privacy CA and ITC manage TNs together. And this article adopted multiple parallel Privacy CA Settings of the regional administration scheme and VM migration prefer to adopt the same Privacy CA management area TN for destination node, it can largely improve the efficiency of management and VM migration TNs. But the TCCP model is not able to run after the entire Zone power on, it must take some time to establish a stable TMR, at this time it need to adopt the Privacy CA strategy. TCCP improved model on the performance of ascension is mainly on the reliability, the TC paralyze, it does not affect the migration of VM ; if the ITC paralyze or lose the platform integrity reliability ,VM migration in the Privacy CAs managing area will not be affected, TNs under its management area TC will re-apply for the new PCA, even if all TNs under the Privacy CA management area paralyze or lose platform integrity reliability, it does not affect other areas.

Through the way of combination of active and two-way authentication can see: Shortened the time needed for node transition, because the source node and destination node credible validation process has been interrupted completion in the active validation. Second, when the transition occurs, the source node does not need to send more parameters which needed to be verified, so that the complexity of the authentication is reduced, saving a certain amount of pay expenses.

3.2 Safety analysis model of TCCP improved model and improved the way of validation

As the performance of the original TCCP model was substantially increased, the model security of the TCCP also needs to be guaranteed. In the original TCCP model, TC is responsible for certification TNs and participate in the creation and migration of the VM, If TC breached then TN identity associated with it activities take in everything at a glance. In this article the TC only involved in the certification TNs not involved in the management process of VM, therefore, cannot be informed of the internal operation of the cloud (which is also CSPs expected). This article sets, lots of Privacy CAs isolation mechanism, it can not only improve the performance of TCCP model, but also can improve the safety of TCCP model. If the Privacy CA is compromised, its influence is limited

to its management of the area. Privacy CAs' electing TC used an internal selection algorithm which can avoid malicious Privacy CAs manipulate election results in some degree. By introducing the iTurtle concept (McCune J. M. et al. 2011), allowing users to understand the state of their own VM platform which become possible, which is one of the keys to solving the problem of user and CSPs trust. By way of a combination of active and two-way authentication can timely and effective to eliminate the threat to the platform of malicious nodes, so as to insure the safety of the platform.

4 CONCLUSION AND OUTLOOK

Based TCCP model, improved coordination of the trusted third party, they will be moved to the part of the management of TC on the node TN, thus preventing accidents caused by the breaking of the security of the entire model due to the leakage of TC. Meanwhile, more conducive to the node management and supervision. Virtual machine migration and creation under the improved scheme, can ensure that the virtual machine is running in a safe environment (Gu X., Z.Q.Xu, & J. Liu 2011). Through two-way authentication and active authentication at the same time make the platform more credible.

In all, there is still much research space of the credible cloud model based on cloud theory. This paper argues that in future studies, focused mainly on the following aspects: weigh the node's different attributes in many ways, In that time, the nodes on the trusted platform are more effective and safer for computing, migration, meanwhile ensure that the node can not threat platform security.

ACKNOWLEDGMENTS

Subject belongs to the national natural fund project. Project number: 2013BAK07B04: Organization code management service cloud platform and security system research.

REFERENCES

Wu J.Y., L.D. Ping & X.Z. Pan (2012). From Concept to The Platform. *Telecommunications Science 12(6)*, 1–2.
Santos N., K. P. Gummadi, & R. Rodrigue (2011). *Towards Trusted Cloud Computing*. In Proc.of the 1st USENIX Workshop on Hot Topics in Cloud Computing, Berkeley, CA, USA.
TowardsTrustedCloudComputing,http://www.bbs. chinacloud.cn.

Zheng L. (2013). Based On The TNC's Trusted Cloud Computing Platplatform Design. *HeNan: ZhengZhou University.* 5–10.

Yang S. (2013). An Improved Virtual Machine Migration Method Based On Trusted Computing Technology. *Computer & Digital Engineering ,41(10),* 2–3.

Sadeghi A.R. & C.Stüble (2014). Property-based Attestation For Computing Platforms: Caring about properties, not mechanisms. *In New Security Paradigms Workshop 2014.*

Wang H.Z. (2011). A Trusted Cloud Computing Platform Model and Its Improvement .*Hefei: University of Science and Technology of China.*

McCune J. M., A. Perrig, A. Seshadri, & L. van Doorn (2011). Turtles all the way down: Research challenges in user-based attestation. *In Usenix Workshop on HotSec'11, 2011.*

Gu X., Z.Q.Xu, & J. Liu (2011). Based on the cloud theory of credible research and prospects. *Journal on Communications 32(7),* 4–6.

Electronic Engineering and Information Science – Wang (Ed.)
© 2015 Taylor & Francis Group, London, ISBN: 978-1-138-02772-5

Research and design of pipe burst analysis algorithms based on Geodatabase data model

Z.B. Yu & W.B. Wang
School of Computer Science and Technology, Harbin University of Science and Technology, Harbin, China

H.P. Xiong
Exploration & Development Research Institute of the Daqing Oilfield, Daqing, China

ABSTRACT: Analysis of pipe burst is the primary function of network maintenance. Aiming at the shortage of the analysis algorithm used by current market, the article used the Depth First Search algorithm (DFS) to replace the Breadth First Search algorithm (BFS), and built the pipe network data model with Geodatabase data model .According to the generated logical network, relying on the principle and topology graph theory, the analysis of pipe burst algorithm are studied and re design, and it overcome the existing algorithms shortcomings of pipe burst analysis, such as low efficiency and not accurate results, improved the efficiency of network maintenance.

KEYWORDS: Water Supply Network; Pipe Burst Analysis; Algorithm Implementation.

1 INTRODUCTION

The existing pipe burst analysis algorithms usually have two kinds, one is based on graph theory, and another one is based on the analysis of pipe burst method using flow. The first use the graph theory and adjacency list to abstract physical network and generate the undirected graph. The second use the GIS component to develop it. It uses a related interface and method of ArcGIS Engine to analysis flow and the pipe, find a valve closed solution (Yang Shanshan 2005, Hu Xinling 2008, Wang Limei 2008).

This article integrates the advantages of two algorithms. Directed graph is used to simulate the network structure and use the direction of graph to represent the direction of the flow, through GIS components to develop the analysis of pipe burst, and it can correct display flow without calculating the angle of pipe and flow, makes programming easier.

2 DESIGNING THE SCHEME

The steps of designing pipe burst analysis algorithms

1. Finish to the topology of network data, and form a directed graph.
2. Find the optional solution from the direct of counter current starting from a certain point on the directed graph with the depth-first traversal.

2.1 Establish logical network topology relation model

ArcGIS Geodatabase database use the geometric network model and logical network model to express the linear network system (Yong Chang & Xing Tao & Liu Dacheng 2012), as shown in Figure 1. Also Geodatabase network model has an important feature; it can describe the flow of resources (such as water, gas) in the geometric network. Therefore, that use Geodatabase network model as the network data model can realize the management of integrated network with spatial data, connectivity data and attribute data (Zhan Liaofang 2013, Zhong Yue 2013, Xing Naying 2009).When need to do the analysis of time burst pipes, just reading a file "node ID of the beginning and end pipe", it stands by the topological relationship between punctuate elements. We can get a Map from the file with the Geodatabase network model.

In geometric network, the property has three values: None, Source and Sink. The Source and Sink mean the source point and ending of the flow. Never generate flow in the network called none. When need analysis of pipe burst, the network resources, supply point (such as water plant, computer, gas station, etc.) is set to the source point, resource using point (such as user valve) is set to end points (sink), then use the IutilytyNetworkGEN (GIS interface) to generate the flow. A Geodatabase network model can form three kinds of flow: determinate flow, indeterminate flow and uninitiated flow. In the case of a point source, all of the tree network structure will generate determinate flow. A part of ring structure will generate determinate flow and some part of ring structure will generate indeterminate flow, while the part that is not connected to the source point generated an uninitialized flow. For cases with multiple source point, most of the backbone network flow generated uninitiated flow (Zhang Lei 2013, Guan Dong & Yin Xiaoming & Zhanhong 2012).

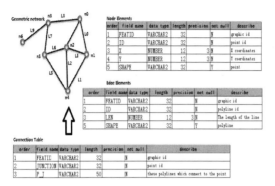

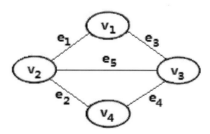

Figure 1.   Geodatabase network data model.

Setting up the topological relations of logical network model, is actually abstract the directed graph on the basis of the logical network model, then generate all topological adjacency relations between point element involved in the analysis of pipe burst in the network, The logical network model is established according to the needs of pipe burst analysis, able to quickly find the valve that it is the nearest location of pipe burst, also can easily find the affected user nodes. In the field of mathematics and computer network is abstracted as figure, so the graph theory is a powerful tool of analysis of the water supply network model. A figure has two parts, one part is a node, and the other part is the edge. The described method of figure has a lot of kinds, such as the adjacency matrix notation of figure, graph adjacency list notation, graph adjacency multiple table notation. In order to analyze pipe burst, adopting the idea of adjacency list to represent the figure when building topology relations. Because it is easy find the first adjacent point and the next adjacent point of one vertex in an adjacency list. When one does not consider the direction of flow, qualitative analysis of the problem, you can think the V_i and V_j segment is undirected, it is mean that the water supply network called $G = (V, G)$, and the V is all dotted elements of that participated in pipe burst analysis in pipe network, called vertex set; The E is all edge of the pipe network, known as the edge set.

In fact, the water has a direction, and there is a pathway between the elements at any point in the pipe network, therefore the water supply pipe network is a directed graph. According to the basic principle of graph theory, for undirected graph, side e has two vertices are U and V, $<u, v>$ express the unordered pair, and the $< u, v >$ and $< v, u >$ are expressed u, v for an undirected edge of the endpoint. Figure 2 can be expressed as $V= \{v_1, v_2, v_3, v_4\}$, $E= \{e_1, e_2, e_3, e_4, e_5\}$, or are represented as $G= (V, E)$, $V = \{v_1, v_2, v_3, v_4\}$, $E= \{<v_1, v_2>, <v_2, v_4>, <v_1, v_3>, <v_3, v_4>, <v_2, v_3>\}$.

For Figure 3, using a ordered pair that directed edge e associated with the vertex U, V to express it, namely $E=<u, v>$, u said starting side of u, v said the endpoint of e. Figure 3 can be expressed as $G= (V, E)$, $V = \{v_1, v_2, v_3, v_4\}$,$E= \{<v_2, v_1>, <v_1, v_3>, <v_4, v_3>, <v_4, v_2>, <v_2, v_3,>\}$

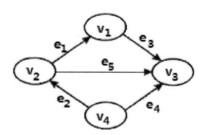

Figure 2.   Undirected graphs.

Figure 3.   Directed graph.

When the need for the analysis of time burst pipes, reading a file "node ID of the beginning and end pipe", it said the topological relationship between punctate elements.

Map is a nonlinear data structure (Liu Zhenxing & Gong Rui & Qin Hua 2012), and used to store it in the adjacency matrix. If there is a group of network instance data, it can use a dimension group $V[n]$ to store the node, and using a two-dimensional array $A(I, J)$ to store the relationship of figure point. In the A, I and J represents the vertex number, the value K of $A(I, J)$ represents the adjacency relationships of vertices. In the following adjacency matrix, Adjacency list value is: $k =0$, it means there is not relationship

314

between vertices; $k = 1$, it said there is. The adjacency matrix of network nodes is show in (1).

$$
\begin{bmatrix}
 & V_0 & V_1 & V_2 & V_3 & V_4 & V_5 \\
V_0 & 0 & 1 & 1 & 1 & 0 & 0 \\
V_1 & 1 & 0 & 1 & 0 & 1 & 0 \\
V_2 & 1 & 1 & 0 & 0 & 1 & 1 \\
V_3 & 1 & 0 & 0 & 0 & 0 & 1 \\
V_4 & 0 & 1 & 1 & 0 & 0 & 1 \\
V_5 & 0 & 0 & 1 & 1 & 1 & 0
\end{bmatrix}
\qquad (1)
$$

2.2 Search the valves which need to be closed

Searching these valves can be divided into two processes: Finding every valve that can make the accident pipe cutoff and the number of pipes is the minimum; Find out the independent valve. First of all, obtain the coordinates of the fault point. See the point as the starting point, and judge the flow direction of the pipe section, and do DFS in a countercurrent flow. Establish a node stack P, and the starting point will be put into the stack P. The stack P pops the node A. Searching the side of a set E which have all of the edges that linked the A, and judge the side whether it is a downstream or upstream for A. Putting all of the un-upstream edges from E into the stack Q. Getting a side from the Q, and gets another node B of the edge which is not in the P stack, and judge B whether it is a valve, if it is, put it in the closing valve array, and pops this edge from the stack Q, Then take the other side from the stack to judge, if it is not then pushed onto the stack P, repeat the above steps until the stack Q is empty. The valve closing array is the effective valve. The structure of the program flow diagram is shown in Figure 4.

3 THE REALIZATION OF THE ALGORITHM

ArcEngine is a GIS component library. Its access interface can support the GeodataBase to check the effectiveness of the network elements and generate a flow in the network flow and getting the flow of elements to realize the pipe burst analysis capabilities which are optimized by DFS.

Pipe burst can be divided into two categories: edge pipe burst and point pipe burst, the position of the pipe burst can use INetFlag interface to set markers. Since the side tube explosion, it can be directly set the edge flag, and then judge the flow of the edge. According to different flow to determine the search direction of the valve, there are three kinds of situations: Known flow uses upstream point to search; the unknown flow

respectively use two endpoints valve to search; uninitialized flow does not search. For point pipe burst, searching directly through this point, namely first determine whether a point is valve type, if it is, you should set firstly this valve as an invalid valve, and then search next point. The valve is set to an invalid valve when it has some valves to be searched which can be closed, (using INetworkFeature interface); finally, restore the effectiveness of the valve, restore the network model to the initial state.

4 THE USE OF ALGORITHM

A part of the pipeline distribution map is shown in Figure 5, and the pipeline model respectively, with DN at the beginning of number, such as: DN100, DN200, But the node about control valves which are numbered starting with SXYC_KZF_ such as: SXYC_KZF_104. If a failure point lies between the valve SXYC_KZF_103 and SXYC_KZF_111 DN200, using traditional algorithms and the algorithm to make an analysis of pipe burst, then get valve results. Experimental results show that the traditional algorithm gets a result is slowly and the need to be close valve tail number which be searched are as follows: 111,104,103,102. From the graph, it can be known that 104 and 102 are independent valves, so the result is not accurate. The number gets by new algorithm is 103. Because the valve number 111 is in the downstream direction of fault point so as to can be regardless of. So the algorithm excludes valve number 111 is correct. From the experimental results by comparison, this algorithm perfectly overcomes the

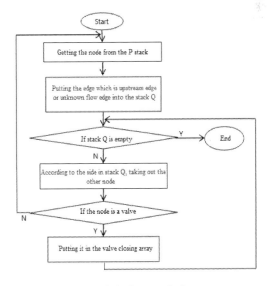

Figure 4. Flow chart of pipe burst analysis.

deficiencies of the traditional algorithm, achieved the expected effect.

5 CONCLUSION

According to the requirement of the analysis of pipe burst, aiming at the shortage of the analysis algorithm used by current market, the article used the depth first search algorithm (DFS) to replace the breadth first search algorithm (BFS), and built the pipe network data model with Geodatabase data model .According to the logical network generation, relying on the principle and topology graph theory, the analysis of pipe burst algorithm are studied and re design, and it overcome the existing algorithms shortcomings such as low efficiency and not accurate results of pipe burst analysis, improved the efficiency of network maintenance. This algorithm is simulated with the actual system, successfully reached the predicted effect, makes the analysis of pipe burst function has been greatly improved, it also shows the feasibility of the algorithm and advanced.

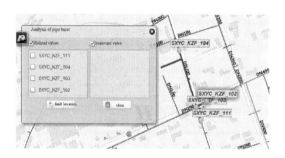

Figure 5.   Water supply pipe burst analysis diagram.

REFERENCES

Guan Dong & Yin Xiaoming & Zhanhong (2012).The application of GIS in urban water supply and drainage pipe network management. *J. Southwest China municipal engineering design & research institute.*

Hu Xinling & Zhang Hongfei (2008). The algorithm study of pipe burst analysis in water supply network geographic information system. *J. Science of Surveying and Mapping*, 33(4): 225.

Liu Zhenxing & Gongrui & Qinhua (2012). Research and Design of Water Resources Information Management System Based on GIS. *J. Computer CD Software and Applications.*

Wang Limei &Wang Pu & Yang Zhaohui (2008). WebGIS - based valve closing analysis in pipe burst emergency of water distribution network. *J. WATER TECHNOLOGY.*

Xing Naying (2009). Cause analysis and preventive measures of pipe burst in water supply pipe network. *J. SCIENCE&TECHNOLOGY INFORMATION.*

Yang Shanshan (2005). The Design and implementation of Analysis of Pipe Burst in Water Supply Network Geographic Information System. D. *Wuhan University Master Thesis,*

Yongchang & Xing Tao & Liu Dacheng (2012). Pipe Burst Detection and Valve-turnoff Algorithm on Water Supply Pipe Network Based on GIS. *J. Computer technology and development.*

Zhang Lei (2013). APPLICATION OF GIS IN THE ENERGY PIPELINE NETWORK. D. *Yanshan University Master Thesis.*

Zhong Yue (2013). Analysis of Coated-after Super heater Pipe Bursting Accident in Utility Boiler. *Proscenia Engineering.*

Zhan Liaofang (2013). Research and Implementation of Water Resources Information Management System Based on GIS. *China University of Geosciences Master Thesis.*

Electronic Engineering and Information Science – Wang (Ed.)
© *2015 Taylor & Francis Group, London, ISBN: 978-1-138-02772-5*

Design and implementation for computer online examination system

H. Li, Y.X. Ling & W.F. Wu
Department of Management Science and Engineering, Officers College of Chinese Armed Police Force

H.Y. Yan
Science and Technology on Information System Engineering Laboratory, National University of Defense

ABSTRACT: In view of the current situation of computer examination system, in the full use of .Net platform, database technology and modern information processing technology, this paper puts forward a technical framework of computer online examination system based on database, and complete the database design. The paper realized computer proposition, automatic generating examination paper, computer examination and automatic scoring. It can meet the needs of paperless examination, and also has a very good reference value and promotion for the other assessment system.

KEYWORDS: Computer examination; Nhibernate; Automatic scoring; Sentence similarity.

1   INTRODUCTION

With the development of computer technology, especially the rapid development of computer network technology, the traditional teaching style has changed greatly, the mode of network teaching has gradually become the main trend(Qin, 2010). At present, most of the computer examination system has adopted the paperless examination, but it still uses artificial browse way to get the score (Jie, Ling, and Wang.2010). Therefore, in the network teaching , it becomes especially important how to achieve efficient and accurate computer examination and automatic scoring. In the full use of .Net platform, database technology and modern information processing technology, this paper builds a computer online examination system based on database. It realizes the computer proposition, automatic generating paper, computer examination and automatic scoring. It can meet the needs of paperless examination, and there is also a very good reference value and promotion for the other assessment system.

2   SYSTEM FRAMEWORK AND FUNCTIONS

The realization of a computer examination system mainly has two modes, C/S (Client/Server) and B/S (Brower/Server), which have their respective advantages and disadvantages. Taking into account the database functions, easy use and safety, the system uses C/S mode, uses SQL Sever 2008 as the database,

uses .Net platform as Microsoft software platform, and uses C# programming language. The basic frame of the system structure is shown in fig.1.

The system is mainly divided into four layers. Access layer mainly realizes the user identity authentication. Storage layer mainly realizes universal access to the database. Application layer is the ultimate function which the system provides. The main function modules of the application layer are shown in fig.2.

Application layer mainly consists of 10 modules, and different module achieves different functions. For example, Item proposition module provides the ability of manager temporary item database to domain

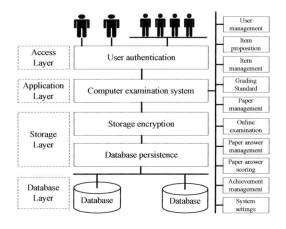

Figure 1.   The basic frame of the system structure.

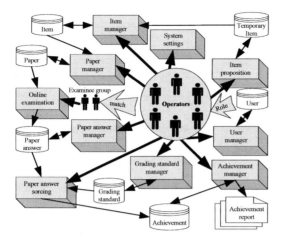

Figure 2. The function modules of the application layer.

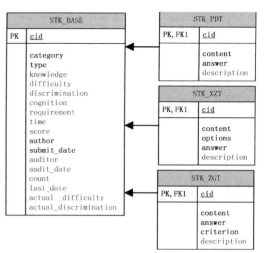

Figure 3. The specific design of item database.

experts, such an adding a new item to temporary item database. Paper answer module provides the ability of generation achievement records from the paper answers to Examiners.

3 DATABASE DESIGN

The database is the entire examination system, which is mainly composed of item database, paper database, paper answer database and user database. Taking item database as an example, we explain the specific design process of the database. Item database is designed for the storage of item content, scoring standard and other information relating to item. Consideration of the great treatment differences between the objective and subjective item, such as the additional options (options) for multiple-choice item, the answer of judgments only two categories, and so on, the property of item is divided into basic attributes and exclusive property. Based on the modern distance education resources construction technology standard, the item database is designed as shown in Fig.3.

Here, public basic information of one item is stored in the basic information table (STK_BASE). There are three exclusive attribute tables (STK_PDT, STK_XZT, STK_ZGT) to store separately the exclusive attribute of judgments, multiple-choice and subjective.

4 DESIGN AND IMPLEMENTATION OF THE KEY TECHNOLOGY

In the automation technology of computer online examination system, there are several key technologies, such as encryption storage for information, strategy for paper generation and automatic scoring for subjective questions. The following were expounded.

4.1 Encryption storage for information

The security storage of network information is the issue which each network system must consider. We realize the security storage process based on the DES and AES algorithms, and the keys will not be visible to any users. In the current production environment, it may be of considerable trouble and wasting time to use the object oriented software and relational database together. The use of Nhibernate could liberate developers from the original boring SQL statement to devote to the business logic implementation. And it can greatly reduce the development time that one uses SQL and ADO.NET directly. In addition, Nhibernate can help users to put the result set from the tabular representation into a series of objects.

Therefore, we use both the information encryption process before persistence and the persistence process based on the Nhibernate toolkit to quickly and reliably realize the information encryption storage module.

4.2 Strategy for paper generation

The strategy for paper generation can be divided into two kinds: automatic mode and manual mode. The manual mode is relatively simple, and automatic mode is a more critical technology, especially to generate a better quality paper for the examination system. The teacher is generally determined by the range of knowledge, the different proportions of items, and paper structure to implement the artificial paper generation process. Therefore, according to the teacher's psychological process, the basic flow of the automatic paper generation for an appraisal system is as show follows:

1 Determine the professional courses, structure of paper and scores of different item types.
2 Determine the external constraints of paper, such as the coverage of knowledge points.
3 Determine the other attributes of paper, such as difficulty distribution, and so on.
4 Finally, get the set of selectable items for each item type against the above constraints, and then sample one item randomly as a result, until meet all the count and score requirements on each item type.

4.3 *Automatic scoring for subjective questions*

Since the 1960s, many foreign scholars committed to the study on the automated scoring technology for subjective questions (Xiao,and Wang 2010). Our automated scoring technology for subjective questions started late, in recent years also has made a lot of researches (Li ,Chai and Zhang. 2010; Liu, and Li 2009)

The automated scoring technology for subjective questions mainly solves the problem that calculates the similarity between the answer and the standard answer. The similarity is higher, the score is more, and vice versa. The current sentence similarity calculation method mainly has three kinds: morphology, word order, word meaning. The main consideration of morphology and word order is the similarity based on word form and word order, and morphology plays a major role in fact. The word meaning method is used to analyze the semantic information of the words, which can handle the similarity calculation on synonyms, synonyms expansion and the deep semantic of words. There are several thesaurus for semantic extension, such as Tongyicicilin, HowNet, Wikipedia entries, and so on.

According to the teacher psychological, there are mainly two factors: point scores and the similarity between the answer and the standard answer. The paper approaches to handling as show follow:

1 According to the scoring standard and standard answers of the subjective question, extract a plurality of scoring points.

2 Use these scoring points to match the answer, and then determine the score based on the similarity level.
3 Finally, according to the weight proportion of the scoring points and the similarity, calculate the final score.

5 CONCLUSION

The paper focuses on the analysis, design and implementation process of a computer online examination system. The goal of the analysis and design is to establish a scalability, usability, security computer online examination system, which realizes the networking and automation on four basic functions (computer proposition, automatic generating paper, computer examination and automatic scoring). This paper proposed the design of system framework and database, and realized the core function of the computer online examination system. Because of the time, technical limitations and other conditions, automatic scoring of subjective question still exist some shortcomings, and need further improvement.

REFERENCES

Qin, X.V. (2010);Subjective marking system design based on similarity computation. *Journal of Anhui Institute of Architecture & Industry.18(4)*, 77–80.
Jie, B., Z.H. Ling, T.C. Wang. (2010) Design and Implementation on Automatic Questions Review System of the Computer Culture Basis Based on Net Platform. *Computer Education.14*, 144–148.
Xiao, M.,&X.L. Wang. (2010) Research on automatic marking of Subjective Item[J]. *China Examinations*. 10, 28–31.
Li, Y.H., L.Y. Chai, Q. Zhang. (2010) Automated correcting algorithm for subjective questions based on segmentation technology and sentence similarity. *Computer Engineering and Design. 31(1)*,2663–2666.
Liu, P.Q., Z.Z. Li (2009),Research on methods of automatic checking over subjective examination based on fuzzy conceptual graphs with weight. *Application Research of Computers, 26(12)*, 4565–4567, 4584.

Electronic Engineering and Information Science – Wang (Ed.)
© 2015 Taylor & Francis Group, London, ISBN: 978-1-138-02772-5

A local multi-resolution discriminant power for face recognition

P. Cui, X.T. Zhang & L.S. Yu
Harbin University of Science and Technology, Harbin, China

ABSTRACT: To make use of label information and enhance recognition accuracy, a local multi-resolution method based on Steerable Pyramid-based Semi-supervised Discriminant Power (SPSDP) is proposed for face recognition. Firstly, Steerable Pyramids (SP) transform is performed to decompose a face image into different sub-bands at different levels and obtain luminance variation in different positions. Then the coefficients are divided into blocks. Feature vectors are combined with statistical measures of blocks. Finally, semi-supervised discriminant power analysis is used for classification. The algorithm is tested on ORL, YALE and FERET face databases. The experimental results show the proposed method obtains higher performance than other local multi-resolution methods..

KEYWORDS: Steerable pyramid, Feature extraction.

1 INTRODUCTION

As high recognition accuracy, traditional multi-resolution recognition methods, the discrete wavelet transform (DWT) and Gabor are very popular (Devi, B.J. 2010, Shen L.L. & Ji Z. 2009). With translational and rotational invariance, Contourlet and Curvelet, steerable pyramid transform (SPT) are similar with the two-dimensional DWT (Boukabou, W.R. & Bouridane, A. 2008, Mandal, T. et al. 2007, Su C.Y, Y.T Zhuang & L. Huang 2005). However, most of traditional multi-resolution methods are unsupervised learning. There are errors once the recognition process deviates the correct direction. Semi-supervised learning is the methods which part labeled samples are used and many unlabeled samples. In the learning process, clustering and classification are performed by constraint condition (Song Y.Q. et al. 2008, Cui P. & R.B. Zhang 2012). To enhance robustness and accuracy, we propose a steerable pyramid-based semi-supervised discriminant power (SPSDP) for face recognition.

2 STEERABLE PYRAMID TRANSFORM

An image contains similarity and feature information different from others. Because of illuminance variation, discriminant information can be obtained from a different position. Discriminant features caused by variation of position and illuminance usually are used for discriminating one from other persons. Steerable pyramid transform (SPT) divides image features into different subbands at different levels.

Steerable pyramid transform (SPT) decomposes a face image into different sub-bands at different levels. In Figure 1, SPT is performed with (128×128, 64×64, 32×32) 3 scales and ($-\pi/4$, 0, $+\pi/4$, $\pi/2$) 4 directions in a face image.

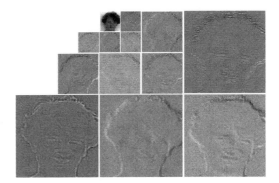

Figure 1. SPT with 3 scales and 4 orientations.

We can see that every direction filter is sensitive to filtering information about perpendicular direction. Multi-scale feature in the space and local feature of multi-direction are combined with steerable pyramid.

High-pass filters and low-pass filters (H_0, L_0) are used to decompose an input image into two subbands. Low-pass subbands are decomposed into K directional band-pass subbands B_0,...,Bk_{-1},and a low-pass subband L_1. Down-sampling and up-sampling corresponding to row and column are carried out by a factor of 2 to obtain lower low-pass subbands (LS), recursively. Each recursive step captures different directional information at a given scale. Considering the polar-separability of the filters in the Fourier domain, the first low-pass, and high-pass filters, are defined as $L_0(r,\theta)= L(r/2,\theta)/2$ and $H_0(r,\theta)= H(r/2,\theta)$. r,θ are the polar frequency coordinates. L, H are raised cosine low-pass, and high-pass transfer functions:

$$B_k(r,\theta) = H(r)G_k(\theta), \qquad k \in [0, K-1) \qquad (1)$$

$B_k(r,\theta)$ means kth directional band-pass filter in recursive stage, radius and angle are defined as

$$H(r,\theta) = \begin{cases} 1 & r \geq \dfrac{\pi}{4} \\ \cos(\dfrac{\pi}{2} \log_2(\dfrac{2r}{\pi})) & \dfrac{\pi}{4} < r < \dfrac{\pi}{2} \\ 0 & r \geq \dfrac{\pi}{2} \end{cases} \qquad (2)$$

$$G_k(\theta) = \begin{cases} \alpha_K (\cos(\theta - \dfrac{\pi k}{K}))^{K-1} & \left| \theta - \dfrac{\pi k}{K} \right| < \dfrac{\pi}{2} \\ 0 & otherwise \end{cases} \qquad (3)$$

where $\alpha_K = 2^{(k-1)} \dfrac{(K-1)!}{\sqrt{K[2(K-1)]!}}$.

Using SP coefficient as feature is simple use of SPT. Multi-scale edges of face image can be locally detected by SPT. Facial contour information exists in lowest spatial-frequency subbands. However tiny contour information occurs in higher spatial-frequency subbands. A face image contains similarity information and discriminant information with respect to other images. Discriminant information is obtained from luminance variation in different face position. Steerable pyramid transform (SPT) decomposes face image into different sub-bands at different levels.

3 SPSDP ALGORITHM

3.1 *SP based feature vector*

Let $M_{i_{pq}}(x,y)$ be the coefficient which image xi are transformed by SPT and are in the position (x,y) at qth block of pth sub-band, the generative feature vector vpq(i)=[$\mu(i)$pq $\sigma(i)$pq^2 $ent(i)$pq], where $\mu(i)$pq, $\sigma(i)$pq^2 and $ent(i)$pq are respectively mean ,variance and entropy of qth block of pth sub-band. The image feature vector $\mathbf{v}(i) = \sum\limits_{p=1}^{k_p} \sum\limits_{q=1}^{k_q} \{v(i)_{pq}\}$ is obtained by connecting each block measure, where kp is sub-band number, kq is of block number of qth sub-band.

Therefore, we can extract features and reduce dimension while maintaining only major discriminating features.

3.2 *Semi-supervised discriminant power*

We define the class-between the variance of part labeled samples to variance of all samples as semi-supervised discriminant power (SDP). The value of SDP can be estimated as following steps:

Step 1: 2D feature is vectorized into 1D, feature matrix of image database can be obtained as

$$\mathbf{T} = \begin{bmatrix} v_{11}(1) & v_{11}(2) & \cdots & v_{11}(l) \\ v_{12}(1) & v_{12}(2) & \cdots & v_{12}(l) \\ \vdots & \vdots & \cdots & \cdots \\ v_{pq}(1) & v_{pq}(2) & \cdots & v_{pq}(l) \end{bmatrix}_{pq \times l} \qquad (4)$$

where T is feature matrix from l samples. c is number of classes. The column vector is a feature vector of each image. Let $\mathbf{T} = \mathbf{T}_{labeled} \cup \mathbf{T}_{unlabeled}$, $\mathbf{T}_{labeled} = \bigcup\limits_{k=1}^{c} \tilde{\mathbf{T}}_k$, $\mathbf{T}_{labeled}$ is feature matrix of labeled samples, and $\mathbf{T}_{unlabeled}$ is feature matrix of unlabeled samples, $\mathbf{T}_k$ reprents kth class samples.

Step 2: variance of class-between is estimated as

$$\tilde{\delta}_{ij}^B = \sum\limits_{k=1}^{c} \tilde{l}_k \left(\frac{1}{\tilde{l}_k} \sum\limits_{v_{ij} \in \tilde{T}_k} v_{ij} - \frac{1}{l} \sum\limits_{i=1}^{l} v_{ij} \right)^2 \qquad (5)$$

where $\tilde{l}_k$ is number of kth class labeled samples; $\dfrac{1}{\tilde{l}_k} \sum\limits_{v_{ij} \in \tilde{T}_k} v_{ij}$ is mean of labeled class-between sample; $\tilde{\delta}_{ij}^B$ is variance of class-between from labeled samples at position (i,j) of feature matrix.

Step 3: variance of all samples is estimated as

$$\delta_{ij}^T = \sum\limits_{v=1}^{l} \left(v_{ij}(a) - \frac{1}{l} \sum\limits_{u=1}^{l} v_{ij}(b) \right)^2 \qquad (6)$$

where $\dfrac{1}{l} \sum\limits_{a=1}^{l} v_{ij}(a)$ is mean of at position (i, j), δ_{ij}^T reprents variance of at position(i,j).

Step 4: SDP at position (i, j) is computed as

$$SDP(i,j) = \frac{\tilde{\delta}_{ij}^B}{\delta_{ij}^T}, \qquad 1 \leq i \leq p, 1 \leq j \leq q \qquad (7)$$

where SDP(i, j) reprents value of SDP at position(i, j).

Step 5: SDP matrix is normalized. Large SDP value means higher discriminant power, which is normalized as 0 to 1. As SDP is a statistical measure, the number of training samples effects its validity. When training samples are enough, SDP is a optimized feature extraction method.

4 EXPERIMENTAL RESULTS

4.1 *Experimental database*

To evaluate the recognition accuracy of SPSDP, experiments were carried out on ORL, and YALE, FERET databases. Figure 2 shows three face databases. In all experiments, the cross-validation method was used. The SPSDP was compared with Contourlet and Curvelet, Gabor, DWT, Curvelet+PCA+ LDA. The results are an average of 20 runs.

(a) ORL

(b) YALE

(c) FERET

Figure 2.   Three face image databases.

4.2 *Analysis on experimental results*

In table1, we select different block sizes and compute recognition accuracy in experiments. Optimal block size of ORL is 32×32. The optimal block size is 8×8 pixel in FERET and 16×16 pixel block size is optimal for YALE. The results show the illuminance and gesture variation can largely affect selection of block size.

Table 1.   Recognition rate of SPSDP with different block size.

Block size	Database		
	ORL	YALE	FERET
8×8	93.6	96.3	87.4
16×16	96.2	97.6	86.3
32×32	98.6	94.7	85.6
64×64	94.2	91.5	74.9
128×128	93.6	81.2	67.5

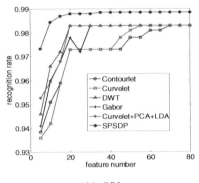

(a) ORL

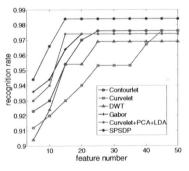

(b) YALE

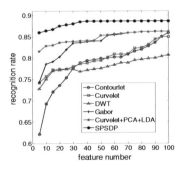

(c) FERET

Figure 3.   Recognition accuracy on three databases.

In Figure 3, we can see SPSDP all reaches the highest recognition accuracy on ORL and YALE, FERET. Because the SPSDP uses some label information and incorporate a statistical measure. As rotation invariation, SPSDP can overcome variation of illuminance and gesture.

5 CONCLUSIONS

We propose a semi-supervised discriminant power for identifying static face images. Each image is decomposed into sub-bands by SPT, which are divided into blocks. Mean, variance and entropy are extracted from blocks to combine feature vector. In the use of labeled information, semi-supervised discriminant power is computed. Experiments are performed on ORL, YALE and FERET face databases. In comparison with others multi-resolution methods, SPSDP obtains best recognition accuracy. However, there also are some problems, such as the computation cost, selection of label number, block size and so on.

ACKNOWLEDGMENT

This research was funded by science technology research of the Heilongjiang provincial education Department under Grant No. 11551086.

REFERENCES

Devi, B.J., Veeranjaneyulu N. & Kishore, K.V.K. (2010). A novel face recognition system based on combining eigenfaces with fisher faces using Wavelets. *Proc. ICEBT2010(2),* 44–51.

Shen L.L. & Ji Z. (2009). Gabor wavelet selection and SVM classification for object recognition. *Acta Automatica Sinica 35(4),* 350–355.

Boukabou, W.R. & Bouridane, A. (2008). Contourlet-based feature extraction with PCA for face recognition. *Proc. AHS08,* 482–486.

Mandal, T., Majumdar, A. & Wu, Q.M.J. (2007). Face recognition by curvelet based feature extraction. *Proc. ICIAR* 4633: 806–817.

Su C.Y, Y.T Zhuang & L. Huang (2005). Steerable pyramid-based face hallucination. *Pattern Recognition* 38(6), 813–824.

Song Y.Q., Nie F.P. & C.S. Zhang (2008). Semi-supervised sub-manifold discriminant analysis. *Pattern Recognition* 29, 1806–1813.

Cui P. & R.B. Zhang (2012). A semi-supervised coefficient select method for face recognition. *Journal of harbin engineering university* 33(7), 855–861.

Electronic Engineering and Information Science – Wang (Ed.)
© 2015 Taylor & Francis Group, London, ISBN: 978-1-138-02772-5

The parallel processing for promoter data base on OpenMP

Y. Shi
College of Computer Science and Technology, Heilongjiang University

J. Lu
College of Computer Science and Technology, Heilongjiang University
Key Laboratory of Database and Parallel Computing of Heilongjiang Province Harbin, China

J.J. Zheng
College of Computer Science and Technology, Heilongjiang University

ABSTRACT: The promoters of Arabidopsis and the Poplar are parallel processed and analyzed in this paper. The mainly works include three parts: First, The Motif data are parallel matched with both plant promoter data based on OpenMP. Then, the matching result is processed by frequent mining technology. The experimental result shows that the parallel processing time based on OpenMP is far less than the serial time. Second, to increase the calculation accuracy, the algorithm of solving the P value is improved by the prime split method. At last, according to the frequent process result, the shared frequent items of the two plants are analyzed.

KEYWORDS: OpenMP; Motif; P-value; Promoter.

1 INTRODUCTION

With the rapid development of science and technology, the technology used in the field of gene research is developed constantly. The research on genetically modifies (GM) crops have a great progress (Jun Ding 2012). Many researchers, research on specific gene expression in plants. The gene expression is adjusted and controlled by all kinds of physical and chemical factors in its body. The gene product is expressed with quantitative timing and location based on its own needs and the changing of the environment. Among the regulatory elements, the function of the promoter is particularly important.

The bottleneck in the research on plant promoters is the unlimited increasing of the data. It plays an important role in the area of Bioinformatics analysis research (Yu Wang 2014). The research purpose of this paper is to make full use of multi-core parallel technology and design more efficient parallel algorithm (Shanghong Zhang 2014).

The main contents of this paper are followed. The first step is the localization of Motif. The matching Motif is processed for duplicate removal by the grading rules. Then, the effective matching components are processed by frequent mining software (Wang Xiaoping 2014). The second step is to improve the

P-value computing mode based on the used parameters in selecting the frequency combinations. So the calculation precision is improved.

2 MOTIF PROCESS

2.1 *Motif matching*

In the procedure for promoter data processing, some fixed elements often appear. The fixed elements are called modulation components, also called Motif. The Motif is a polymer with the certain function. It is often called the functional motif or structural motif (NIE Li-na 2008). In this module, the 1 KB upstream area promoter data is captured as the experimental data called the sequences. The Motif also is a promoter sequence whose length is smaller. In this paper, the 469 fixed motifs are from the PLACE database (Aravind Madhavan 2014). The work of matching the 469 fixed motifs with the promoter sequences is called motifs location. Then the matching time and the matching position will be marked. The fixed motifs are matched with the sequences to test whether the motifs exist in the sequences. The matching method is to test one by one. When the motif is matched with the sequences, the number and the position are recorded (Ananth Krishnamurthy 2004).

2.2 Matching motif selecting

The matching motifs are removed by the grade rule in the first time. In this paper, the cut value is set 4. That is to say, the adjacent position distance of the matching motif in the sequence cannot be less than 4. The cutoff value is set 0.85, that is to say, when the similarity matching degree is higher than 0.85, the motif matches successfully in the sequence. Otherwise, this matching cannot be recorded in the matching file. The matching motifs are selected once again for the sake of the subsequent frequent mining processing. The effective matching motifs are processed by the frequent mining software. The algorithm of the mining frequent is the FP-Growth (Lijuan Zhou 2014). The support value is set 100, 500, 1000 or 2000. Different support value will get the different frequent mining result. At the last, the shared frequent combinations are fined in Arabidopsis and Poplar.

2.3 The parallel algorithm

In the process of the matching motifs, because the plant promoter data is very large, the matching work consumes a lot of time. In this paper, the OpenMP parallel technology is adopted to process the motif matching. The task of motif matching is same and there is no correlation in the motifs. It can be implemented completely in parallel. The OpenMP is used the Fork/Join parallel mode (Xiaodong Fu 2014). The main thread completes the input and the output task. The matching work between motifs and promoter data will be completed in the parallel area.

2.4 The pseudo code of the algorithm implementation

```
Input: sequences; motifs; cutoff; cut.
Output: hit-index, hit-position
Begin:
Set: Read sequences & motifs from the files;
     cutoff=0.85; cut=4;
#pragma omp parallel num_threads (value)
  { #pragma omp for private (m) schedule (static)
    Loop m=0 to motif-number
       Hit_Motif_Scan(sequences, motifs, cut,
cutoff);
  }
Write the results (hit-index & hit-position) into the
file
End
```

2.5 Hardware configuration

In this experiment, the hardware configuration is described as follows: The CPU model is Intel(R) Xeon(R) CPU E5-2650 0 @ 2.00GHz; The CPU op-mode(s) is 64-bit; The CPU(s) is 32; The CPU MHz is 1200 MHz; The hardware Memory is 2 TB; The Linux is cu01 2.6.32-220.el6.x86_64 NUG/Linux.

2.6 Experiment results

In the paper, the Arabidopsis and Poplar promoter are respectively: Athaliana_167 promoter data has 27415 items; Poplar_210 promoter data has 41377 items. After parallel processing based on two kinds of plant promoter data, the execution time is compared between the serial and the parallel. It includes CPU, OpenMP 4 threads, OpenMP 8 threads and OpenMP 16 threads. In each model, the promoter data is tested five times. At last, the average time is as comparison result. It can be shown in Table 1.

Table 1. Serial/parallel contrast time of Poplar promoter processing.

Time	Serial	Thread4	Thread8	Thread16
1	3.719	2.516	1.267	0.857
2	3.694	2.607	1.271	0.869
3	3.706	2.504	1.277	0.861
4	3.816	2.547	1.247	0.855
5	3.774	2.604	1.276	0.869
Average	3.7418	2.5556	1.2676	0.8622

At the same time, the speed-up value is calculated to analysis the Arabidopsis and Poplar processing effect based on OpenMP parallel. The result is shown in figure.1. From the experimental result, there is a big advantage for promoter gene research based on OpenMP technology.

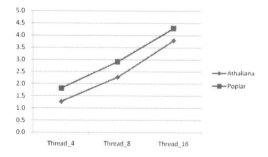

Figure 1. The Speed-up char.

3 P-VALUE ALGORITHM IMPROVEMENT

3.1 P-value

The motif matching result can be mined by the existed frequent mining software. Then the frequent mining result can be selected in the references (Yan Shi 2014). That P value is an important factor for selecting the frequent motif combinations. The P value can be calculated by Equation 1.

$$P_i = \sum_{i=g}^{N} C(\mathrm{N}, i) \mathrm{p}^i (1-p)^{N_i} \qquad (1)$$

In the Equation1, the N represents the number of motifs. The g represents the number of the frequent motifs. The i represents the number of motifs which appears frequently in the sequences. In the implementation process of the algorithm in the paper, the once multiplication and once division way is adopted. But, along with the increase of the i, the calculating precision of the $0.05 / C(N, i)$ will become lower and lower. Aiming at this point, the new algorithm is proposed—the prime split algorithm.

3.2 Prime split algorithm

The main principle of the prime split algorithm is that the ordinal number is split. By looking for the prime splitting items in the file, the multiplier and the divisor involved in prime decomposition are stored respectively in molecular and the denominator (Zhang Zhuo 2014). At this time, the calculation is converted to the product of prime power form. Then, when the base number of the molecules and denominator is same, the exponent offsets to simplify the calculation way. This algorithm can reduce the time consumed by multiplication and division. The more important thing is that the accuracy of the calculation can be improved.

3.3 Prime split algorithm described

When listing facts use either the style tag List signs or the style tag List numbers.

Begin:
Input: C (N, i)
Output: p-value.
1、 Read prime splitting items from the file to the FenJie-items array;
2、 Loop i=0 to N
    {
    Call the prime split function
    FenJie-Func (int item)
        //to search the splitting items of the data;
    Add the splitting items
        into the Fenjie_up[] or Fenjie_down[];
    }
3、 The items in Fenjie_up[] subtract the corresponding items in Fenjie_down[], then the items in Fenjie_up[] are multiplied and are saved in value f.
4、 0.05 divides f and the result is stored in p-value;
End

3.4 Optimization results

By prime split algorithm, the result of the improved P value calculation is shown in Table 2.

Table 2. The P value calculation in different algorithm.

Multiplication & division	Prime split algorithm
2.701324790072692e-010	2.701378924514735e-010
1.684715947845725e-011	1.684734879215485e-011
2.384977576881468e-010	2.384975986214564e-010
2.485975154594657e-011	2.485926748154577e-011

It can be seen from Table 2, the P-value calculation result of multiplication and division algorithm can only be accurate to 15 decimal after the point, and then followed by using random numbers to fill. By contrast, the precision of P-value calculation by the prime split algorithm is determined 18 decimals. It is effective to improve the calculation precision of the P-values.

4 SHARING ANALYSIS

In this task, the shared promoters are calculated between Arabidopsis thaliana and Poplar based on the frequent regulatory motif combinations. The frequent regulatory motifs can obtain by the frequent mining technology. Then, the frequent motifs combinations identified by the plant gene ID. At last, the shared frequent motif combinations are obtained. In the experiment, 40165 effective Arabidopsis combination items and 72962 effective Poplar combination items are found (Wen Yang 2008). The frequent combination items of the two plants are processed and analyzed.

The shared combination items in Arabidopsis and Poplar are fined. It is to say that all items of combination items in one plant data can be searched in the other plant data (Acharya Sefali 2013). When a combination item is searched, the shared combination accumulator will be counted. The searching process in the total items is the same and there is no correlation between different combination items. Then the searching process can be implemented in parallel. In searching task, the OpenMP technology is adopted to find the shared combination items of two plants. The time of the serial algorithm and parallel algorithm with different thread numbers are contrasted. The result is shown in Figure2.

At last, the frequent combination items of the two plants are analyzed. There are 17013 shared items in Arabidopsis and Poplar. The result is shown in Figure3.

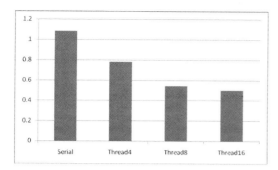

Figure 2. The contrast time of the series/parallel for selecting the shared combinations.

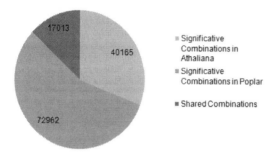

Significative Combinations in Athaliana

Significative Combinations in Poplar

Shared Combinations

Figure 3. The shared combinations between Arabidopsis and Poplar.

5 CONCLUSION

In this paper, the Arabidopsis and Poplar promoter data are processed by OpenMP technology. It has completed the motif matching work in parallel. Compared with serial algorithm, the parallel algorithm on the promoter data processing has obvious advantages. It is significantly improved the efficiency of the promoter data analysis and processing. At the same time, the prime split algorithm is proposed for P-value to improve the calculation accuracy. Finally, the shared combination items in two plants data are searched and analyzed.

ACKNOWLEDGMENT

This research is funded by a grant (No. YJSCX2014-064HLJU) from the Postgraduate Innovation Researching Projects of Heilongjiang University.

REFERENCES

Jun Ding (2012) Thousands of Cis-Regulatory Sequence Combinations Are Shared by Arabidopsis and Poplar1.C.J. EN,1–12.
Yu Wang (2014). Connexin 32 and 43 promoter methylation in Helicobacter pylori-associated gastric tumorigenesis .J. World Journal of Gastroenterology, 17–19.
Shanghong Zhang (2014). Parallel computation of a dam-break flow model using OpenMP on a multi-core computer .J. Journal of Hydrology, 1–4.
Wang Xiaoping (2014). The OpenMP application research on multi-core processors .J. Computer science and application. 0409: 1–5.
NIE Li-na (2008). Progress on C loning and Functional Study of Plant Gene Promoters .J. Chinese Academy of Agricultural Sciences, 3–9.
Aravind Madhavan (2014). Promoter and signal sequence from filamentous fungus can drive recombinant protein production in the yeast Kluyveromyces lactis .J. Bioresource Technology, 41–67.
Lijuan Zhou (2014). Research of the FP-Growth Algorithm Based on Cloud Environments .J. Journal of Software, 4–15.
Ananth Krishnamurthy (2004). Analysis of a Fork/Join Synchronization Station with Inputs from Coxian Servers in a Closed Queuing Network .J. Annals of Operations Research,:5–11.
Xiaodong Fu (2014). The discontinuous deformation analysis of parallel computing method based on OpenMP .J. Rock and soil mechanics, 24–27.
Yan Shi (2014). The Key Algorithms of Promoter Data Parallel Processing based on OpenMP .D. College of Computer Science and Technology, Heilongjiang University,1–3.
Zhang Zhuo (2014). The General parallel combination coding algorithm .D. Heilongjiang University, 15–24.
Wen Yang (2008). GO in the application of the biological data integration .J. Books intelligence work, 124–127.
Acharya Sefali (2013). Efficient chimeric plant promoters derived from plant infecting viral promoter sequences. J. Planta, 6–10.

Electronic Engineering and Information Science – Wang (Ed.)
© *2015 Taylor & Francis Group, London, ISBN: 978-1-138-02772-5*

Research and implementation of latent semantic analysis based on UMLS and biomedical pathway databases

J.Q. Ma
Mutimedia Technology Laboratory, Institute of Information Technology, Heilongjiang University, China

M.C. Shang
Computational Biology Laboratory, Institute of Information Technology, Heilingjiang University, China

ABSTRACT: In today's world, the rapid development of biotechnology to bring a huge challenge, many biology literature based on different ways of expression can be seen on the network. There is a lot of latent semantic hide in literature, if scientists can dig out the latent semantic behind in literature effectively, they can explore biology more accuracy in the future. In this paper, we proposed a latent semantic analysis method based on UMLS and biomedical pathway databases. The experiment shows that this method can be used to identify the degree of semantic relationship between genes and biological processes.

KEYWORDS: latent semantic analysis; biomedical pathways; UMLS.

1 INTRODUCTION

The 21st century is an information age, so much biology field information emerged in literature, however, the heterogeneity of data organization and different expression forms of knowledge put forward the challenge to correct explanation of biological experiment results (Plake C, Royer L, Winnenburg R 2009), especially in the terms of integration and extraction of biological information. In order to mining the latent relationship in different areas of biology literature, the latent semantic mining technology based on biology developed quickly.

UMLS (Unified Medical Language System) is the U.S. national library of medicine (NLM) research and development of the integration of Medical Language System since 1986 (Bodenreider O2004). This system is a computerized information retrieval language integrated system, it is not only a language translation, natural language processing and language standardization of tool, but also the realization of cross-database retrieval vocabulary translation system, it can help user contact with intelligence sources, including computerized medical records, bibliographic database, fact database and expert system, retrieving among the electronic integration of biomedical information.

Based on keyword-document vector space model (VSM) is one of the traditional space models, the unstructured text transformed into vector model, and this model makes it possible to process various mathematical problem. Its advantage is that processing logic simply and quickly. However, the basic hypothesis of vector space model about the relationship between words is independent seems hard to achieve in a real environment, terms appear in the text often has some relevance, in some extent affect the results of the experiment.

At the same time, this text processing method based on keyword, mainly based on word frequency, two text similarity degree depends on the number of common words which they have, so this method can't distinguish the semantic fuzziness of natural language. In addition, there are a lot of synonyms and polysemy in natural language, the accurate expression of semantic doesn't only depend on the appropriate use of vocabulary, but also depends on the meaning limited in the context. If we ignore the limitation of context, only use keyword frequency to indicate the contents of the text, with no doubt affect the accuracy and completeness of experiment results.

Initially, LSA applied to retrieve document information which solves the synonyms and polysemy problem effectively, by identifying synonyms, the accuracy of information retrieval increased by 10% to 30%. As applications continue to expand, LSA applied to many areas such as information filtering, information classification / clustering, cross-language retrieval, information understanding, judgment and prediction and so on.

In China, most of latent semantic analysis studies focused on screening and collecting network information areas. In terms of the Internet, latent semantic analysis not only determines whether a website and a keyword which user searches are relevant, but also decided website ranking about this keyword. LSA algorithm through SVD algorithm calculates the relevant score in website - keyword matrix, the higher score the website gets, the more relevant between the website with a keyword, it also means they are strong correlated in this matrix. In this paper, we use biological processes corresponding pathways database to improve the accuracy of the calculation, however, between the genes and biological pathways database there is no clear analysis. Furthermore, the document set and the retrieved pathways present a different structure of the information. The document set's decomposition is suitable for the LSA, while the pathways database are represented as a connected graph, they are two different structures. In this article we will use the LSA to analysis this case.

The main content of the project is to use UMLS integration a set of genes and biological processes which user input to locate CUIs, and find the relevant AUIs according each CUI, all the CUIs and AUIs and user input biological process considered as a set of keywords, and then document abstracts in nature form language according to each keyword in the keyword set retrieved through PubMed web services (Doms, A., Schroeder, M2005). In addition, based on user input biological processes access to relevant pathways, all of which genes are considered as trusted information. A term-document matrix has built up, then it use the matrix Singular Value Decomposition, after rank-validation, make cosine similarity calculation among the vector in the new matrix .finally, according similarity score to evaluation the relevance between gene and biology process in descending order.

2 BIOLOGICAL DATA INTEGRATION

The so-called biological data integration refers to unify a number of biological semantic information and data processing. Here, we take advantage of Metamap, it's an algorithm of the UMLS. Therefore the relevant biological terms in biology document abstracts are screened, UMLS solves two main aspects: (1) the same concept by different people or in different databases may have different expression; (2) retrieval incomplete problems caused by the dispersion database system. In order to avoid interference of unrelated words in the literature, Figure.1 shows the system processes document abstracts which the user retrieved, this step include: removal of punctuation, remove extra spaces, the spaces will be split according to the document abstract ,remove function words (articles, prepositions, conjunctions, etc.), and then these terms send into the UMLS localization interface to integrate semantic, after integration remove duplicate keywords, these keywords are the final terms.

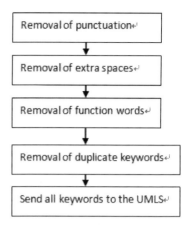

Figure 1. Word filtering process.

3 SPRINKLING TECHNOLOGY

Genetic pathway is considered as a trusted information source. In fact, genetic pathways express each relevant gene based on connect-graph. Thus, it provides a straight-chain between a gene and its corresponding pathways. In addition, genetic pathway is constantly checked, so it is very credible. A server client software called Pathway Commons used to query a list of pathways has been developed (Romero, P., Wagg, J., Green, M.L., Kaiser, D., Krummenacker, M., Karp, P.D2004). We make the terms corresponding to the first pathway and apply to LSI matrix of the column direction, and replication. We refer at this procedure as document padding, to improve the accuracy of the integral matrix calculation.

$$NPD = \lceil p \circ PAD \rceil \qquad (1)$$

According to the Equation 1, the increasing number of columns is the number of pathway documents called NPD, NAD represents all the retrieved article abstracts from PubMed, PubMed is a search engine for life science journals, play a fundamental role for the scientist because they the main source of up-to-date information, p is a padding factor spanning from 0 to 1. In the LSI algorithm, the first step is to establish the term-document matrix based on keywords and retrieved literature, after building the original matrix, the system will sprinkling by p value and biological process which user input. No doubt, Sprinkling technology has also changed the correlation scores between keywords, for example: keyword A and keyword B's correlation in abstract document is very weak, because they are linearly independent, but they are strong relevant in the pathway documents, so their overall correlation score higher, that means the degree of correlation increases.

4 LATENT SEMANTIC ANALYSIS

Latent Semantic Analysis is based on the principle of vector space model, it is used to calculate the correlation between terms, texts, even between term and text. In our system, abstract documents, terms, and pathway documents which result from user input biological processes were constructed term-document matrix, after establishing the matrix, the Singular Value Decomposition (SVD) allows to decompose the matrix into a new matrix.

Firstly, we should construct an m×n term-document matrix A, there are two dimensional occurrence matrix A in which the columns represent the dimensional space of a corpus of documents and the rows represent the dimensional space of the terms occurring into the documents. In the matrix A, aij represents the frequency of the term i into the document j, m is the number of terms, n is the number of documents, they include abstract documents and pathway documents. Then, we used Singular Values Decomposition, it results in an approximate matrix named A_k. Actuary, $A=A_k=U_k\Sigma_k V_k^T$, U_k and V_k referred to the left and right singular vectors of the matrices A_k, respectively. Σ_k is a diagonal matrix, diagonal elements are called the singular value of matrix A_k. A_k the final matrix which we need, calculate cosine angle between the row vector of the matrix A_k, according the following Equation 2:

$$sim(A,B) = \frac{A \times B}{\sqrt{A^2 + B^2}} \tag{2}$$

A represent a term, and B represents a biological process, By adding some pathway documents the system calculates genes and biological process to obtain a similarity score, the higher the score, the more strong relationship between this gene and the biological.

5 EXPERIMENT

Firstly, as the information is inputted by the user, we integrate the biological process and genes by the use of the UMLS. After the integration, we collect the abstract of that by sending terms of the biological process and genes into PubMed. Then, we can get the Literature - key word matrix by sending the abstract into UMLS. On the condition that p=0.5, we use sprinkling to augment the Matrix. After that, our work is calculating the cosine value of vectors of the biological process and genes and making an ordered list of all the values. It is shown in Figure.2.

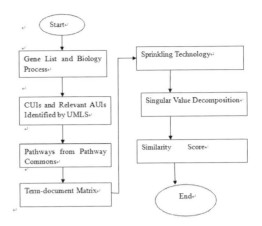

Figure 2. Complete semantic analyses flow.

The following experiment tests a similarity of biological process (Angiogenesis) and 5 genes (VEGFA, ANGPT2, ANGPT1, P53, PTEN) (Nguyen V P K H, Chen S H, Trinh J2007). All the data is coming from the documents and pathway database of PubMed. In the experiment, we set the value of P as 0.5 and the pathway data is half of the original document. What we care about is if the system can correctly generate the matrix and finish the calculation on condition that the biological process and genes have been correctly inputted. To build a test, we design the case as is shown in Table 1.

Table 1. Compared experiment result.

Biology Process	Gene	Score Similarity	Degree	Result
Angiogenesis	VEGF-A	1.0	100%	TRUE
Angiogenesis	ANGPT2	0.95	97%	TRUE
Angiogenesis	ANGPT1	0.95	96%	TRUE
Angiogenesis	P53	0.49	46%	TRUE
Angiogenesis	PTEN	0.52	47%	TRUE

After a series of tests, we can conclude that: every part of the system works well, the core question is properly solved- Latent semantic analysis module can calculate the relationship between biology process and genes. The last but not the least, the error's range is acceptable.

6 CONCLUSION

We propose a new way to use Latent semantic analysis algorithm when dealing with the biology process and genes technologies. The new way can calculate the similarity between terms of biomedical articles. After pre-processing of the biological information, the system based on LSA adds reliable data, that biology pathway data, to the article. By the use of extending matrix, the result of the experiment shows the comparison of calculated correlation and known correlation. It proves adding pathway data can improve the accuracy of solved- Latent semantic analysis.

REFERENCES

Plake C, Royer L, Winnenburg R (2009). GoGene: gene annotation in the fast lane. Nucleic acids research, 37(suppl 2): W300–W304.

Bodenreider O(2004). The unified medical language system (UMLS): integrating biomedical terminology. Nucleic acids research, 32(suppl 1): D267–D270.

Doms, A., Schroeder, M(2005). GoPubMed. exploring PubMed with the Gene Ontology. NucleicAcids Research.

Romero, P., Wagg, J., Green, M.L., Kaiser, D., Krummenacker, M., Karp, P.D(2004). Computational prediction of human metabolic pathways from the complete human genome. GenomeBiology.

Nguyen V P K H, Chen S H, Trinh J(2007). Differential response of lymphatic, venous and arterial endothelial cells to angiopoietin-1 and angiopoietin-2. BMC cell biology, 8(1): 10.

Electronic Engineering and Information Science – Wang (Ed.)
© 2015 Taylor & Francis Group, London, ISBN: 978-1-138-02772-5

Distributed neural network self-learning algorithm

J.L. Cui, Y. Zhao & P.L. Qiao

Computer Science and Technology, Harbin University of Science and Technology, Harbin, P.R.China

ABSTRACT: For the problem of selecting the learning rate and initial connection weights, it improves BP algorithm of artificial neural network. The distributed neural network self-learning algorithm has been proposed. In the course of training, it implements the selection of input units, hidden units and learning rate and improves the learning rate of the algorithm. The conclusion shows that the distributed neural network self-learning algorithm is superior to the traditional BP algorithm. The speed of the convergence is faster and the detection rate is higher. It not only clearly reduces the complexity of the algorithm, but also ensures the integrity of the result of the study.

KEYWORDS: Neural network; Back propagation algorithm; Self-adapting; Distributed.

1 INTRODUCTION

The algorithm model as a distributed information parallel processing, artificial neural network, according to the relationship between the complexity of adjusting internal node system connection, through this way to process information. Neural network and its behavior of this kind of behavior and human or animal are similar, so named for the neural network. The neural network has a known as "training" learning process analysis, it analyzes the input and output data analysis, we find some rules of the two mutual, finally, when the input data, through these rules prior to calculate the output results (Qi D. & Kang J.C. 2012).

In the neural network model, BP neural network model is one important model. In general, the connection weights between neurons, using BP algorithm and some learning rules to adjust, through this learning process, topology structure and learning the rules of the network is the same. Therefore, the neural network model of adaptability and learning ability need to further strengthen them (Zhang D.Y. 2011, Zhang W.M. et al. 2010).

In practical application, using the BP algorithm can solve the problem well, but the BP algorithm has some deficiencies, such as: the emergence of local minimal and lead to the global optimal solution can be obtained, low learning efficiency, slow convergence speed, a new sample learning, will produce disturbance to the old samples. In view of the learning efficiency and convergence, many domestic and foreign scholars have put forward the following improvement scheme: adjust the gradient direction, the momentum term to adjust the

speed of convergence, adaptive step size increases or decreases the learning rate BP algorithm (Zhang D.Y. 2010).

In this paper, based on the above situation, further optimization of the traditional BP algorithm, presents a self distributed neural network learning algorithm, can not only speed up the network convergence speed, but also improve the learning speed, good optimization of the topology of the network, so as to improve the BP neural network adaptive ability.

2 BP NEURAL NETWORK MODEL

The BP neural network is a multilayer feed forward network, its structure is a hierarchical definition, the output neurons of each layer can only be connected to the neurons in the next layer. It consists of three parts: input layer, hidden layer and output layer, the network structure as shown in figure1.

BP neural network algorithm with error back-propagation algorithm as the core algorithm, forward and backward respectively, two kinds of learning process of the algorithm. Forward propagation, the

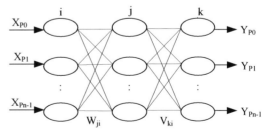

Figure 1. The structure of BP neural network.

so-called no doubt, is according to the output process input vectors were calculated at each successive layer node; while the back propagation, refers to the output node using the ideal error with the actual, from the output layer to the input layer of the successive change weight matrix V and W. The algorithm after a few repetitions, successful converges to a value (Liu Y.H. 2013).

By using the gradient descent algorithm, combined with the traditional BP neural network model, the optimization of sample input and output, this process is nonlinear. Nonlinear simulation model error minimum defects are as follows:

1   The traditional BP network, the extreme value gradually correction network, so as to minimize the error function, the local minima problem, nonlinear optimization, get the advantages of local.
2   The learning speed is slow, the extreme drop is slow, in the learning process is prone to an error in flat regions, and will continue for a long time.
3   Due to the over fitting the training, causes the network efficiency is low, not only the network structure, network performance is poor, fault tolerance is low, floating point overflow problems, are due to human factors.

3   IMPROVED BP NEURAL NETWORK MODEL

3.1   Select the initial weight

Based on the idea of random, statistical principles, choosing the initial weights algorithm, a large number of generating random, initial points, and the iterative optimization of initial weights. Although, the error function E through the initial weights to obtain the global optimal solution, in order to increase the probability of randomly generated, but it inevitably the blindness and randomness. Therefore, generally in the [-1, 1] range selection of initial weights, and cannot be too large. On the basis of the principle of distributed parallel, by the method of stepwise search, in order to solve the problem of blind choice of initial weights, the initial weights of equal area H N, and then select the minimum error function E to should be regional, then carries on the N parts, continue to repeat the previous steps, so that the error function E is no longer reduced, until the cessation of iteration finally, the best result.

3.2   To determine the number of input layer nodes

Research on the problem of traditional BP network modeling, the key is to determine the dimension of input, output layer. So, as much as possible in order to ensure that those important factors have not been missed election, we must first select some factors (independent variables), because if the variables are affected by these variables influence the number of independent variables is large and many, it is necessary to establish the system model, the qualitative analysis method this method is called.

Let O_{pi} represent the output, the output generated by the hidden node I is learning the first p samples values; Opj also represents the output, the output generated by the hidden node j is learning the first p samples of the value, the total number of N as the sample learning, then

$$\bar{O}_i = \frac{1}{N}\sum_{P=1}^{N} O_{pi} \tag{1}$$

$$\bar{O}_j = \frac{1}{N}\sum_{P=1}^{N} O_{pj} \tag{2}$$

Make

$$x_p = O_{pi} - \frac{1}{N}\sum_{P=1}^{N} O_{pi} = O_{pi} - \bar{O}_i \tag{3}$$

$$y_p = O_{pj} - \frac{1}{N}\sum_{P=1}^{N} O_{pj} = O_{pj} - \bar{O}_j \tag{4}$$

Opi and O_{pj} is defined as follows:

$$R_{ij} = \frac{\sum_{P=1}^{N} x_p y_p}{\sum_{P=1}^{N} x_p^2 \cdot \sum_{P=1}^{N} y_p^2} \tag{5}$$

Easy to get:

$$\left| R_{ij} \right| \leq 1 \tag{6}$$

If R_{ij} is infinitely close to ± 1, then the linear correlation of O_{pi} and O_{pj}, then the two sequences are regression dispersion is smaller; on the other hand, is the linear correlation of O_{pi} and O_{pj} is small, so the two sequences are return dispersion is greater.

Using the adaptive method of selecting learning parameter, the main idea is: when the w_{ji} from the stationary point when further, ETA greater; when it is infinitely close to a stable point (E, 0) when the smaller, eta. Specific practices are as follows:

$$\eta(t+1) = \eta(t) \cdot E(t) / E(t-1) \tag{7}$$

Among them

$$E = \sum_{P=1}^{N} E_P / N \tag{8}$$

The design of this method is simple, is relatively easy to implement, not only can fully solve the problem of local minima, but also to speed up the convergence speed, reduces the training time, thereby improving the resilience of the network.

4 THE DISTRIBUTED NEURAL NETWORK LEARNING ALGORITHM

Through those divided by the results of the data, if this study once finished, so some of the more complete network is contained in each partition of the data, there is some network behavior that is not complete. Therefore, we need to learn the results of the sample space, so as to achieve a form a complete network behavior the goal of knowledge. In this process, some of the features of network behavior may be due to some knowledge cannot be divided by the formation, this problem has been resolved, but also reduces the generalization ability of the learning results

Definition 1 The correlation coefficient P_{ij} represents the same hidden layer nodes i and j, gives the following definition:

If P_{ij} is too large, then repeat the same nodes i and j function, needs to carry on the compression and consolidation.

Definition 2 S_i express sample divergence, gives the following definition:

If S_i is too small, then the output of hidden node i value changes little, no network training, they can be deleted.

Regulation 1 If $\left| \rho_{ij} \right| \geq C_1$ and $S_i, S_j \geq C_2$, so i and j, the two of the same layer hidden nodes can be merged into one. Where C_1 and C_2 to a predetermined lower limit, the general C_1 take 0.8~0.9, C_2 take 0.001~0.01.

Regulation 2 If $S_i < C_2$, indicates that the effect of node i with the same threshold, can be merged with the threshold, that is to say it can delete a node i.

4.1 Step of the algorithm

Following the steps of an algorithm of distributed neural network self learning:

Step 1. Structuring an initial structure of the BP neural network model (an input layer, a hidden layer, an output layer and more than enough hid nodes).

Step 2. According to the practical problems, include in the variable to affect the dependent variable (independent variable). Use the adaptive method to select the dependent variable of the most influential variables, in order to determine the number of input nodes.

Step 3. According to the practical problems, determine the number of output layer nodes. To initialize the modifying algorithm (including a given learning accuracy, iteration step M_0, maximum number of hidden nodes the R, the initial value of the learning parameters η, the momentum coefficient a, other constants C_1 and C_2).

Step 4. According to the principle of parallel, making the tolerance region H of the initial weights becoming N aliquots, marked as β_1, β_2, ..., β_N.

Step 5. Enter the learning samples. Make the sample parameters between 0 and 1.

Step 6. Randomly generate initial weights on the selected small area (i=1,2,..., N).

Step 7. Learn network, according to the BP algorithm.

Step 8. To determine whether an iterative steps in excess of the prescribed steps or whether learning precision meets the requirement, if it is yes, go to Step9, if it is not, then go to the Step 7.

Step 9. Parallel computes the related parameters and the remittance of hidden nodes. Firstly, delete the node according to rule 2. If neither of the two rules is satisfied, merge the node according to rule 1 again. If two rules are both satisfied, then delete the node according to rule 2 only. If both rules are not satisfied, then do not merge and delete the node. If no node is merged and deleted, then transferred to Step10; otherwise, returns the Step 6.

Step 10. Whether learning accuracy meets the requirements or iterative steps in excess of the prescribed steps, if so, the algorithm terminates; if not, then return to the Step 6.

5 ANALYSIS AND VERIFICATION OF EXAMPLES

5.1 Using the adaptive method for selecting input variables

We select some key factors affecting the level of consumption of Chengdu City to establish the model; they have a direct relationship with the Chengdu total retail sales of social consumer goods, used to verify the feasibility of the algorithm. The sample interval for the 1997 January to 1998 September data. Then used to establish the metrical model of a distributed neural network learning algorithm.

The principle and method of computation and the use of adaptive selection, BP neural network input variables, these variables are variable, associated with the Chengdu City, total retail sales of social consumer goods.

The input variables:

Y Chengdu City, the total retail sales of consumer goods (100 million Yuan)

P the one-year deposit rate (%)

M_0 cash circulation

V_1 the national consumer price index (%)

V_2 national fiscal deficit (100 million Yuan)

V_3 the RMB dollar exchange rate ($one hundred)

V_4 fixed asset investment growth (100 million Yuan)

V_5 the national export growth rate (%)

V_6 measure of money supply

We chose 17 of the sample data, respectively from 1997 January to 1998 May, data selection, and through the simulation optimization, success-fully selected 6 variables V_1, V_2, V_3, V_4, V_5, V_6, the obtained model is shown as follows:

$$y = a_0 + \sum_{i=1}^{6} a_i v_i \qquad (9)$$

Among them, a_0=471.76831, a_1=-0.60538, a_2=0.69255, a_3=-0.44378, a_4=-0.35960, a_5=-0.00006, a_6=-0.59829.

5.2 Analysis and forecast of the model

In order to verify the practicability of BP algorithm, we selected in June 1998 to September 1998 data as forecast data, distributed neural network self-learning algorithm is adopted to establish the model, and total retail sales of social consumer goods for prediction of Chengdu. We choose the structure of BP network is as follows: 8 input nodes and one output node. Learning samples to choose from January 1997 to May 1997, the data between the W and V is the model is obtained by studying the optimal connection weights matrix, the predicted results and error analysis as shown in table 1.

Table 1. The predicted results and error analysis of each model.

type / date	the total retail sales of social consumer goods of Chengdu (one hundred million Yuan)	distributed neural network self-learning algorithm		BP neural network adaptive learning algorithm	
		predictive value (one hundred million Yuan)	Relative error (%)	predictive value (one hundred million Yuan)	Relative error (%)
1998.6	34.21	34.29	0.24	34.29	0.56
1998.7	33.87	33.65	−0.65	33.57	−0.89
1998.8	34.38	34.32	−0.17	34.42	0.12
1998.9	35.27	35.47	0.57	35.51	0.69

From the table above can be concluded that using distributed neural network self-learning algorithm to solve the predicted results of model with high accuracy, and the maximum relative error less than 0.65%. Compared with the BP neural network adaptive learning algorithm, the traditional BP network learning algorithm has better advantage. Because it is distributed, running time of the algorithm is shorter than other algorithms, and improves the efficiency of the algorithm.

6 CONCLUSIONS

In this paper, the traditional BP algorithm is improved, and the distributed neural network self-learning algorithm is proposed, so enhancing the adaptability of the BP network and improving the rate of learning. Combined with the adaptive algorithm to determine the input node, make the network structure more reasonable. This new algorithm not only accelerates the convergence speed of the network, but also can optimize the topology of the network, and improves the ability of the BP neural network to adapt to.

ACKNOWLEDGMENTS

This research is supported by research project of Harbin city(2011AA1CG063).

REFERENCES

Qi D. & Kang J.C. (2012). On Design of the BP Neural Network. *Computer Engineering and Design 43(2)*, 115–120.

Zhang D.Y. (2011). Neural Network Algorithm Based on Distributed Parallel Computing. *Systems Engineering and Electronics 32(2)*, 386–391.

Zhang W.M., Luo J.Y & Wang Q.X. (2011). Survey on Network Topology Visualization. *Application Research of Computers 25(6)*, 1606–1610.

Zhang D.Y. (2010). Neural Network Algorithm Based on Distributed Parallel Computing. *Systems Engineering and Electronics 32(2)*, 386–391.

Liu Y.H., D.X. Tian, X.G.Yu & J. Wang (2013). Arge-Scale Network Intrusion Detection Algorithm Based on Distributed Learning. *Journal of Software 4(19)*, 993–1003.

Electronic Engineering and Information Science – Wang (Ed.)
© 2015 Taylor & Francis Group, London, ISBN: 978-1-138-02772-5

The wireless sensor network routing choice model based on game theory

R.Y. Qiao
Department of Postgraduate of the Academy of Equipment, China

X.G. Zhao
Department of Aerospace Commanding of the Academy of Equipment, China

ABSTRACT: The shortest routing choice model was easy to cause the imbalance of network resource utilization and can not effectively improve the SNR, in order to solve these problems, according to the game theory, it was built that the wireless sensor network routing choice model based on game theory, and the existence of the equilibrium strategy was proved, and the equilibrium strategy was solved, through the simulation, the model had a certain advantage in reducing the "hot spots" of the network and improving the SNR.

KEYWORDS: Wireless sensor network; Routing choice model; Game theory.

1 INTRODUCTION

In a static wireless sensor network (Shu Wang & Yujie Yan & Fuping Hu & Xiaoxu Qu 2007), each node has a routing table, the source node generally chooses the shortest routing to transmit the data. When some nodes are in the Intersection of many the shortest routing of some source nodes, it is easy to form the "hot spot" (Qiao Rongyan 2010) of the network that the shortest routing choice model is used. And in the case of no incentive, the transmission power of node will be constant, when the environmental noise becomes large, the SNR of the information transmission will decline, the shortest routing choice model can not solve this problem, so that model is not completely suitable for the wireless sensor network. So, according to the characteristics of the wireless sensor network and the mechanism of competition, the paper sets up the wireless sensor network routing choice model based on Game Theory (Drew Fudenberg & Jean Tirole 2006, Li Guangjiu 2004).

2 THE RELEVANT DEFINITIONS

Definition 1: $N = \{n_1, n_2, \cdots, n_m\}$ is a set of nodes in a static wireless sensor network, and n_i is a certain node.

Definition 2: $L = \{1, 2, \cdots, k\}$ is a set of the routing of n_i, when the routing j $(j \in L)$ transmits the information, the SNR of j is $r_j \in [r_{min}, r_{j\max}(t)]$, and r_{min} is the minimum SNR that is required by the network; the SNR $r_{j\max}(t) = \dfrac{p_{\max}}{VN_0^{(j)}(t)}$ (Bernard Sklar

2007), $p_{\max}$ is the maximum transmission power of node, the transmission rate is V, $N_0^{(j)}(t)$ is the environment noise power of the routing j at the moment t.

Definition 3: m_i is the contribution value of n_i, m_i is proportional to the amount of forward information, the greater m_i, the higher the priority level of n_i uses the network resources.

3 THE ESTABLISHMENT OF THE MODEL

$L = \{1, 2, \cdots, k\}$ is the set of participants that the set of routing of node n_i. The strategy of any participant $j \in L$ is to provide a reasonable SNR r_j. $r_j = f(r_{j\max}(t))$ is the inference function of any participant j, the function is a strictly increasing and differentiable, the probability density distribution of $r_{j\max}(t)$ is $g(x)$, and $x \in [0, +\infty)$, $g(x)$ is known by all participants. The income function of any participant j:

$$y_j = \begin{cases} 1 - \dfrac{r_j}{r_{j\max}(t)}, & r_j = \max(r_1, r_2, \cdots r_k) \\ 0, & r_j \neq \max(r_1, r_2, \cdots r_k) \end{cases} \qquad (1)$$

When r_j of the participant j is the greatest, the participant j will receive the contribution value is 1, or the contribution value is 0, and $\dfrac{r_j}{r_{j\max}(t)}$ is the price; according to probability theory (Sheng Zhou & Xie

Shiqian & Pan Chengyi 2001), so the best strategy of participant j should be such as type (2):

$$\max(1 - \frac{r_j}{r_{j\max}(t)}) \prod_{i=1}^{k} P(r_j > r_i) \qquad (2)$$

In the type (2), $P(r_j > r_i)$ is the probability of $r_j > r_i$, and $i \neq j$.

4 THE SOLUTION OF THE MODEL

$$\frac{\partial^2 y_i}{\partial r_j^2} = \frac{(1-k)g(r_{j\max}(t))(\int_{r_{\min}}^{r_{j\max}(t)} g(x)\,dx)^{k-2}}{f'(r_{j\max}(t))} \qquad (3)$$

$$f'(r_{j\max}(t)) < 0 \Rightarrow \frac{\partial^2 y_i}{\partial r_j^2} < 0 \qquad (4)$$

So the function y_j is quasi concave, and the game has the equilibrium strategy (Li Guangjiu 2004). Set

$$\frac{\partial y_j}{\partial r_j} = 0 \qquad (5)$$

to obtain

$$r_j = f(r_{j\max}(t)) = \frac{k-1}{k} r_{j\max}(t) + \frac{1}{k} r_{\min} \qquad (6)$$

And the equilibrium strategies of k alternate routing of the node n_i:

$$\begin{cases} r_1 = \frac{k-1}{k} r_{1\max}(t) + \frac{1}{k} r_{\min} \\ r_2 = \frac{k-1}{k} r_{2\max}(t) + \frac{1}{k} r_{\min} \\ \vdots \\ r_k = \frac{k-1}{k} r_{k\max}(t) + \frac{1}{k} r_{\min} \end{cases} \qquad (7)$$

5 THE SIMULATION OF THE MODEL

24 nodes (ID1 ~ ID24) and a base station are random distributed, using the MATLAB to simulate, $V = 250kbit/s$, $r_{\min} = 10$, $p_{\max} = 0.25 \times 10^{-3}W$

$N_0 \in [10^{-11}, 10^{-12}]$. In this paper, if the amount of forwarding of some node is 2 times greater than the average amount of forwarding of the network, the node is called the "hot spot", as shown in Figure 1:

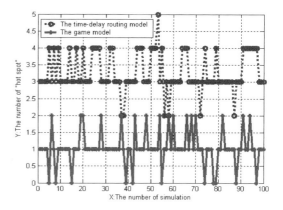

Figure 1. The number of "hot spot".

The number of "hot spot" produced by the game mode is basically less 2, in most cases, to keep 0 ~ 1; and the number of 'hot spots' produced by the shortest delay routing basic in 3 ~ 4, is 3 ~ 4 times of the game routing.

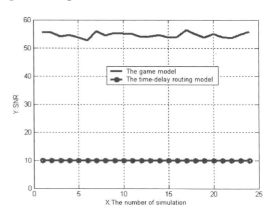

Figure 2. The SNR of two models.

From Figure 2, when each node builds two routing, the SNR of the game routing choice model is greater 5 times than the SNR of the shortest routing model.

6 THE CONCLUSION

This paper analyzes the problems in using the shortest routing choice model, the competition mechanism is introduced, and the equilibrium strategies is solved, through comparing the simulation results, it shows that the game choice routing model can reduce the number of the "hot spot" and increase the SNR.

REFERENCES

Shu Wang & Yujie Yan & Fuping Hu & Xiaoxu Qu (2007). The Theory and Application of Wireless Sensor Network. *Beihang University Press*, pp.1–94.

Qiao Rongyan (2010). Research on the routing algorithm of wireless sensor network based on the game theory. *Institute of Command and Technology of Equipment Master Thesis,* pp. 67–50.

Drew Fudenberg & Jean Tirole (2006). Game Theory. *China Renmin University Press*, pp. 1–264.

Li Guangjiu (2004). The game theory. *The chemical industry press*, pp.107–147.

Bernard Sklar (2007). Digital Communications Fundamentals and Applications. *Beijing: Electronic Industry Press*, 172–173,426–427,90–91.

Sheng Zhou & Xie Shiqian & Pan Chengyi (2001). Probability and Mathematical Statistics. *The Higher Education Press*, pp. 37–62.

Electronic Engineering and Information Science – Wang (Ed.)
© 2015 Taylor & Francis Group, London, ISBN: 978-1-138-02772-5

Calculation model and numerical simulation of positive pressure dense phase pneumatic ash removal system

W.B. Wang & L. Shi
School of Computer Science and Technology, Harbin University of Science and Technology, China

W.W. Liu
Exploration & Development Research Institute of the Daqing Oilfield, Daqing, China

ABSTRACT: In this paper, we analysis the current design status of positive pressure dense-phase pneumatic ash removal system, and promulgates its unreasonable parameters calculation. We put forward parameters calculation method, which minimizes the ash removal system power index, and establish a calculation model based on the critical gas flow velocity of pneumatic conveying pipe V_c, diameter D_c, economic ash and gas conveying ratio μ_c. In addition, we compare our experimental results with the he economic flow velocity and the bin pump pressure, and use the dynamics software CFX to simulate the high concentration of gas solid with two phase flow through the straight pipe section. The simulation result indicates that the minimum energy consumption, economic velocity can be confirmed in certain conditions which show our calculation model is reliable, reasonable and effective.

KEYWORDS: Pneumatic Ash; System Optimization; Numerical Simulation.

1   INTRODUCTION

There are advantages of positive pressure dense phase fluidization storehouse pump pneumatic ash removal system to other conventional dilute phase bin pump system with low flow rate, ash gas ratio (BROCKS M 2005, DING Yanfeng 2006, HANG Biao 1984). The parameters of positive pressure dense phase fluidized pneumatic ash removal system is determined the calculation method based on suspension type of dilute phase positive pressure pneumatic conveying system, Then flow rate, ash gas ratio and other important parameters selected are still dependent on experience, so calculation's error relatively large makes engineering design and operation management uncertain (LIN Zhaofu 2002). In this paper, the calculation model of positive pressure dense phase fluidization storehouse pump pneumatic ash removal system was established. It gives the system design parameters of power system economic conditions in the pneumatic conveying, and through the CFX software to do a numerical simulation for the high concentration gas solid two phase flow, obtains the optimal economic speed under certain conditions(LIU Zongming & YUE Yunlong & LI Haidong,et al 2003).

2   CALCULATING PRESSURE LOSS OF PNEUMATIC ASH REMOVAL SYSTEM

The theory pressure loss of positive pressure dense phase fluidization storehouse pump pneumatic ash removal pipe ΔP can be calculated by formula (1).

$$\Delta P = \left(\sqrt{P_e^2 + \frac{2P_e \lambda L \gamma_e V_e^2}{2Dg}} - P_e \right)(1 + n\mu) \qquad (1)$$

In formula (1), p_e is the absolute pressure of the pipe section terminal in kg/m². λ is air friction coefficient, L is the ash conveying pipe section equivalent length. D and m are the pipe section diameter. γ_e is the pipe terminal air density in kg/m³. V_e is the pipe terminal velocity in m/s. μ is the ash gas ratio, n is the two phase flow coefficient, and g is the acceleration of gravity in m/s².

However, in the actual operation process, pneumatic ash removal system turn from dilute-phase to dense-phase, the material in the pipeline is no longer uniform distribution, material particles will separate from the gas stream, conveyed with a dune shape (MALLICK S S 2009). In addition to, the range of the cooling caused by that air expands in the transfusion is not equal to the extent of warming caused by that the frictional heating created by while air is mixed with ash (WANG Qigai & WU Chengguang & TU Qiu 2005).

2.1 Calculation model

Economic speed V_c is when Ash grain sinks down pipe bottom gradually and form the stillness accumulation, namely dense phase turn to dilute phase. For dilute phase conveying, $V > V_c$, for dense phase pneumatic conveying, $V \leq V_c$;

The pressure loss of positive pressure dense-phase bin pump pneumatic ash removal pipe is formed by formula (2).

$$\Delta P_m = \Delta P_a + \Delta P_s \qquad (2)$$

In the formula (2): ΔP_a is pure movement of air pressure loss, kg/m2; ΔP_s is the additional pressure loss of the material movement, kg/m^2.

According to Darcy's law:

$$\Delta P_a = \frac{\lambda \Delta L \rho V^2}{2D} \qquad (3)$$

$$\Delta P_m = \frac{\lambda \Delta L \rho V^2}{2D} + \frac{\lambda_m \Delta L \delta_m \mu \rho V^2}{2D} + \frac{\mu V_1 \Delta L \gamma}{\delta_m V} \qquad (4)$$

In the formula (4): γ is air density, kg/m3; V_1 is the suspension velocity of material m/s; ΔL is ash velocity ratio; δ_m is selection of pipe length, m; ρ is the air density, kg/m3; λ_m is the friction pressure loss coefficient of the material to the pipe wall.

2.2 Establishing the objective function

Taking the minimum power exponent in a pneumatic ash system as optimization objective, and the objective function is established. The physical significance of dynamic index is the power that 1t material conveying distance is 1m needs.

$$K = \frac{Q \Delta P_m}{2G_h \Delta L} \qquad (5)$$

In the formula (5): Q is the volume of air, m^3; G_h is the amount of ash, t/h.

Q, G_h, ΔL are all assumed constants. To make K to be the smallest, ΔP_m must be the smallest.

According to (5), take V for the partial differential to make $\dfrac{\partial(\Delta P_m)}{\partial(V)} = 0$:

$$\frac{\partial(\Delta P_m)}{\partial(V_c)} = \frac{2 \Delta L \rho (\lambda + \lambda_m \mu \delta_m) V_c}{2 D_c} - \frac{\mu V_1 \Delta L \gamma}{\delta_m V_c^2} \qquad (6)$$

K is smallest when pressure loss is the minimum:

$$V_c = \sqrt[3]{\frac{g D_c \mu V_1}{\delta_m (\lambda + \lambda_m \mu \delta_m)}} \qquad (7)$$

In the formula (7): D_c is the economic diameter when pressure loss is the minimum, m.

When dilute phase to dense phase, $\lambda_m = 0$

$$V_c = \sqrt[3]{\frac{g D_c \mu V_1}{\delta_m \lambda}} \qquad (8)$$

3 INTERPRETATIONS

Fluidization storehouse pump simulation test ratio is 1:1, adopts the seamless steel pipe, experimental data are shown in Tables 1 and 2.

Table 1. The transportation distance L=1000m, D133mm× 4.5mm.

Conveying capacity t/h	Gas consumption, m^3/min	Pump-pressure MPa
19.3	7.5	0.51
18.1	8.6	0.52
18.5	8.0	0.52
18.2	8.1	0.52

Terminal velocity m/s	Ash gas ratio μ	Dynamic index kwh/tkm	Gas proportional %
19.13	35.7	2.0	0.49
21.30	29.0	2.3	0.46
20.71	32.0	2.2	0.48
20.31	31.2	2.2	0.48

4 ECONOMIC FLOW VELOCITY CONTRACTION

1 Comparing with the experimental data in Table 1, the average particle size of ash d=0.3mm, V_1 =3.5m/s, mixture ratio $\mu = 35.7$ .Using formula(8) to calculate the economic flow velocity $V_c = 19.40 m/s$, the test value $V'_c = 19.13 m/s$, error is 1.41%.

Table 2. The transportation distance L=1000m, D108mm× 4.0mm.

Conveying capacity t/h	Gas consumption, m³/min	Pump-pressure MPa
17.5	9.4	0.51
20.4	6.6	0.47
19.7	10.6	0.48
20.1	10.1	0.48

Terminal velocity m/s	Ash gas ratio μ	Dynamic index kwh/tkm	Gas proportional %
25.89	25.8	3.3	0.78
19.97	43.0	1.8	0.55
34.09	25.9	2.9	0.66
33.60	27.6	2.9	0.68

2   Comparing with the experimental data in Table 2, the average particle size of ash d=0.3mm, V_1 =3.5m/s, mixture ratio $\mu = 43.0$ .Using formula(8) to calculate the economic flow velocity $V_c = 19.39 m / s$, the test values $V'_c = 19.97 m / s$, and the error is -2.90%.

The error is within the control, suggesting that the economy speed formula is effective.

5   DENSE PHASE GAS-SOLID TWO-PHASE FLOW OPTIMAL ECONOMIC VELOCITY

In order to verify the accuracy of the test and the result, the straight segment of gas-solid two-phase flow is simulated by using the computational fluid dynamics software CFX. The Euler Lagrange model is established for choosing a numerical simulation object according to the working conditions, determine the software simulation conditions. Through numerical simulation of the distribution of the resistance of the system and the particle deposition amount under different conditions, the minimum energy consumption economic velocity under certain conditions is simulated (YANG Xiuli &YANG Xiaohui & ZHOU Qing-liang,et al 2008).

5.1   Geometric model and mesh

We use Solid Works software to draw a geometric model of straight pipe as Figure 1 and ICEM-CFD to deal with the grid division of the 3D geometric model as Figure 2.

In order to ensure the numerical simulation accuracy and acceptability of computational time at the last, analyze the quality of and total mesh the mesh

partition as Figure 2, the total number of grid is 453255, grid quality minimum is 0.349, grid quality maximum is 0.9997, grid quality mean is 0.744, and it is reasonable grid.

Simulation pipe segment is a horizontal straight tube that inner diameter 80mm, length 8m, the density of solid particles is 1020 kg/m³. The working condition 1 is described as below. The diameter size of solid particle is 200μm. The speed of mass flow is 200kg/(m², s). The gas phase velocity are 7m/s, 14m/s, 16m/s, 18 m/s, 20 m/s and 22m/s. Working condition 2: solid particle diameter is 200μm, mass flow rate is 200 kg/ (m². s), and gas phase velocity are 10m/s, 15m/s, 18m/s, 20m/s, 22 m/s and 24 m/s. The results showed that: The optimal economic velocity of 1 is about 11.5 m/s, the corresponding unit pressure drop value is 135 Pa; the optimal economic velocity of 2 is about 15.5 m/s, the unit pressure drop value is 220Pa.

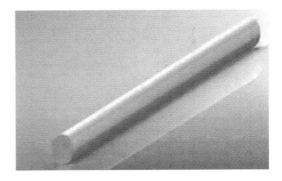

Figure 1.   A three-dimensional geometric model of the straight section.

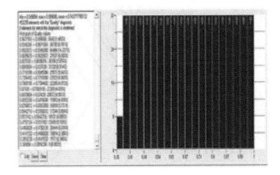

Figure 2.   Division of the regional grid computing.

6   CONCLUSIONS

Based on the relationship analysis, among the characteristics of the system of positive pressure dense-phase, the parameters of storehouse pump ash pneumatic removing mechanism are calculated. These parameters include a dynamic index of pneumatic ash removal system, pipe gas flow velocity, pipe diameter, ash and gas conveying ratio. According to the goal of the minimum power index of a pneumatic ash removal system, the optimization model is established. It gives out the method that calculates the critical gas flow velocity V_c and diameter D_c and economic ash gas transmission ratio μ_c when the power index is the minimum. The simulation result shows that once the solid particle concentration reaches a certain value, the pressure loss per unit length increases with the gas velocity first and then decreases and increases finally, additionally, there exists an optimal speed which gives the system the lowest energy consumption. Therefore, our model is feasible and economical to obtain the optimum speed under certain conditions, and contribute an important reference value for the pneumatic conveying system designed to ensure the safety and economy.

REFERENCES

BROCKS M (2005).Reliable and economical ash disposal in coal fired power stations using pneumatic conveying systems.*Bulk Solids Handling*, 25(1):38–42.

DING Yanfeng (2006).Retrofit commission and analysis on the pneumatic ash conveying system in Yueyang Power Plant .*Electric Power*,39(10):69–72.

HANG Biao (1984). Pneumatic conveying. *Shanghai: Shanghai science and technology.*

LIN Zhaofu (2002).Debugging of fly ash conveyor system with a fluidized pump and existing problems.*Electric Power Environmental Protection*, 18(1):45–49.

LIU Zongming & YUE Yunlong & LI Haidong,et al (2003).An experimental study on the fly ash removal by dense phase pneumatic conveying. *Proceedings of the International Conference on Energy and the Environment.Shanghai*, 2(1): 1288–1291.

MALLICK S S (2009).WYPYCH P W.Modeling solids friction for dence phase pneumatic conveying of powders. *Particulate Science and Technology*,27(5):444–455.

WANG Qigai & WU Chengguang & TU Qiu (2005).Design and engineering application of duc-t in pneumatic ash disposal system with positive pressure. *Electric Power Environmental Protection*, 21(1):40–42.

YANG Xiuli &YANG Xiaohui & ZHOU Qing-liang,et al (2008).Failure analysis of pumps for dense phase and positive pressure fly ash air conveying system. *CoalMineModernization*:77–78.

Electronic Engineering and Information Science – Wang (Ed.)
© *2015 Taylor & Francis Group, London, ISBN: 978-1-138-02772-5*

Experimental research on spectral evolution of the stokes pulse via Brillouin amplification in SMF

H.Y. Zhang, S.P. Zhang & Z.J. Yuan
Department of Optoelectronic Information Science and Engineering, Harbin University of Science and Technology, Harbin, China

ABSTRACT: We report on the spectral evolution of the Stokes pulse via Brillouin amplification in a single-mode optical fiber. The resonant frequency components of the Stokes pulse experience higher Brillouin gains than the non-resonant ones due to the narrowband gain characteristic of stimulated Brillouin scattering. 10 and 100 ns Stokes pulses are examined for Brillouin amplification, with the results showing that under the condition of a high Brillouin gain the spectrum narrowing and frequency pulling effects of the Stokes pulse become more obvious for a shorter pulse with a spectrum broader than the Brillouin gain bandwidth.

KEYWORDS: stimulated Brillouin scattering; spectral evolution; spectrum narrowing; frequency pulling.

1 INTRODUCTION

Stimulated Brillouin scattering (SBS) has been a subject of interest because of its ability to achieve distributed Brillouin optical fiber sensors (Diaz et al. 2008, Soto et al. 2010, Dong et al. 2009, Dong et al. 2011), Brillouin slow light (Okawachi et al. 2005, Lu et al. 2007, Dong et al. 2008, Song et al. 2005, Zhu et al. 2005) and amplification of weak signals (Gao et al. 2011, Gao et al. 2012). SBS is classically described as a nonlinear process involving a pump wave, a Stokes wave and an acoustic wave, where the angular frequencies ω_p of the pump wave, ω_s of the Stokes wave, and Ω of the acoustic wave satisfy $\Omega = \omega_p - \omega_s$ (Agrawal 2001, Boyd 2010). Because of damping of the acoustic wave, the bandwidth of SBS amplification is spectrally broadened around the Brillouin resonant frequency, resulting in a Brillouin gain bandwidth of a few tens of MHz in a silica single-mode optical fiber. This resonant characteristic has been exploited for narrowband filter and amplifier (Tanemura 2002, Lee 2007), and spectrum evolution of amplified spontaneous Brillouin scattering has also been investigated in previous studies (Gaeta 1991, Yeniay 2002), which show that the Stokes wave of the spontaneous Brillouin scattering is characterized by spectrum narrowing as pump power is increased.

In this paper, we investigate the spectral evolution of the Stokes pulse via Brillouin amplification in a single-mode optical fiber. The spectrum narrowing of the Stokes pulse is observed, agreeing with previous studies (Gaeta 1991, Yeniay 2002). Under the condition of detuning, the central frequency of the amplified Stokes pulse is pulled towards the Brillouin resonant frequency because of the considerable narrow bandwidth Brillouin amplification, which we name the frequency pulling effect in this paper.

2 EXPRIMENTAL RESULTS AND DISCUSSION

The experimental setup is shown in Figure 1. The pump and Stokes waves were provided by two narrow linewidth (3 kHz) fiber lasers operating at 1550 nm, respectively; with their frequency difference being locked by a microwave frequency counter. The output of laser 2 was launched into an electro-optic modulator to create a Stokes pulse and then entered a 2 km standard single-mode optical fiber, where it experienced Brillouin amplification at the expense of the counter-propagating cw pump wave. Because of the random polarization variation in a single-mode optical fiber, a polarization controller was used to achieve the maximum amplification for the pulse. The gain parameter G was about 1.5 in the experiment. The spectrum of the Stokes pulse was analyzed by using a 1-GHz bandwidth photo-detector and a fiber scanning Fabry-Perot interferometer with a free spectral range of 2 GHz and a resolution of 3 MHz.

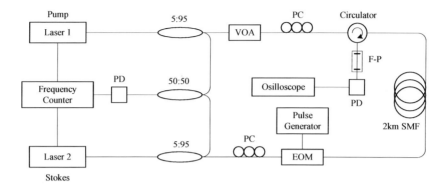

Figure 1. Experimental setup. PD: photodiode, PC: polarization controller, EOM: electro-optic modulator, F-P: scanning Fabry-Perot interferometer, VOA: variable optical attenuator.

The impact of pump power on the Stokes pulse spectrum was studied. A 10 ns pulse with a peak power of 5 mW was used, and the pump was tuned to make a 30 MHz detuning. The spectra of the amplified pulse with pump power of 4, 7 and 10 mW are shown in Figure 2. It can be seen that as the pump power increases, the power of the resonant frequency component increases dramatically, while the pulse spectral width (FWHM) decreases to 57, 29 and 24 MHz, respectively.

the amplified pulse. Moreover, for the 100 ns Stokes pulse, an oscillation on the top of the output pulse is also observed when the detuning changes from 10 to 60 MHz. Due to the short rising time of 2.5 ns, which contains spectrum components with frequencies of several hundreds of MHz, interference between the amplified resonant frequency and the carrier frequency of the Stokes pulse yields the oscillation.

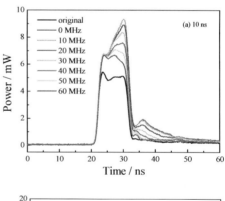

Figure 2. Spectral evolution of a 10 ns Stokes pulse with a detuning of 30 MHz at different pump powers.

We also studied the waveform evolution of the Stokes pulse. The output waveforms of 10 and 100 ns pulses at different detunings are plotted in Figure 3, respectively. In both cases, there is a ~10 ns relaxation time following the trailing edge. Because of the SBS-induced narrowband phase modulation to the resonant components of the pulse spectrum, a chirp is induced in the output pulse, and the scattered light during the relaxation time contributes to the resonant component and is independent of detuning, which explains the spectrum narrowing of resonant component of

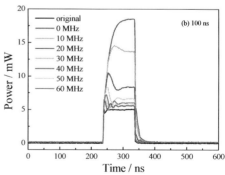

Figure 3. Waveform evolution at different detunings for (a) 10 ns and (b) 100 ns Stokes pulses.

346

3 CONCLUSIONS

To summarize, we have demonstrated the spectral evolution of the Stokes pulse via Brillouin amplification in a single-mode optical fiber, and studied the spectrum narrowing and frequency pulling effects with different pulse widths and pump powers. The results show that these two effects become more obvious for a shorter pulse and a higher pump power, and the frequency pulling effect may induce a distortion on the output waveform of the Stokes pulse.

ACKNOWLEDGMENT

The authors would like to thank Prof. Yongkang Dong from Harbin Institute of Technology for the helpful discussion. This project is supported by Scientific Research Fund of Heilongjiang Provincial Education Department (NO: 12531093).

REFERENCES

Agrawal, G.P. (2001). *Nonlinear Fiber Optics*, third ed., San Diego: Academic.

Boyd, R.W. (2010). *Nonlinear Optics*, third ed., Singapore: Academic.

Diaz, S., Mafang, S. F., Lopez-Amo, M. & Thévenaz, L.(2008). A high-performance optical time-domain Brillouin distributed fiber sensor. *IEEE Sensors Journal* 8 (7): 1268–1272.

Dong, Y., Bao, X. & Li, W. (2009). Differential Brillouin gain for improving the temperature accuracy and spatial resolution in a long-distance distributed fiber sensor. *Appl. Opt.* 48 (22): 4297–4301.

Dong, Y., Chen, L. & Bao, X. (2011). Time-division multiplexing-based BOTDA over 100km sensing length. *Opt. Lett.* 36 (2): 277–279.

Dong, Y., Lu, Z., Li, Q. & Liu, Y. (2008). Broadband Brillouin slow light based on multifrequency phase modulation in optical fibers. *J. Opt. Soc. Am. B-Opt. Phys.* 25 (12): C109-C115.

Gaeta, A. L & Boyd, R. W. (1991). Stochastic dynamics of stimulated Brillouin scattering in an optical fiber. *Phys. Rev. A* 44: 3205–3209.

Gao, W., Lu, Z., He, W. & Zhu, C. (2011). Investigation on competition between the input signal and noise in a Brillouin amplifier. *Appl. Phys. B* 105: 317–321.

Gao, W., Hu, X., Sun, D. & Li, J. (2012). Simultaneous generation and Brillouin amplification of a dark hollow beam with a liquid-core optical fiber. *Opt. Express* 20 (18): 20715–20720.

Lee, K.-H. & Choi, W.-Y. (2007). Harmonic signal generation and frequency upconversion using selective sideband Brillouin amplification in single-mode fiber. *Opt. Lett.* 32 (12): 1686–1688.

Lu, Z., Dong, Y. & Li, Q. (2007). Slow light in multi-line Brillouin gain spectrum. *Opt. Express* 15 (4): 1871–1877.

Okawachi, Y., Bigelow, M. S., Sharping, J. E., Zhu, Z., Schweinsberg, A., Gauthier, D. J., Boyd, R. W. & Gaeta, A. L. 2005. Tunable all-optical delays via Brillouin slow light in an optical fiber. *Phys. Rev. Lett.* 94 (15): 153902-01-04.

Song, K.Y., M.G. Herráez, & Thévenaz, L. (2005). Gain-assisted pulse advancement using single and double Brillouin gain peaks in optical fibers. Opt. Express 13(24): 9758–9765.

Soto, M. A., Bolognini, G., Pasquale, F. D. & Thévenaz, L. (2010). Simplex-coded BOTDA fiber sensor with 1 m spatial resolution over a 50 km range. *Opt. Lett.* 35(2): 259–261.

Tanemura,T., Takushima, Y.& Kikuchi, K.(2002). Narrowband optical filter, with a variable transmission spectrum, using stimulated Brillouin scattering in optical fiber. *Opt. Lett.* 27(17): 1552–1554.

Yeniay, A., Delavaux, J.-M.,&Toulouse, J. (2002). Spontaneous and stimulated Brillouin scattering gain spectra in optical fibers. *IEEE J. Lightwave Technol.* 20(8): 1425–1432.

Zhu, Z., , Gauthier, D. J., Okawachi, Y., Sharping, J. E., Gaeta, A. L., Boyd, R. W. & Willner, A. E.(2005). Numerical study of all-optical slow-light delays via stimulated Brillouin scattering in an optical fiber. *J. Opt. Soc. Am. B-Opt. Phys.* 22: 2378–2384.

Electronic Engineering and Information Science – Wang (Ed.)
© 2015 Taylor & Francis Group, London, ISBN: 978-1-138-02772-5

Application of evolution in path searching and program generation

P. He
School of Computer Science and Educational Software, Guangzhou University, Guangzhou, China

A.C. Hu
School of Computer and Communication Engineering, Changsha University of Science and Technology, Changsha, China

M.B. Tang
Key Laboratory of High Confidence Software Technologies (Peking University), Ministry of Education, Beijing, China

X.R. Chen
South China Institute of Software Engineering, Guangzhou University, Guangzhou, China

ABSTRACT: The problem with performance of Genetic Programming (GP) comes in part from what description tool we use and what convenience it may offer. As random search technologies, a major challenge GP must face is to get ideal approaches for depicting the search space and evolution rules. To this end, model approaches aiming to delineate relationships among given components or constructors is initiated under finite state transition systems, and a deep investigation into efficient implementation of genetic operators is carried out. To make it more convincing, we also conduct experiments with classical regression problems, obtaining positive result from comparisons between the present approach and an important GP variant like grammatical evolution.

KEYWORDS: Genetic programming; Grammatical evolution; Finite state transition; Genetic operators.

1 INTRODUCTION

Genetic programming (GP) (Koza 1992) was first proposed by John R. Koza to construct programs solving given problems. Although there appear many GP variants (Oltean et al. 2009, O'Neill & Ryan 2001, Byrne et al. 2014 in press, Montana 1995, Sabar et al. 2013) such as Gene Expression Programming (GEP), Grammatical Evolution (GE), Strongly Typed Genetic Programming (STGP), etc. since then, their way to construct expected programs from given sets of components are almost the same. They all solve optimization problems with random search technologies. This means few knowledge or constrain can be utilized to optimize the evolution process, therefore encountering difficulty in narrowing down the big search space.

In view of this, model approaches (He et al. 2011a, b) aiming to describe relationships among given components or constructors is initiated under finite state transition systems (Huth & Ryan 2004). The work of He et al. (2011a, b) indicates all possibly valid programs can be well summarized with models or transition systems, and every desired program has its path counterpart in the concerned model by which to reconstruct it. In this paper, much of our focus will center on evolutionary techniques beneficial to path searching and program generations.

In the following sections, we first introduce transition system, the way to represent programs, then the evolutionary mechanism of programs and applications respectively.

2 TRANSITION SYSTEM AS MODEL OF PROGRAMS

By transition system, we mean a mathematical model or a labeled directed graph (Huth & Ryan 2004) defined on: i) three sets of states, input symbols, and transitions over them; ii) an initial state $s_0 \in S$ and a set of final states. These systems can be widely applied in compiler designs and protocol verifications, etc.

Most existing GPs are automatic program approaches devised from random search technologies like genetic algorithm (GA). These evolutionary frameworks are too simple to employ knowledge or constrain in deducing optimal solution for given problems, therefore facing a challenge to check the huge search space.

To overcome this shortcoming, we make a deep investigation into model-based evolutionary approach, obtaining both Hoare logic-based genetic programming (HGP) and model-based grammatical evolution (MGE). The present approach is divided into three steps:

a. Define search space for a set of generalized Hoare formulae (in the case of HGP) or a set of grammatical inferences (in the case of MGE);
b. Model search space in the context of relations of subsets by transition diagram;
c. Search for the desired gene or solution under the concerned model through a well defined heuristic algorithm . For the case of HGP, gene is a program represented by certain path of the diagram; for the case of MGE, gene represented also by path is a sequence of production rules which designates the grammatical inference steps for the expected program.

For instance, Figure 1 is a transition system covering all sentential deductions of the following context-free grammar. In this case, the search space $(O(256^n)$ of standard GE can be narrowed down dramatically to $O(k^{\frac{n}{2}})$. For the details, one can refer to He et al. (2011b).

The grammar used in Figure 1 is $G = (V_N, V_T, S, P)$, where $V_N = \{$expr, var, op, pre_op$\}$, $V_T = \{$sin, cos, exp, log, +, -, *, /, y, 1.0, (,)$\}$, $S = $<expr>, and P are the following productions:

1 <expr>::= <expr> <op><expr>   (1a)

   | (<expr><op><expr>)   (1b)

   | <pre_op> (<expr>)   (1c)

   | <var>   (1d)

2 (<op>::= +   (2a)

   | -   (2b)

   | *   (2c)

   | /   (2d)

3 <pre_op>::= sin   (3a)

   | cos   (3b)

   | exp   (3c)

   | log   (3d)

4 <var>::= y   (4a)

   |1.0   (4b)

3 MAJOR ISSUE WITH EVOLUTIONARY MECHANISM OF PROGRAMS

Obviously, GP framework (Koza 1992, O'Neill & Ryan 2001, He et al. 2011a, b) can be employed to construct programs for given problems, but evolutionary results, i.e. strings of deduction rules, may be inconsistent with the concerned model like Figure 1. This casts a shadow over their implementation and performance. Hence, one of the major issues to be addressed urgently has a close relationship with genetic operators. To solve this problem, novel genetic operators must be provided for evolving only valid chromosomes (i.e. there is no breakpoint among adjacent genes of the chromosome). The method is given below. It has similarity with the work of He et al. (2009), and can be used to search for desired paths in all model-based evolutionary frameworks. Paths in that case stand for sequences of program components, but in this case, they are sequences of production rules to be paraphrased into programs of some language.

Definition 1 Supposing $G = (V, E)$ is a model with V for states, and $E = \{e_i \mid 1 \leq i \leq n\}$ for edges. $Succ$: $E \to 2^E$ is a mapping defined for $f \in E$ by: $succ(f) = \{g \in E \mid$ exist ε edges lying between f and g $\}$. Note that $succ$ (ε) is E.

Similarly, we will define the following $Pred$.

Definition 2 Given $G = (V, E)$ as above, $Pred$: $E \to 2^E$ is a mapping defined for $g \in E$ by $pred(g) = \{f \in E \mid$ exist ε edges lying between f and g $\}$. Note that $pred(\varepsilon) = E$.

Definition 3 Given the above mentioned $G = (V, E)$, $C(\alpha, \beta)$ is a function of $\alpha, \beta \in E^*$ defined by: $C(\alpha, \beta) = \{e \in E \mid e$ appears in α and $\beta\}$.

Up to now, the novel genetic operators mutation and crossover can be implemented with $Succ$, $Pred$, and $C(\alpha, \beta)$ as well. As far as the initialization is concerned, valid chromosomes are constructed similarly from the above described functions.

Design of mutation:

a. Suppose p is an individual taking of the form $P = f_1 f_2 \cdots f_m$;
b. Select a mutation position i from p;
c. Select some f from $succ(f_{i-1}) \cap pred(f_{i+1})$ randomly;
d. Replace f_i with the above obtained f.
   Particularly, $f_i = \varepsilon$ for either $i \leq 0$ or $i > m$.
   Design of crossover:

a. Suppose p_1 and p_2 satisfying $C(P_1, P_2) \neq \varnothing$ are two chromosomes of the forms $P_1 = f_1 f_2 \cdots f_m$, $P_2 = h_1 h_2 \cdots h_n$;
b. Select some f from $C(P_1, P_2)$;
c. Select crossover points i and j within P_1, P_2 in terms of the above obtained f;

d. Interchange $h_{j+1} \cdots h_n$ with $f_{i+1} \cdots f_m$, obtaining two non-conflict or valid chromosomes.

In view of the above discussions, the present approach takes into account the restriction of evolutionary models. The fitness applied in the experiments is the least square error.

4 APPLICATIONS

Up to now, we have applied the above-mentioned technologies to solve many regression problems. Because of limited space, we will take for example only the improved grammatical evolution.

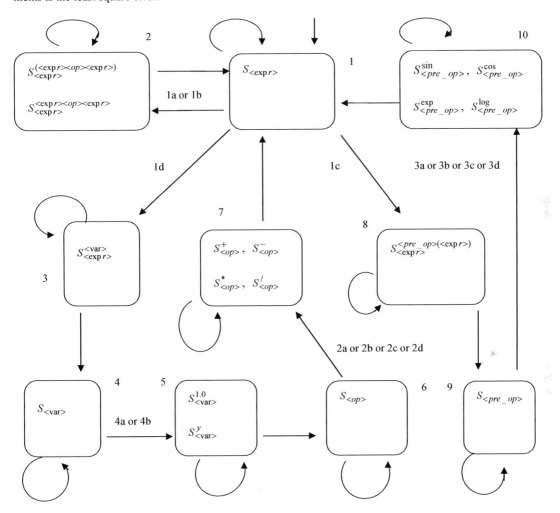

Figure 1. Model (with start state 1) of the above mentioned grammar. Unlabeled arrows are ε arrows. Edge labels in the form of disjunctions stand for sets of productions. For instance, 4a or 4b corresponds to {4a, 4b}.

Table 1. Survey of the three experiments.

	Solving of $f(y)$ for 100 times		Solving of $g(y)$ for 100 times		Solving of $h(y)$ for 100 times	
	Average time	Average error	Average time	Average error	Average time	Average error
MGE	3.46826	0.63864	2.04116	0.15309	2.22013	0.11873
GE	11.28804	0.57576	3.54410	0.18394	2.86162	0.16070

Table 2. Wilcoxon-Mann-Whitney test on distribution of the experimental data.

		$H_0: \mu_1 = \mu_2 \leftrightarrow H_1: \mu_1 \neq \mu_2$		
		Risk level $\alpha = 0.05$	P-value	Report on the efficiency
Experiment 1	GE MGE	Accept H_1 in light of α	<2.2e-16	MGE is significantly better than GE in the experiments.
Experiment 2	GE MGE	Accept H_1 in light of α	9.027e-09	
Experiment 3	GE MGE	Accept H_1 in light of α	0.001046	

We make comparisons between traditional grammatical evolution and the present approach in terms of the following target functions 5-7 on [-1, 1].

$$f(y) = y^4 + y^3 + y^2 + y \qquad (5)$$

$$g(y) = \sin(y^4 + y^2) \qquad (6)$$

$$h(y) = \sin(\exp(\sin(\exp(\sin(y))))) \qquad (7)$$

The parameters employed in the experiments can be briefly summarized as: size of generation: 100, probability of crossover and mutation are 0.9 and 0.15 respectively, crossover model: single-point, and the like. Comparisons shown in tables 1-2 are obtained from running each method over the same sample dataset of [-1, 1] for 100 times. It follows MGE is significantly better in time complexity than classical GE.

5 CONCLUSIONS

To improve the performance of GP, we make an attempt at reducing the search space through using transition systems. The method can be summarized as defining search space of all possible solutions first, then modeling solutions according to their relationships, and finally designing efficient genetic operators in light of solution model. Experiments indicate that the present approach is significantly better than existing GEs in performance. The future work will focus on novel modeling techniques, more efficient search approaches, and structural representations, etc.

ACKNOWLEDGMENTS

The research work was supported by National Natural Science Foundation of China (Grant NO. 61170199), the Scientific Research Fund of Education Department of Hunan Province, China (Grant NO. 11A004) and Guangzhou Teaching Reform Project (2013A022).

REFERENCES

Aho A.V., M.S. Lam, R. Sethi & J.D.Ullman (2007). Compilers: principles, techniques, and tools (2nd edition). *Pearson education, Inc.*

Byrne J. M. Fenton, E. Hemberg, J. McDermott & M. O'Neill (2014). Optimising complex pylon structures with grammatical evolution. *Information Sciences.* (In press).

He P. L.S. Kang & D.C. Huang (2009). Construction of Hoare Triples under Generalized Model with Semantically Valid Genetic Operations, *Lecture notes in computer science*, 5821, 228–237.

He P., L.S. Kang, C.G. Johnson & S. Ying (2011). Hoare logic-based genetic programming. *Science china information sciences 54(3)*, 623–637.

He P., C.G. Johnson & H.F. Wang (2011). Modeling grammatical evolution by automaton. *Science china information sciences 54(12)*, 2544–2553.

Huth, M. & M. Ryan (2004). Logic in computer science: modelling and reasoning about system. England: Cambridge university press.

Koza, J.R. (1992). Genetic programming: on the programming of computers by means of natural selection. Cambridge, MA:The MIT Press.

Montana, D.J. 1995. Strongly typed genetic programming. *Evolutionary computation 3(2)*, 199–230.

Oltean, M., C. Grosan, L. Diosan & C. Mihaila (2009). Genetic programming with linear representation: a survey. *International journal on artificial intelligence tools 19(2)*, 197–239.

O'Neill M. & C. Ryan (2001). Grammatical evolution. *IEEE transactions on evolutionary computation 5(4)*, 349–358.

Sabar N.R., M. Ayob, G. Kendall & R. Qu (2013). Grammatical evolution hyper-heuristic for combinatorial optimization problems. *IEEE trans on evolutionary computation 17(6)*, 840–861.

Electronic Engineering and Information Science – Wang (Ed.)
© 2015 Taylor & Francis Group, London, ISBN: 978-1-138-02772-5

Library electronic resources based on marketing theory

P. Liu
Library of Harbin University of Science and Technology, Heilongjiang Province, China

L. Liu
Library of Heilongjiang University of Chinese Traditional Medicine

ABSTRACT: This article introduces the basic content of marketing theory, analyses the problems of taking marketing theory into library service, the psychology of using electronic resources of readers, and at last tries to promote library service by marketing theory with certain methods.

KEYWORDS: Library; Electronic Resource; Marketing theory.

1 INTRODUCTION

University Library as an organic part of colleges and universities are responsible for the preservation of literature information and transfer of education. Broadly speaking, university libraries are institutions for information transmission, spreading, processing, research and development. University libraries in the environment of information society are faced with unprecedented competition and challenges, and how making customer satisfied and attract more users to use library resources is an urgent problem that many libraries and librarians are facing.

The application of library information service mainly relies on the electronic resources. But many library users are not accustomed to using electronic resources (including electronic journals, electronic books, dissertations, conference papers) and there are a few users who do not understand the value of electronic resources for their scientific research and teaching. For enhancing the usage of electronic resources as well as the understanding of electronic resources, this paper takes an attempt to introduce marketing theory to the library.

2 LIBRARY MARKETING

Library Services Promotion Strategy should integrate various marketing theories, because library resources and services are different from the sales of products on the market. The resources and services that the library provides are free of charge, and the library has a relatively stable, user group, such as the school teachers and students. So we should be aware of this point on the process of introducing marketing theory into the electronic resources.

2.1 Electronic resources marketing concepts

Library marketing is a series of purposeful activities that librarians carry out to the current and potential users, including the service product, methods, techniques and service costs. The library's electronic resources in terms of content, including foreign electronic journals, electronic books, conference papers, electronic newspapers, catalog query system and self-built features databases. From deepest sense, the library's electronic resources include user training, reference services and technical support network of a variety of intangible services related to electronic resources. Marketing of electronic resources should be user-oriented, respect each user's information needs, and use a variety of electronic resources marketing mode, so that libraries can provide more perfect service. The Electronic Resource Management Initiative (ERMI) Report (2004) stated what had become obvious to librarians, that "as libraries have worked to incorporate electronic resources into their collections, services, and operations, most of them have found that their existing integrated library systems (ILSs) are not capable of supporting these new resources."

2.2 Marketing strategy

Library as a non-profit organization is not for profit, but for meeting the needs of users. The godfather of economics Jerry McCarthy was the first to propose the theory of 4P, namely: Product, Price, Place, and Promotion.

From the marketing point of view, the product is everything that provides the market to use and consume and then to meet needs, including tangible products, services, personnel, organizations, ideas or a combination of them; There are four main factors affecting the price: cost, demand, channel, competition. The highest price depends on market demand; the lowest price depends on the cost of the product. Channel, the so-called sales channel is the various links and driving forces, referring to the flow of goods from manufacture enterprises to producers. Sales promotion is used to give products, services, image and concept notices to the target market mechanism, and is used to persuade and remind to trust product and support.

2.3 Library marketing theory

Libraries in order to form a mutual aid with the user, mutual demand, mutual need, libraries have to tie the end users together and reduce customers losing, they must adopt a number of effective ways so as to make the law in the service, demand and other aspects associated with users, to increase user confidence of the library, to win the long-term access and utilization improving rate. For libraries, the more realistic question is not how to develop systems and to make measures, but rather to listen to the users' hopes, desires and needs with timely manner. Libraries emphasis on the interaction with the user, and on the respond of users' needs, which is necessary to develop library's electronic resources.

In recent years, a lot of people are talking about that library facing with competition has to lead the marketing strategy in the library work in order to survive. The dynamics of marketing can be simply expressed: (1) identify your product, (2) identify your market, and (3) bring product and market together. In China, Xie Chunzhi (2003) considers that if digital library wants to occupy the advantage in the fierce market competition, to attract the attention of the public, it is necessary for library to introduce the marketing idea. Li Li (2005) proposed that the development of digital libraries lateral marketing theory that is personalized service to users in the center. Liu Kunxiong (2007) considered that the library information marketing is not only a resource of production resource marketing, but also a resource of productive forces and production relations marketing. From the perspective of marketing information, the basic strategies of library information resources development are: product strategy, pricing strategy, channel strategy. Fully understand and grasp this relationship between the information marketing and library information resources development is necessary. Feng Xiangjun (2009) proposed that library experience marketing strategy development ideas are based on

the user information from the perspective of behavioral research. Wu Junying (2009) proposed a strategy of an integrated marketing mix with digital library marketing. The research on the electronic resources, marketing of many experts emphasizes on its importance in theory and on putting forward any specific marketing strategy that can be taken.

In western nations, the four most popular techniques among all kinds of libraries are patron training (group), flyers/brochures, e-mail (external), and surveys. In China, the most popular techniques among various libraries are face-to-face communication, telephone, e-mail, and chatting tools, such as QQ.

3 THE PROBLEMS OF LEADING MARKETING INTO LIBRARY

The marketing is leaded into electronic resources began in the library having the market investigation and analysis of users, ended with meeting users' specific information needs and completing the library 's value exchange with the user.

3.1 Users' requirements

As the number and variety of electronic resources is increasing, the cost of electronic resources has been increasing; and the library is in dilemma of providing limited capacity and users' unlimited needs every year. The users of university library electronic resources are mainly university teaching and research staffs, graduate students and advanced undergraduates. This is the three different types of user groups. Their needs and preferences of resource services have obvious differences. Such as university researchers timely need access to information resources, they also need strong academic information and track the latest trends in the research process. Graduate hand need teachers designated professional reference materials, on the other hand also need to look classic and professional-related and read a lot of domestic and international academic literature to complete the dissertation writing and research. Undergraduate find the relevant information in order to complete the job one hand, on the other hand they do simulation title in order to get CET, computer certification. There is also a difference in the demand and method of use of different disciplinary background users of electronic resources service. As the author's school, the concerns and satisfaction are generally low on electronic resources to the teachers and students in liberal arts as the background, the reason is the purchase of library electronic resources, especially foreign electronic resources for liberal arts students is small, another reason is liberal arts students are not used to find electronic resources literature. As well as the science and engineering students have a high attention to electronic resources and utilization, but the existing electronic resources cannot meet the needs of all science and engineering students and teachers of literature.

3.2 *Two-way communication*

In general, the university library all attach great importance to the construction of electronic resources, provides many kinds of electronic resources in the campus network, but the promotion and publicity of electronic resources is slightly weak. Namely, it's the lack of communication between user and library. The exchange of library and users can be said one of the library services. Such as the establishment of the library desk, as well as reader service department, as well as volunteer recruitment activities organized by the library. For users in the use of electronic resources when they encounter problems, such as the need to answer the library and they need to give feelings back to the library after using electronic resources, etc. They have almost no communication with library of these problems.

So it's rare that the real user communicates with library, and the communication opened by the library is not fully play its role. Many users have experienced problems used to Baidu, or ask colleagues, classmates, or post on the forum for help, and not willing to try to communicate with the library to obtain the solution to the problem. The library had difficulty in receiving feedback and consulting from users, and library understands the use of electronic resources in library user has been hindered, meet the user's problem is not timely and professional solutions.

3.3 *Knowledge services*

There is a certain gap between expected electronic resources service and realistic electronic resources service for academic staffs and undergraduates of colleges. The most critical step to narrow the gaps is to increase the knowledge service. Knowledge service is based on information and knowledge search, organization, analysis, the reorganization of the knowledge and ability. According to the user's question and environment, integrate into the process of user's solving problems and provide services to effective support for knowledge application and knowledge development. Now, although the university library has been aware of the importance of knowledge services, and has gradually carried out such service, the knowledge service range has no unified standard and requirement, and service object localization service mode is still passive in waiting for user service. At the same time, which has been carried out in the knowledge services is narrow, shallow-level, in-depth narrower, the users' acceptance of knowledge services is not high, and

they don't get enough from the knowledge service. How to change service mode and amplify scope of knowledge, change the traditional passive approach to service, according to the needs of different users and professional conduct in-depth knowledge of services, making it the extension user group service, the above is worth of exploring.

3.4 *Personalized service*

Personalized service is a process of collecting users' preference, needs between digital libraries and users, and then transfer users' information and libraries' service according to collected needs. Expanding library services as a means to win the competition in the market, the implementation of personalized service has become an inevitable trend of development of the library. We should do a good job for personalized service from the aspect of the personalized service object, users' needs, and users' receiving mode. At the same time, we also need the help of the network technology, push technology, web page generation technology, database technology and a series of personal information service technology, in order to solve practical problems for users, to achieve the efficient knowledge transfer.

3.5 *Marketing effectiveness evaluation*

Marketing effectiveness evaluation is the most important link for commodity marketing. A good marketing effectiveness evaluation has definite object, foster strengths and circumvent weaknesses, which can prolong the life of goods market and expand market share. Evaluation of the good marketing effect can intuitively reflect the stage of the marketing effect; To the library, it is an imminent task to set up the electronic resources service. Because the library situation is different, the resources and services are different; the library can establish a set of suitable electronic resources for its own marketing effectiveness evaluation system. Each library has to change its own marketing strategies in order to develop in a sustainable way.

4 USER'S PSYCHOLOGY OF ELECTRICAL RESOURCES

With electronic resources continue to enrich and the user's search continued to improve, how to attract and retain existing users as well as the development of new electronic resources users of university library electronic resources has become the most important issue of marketing electronic resources. It is the basis of marketing activities to aware and to understand the target user in the use of electronic resources.

The users' behavior in the use of electronic resources can be divided into: willingness to use, and usage behavior. Users could use requirements in the use of electronic resources, and have a willingness to use. The use of decision is the user's tendency in the selection and use of electronic resources before the psychological activity and behavior, which belongs to the users' attitude. And the usage behavior is the process of the use of the decision.

5 CONCLUSIONS

This paper provides a reference for promotion and publicity of library electronic resources by introducing the marketing theory and problem in the process of marketing. With the developing of network and information, more and more new technologies and new methods are used on the product of electronic resources, not only on limit access to electronic resources within campus, off campus and throughout the Internet environment. Users who use electronic resources are watching all the time. University libraries need to consider the long-term development. They need to try to put service into marketing electronic resources among users and to consider the long-term development of electronic resources.

REFERENCES

Zhou, D.F. (2007). Customized marketing: new ideas for library service promotion. *Modern Information*. 3, 115–116.

Wang, R. (2008). Analysis on the Service Promotion Strategy in University Libraries of Guangdong *Library Tribune*, 6, 34–36.

Feng, X.J.,(2009). Analysis on Experiential Marketing Strategy of Library Based on User Information Behavior. *Library Work and Study*. 12, 3–6.

Elliott Ettenberg.(2006). *4R marketing: New Theory subversion 4P marketing*. The Second Edition. Beijing: Enterprise Hall Lane Press.

Liu. K. X., & H. L. Li.(2007). Library Information Marketing and Information Resource Development. *Journal of Library Science In China*, 2, 47–51.

Ma. J.T.& H.Maslow. (2002). *Classics in Humanity Management*. Beijing: Beijing Industry University Press.

Lindsay, A.R.(2004). *Marketing and Public Relations Practices in College Libraries*, CLIP Note, ALA, Chicago, IL.

Liu. K.X., & C.P, Hu.(2007). Discusses on the Management Science Attribute of Library Information Marketing. *LIBRARY AND INFORM ATION SERVICE*, 1, 65–68.

Zhang, W.Y.(2007). The Application of New Media in Library Outreach Service. *Library Journal*, 6, 19–21.

Turner, A., Wilkie, F. &Rosen, N. (2004). Virtual but visible: developing a promotion strategy for an electronic library, *New Library World*.105(1), 262–8.

Electronic Engineering and Information Science – Wang (Ed.)
© *2015 Taylor & Francis Group, London, ISBN: 978-1-138-02772-5*

A flat micro heat pipe with fiber wick and mathematical model

T. Han, Z. Zhang & Y.D. Shen
Harbin University of Science and Technology, Harbin, China

ABSTRACT: A novel flat micro heat pipe with glass fiber wick structure is introduced in this paper. Fibers are used as the wick structure of the micro heat pipe. The fibers are pushed closely and the space among the fibers forms the liquid channel. A mathematical model is founded by finite element method to calculate this kind of micro heat pipe. Governing equations are formulated for each finite element and all the equations are calculated by iteration. For the structure introduced in this paper, 60 steps of iteration are needed to get steady results of the model. The iteration processes of the meniscus radius and temperature difference are presented. And the theoretical results of the model can be used to predict the thermal resistance under different input powers. The thermal resistance profile is validated by experimental testing. The theoretical maximum heat transport is 2w which is in good accordance with the testing result.

KEYWORDS: micro heat pipe; fiber wick; maximum heat transfer.

1 INSTRUCTIONS

The amount and density of electronic components in VLSI chips sharply increases on modern integrated circuits. The heat generated in these chip results in high-operating temperature if not removed, which will degrade the speed and the stability of the electronic devices(Suma&De 2005). Furthermore, the thermal stresses inherent within a chip during operation typically result in failure caused fatigue of the mechanical devices or connections (Pandraud,& Martine 2006). Cotter first introduced the concept of micro heat pipe for the dissipation and removal of heat. Then micro heat pipe is used in many fields because it has the merit of high efficiency, no noise, and without power(Jiao, and Ma 2007, Suman&Hoda. 2005). Recently many micro heat pipes use fibers as wick structure to increase the characteristic, and fiber wick becomes one of the significant researches (Peterson and Ma 1996, Suh, Greif and Grigoropoulos.2001))

Precious mathematical models are mostly designed for micro grooved heat pipe, so developing a new mathematical model to predict the characteristic of the micro heat pipe with fiber wick is necessary (Suh, Greif, and Grigoropoulos. 2001). A three-dimensional mathematical model is developed by finite element method in this paper. The effects of phase changing, the contact angle, gravity, and heat conducting between the fibers are accounted in the model. The governing equations are formulated by conservation of mass, momentum and energy in three dimensions. The model is calculated by iteration and the results can be used to predict the thermal

resistance of the micro heat pipe under different input powers. The theoretical results are analyzed and compared with experimental testing data.

2 STRUCTURE OF THE MICRO HEAT PIPE

Micro heat pipe is usually divided into three parts, evaporator condenser and adiabatic region which are shown in Figure 1. In evaporator the liquid absorbs heat and changes into vapor. In condenser the vapor releases heat and changes into liquid. The liquid flows back to the evaporator by capillary force which is produced by the wick structure. This is the material circle in micro heat pipe.

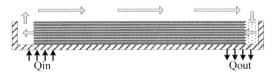

Figure 1. The shematic diagram of the mciro heat pipe with fiber wick.

This kind of micro heat pipe is designed to use glass fibers as the wick structure. Glass fibers are pushed closely, and the space among the fibers forms the liquid channel. The cross-sectional area is an analogous triangular which is surrounded by three tangent circles. The cross-sectional area is shown in Figure 2. The sharp corner in the cross-sectional area can produce strong capillary force for the liquid flowing back to the evaporator.

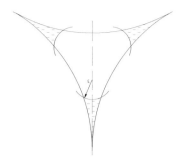

Figure 2.　The cross-sectional area of the liquid channel.

In order to fix the fibers in the micro heat pipe, two copper plates are designed. The fibers are pushed closely in the back plate and front plate is packaged on the fibers. The packaging process is presented in Figure 3. Four beams are designed in the front plate and the position of the four beams is shown in Figure 4. The front plate can not only push the fibers closely but also make the vapor can flow from the evaporating part to the condensing section.

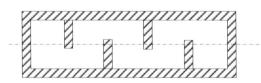

Figure 3.　Structure of the flat micro heat pipe.

Figure 4.　The shape of the front plate.

3　MATHEMATICAL MODEL AND CALCULATION

In order to predict the characteristic of the micro heat pipe, a mathematical model is founded by finite element method. The micro heat pipe is divided into many elements in three dimensional. First the direction along the length of the micro heat pipe is denoted z axis, and the cross sectional area is denoted x and y axis. Along the z direction the micro heat pipe is divided into many small elements, the length of which is dz. In x and y direction, the cross sectional area is

meshed into several parts and there is key point in each part. The governing equations are formulated for each key point in the model. The key points are shown in Figure 5.

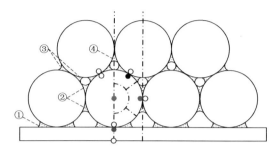

Figure 5.　The key points in the cross-sectional area.

In this mathematical model, the vapor flow, the liquid flow and the heat conduction in the solid is regarded. The mathematical model is founded under the following assumptions: 1) The micro heat pipe works in steady state. 2) The vapor and liquid is incompressible. 3) The pressure and temperature of the vapor is homogeneous in the heat pipe. 4) The heat conduction in the liquid along the z direction is neglected. In order to formulate the governing equations, the heat pipe is divided into a series of small control volumes of length dz. In a control volume, the continuity equation for the liquid is expressed as follow:

$$\rho u_1 A_1 + m_{ph} = \rho u_2 A_2 \tag{1}$$

Where u_1, u_2, A_1, A_2, ρ, and m_{ph}, denote the velocity of the liquid flowing into the control volume, the velocity of the liquid flowing out of the control volume, the cross-sectional areas of the analogous triangular, the density of the liquid and the mass of the liquid changing phase on the interface. For the incompressible flow in the micro heat pipe, Bernoulli equation is used to formulate the governing equations. And in the Bernoulli equations, the flowing resistance and the gravity are regarded and the conservation equation of the axial momentum in the control volume is written as:

$$\frac{p_1}{\rho g} + \frac{u_1^2}{2g} = \frac{p_2}{\rho g} + \frac{u_2^2}{2g} + \frac{u_2^2}{2g} \cdot f + l \sin \phi, \ f = \frac{C_f \cdot l}{D_h} \tag{2}$$

Where P_1, P_2, f, φ, C_f, and D_h is the pressure of the liquid in the control volume, friction between the liquid and the channel, the included angle between the micro heat pipe and the horizontal line, friction factor of the flowing liquid and hydraulic diameter of the cross-sectional area of the liquid flow.

The energy equations of the mathematical model are made of two parts. The first part is the phase changing process which occurs on the interface of the liquid and vapor, and the second part is the heat conduction in the wall of the micro heat pipe, between the wall and fiber, and between the fibers.

The temperature of interface between liquid and vapor is expressed below:

$$T_{iv} = T_v(1 + \frac{P_c}{h_{fg}\rho_l}), \; P_c = \frac{\sigma}{r} \tag{3}$$

So the density of the heat flow on the interface is written as follow:

$$q_{flus} = \frac{T_l - T_{iv}}{R_i}, \; R_i = \frac{T_v\sqrt{2\pi\delta T_v}}{2h_{fg}^2\rho_v} \tag{4}$$

Where h_{fg}, ρ_l, ρ_v, δ and σ are the latent heat of vaporization, the density of the liquid, the density of the vapor, gas constant, and the surface tension.

The heat conduction in the solid is described by the energy conversation equations for three kinds of nodes, which is shown in Figure 5. For different kind of nodes different equations is formulated to calculate the heat conduction. In order to distinguish the different node, the three kinds of nodes are numbered by the following rules: Num=4n-1, where n is the number of the fiber levels. The number along the z direction is 1-m, where m is the number of the meshing point along z axis. The governing equations are written as follow.

For the first kind of nodes, which is in the wall of the micro heat pipe and is shown in Figure 5 in blue, the governing equation is written as:

$$\frac{1}{2}\lambda_m dR(T_{2,i-1} - T_{2,i})/dz + 2\lambda_m R(T_{1,i} - T_{2,i})dz/d$$

$$= \frac{1}{2}\lambda_m dR(T_{2,i} - T_{2,i+1})/dz + \lambda_{mf}(T_{2,i} - T_{3,i})dz \tag{5}$$

$$+ \frac{1}{2}q_{flux}A_{iv}$$

For the second kind of nodes, which is shown in Figure 5 in red, the governing equation is written as:

$$\frac{\pi R^2}{8dz}\lambda_m(T_{4n,i-1} - T_{4n,i}) + \lambda_m\frac{dz}{2}(T_{4n-1,i} - T_{4n,i})$$

$$+ \lambda_m\frac{\pi dz}{12}(T_{4n+1,i} - T_{4n,i}) + \lambda_m\frac{dz}{2}(T_{4n+2,i} - T_{4n,i})$$

$$= \frac{\pi R^2}{8dz}\lambda_m(T_{4n,i} - T_{4n,i+1}) + \frac{1}{2}q_{flux1}A_{iv1} \tag{6}$$

$$+ \frac{1}{2}q_{flux2}A_{iv2}$$

For the third kind of nodes, which is shown in Figure 5 in green, the governing equation is written as:

$$T_{4n-1,i} + T_{4n,i} + T_{4n+1,i} = 3T_{4n+1,i} \tag{7}$$

These governing equations can be written for each finite element, and all the equations can be calculated by iteration in computer program. A first value is given for the equations, and the iteration process is shown in Figure 6 which shows the calculation process of the meniscus radius. During the iteration, the value is stable after about 60 steps of calculation.

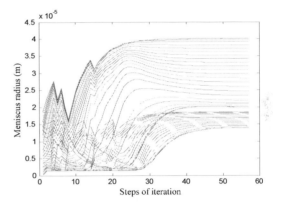

Figure 6. Iteration of the meniscus radius.

4 RESULTS AND DISCUSSION

The mathematical model can be calculated by iteration and the characteristics of the micro heat pipe can be analyzed by the calculation results. Figure 7 shows the iteration process of temperature difference between the evaporator and condenser. The thermal

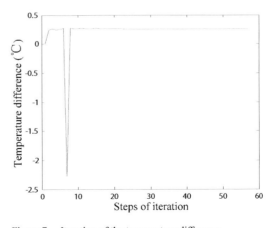

Figure 7. Iteration of the temperature difference.

resistance of the micro heat pipe can be calculated by the temperature difference and the input power. So the mathematical model can be used to predict the thermal resistance of the micro heat pipe.

In order to validate theoretical results, several experiments are performed. The micro heat pipes with fibers wick are fabricated and tested in vacuume. The experiments are performed in vacuum because the heat transfer between the micro heat pipe and air is ignored. The testing results are presented in Figure 8. The temperature profiles along the z direction at different input powers are tested. So the experimental thermal resistance is calculated by the testing results. Figure 9 suggests that the micro heat pipe provides well isothermal ability when the input power is 1w and 2w, and if the heat input increases to 3w or 4w the thermal resistance of the micro heat pipe rise rapidly which is shown in Figure 9. Therefore, it is determined from the experimental results that the maximum heat transport of the micro heat pipe is about 2 w. The maximum heat transport per unit area is 1.3×10^5 w/m^2 and the coefficient of thermal conductivity is 6500 w/m/k which is 15 times bigger than copper.

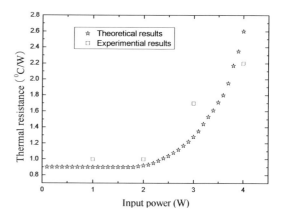

Figure 9.   Thermal resistance for different heat input.

5   CONCLUSTION

A mathematical model is introduced for a kind of flat micro heat pipe with fiber wick. The mathematical model is founded by finite element method and calculated by iteration. The thermal resistance of the micro heat pipe is calculated by the mathematical model and the results are compared with the experimental results. The testing results are in good accordance with the theoretical results when the micro heat pipe works normally. When the input power increases over the maximum heat transport of the micro heat pipe, there are some errors between the model and experimental results. The maximum heat transport can be obtained from the theoretical thermal resistance profile, and the results are validated by the testing results.

ACKNOWLEDGMENT

This work is supported by Heilongjiang Provincial Natural Science Foundation (No. F201315).

This work is supported by Harbin Research Fund for Technological Innovation (No. 2013RFQXJ104).

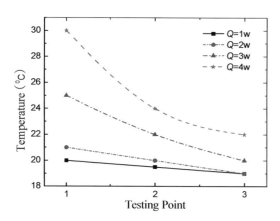

Figure 8.   Temperature profile alone the heat pipe.

The experimental thermal resistance and the theoretical thermal resistance are compared in Figure 9. The theoretical results can predict the variation of the thermal resistance when the input power increases, and the theoretical results also can point out the maximum heat transports is about 2w, because the thermal resistance begins to rise at 2w. The agreement between the model and the experimental results is excellent when the input power is 1w or 2w. when the input power increases to 3w, the balance circle of the working material in the micro heat pipe may be broke out, and the micro heat pipe is not in steady state. So there are some errors between the model and experimental results when the input power increases to 3w and 4w.

REFERENCES

Suman, B., &S. De. (2005)*International Journal of Heat and Mass Transfer. 48,* 1633.

Pandraud, G. &L. B. Martine. (2006) *Journal of Electronic Packaging. 128,* 294.

Jiao, A. J. Ma, H. B. & J. K. (2007)*International Journal of Heat and Mass Transfer. 50,* 2905.

Suman, B.&Hoda. N. (2005) *Journal of Heat and Mass Transfer. 48,* 2090.

Peterson, G. P. & Ma. H. B. (1996)*Journal of Heat Transfer. 118,* 731.

Suh, J. ,Greif, R.& Grigoropoulos. C. P. (2001) *International Journal of Heat and Mass Transfer. 48,* 4498.

Suh, J. ,Greif, R. & Grigoropoulos. C. P. (2001) *International Journal of Heat and Mass Transfer. 10,* 3103.

Electronic Engineering and Information Science – Wang (Ed.)
© *2015 Taylor & Francis Group, London, ISBN: 978-1-138-02772-5*

A novel method of correlated jamming suppression for DS-CDMA system based on ICA

M. Yu & J.Z. Chen
Nanjing Telecommunication Technology Institute, Nanjing, China

Y.Y. Qi
Institute of Communication Engineering, PLA University of Science and Technology, Nanjing, China

ABSTRACT: When correlated jamming exists, the performance of the Direct Spread Code Division Multiple Access (DS-CDMA) system declines rapidly and even turns into disabled. Because of the similarity between the useful DS-CDMA signal and correlated jamming, common technology is difficult to deal with such jamming signal. A novel method of correlated jamming suppression in a DS-CDMA system based on Independent Component Analysis (ICA) is proposed in this paper, which utilizes the independence between correlated jamming and DS-CDMA signal in the baseband. The validity of the proposed method is proved by the simulation results. Even if the power of the correlated jamming is much higher than that of the communication signal, the proposed method can also improve the bit error rate performance of a DS-CDMA system greatly.

KEYWORDS: DS-CDMA; Correlated Jamming; Jamming suppression; ICA; BSS.

1 INTRODUCTION

The Direct Sequence Spread Spectrum (DSSS) signal has been used as secure communication for a long time. Combining the double merits of DSSS and multiple access, Direct Spread Code Division Multiple Access (DS-CDMA) has been used widely in military and civil communications (Lee & Miller 1998).

DSSS signal has good ability of resisting narrow band jamming. Broad band jamming is not practical for DSSS because the energy consumption cannot be endured. With the development of technology, some effective detection methods of DS-CDMA signal have been proposed. These methods include energy radiometer (Medley et al. 1994), correlation function (Polydoros & Weber 1985), cyclostationary analysis (Jin, & Ji 2005). Several promising approaches to estimate the parameters of DS-CDMA signal such as carrier frequency (Hill & Bodie 2001), chip rate (Yu et al. 2008), symbol rate (Burel et al. 2001) have also been developed. Even the critical pseudo noise (PN) sequence of DS-CDMA signal, could be estimated by triple correlation function (Adams et al. 1998), principal component analysis (PCA) (Nzèza 2004) or independent component analysis (ICA) method (Shen et al. 2007). Thus, it is reasonable to suppose the interceptor have acquired most prior information about the transmitted DS-CDMA signal. Then, the intentional interceptor is able to generate correlated jamming with the same parameters and pseudo noise sequence as the victim DS-CDMA signal advisedly, which is much more efficient than broad band jamming and narrow band jamming. For the similarity between the jamming and DS-CDMA signal, common technology such as filter could not deal with correlated jamming efficiently.

If the correlated jamming is only the duplicate of the object signal with certain delays, it degenerates into repeater jamming. Such repeater jamming could be regarded as the multipath form of the original DS-CDMA signals, which could be diversely combined with a RAKE receiver efficiently. So, to be effective against a spectrum communication system, the payload of the correlated jamming is usually a noise-modulated signal (Torrieri 1989). However, the payload of the DS-CDMA signal is information sequence. Thus, the baseband signals of correlated jamming and the DS-CDMA signal are quite different, and they remain statistically independent indeed. According to this, Independent Component Analysis (ICA) could be introduced to separate the correlated jamming and the DS-CDMA signal in the baseband.

For the independent source signals which are linearly mixed instantaneously, ICA achieves the blind separation of the mixed signals without any prior knowledge. Ever since the theory of the ICA was first established, many algorithms based on different object functions have been developed (Lathauwer et al. 1996, Cardoso 1997). Among these algorithms, fastICA has become the most popular for its good convergence and computational simplicity (Hyvärinen & Oja 1997).

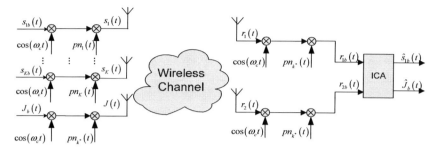

Figure 1. System chart of the proposed method.

ICA has been introduced in DS-CDMA signal processing since 1999 (Ristaniemi & Joutsensalo 1999) based on the independence between the signals of different users in DS-CDMA system. But the statistical distribution of correlated jamming signal is almost the same as that of the victim DS-CDMA signal, so original ICA method could not be adopted to suppress the correlated jamming directly.

A novel method of correlated jamming suppression for DS-CDMA signal with ICA is proposed in this paper. The method achieves the blind separation between the correlated jamming and the DS-CDMA signal in base-band. The validity of the proposed method is proved by the simulation results. Even if the correlated jamming is much stronger than the DS-CDMA signal, the method is still very efficient.

2 BLIND SEPARATION BETWEEN CORRELATED JAMMING AND DS-CDMA SIGNAL IN BASEBAND

2.1 The method of correlated jamming suppression in baseband

Figure 1 is the system chart of the blind separation between the correlated jamming and the DS-CDMA signal in base-band.

K users' DS-CDMA signals and one correlated jamming which are linearly mixed instantaneously through a wireless channel are mainly considered here. s1(t), s2(t),..., sK(t) are K users' DS-CDMA signals whose information sequence are s1b(t), s2b(t), ..., sKb(t). The pseudo noise sequence of the kth user's DS-CDMA signal can be written as

$$pn_k(t) = \sum_{p=1}^{P} c_{k,p} \cdot g_T(t - pT_c) \qquad (1)$$

where $c_{k,p}$ is the pth chip code of the kth user. As a correlated jamming, $J(t)$ uses the same carrier frequency and the same pseudo noise sequence as the object user $s_v(t)$, $1 \le v \le K$.

To implement the ICA algorithm, two antennae are needed at the receiver end. Two received signals from these two antennae could be represented as

$$\begin{cases} r_1(t) = b_{11}s_1(t) + \cdots + b_{1K}s_K(t) + b_{1K+1}J(t) + n_1(t) \\ r_2(t) = b_{21}s_1(t) + \cdots + b_{2K}s_K(t) + b_{2K+1}J(t) + n_2(t) \end{cases} \qquad (2)$$

Then (2) could be written in a matrix form

$$r(t) = \mathbf{B} \cdot \begin{bmatrix} s(t) \\ J(t) \end{bmatrix} + n(t) \qquad (3)$$

where $\mathbf{B}$ is the mixing matrix, for the simplicity, the classical model of linear instantaneously mixture is mainly considered here. $J(t)$ is the correlated jamming and $s(t)$ are the DS-CDMA signals

$$s(t) = [s_1(t), s_2(t), \cdots, s_K(t)]^T \qquad (4)$$

$[\bullet]^T$ means the transpose of matrix. When demodulated and despreaded by the vth user's pseudo noise sequence, two recovered baseband signals could be represented as

$$\begin{cases} r_{1b}(t) = r_1(t) \cdot z(t) \\ r_{2b}(t) = r_2(t) \cdot z(t) \end{cases} \qquad (5)$$

where $z(t)$ is proportional to $\cos(\omega_c t) \bullet pn_v(t)$, with the influence of low pass filter.

In DS-CDMA system, the pseudo noise sequences of different users are uncorrelated, when (3) is imported, (5) turns into

$$\begin{cases} r_{1b}(t) \approx \tilde{b}_{1v}s_{vb}(t) + \tilde{b}_{1K+1}J_{cb}(t) + n_1(t)z(t) \\ r_{2b}(t) \approx \tilde{b}_{2v}s_{vb}(t) + \tilde{b}_{2K+1}J_{cb}(t) + n_2(t)z(t) \end{cases} \qquad (6)$$

which could be written in a matrix form

$$r_b(t) = \begin{bmatrix} r_{1b}(t) \\ r_{2b}(t) \end{bmatrix} = \tilde{B} \cdot \begin{bmatrix} s_{vb}(t) \\ J_b(t) \end{bmatrix} + \tilde{n}(t) \qquad (7)$$

362

From the analysis above, the statistical distribution of svb(t) and Jb(t) are quite different and they remain independent in fact. So (7) is just the mathematical model of ICA with noise, it is appropriate for ICA to separate the useful signal and correlated jamming in baseband.

2.2 Calculation of the unmixing matrix

Under the assumption of statistical independent, the calculation of unmixing matrix **W** could be regarded as a dynamic optimizing process with certain object function, which is depicted in Fig.2.

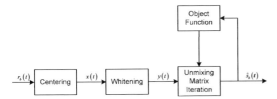

Figure 2. Flow chart of ICA algorithm.

The received signal rb(t) is not suitable to be processed by ICA directly, it should go through two preprocessing steps, centering and whitening[18]. Then y(t) is the stochastic variable of unit variance and zero mean. The unmixing matrix W could be estimated by maximizing certain object functions. From the central limit theorem, non-Gaussianity could be chosen as the object function of ICA. In (Hyvärinen & Oja 2000) negentropy and High Order Statistics (HOS) are used to measure the non-Gaussianity of the separated signal. From our experience, ICA based on negentropy is more robust than that of HOS. So the ICA algorithm with negentropy is adopted in this paper to achieve the blind separation between the correlated jamming and the DS-CDMA signal in base band.

Negentropy could be represented as

$$f(y) = H(y_g) - H(y) \qquad (8)$$

where $H(y)$ is the differential entropy and y_g is the Gaussian stochastic variable with the same power as y. From the knowledge of information theory, negentropy is non-negative, and it is zero if and only if y is a Gaussian stochastic variable. Negentropy can be approximated by nonlinear functions (Hyvärinen & Oja 2000),

$$f(y) \approx \sum_{i=1}^{p} k_i \left[E\{G_i(y)\} - E\{G_i(v)\} \right]^2 \qquad (9)$$

The demand on $G(\cdot)$ is not growing too fast, two functions are recommended in (Hyvärinen 1999)

$$G_1(u) = \frac{1}{a} \log \cosh(au), G_2(u) = -\exp(-u^2/2) \quad (10)$$

where $1 \leq a \leq 2$ is some suitable constant. The derivatives of the functions in (10) are (Hyvärinen et al. 2001)

$$g_1(u) = \tanh(au), \quad g_2(u) = u \exp(-u^2/2) \qquad (11)$$

Then the fixed point ICA algorithm based on the negentropy calculates the unmixing matrix through the following steps (Hyvärinen 1999, Hyvärinen & Oja 2000, Hyvärinen et al. 2001)

step1. Let k = 1, k is the row index of the unmixing matrix

step2. Choose an initial random unit vector $\vec{w}_k$

step3. Update the unmixing vector $\vec{w}_k$ in the fixed point direction of negentropy

$$w_k = E\{xg(w^T x)\} - E\{g'(w_k^T x)\} w_k \qquad (12)$$

step4. Normalize $\vec{w}_k$ by Gram-Schmidt orthogonalization procedure

$$\vec{w}_k = \vec{w}_k - \sum_{i=1}^{k-1} (\vec{w}_k^T \vec{w}_i) \vec{w}_i, \vec{w}_k = \vec{w}_k / \|\vec{w}_k\| \qquad (13)$$

step5. If $\vec{w}_k$ does not convergence, go back to step3
step6. If $k < K$, $k = k+1$, go back to step2; else, iterative ends.

After the steps above, the unmixing matrix $W \approx \tilde{B}^{-1}$ is calculated out. When $r_b(t)$ is multiplied by W, the estimation of source signal $\hat{s}_{vb}(t)$ could be acquired accurately.

$$\mathbf{W} \cdot r_b(t) \approx \begin{bmatrix} s_{vb}(t) \\ J_b(t) \end{bmatrix} + \tilde{B}^{-1} \cdot \tilde{n}(t) \qquad (14)$$

3 NUMERICAL SIMULATION

Several numerical experiments are completed in this section to testify the performance of the correlated jamming suppression receiver. FastICA algorithm is adopted in numerical simulation for its good computational efficiency and fast convergence speed.

The complex situation of six asynchronous users and one correlated jamming are considered in the experiments. The object of correlated jamming is randomly chosen from total six users in each simulation. The attenuation factor of each user, $0 < a_k < 1$, is randomly chosen. The delay factors, τ_k, are uniformly randomly chosen from $\{1, 2, \cdots, P\}$, $P = 63$ is the period of the pseudo noise sequence. It is supposed that the signal of 5000 bits length is acquired at the receiver part. No coding is considered in the following experiments, and thus the results are expressed in raw Bit Error Rate (BER). The following measure quantities are the averaged results over 1000 Monte Carlo experiments. None channel code is adopted in the simulations. The BER performance of the received signal is affected by both SNR and SIR. The relationship between BER of the received signal with SNR and SIR is revealed in Figue 4. The minimum BER value of $r_{1b}(t)$ and $r_{2b}(t)$ is counted in the experiment.

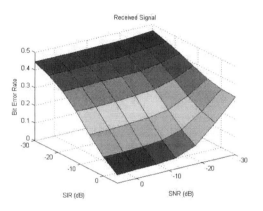

Figure 4. Relationship of the received signal between BER with SNR and SIR.

In Figure 4, the BER surface gets its minimum value, only when noise and jamming are both very weak. With the decrease of SNR or SIR, the BER increases greatly. Figure 5 reveals the relationship between BER of the separated base band signal with SNR and SIR.

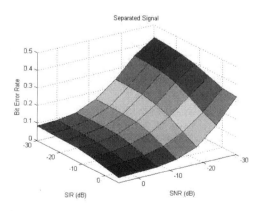

Figure 4. Relationship of the separated signal between BER with SNR and SIR.

Seen from Figure 5, the BER of $\hat{s}_{k,b}(t)$ increases with the decrease of SNR only. When the strength of noise is fixed, BER hardly changes with SIR. When SIR = −30dB and SNR = 6dB, BER of the received signal in Figure 4 is 0.4467, while BER of the separated signal from ICA under the same condition in Figure 5 is only 0.0838. Compared with Figure 4, the advantage of the proposed method is very obvious when the strength of correlated jamming is much higher than the victim DS-CDMA signal.

4 CONCLUSIONS

According to the independence between the correlated jamming and the DS-CDMA signal in base band, a novel method of correlated jamming suppression for DS-CDMA signal is proposed in this paper. The method utilizes independent component analysis algorithms to separate the DS-CDMA signal and the correlated jamming in baseband, which improves the BER performance of the victim DS-CDMA signal greatly. The validity of the proposed method is proved by numerical simulation results. The proposed method is robust to the jamming signal, whose separation performance hardly changes with the change of SIR. Even when the power of the correlated jamming is much higher than that the victim DS-CDMA signal, the proposed method is still very efficient.

ACKNOWLEDGEMENTS

This work is supported by the National Natural Science Foundation of China (No. 61179006) and the Natural Science Foundation of Jiangsu Province (No. BK20141068).

REFERENCES

Lee, J. S. & L. E. Miller (1998). *CDMA systems Engineer Handbook*. Artech House, USA:New York.

Medley, M., G. J. Saulnier & P. Das (1994). Radiometric detection of direct-sequence spread spectrum signals with interference excision using the wavelet transform. *SuperComm/ICC'94*, 1648–1652.

Polydoros, A. & C. Weber(1985). Detection performance considerations for direct-sequence and time-hopping LPI waveforms. *IEEE Journal on Selected Areas in Communications.5*, 727–744.

Jin, Y. & H. B. Ji (2005). A cyclic-cumulant based method for DS-SS signal detection and parameter estimation. *IEEE MAPE 2005. 2*, 966–969.

Hill, D. A. & J. B. Bodie (2001). Carrier detection of PSK signals. *IEEE Trans. on Communications. 49*, 487–496.

Yu, M., S. J. Li, H. Feng, & Z. M. Yang (2008). Blind detection and parameter estimation of multiuser and multipath DS-CDMA signal using cyclostationary statistics. *wicom2008*(2008).

Burel, G., C. Bouder, & O. Berder (2001). Detection of direct sequence spread spectrum transmissions withoutprior knowledge. *Globecom'01*. *1*, 1–5. Adams, E. R., M. Gouda, & P. C. J. Hill (1998). Statistical techniques for blind detection & discrimination of m-sequence codes in DS/SS systems. *IEEE 5th International Symposium on Spread Spectrum Techniques and Applications*. *3*, 853–857.

Nzèza, C. N., R. Gautier, & G. Burel(2004). Blind synchronization and sequences identification in CDMA transmissions. *Milcom2004*. *3*, 1384–1390.

Shen, L., S. J. Li, Y. B. Wang, & F. N. Chen (2007). Blind estimation of pseudo-random sequences of spread spectrum signals in multi-paths. *Journal of Zhejiang University (Engineering Science)*. *41*, 1828–1833.

Torrieri, D. J. (1989). Fundamental limitation on repeater jamming of frequency-hopping communications, *IEEE Journal on selected areas in communication*. *7*, 569–575.

Lathauwer, L. D., B. D. Moor, & J. Vandewalle (1996). Independent component analysis based on higher-order statistics only. *Proc. IEEE SSAP workshop*, 356–359.

Cardoso, J. F. (1997). Infomax and Maximum Likelihood for Blind Source Separation. *IEEE Signal Processing Letters*. *4*, 112–114.

Hyvärinen, A., & E.Oja (1997). A fast fixed point algorithm for indepdent component analysis. *NeuralComputation*. *9*, 1483–1492.

Ristaniemi, T., & J. Joutsensalo (1999). On the performance of blind source separation in CDMA downlink. In Proc. *Int. Workshop on Independent Component Analysis and Signal Separation (ICA'99)*, 437–441, Aussois, France.

Hyvärinen, A., & E.Oja (2000). Independent component analysis: algorithms and applications. *Neural Networks*. *13*, 411–430.

Hyvärinen, A. (1999). Survey on independent component analysis. *Neural Computing*. *2*, 94–128.

Hyvärinen, A., J. Karhunen, & O. Oja (2001). Independent component analysis. *John Wiley & Sons*, United States: New York.

Electronic Engineering and Information Science – Wang (Ed.)
© *2015 Taylor & Francis Group, London, ISBN: 978-1-138-02772-5*

Research of Data Mining based on clustering model

Y. M. Wang
Faculty of Management and Economics, Kunming University of science & Technology, Kunming, China

H. L. Yin
School of Computer Science and Information Technology, Yunnan Normal University, Kunming, China

ABSTRACT: In recent decades, with the rapid development of network technology and information technology, the accuracy of the information people master is higher and higher. At the same time, the output of information is also increasing, and much of the information is useless for people, or even spam. How to find the useful information we really need has become an urgent problem to resolve for the majority of computer workers. The clustering model in Data Mining is an important topic in current researches. It is extremely helpful for us to find deeper information, and it can also be used as an important step in the process in Data Mining – pre-processing step to provide valuable assistance for people's decision.

KEYWORDS: Genetic Algorithm; Clustering Analysis; Data Mining.

1 INTRODUCTION

What is Data Mining? What is the purpose of Data Mining? The answers are different, but the goal of Data Mining we all agree is extracting the useful and relevant information for people from the vague, massive, dubious and limited data, no matter which kind of methods it uses (Jain 1988, Mitra 2002).

Data mining is a crosscutting discipline instead of a simple technology, and its development is influenced by the constraints of multiple disciplines. These subjects include traditional subjects such as Statistics and Mathematics, new subjects appeared in recent decades, such as Artificial Intelligence and Pattern Recognition, and also new technology with the continuous development of computer such as Database technology, Visualization technology, Information technology and other technical disciplines. Figure 1 shows the whole process of Data Mining (Shaw 2001).

Figure 1. Whole process of Data Mining.

Here we describe in details the action each step plays in the Data Mining process in Figure 1:

- Cleanness and Integration. It is a preliminary investigation of data in the database, cleaning out the obviously irrelevant data and sending the remaining data are being combined into the data warehouse.
- Selection and Exchange. The data in the data warehouse are further converted into pending data set, which can make the next step of Data Mining facilitate.
- Data Mining. This is a very important step for us to find useful information, arranging data in patterns through the use of various intelligent algorithms.
- Assessment and Expression. Find out the real useful information for people from the models by assessing and representing it in a reasonable manner, which is called knowledge.

Data Mining is mainly divided into two functions from the current applications. One is the description of the data that is used to understand and interpret data. The other is the prediction/verification, which is mainly used to predict the data and give a rough description of the unknown property values of the data. Figure 2 clearly describes the current use of Data Mining functions.

2 CLUSTER ANALYSIS

Each class of cluster analysis has a specific nature. There are three general forms to indicate the natures of the classes: class center, visualize cluster nodes in the fast search tree and logical expression of the sample properties.

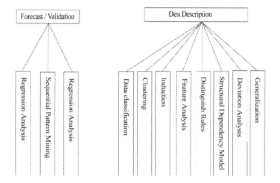

Figure 2.   Current use of Data Mining functions.

2.1 Main ideas

Here gives a data set, which contains n objects. We divide this data set into k small data sets, which is called a small cluster. It should be noticed that the division is not just simply dividing the data set on the principle of equal portions, and each small data set does not necessarily contain a n/k of the object. The data set is divided in accordance with clustering technology, and k is less than or equal to n. Each cluster needs to meet the following two conditions:

- A cluster contains at least one object.
- Each object in the data set must belong to a cluster, and it cannot belong to other clusters at the same time.

Here we need to mention in particular that: we need to send the object into different clusters through iterative re-positioning technology. Objects in the same cluster should be related to each other or as similar as possible while objects in the different clusters should be as different as possible.

2.2 Evaluation functions

We have to evaluate its merits after the division of a large data set. It general references to the two main aspects: One is whether the similar degree of the objects in the same cluster is in the standard level; the other is the difference of the objects in different clusters should be large enough. Dai (2008) has introduced the evaluation function for how to find a reasonable assessment of these two aspects, giving a scientific answer with a function.

The easiest way for the degree evaluation of the objects in the same cluster is calculating the distance from each object to the cluster center in the data cluster. The formula is as the following:

$$w(C) = \sum_{i=1}^{k} w(C_i) = \sum_{i=1}^{k} \sum_{x \in C_i} d(x, \overline{x_i})^2 \qquad (1)$$

The difference between the clusters is evaluated by the distance of their centers. The formula is as the following:

$$b(C) = \sum_{1 \le j < i \le k} d(\overline{x_i}, \overline{x_j})^2 \qquad (2)$$

Results of W(c) and b(c). There are many methods to make the comparison, and the simplest is to get their ratio, which is B(c)

The overall effect of the division can be got from the comparison of the /W(c).

3 PAM ALGORITHM

This is a division method around the center of the clusters. The principle is that we should firstly select k objects randomly as the centers of the division for the data set. Then we analyze all objects in the data set and find out the best k objects at the center point through constant iteration. We use capital letters O standing by the object, and O_i standing by a particular object. The following four cases may occur in the dealing process.

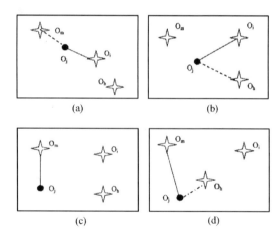

Figure 3.   Four cases of PAM algorithm.

The first case: Oi is the center of the cluster, and Oj is in this cluster. After the iterations, if Oi is replaced by the new center Oh. Then, after the calculating, Oj may be closer to another center Om, so Oj will be re-assigned to the cluster that has Om as its center.

The second case: Oi is the center of the cluster, and Oj is in this cluster. After the iterations, if Oi is replaced by the new center Oh. Then, after the calculating, Oj may be closer to another center Oh, so Oj will be re-assigned to the cluster that has Oh as its center.

368

The third case: Om is the center of the cluster, and Oj is in this cluster. After the iterations, if Om is not replaced, and there appear other two centers Oh andOi. Then, after the calculating, Oj may be still closest to Om, so Oj will not be re-assigned.

The fourth case: Om is the center of the cluster, and Oj is in this cluster. After the iterations, if Om is not replaced, and there appear other two centers Oh and Oi.Oj may be closer to another center Oh, so Oj will be re-assigned to the cluster which has Oh as its center.

Figure 3 describes clearly the four cases of PAM algorithm in continuous iterative process with graphical form.

4 DESIGN AND IMPLEMENTATION OF DATA MINING SYSTEM

It is a very complex process to design a Data Mining system. There are so many algorithms for Data Mining, and we can't say surely that which one is not good and which one is good, because the methods of data collection and analysis are not the same in different industries (Meo 1998 Zimmermann & Surcka 2008).

The objectives of the Data Mining systems are consistent, no matter which type they are. So we can design a suitable architecture for all users. However, users need to conduct the internal specific algorithms according to their own choices and designs. In the data mining system, there are three most important aspects as the database connection configuration, data acquisition and data conversion, besides the data mining feature(Wang 2013). Figure 4 shows an architecture diagram of Data Mining system.

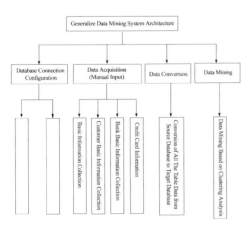

Figure 4. Architecture diagram of Data Mining system.

Database connection configuration is an essential function in a data mining system. Almost all the programming languages use database connectivity, and

many papers and books also describe it. Here, we no longer describe it as it is relatively simple. We mainly introduce the data mining algorithm and data collection in the follows.

Here, we give the main core code based on a PAM clustering analysis algorithm talked about before. C# is used to implement the function of clustering analysis algorithm.

Private void InitJlmatrix()//Initialize the distance matrix

```
{
This.Jlmatrix=new double[this.Mydistance.Tables
[0].Rows.Count;
This.Mydistance.Tables[0].Rows.Count];
Int j;
Int i;
for(i=0;i<=this.Mydistance.tables[0].Rows.
Count-l; i++)
{
For(j=0;j<=this.Mydistance.tables[0].Rows.
Count-l;j++)
{
This.Jlmatrix[i,j]=Convert.ToDouble(this.
Mydistance.tables[2].Rows[i][j]);
}
}
}
```

In the data mining system, it plays an important role how to collect the data. So here we design a class for data collection. Readers can use the corresponding function as long as instantiate it in the application process. It will reduce the work of the Development. The specific code of the class for data collection is as follows:

```
ClassShouJiShuJu:
Public void BaoCunShuJu()
{
If (this.CunshuKu.Equals(string.empty)
{
This.CunshuKuMingZi=this.genXinBiaoMingZi;
}
Else
{
This.CunshuKuMingZi=this.BaocunObjectMain;
}
}
Public void ShanchuShuJu(object MainID)
{
This.genxinStatus(MainID,int.Mainzhi,-1);
}
```

5 CONCLUSION

This paper first analyzes the Data Mining, highlighting the clustering patterns and giving a detailed analysis of PAM. Finally, it gives a structured design of the Data Mining system and code of the initialization for the Matrix. The code of data collection base is realized with C # language. It does not give the specific code of the data conversion function and other functions due to the limited space, so readers can write the code by themselves based on the description of the algorithm.

ACKNOWLEDGMENT

The work is supported by Natural Science Foundation of China (NO. 71262029, NO. 71362030, NO. 71362025, NO. 71362024), Natural Science Foundation of Yunnan (NO. 2013FB029, NO. 2013FZ048). Foundation for Food Safty of Yunnan Province (NO. 2013SY05).

I would express special thanks and gratitude to reviewer and editor for their efforts in reviewing and correcting this paper manuscript.

REFERENCES

Dai, S. P. & X. J. Zhang(2008). A Quantum Genetic Algorithm for data mining on E-business, *7th Wuhan International Conference on E-Business,* 1046–1050.

Jain, A. K. & R. C. Dubes(1988). *Algorithms for Clustering Data,* NJ: Prentice-Hall.

Mitra, S.,S.K. Pal, &P. Mitra(2002). Data mining in soft computing framework: A survey, *IEEE Transactions on Neural Networks,* 3–14.

Meo, R., G. Psila, & S. Ceri(1998). A tightly-coupled architecture for data mining, IEEE Computer-Society *14th International Conference on Data Engineering.* 316–323.

Shaw, M. J., C. Subramaniam, & G. W. Tan(2001). Knowledge management and data mining for marketing, *Decision Support Systems,* 127–137.

Surcka, A., & K. V. Indukuri(2008). Using Genetic Algorithms for Parameter Optimization in Building Predictive Data Mining Models", *Advanced Data Mining and Applications, Proceedings,* 5139:260–271.

Wang, Y.M., & H.L. Yin(2013). A Novel Genetic Algorithm for Flexible Job Shop Scheduling Problems with Machine Disruptions, *International Journal of Advanced Manufacturing Technology, Springer London ,* 1317–1326.

Zimmermann, H. J., "Knowledge management, knowledge discovery, and dynamic intelligent data mining", *Cybernetics and System,:* 509–531.

Electronic Engineering and Information Science – Wang (Ed.)
© 2015 Taylor & Francis Group, London, ISBN: 978-1-138-02772-5

Influences of different ways of communication over team innovation efficiency

Y. Zhao & L. F. Zhang
Zhejiang University of Technology, Hangzhou, China

X.X. Qiao
Zhejiang University of Technology, Hangzhou, China

ABSTRACT: Design communication is an important means of resource coordination in product concept design. This paper focuses on the influences of different ways of communication over design innovation efficiency, divides the innovation process into three stages of design analysis, idea generation and solution decision for study, proposes eight hypotheses, conducts a scientific situational experiment participated by 42 industrial designers, analyzes the statistics of the experiment outcome and verifies the hypotheses. The experiment verification outcome is of certain instructive significance to the improvement of the team innovation efficiency.

KEYWORDS: Design communication, communication efficiency, concept design, innovation evaluation.

1 GENERAL INSTRUCTIONS

In terms of information, the design innovation process concentrates on the collection and analysis of design information, including collection of demand information at the pre-design stage, opinions generated in the analysis process, visual design, information formed in the sketch conception, solution decision and discussed skills, man-machine and feasibility information in the evaluation. This study is aimed to explore influences of different ways of communication over the design innovation efficacy in the form of experiment, analyze the innovation achievements at different stages through comparison, and study how ways of communication influence the design innovation progress and the influences of design innovation achievements.

2 THEORETICAL BASIS AND RESEARCH HYPOTHESIS

2.1 *Relations between different ways of communication and innovation efficacy at design analysis stage*

The design analysis stage is the process where the design team discusses and analyzes the design tasks to eventually reach consistent design goals. With the approach of objective measurement, this paper categorizes design report completed by subjects at the

first stage, including target groups, operating environment, demand analysis, behavioral analysis, market analysis and design orientation. Then key words corresponding to each category are selected and effective information about each subject is gathered. Hence, this study proposes the following hypotheses: H1a: Different ways of communication cause significant differences in communication efficacy of the design team. H1b: Face-to-face communication generates more effective information than instant messaging does.

2.2 *Relations between different ways of communication and innovation efficacy at idea generation stage*

All sketches in the process will be deemed as the measurement objects during the design information review. Studies show that obvious differences exist in the sketch interactions between close distance and long distance. Sketches can be better completed with face-face communication. Instant messaging usually leads to inadequate interaction and even zero interaction. However, some studies demonstrate that instant messaging is better than face-to-face communication for the innovative task of "brainstorming". As a result, the following hypotheses are put forward: H2a: Different ways of communication cause significant differences to the number of sketches completed by the design team. H2b: Different ways

371

of communication cause significant differences in refinement degree of sketches.

2.3 *Relations between different ways of communication and innovation efficacy at solution decision stage*

According to previous studies, face-to-face communication is obviously more effective than instant messaging [7]. In this study, design innovation evaluation is regarded as the standard for measuring design innovation achievements. As a result, the following hypotheses are put forward:

H3a: Different ways of communication cause significant differences to the originality and feasibility of the design team's final solution.H3b: Different ways of communication cause significant differences to the feasibility of the design team's final solution.

3 EXPERIMENTAL PROCESS AND DATA ANALYSIS VERIFICATION

3.1 *Experiment sample*

The subjects of the study are industrial design juniors of a university who have taken the same design courses for training. They possess the basic ability of freehand sketching, learn about the product development process and have made design experiments themselves. There are 42 subjects in total including 18 boys and 24 girls with an average age of 22 years old, and are randomly allocated into 14 groups, each of which includes three subjects: seven groups of face-to-face communication numbered A and seven groups of instant messaging numbered B. See Table 3-1. The samples in the experiment are grouped as follows:

Table 3-1. Sample grouping.

Experimental Group	No.
Face to face group（FTF）	A0 A1 A2 A3 A4 A5 A6
Instant messaging group（IM）	B0 B1 B2 B3 B4 B5 B6

3.2 *Experiment process*

Phase I: The time duration is set: 35m for FTF group and 45m for IM group. Discussion of the design tasks follows such train of thought: crowd → environment → demand→ behavior → market (products of the same kind) → orientation (function and structure). The survey results for user demand are added at the earlier stage so as to make discussions consistent and not to deviate from the focus.

Phase II: The duration for FTF group is changed to 35m and 50m for IM group. Serial numbers and text description are required in sketches. Team members take photos of sketches and upload them to the discussion group to prepare for the next step.

Phase III: Interaction is made about sketches with the help of file transfer and screenshot functions of instant messaging. Members give suggestions, but the final decision is up to the designers and must be accompanied by design description.

3.3 *Statistical approach of the experiment*

The creative design is divided into three parts in the entire study so different statistical approaches are adopted for each part:

a. Objective measurement is employed at the design analysis stage where each subject needs to complete a discussion report which covers detailed design orientation, design goal and design strategy. Key words are collected. The greater the matching between key words and design orientation, the better the information quality. Quantity means the number of key words of better information quality, that is, quantity of effective information.

b. At the idea generation stage, subjects are required to complete the creation sketches within the scheduled time, further express functions and structures that meet with requirements and will eventually finish a number of sketches. Then, each solution is scored in terms of refinement degree and sketch details with the method of Likert's five-point scale, and total scores are calculated.

c. At the solution evaluation stage, the final solution is finalized as long as it's clearly delivered. Contents mainly cover creation subject, interpretation of the body and symbols and text description.

The evaluation goes like this: three designers with over five years' experience of product design are invited to score according to random sequence and they don't know corresponding participants for relevant solutions. Likert's five-point scale is adopted: 1 means completely inconsistent; 2 represents inconsistent; 3 indicates noncommittal and 4 means consistent while 5 means completely consistent. The final solution is evaluated from two perspectives of originality (innovation) and feasibility (practice). (Finke, 1990 Goldschmidt and Smolkov, 2006).

3.4 *Analysis and verification of the experiment outcome*

Quantity of effective information in the design report completed by each subject at the first stage is calculated. 36 effective samples are obtained after removing samples that are not in line with experimental requirements, including 18 from FTF group and 18 from IM group. The statistics are as follows: it can be seen from table 3-2 that the quantity of effective information from FTF group is greater than IM group.

Table 3-2. Basic description of effective information quantity of FTF and IM groups.

	Experimental Group	Sample Number	Mean Value	Standard Deviation	Std.Error
The quanity of Effective information	FTF	18	16.52	1.602	0.377
	IM	18	15.44	1.753	0.413

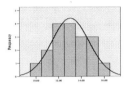

Figure 3-3. Frequency distribution of FTF group's effective information quantity.

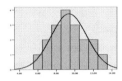

Figure 3-4. Frequency distribution of im group's effective information quantity.

Fig. 3-3 and Fig. 3-4 respectively demonstrate the frequency data of total effective information of the two groups. It can be noted from the table that the two groups' data distribution basically takes on a normal trend and the average value of FTF is greater than IM. Independent-sample t-test is conducted on the two groups in SPSS 13.0 to examine whether data differences between the two groups are worth calculating or not. The test result is as follows:

Table 3-5. Result of independent-sample t-test on effective information quantity.

	Levene's Test		t-test for Equality of Means		
	F Value	Sig.Value	t Value	Sig. Value	Mean Difference
The quanity of Effective information	0.570	0.456	6.194	0.000	3.66667

Levene's Test shows that F is 0.570. In terms of significance probability, when p>0.05, it means there's no significant difference between the two groups at the variance of $\alpha=0.05$, that is, the variance is homogeneous. $\alpha=0.05$ is adopted. After the analysis

of data in the t-test comparison table, it can be seen from the outcome (table 3-5) that significance differences are displayed in the two groups' data, t (34) =6.194, p<0.001 (two-tailed test). In other words, the effective information quantity produced from FTF is significantly higher than IM within scheduled time. Moreover, different outcome results from the communication process of the two groups. t =6.194, P=0.000, P<0.05. H1a is accepted at the α level. Moreover, P<0.001 and differences are extremely significant so H1b is accepted as well.

a. Analysis of sketch quantity. According to the standard, the creation and feasibility of concept sketchs related with the design theme are not considered. Table 3-6 shows the basic data on the variable number of the two groups' sketchs. It can be seen from the statistics result that the number of sketchs completed by FTF group is greater than IM group within scheduled time (respectively M=2.61 SD=0.45; M=2.11 SD=0.83).

Table 3-6. Statistics result of the sketch quantity.

	Experimental Group	Sample Number	Mean Value	Standard Deviation	Std.Error
Sketch quantity	FTF	18	2.61	0.849	0.200
	IM	18	2.11	0.832	0.196

Independent-sample t-test is conducted on the two groups' sketch quantity in SPSS 13.0. According to the Levene's Test result (table 3-7), the sketch quantity of the two groups F is 0.521 and the significance probability p=0.475, which means there's no significant difference between the two groups at the variance of $\alpha=0.05$, that is, the variance is homogeneous. $\alpha=0.05$ is adopted. The data in the table (table 3-6) is calculated with independent-sample t-test, the outcome (table 3-7) shows that there's no obvious difference between the two groups' sketch quantity within scheduled time. t (34) =1.783 p=0.083 (two-tailed test), Analysis of the verification: Unproven H2a.

Table 3-7. Result of independent-sample t-test on sketch quantity.

	Levene's Test		t-test for Equality of Means		
	F Value	Sig.Value	t Value	Sig. Value	Mean Difference
Sketch quantity	0.521	0.45675	1.784	0.083	0.50000

b. Evaluation of the sketch refinement degree. Sketches of each subject completed within scheduled time are gathered to score the sketch refinement based on relevant evaluation standards. Table 3-8 showcases the basic data on the quality of completion. On the whole, the score for the refinement degree of FTF group is higher than

IM group within scheduled time. (respectively M=27.89, SD=8.17; M=22.33, SD=6.43).

Table 3-8. Basic data on sketch quality score at the second stage.

	Experimental Group	Sample Number	Mean Value	Standard Deviation	Std.Error
Sketch quality	FTF	18	27.89	8.174	1.927
	IM	18	22.33	6.426	1.515

According to Levene's test result (see table 3-9), F is 0.622. In terms of significance probability, when p>0.05, it means there's no significant difference between the two groups at the variance of α=0.05, that is, the variance is homogeneous. α=0.05 is adopted. After the analysis of data in the t-test (table 3-8) comparison table, it can be seen from the outcome (table 3-9) that t (34) =2.267, p=0.03 (two-tailed test). The sketchy quality of FTF group is obviously better than that of the IM group within scheduled time. Analysis of the verification: Proven H2b.

Table 3-9. Result of independent-sample t-test on sketch quality.

	Levene's Test		t-test for Equality of Means		
	F Value	Sig. Value	t Value	Sig. Value	Mean Difference
Sketch quality	0.622	0.436	2.267	0.030	2.45068

Evaluation of originality and feasibility of solutions. In order to further analyze details, contents are evaluated in two dimensions of originality (innovation) and feasibility (practice). First, the total of each expert's score is calculated for the two groups and then the average score of each subject is concluded in terms of originality and feasibility. Statistical analysis is further made for the two factors. Table 3-10 is about basic data on evaluation of originality. On the whole, the FTF group has greater originality than the IM group (respectively M=13.87, SD=1.65; M=12.24, SD=1.48).

Table 3-10. Basic data on originality evaluation.

	Experimental Group	Sample Number	Mean Value	Standard Deviation	Std. Error
Originality Evaluation	FTF	18	13.87	1.654	0.390
	IM	18	12.24	1.481	0.349

According to Levene's test result (see table 3-11), F is 0.216. The significance probability is 0.645 and p>0.05, which means there's no significant difference between the two groups at the variance of α=0.05, that is, the variance is homogeneous. α=0.05 is adopted. After the analysis of data (table 3-11) in the independent-sample t-test, it can be seen from the outcome (table 3-10) that the originality of FTF group is obviously greater than

IM group. t (34) =3.115, p=0.004 (two-tailed test). Analysis of the verification: Proven H3a.

Table 3-10. Result of independent-sample t-test on originality.

	Levene's Test		t-test for Equality of Means		
	F Value	Sig. Value	t Value	Sig. Value	Mean Difference
Originality	0.216	0.645	3.115	0.004	1.62978

b. See table 3-11 for basic data on feasibility evaluation of solutions. On the whole, the feasibility (practice) degree of FTF group is higher than IM group (respectively M=16.52, SD=1.60; M=15.44, SD=1.75).

Table 3-11. Basic data on feasibility evaluation.

	Experimental Group	Sample Number	Mean Value	Standard Deviation	Std. Error
Feasibility	FTF	18	16.52	1.602	0.377
	IM	18	15.44	1.753	0.413

According to Levene's test result (see table 3-12), F is 0.139. The significance probability is 0.667 and p>0.05, which means there's no significant difference between the two groups at the variance of α=0.05, that is, the variance is homogeneous. After the analysis of data (table 3-11) in the independent-sample t-test, it can be seen from the outcome (table 3-15) that there is no significant difference between the two groups' feasibility (practice). t (34) =1.919, the significance probability is 0.063 (two-tailed test) and p>0.05. Analysis of the verification: Unproven H3b.

Table 3-12. Result of independent-sample t-test on feasibility.

	Levene's Test		t-test for Equality of Means		
	F Value	Sig. Value	t Value	Sig. Value	Mean Difference
Feasibility	0.139	0.667	1.919	0.063	1.07411

4 EXPERIMENT CONCLUSION

In accordance with the analysis above, the experimental outcome shows that the information analysis degree and information extraction rate of the design team are influenced by interaction and richness of communication, thus affecting the decisions on the final representative solution. Face-to-face communication can generate fuller information to share and higher information extraction rate, and facilitates the transformation of concept sketches into final design solutions. The auxiliary functions of instant messaging software are not able to well satisfy the demands for effective communication. Analyzed from the two aspects of originality and feasibility, the influences of ways of communication over innovation are mainly reflected in the solution originality which, in addition to the influences of ways of communication, is related to some extent with the design analysis outcome and the concept of the first two stages. The study outcome in this paper is of favorable, inspiring significance to the follow-up product concept design and in particular, it serves as considerably good reference for the team communication in collaborative innovation design.

REFERENCES

Maher. M. L. (2001).A model of co-evolutionary de sign. *J. Engineering with Computers.* 16(2), 195–208.
Goldschmidt G. Smolkov M. (2006) Variances in the impact of visual stimuli on design problem solving performance. *J. Design Studies.* 27(5), 549–569.
Garner, S. (2001). Comparing graphic actions between remote and proximal design teams. *J. Design Studies.* 22(04):365–376.
Maher M L. (2001).A model of co-evolutionary de sign. *J. Engineering with Computers.* 16(2), 195–208.
Liu Zheng. (2008).Study on the Earlier Development Stage of Product Design in the Matching Process *D. Zhejiang University.*
Menezes A, Lawson B. (2003) How designers per- ceive sketches. *J. Design Studies.* 13(3), 192–200.
Finke, R. Creative imagery.(1990). Discoveries and inventions in visualization. *NJ: Erlbaum, Hillsdale, Mech. 40,* 223–238.

Electronic Engineering and Information Science – Wang (Ed.)
© 2015 Taylor & Francis Group, London, ISBN: 978-1-138-02772-5

Design of controller of automatic variable rate spraying based on ARM9

J.X. Huang, G.Y. Song & L.M. Li
Institute of Information Electronic Technology, Jiamusi University, China

Y. Hou*
College of Clinical Medical Laboratory; Jiamusi University, China

ABSTRACT: This paper, based on the design of the ARM9 automatic variable spraying control system, completed the experimental hardware design, automatic variable spraying controller software design and system. The control system uses GPS (Global Positioning System) to determine the spraying machine seat block grid coordinates, through the spray volume of prescription map information read U disk memory, get the current block required injection dosage. The combination of pressure, velocity and flow rate sensor information, control the executing mechanism, implementation of automatic control in the best condition of variable spraying atomization. Through the analysis of the experimental results show that the actual accuracy meets the requirements of production, has a strong practical value.

KEYWORDS: variable spraying; control system; ARM9; GPS.

1 THE PUTTING FORWARD OF THE PROBLEM

The general level of china's plant protection equipment, for more than a score of years, remained in a relatively low position. Ideal effect can hardly be achieved on nebulization, adhesion and uniformity for spraying, which induces not only the pesticide squander, but the environmental pollution as well. The traditional agricultural sprayer with fixed droplet diameter is unsuitable for the changing working environment according to the factors of density, winds, parts of injury by pest, etc. Then we bring forth a way of variable spraying, which is used to control the dosage and manner of spraying according to the characters of changing working environment, and also is definitely what this paper dealt with—the design of the controller of automatic variable rate spaying.

2 PRINCIPLE OF THE CONTROL SYSTEM

Generating the digital map of plot partition and mesh the plot, before the equipment start working, by using GPS and GIS information. The a real time monitoring for diseases and pests is performed by means of advanced Remote Sencing technoligy, When diseases and pests is found, data collection is done combined with the GPS imformation again ,and the analysis and process is maded out by the decision-making sustain system which is consist of agricultural specialists in Lab, the spraying prescription chart is given for every small grid and is saved to USB flash disk. When the spraying machine, the system first location information to read GPS to the controller, the calculated by calculating the working length and area of operation unit grid spraying machine where the decision application dosage and then queries the corresponding to the operation unit, and the LCD screen real time display of the current position and application amount, at the same time the loop pressure signal read variable spraying system pressure sensor system came to the controller, and provide the prescription map information in the system to achieve the regulation of pressure atomization quantity according to the optimal.

At the same time the speed sensor speed spraying machine signal, system flow signal reading to the controller, the extraction of the spraying machine speed of advance information and system flow of information. Finally, through the spray volume formula to calculate the nozzle solenoid valve required frequency, the final realization of the variable spraying.

Every hectare of crop required amount of spraying process for variable spraying.

$$Q = \frac{60000 \times q}{V \times L} \tag{1}$$

Corresponding author, Address: College of Clinical Medical Laboratory; Jiamusi University.

Where Q—spraying prescription volume per hectare of crop pesticides (L/km2);q—Single nozzle flow (L/min);V—Walking speed spraying machine (km/h);L—Sprinkler spacing (cm).

Based on the formula one, if you keep the system pressure in the nozzle entrance, the solenoid valve is installed, you can change the nozzle flow. Nozzle flow is adjusted by the intermittent on-off solenoid valve, the dosage change open time by electromagnetic valve to control the length of, so that you can achieve a larger range of control variables, and the droplet velocity than traditional index of small changes, based on the pressure change regulation mode fast.

3 THE HARDWARE DESIGN OF CONTROL SYSTEM

The control system consists of the controller, the GPS positioning system, detection device (speed sensor, flow sensor, pressure sensor), executive device (solenoid valve, electric valve); storage equipment, man-machine interface (keyboard, display) and power supply, the system hardware structure as shown in figure 1.

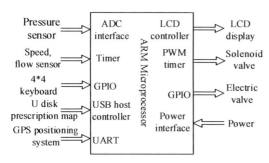

Figure 1. Schematic diagram of the hardware structure of the control system.

3.1 The choice of the microprocessor

The microprocessor is mainly used to receive GPS positioning system spraying machine, transmission speed, position of pressure sensor data and flow sensor data, and real-time processing of the data, the output control information, control devices work. Therefore, the microprocessor should have data processing, real-time, accurate and flexible, able to carry out serial and parallel communication, convenient extension of the keyboard input device, output device display, storage equipment, excellent reliability characteristics. Therefore, this design uses the S3C2410 processor developed by Samsung company as the system control center. It is a 32 bit ARM920T microprocessor

CPU core and 0.18um based on CMOS technology, frequency reaches 203MHZ, the realization of the MMU, AMBA bus and Harvard cache structure, the structure has a 16KB command Caehe and 16KB data Cache independent. The processor has the advantages of low power consumption, high degree of integration of multiple advantages.

3.2 GPS global positioning system

The selection of the purpose of the GPS is to accurately determine the location and spraying machine, automatic variable spraying process using multiple positioning machine, in order to facilitate, accurate block grid map, prescription map, positioning the spraying machine when working on real time range, comprehensive consideration, this design choice of American Trimble company production of AgGPS132 differential receiver as a positioning device, AgGPS132 is used for the decimeter precision agriculture GPS receiver, embedded monitor and keyboard, small volume, low power consumption, strong function, easy to install. GPS interface circuit as shown in figure 2.

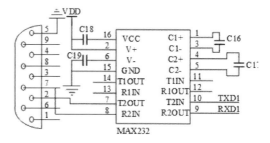

Figure 2. GPS interface circuit.

3.3 Sensors and executive device

The speed of travel speed sensor to collect information to determine sprayer, spraying frequency according to the size of the spray machine speed; the pressure sensor returns the ARM micro controller system pressure, by controlling the return to reach the appropriate pressure to achieve the best effect of atomization. Flow sensor is used to give feedback to the ARM micro spraying amount of the total system controller, electric valve actuator is used to control the system flow, through the control of electric valve opening angle to control the size of the spraying pipe system flow, change the droplet size, the realization of different weather conditions, different spray effect. Electromagnetic valve is installed in the nozzle of the spraying machine, is the actuator control each nozzle opening and closing frequency, pulse width control by PWM, the length of time the opening and closing of each nozzle to control the amount of spraying. Due to the vibration, the noise field in the process of operation are relatively large, considering the accuracy and precision of measurement, this design adopts DPL-A type radar sensors, LW-50 type turbine flowmeter and piezoresistive pressure sensor are respectively to measure the velocity, flow rate and pressure signal; the use of electric valve direct acting solenoid valve and DC0-10V standard voltage signal as the actuator. The circuit in Figure 3, 4 and 5 below its signal conditioning.

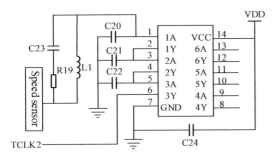

Figure 3.   Conditioning circuit of speed sensor.

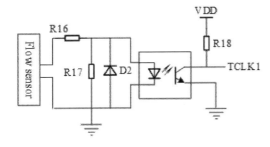

Figure 4.   Flow sensor conditioning circuit.

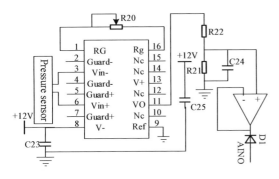

Figure 5.   Conditioning circuit of pressure sensor.

3.4   *The man-machine interactive equipment*

A human–computer interaction device is mainly composed of input and output (keyboard, display) is composed of two parts. The LCD display has the characteristics of small volume, light weight, low power consumption, long service life, at the same time, S3C2410 is integrated with LCD controller, the LCD screen can be very convenient to control various types of Samsung Corp, it is selected in the design of 320 * 240TFT, its model for LTV350QV. Input device selection of cross type 4 * 4 keyboard interface, different keys corresponding to different operation.

3.5   *Power supply*

Power is the power source of the system, as the voltage of spraying machine control center and peripheral equipment to provide the needed, both to meet the different module voltage different problems, but also can reduce the power consumption of the system. The design of direct power source from vehicle 12V DC voltage, and then voltage rectifier, filter, voltage conversion, obtain power supply corresponding components required by the control circuit.

4   SYSTEM SOFTWARE DESIGN

Completed the software design of this system of the main tasks are: to receive GPS positioning system by latitude and longitude of the location information of the serial port to send the prescription map information, read U disk through the USB interface transmission, obtain the transmission pressure, velocity and flow sensor information, feedback of the running state of the system through the man-machine interface, finally complete the control the operation of the solenoid valve, electric valve. Therefore, the application of the whole spraying system can be divided into three modules: one is the information access and processing module, the two is the control module, human-computer interaction module three is.

5   CONTROL SYSTEM TEST

5.1   *Test location and equipment*

This system in Jiamusi University four Feng Shan Nong Chang, maize emergence after the 40 day of the corn of variable spraying test, operation point coordinates (96.45629 degrees E, 29.58643 degrees N), the grid size of 15 m * 15 m, operating area of 1 Hm2, according to the operation block ridge distance adjusting nozzle spacing is 0.55 M. Control spray prescription volume set is given by the agriculture

379

expert system, variable rate spraying prescription as shown in figure 6.

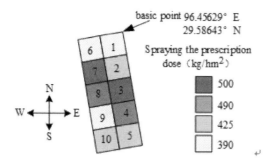

Figure 6.   Prescription map for variable spraying.

5.2   *Test method*

Spraying amount set in advance of the prescription writing U disc. The spraying machine from the point of departure, with 3.5km/h speed along the length of ridge direction, ARM processor receives the GPS information for locating and speed of the machine, read U disk memory put spray prescription volume at the same time, the data of the flow sensor and the pressure sensor detects the signal, control the electric valve and solenoid valve work, completes the automatic control of variable spraying.

5.3   *The results and analysis*

Spraying prescription volume set and measured the amount of spraying prescription data, such as shown in table 1.

Table 1.   Test data and results of control system.

Grid label	Spraying prescription volume set (kg/hm^2)	Spraying prescription measured (kg/hm^2)	Relative error %
1	390	406	4.10
2	425	431	1.41
3	500	487	0.26
4	490	472	3.67
5	425	439	3.29
6	390	408	4.61
7	500	483	3.40
8	490	481	1.83
9	390	403	3.33
10	425	413	2.82

Controlled by the spray volume error test that: 15m * 15 m operations within the grid work, the spraying machine with 3.5km/h speed, spraying the prescription dose in the range of 390 to 500kg/hm2 changes, using the system to complete the variable spraying system operation, maximum error within 5%, show that the system hardware and software design is reasonable.

6   CONCLUSION

The design of the ARM embedded system is applied to the agricultural production process, through the realization of variable spray control system software and hardware design purpose. The design of the control system to improve the utilization rate of the pesticide, reducing pesticide residue and pollution of the environment. The tests show that the control system is reasonable in design, simple in operation, has a potential application prospect in the market.

FUNDED PROJECTS

This work is supported by Plan Project of Youth Foundation of jiamusi University (Lq2013-032); Teaching key scientific research project of Jiamusi University (JKA2013-047).

REFERENCES

Zhao, H.Q., Zhang Y.X.& Chen P.Y. et al. (2006) The problems and measures of our country exists in the use of pesticides in the. *Hebei fruits.6*, 17–19.
Gerhards R,Oebel & H.Practical (2006)experiences with a system for site-specific weed control in arable crops using real-time image analysis and GPS-controlled patch spraying. *Weed Research, ,46(3),*185–193.
Tang G Y,Li J. (2008)Optimal fault diagnosis for systems with delayed measurements.*Control Theory&Applications,* 11, 990–998.
Ma Z.M., Ma halo& Xu Y.H. (2002).*ARM embedded processor architecture and application foundation.* Beijing: Beihang University press, 9~11.
Jia Z.P. (2005). *Embedded system principle and interface technology.* Tsinghua University press.,7.
Shi W.P., X. Wang.(2007) Research on Variable Rate Spraying Technology of agricultural mechanization. *GPS based on GIS.* 2, 19–21.
Qi,J.T., S.H.Zhang,Y.J. & Yu,et al (2009). Development of a Ground Speed Collecting System for the Variable Rate Fertilizer Machine Based on Bluetooth.*Transactions of the CSAM,* 40(1), 200–204.

Electronic Engineering and Information Science – Wang (Ed.)
© 2015 Taylor & Francis Group, London, ISBN: 978-1-138-02772-5

Electrotechnics curriculum reform aiming at cultivating innovation ability

J.X. Huang & Y.Hou
College of Information-electronic Technology, Jiamusi University & College of Clinical Medical Laboratory; Jiamusi University, China

J.Li*
Professor, Institute of Information-electronic Technology, Jiamusi University, China

ABSTRACT: Improving college students' comprehensive quality, practical ability, especially innovation ability, is of important significance, and must be implemented in the specific process of teaching. According to the practical situation of our school and the teaching experience of many years, the course group went on active reform and practice of the content of 《Electrotechnics》 , teaching method and approach, experimental teaching and practical teaching, web-based instruction, and the construction of the teachers' time. The course group has accumulated some profitable experience.

KEYWORDS: Electrotechnics; innovation abilities; teaching reform.

1 INTRODUCTION

Innovation sustains the progress, and is the inexhaustible motive force of a county's prosperity. One of the missions of the college students' education is to make the students have a certain innovation ability. The training of students' innovation ability is the core of the training of college students' integrated qualities, and it sustains the development of the country and the future of the nation. So how to cultivate talents of innovation ability better and faster becomes an important goal of university education.

Electrotechnics is an important technical basic course of enormous quantity of range of the non electricity specialty of higher engineering college. It plays an important role of the link in the whole non electricity specialty talent training programs and curriculum system, with features of fundamentality, applicability and advancement (Wang, 2012). Electrotechnics and all aspects of electronic field are involved in its teaching content, so it is an indispensable and an important part in the knowledge structure of contemporary college students. But in recent years, due to school compression and the constantly emerging of new knowledge and new technology, the contradiction of less class hours and more content is very prominent (Huang, 2008). This contradiction cannot be released, and innovative talents who can meet the need of the society cannot be developed, unless a comprehensive

reform of the content of textbook, teaching method and approach, practical teaching, and the construction of the teachers' team, and so on. For this reason, we intensify the teaching reform of electrotechnics course, in order to strengthen the effect of class-teaching, improve the path of practice, and improve students' comprehensive quality and innovation ability.

2 REFORMING OF CLASS-TEACHING MODE

2.1 *Integrating and updating the teaching content*

Referencing to the teaching basic requirement of electrical engineering courses that made by the Ministry of Education Teaching Steering Committee, and combining with the requirement of the actual situation of each major and follow-up course on electrical engineering courses, and in line with the thought of improving students' comprehensive quality and innovation ability, the task group published "electrical technology" and "electronic technology" after more than two years' time in 2011. According to many years' teaching experience, the task group set facilitating teachers to teach and students to understand and master as the principle, and deleted some old content in the writing process, concentrated on basic content, highlighted the main line. They also properly adjusted some content and arrangement order of electrician technology, carefully selected examples and problem sets, and strengthened the combination of theory and application. What's more, they added the PLC control, MULTISIM2011 and Mat lab circuit simulation technology, and micro-controller hardware circuit and other advanced, practical contents, that expanded the horizon of students and contact of subsequent course (Xu, 2011). At present in our school there are more than 1000 students (contains secondary college students) of twenty classes that in nine electricity class specialized learning the electrical course. Owing to different training scheme of every major and the poor basis of secondary college students, there are different requires of electrical course, and this makes a big difference in teaching content and teaching depth, so the curriculum implementation has certain difficulties. In view of the above situation, the course group visited each lecture institute in the specialty, and established electrotechnics teaching plan for every major and secondary college students。 After two years' teaching test and verify, this plan turned out to be effective and gain the approval of experts and every teaching accepted college.

2.2 *The reform of the teaching methods and means*

Firstly, reforming the traditional teaching mode that teacher does the most talking and the cultivation of students'conscious and ability of innovation are ignored. Then let students play an important part in teaching and ensure students participate in the whole process of teaching and learning. Meanwhile, teachers should be the leading factor and work as the instructor. Other than the classroom teaching mode that teachers give lessons, other teaching modes should be concerned (Qi, 2010). Let students study independently, such as, photoelectric device, multiplexer and demultiplexer, and so on; Teachers tutor students when they study content has certain difficult and clear logic, for example, multi-level amplifier circuit, field-effect transistor amplifier circuit, and so on. Firstly, teacher raises some questions to let students study independently and find the answers, then, according to the situation of the self-study, prepare lessons flexibly; Employing an exploratory learning method when study RL transient. Let students raise questions under teacher's guide, and then analysis and solve the questions themselves. The cultivated students' ability of pondering independently. At the same time, we put the heuristic teaching through all of teaching, and stimulate students' thirst for knowledge, active their academic thought, and cultivate their innovative ability.

According to the practical need of teaching, the research group developed hypermedia CAI that contains words, pictures, sound and animation and supports the electrotechnics teaching material. In the production process, we added the demonstration of EMW that show the related content such as changing parameters' effect on the circuit working condition in a circuit vividly and visually. At the same time, we added some simulation, animation, live-scenes, and so on. Therefore, when introduce motor, we can let the students understand the structure and composition of motor through watching animation there-dimensional simulation model, and explain the manufacturing operation of the motor through live documentary. In this way we can not only transfer richer and more visually knowledge and information to students, but also arouse students' studying enthusiasm and develop and improve their cognitive structure. Meanwhile the classroom teaching information is increased and the contradiction of less learning time and more teaching content are solved. The teaching efficiency is also improved greatly.

3 THE REFORM OF PRACTICING TEACHING MODE

The overall goal of practicing, teaching reform is to constantly adapt to the situation of higher education reform and development, and improve students' comprehensive quality, practical ability and Innovation ability. Since the electrotechnics course is of strong practice, it is necessary to deepen the students' comprehending of theory from practice, and cultivate their innovation consciousness, and improve their practical ability and creative ability. This course is not only the auxiliary and supplement of theoretical teaching, but also an important approach to cultivate the students' practical ability and improve their skill levels. According to 'the actual situation of students, we arrange the experiment classes of 30 hours, which contents basic experiment of 10 hours and new technology experiment of 20 ours. To guarantee that the basic skill of students can be trained, the replication experiment reduced to two or three from the original six, such as superposition theorem and Thevenin's theorem. At the same time, the number of designing experiments and comprehensive experiment increased from four to seven or eight to prepare well for the subsequent experiment, such as amplifier circuit design and the design of the voltage. Students are only given the aim of the experiment and the requirement of design in designing experiments and comprehensive experiment, and they have to complete the experimental plan and scheme on their own. Through this way can make students the real protagonist during the experiment, and improve their abilities of designing and experiment independently, and cultivate their engineering consciousness, innovative spirit and the ability of applying knowledge comprehensively. To cultivate the students' innovation ability, in 2010 the course group sets new course teaching platform and established the new technology of electrotechnics experiment course for the non electricity specialty for example major of mechanics and material. This course establishes after the electric technology and electronic course and combine with the technology comprehensive practice curriculum such as computer simulation technology, PLC control and the SCM control. Through the effort of teaching-research office and teachers of electrical laboratory, and adjust the practice content as well as assessment methods such as grade on the spot, summary after class and comprehensive assessment, this course has been developed well, and deeply popular with students.

In respect of the cultivating the innovation ability of students and the goal of project training, experiment in the classroom is far from sufficient. It is also necessary to provide students with more chances of independent practice. So in 2010 we efficiently integrated the electrical and electronic laboratory with other equipment of the laboratory, and established Electric and Electronic, Experimental Center which open all day (Liu, 2011). Since its opening, there are twenty percent of students join in the comprehensive experiment and innovative experiment in two years, among whom some students attended Electronic technology contest and design competition in 2011. Students' theoretical knowledge and practice are connected, innovation, consciousness are strengthened, and innovation ability is improved. In addition, we also guide and encourage students who have more abilities to attend our scientific research project. In 2010 there were ten percent to twenty percent students of mechanical engineering major took part in the work of scientific research item such as investigation, experiment and so on which processed by teachers in course group. During the work the teachers actively guided while students actively pondered. Thus, they broadened the horizon, know about the frontier knowledge, and improved innovation ability. In 2011 there were eight percent students of materials major and mechanical major attended the University Innovation Fund project set up by our school. In the program implementation process the teachers' guidance is complementary, while students' practical activities are the main principle. The survey shows that students who attended the University Innovation Fund project had a great improvement in abilities of finding, analyzing, and solving practical problems, and they can combine theory knowledge and practice well. Thus, it cultivates students' innovation ability and also help them successfully complete the transition from basic course to specialized course.

4 SET UP NETWORK TEACHING PLATFORM

Electrotechnics course is the exquisite course of our school. All the teaching resources are shared on the network, and the course group puts the whole teaching materials on the network, such as courseware, electronic teaching plan, the video of the teachers' curriculum, simulation test paper database, reference material, experimental teaching, problem sets and solutions, and so on. Students can study curriculum theory and experiment without the limit of time and space. At the same time there is an online FAQ system on the platform, and teachers will solve various problem students encounter during study at a certain time. And this will strengthen students'-interest and initiative of studying. Updating advanced knowledge that is related to Electrotechnics extents students' knowledge. The network teaching, classroom teaching and practical teaching combine efficiently and complement each other. After more than three year' research, construction and use, Electrotechnics curriculum network teaching platform get good effect and praise from teachers and students.

5 THE CONSTRUCTION OF TEACHERS' TEAM

The level of teachers'qualities affects directly the study and the cultivating of innovation ability of students. Nowadays the science and technology develops rapidly, and the development of new methods, new technology and new equipment of electronic and electric change with each passing day. This requires a high level of teachers' ability structure and knowledge structure, solid theoretical knowledge rich practical experience, proficient practical ability, and innovation ability. In form of all the new technology and situation and so on which appear in teaching process to students so that extends their horizons, stimulates them to learn new knowledge, master and use new knowledge actively, and achieve the goal of training their innovation consciousness. Therefore, it needs to take on proper training of university teachers through various ways and improving teachers' abilities and professional skill through developing visit and studying from domestic famous universities, taking part in teaching seminars exchange, and learning technology and training skills at the frontline of scientific research and production. Since 2009, electrotechnics teaching-research office of our school sent one teacher to each famous university and enterprise that related to profession to learn half a year to carry on double teacher education. Through studying, teachers improved and developed themselves on theory and practical, mastered the development trend of new theory new technology and new production, enhanced their innovative ability, and edified and cultivated innovative students as innovation type teachers.

6 EPILOGUE

Through several years' practical exploration, the course group has a further experience of how to mobilize students' learning, enthusiasm, create a good atmosphere in the class, selecting course content, improving teacher efficiency and effectiveness, strengthening experimental teaching and network teaching, cultivating students'innovation ability. Nowadays the development of electrotechnics and electronic techniques change with each passing day, so the reform of electrotechnics curriculum is a long-term and complex system engineering. We need to update the idea, active in the teaching reform and bold innovation, and constantly improve teaching quality, and make due contribution of cultivating the talents of innovation ability in a new age.

FUNDED PROJECTS

This work is supported by Plan Project of Youth Foundation of jiamusi University (Lq2013-032); Teaching key scientific research project of Jiamusi University (JKA2013-047).

REFERENCES

Wang, Z. (2012) Teaching mode design undergraduate innovation ability training. *Higher education BBS* 7 102.

Huang, Q. Z. (2008) *Electrotechnics* (Bei Jing: higher education press), 132–136.

Xu, Y .G. (2011)The electric professional teaching reform of electrotechnics. *Higher education BBS* 5 88.

Qi, J. (2010) college students'innovation ability training research. Yan shan university.

Liu, X. J. (2011) The though of the construction of electrotechnics'selected course. *Journal of Peking University* 3 23.

Electronic Engineering and Information Science – Wang (Ed.)
© 2015 Taylor & Francis Group, London, ISBN: 978-1-138-02772-5

Fault forecasting of numerical control machine by improved grey neural network GNNM(1, 1) modelbased on momentum

M. Bao
College of Electrical Engineering, HeiLongJiang Vocational College, Harbin, China

J. Bao
Public Foundation Institute, HeiLongJiang Vocational College, Harbin, China

J.W. Yang
College of Electrical Engineering, HeiLongJiang Vocational College, Harbin, China

ABSTRACT: To solve the economic loss of the numerical control machine fault, a new model based on grey neural network GNNM(1,1) is proposed. To improve the model by adding quadratic momentum and the convergence of grey neural network GNNM(1,1) model is proved. The experiment results show that this model by adding quadratic momentum have a better prediction results and iteration number than traditional GNNM(1,1) model.

KEYWORDS: Grey System; Neural Network; Quadratic Momentum; Fault Forecast.

1 INTRODUCTION

The numerical control machine is the major cornerstone of the development of modern Mechanical Manufacturing Industry. With the rapid development of information technology, the enterprise depends on the numerical control machine more and more (Shen et al. 2012).The result showed that the total economic loss of fault forecast reached 1×10^{11} Yuan. Therefore, fault forecast of numerical control machine has become the problems that brook no delay. Among the fault forecast methods, the grey system is widely to solve a incertitude system whose part information is known, part information is unknown (Sun 2009). A new grey forecasting model based on neural network named GNNM(1,1) was demand.

2 ESTABLISHING GNNM(1,1) MODEL

Assume that $x^{(0)} = \left\{ x^{(0)}(1), x^{(0)}(2), \cdots, x^{(0)}(k) \right\}$

Establishing the 1-AGO sequence

$$x^{(1)}(k) = \sum_{m=1}^{k} x^{(0)}(m)$$

Let $x^{(1)}(k) = y(t)$

Grey differential equation GM(1,1) can be expressed in following form:

$$\frac{dy(t)}{dt} + ay(t) = u \tag{1}$$

The time response model is:

$$y(t) = \left(y(0) - \frac{u}{a} \right) e^{-at} + \frac{u}{a} \tag{2}$$

Let Eqn.(3) is transform as follows:

$$
\begin{aligned}
y(t) &= \left[(y(0) - \frac{u}{a})(\frac{e^{-at}}{1+e^{-at}}) + \frac{u}{a}(\frac{1}{1+e^{-at}}) \right](1+e^{-at}) \\
&= \left[(y(0) - \frac{u}{a})(1 - \frac{1}{1+e^{-at}}) + \frac{u}{a}(\frac{1}{1+e^{-at}}) \right](1+e^{-at}) \\
&= \left[(y(0) - \frac{u}{a}) - y(0)(\frac{1}{1+e^{-at}}) + \frac{2u}{a}(\frac{1}{1+e^{-at}}) \right](1+e^{-at})
\end{aligned}
\tag{3}
$$

Then, the corresponding BP network weights can be assigned as follows:

$$w_{11} = w, \qquad w_{21} = -y(0), \qquad w_{31} = w_{32} = 1 + e^{-wt}$$

Among them: $w_{11} = w = a$, $\qquad w_{22} = v = \frac{2u}{a}$

Note:
(a) The weight w_{21} kept invariant during the network.
(b) w_{31}, w_{32} are directly obtained from the inputs t and w_{11}.

385

According to Eqn.(5), Grey neural network GNNM(1,1) model is shown in Fig.1 From the figure 1, we can see that GNNM(1,1) model is a feed forward neural network. The threshold of layer LD is: $\theta_y = (1 + e^{-wt})(\frac{v}{2} - y(0))$. The activation function of neural network layer LB is sigmoid function: $f(x) = \frac{1}{1 + e^{-x}}$; The activation function of others layers is : $f(x) = x$.

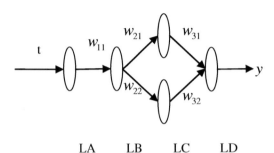

LA   LB   LC   LD

Figure 1.   Grey neural network GNNM(1,1) model.

3   INTRODUCING QUADRATIC MOMENTUM INTO GNNM(1,1) MODEL

At the assumption of training sample set by $\{t, y(t)\}_{t=1}^{T} \subset N \times R$, the actual output of the network is computed by

$$z(t) = \left[\left(y(0) - \frac{v}{2} \right) - y(0) \times (\frac{1}{1 + e^{-wt}}) + v \times (\frac{1}{1 + e^{-wt}}) \right](1 + e^{-wt})$$

$$= \left(\frac{v}{2} - y(0) \right)(1 + e^{-wt}) + y(0) \tag{4}$$

where t is input

Let $X = (w, v)^T$, $X_n = (w_n, v_n)^T$
Then the error function is

$$E(X) = \frac{1}{2} \sum_{t=1}^{T} f^2(X) \tag{5}$$

Among them :

$$f(X) = z(t) - y(t) = (\frac{v}{2} - y(0))(1 - e^{-wt}) + y(0) - y(t) \tag{6}$$

Find a X^* to meet $E(X^*) = \min E(X)$ \tag{7}
The iteration formula of grey neural network is

$$w_{n+1} = w_n + \Delta w_n, \quad v_{n+1} = v_n + \Delta v_n$$

and $n = 0, 1, 2, \cdots$
Among them:

$$\Delta w_n = -\eta E_w(X_n) + \tau_n \Delta w_{n-1} + \frac{\tau_{n-1}}{L_n} \Delta w_{n-2}$$

$$= -\eta \left[\sum_{t=1}^{T} (\frac{v_n}{2} - y(0)) t e^{-w_n t} f(X_n) \right] \tag{8}$$

$$+ \tau_n \Delta w_{n-1} + \frac{\tau_{n-1}}{L_n} \Delta w_{n-2}$$

$$\Delta v_n = -\eta E_v(X_n) + \tau_n \Delta v_{n-1} + \frac{\tau_{n-1}}{L_n} \Delta v_{n-2}$$

$$= -\frac{1}{2} \eta \left[\sum_{t=1}^{T} (1 - e^{-w_n t}) f(X_n) \right] \tag{9}$$

$$+ \tau_n \Delta v_{n-1} + \frac{\tau_{n-1}}{L_n} \Delta v_{n-2}$$

The constant $\eta > 0$ is the quadratic momentum and learning rate can be assigned as follows:

$$\tau_n = \begin{cases} \dfrac{\tau \|E_x(X_n)\|}{\|\Delta X_{n-1}\|} & as \quad if \quad \|\Delta X_{n-1}\| \neq 0 \\ \\ 0 & otherwise \end{cases} \tag{10}$$

$$L_n = \begin{cases} L_{n-1} + 1 & as \quad if \quad E(X_n) > E(X_{n-1}) \\ \\ L_{n-1} & as \quad if \quad E(X_n) \leq E(X_{n-1}) \end{cases}$$

Moreover: $10 \leq L_n \leq 30$, constant $\tau \geq 0$

4   CONVERGENCE PROOF OF MGNNM(1,1) MODEL

The main assumptions are as follows:

(c): $\left\{ \|X_n\| \ \middle| \ n = 1, 2, \cdots \right\}$ is bounded.

(d): The set $\Omega_0 = \left\{ X \in \Omega \ \middle| \ E_x(X) = 0 \right\}$ is a finitude point set.
Denote

$$\Delta X_n = X_{n+1} - X_n \tag{11}$$

$$\Delta \tilde{X}_n = \theta \tilde{X}_n + (1 - \theta) \tilde{X}_{n+1} \tag{12}$$

Let $\theta \in (0, 1)$
From (7) and (8), we have

$$E_{ww}(X) = \sum_{t=1}^{T}\left[\left(y(0)-\frac{v}{2}\right)t^2e^{-wt}f(X)+\left(y(0)-\frac{v}{2}\right)^2t^2e^{-2wt}\right]$$

$$E_{ww}(X) = \sum_{t=1}^{T}\left[\left(y(0)-\frac{v}{2}\right)t^2e^{-wt}f(X)+\left(y(0)-\frac{v}{2}\right)^2t^2e^{-2wt}\right]$$

$$E_{vv}(X) = \frac{1}{4}\sum_{t=1}^{T}(1-e^{-wt})^2$$

By assumption c, there exist a constant number $C>0$ such that

$$\max\left\{\left|E_{ww}(\tilde{X}_n)\right|,\left|E_{vv}(\tilde{X}_n)\right|,\left|E_{vv}(\tilde{X}_n)\right|\right\}<C \quad (13)$$

And $\tau = s\eta,\ 0<s<1$ (14)
From (8) (9) (10), we have

$$\left\|\Delta X_n\right\| \leq \left\|-\eta E_X(X_n)+\tau_n\Delta X_{n-1}+\tau_{n-1}\Delta X_{n-2}\right\|$$

$$\leq \eta\left\|E_X(X_n)\right\|+\tau_n\left\|\Delta X_{n-1}\right\|+\frac{\tau_{n-1}}{8}\left\|\Delta X_{n-2}\right\|$$

$$\leq \eta\left\|E_X(X_n)\right\|+\tau\left\|E_X(X_n)\right\|+\frac{\tau}{8}\left\|E_X(X_{n-1})\right\| \quad (15)$$

$$\leq (2\eta+\frac{\eta}{8})\left\|E_X(X)\right\|$$

Among them:

$$\left\|E_X(X)\right\| = \max\left\{\left\|E_X(X_n)\right\|,\left\|E_X(X_{n-1})\right\|\right\} \quad (16)$$

Using the Taylor's expansion of two-variable function and (11)-(16), we have

$$E(X_{n+1})-E(X_n) = E_X(X_n)\Delta X_n$$
$$+\frac{1}{2}\left[\Delta w_n^2 E_{ww}(\tilde{X}_n)+2\Delta w_n \Delta v_n E_{wv}(\tilde{X}_n)+\Delta v_n^2 E_{vv}(\tilde{X}_n)\right]$$
$$=-\eta\left[E_X(X_n)\right]^2+\tau_n E_X(X_n)\Delta X_{n-1}+\frac{\tau_{n-1}}{8}E_X(X_n)\Delta X_{n-2}$$
$$+\frac{1}{2}\left[\Delta w_n^2 E_{ww}(\tilde{X}_n)+2\Delta w_n \Delta v_n E_{wv}(\tilde{X}_n)+\Delta v_n^2 E_{vv}(\tilde{X}_n)\right]$$
$$\leq -\eta\left[E_X(X_n)\right]^2+\tau_n E_X(X_n)\Delta X_{n-1}$$
$$+\frac{\tau_{n-1}}{8}E_X(X_n)\Delta X_{n-2}+C(\Delta w_n^2+\Delta v_n^2)$$
$$\leq -\eta\left[E_X(X_n)\right]^2+\tau_n\left\|E_X(X_n)\right\|\left\|\Delta X_{n-1}\right\| \quad (17)$$
$$+\frac{\tau_{n-1}}{8}\left\|E_X(X_n)\right\|\left\|\Delta X_{n-2}\right\|+C\left\|\Delta X_n\right\|^2$$
$$\leq (-\eta+s\eta+\frac{s\eta}{8})\left\|E_X(X)\right\|^2+(2\eta+\frac{\eta}{8})^2 C\left\|E_X(X)\right\|^2$$
$$\leq (-\eta+\frac{(8+1)}{8}s\eta+5C\eta^2)\left\|E_X(X)\right\|^2$$
$$\leq 5C\eta\left(\eta-\frac{(1-2s)}{5C}\right)\left\|E_X(X)\right\|^2$$

Choose $0<\eta<\dfrac{(1-2s)}{5C}$ Therefore, the error function is decreasing function.
For any $n=1,2,\cdots,\ E(X_n)\geq 0$, so there exists $E^*\geq 0$

such that $\lim_{n\to\infty}E(X_n)=E^*$ (18)

From (17) (18) and the application of "Double Sides Approximate Process", we have

$$\lim_{n\to\infty}\left\|E_X(X_n)\right\|=0 \quad (19)$$

From (15) and (19), we have $\lim_{n\to\infty}\left\|\Delta X_n\right\|=0$
Thus, $\lim_{n\to\infty}X_n=X^*$
This completes convergence of GNNM(1,1) based on quadratic momentum proof.

5 SIMULATION INSTANCE ANALYSIS

Spindle drive system is one of the key subsystem of the numerical control machine, playing an important role in numerical control machine of normal operation. Model taking date1-19 as input sample of GNNM(1,1) model, as shown in TABLE.1.

Table 1.   Sample date.

Number	1	2	3	4	5
Time(hour)	131.74	378.74	395.21	675.14	823.32
Number	6	7	8	9	10
Time(hour)	1020.95	1053.88	1267.95	1805.87	2041.89
Number	11	12	13	14	15
Time(hour)	2102.27	2239.5	2552.37	2695.08	2865.24
Number	16	17	18	19	20
Time(hour)	3342.78	3589.78	3787.38	4923.60	5121.20

Introducing one and quadratic momentum into Grey neural network GNNM(1,1) which were named as $GNNM_1(1,1)$ and $GNNM_2(1,1)$. The paper compared $GNNM_2(1,1)$ with traditional GNNM(1,1) and $GNNM_1(1,1)$ model, as shown in TABLE.2.

Table 2.   Model date.

Model	$GNNM(1,1)$	$GNNM_1(1,1)$	$GNNM_2(1,1)$
Prediction Data	5428.88	5424.63	5420.03
Real Data	5121.20	5121.20	5121.20
Relative Error	6.01%	5.92%	5.83%
Iteration Number	239	196	186

6 CONCLUSIONS

In order to solve the economic loss of the numerical control machine fault, a model of adding quadratic momentum GNNM(1,1) (named $GNNM_2(1,1)$) is presented. The experimental results of three models showed that the iteration number and prediction accuracy of adding quadratic momentum GNNM(1,1) models was much higher than that of GM(1,1) model and $GNNM_1(1,1)$ model. Thus, the proposed model has more realistic applications.

ACKNOWLEDGMENT

The paper thanks the education department of Hei Long Jiang Province12535164 for the development of opening motion controller based on USB bus.

REFERENCES

Shen, Y., J. Bao, & H. Sun(2012). A Modified Grey Neural Network GNNM(1,1) Model Based on Momentum and Its Application, *2012 Third Global Congress on Intelligent Systems*, 405–408.

Sun, Y. W.(2009). Convergence of a batch method for grey neural network GNNM(1,1), *Dalian University of Technology, 20090601*.

Electronic Engineering and Information Science – Wang (Ed.)
© 2015 Taylor & Francis Group, London, ISBN: 978-1-138-02772-5

Numerical simulation of the laser transmission soliton initial stage

F.X. Li, J. Han, Z.F. Fei & J. Wei
School of Applied Sciences, Harbin University of Science and Technology, Harbin, P. R. China

ABSTRACT: In order to improve the convergent rate of the traditional reproducing kernel method for solving the nonlinear Schrodinger equation in the laser transmission initial stage, we presented an improved reproducing kernel method. In the first, we gave the error estimation of the traditional reproducing kernel method. Secondly, using the properties of the error estimation and combining extrapolation method, we raised the convergent order of the traditional reproducing kernel method from 1st-order to 4th-order. All test problems show the 4th-order convergence of our method.

KEYWORDS: reproducing kernel; extrapolation method; singular; initial value; numerical simulation; laser.

1 INTRODUCTION

Laser is the carrier of strong energy with the excellent directivity and stability. People are interested in the formation of soliton. In view of it's special properties, laser plays an important role in modern science and technology. The intensity (E) of the electric field in the emission of laser satisfies the nonlinear Schrodinger equation

$$2iK(\frac{\partial}{\partial z}+\frac{1}{v}\frac{\partial}{\partial t})E+\frac{\partial^2 E}{\partial x^2}+\frac{\partial^2 E}{\partial y^2}-K\frac{\partial^2 K}{\partial \omega^2}\frac{\partial^2 E}{\partial t^2}$$
$$+2K^2\frac{n_2}{n_0}|E^2|E=0 \tag{1}$$

It changes to (2) with the 2-order ordinary differential problem with singular coefficients through evolution and by using the special solution $U = U(\rho)$ $exp(i\beta\zeta)$

$$\frac{1}{2}(U''+\frac{2}{\rho}U')-\beta U+U^3=0, U(0)=1, U'(0)=0 \tag{2}$$

According to (Lv et al. 2004), it's electric field intensity satisfies nonlinear Schrodinger equation in the laser emission mechanism, which can be reduced to singular differential equation. And nonlinear Schrodinger equation is an important model in physics. Hence, the problem can be transformed into solving the second-order singular nonlinear differential equations with initial value condition.

A detailed argument for the existence and the uniqueness of the solution to Eq. 3 is proved in (Wang 2009, Agarwal, 1999). For the second-order singular nonlinear differential equations, a few methods are presented to solve its approximate solutions. Twofold Spline Chebyshev method and a backward Euler method are used to solve the numerical approximation of solutions to the singular second-order differential equations in (Khuri 2013, Benko 2008). The numerical solution to a singular second-order initial value problems is obtained by Nystrom method and the order of convergence is determined in (Benko 2009). The Laplace decomposition method for solving the second-order singular nonlinear differential equation was given in (Ahmad 2014), combining Laplace transform with Adomian's decomposition method. In this paper, numerical simulation of Eq. 2 is given by using an improved reproducing kernel method.

In this paper, we consider the following second-order singular nonlinear differential equations with initial value conditions:

$$\begin{cases} u''+p(x)u'=f(x,u(x)), & x\in(0,1] \\ u(0)=0, u'(0)=0 \end{cases} \tag{3}$$

where $\int_0^1 p(x)dx=+\infty, f(x,y)$ is a continuous function. Eq. 3 can be expressed as follows:

$$Lu = f(x,u(x)), u(0)=0, u'(0)=0 \tag{4}$$

where, $f(x,y)$ is a continuous function, $Lu \overset{\Delta}{=} u''(x)$ $+p(x)u'(x), x\in(0,1]$

2 REPRODUCING KERNEL METHOD (RKM)

First of all, two reproducing kernel spaces are introduced below.

Reproducing kernel space $W_1[0,1]$ is defined by $W_1[0,1]=\{u(x)|\ u$ is an absolutely continuous function $u' \in L^2[0,1]\}$. The inner product and the norm of space $W_1[0,1]$ is defined as follows:

$$\langle u(x), v(x) \rangle_{W_1} = u(0)v(0) + \int_0^1 u'(x)v'(x)\mathrm{d}x,$$

$$\|u(x)\|_{W_1} = \sqrt{\langle u(x),v(x)\rangle_{W_1}} \text{ for } u(x),v(x) \in W_1[0,1]$$

$W_1[0,1]$ is proved a reproducing kernel space equipped with reproducing kernel function

$$R_x^{\{1\}}(y) = \begin{cases} 1+y, y \le x \\ 1+x, y > x \end{cases}$$

The reproducing kernel space $W_3[0,1]$ is defined by $W_3[0,1]=\{u(x)|\ u$ is an absolutely continuous function $u''' \in L^2[0,1], u(0)=u'(0)=0\}$. The inner product of space $W_3[0,1]$:

$$\langle u(x), v(x) \rangle_{W_3} = u''(0)v''(0) + \int_0^1 u'''(x)v'''(x)\mathrm{d}x$$

for $u(x),v(x) \in W_3[0,1]$, the norm

$$\|u(x)\|_{W_3} = \sqrt{\langle u(x),v(x)\rangle_{W_3}} \ .$$

The reproducing kernel function of $W_3[0,1]$ can be written as

$$R_x^{\{3\}}(y) = \begin{cases} \dfrac{x^2y^2}{4} + \dfrac{x^2y^3}{12} - \dfrac{xy^4}{24} + \dfrac{y^5}{120}, y \le x \\ \dfrac{x^5}{120} - \dfrac{x^4y}{24} + \dfrac{x^2(3+x)y^2}{12}, y > x \end{cases}$$

Secondly, the reproducing kernel method to solve Eq. 4 is presented.

It is easy to infer that $L: W_3[0,1] \to W_1[0,1]$ is a bounded linear operator and the operator L^{-1} exists by using $Lu = u''(x) + p(x)u'(x)$ in Eq. 4. Suppose $\{x_i\}\infty$ i=1 is dense on [0,1], let $\varphi_i = R\{1\}\ x\ (x)$, $\psi_i(x) = L^*\varphi_i(x)$, where L^* is the conjugate linear operator of L. Dealing with $\{\psi_i\}\infty$ i=1 by means of Schmidt orthogonalization and normalization, we obtain $\{\psi_i\}\infty$ i=1, where $\psi_i(x) = \Sigma\beta_{ik}\ \psi_k(x)$, $(\beta_{ii}>0,\ i=1,2,\dots)$. We construct an iterative scheme:

$$w_n(x) = \sum_{i=1}^n \sum_{k=1}^i \beta_{ik} f(x_k, u_{k-1}(x_k))\overline{\psi}_i(x)$$

where $u_0(x) \in W_3[0,1]$, $u_n = P_n u$, $P_n: W_3[0,1] \to span$ $\{\psi_1, \psi_2,\dots,\psi_n\}$ is an orthogonal projection operator, $u(x)$ is the exact solution to Eq. 4.

Theorem 2.1[9]. Assume that $f(x, y)$ is a continuous function about x, y and $\{x_i\}\infty$ i=1 is dense on [0,1]. Then $w_n(x)$ converges to the exact solution $u(x)$ of Eq. 4, denoted by $w_n(x) = u_n(x)$ and

$$u(x) = \sum_{i=1}^\infty \sum_{k=1}^i \beta_{ik} f(x_k, w_{k-1}(x_k))\overline{\psi}_i(x)$$

where $w_0(x) = u_0(x)$.

Theorem 2.2. Let $n \in \mathbb{N}^+$, $x_i \in [0,1]$, $h = 1/n$, $x_i = (i=1,2,\dots,n+1)$. Suppose $u(x)$ is the solution to Eq. 4, the conditions of Theorem 1 hold, and $Lu=u''(x)+p(x)u'(x)$, $|w_n'''(x)| \le K$, then $|w_n(x)-u(x)| = O(h)$, where K is a positive constant.

Let $x_1=0$, $w_0(x)=v(x)$, for any function $v(x) \in W_3[0,1]$. From the initial condition, we have $w_0(x_1) = u(x_1)$, $w_0'(x_1) = u'(x_1)$, and then

$$w_1(x) = \beta_{11} f(x_1, w_0(x_1))\overline{\psi}_1(x) = u_1(x)$$

Therefore, we obtain a convergent projection iterative sequence:

$$\begin{cases} w_0(x) = v(x), \forall v(x) \in W_3[0,1] \\ w_1(x) = \beta_{11} f(x_1, w_0(x_1))\overline{\psi}_1(x) \\ \dots \\ w_n(x) = \sum_{i=1}^n \sum_{k=1}^i \beta_{ik} f(x_k, w_{k-1}(x_k))\overline{\psi}_i(x) \end{cases} \quad (5)$$

Let $xi=ih$, $h=1/n$, $(i=0,1,2,\dots,n)$. We get the approximate solution to $u(x)$, denoted by $w_n(x)$ and $e_n = u(x)-w_n(x) = O(h)$. This is the traditional reprod-ucing kernel method (RKM). In order to get better accuracy of the approximate solutions $w_n(x)$, n must be sufficiently large, thereby increasing the calculation of the orthogonalization and the number of iterations, which restricts the application of RKM. In this paper, combining RKM with extrapolation method, we give a more effective method, called an improved reproducing kernel method (RKEM), to solve the Eq. 4.

3 AN IMPROVED REPRODUCING KERNEL METHOD(RKEM)

In formula (5), $w_n(x)$ is obtained by letting $x_i = ih$, i = 0,1,2,…,n, with $h = 1/n$, and then we get $F_1(h)= w_n(x)$. $w_m(x)$ is generated by letting $y_i = iqh$, i = 0,1,2,…,m, here $m= 1/qh$, $q \in (0,1)$. Then we obtain $F_1(qh) = w_m(x)$. According to extrapolation method, we get

$$F_2(h) = \frac{F_1(qh) - q^{p_1}F_1(h)}{1 - q^{p_1}}, \cdots,$$

$$F_m(h) = \frac{F_{m-1}(qh) - q^{p_{m-1}}F_{m-1}(h)}{1 - q^{p_{m-1}}}$$

Suppose $p_{m-1} = m-1$, for any $m = 2,3,\ldots$, then we know that the error of $F_m(h)$ is at least $O(h^m)$. According to the correlation between $F_m(h)$ and $F_{m-1}(h)$, we can infer $F_m(h)$ goes near to the total amount of calculation of $F_{m-1}(h)$ and $F_{m-1}(h)$. However, the convergent rate of $F_m(h)$ may be faster than that of $F_{m-1}(h)$.

4 NUMERICAL EXAMPLES

In this section, we apply the method given in this paper to some numerical examples. By software Mathematica7.0, the concrete numerical computations are completed.

Example 1. The initial value problem

$$u''(x) + \frac{2u'(x)}{\sin(x)} + u^3(x) = f(x), u(0) = 0, u'(0) = 0$$

where

$$f(x) = 4\cos(2x) - 4x\sin(2x) + x^3\sin^3(2x)$$
$$+ 2\csc(x)(2x\cos(2x) + \sin(2x))$$

the exact solution is $u(x) = x\sin(2x)$. Let $q=1/2$ and interval $[0,1]$ is divided into n subintervals. When n is equal to 10, 20, 40, 80 or 160, it implies that the approximate solutions $w_{10}, w_{20}, w_{40}, w_{80}$ and w_{160}. We get an approximate solution $F_{3,q}(h)$ by using twice extrapolation with w_{20}, w_{40} and w_{80}. Meanwhile, an approximate solution $F_{4,q}(h)$ is obtained by using three times extrapolation with w_{10}, w_{20}, w_{40} and w_{80}. The maximums of absolute errors obtained by w_{80}, $w_{160}, F_{3,q}(h)$ and $F_{4,q}(h)$ are compared, and the numerical results are shown in Table 1.

Table 1. Numerical results for $q=1/2$.

Method	w_{80}	w_{160}	$F_{3,q}(h)$	$F_{4,q}(h)$
The error	2.13×10^{-5}	5.30×10^{-6}	3.04×10^{-6}	4.38×10^{-7}
Time[s]	67.268	662.365	76.972	77.237

Example 2. The initial value problem

$$u''(x) + \frac{u'(x)}{\sin x} + u^2 = f(x), u(0) = 0, u'(0) = 0$$

where

$$f(x) = 2\cos x - x\sin x$$
$$+ \csc x(\sin x + x\cos x) + x^2\sin^2 x$$

the exact solution is $u(x) = x\sin x$. According to the process of Example 1, the numerical results obtained by the RKEM are shown in the following Table 2.

Table 2. Numerical results for $q=1/2$.

Method	w_{80}	w_{160}	$F_{3,q}(h)$	$F_{4,q}(h)$
The error	8.48×10^{-6}	2.11×10^{-6}	5.27×10^{-7}	8.22×10^{-8}
Time[s]	65.271	643.441	75.151	75.416

Two numerical examples show that $F_{3,q}(h)$ and $F_{4,q}(h)$ go near to the calculation of w_{80}, while $q=1/2$ and the accuracy of $F_{3,q}(h)$ and $F_{4,q}(h)$ are higher than that of w_{80}. Meanwhile, compared to w_{160}, $F_{3,q}(h)$ and $F_{4,q}(h)$ need fewer calculations and get better accuracy.

5 CONCLUSIONS

In the paper, the reproducing kernel method to solve the nonlinear Schrodinger equation in the laser transmission initial stage and the error estimation are given. Utilize the error estimation and combine extrapolation method, and then we present an improved reproducing kernel method for the solutions to second-order singular nonlinear differential equations with initial value conditions. This method can improve the convergent order of the reproducing kernel method. The numerical examples show that the improved reproducing kernel method, compared to the traditional reproducing kernel method, has the superiority for solving the nonlinear Schrodinger equation. Since the method improves the convergent order effectively and reduces calculation greatly, it contributes to the application of the reproducing kernel method to solve nonlinear problems.

ACKNOWLEDGMENTS

This research was funded by Natural Science Foundation of Heilongjiang Province (No.A200811) and by the Educational Department Scientific Technology Program of Heilongjiang (No. 11521045).

REFERENCES

Lv, Y. & Chen, Ch.M. (2004). The Finite Element Solutions of the 2-order Ordinary Differential Problems with Singular Coefficients. *Journal of Zhuzhou Institute of Technology.* 18(2), 33–35(in Chinese).

Wang, X.C. & Zhou, M.X. (2009). The uniqueness of solution for a class of second order singular initial value probelm. *Journal of Zhejiang normal university (Nat. Sci)*. 32 (3), 287–292.

Agarwal, R.P. & O'Regan, D. (1999). Second-order initial value problem of singular type. *Journal of Mathematical Analysis and Applications*. 229 (2), 441–452.

Khuri, S.A. & Sayfy, A. (2013). A Twofold Spline Chebyshev linearization approach for a class of singular second-order nonlinear differential equations. *Results in Mathematics*. 63 (3-4), 817–835.

Benko, D., Bykes, D.C., Robinson, M.P. & Spraker, J.S. (200 8). Nystrom methods and singular second-order differential equations. *Comput. Math. Appl.* 56, 1975–1980.

Benko, D., Biles, D.C., Robinson, M.P. & Spraker,J.S. (2009). Numerical approximation for singular second-order differe-nttial equations. *Math. Comput. Model*. 49 (5-6), 1109–1114.

Ahmad, J., Hussain, F., & Naeem, M. (2014). Laplace decomp-osition method for solving singular initial value problems. *AJMP*. 5 (1), 1–15.

Electronic Engineering and Information Science – Wang (Ed.)
© *2015 Taylor & Francis Group, London, ISBN: 978-1-138-02772-5*

Existence and stability of almost periodic solution to a class of neutral tape of integro-differential equation

Y.F. He & W.B. Ma
College of sciences, Inner Mongolia Agriculture University, Huttot, China

Z.B. Cao
Harbin University of Science and Technology, Harbin, China

ABSTRACT: In recent years, Integro-Differential equation is applied in lots of fields. In this paper, Existence and university and stability of almost periodic solution to a class of neutral type of Integro-Differential equation are obtained by applying exponential dichotomy and fixed point method.

KEYWORDS: neutral tape, Integra-differential equation, almost periodic solution, existence, uniqueness, stability.

1 INTRODUCTION

Neutral Type of Integra-Differential Equation have been extensively studied in past years and some important results have been reported.Paper (Wang, 2002)and (Wang, 2003)discussed existence and uniqueness stability of almost periodic solution about the following Integra-Differential equation:

$$\frac{d}{dt}\left(x(t)-\mathrm{E}(t)x_t-\int_{-\infty}^t B(t,s)x(s)ds\right)=\mathrm{A}(t)x(t)+\int_{-\infty}^t C(t,s)x(s)ds+\ g\left(t,x(t),x_t\right) \quad (1)$$

Where $x_t = x(t+\theta)$.
$$-\infty < \theta \le 0$$

Paper (Wang, 2002) and (Wang, 2003)discussed existence and uniqueness stability of almost periodic solution about Eq. (1) under the circumstance of E (t) = 0, $g\left(t,x(t),x_t\right)=f(t)$ and $g\left(t,x(t),x_t\right)=\cdot g\left(t,x(t)\right)$. Paper (Fang, Wang2004) discussed existence and uniqueness stability of almost periodic solution about Eq.(1) under the circumstance of B(t,s)≡C(t,s)≡0, $g\left(t,x(t),x_t\right)=f(t,x_t)$,Has been inspired, The paper popularized their conclusion and gained existence and uniqueness stability of almost periodic solution about Eq.(1)

2 MAIN RESULTS

Assume Eq. (1) satisfy the following conditions:

(A_1) A(t), E(t) are almost periodic function matrix; B(t,t+s), C(t,t+s) are uniform almost periodic function matrix for s∈D, (D is compact subset of R); $g\left(t,x(t),x_t\right)$ is almost periodic function vector in $R\times R^n\times C\to R^n$, where $C=\left((-\infty,0]\to R^n\right)$

(A_2) b (t)is the average value of almost periodic function, define

$$\mathrm{M}[b]\underset{=}{\Delta}\lim_{t-s\to\infty}\frac{1}{t-s}\int_s^t b(r)dr=-a<0,\ \ \text{Where}\ \ b(t)=\max_{1\le j\le n}$$
$$\left\{a_{jj}(t)+\sum_{i=1,i\ne j}^n\left|a_{ij}(t)\right|\right\}$$

(A_3) There exist constant k_1 such that $\int_{-\infty}^t \|B(t,s)\|ds\le k_1$ and $\int_{-\infty}^t \|C(t,s)\|ds$ is bounded for $\forall t\in R$, where $0\le k_1<1$

(A_4) Given $\forall \varepsilon>0$, there exist $L=L(\varepsilon)>0$ such that $\int_{-\infty}^{t-L}\|B(t,s)\|ds<\varepsilon$, $\int_{-\infty}^{t-L}\|C(t,s)\|ds<\varepsilon$, for $\forall t\in R$

(A_5) There exist Non-negative almost periodic function $\alpha_1(t)$ and $\alpha_2(t)$ such that $\|g(t,x(t),x_t)-g(t,y(t),y_t)\|\le\alpha_1(t)\|x(t)-y(t)\|+\alpha_2(t)\|x_t-y_t\|$ for $\forall x,y\in R^n$.

(A_6) There exist constant $k>\dfrac{1}{1-k_1-E}$ such that
$$b(t)+k[\int_{-\infty}^t\|G(t,s)\|ds+\|\mathrm{A}(t)\mathrm{E}(t)\|+\alpha_1(t)+\alpha_2(t)]\le 0 \quad \text{for}$$
$\forall tx,y\in R^n$

Where $b(t)$, $\alpha_1(t)$ and $\alpha_2(t)$ are listed in (A_2) (A_5), $\mathrm{E}=\sup_{t\in R}\left\{\|\mathrm{E}(t)\|\right\}$, $k_1+E<1$, $G(t,s)=\mathrm{A}(t)B(t,s)+C(t,s)$

Let $BC(-\infty,t_0]=\left\{\phi(t)\middle|\phi:(-\infty,t_0]\to R^n\right\}$ is bounded and continous}, we define its norm as $\|\phi\|=\sup\left\{\|\phi(t)\|,t\in(-\infty,t_0]\right\}$ for $\forall \phi\in BC(-\infty,t_0]$ the solution of Eq. (1) will be denoted as $x(t,t_0,\phi)$, $x(t,\phi)$ or $x(t)$ with bounded and continuous initial function $\phi\in BC(-\infty,t_0]$ under the circumstance of not being confused.

$$\|\mathrm{X}(t)\mathrm{X}^{-1}(r)\|\le\exp(\int_s^t b(r)dr) \quad (3)$$

Next, we can gain following theorem.

Theorem 1. if Eq. (1) satisfy conditions $A_1 \sim A_6$, then Eq. (1) exists unique and uniformly stable almost period solution.

Proof. Let $B_1 = \{u_1(t) | u : R \to R^n \text{is almost periodic function}\}$, then B_1 is a Banach space under norm $\|u\| = \sup\{\|u(t)\| : t \in R\}$. Denote $G(t,s)=A(t)B(t,s)+C(t,s)$, according to the condition of the theorem, we know $G(t,s)$ have property of $C(t,s)$. for $\forall u \in B_1$, it is easily to find $E(t)u_t$ is almost periodic function by the property of almost periodic function. By lemma 2 we know $\int_{-\infty}^t B(t,s)u(s)ds$ and $\int_{-\infty}^t G(t,s)u(s)ds$ are almost periodic function. By lemma4 and property of almost periodic function, we can deduce $\int_{-\infty}^t X(t)X^{-1}(r)\left[\int_{-\infty}^r G(r,s)u(s)ds + A(r)E(r)u_r + g(r,u(r),u_r)\right]dr$ is almost periodic function too.

Let $x_u(t) = E(t)u_t + \int_{-\infty}^t B(t,s)u(s)ds + \int_{-\infty}^t X(t)X^{-1}(r)\left[\int_{-\infty}^r G(r,s)u(s)ds + A(r)E(r)u_r + g(r,u(r),u_r)\right]d$ then $x_u(t)$ also is almost periodic function.

Where $u_r = u(t+\theta)$. $_{-\infty \leq \theta \leq 0}$

Let $Fu(t) = x_u(t)$ then

$\| Fu_1(t) - Fu_2(t)\| \leq E \|u_{1_r} - u_{2_r}\| + \int_{-\infty}^t \|B(t,s)\| \|u_1(s) - u_2(2)\| ds \cdot$

$+ \int_{-\infty}^t \|X(t)X^{-1}(r)\| \left[\left(\int_{-\infty}^r \|G(r,s)\| ds + \alpha_1(r)\right) \|u_1(r) - u_2(r) + (A(s)$

$A(s)E(r)\| + \alpha_2(r)\|u_{1r} - u$

$\leq (E + k_1)\|u_1 - u_2\| +$

$\int_{-\infty}^t \exp\left(\int_r^t b(\tau)d\tau\right)\left(\int_{-\infty}^r \|G(r,s)\| ds + \|A(r)E(r)\| + \alpha_1(r) + \alpha_2(r)\right)$

$d \cdot \|u_1 - u_2\|$

$\leq \int_{-\infty}^t \exp\left(\int_r^t b(\tau)d\tau\right) \cdot \left(-\frac{b(r)}{k}\right)dr \cdot \|u_1 - u_2\| \leq \left(k_1 + E + \frac{1}{k}\right) \cdot \|u_1 - u_2\| \leq$

By condition (A_6), we can know $k_1 + E + \frac{1}{k} < 1$.

Since F is compressed in B_1, so F exists unique fixed point in B_1. That is exist to unique $x \in B_1$ such that

$x(t) = E(t)x_t + \int_{-\infty}^t B(t,s)x(s)ds + \int_{-\infty}^t X(t)X^{-1}(r)\left[\int_{-\infty}^r G(r,s) \cdot x(s)\right.$

$ds + A(r)E(r)x_r + g(r,x(r),x_r)\left.\right]dr$. So,

$x(t) - E(t)x_t + \int_{-\infty}^t B(t,s)x(s)ds = \int_{-\infty}^t X(t)X^{-1}(r)\left[\int_{-\infty}^r G(r,s) \cdot\right.$

$x(s)ds + A(r)E(r)x_r + g(r,x(r),x_r)\left.\right]dr$ (4)

By doing derivation for Eq. (4), we get $x(t)$ to satisfy Eq. (1), namely, $x(t)$ is Eq.(1) unique almost periodic solution.

Finally, we prove that any solution of Eq. (1) is uniformly stable.

Eq. (1) is transformed the following form:

$\frac{d}{dt}Dx_t = A(t)x(t) + \int_{-\infty}^t G(t,s)x(s)ds + A(t)E(t)x_t + g(t,x(t),x_t)$ $1^{(/)}$

Where $Dx_t = x(t) - \int_{-\infty}^t B(t,s)x(s)ds - E(t)x_t$

According to the constant variant method, we can get:

$Dx_t = X(t)X^{-1}(t_0)Dx_{t0} + \int_{t_0}^t X(t)X^{-1}(r) \cdot \left[\int_{-\infty}^r G(r,s)x(s)ds + A(r)\right.$

$E(r)x_r + g(r,x(r),x_r)\left.\right]dr$ $(t \geq t_0)$ (5)

$Dy_t = X(t)X^{-1}(t_0)Dy_0 + \int_{t_0}^t X(t)X^{-1}(r)\left[\int_{-\infty}^r G(r,s)x(s)ds + A(r)E(r)y_r + \right.$

$g(r,y(r),y_r)\left.\right]dr$ $(t \geq t_0)$ (6)

$x(t,t_0,\phi_1) = X(t)X^{-1}(t_0)Dx_{t_0} + E(t)x_t + \int_{-\infty}^t B(t,s)x(s)ds +$

$\int_{t_0}^t X(t)X^{-1}(r)\left[\int_{-\infty}^r G(r,s)x(s)ds + A(r)E(r)x(r) + g(r,x(r),x_r)\right]dr$ $(t \geq t_0)$

$y(t,t_0,\phi_2) = X(t)X^{-1}(t_0)Dy_{t_0} + E(t)y_t + \int_{-\infty}^t B(t,s)y(s)ds$

$\int_{t_0}^t X(t)X^{-1}(r)\left[\int_{-\infty}^r G(r,s)y(s)ds + A(r)E(r)y(r) + g(r,y(r),y_r)\right]dr$ $(t \geq t_0)$

Therefore, for $\forall \varepsilon > 0$, I et $\delta = \frac{\varepsilon}{k(1+k_1+E)}$ then we have

$\|x(t,t_0,\varphi_1) - y(t,t_0,\phi_2)\| < \varepsilon$ (7)

When $\|\phi_1 - \phi_2\| < \delta$.

Otherwise it must exist $t_1 > t_0$ such that

$\|x(t,t_0,\phi_1) - y(t,t_0,\phi_2)\| < \varepsilon$ $(t_0 < t < t_1)$

However, $\|x(t_1,t_0,\phi_1) - y(t_1,t_0,\phi_2)\| = \varepsilon$

$\therefore \varepsilon = \|x(t_1,t_0,\phi_1) - x(t_1,t_0,\phi_2)\| \leq \|X(t_1)X^{-1}(t_0)\| \cdot \|Dx_{t_0} - Dy_{t_0}\|$

$+ \|E(t)\| \|x_{t_1} - y_{t_1}\| +$

$\int_{-\infty}^{t_1} \|B(t_1,s) \cdot \| \|x(s) - y(2)\| ds +$

$\int_0^{t_1} \|X(t_1)X^{-1}(r)\| \left[\int_{\aleph} \|G(r,s)\| \|x(s) - y(s)\| ds\right.$

$+ A(r)E(r)\| \|x_t = y_r\| + \alpha_1(r)\|x(r) - y(r)\| + \alpha_2(r)\|\cdot$

$+\alpha_2(r)\|x_r - y_r\|\left.\right]d$

$\leq \exp\int_{t_0}^{t_1} b(r)dr \cdot \left[\|x(t_0) - y(t_0)\| + \int_{-\infty}^{t_0} \|B(t_0,s)\| \cdot \|\right.$

$\phi_1(s) - \phi_2(s)\|\left.\right]ds$

$\|E(t_0)\| \|x_0 - y_0\| + k_1\varepsilon + \varepsilon\int_0^1 \exp\left(\int_r^1 b(\tau)dr\right) \cdot \Big|$

$\left[\int_{-\infty}^r \|G(r,s)\| ds + A(r)E(r)\| + \alpha_1(r) + \alpha_2(r)\right]dr$

$\leq \exp\int_{t_0}^{t_1} b(r)dr(1+k_1+E)\delta + (k_1+E)\varepsilon + \varepsilon\int_{t_1}^{t_0}\exp$

$\left(\int_r^{t_1} b(\tau)dr\right) \cdot \left(\frac{b(r)}{k}\right)dr$

$\leq \exp\int_{t_0}^{t_1} b(r)dr(1+k_1+E)\delta + (k_1+E)\varepsilon +$

$\frac{\varepsilon}{k}\left(1 - \exp\int_{t_0}^{t_1} b(r)dr\right)$

$$< \exp \int_{t_0}^{t_1} b(r)dr \cdot \frac{\varepsilon}{k} + \left(1 - \frac{1}{k}\right)\varepsilon + \frac{\varepsilon}{k}\left(1 - \exp \int_{t_0}^{t_1} b(r)dr\right) = \varepsilon$$

This contradicts the hypothesis.
So Eq. (7) is true, namely, the solution of Eq. (1) is uniformly stable. The proof of theorem1 is over.

REFERENCES

Wang, Q. Y. (2002): Existence Uniqueness and Stability of Almost Periodic Solution to a Class of Neutral Tape of Functional Differential Equation. *Journal of HuaQiao university natural science.*23(3):222–228.

Wang, Q. Y. (2003): Almost Periodic Solution a Class of Neutral Tape of Functional Differential Equation. *Journal of HuaQiao university natural science.* 24(4):349–353.

Fang, C.N., W.Y. Wang(2004): Existence Uniqueness and Stability of Almost Periodic Solution to a Class of Neutral Tape of Functional Differential Equation. *Journal of JiMei university natural science.* 9(1):82–86.

Wang, Q. Y. (1996).: Existence and Uniqueness Almost Periodic Solution with infinite delay. *Journal of HuaQiao university natural* science,17(4)336-340.

Wang, Q. Y. (1997). Existence and Uniqueness and Stability of Almost Periodic Solution. *Acta Mathematical Sinica.*, 40(1):80–89.

Electronic Engineering and Information Science – Wang (Ed.)
© 2015 Taylor & Francis Group, London, ISBN: 978-1-138-02772-5

Homotopy iteration method for nonlinear system

J. Wei, Z.F. Fei, J. Han & F.X. Li
School of Applied Sciences, Harbin University of Science and Technology, Harbin, P. R. China

ABSTRACT: Basing on the 1st-order deformation equation of Homotopy-Analysis Method(HAM), an itera-tion algorithm is presented to solve the nonlinear equations in this paper. The new algorithm is called Homotopy Iteration Method (HIM). The solutions gotten by HIM are better than the ones gotten by HAM in some cases, avoiding the inherent huge computation of HAM. What's more, the advantage of HAM will not be influenced at all.

KEYWORDS: nonlinear equation; homotopy analysis method; homotopy iteration method.

1   INTRODUCTION

In the early 1990s of 20th century, Doc. Liao pro-posed the HAM (Liao 1992) based on homotopy, a fundamental conception in topology and differen-tial geometry. The method offers great freedom in choosing initial approximate solution, auxiliary lin-ear operators. At the same time, it has the function to regulate the region and the convergent rate of the series. Because of the freedom and the advantages mentioned above, it was applied to solve practical problems more and more widely. But the existence of huge computation makes the procedure complex in practical operation. To simplify the process of com-putation, this paper applies the iteration to the HAM, the advantage of HAM will not be influenced at all.

2   THE IDEA OF HAM AND HIM

Let's consider a nonlinear equation

$$N[u(t)] = 0 \qquad (1)$$

For simplicity, we ignore the boundary conditions and initial conditions and construct the zero-order deformation equation based on the HAM

$$(1-q)L[\Phi(t,q) - u_0(t)] = qhH(t)N[\Phi(t,q)] \qquad (2)$$

L is the auxiliary operator with the property

$$L[f(t)] = 0 \text{ if } f(t) = 0 \qquad (3)$$

$u_0(t)$ is the initial guess solution, $q \in [0, 1]$ is embed variable. Obviously, at $q = 0$ the Eq. 2 becomes $L[\Phi(t, q) - u_0(t)] = 0$. In the light of the property (3), we can get $\Phi(t,0) = u_0(t)$. At $q = 1$, together with $h \neq 0$ and

$H(x) \neq 0$, $N[\Phi(t,1)] = 0$ is obtained. It is exactly the same as the original equation, thus

$$\Phi(t,1) = u(t) \qquad (4)$$

Therefore, the process of the embedded variable q varying continuously from 0 to 1 is just the continu-ous variation of $\Phi(t, q)$ from the initial guess $u_0(t)$ to the solution $u(t)$ of original Eq. 1.

Because $\Phi(t, q)$ is the function of the embed varia-ble q, we can expand it into Maclaurin series

$$\Phi(t,q) = \sum_{m=0}^{+\infty} u_m(t)q^m \qquad (5)$$

of which $\Phi(t,0) = u_0(t)$ and

$$u_m(t) = \frac{1}{m!} \frac{\partial^m \Phi(t,q)}{\partial q^m}\bigg|_{q=0} \quad (m \geq 1)$$

are attached. Here, the series (5) is called homotopy-series, $um(t)$ is called the mth-order homotopy-derivative of $u(t)$. Obviously, the convergence of series (5) is dependent on h. Assuming that h is so properly selected that series (5) is convergent at $q = 1$ (Liao 2008, Liao 2009), then setting $q = 1$ in (5), we have

$$u(t) = u_0(t) + \sum_{m=1}^{+\infty} u_m(t) \text{ by (4).}$$

The above formula connects the initial approx-imate solution $u_0(t)$ with the solution $u(t)$ of (1) by means of homotopy-derivative $um(t)$. The mth-order homotopy-derivative can be obtained by differenti-ating the Eq. 2 for m times with respect to q, then dividing by $m!$, and seting $q = 0$, i.e.

$$L[u_m(x) - \chi_m u_0(x)] = hH(x)R_m(u_{m-1},t)$$

where

$$
\begin{cases}
\chi_m = \begin{cases} 0, m \le 1 \\ 1, m > 1 \end{cases} \\
R_m(u_{m-1},t) = \dfrac{1}{(m-1)!} \dfrac{\partial^{m-1} N[\Phi(t,q)]}{\partial q^{m-1}} \Big|_{q=0}
\end{cases}
\tag{6}
$$

We can find that the mth-order deformation Eq. 5 is linear with respect to mth-order homotopy-derivative $u_m(t)$ $(m \ge 1)$. These linear differential equations can be solved one after another in order by a lot of traditional approaches. So the mth-order HAM approximate solution is

$$u(t) \approx u_0(t) + \sum_{k=1}^{m} u_k(t).$$

To overcome the huge computations of HAM, this paper introduces a new method named homotopy iteration (HIM) which is based on the 1st-order deformation equation of HAM.

We here differentiate the zero-order deformation Eq. 2 once with respect to q and set $q = 0$, i.e.

$$L\left[\frac{\partial \Phi(t,q)}{\partial q}\right] = hH(t)N[\Phi(t,q)]\Big|_{q=0} \tag{7}$$

Take $U_0(t) = u[1]_0(t)$ as the initial guess solution, Eq. 7 becomes

$$L[u_1^{[1]}(t)] = hH(t)N[u_0^{[1]}(t)] \tag{8}$$

We can get the expression of $u[1]_1(t) = u_1(t)$ from (8), the 1st-order approximate solution of (1) is $U_1(t) = u[1]_0(t) + u[1]_1(t)$, $\Phi(t, q)$ is a function of the parameter q $(q \in [0,1])$, we here take $U_1(t)$ as the new initial guess solution, the Maclaurin series of $\Phi(t, q)$ is

$$
\begin{aligned}
\Phi(t,q) = U_1(t) + u_2^{[1]}(t)q + u_2^{[2]}(t)q^2 + \cdots \\
+ u_2^{[m]}(t)q^m + \cdots
\end{aligned}
\tag{9}
$$

Substituting (9) into (7), it reads

$$L[u_2^{[1]}(t)] = hH(t)N[U_1(t)] \tag{10}$$

From the equation mentioned above, we can get the expression of $u[1]_2(t)$.

In conclusion, enforcing

$$U_{m-1}(t) = \sum_{k=0}^{m-1} u_k^{[1]}(t),$$

it means that the $(m-1)$th-order approximate solution of nonlinear Eq. 1 is $Um-_1(t)$, we can expand $\Phi(t,q)$ into Maclaurin series with $Um-_1(t)$ as initial guess solution

$$\Phi(t,q) = U_{m-1}(t) + u_m^{[1]}(t)q + u_m^{[2]}(t)q^2 + \cdots \tag{11}$$

When associate with (7)

$$L[u_m^{[1]}(t)] = hH(t)N[U_{m-1}(t)] \tag{12}$$

is obtained. In this way, we obtains $u[1]_m$ one by one in the order $m = 1,2,3\ldots$. So the mth-order approximation of nonlinear Eq. 1 with HIM is

$$u(t) \approx U_m(t) = \sum_{k=0}^{m} u_k^{[1]}(t) \tag{13}$$

Notice that the rate and the region of convergence of series (13) are determined by the value of h which plays a same role on HAM (Liao 1997, Liao 2004).

Theorem. As long as the series

$$\sum_{m=0}^{+\infty} v_m(t)$$

is convergent, it must be the solution of the nonlinear equation $N[v(t)] = 0$, where $v_m(t)$ meets the deformation equation

$$L[v_m(t)] = hH(t)N\left[\sum_{k=0}^{m-1} v_k(t)\right]$$

and subjects to the initial condition, in addition, the function N is continuous.

Therefore, as long as L, $H(x)$, $u_0(x)$ and h are chosen properly, making that the series gotten by HIM is convergent, the series must be convergent to the solution of the nonlinear equation.

As we have seen, the computation process of HIM is based on the 1st-order deformation equation of HAM, it needs take 1st-order derivative only, compared with differentiating the equation for m times, it reduces the computation effectively.

Example. Consider the nonlinear equation (Liao 1997)

$$v'(t) + v^2(t) = 1, \; t \ge 0, \; v(0) = 0 \tag{14}$$

whose exact solution is $v(t) = \tanh(t)$. Choosing $\{t^{2k+1} \mid k = 0,1,2,3,4\ldots\}$ as the basic functions to express the solution of (14), that is

$$v(t) = \sum_{k=0}^{+\infty} a_k t^{2k+1} \tag{15}$$

which a_k is the coefficient.

According to the expression of solution, along with the initial condition it's obvious to choose $v_0(t) = t$ as the initial guess solution. Choosing

$$L[\Phi(t,q)] = \frac{\partial \Phi(t,q)}{\partial t}$$

as the auxiliary linear operator, obviously, the constant c is a solution of the linear equation $L[v] = 0$, and defining the nonlinear operator

$$N[\Phi(t,q)] = \frac{\partial \Phi(t,q)}{\partial t} + \Phi^2(t,q) - 1$$

We construct such a zero-order deformation equation

$$(1-q)L[\Phi(t,q) - v_0(t)] = hH(t)N[\Phi(t,q)]$$

where $u0(t)$ is an initial guess meeting the initial condition. In this case, we choose $H(t) = 1$ as the auxiliary function (Liao 2004) according to the expression of solution and the rule of ergodicity for coefficient.

Using HIM, one has the corresponding solutions

$$v_1(t) = \frac{1}{3}ht^3$$

$$v_2(t) = \frac{1}{3}h(1+h)t^3 + \frac{2}{15}h^2t^5 + \frac{1}{63}h^2t^7$$

$$v_3(t) = \frac{h}{3}(1+h)^2t^3 + \frac{h^2}{15}(2+3h)t^5 + \frac{h^3}{315}(5h^2 + 25h + 37)t^7$$

So the mth-order approximate solution of (14) is

$$v(t) \approx \sum_{k=0}^{m} v_k(t).$$

Similarly, by the HAM one can get that

$$v_1'(t) = \frac{1}{3}ht^3$$

$$v_2'(t) = \frac{1}{3}h(1+h)t^3 + \frac{2}{15}h^2t^5$$

$$v_3'(t) = \frac{1}{3}h(1+h)^2t^3 + \frac{4}{15}h^2(1+h)t^5 + \frac{17}{315}h^3t^7$$

The corresponding mth-order approximate solution of (14) is

$$v'(t) \approx \sum_{k=0}^{m} v_k'(t).$$

Table 1. The comparison between 5th-order approximate solution and the exact solution at $h = -1$.

	HIM		HAM		
x	Approximate solution	Absolute error	Approximate solution	Absolute error	Exact solution
0.1	0.099668	2.78×10^{-17}	0.099668	3.47×10^{-16}	0.099668
0.3	0.291313	3.43×10^{-11}	0.291313	5.53×10^{-10}	0.291313
0.5	0.462117	2.23×10^{-8}	0.462117	3.98×10^{-7}	0.462117
0.7	0.604366	1.41×10^{-6}	0.604339	2.90×10^{-5}	0.604368
0.9	0.716270	2.78×10^{-5}	0.715610	6.87×10^{-4}	0.716298

Table 2. The comparison between 10th-order approximate solution and the exact solution at $h = -1$.

	HIM		HAM		
x	Approximate solution	Absolute error	Approximate solution	Absolute error	Exact solution
0.1	0.099668	0	0.099668	0	0.099668
0.3	0.291313	5.55×10^{-17}	0.291313	0	0.291313
0.5	0.462117	2.78×10^{-16}	0.462117	4.25×10^{-12}	0.462117
0.7	0.604368	3.74×10^{-13}	0.604368	8.97×10^{-9}	0.604368
0.9	0.716298	7.36×10^{-11}	0.716300	2.62×10^{-6}	0.716298

It is obvious that the solution of HIM is much closer to the exact solution compared with HAM when we choose the same auxiliary linear operator, auxiliary function, basic function and value of h.

3 CONCLUSIONS

The method of HIM proposed in this paper avoids the inherent huge computation of HAM effectively, what's more, it is turned out that the solution of the HIM is much closer to the exact solution compared with the one gotten by HAM in some cases.

ACKNOWLEDGMENTS

This research was funded by Natural Science Foundation of Heilongjiang Province (No. A200811) and by the Educational Department Scientific Technology Program of Heilongjiang (No. 11521045).

REFERENCES

Liao, S.J. 1992. The Propose Homotopy Analysis Technique for The Solution of Nonlinear Problem. Shanghai: Shanghai Jiao Tong University.

Liao, S.J. 2008. Beyond Perturbation: The Basic Concepts of The Homotopy Analysis Method and Its Applications. *Advances in Mechanics* 38(1): 1–34.

Liao, S.J. 2009. Analysis of Nonlinear Fractional Partial Differential Equations with The Homotopy Analysis Method. *Commun. Nonlinear Sci. Numer. Simulat.* 14: 983–997.

Liao, S.J. 1997. A Kind of Approximate Solution Technique Which Does Not Depend upon Small Parameters — II. An Application in Fluid Mechanics. *Int. J. Non-Linear Mechanics s* 32: 815–822.

Liao, S.J. 2004. Beyond Perturbation: Introduction to the Homotopy Analysis Method. New York: Chapman & Hall/CRC.

Liao, S.J. 1997. Numerically Solving Non-linear Problems by The Homotopy Analysis Method. *Computational Mechnics Vol* 20(6): 530–540.

Liao, S.J. 2004. On The Homotopy Analysis Method for Nonlinear Problems. *Applied Mathematics and Computation* 147(2): 499–513.

Electronic Engineering and Information Science – Wang (Ed.)
© 2015 Taylor & Francis Group, London, ISBN: 978-1-138-02772-5

Integral equation method of mixed crack problem

W. Zheng, W. Sun & C.Z. Bao
Department of Applied Sciences, Harbin University of Science and Technology, Harbin, China

ABSTRACT: In this paper, we discuss a mixed crack problem defined in the exterior of an open arc (crack). It is a boundary value problem with two different boundary conditions. We construct the solution to the problem by a linear combination of an angular potential and a single potential and deduce an equivalent integral system. The system has no super-singular integrals, but integral equations which are deduced by traditional method contain them. Existence and uniqueness of the equations to the system are considered.

KEYWORDS: Laplace equation; angular potential; double layer potential; integral equation.

1 INSTRUCTIONS

Crack problem has attracted more and more attentions because of considerable interest for nondestructive testing in material sciences and the pure mathematical interest. The problem can be governed by Laplace equation with different boundary conditions. The difficulty of the crack problem is the presence of the tips of the crack.

In this paper, we discuss a mixed crack problem. It is a mixed problem value problem for Laplace equation defined in the complementary set of an open arc. The purpose of the paper is to establish the integral equations which are equivalent equations of the problem and to prove that there is a solution to the integral equations in Sobolev spaces.

The results and methods are closed to those of Cakoni & Colton (2003) and Liu & Krutitskii et al. (2011). In Cakoni & Colton (2003), the authors study Helmholtz equation with different boundary conditions in the exterior of an open arc. They use Green formula to obtain the representation of the solution and deduce existence and uniqueness of the solution in Sobolev space. In Liu & Krutitskii et al. (2011), the authors consider the similar problem in Cakoni & Colton (2003). They construct the solution by a linear combination of an angular potential and a single layer potential and prove there is only one so-lution to the problem in Hölder spaces.

Angular potential is considered in Liu & Krutitskii et al. (2011), Krutitskii (2007) and Krutitskii & Kolybasova (2005). It can be considered as a double layer potential under some condition. But compared with the angular potential, double layer potential has more strong singularity near the crack. The integral equations which are obtained with angular potentials

and single layer potentials contain a Cauchy singular integral. But if the solution to the problem is constructed by classical potentials, the equivalent system will have super-singular integrals.

The article is organized as follows. In Section 2, the mathematical model of the problem in two dimensions is presented and some spaces which are used in this paper are listed. In Section 3, we construct the solution by a linear combination of an angular potential and a single layer potential and deduce the equivalent integral systems. In Section 4, existence and uniqueness of the solution to the integral equations is discussed.

2 STATEMENT OF THE PROBLEM

We consider a simple open arc $\Gamma := \{x := (x_1(s), x_2(s)), s \in [a,b]\}$, which is oriented smooth non-intersecting. Γ^+ is the left-side of Γ as the parameter s increases. Thus the other side is denoted by Γ^-. The normal unite vector n of Γ is denoted into Γ^-.

First, we extend Γ to an arbitrary closed curve ∂D such that the normal vector on ∂D coincides with the normal vector on Γ. Define

$$L^2(\Gamma) := \{u|_\Gamma : u \in L^2(\partial D)\},$$

$$H^{\frac{1}{2}}(\Gamma) := \{u|_\Gamma : u \in H^{\frac{1}{2}}(\partial D)\},$$

$$\tilde{H}^{\frac{1}{2}}(\Gamma) := \{u \in H^{\frac{1}{2}}(\Gamma) : supp\, u \subseteq \bar{\Gamma}\}.$$

If $u \in \tilde{H}^{\frac{1}{2}}(\Gamma)$, denote its extension by zero to

$$\tilde{H}^{\frac{1}{2}}(\Gamma) := \{u \in H^{\frac{1}{2}}(\Gamma) : supp\, u \subseteq \bar{\Gamma}\}.$$ the whole boundary ∂D by $\tilde{u}$, $\tilde{u} \in H^{\frac{1}{2}}(\partial D)$. We denote the dual space

of $\tilde{H}^{\frac{1}{2}}(\Gamma)$ by $H^{\frac{1}{2}}(\Gamma)$ and the dual space of $H^{\frac{1}{2}}(\Gamma)$ by $\tilde{H}^{-\frac{1}{2}}(\Gamma)$. Hence it is easy to see $\tilde{H}^{\frac{1}{2}}(\Gamma) \subset H^{\frac{1}{2}}(\Gamma) \subset L^2(\Gamma)$ $\tilde{H}^{-\frac{1}{2}}(\Gamma) \subset H^{-\frac{1}{2}}(\Gamma)$. From Cakoni & Colton (2003), $\tilde{H}^{-\frac{1}{2}}(\Gamma)$ is the same as $H_{\Gamma}^{-\frac{1}{2}}(\partial D)$ $:= \{u \in H^{-\frac{1}{2}}(\partial D): \text{supp } u \in \overline{\Gamma}\}$.

Let us consider the mixed crack problem: Given $g \in H^{\frac{1}{2}}(\Gamma)$, $f \in H^{-\frac{1}{2}}(\Gamma)$, find $u \in H_{loc}^1(R^2 \setminus \Gamma)$ such that

$\Delta u = 0 \quad in \ R^2 \setminus \Gamma$,

$\dfrac{\partial u}{\partial n}\Big|_{x \in \Gamma^+} = f$,

$u\big|_{x \in \Gamma^-} = g$,

$u = O(1) \quad as \ x \to \infty$.

Theorem 1.(Cakoni & Colton (2003)) The mixed problem has at most one solution.

3 REDUCTION OF THE PROBLEM TO INTEGRAL EQUATIONS

Consider an angular potential

$v[\mu](x) = -\dfrac{1}{2\pi} \int_{\Gamma} \mu(y(\sigma)) V(x, y(\sigma)) ds_y$,

where the kernel is defined by

$\cos V(x, y) = \dfrac{x_1 - y_1(\sigma)}{|x - y(\sigma)|}$,

$\sin V(x, y) = \dfrac{x_2 - y_2(\sigma)}{|x - y(\sigma)|}$,

$y(\sigma) = (y_1(\sigma), y_2(\sigma)) \in \Gamma$,

$|x - y(\sigma)| = \sqrt{(x_1 - y_1(\sigma))^2 + (x_2 - y_2(\sigma))^2}$.

It is easy to see $V(x, y)$ is the angle between the vector $y(\sigma)x$ and x-axis. $V(x, y)$ is a many-value function of x. Because of the definition of $V(x, y)$, $v[\mu](x)$ is a many-value function. We add an additional condition which makes $V(x, y)$ be a single-value function:

$\int_{\Gamma} \mu ds = 0$.

Integrating it by parts, the angular potential becomes a double layer potential:

$v[\mu](x) = \dfrac{1}{2\pi} \int_{\Gamma} \rho[\mu] \dfrac{\partial}{\partial n_y} \ln |x - y(\sigma)| ds_y$,

where $\rho[\mu] = \int_{\Gamma_\sigma} \mu ds$ and $\Gamma_\sigma := \{y = y(s) = (y_1(s), y_2(s)), s \in [a, \sigma]\}$.

We construct the solution of the mixed crack problem by sum of an angular potential and a single layer potential.

$u[\phi_1, \mu] = v[\mu] - w[\phi_1]$, \hfill (1)

where

$w[\phi] = -\dfrac{1}{2\pi} \int_{\Gamma} \phi(y) \ln |x - y(\sigma)| ds_y$.

Theorem 2. If $g \in H^{\frac{1}{2}}(\Gamma)$, $f \in H^{-\frac{1}{2}}(\Gamma)$, $\varphi_1 \in \tilde{H}^{-\frac{1}{2}}(\Gamma)$, $\mu \in \tilde{H}^{-\frac{1}{2}}(\Gamma)$ satisfy the following system:

$$
\begin{cases}
-\dfrac{1}{2\pi}\int_{\Gamma}\phi_1(y)\ln|x-y(\sigma)|ds_y - \dfrac{1}{2\pi}\int_{\Gamma}\mu(y)V(x,y)ds_y \\
\qquad\qquad\qquad\qquad -\dfrac{1}{2}\rho[\mu] = g, \\[2mm]
-\dfrac{1}{2\pi}\int_{\Gamma}\phi_1(y)\dfrac{\partial \ln|x-y(\sigma)|}{\partial n_x}ds_y + \dfrac{1}{2}\phi_1(x) \\
\qquad\qquad -\dfrac{1}{2\pi}\int_{\Gamma}\mu(y)\dfrac{\sin\varphi(x,y)}{|x-y|}ds_y = f, \\[2mm]
\int_{\Gamma}\mu ds = 0.
\end{cases}
$$

where $\phi(x, y)$ is the angular between the vector $\vec{xy}$ and $n(x)$.

Proof. Put (1) into the boundary condition of the problem. Because of the jump relation of the surface potential and

$\lim\limits_{x \to \Gamma^{\pm}} \dfrac{\partial v[\mu]}{\partial n} = -\dfrac{1}{2\pi} \int_{\Gamma} \mu(y) \dfrac{\sin(\phi(x, y))}{|x - y|} ds_y$,

we can get the conclusion.

4 EXISTENCE AND UNIQUENESS OF THE INTEGRAL EQUATIONS

We define some boundary integral operators

$S: \tilde{H}^{-1/2}(\Gamma) \to H^{1/2}(\Gamma)$,

$K: \tilde{H}^{1/2}(\Gamma) \to H^{1/2}(\Gamma)$,

$K': \tilde{H}^{-1/2}(\Gamma) \to H^{-1/2}(\Gamma)$,

$T: \tilde{H}^{1/2}(\Gamma) \to H^{-1/2}(\Gamma)$,

$M: W \to H^{1/2}(\Gamma)$.

where

$$W := \{\mu \in \tilde{H}^{-1/2}(\Gamma): \int_\Gamma \mu ds = 0\}, \quad M\mu = \int_{\Gamma_\sigma} \mu ds,$$

$$S\phi = -\frac{1}{\pi}\int_\Gamma \phi(y)\ln|x-y|ds_y,$$

$$K\phi = -\frac{1}{\pi}\int_\Gamma \phi(y)\frac{\partial \ln|x-y|}{\partial n_y}ds_y,$$

$$K'\phi = -\frac{1}{\pi}\int_\Gamma \phi(y)\frac{\partial \ln|x-y|}{\partial n_x}ds_y,$$

$$T\phi = -\frac{1}{\pi}\frac{\partial}{\partial n_x}\int_\Gamma \phi(y)\frac{\partial \ln|x-y|}{\partial n_y}ds_y.$$

We use these operators to rewrite the system. Let us put the solution (1) into the two boundary conditions of the problem). We obtain

$$\begin{pmatrix} S & -K-I \\ K'+I & -T \end{pmatrix}\begin{pmatrix} \varphi_1 \\ \varphi_2 \end{pmatrix} = \begin{pmatrix} 2g \\ 2f \end{pmatrix}, \quad (2)$$

where

$$\begin{pmatrix} \phi_1 \\ \phi_2 \end{pmatrix} = \begin{pmatrix} 1 & 0 \\ 0 & M \end{pmatrix}\begin{pmatrix} \phi_1 \\ \mu \end{pmatrix}.$$

We denote the matrix operator in (2) by A. It is easy to see that A maps $H = \tilde{H}^{-\frac{1}{2}}(\Gamma)\times\tilde{H}^{\frac{1}{2}}(\Gamma)$ into $H^* = H^{\frac{1}{2}}(\Gamma)\times H^{-\frac{1}{2}}(\Gamma)$.

Theorem 3. A is Fredholm with index zero.

Proof. From Mclean (2000), there exits two compact operators L_S and L_T,

$$L_S: \tilde{H}^{\frac{1}{2}}(\Gamma)\rightarrow H^{\frac{1}{2}}(\Gamma),$$

$$L_T: \tilde{H}^{-\frac{1}{2}}(\Gamma)\rightarrow H^{-\frac{1}{2}}(\Gamma)$$

such that

$$\mathrm{Re}\langle(S+L_S)\varphi,\overline{\varphi}\rangle \geq C\|\varphi\|_{\tilde{H}^{-1/2}(\Gamma)}^2 \text{ for } \varphi \in \tilde{H}^{-1/2}(\Gamma), \quad (3)$$

$$\mathrm{Re}\langle-(T+L_T)\phi,\overline{\phi}\rangle \geq C\|\phi\|_{\tilde{H}^{1/2}(\Gamma)}^2 \text{ for } \phi \in \tilde{H}^{1/2}(\Gamma), \quad (4)$$

Let us define $S_0 := S + L_S$ and $T_0 := -(T + L_T)$.

We rewrite $A = A_0 + L$, where

$$A = \begin{pmatrix} S_0 & -K-I \\ K'+I & T_0 \end{pmatrix},$$

$$L = \begin{pmatrix} -L_S & 0 \\ 0 & L_T \end{pmatrix}.$$

Obviously, $L: H \rightarrow H^*$ is compact. Let $\zeta = (\phi,\varphi)^T$.

$$\langle A_0\zeta,\overline{\zeta}\rangle = (S_0\varphi,\varphi) + (-K\phi,\varphi) - (\phi,\varphi) + (K'\varphi,\phi)$$
$$+ (\varphi,\phi) + (T_0\phi,\phi)$$

where $(\cdot,\cdot)$ denotes the scalar product on $L^2(\Gamma)$. According to (3) and (4),

$$\mathrm{Re}[(S_0\phi,\phi) + (T_0\varphi,\varphi)] \geq C[\|\phi\|_{\tilde{H}^{-1/2}}^2 + \|\varphi\|_{\tilde{H}^{1/2}}^2] = \|\zeta\|_H^2.$$

Because K and K' are adjoint,

$$\mathrm{Re}[(-K\phi,\varphi) + (K'\varphi,\phi)] = \mathrm{Re}[(-K\phi,\varphi) + (\varphi,K\phi)] = \mathrm{Re}[(-K\phi,\varphi) + \overline{(K\phi,\varphi)}] = 0.$$

Thus,

$$\mathrm{Re}\langle(A-L)\zeta,\overline{\zeta}\rangle_{H,H^*} \geq C\|\zeta\|_H^2 \quad \text{for } \zeta \in H.$$

So A is Fredholm with index zero.

Theorem 4. Equation (2) has only one solution.

Proof. From Krutitskii (2009), homogeneous equation (2) has only a trivial solution. It is easy to prove that M has a bounded inverse operator. According to Theorem 4, we get (2) has only one solution.

5 CONCLUSION

This paper has derived the equivalent integral system of the mixed crack problem, where the integral representation for the solution is obtained in the form of a linear combination of an angular potential and a single layer potential. Since the angular potential can be considered as a double layer potential under some condition, existence and uniqueness of the system can be proved. Because the integral equations contain no super-singular integrals, it is possible to compute the problem numerically better.

ACKNOWLEDGEMENTS

The work was supported by Education Department of Heilongjiang No. 12531136.

REFERENCES

Cakoni F. & Colton D. 2003. The linear sampling method for cracks, *Inverse Problem*, 19: 279–295.
Liu J. & Krutitskii P.A. & Sini M. 2011. The linear sampling method for cracks, *J. Comput. Math.*, 29: 141–166.

Krutitskii P.A. 2007. Properties of solutions of the Dirichlet problem for Helmholtz equation in a two-dimensional domain with cuts, *Diff. Equ.* , 43: 1200–1212.

Krutitskii P.A. & Kolybasova V.V. 2005. V.V.A generalization of the Neumann problem for the Helmholtz equation outside cuts on the plane, *Diff. Equ.* , 43:1213–1224.

Mclean W. 2000. Strongly Elliptic Systems and Boundary Integral Equations, *Cambridge University*, Cambridge.

Krutitskii P.A. 2009. Boundary Value Problem for the Laplace Euqation Outside Cuts on the Plane with Different Conditions of the Third Kind in Opposite Sides of the Cuts, *Diff. Equ.* 45:86–10.

Electronic Engineering and Information Science – Wang (Ed.)
© 2015 Taylor & Francis Group, London, ISBN: 978-1-138-02772-5

Research on parameterizations of stable distribution

H.L. Chen
School of Computer Science and Technology, Harbin University of Science and Technology, Harbin China
College of Computer Science and Technology, Harbin Engineering University, Harbin China

J.T. Wang, Y. Zhang & C.L. Liu
School of Computer Science and Technology, Harbin University of Science and Technology, Harbin China

ABSTRACT: Non-Gaussian signal processing is a new signal processing field with the development of signal processing techniques in recent years. Alpha stable distribution is a useful tool to study non-Gaussian signal, however, there are multiple parameterizations, and various parameterizations make researchers confused. This paper sets forth four representations of stable random variables characteristic function, clears the meaning of each parameter and finds relations between parameters in these representations. Secondly, we focus on researching the relevant properties in different parameterizations and obtaining the change of the related parameters, which are convenient to calculate in the future. Finally, simulate the probability density function of alpha-stable distribution in different parameterizations, and analyze advantages of the parameterization.

KEYWORDS: stable distribution; parameterization; characteristic function; probability density function.

1 INTRODUCTION

Probability Density Function (PDF) closed expression of stable distribution does not exist, other than Gaussian distribution, Cauchy distribution and Levy distribution, so stable distribution is usually characterized by characteristic function. Mathematicians were interested in parameterizations that the characteristic function has the simplest form, namely, standard parameterization. However, standard parameterization used to stable distribution is not suitable for numerical calculation and modeling. In addition, there are multiple parameterizations and various parameterizations make researchers confused. For example, Chambers et al. (1976) study the stable random number generator, which has a slightly incorrect form: Most researchers expect to get a $S_\alpha(1, \beta, 0)$ result, however, the routine actually returns random variables with a $S_\alpha(1, \beta, -\beta \tan \pi\alpha/2)$ distribution when $\alpha \neq 1$, and a $S_\alpha(1, \beta, 0)$ distribution when $\alpha = 1$. And another example is that Hall (1981) describes a "comedy of errors" caused by parameterization choices. The most common mistake concerns the sign of the skewness parameter when $\sim= 1$ (Holt & Crow 1973).

After analyzing the use of various parameterizations, we decided to define some parameters with more intuitive meanings which will cause the expression of characteristic function more complicated,

while the connection between the parameters is more intuitive. This paper will preferably help researchers make suitable choices.

2 CHARACTERIZATIONS OF ALPHA STABLE DISTRIBUTION PARAMETERS

From now on in this paper superscripts shall specify the type of the parameterization.

Parameterization 1. (Zolotarev 1986) (Lévy-Khintchine representation) We will write that $X \sim S_\alpha^{(1)}(\sigma^{(1)}, \beta^{(1)}, \mu^{(1)})$, if the characteristic function of X random variable has the following representation

$$\phi_X(t) = \begin{cases} \exp\{-(\sigma^{(1)}|t|)^\alpha[1 - i\beta^{(1)}\mathrm{sign}(t)tg\frac{\pi\alpha}{2}] + i\mu^{(1)}t\}, \alpha \neq 1, \\ \exp\{-\sigma^{(1)}|t|[1 + i\beta^{(1)}\frac{2}{\pi}\mathrm{sign}(t)\ln(|t|)] + i\mu^{(1)}t\}, \alpha = 1, \end{cases} \quad (1)$$

where $\alpha \in (0, 2]$, $\beta^{(1)} \in [-1, 1]$, $\mu^{(1)} \in \mathrm{R}$, $\sigma^{(1)} > 0$.

Canonical representation (1) has one peculiarity, namely, the function has discontinuities in all points for $\alpha = 1$, $\beta^{(1)} \neq 0$. V.M. Zolotarev in his study (Liu 2014) suggested the following parameterization:

Parameterization 2. (Chen 2014) We will write that $X \sim S_\alpha^{(2)}(\sigma^{(2)}, \beta^{(2)}, \mu^{(2)})$, if the characteristic function of X random variable has the following representation

$$\phi_X(t) = \begin{cases} \exp\{-(\sigma^{(2)}|t|)^\alpha[1+i\beta^{(2)}\operatorname{sign}(t)tg\frac{\pi\alpha}{2}((\sigma^{(2)}|t|)^{1-\alpha}-1)]+i\mu^{(2)}t\}, \alpha \neq 1, \\ \exp\{-\sigma^{(2)}|t|[1+i\beta^{(2)}\frac{2}{\pi}\operatorname{sign}(t)\ln(\sigma^{(2)}|t|)]+i\mu^{(2)}t\}, \alpha = 1, \end{cases} \quad (2)$$

where $\alpha \in (0,2]$, $\beta^{(2)} \in [-1,1]$, $\mu^{(2)} \in R$, $\sigma^{(2)} > 0$.

Expression (2) of the characteristic function is continuous in α and $\beta^{(2)}$. The peculiarity of this form resides in the fact that $\mu^{(1)}$ has no natural interpretation as the positional parameter. That is why the majority of authors uses (1) form to express the characteristic function of α - stable random variable.

There are the following relations between parameters in (1) and (2):

$$\beta^{(2)} = \beta^{(1)}, \ \sigma^{(2)} = \sigma^{(1)}, \ \mu^{(2)} = \begin{cases} \mu^{(1)} + \beta^{(1)}\sigma^{(1)}tg\frac{\pi\alpha}{2}, \alpha \neq 1, \\ \mu^{(1)} + \beta^{(1)}\sigma^{(1)}\frac{2}{\pi}\ln\sigma^{(1)}, \alpha = 1. \end{cases} \quad (3)$$

Parameterization 3. (Janicki & Weron 1994) We will write that $X \sim S_\alpha^{(3)}(\lambda^{(3)}, \beta^{(3)}, \gamma^{(3)})$, if the characteristic function of X random variable has the following representation

$$\phi_X(t) = \begin{cases} \exp\{\lambda^{(3)}[it\gamma^{(3)} - |t|^\alpha + it|t|^{\alpha-1}\beta^{(3)}tg\frac{\pi\alpha}{2}]\}, \alpha \neq 1, \\ \exp\{\lambda^{(3)}[it\gamma^{(3)} - |t| - it\beta^{(3)}\frac{2}{\pi}\ln|t|]\}, \alpha = 1, \end{cases} \quad (4)$$

where

$$\alpha \in (0,2], \ \beta^{(3)} \in [-1,1], \ \gamma^{(3)} \in R, \ \lambda^{(3)} > 0.$$

Parameters in (4) are related to parameters of representation (1) as follows:

$$\lambda^{(3)} = (\sigma^{(1)})^\alpha, \ \gamma^{(3)} = \frac{\mu^{(1)}}{\lambda^{(3)}}, \ \beta^{(3)} = \beta^{(1)}.$$

Parameterization 4. (Uchaikin & Zolotarev 1999). We will write that $X \sim S_\alpha^{(4)}(\lambda^{(4)}, \beta^{(4)}, \gamma^{(4)})$, if the characteristic function of X random variable has the following representation

$$\phi_X(t) = \begin{cases} \exp\{\lambda^{(4)}[it\gamma^{(4)} - |t|^\alpha \exp(-i\frac{\pi}{2}\beta^{(4)}K(\alpha)\operatorname{sign}(t))]\}, \alpha \neq 1, \\ \exp\{\lambda^{(4)}[it\gamma^{(4)} - |t|(\frac{\pi}{2} + it\beta^{(4)})\operatorname{sign}(t)\ln|t|]\}, \alpha = 1. \end{cases} \quad (5)$$

where

$$\alpha \in (0,2], \ \beta^{(4)} \in [-1,1], \ \gamma^{(4)} \in R, \ \lambda^{(4)} > 0,$$

$$K(\alpha) = \alpha - 1 + \operatorname{sign}(1-\alpha).$$

Parameters in (5) are related to parameters of representation (4) as follows:

if $\alpha \neq 1$, then

$$\lambda^{(4)} = \lambda^{(3)}\left[\cos\left(\frac{\pi}{2}\beta^{(4)}K(\alpha)\right)\right]^{-1},$$

$$\gamma^{(4)} = \gamma^{(3)}\cos\left(\frac{\pi}{2}\beta^{(4)}K(\alpha)\right),$$

$$tg\left(\frac{\pi}{2}\beta^{(4)}K(\alpha)\right) = \beta^{(3)}tg\left(\frac{\pi\alpha}{2}\right);$$

if $\alpha = 1$, then

$$\lambda^{(4)} = \frac{2}{\pi}\lambda^{(3)}, \ \gamma^{(4)} = \frac{\pi}{2}\gamma^{(3)}, \ \beta^{(4)} = \beta^{(3)}. \quad (6)$$

3 PROOF OF RELEVANT PROPERTIES

Characteristic function is a useful tool to study stable distribution, so we should use it to get some basic properties of stable random variables. We know that some properties are usually studied in the parameterizations 1, however, if these properties are equally applicable to other parameterizations, which will be researched below.

Theorem 1. If $X_1, X_2, ..., X_n$ – independent random variables, $X_j \sim S_\alpha^{(1)}(\sigma_j, \beta_j, \mu_j)$, $j = \overline{1,n}$, then

$$X_1 + X_2 + ... + X_n \sim S_\alpha^{(1)}(\sigma, \beta, \mu),$$

where

$$\sigma = (\sigma_1^\alpha + ... + \sigma_n^\alpha), \ \beta = \frac{\beta_1\sigma_1^\alpha + ... + \beta_n\sigma_n^\alpha}{\sigma_1^\alpha + ... + \sigma_n^\alpha},$$

$$\mu = \mu_1 + ... + \mu_n.$$

If $X \sim S_\alpha^{(1)}(\sigma, \beta, \mu)$, $a \in R$, $a \neq 0$, then aX is also an α-stable random variable, and we will write that $aX \sim S_\alpha^{(1)}(\sigma_0, \beta_0, \mu_0)$

where

$$\sigma_0 = |a|\sigma, \ \beta_0 = \beta\operatorname{sign}(a),$$

$$\mu_0 = \begin{cases} a\mu & \text{when } \alpha \neq 1, \\ a\mu - \beta\frac{2}{\pi}\sigma a\ln|a| & \text{when } \alpha = 1. \end{cases}$$

By generalization of theorem 1, we get the following conclusion.

Theorem 2. If $X_1, X_2, ..., X_n$ are independent random variables, $X_j \sim S_\alpha^{(2)}(\sigma_j, \beta_j, \mu_j)$, $j = \overline{1,n}$, then

$$X_1 + X_2 + ... + X_n \sim S_\alpha^{(2)}(\sigma, \beta, \mu),$$

where

so that

$$\sigma = (\sigma_1^\alpha + ... + \sigma_n^\alpha)^{1/\alpha}, \quad \beta = \frac{\beta_1\sigma_1^\alpha + ... + \beta_n\sigma_n^\alpha}{\sigma_1^\alpha + ... + \sigma_n^\alpha},$$

$$\mu = \begin{cases} \sum_{j=1}^n \mu_j + tg\frac{\pi\alpha}{2}\Big(\beta\sigma - \sum_{j=1}^n \beta_j\sigma_j\Big), \alpha \neq 1, \\ \sum_{j=1}^n \mu_j + \frac{2}{\pi}\Big(\beta\sigma\ln\sigma - \sum_{j=1}^n \beta_j\sigma_j\ln\sigma_j\Big), \alpha = 1. \end{cases}$$

If $X \sim S_\alpha^{(2)}(\sigma,\beta,\mu)$, $a \in R, a \neq 0$, then

$aX \sim S_\alpha^{(2)}(\sigma_0,\beta_0,\mu_0)$, where $\sigma_0 = |a|\sigma, \beta_0 = \beta\text{sign}(a)$,

$\mu_0 = a\mu$.

By generalization of theorem 2 we draw the following conclusion.

Theorem 3. If $X_1, X_2, ..., X_n$ are independent random variables, $X_j \sim S_\alpha^{(3)}(\lambda_j,\beta_j,\gamma_j)$, $j = \overline{1,n}$, then

$$X_1 + X_2 + ... + X_n \sim S_\alpha^{(3)}(\lambda,\beta,\gamma),$$

where

$$\lambda = \lambda_1 + ... + \lambda_n, \quad \beta = \frac{1}{\lambda}\sum_{j=1}^n \beta_j\lambda_j, \quad \gamma = \frac{1}{\lambda}\sum_{j=1}^n \lambda_j\gamma_j.$$

If $X \sim S_\alpha^{(3)}(\lambda,\beta,\gamma)$, $a \in R, a \neq 0$, then

$aX \sim S_\alpha^{(3)}(\lambda_0,\beta_0,\gamma_0)$ is true, where

$$\lambda_0 = |a|^\alpha \lambda, \quad \beta_0 = \beta\text{sign}(a),$$

$$\gamma_0 = \begin{cases} |a|^{-\alpha}a\gamma & \text{when } \alpha \neq 1, \\ \text{sign}(a)[\gamma - \beta\frac{2}{\pi}\ln|a|] & \text{when } \alpha = 1. \end{cases}$$

By generalization of theorem 3 we get the following conclusion.

Theorem 4. If $X_1, X_2, ..., X_n$ are independent random variables, $X_j \sim S_\alpha^{(4)}(\lambda_j,\beta_j,\gamma_j)$, $j = \overline{1,n}$, then

$$X_1 + X_2 + ... + X_n \sim S_\alpha^{(4)}(\lambda,\beta,\gamma),$$

where

$$\gamma = \frac{\lambda_1\gamma_1 + \lambda_2\gamma_2 + ... + \lambda_n\gamma_n}{\lambda} = \frac{1}{\lambda}\sum_{j=1}^n \lambda_j\gamma_j, \quad \beta \in [-1;1],$$

$$\lambda = \begin{cases} \{[\sum_{j=1}^n \lambda_j\cos(\frac{\pi}{2}\beta_j K(\alpha))]^2 + [\sum_{j=1}^n \lambda_j\sin(\frac{\pi}{2}\beta_j K(\alpha))]^2\}^{1/2}, \alpha \neq 1, \\ \lambda_1 + \lambda_2 + ... + \lambda_n \quad\quad\quad\quad , \alpha = 1, \end{cases}$$

$$\cos(\frac{\pi}{2}\beta K(\alpha)) = \frac{1}{\lambda}\sum_{j=1}^n \lambda_j\cos(\frac{\pi}{2}\beta_j K(\alpha)), \alpha \neq 1$$

$$\beta = \frac{\lambda_1\beta_1 + \lambda_2\beta_2 + ... \lambda_n\beta_n}{\lambda} = \frac{1}{\lambda}\sum_{j=1}^n \lambda_j\beta_j, \alpha = 1.$$

If $X \sim S_\alpha^{(4)}(\lambda,\beta,\gamma)$, $a \in R, a \neq 0$, then $aX \sim S_\alpha^{(4)}(\lambda_0,\beta_0,\gamma_0)$ is true, where

$$\lambda_0 = |a|^\alpha \lambda, \quad \beta_0 = \beta\text{sign}(a),$$

$$\gamma_0 = \begin{cases} |a|^{-\alpha}a\gamma & \text{when } \alpha \neq 1, \\ \text{sign}(a)[\gamma - \beta\frac{2}{\pi}\ln|a|] & \text{when } \alpha = 1. \end{cases}$$

By generalization of theorem 4 we get the following conclusion.

4 SIMULATIONS AND ANALYSIS

Combining with the conversion relationship between parameters of four parameterizations described above and demonstrating the relevant properties. Fig. 1 and Fig. 2 depicts the curve changes of PDF in the parameterization 1 and parameterization 2. Compared to the parameterization 1, only the meaning of μ has changed in the parameterization 2. We can see that $\mu^{(2)}$ does not indicate median of alpha stable distribution any more.

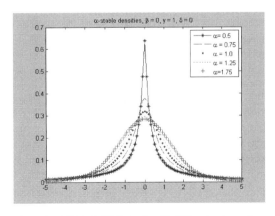

Figure 1. PDF of $S_\alpha^{(1)}(1,0,0)$ with different values of α.

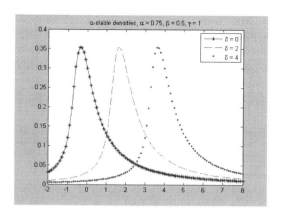

Figure 2. PDF of $S_{0.75}^{(2)}(1,0.5,\mu^{(2)})$ with different values of μ.

With the same principle of obtaining probability density function of Alpha stable distribution in the parameterization 2, compared to the standard parameterization, the meaning of β,γ have changed in the parameterization 3. Fig. 3 depicts the curve changes of $S_\alpha^{(3)}(1,0,0)$ probability density function with α. Combined Fig. 1 with Fig. 3, we can see that when $\beta = 0, \lambda = 1, \gamma = 0$, the probability density function will appear the phenomenon in different parameterizations.

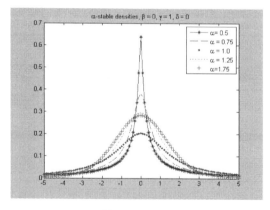

Figure 3. PDF of $S_\alpha^{(3)}(1,0,0)$ with different values of α.

5 CONCLUSIONS

This paper clarifies the relationships between four common parameterizations and gives a unified description of each parameterization. The clearer the meaning of parameters in the common parameterizations, and draw the conversion relationships between various parameters. Study the properties in different parameterizations, and analyze the changes in their parameters further. Finally, simulate the probability density function of alpha-stable distribution in different parameterizations.

ACKNOWLEDGMENT

This work was financially supported by Natural Science Foundation of Heilongjiang Province of China (Grant No. A201301), Doctoral Fund of Ministry of Education of China (Grant No. 20122303120005), The 51th class General Financial Grant from the China Postdoctoral Science Foundation (Grant No. 2012M510926), Heilongjiang Postdoctoral Financial assistance (Grant No. LBH-Z12069).

REFERENCES

Chambers, J. M., C. L. Mallows, & B. W. Stuck(1976). A method for simulating stable random variables, *Journal of the American Statistical Association. 71,* 340–344.
Hall P. (1981). A comedy of errors: the canonical form for a stable characteristic function, *Bulletin of the London Mathematical Society. 13,* 23–27.
Holt D. R., & E. L. Crow (1973). Tables and graphs of the stable probability density functions, *Journal of Research of the National Bureau of Standards B, Vol. 77,* p. 143–198.
Zolotarev V. M. (1986). One-dimensional stable distributions, *American Mathematical Soc.*
Liu C. Li., H. L. Chen, L. Shao, J. T. Wang, & C. J. You (2014). Parameterizations and parameters relations of stable distribution. *VMEIT 2014. 543–547,* 1721–1727.
Chen H. L., J. Yang, C. L. Liu, & J. T. Wang (2014). Parameters estimations for autoregressive process. *VMEIT 2014, 543–547,* 1711–1716.
Janicki A., & A. Weron (1994). Can one see α-stable variables and processes?, *Statistical Science,* 109–126.
Uchaikin V. V., & V. M. Zolotarev (1999). Chance and Stability: Stable Distributions and their applications, *Walter de Gruyter.*

Electronic Engineering and Information Science – Wang (Ed.)
© 2015 Taylor & Francis Group, London, ISBN: 978-1-138-02772-5

Spam filtering based on AdaBoost and active learning methods

X.Y. Liu & J.H. Liu
College of Computer Science and Technology, Harbin University of Science and Technology, Harbin, China

ABSTRACT: In this paper, a new method is proposed to enhance the spam filtering performance. In the spam filter of the past, an over-fitting situation occurs frequently. AdaBoost method is used to solve the problem, and through the weighted weak classifier finally improves the classification performance. We select a suitable model as a weaker classifier to solve different data. Labeling samples require high cost in spam filtering. In order to solve this problem, the active learning methods are introduced to reduce the number of label feedback and the costs. The active learning method selects samples which is the most informative to improve the classification performance. We improve the performance of a spam classifier effectively through the AdaBoost method and the active learning method. Our experiments prove that the combination of these methods can reduce a lot of label feedback effectively and enhance the classification performance.

KEYWORDS: AdaBoost; over-fitting; active learning; spam filtering.

1   INTRODUCTION

Traditional spam classification has a phenomenon of over-fitting. The AdaBoost method can avoid over-fitting situations arise, and by the weighted number of soft classifiers can improve classification performance. To obtain the true labels may require major effort and incur excessive costs (Zliobaite & Bifet 2014). We reduce the number of labeling by active learning method (Cohn 1994, Attenberg 2011).

We propose a method by using a combination of AdaBoost and active learning. The method can avoid over-fitting phenomenon, it also can mark a lower cost to get a higher classification performance. In this paper, we will explain how the use of a specific combination of AdaBoost and active learning to solve the above problems, and the experiments show that our method is effective.

Current spam filtering systems use a simple linear classifier, such as Logistic Regression classifier (Klinkenberg 2001). Although this type of classification is simple and fast, the classification results are often not ideal. It's a great impact on people if the ham has been identified as spam. They may miss some important e-mail messages. Therefore, how to improve the accuracy of classifiers becomes a focus on the moment.

Our AdaBoost classifier is based on some weak classifiers such as a Logistic Regression classifier and Naive Bayesian classifier. Logistic Regression classifier and Naive Bayesian classifier are very suitable as weak classifiers. We need to select a suitable model as a weaker classifier to solve different data.

We raise the weight of the classifier if it gives a right answer, reduce the weight of classifier if it does wrong. Repeat the process again and again to get a strong classifier.

Our method can reduce the number of labels, though the active learning methods (Wang & Hua 2011). This saves a huge cost of labeling and improves the performance of the classifier effectiveness.

We combine the two methods to get a better classifier and overcome some problems. And we prove it is effective by experiments.

We talk about the AdaBoost method firstly. Then we describe the active learning method. Our experiments prove the combination of these methods is effective. At last, we give conclusions of this paper.

2   ADABOOST METHOD

AdaBoost is a representative boosting method, in which an effective statistical method. Spam filter changes the weights of training samples to learn some weak classifiers. We need select suitable models as weak classifiers to solve different data. Through the linear combination of these classifiers, we can improve spam classifier performance.

Spam-filter predicts the coming messages. We mark the predictions and add the feature to the library. Finally the prediction model with the new feature library has been updated.

The feature library is very important. It contains many features of the messages. Every message includes different features. We should do feature selection sometimes, because some features are not useful.

Boosting is a commonly used method for statistical learning method, which is widely used. In the spam classification problem, boosting learns multiple classifiers by changing the right weight of training samples. A linear combination of these classifiers can improve the performance of garbage classification.

For the purposes of classification, seeking a weak classifier is much easier than a strong classifier. Boosting is starting from a weak learning algorithm to get a series of based classifiers, and then combine them into a strong classifier. Most of the boosting methods change the weight distribution of the training data. A series of weak classifiers have been learned to improve the classification accuracy.

Then let's talk about the AdaBoost algorithm, here is a training data set T, $T = [(x_1, y_1), (x_2, y_2),\ldots, (x_n, y_n)]$.

The process of the AdaBoost algorithm is as follows.

Input: a training data set T.

Output: the final classifier $G(x)$.

Initialization the weight distribution of the training data: $D_1 = \{w_{11},\ldots, w_{1i},\ldots, w_{1n}\}$, $w_{1i} = 1/x$, $i=1, 2,\ldots, N$.

We get a basic classifier by using the training data:

$$G_m(x): \chi \to \{-1,+1\} \tag{1}$$

Calculate the classification error rate of $G_m(x)$ on the training data set:

$$e_m = P\left(G_m\left(x_i\right) \ne y_i\right) = \sum_{i=1}^{N} \omega_{mi} I\left(G_m\left(x_i\right) \ne y_i\right) \tag{2}$$

Calculate the coefficient of $G_m(x)$:

$$G_m(x): \alpha_m = \frac{1}{2}\log\frac{1-e_m}{e_m} \tag{3}$$

The weight distribution of training data is updated. Construct a linear combination of basic classifiers to get the final classifier.

Calculate the coefficient of:

$$G(x) = \sin\left(f(x)\right) = \sin\left(\sum_{m=1}^{M}\alpha_m G_m\left(x_i\right)\right) \tag{4}$$

There is a large amount of boosting research, many algorithms have been proposed. We use a representative AdaBoost method to improve the learning accuracy of the spam filter.

AdaBoost raises the weights of samples if they are in the wrong classification last time. As a result of these samples will get more attention next time. Then increase the weights whose classification error rate is small, on the contrary, reduce the weights whose classification error rate is big.

In the spam filtering through the above methods, the training data set can divide and rule, each part is a weak classifier, thus avoiding the occurrence of over-fitting.

AdaBoost method is an effective way to solve many machine learning problems.

3 ACTIVE LEARNING METHOD

The active learning method selects samples which would be informative to improve the classification performance. This method also saves a lot of time. The samples we labeled are obtained from an expert.

Active learning is a way to reduce the cost of labeling. Through different sampling methods, it reduces the number of tags without reducing the accuracy of the classifier. In spam filtering, the use of active learning methods can greatly reduce the cost of training time and the cost of labeling.

An active learning does not casually reduce the number of tags, but by marking some samples, these samples are unsure. According to the strategy of selecting data, it can be roughly divided into two types of active learning algorithms.

Committee-based Method is the one of active learning algorithms. Using a variety of different learning devices for sample annotation marked by the staff and then to make a final judgment on the results of the disputed mark.

The other one is Certainty-based Method.

First, giving a sample with a lower degree of confidence by model, than people labeled selectively.

There are three active learning sampling methods: Label Efficient b-sampling, Logistic Margin Sampling and Fixed Margin Sampling method.

The method we are using belongs to the certainty-based method. It can be combined to AdaBoost more effective and improve the performance of our spam filter.

4 EXPERIMENTAL ANALYSIS

To fully prove the effectiveness of our approach, we use four benchmark data sets to validate our approach. The first is trec06p from TREC spam filtering competitions. It contains 37822 total messages. The second data set is sewm2010 containing 75000 total messages from SEWM publicly available benchmark datasets. The third data set is trec06c with 64320 total messages. The last data set is trec07p with 75419 total messages.

In the graph, we can see the results of our experiment, our classification becomes better and better by the time. ROC is a characteristic curve. The larger value of ROC, the better performance of spam filter.

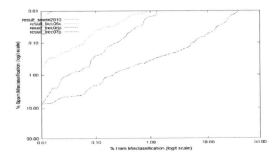

Figure 1. The results of ROC.

From the Fig. 1, we can see that ROC is growing, which means getting better classification results. ROCA is the area of ROC and x-axis.

Therefore, the classification performance is good when the value of (1-ROCA)% is small. We use ROC and (1-ROCA)% as the standard performance measures. They are important indicators to evaluate the performance of a classifier (Qin & Yang 2006).

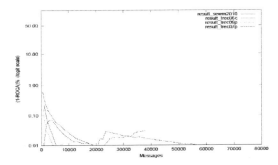

Figure 2. The results of (1-ROCA)%.

Fig. 2 shows that (1-ROCA)% becomes smaller and smaller, and very close to zero by the time.

We observe that (1-ROCA)% finally tends to converge over time, the number of labels requested by our AdaBoost and active learning methods tend to decrease.

For Logistic Regression and Naive Bayesian model, the training data for classification works well, but the effect is not good markers for predicting new data, which is the over-fitting phenomenon, and AdaBoost method for the prediction of new data still with higher accuracy.

5 CONCLUSIONS

Traditional Naive Bayesian or Logistic Regression model has a good performance for the small amount of training data, but there is the phenomenon of over-fitting. When they are dealing with a large number of data, the performance of spam-filter reduces significantly. We prove that AdaBoost method and active learning method can be combined effectively by experiments. In AdaBoost method, we need select a suitable model as the weak classifier to solve different data. Our method can reduce a lot of label feedback and enhance the classification performance.

ACKNOWLEDGEMENT

This paper is supported by Natural Science Foundation of Heilongjiang Province of China (No. F201304).

REFERENCES

Zliobaite I.& A. Bifet(2014). Active learning with drifting streaming data, *IEEE transactions on neural networks and learning systems*, 27–39.

Cohn D., L. Atlas, & R. Ladner(1994), Improving generalization with active learning, *Mach. Learn.*, 201–221.

Attenberg J. F. Provost(2011). Online active inference and learning, *in Proc. 17th ACM SIGKDD Int. Conf. Knowl. Discovery Data Mining*, 186–194.

Klinkenberg R. (2001). Using labeled and unlabeled data to learn drifting concepts, *in Proc. Workshop Notes IJCAI'01 Learn. Temporal Spatial Data*, 16–24.

Wang M. & X. S. Hua (2011). Active learning in multimedia annotation and retrieval, *ACM Transactions on intelligent System and Technology*, 1–3.

Qin F. & B. Yang (2006). Classifier Performance Evaluation Standards, *in Computer Technology and Development*, 1–3.

Electronic Engineering and Information Science – Wang (Ed.)
© 2015 Taylor & Francis Group, London, ISBN: 978-1-138-02772-5

An improved design and realization of the AES encryption algorithm

J.L. Cui, G.L. Huang & P.L. Qiao
Computer science and technology, Harbin University of Science and Technology, Harbin, China

ABSTRACT: The safety performance of S box plays a decisive role in the property of block cipher, so this paper improves the capability of the algorithm against differential cryptanalysis and algebraic attacks through the improved S box which is a nonlinear component in AES encryption algorithm. In comparison with traditional AES encryption algorithm, the improved S box has better algebraic properties.

KEYWORDS: S-box; block cipher; differential cryptanalysis; algebraic attack; nonlinear components.

1 INTRODUCTION

C.E.Shannon introduced two designs Cryptography basic methods: diffusion and confusion which are to resist the invaders on the statistical analysis of cryptosystems. We want to take full advantage of both, especially in block cipher designs which can effectively resist the invaders through the statistical characteristics of the cipher inferred key or plaintext. Diffusion and confusion are the basis of design in modern block cipher, and AES is encryption and decryption algorithm based on a block cipher.

In the actual process of the application AES, we achieve the features of confusion in the password guidelines through SubBytes. Because it is the only nonlinear parts in AES, algorithm of performance is decided by S box. SubBytes transform is realized through the replacement value table of S box. The entire elements in this replacement value table which is in limited domain GF (2^8) have already been prepared. The complexity of AES algebraic expressions is related to the ability to resist attack (Feng 2000, Feng 2002, He 2003, Hu 1999). Meanwhile, it is also an important basis of their safety performance.

We generally adopt mathematics-based method in GF(2^8) domains to solving algebraic expressions of S box and inverse S box (Wei 2005, Wei 2006). Algebraic expression of S box is:

$$
\begin{aligned}
y = {}&'05'x^{-1} + '09'x^{-2} + 'F9'x^{-4} + '25'x^{-8} \\
&+ 'F4'x^{-16} + '01'x^{-32} + 'B5'x^{-19} + '8F'x^{-127} + '63' \\
= {}&'05'x^{254} + '09'x^{253} + 'F9'x^{251} + '25'x^{247} \\
&+ 'F4'x^{239} + '01'x^{223} + 'B5'x^{191} + '8F'x^{127} + '63'
\end{aligned} \tag{1}
$$

S Box algebraic expression is only nine, even though it has a higher frequency, its simplicity is self-evident. The inverse of multiplication cannot increase the number of algebraic expressions, but increase their frequency, and GF (2) affine matrix operations cannot increase coefficient of the expression. It is known that algebraic expression of the inverse S box has 255 non-zero entries, with good algebraic properties, satisfying the security requirements of random S-box and the S-box algebraic expression is too simple.

2 NATURE OF S-BOX

Affine transformation cycle of S box: the length of cycle affects the S-box security in the certain extent. Generally speaking, the longer cycle time is, the higher security will be. According to the construction methods of the S box, it adopts the radiation transformation cycle is 4, but the maximum cycle can reach 16, and it has 5 optional cycles, they are 16,8,4,2,1, so we'd better make the cycle reaching 16.

S-box iterative output cycle: Iteration output cycle is the number of elements in this period, which the element returns to the replacement of the element itself after a series change of S-box. Each element of GF(2^8) is substituted by S-box, namely: $L_{u,v}^{k}(a(x))$ k=1,2..., n; a= 0,1, ..., 255, The entire S-box space capacity is 256, but the cycle of each element is less than 88, even some elements of the cycle are 2. This makes that the differential output of S-box cannot traverse all the values, easily lead to a strong differential attack, so this weakness of short-cycle restricts the iterative output of S-box.

S-box Avalanche probability: An important measure of block ciphers resisting various attacks is the avalanche probability of S-box, meanwhile indirectly reflects the principle of diffusion and confusion-on. Here, we refer to the probability is the probability of the output bits when we change an entering bit. So the best probability is 0.5.

3 IMPROVED DESIGN OF S-BOX

This paper presents the S-box structure improvement program, using three steps: firstly to make once affine transformation of the elements, secondly to reach the conclusion of multiplicative inverse, finally to repeat the first step.

Get the pair of(u,v),that derived by the thought and principles of AES,we used to write by decimal:(254,141)(253,28)(1,4)(2,109)(7,156)(8,111) (11,219)(22,39)(25,238)(31,213)(32,91)(35,85) (37,61)(38,43)(41,49)(42,2)(44,80)(47,156)(50,9) (52,139)(55,110)(61,72)(62,9)(64,2)(67,147) (69,89)(73,6)(76,180)(79,51)(81,53)(87,112) (88,20)(91,178)(94,228)(97,21)(100,147)(104,102) (107,57)(109,6)(112,28)(115,30)(117,73)(118,63) (121,106)(127,3)(128,2)(131,29)(137,35)(138,154) (143,21)(148,126)(151,69)(152,8)(155,154) (157,27)(161,18)(167,111)(168,149)(171,197) (173,175)(174,164)(176,83)(179,56)(181,134) (182,118)(185,197)(188,65)(191,72)(193,95) (194,1)(200,57)(203,57)(206,133)(208,53)(211,53) (213,40)(214,70)(218,32)(220,1)(223,88)(227,219) (229,18)(233,40)(239,3)(241,161)(242,89) (247,126)(248,3)(251,9).

In general,we can use any pair of the coefficient as the affine transformation in the entire space, meanwhile meeting the design guidelines for AES affine transformation. These 91 pairs of the affine transformation are calculated one by one,finding: the avalanche distance of (233,40)(61,72)(79,51) (167,111)(211,53) is 304,the avalanche distance of (213,40)(1,4)(2,109)(8,111)(25,238)(32,91)(35,85) (50,9)(64, 2)(100,147)(128,2)(200,57) is 516,the avalanche distance of (218,32)(91,178)(107,57) (109,6)(173,175)(181,134)(182,118)(214,70) is 488,the avalanche distance of (168,149)(42,2) (69,89)(81,53) (138,154)is 468,the avalanche distance of (248,3) (7,156)(11,219)(22,39)(31,213) (37,61)(38,43)(41,49) (62,9)(73,6)(76,180)(88,20) (97,21)(112,28)(131,29)(137,35)(143,21)(148,126) (152,8)(176,83)(193,95)(194,1)(227,219)(241,161) is 432,the avalanche distance of (206,133) is 444,the avalanche distance of (220,1)(55,110)(115,30) (155,l54)(185,197) is 428,the avalanche distance of (208,53)(52,139)(67,147) (104,102)(161,18) is 424,the avalanche distance of (174,164)(87,112) (117,73)(171,197) is 412, the avalanche distance of (242,89)(47,156)(94,228)(118,63) (121,106) (151,69)(157,27)(179,56)(188,65)(203,57) (229,18) is 368, the avalanche distance of (254,141) (127,3) (191,72)(223,88)(239,3)(247,126)(251,9)(253,28) is 348,according to the definition of avalanche distance,the shorter the avalanche distance,the better the diffusion algorithm, so randomly selecting a pair

from (233,40)(61,72)(79,51)(167,111)(211,53) the operation steps are as follows:

1 We select (167,111)as affine transformation,the hexadecimal representation is ('A7','6F'),draw conclusion:

$$x' = Lb \times x + '6F' = \begin{bmatrix} 1 & 1 & 1 & 1 & 0 & 0 & 1 & 0 \\ 0 & 1 & 1 & 1 & 1 & 0 & 0 & 1 \\ 1 & 0 & 1 & 1 & 1 & 1 & 0 & 0 \\ 0 & 1 & 0 & 1 & 1 & 1 & 1 & 0 \\ 0 & 0 & 1 & 0 & 1 & 1 & 1 & 1 \\ 1 & 0 & 0 & 1 & 0 & 1 & 1 & 1 \\ 1 & 1 & 0 & 0 & 1 & 0 & 1 & 1 \\ 1 & 1 & 1 & 0 & 0 & 1 & 0 & 1 \end{bmatrix} \begin{bmatrix} x_7 \\ x_6 \\ x_5 \\ x_4 \\ x_3 \\ x_2 \\ x_1 \\ x_0 \end{bmatrix} + \begin{bmatrix} 0 \\ 1 \\ 1 \\ 0 \\ 1 \\ 1 \\ 1 \\ 1 \end{bmatrix} \quad (2)$$

2 Multiplicative inverse:

$$x' = (x^n)^{-1} = \begin{cases} (x'')^{254}, x'' \neq 0n \\ 0, x'' = 0n \end{cases} \quad (3)$$

3 In the last,make affine transformation again,y: $y = Lb \times x' + '6F'$;Substitution table (hex) of S-box is shown in Table 1:

We use ('A7','6F') as an affine transformation pair, and have to choose ('D0','35') as the inverse S-box transformation pair:

1 Make affine transformation once of ('D0','35'), the result is:

$$x' = Lbx^{-1} \times y + '35' = \begin{bmatrix} 1 & 0 & 0 & 0 & 0 & 1 & 0 & 1 \\ 1 & 1 & 0 & 0 & 0 & 0 & 1 & 0 \\ 0 & 1 & 1 & 0 & 0 & 0 & 0 & 1 \\ 1 & 0 & 1 & 1 & 0 & 0 & 0 & 0 \\ 0 & 1 & 0 & 1 & 1 & 0 & 0 & 0 \\ 0 & 0 & 1 & 0 & 1 & 1 & 0 & 0 \\ 0 & 0 & 0 & 1 & 0 & 1 & 1 & 0 \\ 0 & 0 & 0 & 0 & 1 & 0 & 1 & 1 \end{bmatrix} \begin{bmatrix} y_7 \\ y_6 \\ y_5 \\ y_4 \\ y_3 \\ y_2 \\ y_1 \\ y_0 \end{bmatrix} + \begin{bmatrix} 0 \\ 0 \\ 1 \\ 1 \\ 0 \\ 1 \\ 0 \\ 1 \end{bmatrix} \quad (4)$$

2 Multiplicative inverse:

$$x' = (x'')^{-1} = \begin{cases} (x'')^{254}, x'' \neq 0n \\ 0, x'' = 0n \end{cases} \quad (5)$$

3. In the last, make affine transformation of ('D0', '35') again,the result of x is: $x = Lbx^{-1} \times x' + '35'$, The substitution table (hex) of inverse S-box is shown in Table 2:

Table 1. Substitution table of improved S-box.

s		y															
		0	1	2	3	4	5	6	7	8	9	a	b	c	d	e	f
x	0	6f	20	c0	a5	b0	d6	0a	a0	88	2a	b3	1b	d5	bd	80	f7
	1	9c	e8	c5	17	09	2d	55	de	32	e4	0e	5c	90	bc	2b	4b
	2	96	8c	ac	49	3a	b4	5b	2e	54	8b	46	e9	72	81	bf	85
	3	c1	ce	a2	12	d7	47	f6	22	98	00	86	c3	4d	7e	7d	ae
	4	93	df	9e	b9	8e	3d	74	60	cd	68	8a	4a	75	f4	c7	5a
	5	fa	0f	1d	ff	fb	bb	24	43	e1	ab	18	06	07	b5	1a	77
	6	38	91	b7	3e	89	82	d1	6d	3b	50	73	79	a3	ed	c9	d9
	7	94	39	d0	ba	9b	a7	31	35	76	a1	ef	cc	6e	e2	87	56
	8	19	78	37	c8	9f	70	0c	eb	97	e5	4e	da	ea	f1	e0	1c
	9	36	b1	ec	5d	95	2c	f5	3f	62	a6	aa	b2	33	a8	fd	d8
	a	ad	e6	5f	ca	5e	af	27	c6	25	61	05	58	c2	1e	71	cb
	b	28	f2	0d	64	d4	d3	db	ee	53	11	02	7a	dd	92	6b	a4
	c	c4	fc	10	f0	0b	d2	cf	f9	14	6a	99	52	30	41	66	51
	d	45	83	f8	84	69	44	6c	14	01	63	26	9d	34	7c	3c	fe
	e	9a	67	4c	59	b8	b6	8d	29	15	65	03	23	40	48	42	04
	f	e3	57	08	4f	2f	1f	be	8f	e7	7b	a9	21	13	7f	f3	dc

Table 2. Substitution table of improved inverse S-box.

s⁻¹		x															
		0	1	2	3	4	5	6	7	8	9	a	b	c	d	e	f
y	0	00	01	8d	f6	cb	52	7b	d1	e8	4f	29	c0	b0	e1	e5	c7
	1	74	b4	aa	4b	99	2b	60	5f	58	3f	fd	cc	ff	40	ee	b2
	2	3a	6e	5a	f1	55	4d	a8	c9	c1	0a	98	15	30	44	a2	c2
	3	2c	45	92	6c	f3	39	66	42	f2	35	20	6f	77	bb	59	19
	4	1d	fe	37	57	2d	31	f5	69	a7	64	ab	13	54	25	e9	09
	5	ed	5c	05	ca	4a	24	87	bf	18	3e	22	f0	51	ec	61	17
	6	16	5e	af	d3	49	a6	36	43	f4	47	91	df	33	93	21	3b
	7	79	b7	97	85	10	b5	ba	3c	b6	70	d0	06	a1	fa	81	82
	8	83	7e	7f	80	96	73	be	56	9b	9e	95	d9	f7	02	b9	a4
	9	de	6a	32	6d	d8	8a	84	72	2a	14	9f	88	f9	dc	89	9a
	a	fb	7c	2e	c3	8f	b8	65	48	26	c8	12	4a	ce	e7	d2	62
	b	0c	e0	1f	ef	11	75	78	71	a5	8e	76	3d	bd	bc	86	57
	c	0b	28	2f	a3	da	d4	e4	0f	a9	27	53	04	1b	fc	ac	e6
	d	7a	07	ae	63	c5	db	e2	ea	94	8b	c4	d5	9d	f8	90	6b
	e	b1	0d	d6	eb	c6	0e	cf	ad	08	4e	d7	e3	5d	50	1e	b3
	f	5b	23	38	34	68	46	03	8c	dd	9c	7d	a0	cd	1a	41	1c

4 CONCLUSIONS

Detailed comparison is shown in Table 3:

As Table 3 shows, the avalanche probability of AES algorithm and the last literature (BAUDRON & GIBERT 1999) is 408, below improved S box in this article and the avalanche probability which is 304; and also resolving the S box in AES and the inverse S-box expressions in the last literature is an overly simple question. Since the number of algebraic expressions of the improved box S and inverse S-box are 255 in this article; The iterative output cycle of S Box in this paper is 256, while iterative output cycle of S box constructed in other two programs does not exceed 88; This article improves program by using affine transformation period which is 16, was superior to affine transformation that the cycle of the AES algorithm and S box in the last literature which is 4. In short, the S box has better algebraic properties.

415

Table 3. Improved S-box compare with traditional S-box on algebraic properties.

Property	Traditional S-box	Improved S-box	Improved S-box in this article
Balance	Balance function	Balance function	Balancefunction
Orthogonality	Yes	Yes	Yes
Nonlinearity	112	112	112
Differential uniformity	4	4	4
Nonzero linear structure	None	None	None
Iterative output cycle	Less than 88	Less than 88	256
Affine transformation cycle	4	4	16
Strict probability of Avalanche Criterion	432	408	304
Number of S-boxes algebraic expression	9	255	255
Number of inverse S-boxes algebraic expressions	255	9	255

ACKNOWLEDGMENTS

This research is supported by Foundation Program named research project of Harbin city (2011AA1CG063).

REFERENCES

Feng D. G., & W. L. Wu (2000). Design and analysis of block cipher. *Tsinghua University press*, 25–45.

Feng D. G. (2002). the Domestic research present situation and development trend of cryptography. *Computer application.*

He D. Q., G. Z. Xiao, & Y. Bo (2003). Modern cryptography *Tsinghua University press.*56–60.

Hu Y. H., & G. Z. Xiao (1999). Design and security analysis of block cipher. *Doctoral dissertation of Xi'an Electronic and Science University.*

Wei B. D., W. P. Ma, & X. M. Wang (2005). AES Boolean function Walsh spectrum analysis. *Computer engineering and science.*

Wei B. D., W. P. Ma, & X. M. Wang (2006). S-box algebraic expressions of AES. *Journal of Xi'an Electronic and Science University (NATURAL SCIENCE EDITION).*29–38.

BAUDRON O, & H. GIBERT (1999). Report on the AES Candidates. *Second Advanced Encryption Standard Candidate Conference.* 53–67.

Electronic Engineering and Information Science – Wang (Ed.)
© 2015 Taylor & Francis Group, London, ISBN: 978-1-138-02772-5

The application of fuzzy probability on price-assessment of second-hand houses

H. Zhao & H. Sui

Harbin University of Science and Technology, School of Applied Science, Harbin, Heilongjiang, China

ABSTRACT: This paper applies the theory of fuzzy mathematics to the price-assessment of second-hand houses and uses Fuzzy Probability comprehensively assess the price of second-hand houses. Firstly, through making sure that the complement of the price-assessment of second-hand houses and quantitating the qualitative index, that built a fuzzy comprehensive evaluation model about second-hand houses. Finally, verify the correctness of the model using instances.

KEYWORDS: Fuzzy Probability; Comprehensive assessment; Characteristic Factor; Membership Function.

1 INTRODUCTION

Probability theory is employed to research the phenomenon of random number law. From it is proposed firstly in 1654, probability theory has a history of more than three hundred years to date. The fuzzy probabilistic method is one of the mathematical evaluation models. In 1965, the concept of Fuzzy set had been proposed and the theory had founded to research the problem of fuzziness and uncertainty by the American professor, L.A.zdahe, of computer and control theory. After 40 years of researching and development, it becomes a vibrant branch of mathematics. Nowadays, the method of fuzzy probability has widely applied on the heavy metals in the soil and monitoring of water pollution.

The immediate problem is how to correctly determine a reasonable price of second-hand houses. There are many uncertainties in the price assessment of second-hand houses. The accuracy of the assessment is closely related to the methods and the richness of people's experiences and information. "The hardness of assessment" has been a prominent issue in the transaction process. The main factors that affect the assessment of second-hand houses are regional factors, individual factors, market factors and consumers' psychological factors. The outstanding problem to be solved is to determine a scientific and reasonable price evaluation model and accurately estimate the price of second-hand houses. Therefore, it is important to objectively and scientifically assess the price of second-hand houses.

This paper applies the theory of fuzzy mathematics to the price-assessment of second-hand houses and uses Fuzzy Probability comprehensively assess the price of second-hand houses (He 1998). First,

by making sure that the complement of the price-assessment of second-hand houses and quantitating the qualitative index, it is constructed that a fuzzy comprehensive evaluation model about second-hand houses. Finally, the instances are given to verify the correctness of the model. This article can provide analytic methods and tools for government making policies, developers making decision and consumers purchase houses.

2 PRINCIPLE OF FUZZY PROBABILITY AND APPLICATION STEPS

2.1 *Mathematical model (Lu & Wang 2007)*

(Ω, Q, P) is a Fuzzy probability space in which Q is a Fuzzy Borel field. The Fuzzy subset A is called Fuzzy event, refers to the membership function $X_A(\omega)$ of A is Borel measurable, where $\omega \in U_1$. $P(A) = U_2$ called the probability of event A Fuzzy.

If for the countable set, $\Omega = \{ X_i \; i = 1,2 \dots\}$

$$P(X_i) = P_i \; i = 1,2 \dots .$$

Then the above equation can be expressed as:

$$P(A) = \sum_{i=1}^{\infty} X_A(x_i)\rho_i \qquad (1)$$

2.2 *Fuzzy evaluation method estimate of second-hand houses and the basic steps*

To ensure the reliability and practicality of the fuzzy evaluation model, this paper first real estate company to obtain the data in a certain area of the city A sample of real estate analysis. Reference standard from the real estate company database, as shown in Table 1, the text data are measured by assessment expert (Ceng et al 2010).

2.2.1 The sample composition.

The application of Fuzzy probability to comprehensively assess the second-hand house prices is the determination of the comprehensive evaluation factors of houses. There are seven characteristics that can fully reflect the prices of second-hand houses: Lot features, quality characteristics, characteristics of the construction area, residential environment characteristics, the degree of decoration features, floor feature, property management features, etc. That making k monitoring points, and have seven evaluation factors, each indicator has three evaluation criteria levels (Xu 1988).

It is possible to establish a set of evaluation factors:
$U=\{ U_1 , U_2 , U_3 , U_4 , U_5 , U_6 , U_7 \}$

individual characteristics of each membership function $X_A(\chi)$ for each evaluation factor level. Floors elemental characteristic factor, for example, to establish the membership function characteristic factor of floors:

$$\mathrm{I} \; : \; X_A = \begin{cases} 1 & x \le 0.49 \\ \dfrac{0.66 - x}{0.17} & 0.49 < x \le 0.66 \\ 0 & x > 0.66 \end{cases}$$

$$\mathrm{II} \; : \; X_A = \begin{cases} \dfrac{x - 0.49}{0.17} & 0.49 < x \le 0.66 \\ \dfrac{0.89 - x}{0.23} & 0.66 < x \le 0.89 \\ 0 & x > 0.89, x \le 0.49 \end{cases}$$

Table 1. The results of each study area.

Feature \ Sample	X_1	X_2	X_3	X_4	X_5	X_6	X_7	X_8	X_9	X_{10}
U_1	0.875	0.631	0.549	0.482	0.801	0.571	0.737	0.654	0.841	0.756
U_2	0.754	0.901	0.613	0.754	0.588	0.857	0.804	0.651	0.842	0.556
U_3	0.607	0.826	0.771	0.932	0.403	0.685	0.422	0.705	0.954	0.576
U_4	0.851	0.521	0.623	0.852	0.327	0.468	0.687	0.424	0.705	0.956
U_5	0.504	0.831	0.524	0.886	0.474	0.685	0.694	0.779	0.926	0.743
U_6	0.603	0.566	0.807	0.785	0.407	0.852	0.504	0.675	0.514	0.694
U_7	0.902	0.874	0.755	0.537	0.842	0.926	0.837	0.861	0.850	0.821

Rank grade: $V = \{\mathrm{I}, \mathrm{II}, \mathrm{III}\}$ = {low, medium, high}

Which corresponds I from 2000 to 3000 yuan / $\mathbf{m}^2$, corresponding II 3000 to 4000 yuan / $\mathbf{m}^2$, corresponding III 4000 to 5000 yuan / $\mathbf{m}^2$.

$$\mathrm{III} \; : \; X_A = \begin{cases} 1 & x > 0.89 \\ \dfrac{x - 0.66}{0.23} & 0.66 < x \le 0.89 \\ 0 & x \le 0.66 \end{cases}$$

2.2.2 Establish individual characteristic factor for the evaluation level membership function

According to the regional rates valuation grading standards, using piecewise linear transforms the

2.2.3 Determination of the probability of characteristic factor of each interval

In the study area randomly selected k samples monitoring points, then according to the actual situation

418

and the monitoring point of each sample measured the value of each individual sample indicators assessment. And dividing the value of sample indicator assessment into a series of continuous interval, what is more, the point is N_i, at the median segment interval, and the probability of the number of bits for each point within the interval is Pi. Choose 10 sample monitoring stations in A district, i.e. k=10. Measurement point figures and the corresponding probability list in table 2 .

a-- characteristic factor index segment lower limit;
b-- characteristic factor index segment upper limit;
$X_A(x)$ ----Membership function characteristic factor.

The calculations show that characteristic factor of membership evaluation grade { |, ||, |||}, , can be divided into sections, said in floor characteristic factor example.

Table 2.　Table-point median interval of two measured values within each segment, the corresponding probabilities and the corresponding.

Items	Each Index Piecewise Interval	Some Digits	Probability	Corresponding To Each Evaluation Grade Of Membership Degress		
				|	||	|||
Floor Feature	X ≤ 0.49	1	0.1	1	0	0
	0.49 < x ≤ 0.58	2	0.2	0.7353	0.2647	0
	0.58 < x ≤ 0.66	2	0.2	0.2353	0.7647	0
	0.66<x ≤ 0.78	2	0.2	0	0.7391	0.2609
	0.78 < x ≤ 0.89	3	0.3	0	0.2391	0.7609
	x > 0.89	0	0	0	0	1
Qualitative Characteristics	x ≤ 0.4	0	0	1	0	0
	0.4 < x0.6	2	0.2	0.5	0.5	0
	0.6 < x0.7	2	0.2	0	0.75	0.25
	0.7 < x0.8	2	0.2	0	0.25	0.75
	x > 0.8	4	0.4	0	0	1
Lot Characteristics	x ≤ 0.5	2	0.2	1	0	0
	0.5 < x0.65	2	0.2	0.7321	0.2679	0
	0.65 < x0.78	3	0.3	0.23	0.77	0
	0.78 < x0.9	1	0.1	0	0.5	0.5
	x > 0.9	2	0.2	0	0	1
Construction Area Characteristics	x < 0.48	3	0.3	1	0	0
	0.48 < x ≤ 0.63	2	0.2	0.6873	0.3127	0
	0.63 < x ≤ 0.72	2	0.2	0.1876	0.8124	0
	0.72 < x ≤ 0.86	2	0.2	0	0.6143	0.3857
	0.86 < x ≤ 0.90	0	0	0	0.11	0.89
	x > 0.9	1	0.1	0	0	1
Degree Of Decoration Feature	x0.48	1	0.1	1	0	0
	0.48 < x ≤ 0.53	2	0.2	0.8529	0.1471	0
	0.53 < x ≤ 0.65	0	0	0.3529	0.6471	0
	0.65 < x ≤ 0.7	2	0.2	0	0.875	0.125
	0.7 < x ≤ 0.85	3	0.3	0	0.375	0.625
	x > 0.85	2	0.2	0	0	1

2.2.4 *Determine the average degree of membership*
Calculating the average degree of membership for each characteristic factor in each part of the range for the evaluation level { |, ||, |||}, according to the type(1), the results are listed in table 2.

$$\mu(a < \chi \le b) = \frac{1}{b-a}\int_a^b X_A(\chi)d_x$$

$\mu(a < \chi \le b)$ ----Characteristics factor index in a, b segment within the average grade of membership of each evaluation.

$$I : X(x) = \frac{1}{x \le 0.49} + \frac{0.7353}{0.49 < x \le 0.58} +$$

$$\frac{0.2353}{0.58 < x \le 0.66}$$

$$II : X(x) = \frac{0.2647}{0.49 < x \le 0.58} + \frac{0.7647}{0.58 < x \le 0.66} +$$

$$\frac{0.7391}{0.66 < x \le 0.78} + \frac{0.2391}{0.78 < x \le 0.89}$$

419

$$\text{III} : \quad X(x) = \frac{0.2609}{0.66 < x \le 0.78} + \frac{0.7609}{0.78 < x \le 0.89} +$$

$$\frac{1}{x > 0.89}$$

$$W_i = S_i / \bar{c}_i \tag{3}$$

2.2.5 Calculate the probability of each individual characteristic of fuzzy factor index

By (2), the each characteristic factor index is computed that belong to the normalized Fuzzy probability of each evaluation grade {| || ||| } and the results are listed in Table 3:

$$\bar{W}_i = \frac{S_i / \bar{c}_i}{\sum_i S_i / \bar{c}_i} \tag{4}$$

$\bar{c}_i$ -- The average value of the i-th characteristic factors measured;

S_i -- Sample room characteristic factor of the i-th

Table 3. The characteristic factors belong to the probability of each evaluation class.

Project \ probability	P (I)	P (II)	P (III)
Floors feature	0.29412	0.42543	0.28045
Quality Characteristics	0.10	0.30	0.60
Lot Characteristics	0.4154	0.3346	0.25
Construction area characteristics	0.47498	0.34762	0.1774

$$P(A) = \sum_{i=1}^{\infty} X_A(\chi_i) P_i \tag{2}$$

P_i --Probability characteristic factor indicators within each range segment of the point-digit

A--The FUZZY event

$X_A(x_i)$ --membership function

2.2.6 Determine the weights of individual characteristics factors

According to Eq. (3), calculate the characteristics of each individual sub-index factor and the normalized weighted index are computed by (4)

Heres S_i the characteristic indexes of sample room set as secondary X_1 as an example, calculating the evaluation value of the price of the house, the results in table 4.

2.2.7 Comprehensive evaluation

Comprehensive calculate the mount of general characteristics factor in each rank according to $P = \sum P(A) \bar{W}_i$. Arrive at a comprehensive assessment of the optimal choice of housing prices and choose the results of its largest membership as the final result.

Table 4. Calculating the corresponding parameter values in real estate, for example X_1

Project	S_i	c_i	W_i	W_i
Floors Feature	0.875	0.6879	1.2720	0.1750
Quality Characteristics	0.754	0.732	1.0301	0.1418
Lot Characteristics	0.607	0.6881	0.8821	0.1214
Construction Area Characteristics	0.851	0.6414	1.3268	0.1826
Degree Of Decoration Feature	0.504	0.7046	0.7153	0.0984
Community Environmental Characteristics	0.603	0.6407	0.9412	0.1295
Property Management Features	0.902	0.8205	1.0993	0,1513

|P=0.1750×0.29412+0.1418×0.10+0.1214×0.4154+0.1826×0.47498+0.0984×0.27058+0.1295×0.3792+0.1513×0.12391=0.2972 ,

In the same way ||P=0.3743 , |||P=0.3284.

2.2.8 *The largest membership*

Each evaluation level correspond to a membership function value. At last, get the assessment result depend on the principle of maximum membership corresponding to rank that can determine the estimated value of the sample room. Take max (P) = max (0.2972,0.3743,0.3284) = 0.3743. This result corresponds to the evaluation level ||, namely the middle.

As can be seen from the rating, the second-hand housing lots downtown extent and residential environment in general, renovated older, because the quality of housing is OK, the building area is relatively large, property management is better, so the final results of the assessment is the middle of the house, about 3000 ~ 4000 yuan / m². Selling price in line with the real estate agency to verify the correctness of the model.

3 CONCLUSIONS

This paper applies the theory of fuzzy mathematics to the price-assessment of second-hand houses and uses Fuzzy Probability comprehensively assess the price of second-hand houses. Firstly, by making sure that the complement of the price-assessment of second-hand houses and quantitating the qualitative index, it is built that a fuzzy comprehensive evaluation model about second-hand houses. Finally, the instances are given to verify the correctness of the model. Since the second-hand houses assessment is fuzzy, the use of this method of valuation is the mathematical description of the decision-making process, people can avoid the problem of subjective decision-making and reduce the impact assessment that emotionally brought, which not only could objectively and scientifically reflect the authenticity of the object, but also improve the accuracy of the assessment. And the assessment results not only reflect the ambiguity also reflect the probability are close to objectivity. But there is much more subjective in terms of determining the characteristics factors, which needs a number of professionals to value.

REFERENCES

He Z. X. (1998). Fuzzy mathematics and its application. *Tianjin science and technology press.*

Lu F. M. & B. Wang (2007). The appraisal model of second-hand analysis. *Journal of price theory and practice.*

Ceng Z. F., H. B. Tang, & Y. Wang (2010). Changsha fuzzy comprehensive evaluation model of second-hard housing prices. *business research.*

Xu S. B (1988). The theory of analytic hierarchy process (ahp). *Tianjin university press.*

Electronic Engineering and Information Science – Wang (Ed.)
© 2015 Taylor & Francis Group, London, ISBN: 978-1-138-02772-5

Particle Swarm Optimization algorithm based on multi-swarm cooperative evolution

G.Z. Wang, X. Zeng & P.L. Qiao

Computer Science and Technology, Harbin University of Science and Technology, Harbin, P.R.China

ABSTRACT: In order to overcome the standard Particle Swarm Optimization (PSO) algorithm is easy to fall into local extremum of the disadvantages of low precision and optimization, presents a multi population co evolutionary particle swarm optimization (MS-PSO) algorithm. By building a gene pool, make an inferior particle according to the gene pool of genetic operation. At the same time using the adaptive function method, to maintain the population diversity, reduce the probability of falling into a local minimum. With 4 representative functions of the experiment, the results show that the MS-PSO algorithm has the advantages of good global performance and optimization high precision.

KEYWORDS: particle swarm optimization; multi-swarm; cooperative evolution.

1 INTRODUCTION

Particle Swarm Optimization (PSO) algorithm is an evolutionary algorithm based on population, by Kennedv and Fberhart is put forward in 1995. In PSO algorithm, a particle is belong to a solution of search space, and with each potential solution has a strict corresponding relation. With a group of random number as the initial value in a certain range, through iterative search to the optimal value. PSO algorithm has less parameters, advantages of simple structure, fast convergence rate. Due to its easy to understand, easy to implement, been successfully applied in many optimization problems. But the disadvantage is that there is late in the process of PSO algorithm in the optimization of slow convergence speed and easy to fall into local optimal solution of the problem.

Based on the deficiency of PSO algorithm, this paper proposes a new improved algorithm, multi population co evolutionary particle swarm optimization algorithm. It will be a subgroup of the whole population is divided into many, then the iterative evolution algorithm. According to the characteristics of particle swarm optimization, particle swarm is divided into three sub groups, evolution group neutron group is used in the process of different characteristics in the iterative evolution, the advantage of this: one is the particles can have an ability to escape local extreme point; two is the characteristic of the particles flying full application of the evolution; three algorithm diversity of population, and then solve the algorithm premature convergence, the

global convergence performance of the algorithm will be improved.

2 PARTICLE SWARM OPTIMIZATION

Particle swarm optimization algorithm has many advantages, such as less parameters, method has the advantages of simple structure, fast convergence speed, then the algorithm is put forward in various fields has been widely used, and the application effects are very obvious. However, in contrast, advances in particle swarm optimization algorithm in the aspect of theoretical research is particularly lag.

In the particle swarm optimization algorithm, called particle belongs to a search space of solutions, and the problems and potential solutions are closely related to the initial value, it is a set of random numbers. Continuous learning in the environment, and constantly adapt to the surrounding environment, and each particle according to the ability to adapt to the different levels of fitness and have their respective, and the fitness value is a function of the corresponding to the decision by the particle. The particle in the search space velocity depends on the specific form of motion in space, all the particles in the population are to have the optimal solution of the particle aggregation. In the population, each particle always with iteration, each particle's speed and position will change, their value depends on the individual optimal particle and the global best particle, by comparing with the optimal particle to change the flight path of their own.

3 INPROVED PARTICLE SWARM ALGORITHM STEPS

Step 1: Initialization

The individual is first divided into n groups, and then initializes each individual and define each population evolutionary algorithm. For the maximum number of iterations T, weighting value of inertia Solution steps

Step 2: The initialization function

Constitution of the gene pool in the selection of optimal particle, it is necessary to calculate the individual ability to adapt, then according to a certain ratio to choose.

Step 3: Evaluation of population

Must choose an evolutionary standard, using the evolutionary standard, can for each population co evolutionary operation of intra cluster and inter cluster.

Step 4: Construction of gene library

The population is divided into two categories, gene particles and non gene particles, for some specific rules, evolution and operation corresponding particle.

Step 5: To determine whether meet the requirements:

Judging whether the maximum number of iterations is reached, or other cycle control end condition, if reached, go to step six, otherwise go to step 3.

Step 6: Judge whether the algorithm over

If the number of iterations to meet the maximum value or to find high quality solutions, it will output the results, at the same time the algorithm will end, otherwise the algorithm jump to step 3, continue to implement the start from step 3.

4 ALGORITHM FLOW CHART

Multi population co evolutionary particle swarm optimization algorithm as shown in Figure 1 flow execution.

5 PERFORMANCE AND ANALYSIS

Algorithm optimization effect to verify improvement, using several typical functions on the performance of PSO algorithm and MSPSO algorithm was compared with the following functions:

1 spherical mapping function Sphere function:

$$f_1(x) = \sum_{i=1}^{n} x_i^2, -100 \leq x_i \leq 100 \qquad (1)$$

A single peak two functions.

2 Rastrigrin function:

$$f_2(x) = \sum_{i=1}^{n} \left[x_i^2 - 10\cos(2\pi x_i) + 10 \right], -100 \leq x_i \leq 100 \qquad (2)$$

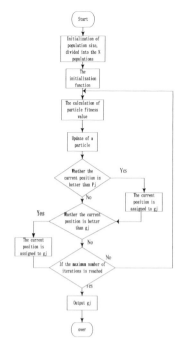

Figure 1. Algorithm flow chart.

With a large number of sinusoidal inflection points and local optimal value points of the regular arrangement.

3 Shaffers F6 function:

$$f_3(x) = \frac{\sin \sqrt[2]{x_1^2 + x_2^2} - 0.5}{[1 + 0.001 * (x_1^2 + x_2^2)]^2}, \qquad (3)$$
$$-100 \leq x_i \leq 100$$

The fluctuation of multi peak function is very intense, very difficult to use the algorithm is difficult to get the optimal solution.

4 Rosenbrock function:

$$f_4(x) = \sum_{i=1}^{n} \left[100 * (x_{i+1} - x_i^2) + (x_i - 1)^2 \right]^2, \qquad (4)$$
$$-100 \leq x_i \leq 100$$

This function is very complex, it is difficult to be minimized.

The experimental parameters are defined as follows:

1 select the number of populations of particles n=50;
2 learning indicator coefficient c1=c2=3;
3 the maximum number of iterations t=60;
4 inertial measure weights, where the minimum value v1=0.4, maximum value v2=0.9;

424

5 The choice of 40 particles used for testing, these particles to choose the best results in the iterative process, as the test particle.

Two kinds of optimization algorithm on average ten times the optimization value as shown in table 1.

Table 1. Comparison of average optimal multiple search results from two optimization algorithms.

Function	Optimal value	PSO	MSPSO
$f_1(x)$	0	6.249578	0
$f_2(x)$	0	2.948736	0
$f_3(x)$	−1	−0.867823	−1
$f_4(x)$	0	0.398746	4.2671e-2

Below is the PSO algorithm, MSPSO algorithm in $f_1(x)$, $f_2(x)$, $f_3(x)$, $f_4(x)$ function to find the optimal value process curve.

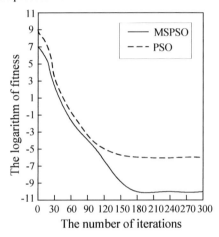

Figure 1. Sphere functions search curve.

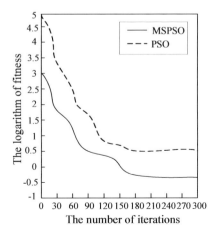

Figure 2. Rastrigrin function search.

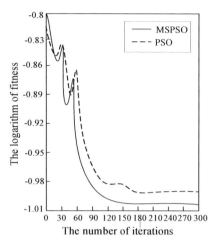

Figure 3. Shaffers f_6 function search curve.

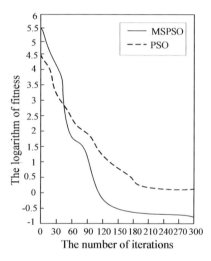

Figure 4. Rosenbrock function search curve.

From figure 1 to figure 4 can be seen in the early to start the search, convergence rate, and function in the MSPSO algorithm has better advantages, the MSPSO algorithm to the optimal solution be found quickly. In the beginning, the convergence process of MSPSO algorithm can get stuck in a local minimum, however, the MSPSO algorithm can quickly find the global optimal value. Therefore, the improved algorithm improves the convergence speed, and ensure the effectiveness of the global optimal value and stability.

Multi population co evolution algorithm has great a improvement in the convergence rate, better than general PSO algorithm. Therefore, multi population co evolutionary particle swarm algorithm has convergence speed and global search performance is very good.

6 CONCLUSIONS

Based on particle swarm optimization algorithm is easy to fall into the shortcomings of local optimum and slow convergence speed, introducing the multiple population coevolution algorithm, is presented based on the multiple population coevolution particle swarm optimization algorithm. Through the test of typical function, it can be seen that the improved particle swarm algorithm has the ability to quickly find the global optimal solution, effectively avoid the premature convergence and into local extremum problems. The future research work is to apply the algorithm to the actual field of have not been used.

ACKNOWLEDGEMENTS

This research is supported by research project of Harbin city (2011AA1CG063).

REFERENCES

MINGWEI L, KANG H. (2013). Hybrid Optimization Algorithm Based on Chaos Cloud and Particle Swarm Optimization Algorithm.*Journal of Systems Engineering and Electronics*. 2(2), 89–94.

RUI W, CHENG Z. (2008). An Effective Immune Particle Swarm Optimization Algorithm for Scheduling Job Shops.*The 3rd IEEE Conference on Industrial Electronics and Applications*. 758–763.

Lutz C, Sattler U, Tender L. (2009). The Complexity of Finite Model Reasoning in Description Logics. *Information and Computation*. 199(1–2). 132–171.

LEE W, STOLFO S J, Shen Z. (2011). With disturbance term improved particle swarm optimization. *Computer Engineering land Applications*. 43 (7), 84–86.

WENKE L. (2011). A New Data Mining Method Based on Multidimensional-data flow Proced Engineering. 5(24), 365–369.

Electronic Engineering and Information Science – Wang (Ed.)
© 2015 Taylor & Francis Group, London, ISBN: 978-1-138-02772-5

An improved orbital sliding algorithm for calculating no-fit polygon

D.Y. Cao & D.C. Liu
Department of Applied Mathematics, Harbin University of Science and Technology, China

H.X. Sun
Department of Computer Science, Harbin University of Science and Technology, China

ABSTRACT: Base on the classical sliding algorithm proposed by Mahadeven for calculating no-fit polygon, an improved algorithm is presented in this paper. The computation complexity of the traditional sliding algorithm has been decreased by reducing the computation of minimum moving distance, which could make it easier to combine many meta-heuristic algorithms with NFP for solving two-dimensional irregular packing problem. The experiments proved that the algorithm is effective and feasible.

KEYWORDS: No-fit polygon; Two-dimensional packing problem; Minimum moving distance.

1 INTRODUCTION

The concept of No fit polygon (No-Fit Polygon, NFP) was proposed in 1976 by Adamowicz, which is an important geometric calculation tool for obtaining relative position of two polygons. NFP could convert the calculation of relative position of two polygons to one point with another polygon, so as to avoid a lot of repeated judgment intersection.

NFP has been widely used in applications such as mechanical CAD/CAM, graphics and image processing and robot around the obstacle path planning and other fields, especially for the two-dimensional packing problem with strong industrial application background, NFP has become an important basic calculation tools (Burke & Kendall 1999, Bennell & Dowsland 2001), but in the actual application process, the calculation of NFP is too complex to hinder that the NFP is used more widely in practice.

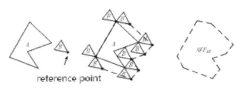

Figure 1. NFP.

For solving NFP of convex-convex polygons and convex-concave polygon, there have been more research results. But for the two polygons are concave, the existing algorithms still have some problems such as low efficiency and difficulty in the realization (Dean & Tu & Raffensperger 2006).

In this paper, based on the orbital sliding method, a new direct and effective process for calculating minimum moving distance is given, which reduces the overall computational complexity effectively, and could be adapted in solving NFP of two simple polygons with random shape.

2 IMPROVED ORBITAL SLIDING ALGORITHM

The orbital sliding method is a classic algorithm for calculating NFP, because of its intuitive and easy to understand, so it is applied widely in the field of application. For the computational complexity, this method complexity is high, in addition, in the treatment of boundary cavity and the internal cavity is still exist some problems have not been solved yet.

The following is the principle of the orbital sliding method: according to the position of *A* and *B* with the connected state in current time and get the moving direction of polygon *B* for the next step, in the direction of movement, calculated the minimum moving distance between *A* and *B*; then *B* will be moved to new location according to the moving direction and moving distance, repeat the movement process until backing to the initial position.

The time consumption of the orbital sliding method mainly concentrated in calculating minimum moving distance.

For convenience of description, we assume that the polygon vertices *A* and *B* are stored counterclockwise, polygon *B* is centered around the *A* with clockwise rotation.

2.1 Select the initial position

Before the polygon B around the polygon A, first of all is to choose the initial position.

Select the vertex of polygon A with minimum y-coordinate, if the total number of vertices is more than one, select the starting point with minimum x-coordinate as a VA for polygon A; for the polygon B, select the vertex with maximum y-coordinate, if the total number of vertices is more than one, then choose one vertex with maximum x-coordinates as a starting point V_B for polygon B.

After selecting the initial position of polygon A and polygon B, moving polygon B to make the vertex V_A and vertex V_B coincidence.

2.2 Determine the moving orbit and moving direction

When the vertex V_A and vertex V_B coincidence, we need determine the moving orbit for moving polygon (the line makes polygon A is connected to polygon B when polygon is moving).

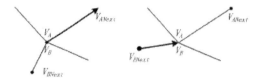

Figure 2.   Determination of moving orbit.

The following is selected method for determining moving orbit:

Assuming for polygon A, the next vertex of vertex V_A (x_1, y_1) is V_{ANext} (x_2, y_2), while the next vertex in polygon B is V_{BNext} (x, y). we only need to judge the position of point V_{BNext} (x, y) and the vector V_A (x_1, y_1) V_{ANext} (x_2, y_2) according to the formula 1), if the V_{BNext} (x, y) lies in the right of vector V_A (x_1, y_1) V_{ANext} (x_2, y_2), then select line V_A $(x_1, y_1)V_{ANext}$ (x_2, y_2) as moving orbit, else select line V_BV_{BNext} as moving orbit (see Figure 2).

$$D = (y_2-y_1)\times x + (x_1-x_2)\times y + x_2\times y_1 - x_1\times y_2 \qquad (1)$$

If $D>0$, then the point (x, y) lies in vector right; if $D=0$, then the point (x, y) on vector; if $D<0$, then the point (x, y) lines in the vector left.

If the V_A V_{ANext} is chosen as the moving orbit R, then moving direction of polygon B and vector R for the direction of the vector V_A V_{ANext} direction, on the other hand, if the vector V_B V_{BNext} is selected as moving orbit, then select the opposite direction vector V_B V_{BNext} as moving direction of polygon B and vector in the direction of R. The starting point if V_A.

2.3 To calculate the minimum moving distance

1 Obtaining the set of candidate intersection vertices

Obviously, when polygon B is moving, we don't need judgment computation for every edge in polygon B to all edges in polygon A, and we only need to calculate some selected edges of the two polygons.

Before choosing candidate intersection vertices from polygon A, computing the maximum distance of vertices in polygon B to moving orbit R, set to S.

The method for selecting a vertex of polygon A can be divided into two categories:

1 Select the vertices in the right side of the R, and the former vertex does not lie in the right side of the R, and the distance to R does not greater than S.
2 Select the vertices in the right side of the R, and the distance L of these vertices to R is greater than the distance between the former vertex with R, at the same time, also L no more than S.

According to these two methods, we could obtain the vertex set A_V.

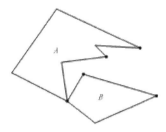

Figure 3.   The candidate points of intersection of polygon A and B.

The selected method for candidate vertices in polygon B is described as following:

Select the vertices in polygon B, which satisfy that the distance to R is less than the distance of the former vertex to R.

These vertices constructing the vertex set B_V, as shown in figure 3.

3 Obtaining the set of candidate intersection edges

The candidate intersection edges in polygon A could be selected according to the following method:

Choose the vertex on the right side of the R, the former vertex lies on the left of the R or lies on the right side, with distance to R is greater than the distance between the former vertex T to R, and the former vertex T' satisfying the same condition, forming the edge set A_E.

The selecting method for candidate intersection edges in polygon B is: the vertex T in B, satisfying that the distance to R is greater than the former vertex T' with distance to R, forming the edge set B_E.

4   Calculating the minimum moving distance.

In the calculation of the minimum moving distance, we need construct lines, which are determined by arbitrary vertex V in vertex set A_V and the slide slope composed of R, and then select one segment with minimum distances constructed by these lines and the vertex in B_V, as shown in figure 4.

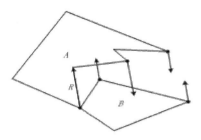

Figure 4.   Calculation of minimum moving distance.

3   EXPERIMENT

The test platform of algorithm is IBM T400 notepad with 2.26G CPU and 2048MB memory, the programming language is the C++. Test data are 3 groups of polygon A and polygon B with variety edge size, the result are given in table 1.

Table 1.   experimental result.

	n=5	n=50	n=500
Traditional algorithm	0.45	25.3	300.21
Improved algorithm	0.43	20.80	220.14

From the numerical results, we could find that the efficiency of the proposed algorithm and the traditional algorithm has little difference when the total number of edges is smaller, but for more large total number edges ($n>=10$), the time consuming of the improved algorithm is far less than traditional algorithms.

For the application of two dimensional irregular problems, the curved row is often regarded as a polygon for approximate processing, so the larger number of NFP for solving boundary polygon is on the practical application background.

4   SUMMARY

For the engineering personnel, orbital sliding method used to calculate the NFP is the most intuitive and easy understanding algorithm. The improved algorithm for solving minimum distance can greatly reduce the computational complexity of the traditional method, which makes the calculation of NFP is more efficient.

REFERENCES

Adamowicz M. & A. Albano (1976). Nesting two dimensional shapesin rectangular modules. *Computer-Aided Design, 8(1)*:27–33.
Burke E. K. & G. Kendall (1999). Evaluation of two dimensional bin packing problem using the no fit polygon. *Proceedings of the 26th International Conference on Computers and Industrial Engineering, Melbourne, Australia.* 286–291.
Bennell J. A. & K. A. Dowsland (2001). Hybridising tabu search with optimisation techniques for irregular stock cutting. *Management Science, 47(8)*: 1160–1172.
Dean H. T., Y. Tu & J. F. Raffensperger (2006). An improved method for calculating the no-fit polygon. *Computers & operations research, 33(6)*: 1521–1539.
Mahadevan A (1984). Optimisation in computer aided pattern packing, *Ph.D. thesis, North Carolina State University.*
Ghosh P. K. (1993). A unified computational framework for Minkowski operations. *Computers and Graphics, 17(4)*: 357–378.
Bennell J. A. & J. F. Oliveira (2008). The geometry of nesting problems: a tutorial. *European Journal of Operational Research, 184(2)*: 397–415.

Electronic Engineering and Information Science – Wang (Ed.)

Construction and performance analysis of a NEMO based IP/LEO satellite network

Z. Zhang, T. Han & X.C. Zhang
Harbin University of Science and Technology, Harbin, China

ABSTRACT: In this paper, a Network Mobility (NEMO) based IP/LEO satellite network structure is established. The IP/LEO satellite network acts as a subnet and connects to the terrestrial Internet network. As an independent subnet, connections can be set up between the satellites using Inter-Satellite Links (ISLs) with IP protocol. One or more mobile routers are set in the satellite network, which work as the interface between the satellite network and the gateway of the terrestrial network. The performances of NEMO based IP/LEO satellite network are simulated. Results show that the NEMO based network performance better in signaling cost and package lost rate compared with the Mobile IPv6 (MIPv6) based network as the number of nodes increases. But in the NEMO based network, packages should be sent through the mobile router which increases the transmission time and leads to the decline of network throughput.

KEYWORDS: IP/LEO Satellite Network; Network Mobility; Mobile IPv6.

1 INTRODUCTION

Internet as a data network has seen explosive growth during recent years. Users expect to be connected to the Internet from anywhere at anytime. So the Low Earth Orbiting (LEO) satellite networks is researching to provide ubiquitous IP connectivity in the world and also in space while guaranteeing short propagation delays. (Carlos Bernardos, Ignacio Soto 2007) Since IP/LEO satellite networks make Internet accesses becoming more and more ubiquitous, demands for mobility are not restricted to single terminals anymore. For example, a person owns more than one mobile device (such as a mobile phone, a laptop, and a personal digital assistant) and each of these devices has multiple network interfaces to connect with each other as well as other networks. When moving together with the owner, the devices constitute a small mobile network. Large scale mobile networks such as aircrafts with mobile devices or airplanes with passengers are more common in IP/LEO satellite networks. [2-4]

Mobile IPv4 (MIPv4) and Mobile IPv6 (MIPv6) have been developed for supporting continuous connections to mobile hosts. While these traditional protocols showed limitations in solving the issue of mobile networks. [5] Each of the mobile devices should be sophisticated enough to run such mobility support protocols, and must be able to attach to the access technology available to connect to the Internet. And the signaling exchanged is limited to a single node sending only one message to avoid signaling overload every time the network moves. [6] Realizing these deficiencies, the Network Mobility (NEMO) Basic Support Protocol is proposed to support network mobility at the IPv6 layer by the IETF NEMO Working Group. [7-8]

2 STRUCTURE DESIGN OF NEMO BASED IP/LEO SATELLITE NETWORK

The limitations of both MIPv4 and MIPv6 protocols in the supporting mobile network provide opportunities for the proposal of NEMO protocol and also the further researches. The combination of NEMO protocol and IP/LEO satellite network improves the utilization rate of satellite channel resource and it also provides a direct connection to the Internet for the devices and platforms without satellite access capabilities.

2.1 *Design requirements of the NEMO based IP/LEO satellite network*

1 Networking scheme

In the space-based satellite constellation network, each satellite has the capability of on-board processing and acts as the switch router, and the Inter-Satellite Links (ISLs) are also needed. In this case, the space-based satellite constellation forms an autonomous domain when operating together with the ground gateway station.

In ground-based satellite constellation network, satellites only act as transponders and repeaters which forward data for ground gateway station and mobile hosts. Without on-board processing capability and ISLs, the coverage area is restricted by the distribution of ground stations. And lots of ground stations are needed for the network to provide services everywhere in the world.

Considering the characteristics of the above two networking schemes and combining the features of the LEO satellite network, space-based constellation network is chosen to construct the NEMO based IP/LEO satellite network in this paper.

2 Types of the mobile networks

Chosen the space-based constellation network, satellites in the network must have the capability of on-board processing and equipped with several devices have independent IP addresses of their own. In this scenario when moving in the satellite network, each satellite forms a mobile network by itself.

The LEO satellites moving in the orbits construct a LEO satellite constellation. Setting one or more mobile router connected with the ground network can set up connections to any satellite in the constellation. This makes the LEO satellite constellation form a distinctive mobile network different from others.

Same as in terrestrial networks, mobile platforms using satellites for communication such as airplanes, ships and also include aircrafts with multiple IP addresses are also mobile networks of the same type.

In NEMO based IP/LEO satellite network, new types of mobile networks are presented because of the movement of the satellites. And the usage of NEMO protocol will also save the network resource during the mobility management processes.

2.2 Structure of NEMO based IP/LEO satellite network

The architecture of NEMO based IP/LEO satellite network with the terrestrial Internet network is shown in Figure 1. In this structure, the LEO satellite network acts as a subnet and connects to the terrestrial Internet network. As an independent subnet, connections can be set up between the satellites using ISLs with IP protocol. One or more mobile routers are set in the satellite network, which work as the interface between the satellite network and the gateway of the terrestrial network. And all the satellites have the capability of access routers, which provide Internet access services to the ground or air users.

In the terrestrial network, one or more gateways should be chosen to work as the HA of the LEO satellite network. That is because the LEO satellite network acts as a mobile subnet of the terrestrial Internet network. As the satellite subnet moving, the satellite

mobile router registered the CoA to the HA and the HA tunnels all the packages on the home network to the CoA. Through this configuration, the changes of the IP routing within the IP/LEO satellite network is isolated with the routing information in the Internet backbone network and the network complexity and signaling load are reduced.

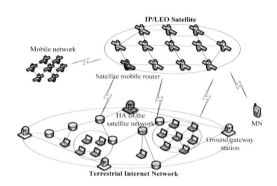

Figure 1. Structure of NEMO based IP/LEO satellite network.

3 THE BINDING UPDATE PROCEDURES USING NEMO PROTOCOL

The same as in the MIP networks, users need to update their address to the HA as they moving. In addition, as a subnet of the Internet backbone network, the IP/LEO satellite network must update IP address to it's HA when leaving home network. The details of the package routing procedures are as follows.

3.1 Binding update procedures of IP/LEO satellite network

Suppose satellite1 (SAT1) with the IP address feca:700:aaaa:1001::/128 is working as the mobile router. As moving into the ground gateway station with the IP address feca:700:aaaa:2000/128, SAT1 should update its IP address to the HA as follows.

1 SAT1 leaves its HA and moves to the ground gateway station (feca:700:aaaa:2000/128) where it gets the new CoA asfeca:700:aaaa:2001::/128.
2 SAT1 sends binding update message to the HA through the gateway station. The message contains the network prefix (feca:700:aaaa:10/56) together with the new CoA (feca:700:aaaa:2001::/128).
3 After receiving the binding update message, the HA establishes a binding between the network prefix and the CoA of SAT1. The binding is feca:700:aaaa:10/56→ feca:700:aaaa:2001::/128 .
4 Then a binding response message is set by the HA.

432

3.2 Binding update procedures of users

As the MNs leaving their HAs, binding update procedures through the IP/LEO satellite network are also needed. Suppose the IP address of the MN's HA is 3ffe:306:1130:1000::/128 and the IP address of the MN is 3ffe:306:1130:1000::1/128, the binding update procedures are as follows.

1. MN leaves its HA and moves to the satellite2 (SAT2) (feca:700:aaaa:1002::/128) coverage area where it gets the new CoA as feca:700:aaaa:1002::1/128.
2. MN sends binding update message to HA through SAT2. The message contains the primary IP address together with the new CoA of MN. Since the message should be capsulated by the satellite router, the encapsulation process should be done by the HA of the satellite and then is forward to the MN's HA.
3. After receiving the binding update message, the HA establishes a binding between the primary IP address and the CoA of MN. That is 3ffe:306:1130:1000::1/128 → feca:700:aaaa:1002::1/128
4. Then a binding response message is set by the HA.

4 PERFORMANCES SIMULATION RESULTS OF THE PROPOSED NETWORK

The performances of the network based on MIPv6 and NEMO protocol are simulated and compared in this section. The results are as follows.

4.1 Signaling cost

In the simulation, the signaling cost is defined as the total number of messages sends from MNs to its HAs. Figure 2 shows the relationship between the number of MNs and the total number of messages under the MIPv6 and NEMO protocol respectively. In the NEMO based network, the number of binding update messages has nothing to do with the number of the nodes in the mobile network, so the signaling cost doesn't increase as in the MIPv6 based network.

4.2 Packet loss rate

In the simulation of MIPv6 based network, the package loss rate is defined as the ratio of the number of packages received by MNs compared to the number of packages sends by the CNs. In the simulation of NEMO based network, the package loss rate is defined as the ratio of the number of packages received by the nodes within the mobile network compared to the number of packages send by the CNs. The simulation result is shown in Figure 3.

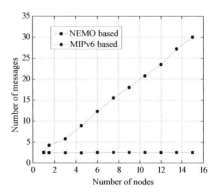

Figure 2. Comparison of signaling cost between NEMO and MIPv6 based network.

When the number of nodes is small, the MIPv6 based network performances better than the NEMO based network in package loss rate. As the number grows, the NEMO based network performances better than the MIPv6 based network. This is because as the number of nodes grows, lots of binding update messages make the handover delay increases, which lead to package loss in MIPv6 based network.

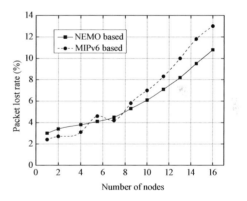

Figure 3. Comparison of package lost rate between NEMO and MIPv6 based network.

4.3 Network throughput

In the simulation, the network throughput is defined as the ratio of the size of the transmitted packet compared with the transmission time. The result is shown in Figure 4. The MIPv6 based network performances better than NEMO based network in network throughput. That is because the throughput is related to the transmission time. In a NEMO based network, packages should be sent through the mobile router that increases the transmission time and leads to the decline of network throughput.

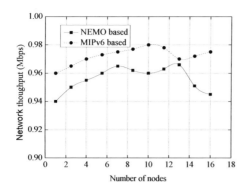

Figure 4. Comparison of network throughput between NEMO and MIPv6 based network.

5 CONCLUSIONS

An IP/LEO satellite network suits for NEMO protocol is presented in this paper. In the proposed network, the IP/LEO satellite network acts as an independent subnet and is able to connect to the terrestrial Internet network through the satellite mobile router by ISLs. Through this configuration, the changes of the IP routing within the IP/LEO satellite network is isolated with the routing information in the Internet backbone network and the network complexity and signaling load are reduced. Simulation results show that the proposed network performances better than MIPv6 based network in signaling cost and package lost rate. But the network throughput declines because the transmission time increases.

ACKNOWLEDGMENT

This work is supported by Heilongjiang Provincial Natural Science Foundation (No. F201422).

This work is supported by the Harbin Research Fund for Technological Innovation (No. 2013RFQXJ104).

REFERENCES

Carlos J.Bernardos, Ignacio Soto(2007)   The Internet Protocol Journal.

Sun,Z.L. (2011) Network Performance Engineering, p.951.

Abu Zafar M, Shahriar, Mohammed Atiquzzaman, William D. Ivancic.(2011) Journal of   Wireless Mobile Networks, Ubiquitous Computing, and Dependable Applications. Vol.46.

Hogie K, Criscuolo, Parise R. (2005)Computer Networks, Vol.47, , p.603.

Kim,H., G. KimC. Kim. (2005)Lecture Notes in Computer Science. Vol.401.

Mohammad H, Z.Zuriati , U.Nur (2012),Wireless Personal Communications, Vol.2, , p.297.

Akbar H, H.Khan (2008) Novel Algorithms and Techniques in Telecommunications, Automation and Industrial Electronics, , p.444.

Ryu,H. D. Kim, Y. Cho, K. Lee, H. Park.(2005)Lecture Notes in Computer Science, Vol.378.

Electronic Engineering and Information Science – Wang (Ed.)
© *2015 Taylor & Francis Group, London, ISBN: 978-1-138-02772-5*

A robust watermarking for 2D CAD graphic based on coordinates modification

T.J. Zhang & H.G. Zhang
School of Software, Harbin University of Science and Technology, China

H.D. Hou
School of Computer Science, Harbin Finance University, China

A.A. Abd El-Latif
Department of Mathematics, Computer Science, Faculty of Science, Menoufia University, Egypt

ABSTRACT: 2D CAD graphic usually contains lots of valuable data, and is of great economic value, but its copyright protection is still a problem. In this paper, a robust watermarking for 2D CAD graphics is proposed, for the ownership declaration and the piratical edition detection. Watermarks are embedded within the tolerance of the graphics, leading to invisible distortions to the relative coordinates of vertices. The method has high capacity to embed enough information and is resilient to partial addition/deletion, cropping and rotation.

KEYWORDS: Blind; Watermarking; Coordinates modification; 2D CAD graphics.

1   INTRODUCTION

Nowadays, CAD graphics are playing an important role in manufacturing, architecture and electrical fields. Most of these graphics are of great worth, because it usually contains special design in it. However, they can be easily duplicated and reused by others, when translated via internet. Robust watermarking can be used to verify the ownership, and receives a lot of attention. Most algorithms focus on applications of watermarking in traditional media such as images, audios, videos and texts. Recently, people have realized the importance of vector graphics, such as vector map and CAD graphic.

Ohbuchi (2003), Voigt (2002) Wang (2007), proposed watermarking algorithms for vector maps in which distances exhibit high correlation. Most of these methods have limited robustness regarding translation, rotation or scaling since such attacks are not usually encountered for GIS data. Thus, these algorithms might not be applicable to engineering graphics where translation, rotation and scaling are commonly manipulated.

Sonnet et al. (2003) proposed an algorithm that changes line attributes, introduces new vertices, and replaces existing stroke segments by new lines. Due to obvious redundancy, the watermarks can be removed easily by attacks designed specifically for reducing redundancy. Solachidis et al (2004) proposed a blind watermarking for vector graphics by polygonal line modification. Properties of the Fourier descriptors ensure that the algorithm can withstand rotation, translation, scaling, reflection and smoothing. But it is fragile to local modification and is not applicable to graphics in which polygonal lines contain few vertices. Similar to the work of Solachidis, Zhang (2005) embedded watermark in the complex wavelet domain of the relative coordinate lines. It is robust as Solachidis' method and is additionally robust to local modification. However, it cannot be applied to graphics in which polygonal lines containing few vertices. Wang (2008), Zhao (2008), Lee (2010) and some other researchers also proposed watermarking methods for 2D CAD graphics respectively, but most of the methods enlarge the data size while embedding the watermarks.

This paper presents a method for robust and blind watermarking of 2D CAD graphic by embedding watermarks in the distance of each two adjacent vertices. It is organized as follows: A detail description of the proposed algorithm is introduced in Section 2. Section 3 analyzes the performance of the algorithm through experimental results and evaluation, followed by conclusions in Section 4.

2   METHOD

2.1   *Pipeline*

In this paper, we define the watermark embedding and extracting procedure as fig.1.

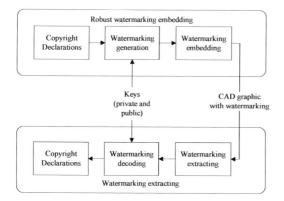

Figure 1. The pipeline of our method.

Firstly, some copyright declarations are defined by the owner, such as the company name, digital fingerprint, ownership and so on. Secondly, this information is encoding into binary data, and then a private key is used to encrypt it to generate the watermark. The watermarking method embeds the watermark into the CAD graphic, and the embedded graphic can be transformed into others. The decoding procedure is mostly the inverse process of the embedding; but changing the key to a public key.

2.2 Embedding method

In our method, the coordinates of vertices in the original graphics are used as the cover data. Given v_i and v_{i+1} are two vertices of the same entity, with coordinates of (x_i, y_i) and (x_{i+1}, y_{i+1}). L_i is the relative distance between v_i and v_{i+1}, which is defined in Eq. 1.

$$L_i = \sqrt{(x_i - x_{i+1})^2 + (y_i - y_{i+1})^2} \tag{1}$$

We can denote the modified L_i as L_i'. The last valid binary bit of L_i' can be used to cover 1 bit watermark. While the relative distance changed, we change the coordinates of the vertex who is farther away to the original point. Supposing that v_{i+1} is farther away to the original point, its coordinates will be calculated by Eq.2.

$$\begin{cases} x_{i+1}' = x_i - \dfrac{L_i'}{L_i}(x_i - x_{i+1}) \\ y_{i+1}' = y_i - \dfrac{L_i'}{L_i}(y_i - y_{i+1}) \end{cases} \tag{2}$$

2.3 Watermarking generation

The watermarking data can be generated by encrypting some ownership information by asymmetric

encryption algorithm. In this paper, we use RSA. After encryption, the watermark is a m-dimensional bit vector $b = (b_1, b_2, ..., b_{mc})$, $b_i \in \{0,1\}$. Repeatedly embedding the same watermarks increases the resiliency of the watermark against cropping, vertex insertion and deletion.

2.4 Watermarking embedding

1 Extraction of the vertices

The vertices are mostly extracted from entities of the lines, positions of texts and etc. N presents the number of the vertex pairs. We mark each farther point between the vertices pair as the embedding vertex and mark it as (x_k, y_k).

2 Embedding watermark into L_i

To embed watermark b_i, the distance L_i between v_i and v_{i+1} are calculated according Eq.1. The L are represented as $L = \{L_1, L_2, ..., L_N\}$ according the length. If $b_i==0$, change the last valid binary bit of L_i to 0. Otherwise, if $b_i==1$, the last valid binary bit of L_i is changed to 1.

3 Changing the coordinates of v_i and v_{i+1}

The watermarked coordinates (x_i', y_i') and (x_{i+1}', y_{i+1}') can then be calculated via Eq.2.

2.5 Watermarking extraction

The algorithm first gets the vertices pairs. Then, the distance L_i' between v_i' and v_{i+1}' is calculated. Denote the last valid binary bit of L_i' as $D_{L_i'}$. The watermark vector can be presented as $b' = (b_1', b_2', ..., b_{mc}')$, $b_i' = D_{L_i'}$.

3 EXPERIMENT RESULTS AND DISCUSSION

We test our method on the DWG format 2D CAD graphics. The graphics contain 14504 entities and 17254 vertices. The length of the watermarks to be embedded is 256(bit). We use firstly use DES to encrypt the data, then use asymmetric encryption algorithm RSA to make a private signature. Most of the evaluate method refers to Hou (2014).

3.1 Capacity

From 2.3, it is clear that the capacity is contributed by vertices of entities in the graphics. Suppose the number of vertex pairs is N, the space provided should be N. The capacity of a graphics is N.

More vertices will result in higher capacity. Most of valuable 2D CAD graphics contain a lot of vertices, so that the capacity of our algorithm is enough.

3.2 Invisibility

Invisibility is another important property of the watermarking. 5 engineering graphics, with the precision of 0.00000000, are used to test the performance of invisibility. Fig. 2 shows details of an original graphic and watermarked graphic, and there is no visible difference between them. Table 1 exhibits the results of experiments on invisibility, in which the mean error (ME) is introduced as the distortion induced by the algorithm. Table 1 shows that MEs of the 5 graphics are all less than the precision, which implies that the distortion induced by the algorithm is low and acceptable.

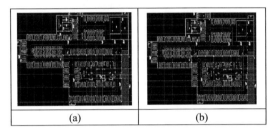

| (a) | (b) |

Figure 2.   (a) An original 2D CAD graphics. (b) The watermarked graphics.

Table 1.   Invisibility of watermarking in the graphic.

Index	File Size	Entities	Vertices	ME
1	1.47MB	12537	16455	1.8112E-10
2	1.73MB	33412	62080	1.6206E-10
3	1.92MB	21334	23461	1.7318E-10
4	2.00MB	35614	63628	1.6802E-10
5	3.29MB	10123	19239	1.8123E-10

3.3 Robustness

Robustness is the ability to keep the integrity of the information in the watermarks, when a 2D CAD graphic is modified by unauthorized user. In this experiment, various attacks are considered. Experiments will be carried out to evaluate the robustness of the algorithm against addition/deletion entities, cropping, translation, and rotation. Graphics G' is the watermarked graphics of G and used to test the performance of robustness. Entity Addition/Deletion

After adding or deleting some entities of the graphics, watermarks embedded will be still decodable since each bit of the watermarks was embedded repeatedly. Here, take deletion for example. While deleting some entities, a number of vertices are deleted randomly. As shown in column 4 of table 2, partial entity deletion will not influence the correctness of the watermarking extraction, which also indicates that the algorithm is resilient against partial entity deletion.

Table 2.   Erroneous caused by entity deletion.

Entities Deleted	Number of Deleted entities	Deleted Vertices	Erroneous Watermarks
Walls	231	413	0
Red Entities	1154	3368	0
Blue Entities	2108	4216	0
Blue and Red Entities	3262	7584	0
Blue, Red and Green Entities	6352	10640	0

1 Cropping

As shown in Table 3, the graphics G' is cropped by various ratios and use the remaining as the test graphics. The results show that the algorithm is robust against cropping.

Table 3.   Erroneous caused by partial cropping.

Ratio of cropping	Removed Entities	Removed Vertices	Erroneous Watermarks
1/14	803	1131	0
5/14	4439	6705	0
9/14	6191	8557	0
27/28	10886	16727	0

2 Translation

Take two adjacent vertices v_i' $\left(x_i',y_i'\right)$ and v_{i+1}' $\left(x_{i+1}',y_{i+1}'\right)$ for example. After translating graphics G' by vector $\left(\Delta x,\Delta y\right)$, coordinates of the two vertices will be $\left(x_i'+\Delta x,y_i'+\Delta y\right)$ and $\left(x_{i+1}'+\Delta x,y_{i+1}'+\Delta y\right)$. $L_i'^T$ is defined as the new distance between v_i' and v_{i+1}', as Eq.3

$$L_i'^T = \sqrt{\left[\left(x_i'+\Delta x\right)-\left(x_{i+1}'+\Delta x\right)\right]^2+\left[\left(y_i'+\Delta y\right)-\left(y_{i+1}'+\Delta y\right)\right]^2}$$
$$= \sqrt{\left(x_i'-x_{i+1}'\right)^2+\left(y_i'-y_{i+1}'\right)^2} = L_i' \qquad (3)$$

Since translation does not affect the distance between two vertices, the algorithm is resilient against translation. Some experiments and the results are shown in the first 2 column in table 4. No matter the graph is translated to any position or rotated to any angle; there is no influence on the watermarking extraction.

Table 4.   Tests of translation and rotation.

$\Delta x, \Delta y$	Erroneous Watermarks	$\Delta\theta$	Erroneous Watermarks
(10, 20)	0	30°	125
(2000, 1000)	0	90°	0
(−5000, −10000)	0	−60°	98
(10000, 20000)	0	−120°	98

ACKNOWLEDGEMENT

This work was supported by Science Foundation of Heilongjiang Province (QC2014C076), China.

CONCLUSIONS AND FUTURE WORK

A robust watermarking method for 2D CAD graphics is proposed in this paper. The algorithm is based on coordinate modification of each two relative vertices and embeds watermarks repeatedly. The watermarking algorithm has high capacity, is invisible and robust to addition/deletion entities, cropping, translation and rotation. It causes little change in the graph and is within the tolerance. In the future, we will focus on lossless CAD watermarking method.

REFERENCES

Ohbuchi, R., H. Ueda and S. Endoh, (2003). Watermarking 2D vector maps in the mesh-spectral domain. *Proceedings of the Shape Modeling International,* May 12–15, Los Alamitos, CA., USA., 216–225.

Voigt, M. and C. Busch, (2002). Watermarking 2D-vector data for geographical information systems. *Proceedings of the Security and Watermarking of Multimedia Contents IV,* 181–184.

Wang, X., Shao, C. Xu X. and Niu, X. (2007). Reversible data-hiding scheme for 2-D vector maps based on difference expansion. *IEEE Trans. Inform. Forensics Security, 2,* 311–320.

Sonnet, H., T. Isenberg, J. Dittmann and T. Strothotte, (2003). Illustration watermarks for vector graphics. *Proceedings of the 11th Pacific Conference on Computer Graphics and Applications,* October 8–10, 2003, Magdeburg, Germany, 73–82.

Solachidis, V. and I. Pitas,(2004). Watermarking polygonal lines using Fourier descriptors. *Comput. Graphics Appl.,24,* 44–51.

Zhang, Q., Xiang H. and Meng, X. (2005). Watermarking vector graphics based on complex wavelet transform. *J. Image Graphics China, 10,* 494–498.

Wang, Y., Xia, K., Fan, Y., Li, Q.(2008). Digital watermarking algorithms for vector images. *IEEE Conference on Intelligent Information Hiding and Multimedia Signal Processing,* 744–747.

Zhao, H., W. Yuan., Z. Wang, (2008). A new watermarking Scheme for CAD engineering drawings. *9th International Conference on Computer-Aided Industrial Design and Conceptual Design,* 518–522.

Lee S H, Kwon K R, (2010). CAD drawing watermarking scheme. *Digital Signal Processing, 20(5),* 1379–1399.

Handan Hou, Jian Li, Jingjia Qi and Junfeng Guo,(2014). A Blind Watermarking for 2D-Vector Engineering Graphics. *Information Technology Journal, 13,* 869–873.

Electronic Engineering and Information Science – Wang (Ed.)
© *2015 Taylor & Francis Group, London, ISBN: 978-1-138-02772-5*

Target detecting method in ballistic midcourse based on minimum mean square error

M. Zhu, J.J. Zhang, H.Z. Liu & S.H. Wang

Naval Aeronautical and Astronautical University, Yantai Shandong, China

ABSTRACT: A new method is proposed to solve the target detecting and tracking problem in the middle trajectory of ballistic missiles. The state formula and measure formula of the target in the middle trajectory are established by applying the minimum mean square error filter to filter and forecast the target detecting data based on serial infrared images. It is difficult to determine the self-covariance matrix of the system noise and measure noise in the minimum mean square error filter. To solve this problem, the parameters setting method of minimum mean square error filter is studied and the filter parameters setting rules are gained by test the different noise parameters condition using the real experiment testing results. This rule can be applied to detect and track the targets in middle trajectory of ballistic missile and provide reference for the target detecting technology.

KEYWORDS: Ballistic Midcourse; Infrared Image; Target Detecting; Minimum Mean Square Error.

1 INTRODUCTION

Image coordinates, target velocity and gray are the main information center detection and tracking of ballistic missile target in the midcourse (Wu, Zhou & Cui2009, Smith M S2005). In order to improve the accuracy of detection of the target to get information and to predict the next frame target information, the need for test results filtering, the classic Wiener filter and Kalman filtering are based on the minimum mean square error method for filtering criteria (Wu, Zhang2005). This paper presents a minimum mean square error filtering method for regulating detection data.

2 METHOD DESCRIPTION

In the warhead target recognition, due to the bullet in the middle of the flight, constantly changing position relative to the stars, bullet coordinates in adjacent frames is changed. We detect warhead target, the target is to obtain accurate information and to predict the next frame of the target information. Forecast the next frame target information can be called a priori information, once the target position is estimated. You can get the coordinates of the target by target detection algorithms. An estimate of the target position there is an error. Similarly, due to the limitations and the effects of the background noise of the instrument itself, and the coordinates obtained by the target detection algorithms also exist errors. Then we can get these two means of data weighted average, get the best estimate. The criteria for the best estimate is that the minimum mean square error.

Suppose by coordinates and gray vector object detection algorithm is $Y_k = (x_k', y_k', g_k')^T$, and the target position and gray once estimated is $\hat{Y}_k = (\hat{x}_k', \hat{y}_k', \hat{g}_k')^T$. x_k', y_k', g_k' respectively represented abscissa, ordinate and the center of gray by the target object detection algorithm. $\hat{x}_k', \hat{y}_k', \hat{g}_k'$ respectively represented a projected of target abscissa, ordinate and center gradation.

If you want to target location and gray once estimated, it must be the introduction of kinetic equations.

Since the target distance can be considered targets of uniform linear motion in image sequences, the coordinates of the target's movement speed and the center of the gray random noise disturbance. Measurement values can be obtained from the target object detection algorithm and center coordinates of gray. Then the state of the object and the observation value can be expressed as follows:

$$\begin{cases} X_k = AX_{k-1} + \omega \\ Y_k = CX_k + v \end{cases} \tag{1}$$

Among them, state vector $X_k = (x_k, y_k, u_k, v_k, g_k)^T$ indicates the status of k time warhead targets, and respectively represented target latitude, longitude, lateral motion speed, vertical motion speed and center gray. Measurement vector $Y_k = (x_k', y_k', g_k')^T$ respectively represented target horizontal, vertical axis and center gray. ω and v are independent zero mean Gaussian white noise. Q and R are self-covariance matrix. State transition matrix A and observation matrix C respectively are:

$$A = \begin{bmatrix} 1 & 0 & T & 0 & 0 \\ 0 & 1 & 0 & T & 0 \\ 0 & 0 & 1 & 0 & 0 \\ 0 & 0 & 0 & 1 & 0 \\ 0 & 0 & 0 & 0 & 1 \end{bmatrix}, \; C = \begin{bmatrix} 1 & 0 & 0 & 0 & 0 \\ 0 & 1 & 0 & 0 & 0 \\ 0 & 0 & 0 & 0 & 1 \end{bmatrix} \quad (2)$$

Among them, T is a unit of time.

As described above, the status of the target can be achieved by two means of obtaining estimates and observations. The difference between the two means of the target state is:

$$\alpha_k = Y_k - \hat{Y}_k \quad (3)$$

It characterizes the improved direction and metrics of the model.

When this improvement becomes optimal, it can be assumed that the measurement means and instruments unbiased. The difference αk not only includes the estimated value and observation value in time k, but also provides new information of yk. We can use a linear combination of the difference between the structure of the state vector sequences directly:

$$\hat{X}_{k+1} = \hat{X}_k | (Y_1 \quad Y_k) = \sum_{i=1}^{k} W_i \alpha_i \quad (4)$$

Then the key problem is transformed the optimal estimate for solving the weights. According to the famous Principle of Orthogonality, the necessary and sufficient condition of minimum mean square error minimization is an estimation error and all input orthogonal. That is, the estimation error is not associated with all inputs. So are:

$$E\left\{ [X_{k+1} - \hat{X}_{k+1}] \alpha_k^H \right\} = 0 \quad (5)$$

Then we can get:

$$E\left\{ X_{k+1} \alpha_k^H \right\} = E\left\{ [\sum_{i=1}^{k} W_i \alpha_i] \alpha_k^H \right\} = W_k E\left\{ \alpha_k \alpha_k^H \right\} \quad (6)$$

It can be drawn:

$$W_k = E\left\{ X_{k+1} \alpha_k^H \right\} \left\{ E\left\{ \alpha_k \alpha_k^H \right\} \right\}^{-1} \quad (7)$$

Then:

$$E\left\{ \alpha_k \alpha_k^H \right\} = E\left\{ [Y_k - \hat{Y}_k][Y_k - \hat{Y}_k]^H \right\}$$
$$= E\left\{ [CX_k + v - C\hat{X}_k][CX_k + v - C\hat{X}_k]^H \right\}$$
$$= E\left\{ \begin{aligned} &vv^H + CA\omega\omega^H A^H C^H \\ &+ CA(X_{k-1} - \hat{X}_{k-1})(X_{k-1} - \hat{X}_{k-1})^H A^H C^H \end{aligned} \right\} \quad (8)$$

If make:

$$E\left\{ (X_{k-1} - \hat{X}_{k-1})(X_{k-1} - \hat{X}_{k-1})^H \right\} = P_{k-1} \quad (9)$$

Then:

$$P_k = AP_{k-1}A^H + Q \quad (10)$$

Formula (8) can be simplified to:

$$E\left\{ \alpha_k \alpha_k^H \right\} = E\left\{ \begin{aligned} &vv^H + C\omega\omega^H C^H + \\ &CA(X_{k-1} - \hat{X}_{k-1})(X_{k-1} - \hat{X}_{k-1})^H A^H C^H \end{aligned} \right\} \quad (11)$$
$$= R + CQC^H + CAP_{k-1}A^H C^H = R + CP_k C^H$$

Left side of Formula (5) is calculated as follows:

$$E\left\{ X_{k+1} \alpha_k^H \right\} = E\left\{ AX_k [Y_k - \hat{Y}_k]^H \right\}$$
$$= AE\left\{ X_k [C(X_k - \hat{X}_k) + v]^H \right\} \quad (12)$$

State vector X_k and the observation noise v are independent. According to Principle of Orthogonality, under the minimum mean square error criterion, optimal estimation $\hat{X}_k$ and prediction error $X_k - \hat{X}_k$ are orthogonal. Therefore, the formula (12) can be written as:

$$E\left\{ X_{k+1} \alpha_k^H \right\} = AE\left\{ \begin{aligned} &[(X_k - \hat{X}_k) + \hat{X}_k] \\ &[X_k - \hat{X}_k]^H \end{aligned} \right\} C^H \quad (13)$$
$$= AP_k C^H$$

Thus, the formula (7) can be written as:

$$W_k = AP_k C^H (R + CP_k C^H)^{-1} \quad (14)$$

Formula (14) into equation (3) can be obtained:

$$\hat{X}_{k+1} = \sum_{i=1}^{k} W_i \alpha_i = \sum_{i=1}^{k} E\left\{ X_{k+1} \alpha_k^H \right\} \left\{ E\left\{ \alpha_k \alpha_k^H \right\} \right\}^{-1} \alpha_i$$
$$= \sum_{i=1}^{k-1} E\left\{ X_{k+1} \alpha_k^H \right\} \left\{ E\left\{ \alpha_k \alpha_k^H \right\} \right\}^{-1} \alpha_i + W_k \alpha_k \quad (15)$$
$$= A\hat{X}_{k-1} + AP_k C^H (R + CP_k C^H)^{-1} (Y_k - CAX_{k-1})$$

Formula (9) and (15) can be recursive target coordinates, speed, and the center gradation of the filtering result and prediction information.

3 THE ANALYSIS OF SIMULATION

The filter parameters setting can be found in the literatures (Li and Guo 2004; Dodgen L J 2006; Wan and Wang 2007). To test the effect of different parameters Q and R on the filtering of results, we have detected a set of experimental data. This set of data is from 72 middle trajectory simulation sequence infrared images. The 72 sets of target coordinates and central gray are calculated (shown in Table 1).

Table 1. The filter experimental data

X coordinate	Y coordinate	Center gray	X coordinate	Y coordinate	Center gray	X coordinate	Y coordinate	Center gray
614.4461	148.7783	1955	584.8911	291.4520	1991	556.1054	434.6016	1976
613.3296	154.5470	1995	583.3891	297.8671	1985	555.0928	440.6342	1988
611.9996	160.4976	2004	582.4988	303.5064	1964	553.8017	446.5046	1989
......	......	......	......	......	......	......	......	......
587.4969	279.5127	1986	558.7676	422.8878	1972	530.7929	566.3016	1911
586.3584	285.7250	1992	557.4968	428.5070	1968	529.2777	572.2705	1927

The xperimental system noise is $Q = I_5$, and the measurement noise is $R = 5I_3$. The pre-filter and filter target coordinates in Fig. 1, Fig. 2 and Fig. 3. After modifying the system noise covariance matrix, target coordinates did not change significantly. The speed change pre-filtering and post-filtering is shown in Fig. 4, Fig. 5 and Fig. 6.

After simulation analysis, filtering target velocity distribution is relatively concentrated, and the X component and Y component of the rate of change is relatively small. This shows that the amount of increase in the measurement noise covariance matrix conditions, filtration filters obvious. Fig. 7 is a diagram of the target center gray before and after filtering.

As can be seen from the figure, in which the parameter settings, the filtered result of the target gradation is more stable, and change magnitude of gray is smaller.

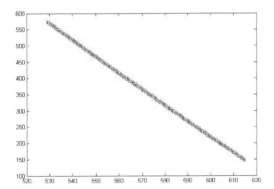

Figure 1. Coordinates before and after filtering, $Q=I_5$, $R=5I_3$.

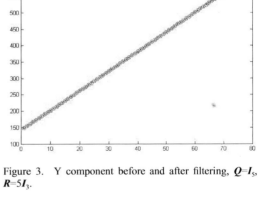

Figure 3. Y component before and after filtering, $Q=I_5$, $R=5I_3$.

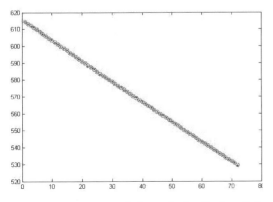

Figure 2. X component before and after filtering, $Q=I_5$, $R=5I_3$.

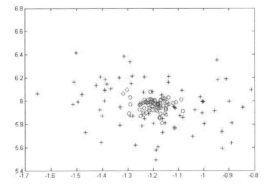

Figure 4. Speed change before and after filtering, $Q=I_5$, $R=5I_3$.

441

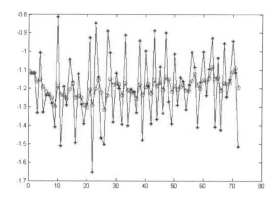

Figure 5. X component of speed beforeand after filtering, $Q=I_5$, $R=5I_3$.

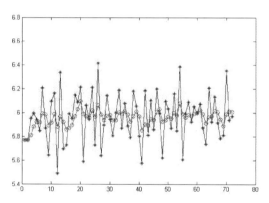

Figure 6. X component of speed before and after filtering, $Q=I_5$, $R=5I_3$ $Q=I_5$, $R=5I_3$.

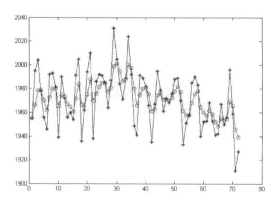

Figure 7. Change before and after filtering, $Q=I_5$, $R=5I_3$.

4 CONCLUSION

In the ballistic midcourse target detection and tracking applications based on infrared image, because the target distance is far, we can believe that the target is in uniform linear motion in the imaging time. Therefore, the system noise is less. The main interference is from the sensor noise and target detection error. Therefore, measurement noise plays a dominant role in the test results. We gain a lot of appropriate measurement noise covariance matrix coefficients, which can make the filtering results more precise.

REFERENCES

Wu,X., Y. Zhou & J. Cui(2009): The Infrared Optical Recognition Technology Analyzes of Missile Defense System. *Infrared and Laser Engineering*, Vol. 38 No.5, p.759.(in Chinese).

Smith M S(2005).: Military Space Programs Issues Concerning, DODs SBIRS and STSS programs, *CRS Report for Congress Aug.*

Wu,C.F., W.Zhang(2005): Problems and Analysis Faced by Target Detection and Recognition of Infrared Imaging Systems Based on Spatial and Temporal Information. *Optical Technology*, Vol. 31 No.2, p.231. (in Chinese).

Li,Q.F. &X.M. Guo(2004): Analysis of Space-based Infrared Sensors Maximum Range. *Electronic Science and Technology Review*, No.3, p.53.(in Chinese).

Dodgen L J(2006): inextricably linked to warfighting[J]. Military Law Review, No.49, p.12.

Wan, Q.,&Y.N. Wang(2007): Moving Target Detection and Tracking based on Kalman Filter. *Hunan University (Natural Science)*,Vol. 34 No.3, p.36. (in Chinese).

Electronic Engineering and Information Science – Wang (Ed.)
© 2015 Taylor & Francis Group, London, ISBN: 978-1-138-02772-5

Read and display of UG models in OpenGL

X.L. Sui, J.T. Chen, X.W. Zhang & C. Hua
Harbin University of Science and Technology, Harbin, Heilongjiang Province, China

ABSTRACT: Aiming at complex three-dimensional objects' animation demonstration problems in the virtual NC milling simulation system, UG model's simulation completed in OpenGL is proposed. Visual C ++ 6.0 is chosen as a development tool, the UG is used as a modeling tool, then UG software's PRT format data files are converted into the STL format data files. 3ds Max is used as a model transformation tool to convert STL files to 3DS files. Ultimately, 3DS models are reappearing in OpenGL by programming and realized the UG files' read and display in OpenGL. The complex problems of UG models' difficulty to realize animation demonstrate and OpenGL software's difficulty to accomplish multiplex modeling of the three-dimensional objects are solved. It is of great significance of improving the simulation speed, real-time simulation as well as the dynamic display process of simulation in the virtual NC simulation system.

KEYWORDS: UG; OpenGL; read; display; 3ds Max.

1 INTRODUCTION

Currently, there are many kinds of three-dimensional geometric modeling software such as UG, 3ds Max, OpenGL, etc. In the aspect of solid modeling, graphics processing and dynamic simulation capabilities has its own advantage each, e.g., UG has a powerful solid modeling, surface modeling, virtual assembly capabilities, but its dynamic motion simulation performance is not very prominent. 3ds Max is a three-dimensional animation production software, it is powerfully modeled, scalable and have a great advantage in terms of animation with rich plugins, but the animation made by 3ds Max has a poor interactivity, so it is hard to realize real-time control. OpenGL is a hardware-independent software interface with good portability, it has modeling, transformation, color mode, texture mapping and powerful double buffered animation function, however, as OpenGL is a 3D graphics underlying graphics library which provide no geometric primitive entity, so it can not be directly used to describe the scene, but through some conversion process to convert 3DS model file into OpenGL software's vertex array could be easily realized, by which method, real-time rendering and interactive control of the established model is convenient to implement (Pan & Mei 2007).

In order to make full use of the respective advantages of UG, 3ds Max, OpenGL, this paper converts UG software's three-dimensional model to 3DS file via 3ds Max, reads an display in OpenGL, implements UG files' read, redrawn, and controls in OpenGL successfully.

2 CONVERSION OF UG AND 3DSMAX MODEL

2.1 *The output format of UG file*

3ds Max is a kind of three-dimensional animation software with strong rendering and animation capabilities, but its three-dimensional solid modeling function is not strong. In order to give full play of UG and 3ds Max respective advantages, it can convert the three-dimensional entity of UG into 3ds Max to complete rendering and animation.

The key technologies of combining UG and 3ds Max is to solve the data exchange between the two software. UG software uses a NURBS model, the output format of UG itself is PRT format, furthermore, it also can output in IGES, STL, DXF, DWG and many other formats, as the model in 3ds Max is mostly polygon model with less NURBS model, thus the graphical output format in IGES, DXF format is not very accurate (You et al. 2008). In this paper, STL format was adopted to accomplish conversion, it can show the real scene of machining in the phase of model design which made the abstract technology visualized, so that the engineering design and layout is more reasonable and effective. But make sure that the exported model's coordinates must be positive, otherwise it may go wrong in exporting.

2.2 *Convert UG file into 3DS file*

Convert UG software's PRT format file into an STL format file in an appropriate accuracy, the exported STL file is a certificate trust list (.stl), STL file is a certificate file which is a common format of three three-dimensional conversion, to open the STL file, it needs to open the three-dimensional software, and then import the corresponding STL file. Pay attention to select the appropriate conversion accuracy, it will affect the quality of the rapid prototyping parts with low precision, and when the precision is too high, if the file size is too big ,it is not easy to transmit data for the large file, and it will also affect the rate of rapid prototyping system of data processing.

The specific steps of converting UG file into 3DS file is: create a three-dimensional solid model in UG or open the model that has been established, then choose "File → Export → STL" ... menu, it will appear "rapid Prototyping" dialog box, choose setting up the option according to need, in the pop-up the "Export Rapid-Prototyping"dialog box input file name and the title, meanwhile, it will pop up a "Class Selection" dialog box, choose the three-dimensional entity which need to be converted into the 3ds Max, at this time it will automatically generate a file with the extension name .STL. Finally, open the 3ds Max software, choose "Import" option in the "File" menu, input the file you want to convert in the pop-up dialog box "Select File to Import" , enter the * .STL file you want to convert and the established entity in UG will open, then it can use the function of rendering and animation in 3ds Max. When import STL files in 3ds Max (Dave 2010), the stitching parameters based on the graphics surface geometry and test, the three-dimensional solid model in figure 2 is a UG model in figure 1 which converted into STL file, then was imported into 3ds Max. It can see that graphic is very precise and the converting speed is fast.

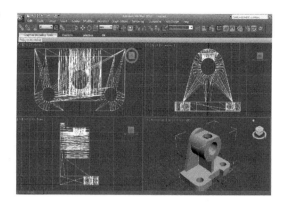

Figure 2. Imported STL file in 3dsMAX.

3 READ AND REDRAW 3DS FILES IN OPENGL

3.1 *Read 3DS files*

This paper mainly introduces the process of OpenGL read 3DS file in Visual C ++ 6.0 environment. OpenGL is a graphic software interface, it is designed as a new kind of interface, its hardware-independent feature make it can be realized on different hardware platforms. OpenGL didn't include window tasks or obtain user input commands, which make it can realize cross-platform performance (Zhou et al. 2009). 3DS is a very common data format, the three-dimensional graphic file saved in 3DS format is very rich, the read and manipulate 3DS files is of great important for three-dimensional graphic application software. The technology of importing the external 3DS file directly to the OpenGL scene can not only shorten the modeling of complex objects, but also lay a solid foundation for the development of visualization systems. The specific steps were shown in figure 3.

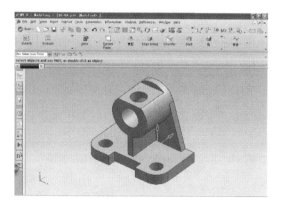

Figure 1. UG model.

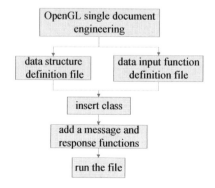

Figure 3. The display steps of 3DS model in OpenGL.

3.2 Show 3DS model in OpenGL

3DS model file is composed of a lot of Chunk, each Chunk includes a head and a body, the Chunk is nested within each other, so it is appropriate to choose to read them in the form of recursive (Gao 2010). First, define two classes "CTriObject" and "CTriList", "CTriObject" mainly used to deal with a variety of objects in 3DS files including geometric coordinates, the object vector, etc. While "CTriList" is mainly used for processing object sequence, which is a collection of the "CTriObject", saved the data form of the three-dimensional model. As the editing Chunk, color Chunk, material Chunk in 3DS model file play a key role, so build read 3DS class is to read into these key Chunk. Definitely a read class "C3dsReader" of 3DS file in the Visual C ++6. 0 MFC framework to read the contents of the 3DS file into the above two classes. There are many read functions which mainly used to read main structure, including color, texture coordinates, the normal vector and coordinate values and 3DS file functions, 3DS model redraw function, 3DS model normalized functions and other major function.

Drawing 3D object is mainly done in "CTriObject", the outer shape is mainly used a triangle to approximate. According to the following steps to import 3DS model After writing "C3dsReader" class (Peace 2006):

Insert a new class"CTriObject" in the project, class type is a generic class, which mainly deals with a variety of objects in3DS file, insert a new class "CTriList" in the project, class type is a generic class, the class mainly processes various display list of 3DS files, using "MFC ClassWizard" to add "OnOpenDocument" "CMy3DSMillDoc" class function, in order to be able to import multiple models and easy to control, it adopts the method of display list to load a model.

Using "MFC ClassWizard" to add member function and member variables for "CMy3DSMillView" class, redrawing the imported model, to add the "Open File" function in the "CMy3DSMillApp" class, then the 3DS model was displayed.

4 FEASIBILITY VERIFICATION OF UG MODELS SHOWM IN OPENGL

In order to verify the feasibility of the program, this paper built a five-axis CNC milling machine used UG, as is shown in figure 4, then exported from 3DS. Compile and run the above program, its operating results are shown in Figure 5, basically realized the model of reading and redraw. From the display effect can be seen that the after-shown graphic on the

shape and color were accurately reflect the information in the original image. Thus it demonstrated the object-oriented programming method was not only accurate but also reliable, and it proved the UG model can be displayed in the OpenGL.

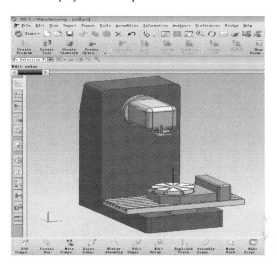

Figure 4. CNC milling in UG.

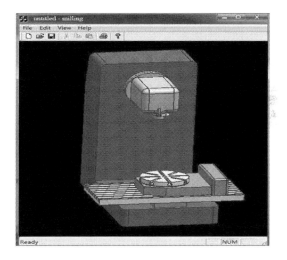

Figure 5. Redrawn CNC milling in OpenGL.

5 CONCLUSION

This paper first proposed that importing UG models into OpenGL to complete virtual CNC milling system simulation in Visual C ++ 6.0 environment. Aiming at the problems that OpenGL is difficult to implement complex three-dimensional object modeling and UG is difficult to realize animation demonstration, the paper combined the two software to get a more realistic, complex model and took a 5-axis CNC milling machine as an example, displayed the complex UG model in OpenGL successfully, and the conversion speed is fast as well as the transformation effect is good, the redraw graphic was basically consistent with the original. This research laid a solid foundation for improving simulation speed, simulation accuracy and real-time simulation which realized by the function of double buffer function in the virtual CNC simulation system.

ACKNOWLEDGMENT

This research was sponsored by Excellent Academic Leaders Project of Harbin science and technology innovation talents of special fund (2013RFXXJ064)

RERERENCES

Pan X. Q. & Mei C. C. (2007). Effective way to get 3ds model data based on OpenGL. *Fujian computer, 3,* 159–160.

You B. D.,Yi C. C. & He Z. R. (2008). OpenGL three-dimensional simulation software based on realistic. *Nanchang University, 12,* 381–383.

Dave S. (2010). OpenGL Programming Guide. *Machinery Industry Press.*

Zhou F., Ni J. F. & Zeng X. Z. (2009). Research on OpenGL read and redraw 3ds model based . *Suzhou University, 29,* 54–56.

Gao H. F. (2010). OpenGL-based import and Control 3DS model. *Wuyi University (Natural Science), 24,* 32–36.

Peace D. S. (2006). OpenGL Programming and Advanced Visualization System. *China Water Power Press.*

Electronic Engineering and Information Science – Wang (Ed.)
© *2015 Taylor & Francis Group, London, ISBN: 978-1-138-02772-5*

Visible camera calibration and registration in the 3D temperature field reconstruction

B.Y. Wang, X.Y. Yu, B.H. He, J.F. Wu, H.B. Wu, X.M. Sun & S.Y. Liu
The Higher Educational Key Laboratory for Measuring & Control Technology and Instrumentations of Heilongjiang Province, Harbin University of Science and Technology, Harbin, China

J.F. Wu
Heilongjiang University of Science and Technology, Harbin, China

ABSTRACT: In order to establish the 3D Temperature Field based on structure light 3D measurement and cross-correlation registration, infrared/visible camera calibration method and heterologous image registration method were studied. First, we simplified the two-step method based on the mature principle of structure light measurement; and the visible camera was calibrated; then, used visible cameras calibrated to form a structured light measurement system; and finally registered the infrared image and visible image in the geometrical position. Experimental results show that: the relative error of measurement system is less than 0.3%;infrared and visible image's alignment error in geometrical position is 0.0015 mm. And measurement systems have good reconstruction results for plane and curved, and cross-correlation method has a good effect of heterologous image registration.

KEYWORDS: Structured light; Infrared camera; Calibration; Registration.

1 INTRODUCTION

The technology of 3D temperature field involved in the field of military, medical and industrial production. In forging industrial production, high temperatures will cause the material to undergo complex physical and chemical changes, the size of the change in the process has a direct impact on the structural performance and accuracy class, constructing 3D temperature field of real-time measurement, in order to achieve online monitoring, has important economic significance for industrial production.

In abroad, Michael Gschwandtner, Roland Kwitt, Andreas Uhl and Wolfgang Pree proposes a self-calibration method including visible camera and infrared camera (Gschwandtner & Witt 2011)in 2011,which is extended on the traditional camera calibration method. The method can be useful for IR camera calibration. In 2012, Stephen Vidas and Ruan Lakemond et al. proposed a calibration method for single infrared camera and multiple infrared camera (Vidas et al. 2012). This method homemade a calibration template, which has been testified that the calibration template is clearer than a heating chessboard in the infrared camera. In China, Li Chenglong proposed an infrared camera self-calibration system (Li 2013). The method uses a multi-camera dynamic calibration method

to calibrate the intrinsic parameters and external parameters of one or more infrared cameras.

These are also a lot of depth and meticulous research in the image registration.

In foreign countries, Moravec presents a gray scale of auto-correlation coefficients to distinguish between a relationship between the area and the area of the pixel, which is a classic feature point detection (Wu & Gong 2011) in 1977. Harris operator (Ma & Zhu 2006) proposed by Harris C. and MJ Stephens in 1988 is a feature point extraction operator, which is the improvement of the Moravec corner detection algorithm, using first-order partial derivatives to describe the brightness changes, this operator inspired by the auto-correlation function in the signal processing, the matrix M is given and linked to auto-correlation function.

David G lowe proposed a local feature description operator based on scale space detecting the feature point, SIFT operator detects the potential feature points for scale and rotation invariant by Gaussian differential functions in scale space. Wang Song proposed a bilateral matching algorithm based on sift operator feature point detection and applying KD-tree and quasi-Euclidean distance. This registration method using two different registration threshold, carrying reverse registration after forward registration.

In China, Zhu Yinghong, Li Junshan et al. proposed a point matching algorithm based on shape context. First, feature point are extracted by the Curvature Scale Space (CSS) (Zhu & Li 2013) corner detector; Ma proposed a pretreatment method that enhancing image edge firstly, and then histogram equalization, finally using matching strategy of gray correlation.

For heterologous image, because of their feature vector's directions are different, the computation of the selected feature points is complexity; while infrared image is generally considered to be the negative image of the visible image, so describing its gray information is more convenient method than feature points after preprocessing visible image, so the text will be taken after gray mutual information method for registration.

2 INFRARED CAMERAS AND VISIBLE CAMERA CALIBRATION THEORY AND EXPERIMENT

In this paper, monocular coplanar lattice produced a calibration plate, the specific calibration step is that: projector projects to template calibrate with the coding pattern and the camera get pictures after rendering. Then, using Matlab writes calibration procedures, including image binarization, calculate the distance between measuring points, the image coordinate calculation and calibration procedures. According to the two-step calibration method we can calculate the internal and external parameters of the camera, camera calibration is complete.

It is assumed to be an ideal environment, so the camera lens hole model for ignoring lens distortion and deformation angle $\alpha = 90°$. Matlab programming calculated calibration results, compared with the standard values listed in Table 1, in order to compare the calibration results with Calib_box.

Through calculating the data in the table, lens focal length maximum relative error of 0.11% maximum relative error of the internal and external parameters is 0.10% can be got. The camera lens in the measurement is approximately holes model, so the lens distortion coefficients are approximately zero.

3 3D MEASUREMENT EXPERIMENTS AFTER VISIBLE CAMERA CALIBRATION

The following will simulate the measurement for flat, to achieve the purpose of validating the calibrated visible camera parameters.

Place a vertical plane at 300×200mm in the measurement system which has been established. The

Table 1. Comparisons between two-step calibration value and the standard value

Calibration parameters	Standard value	Calibration values	Measurement error
f_x	49.455	49.396	0.059
f_y	49.455	49.396	0.059
R_{11}	0.94	0.94	0
R_{22}	0	0	0
R_{33}	0.342	0.342	0
R_{21}	0	0	0
R_{22}	1	1	0
R_{23}	0	0	0
R_{31}	−0.342	−0.342	0
R_{32}	0	0	0
R_{33}	0.934	−0.934	0
T_1	0	0.056	0.152
T_2	50	50.00	0
T_3	−584.76	−584.165	0.595

experimental procedure taking 430mm depth values as a starting point, will place the vertical plane in the measured depth value point, and put eight encoded images sequentially irradiated onto the plane and were taken with the camera, the image was calculated through the measurement procedures written in Matlab and got 3D information. Such objects can be drawn by measuring extreme values on each axis. Reconstruction results and measurement error are shown in Figure 1.

Figure 1. Reconstruction results and measurement error.

At X-axis, maximum of measuring is 149.871mm, minimum is -149.302mm; At Y-axis, the maximum and minimum values are 100.291 mm; Through simple calculation ,we can obtain that the length of plate is 299.173mm, and height is 200.582 mm. And the result of depth error is calculated for each point, and gets that the maximum measurement error is 1.051mm when the depth is 430 mm, the average measuring error is 0.415 mm, the relative maximum error is 0.24%.

4 THE EXPERIMENTAL INFRARED CAMERA CALIBRATION AFTER REGISTRATION

Mutual information refers to any variable information and other variable information has many common parts. That is to say, this principle is statistical correlation corresponds to gray value of visible image and an infrared image. Figure 2 using a Venn diagram to express the relationship between various kinds of entropy and mutual information.

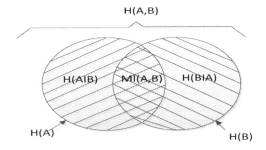

Figure 2. Relationship between entropy and mutual information.

The above analysis briefly explains the principle of mutual information registration. The following work applied it to register the infrared image and visible image.

Figure 3. Pretreatment result.

After a series of treatment, compared to visible light image and the infrared image, the white pixels of the visible image are concentrated within the outline of objects, which is identical with the infrared image. Registration difficulty at this time will be greatly reduced. Pretreatment result is shown in Figure 3. Next, use a mutual information image registrar for two images.

Registration work is still using GUI functionality on the MATLAB platform, simply enter registration and reference images in the interface can automatically complete the search and integration work. After entering the pretreatment visible and infrared image's grayscale, the program will automatically complete the registration, the registration parameters are as follows:

$$X, Y, Angle = [18.033][38.004][-0.015]$$

$$MI_{value} = [0.947]$$

Figure 4 is a fusion of the image registration:

Figure 4. Grayscale image after registration.

5 CONCLUSION

This paper completed the calibration work of visible light camera, and compared with the setting parameters, the experimental results show that: the maximum relative error of the lens focal length is 0.11%, the maximum relative error of internal and external parameters is 0.10%. The experimental results are quickly and accurately. After the completion of the calibration, planar was measured using the 3D measuring system; Finally tried to use the mutual information registering heterologous image.

After more in-depth research, this study still needs further exploration, the next step work should first focus on forming binocular measurement in the application of infrared camera and visible camera, further exploring the registration algorithm of infrared image and visible image, and completing construction of the 3D temperature field.

ACKNOWLEDGEMENTS

This study was supported by the Scientific and Technological Project of Education, Department of Heilongjiang Province with grant number (12541171).

REFERENCES

Gschwandtner M. & R. K. Witt (2011). Infrared Camera Calibration for Dense Depth Map Construction. *IEEE Intelligent Vehicles Symposium* ,857–861.

Vidas, Stephen & Lakemond (2012). A mask-based approach for the geometric. *IEEE Transactions on Instrumentation and Measurement, 61*, 1265–1365.

Li C. L. (2013). Research based on three-dimensional reconstruction method of infrared thermal image. *Anhui University.*

Wu M. & K. Gong (2011). Some discussion on moravec operator. *China science and technology achievements, 16*, 62–65.

Ma L. & L. Zhu (2006). The Approach of Feature Point Detection for Illumination Pattern-based Images under Bias Field.

Zhu Y. H. & J. S. Li (2013). A feature point description and matching algorithm for edge in IR/visual image. *Journal of Computer-Aided Design & Computer Graphics, 25*, 858–864.

Electronic Engineering and Information Science – Wang (Ed.)
© 2015 Taylor & Francis Group, London, ISBN: 978-1-138-02772-5

Graphics processing of CNC cutting based on auto CAD

W.L. Guo & H.J. Xing
College of Computer Science and Technology, Harbin University of Science and Technology, Harbin, China

T. Zhang
Harbin Tianyuan CNC Company, Harbin, China

ABSTRACT: In this article, we will introduce a method for numerical control cutting processing of engineering parts. First, draw the graphics of engineering parts by Auto CAD drawing software, these graphics will save as files of DXF format. These DXF format files contain the necessary data of the parts to be processed by numerical control cutting. This data must be further processed because of it is generated automatically by the Auto CAD drawing, the order of this data cannot in full compliance with the need of engineering, adjust the data sequence in order to form a complete cutting curve. Then, write a program in Visual C++ 6.0 software to deal with the data which saved in DXF files. This program will use b-spline curve fitting method to process the data. Finally, use the processed curves to generate the code files of numerical control cutting machine, and complete the cutting processing of CNC cutting machine with the engine parts.

KEYWORDS: AutoCAD; Drawing Exchange Files; VC++6.0; Curve fitting.

1 INTRODUCTION

Auto CAD is an essential application software to industrial engineering drawings, it is widely used in various engineering and construction fields. DXF, which short for Drawing Exchange Files are ASCII files that write all the information about the graph according to the rule format (Mu et al. 2001). Auto CAD can directly draw the graphics based on the information that provided by the DXF, also can automatically convert into the corresponding DXF by the internal graphics file (the DWG file).

Curve fitting is a data processing method of approximately depicting or assimilating the functional relation of coordinate which be said by a group of discrete points in the plane by the continuous curve (Zhao et al. 2013). Curve fitting technique is a kind of curve surface modeling method, which applied broadly in the field of image processing, reverse engineering, test data processing, and so on (Zhang et al. 2003). Many problems in computer aided design are related to the curve fitting problems.

2 GRAPHIC DRAWING PARTS

The Draw the graphical information of the machining parts which need to process with the AutoCAD software, then store the graphics as DXF format, in order to extract and process the data next step.

Figure 1. The Straight Line Drawn.

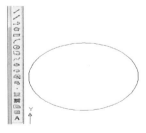

Figure 2. The Circle Drawn.

When drawing graphics with Auto CAD, it can choose a straight line, round and circular, elliptical and elliptic arc to draw (Guan D.Z., 2007). Click the drawing line command button on the toolbar while drawing a straight line, then draw a triangle which as shown in Figure 1.

Using the circle drawing tool can draw a circle as shown in Figure 2. One of the machine parts graphics in this design is shown in Figure 3.

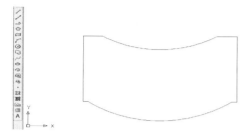

Figure 3.   One of The Machine Parts.

3   DRAWING EXCHANGE FILES CONTENT

The graphic files need to be stored as DXF format after drawing with Auto CAD, which convenient to extract and process the file content by VC++ program.

DXF is the information about graphics of writing as a specified format file. Therefore, DXF is the application's interface files between Auto CAD and high-level programming language, it has now become one of the industry standards in the field of computer aided design (Mu et al. 2001).

DXF include six sections, i.e. the title, class, table, segment, entity, and object.

The title section contains basic information on the graphic, each of the parameters including a variable name and the set of values. The class definition section contains the information about the application definition class. The class definition in the class hierarchy is fixed. The table section contains the definition of the symbol table. The segment includes blocks of each piece of reference in the definition and composition of the graphics primitives.

The entity section contains the graphical object in the graphics, and the block reference. The object section contains non-graphical objects in the graphics. The entity section of the machine part shown as Figure 4.

The graphical geometry information is needed while programming the CNC program of CNC cutting machine, this part of information is included in the entity section of DXF.

Graphics geometry information mainly contains geometric entities and entity label information.

Geometric entities of the entity section are point, line, circle, arc, trace line, broken line, the b-spline curve, form, text, and so on.

The annotation information about the entity section is Line mark information, arc labeling information, round labeling information and angular dimension.

Read and process the geometric entity information by VC++ program, resulting in the G code that can control the CNC cutting machine.

```
      O
  SECTION
      2
  ENTITIES
      O
    ARC
      5
    11E
    330
    1F
    100
  AcDbEntity
      8
    100
  AcDbCircle
     10
  1250. 970172196702
     20
  1503. 465154565568
     30
   0. 0
     40
   1318. 0
    100
  AcDbArc
     50
  232. 0628502518404
     51
  307. 9371497481595
      O
    . . .
  ENDSEC
```

Figure 4.   Part of Entity Section.

4   THE PROCESSING OF GRAPHICAL INFORMATION FILE

The geometric entity information about the DXF entity section extracted by computer is stored in order according to the order in the DXF, which in the order of drawing the graph.

In the actual machining process, the cutting machine was conducted in a certain closed wire-cutting processing order, these two are not exactly the same order. So, the extracted geometric entities must be reorder as the processing order.

According to the requirements of processing technology, the starting point of the section needs to know when the CNC wire-cutting processing, and it did not specify in the DXF. In order to solve this problem, a few auxiliary points need to be created. That is to say, when the paint used in wire-cutting processing, two auxiliary points are added to the ends of the first

452

entity in processing, the turn of these two points decides the order of processing.

As shown in Figue 5, Pl, P2 are the two points. If drawing order is Pl to P2, and the processing order is: 1 to 2 to 3 to 4, so, Pl is the starting point for the processing; if drawing order is P2 to Pl, and the processing order is: 1 to 4 to 3 to 2, so, the starting point is P2 for the processing.

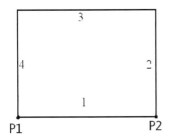

Figure 5.   The Order of Process.

In order to achieve the reorder of geometric entities in processing order, two auxiliary points must be found in the data storage file. The first point is Pl, then the second is found as P2, so that the first geometric entities wait for processing is decided.

Then, the end point P2 of this geometric entity, which connected to the next entity can be found, too. By analogy, the reorder sequence of entities is got. After reordering, the entity files are stored in TXT format files.

5   THE CURVE FITTING AND HANDLING OF GRAPHICS

Because of the data extracted from DXF is discrete geometry data, and the cutting machine is required for the curve graph, the data curve fitting processing for graphics rendering is necessary, then VC++ can program the right code for CNC cutting processing. In this design, the B-spline curve fitting method has been used to restore the data.

B-spline curve fitting method has the powerful features of the free curve and surface design, is one of the most popular mathematical description of the mainstream method.

The given plane or space vertex p_i $(i = 0,..., n)$ is n times parameter curve segment (Sun L.X. 2000).

$$p(t) = \sum_{i=0}^{n} p_i G_{i,n}(t) \qquad 0 \ll t \ll 1 \qquad (1)$$

Function (1) is n times B-spline curve, and the polygon of its vertex p_i $(i = 0, 1,..., n)$ is the characteristics

polygon of the b-spline curve, $G_{i,n}(t)$ is the basis function of n times B-spline function. The basis function is shown as equation (2).

$$G_{i,n}(t) = \frac{1}{n!} \sum_{j=0}^{n-1} (-1)^j c_{n+1}^j (t + n - i - j)^n$$
$$0 \ll t \ll 1, i = 0, 1,..., n \qquad (2)$$

The quadratic B-spline curve, which need to use in this design is shown in Figure 6.

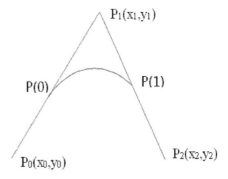

Figure 6.   Quadratic B-spline Curve Be Determined by Three Vertices.

This article using VC++ language writes the program to do the curve fitting processing of the data in DXF and reorder it, and takes the data as the gist of numerical control cutting, then, program the code that CNC cutting machine can identify, to control the cutting of the parts.

6   CONCLUSION

Save the graphic files as DXF after processing parts of the graphics by Auto CAD software, then, program with VC++ language to read the graphics geometry information in the DXF, reorder the data information and use the B-spline curve fitting to process the data, so as to get more accurate numerical control cutting discrete data, and generate the G code for identification of the CNC cutting machine, finally, complete the processing parts of CNC cutting to over this design.

ACKNOWLEDGMENT

This work was supported by Technology innovation fund of Harbin city science and technology oriented small and medium sized enterprises (NO: 2014FF2CJ021).

453

REFERENCES

Mu N. W. , Y. D. Zhao & Z. C. Shan (2001). Automatic Programming of The Edm CNC Wire Cutting Machine Based on DXF. *The Application of Science and Technology*, 28(4), pp. 1–3.

Zhao L.H., J.Q. Jiang, & L.Y. Bao (2013). The Research Status of B-spline Curve Fitting Based on Intelligent Algorithm. *Journal of National University of The Inner Mongolia (Natural science edition)*.

Zhang J.L., S.C. Zhen, P. Cao (2003). The Curve Fitting Method and Its Application of Experimental Data. *Journal of Testing Technology*, 17(3), pp.255–257.

Guan D.Z., Computer Drawing. *Mechanical Industry Press*, pp.27.

Sun L.X. (2000). Computer Graphics. *Harbin Industrial University Press*, pp.135–137.

Electronic Engineering and Information Science – Wang (Ed.)
© *2015 Taylor & Francis Group, London, ISBN: 978-1-138-02772-5*

Application of multi-threshold Otsu algorithm in image edge detection

M.Z. Liu, L. Shi & Q. Wu
The Higher Educational Key Laboratory for Measuring & Control Technology and Instrumentations of Heilongjiang Province, School of Measurement and Control Technology and Communications Engineering, Harbin University of Science and Technology, Harbin, China

ABSTRACT: In the image edge detection, the threshold of original Canny operator relays on the statistics of the image global gradients. However, because the distribution of the image gradient information exists differences, it might lead to some unexpected issues such as local edge information being missed and the edge details being lost. In this paper, a new edge detection algorithm has been provided, which is to divide the whole image into several parts based on the number of peaks on its grayscale histogram, and then uses Otsu algorithm to select the multi-pair thresholds of the Canny operator for every sub-image. Experimental results show that when the number of peaks in the grayscale histogram matched with the number of image partition number, the image edge can be detected and recognized effectively, and the detection effect is superior to the original Canny operator edge detection algorithm.

KEYWORDS: Canny operator; gray statistic; edge detection; image partition; multi-threshold.

1 INTRODUCTION

Edge detection plays an important role in image processing. One of the main goals of the edge detector is to select the appropriate detection thresholds to sharpen the image edges and benefit to the information extraction. Using different edge detection operators can get different detection effects. The traditional Canny operator edge detection is to deal with the whole image directly, whether using the genetic algorithm to find the double threshold of Canny operator, or using the adaptive method to change the weights of the Gaussian filters, both of them are aimed to find the optimal value for processing objects from the perspective of determining parameters (He 2013, Deng 2013). But for the whole image, the features of every part of the image are different, and its gray scales will change in a large range. If we directly use the traditional edge detection method for the whole image, it will miss some edge information. In this paper, a block method for the whole image has been used in the edge detection process, and using the gray level statistic values to select the optimum thresholds have been introduced in the process of edge detection.

2 THE TRADITIONAL CANNY OPERATOR

Using the finite difference approximate differential method can realize the derivation of the image. According to the algorithm is different, can be divided into edge detection based on the first derivative operator and the second derivative operator. Robert operator, Sobel operator and Prewitt operator all belong to the first derivative operator. Based on the second derivative of the edge detection operator have Laplacian operator, Marr-Hildreth operator and Canny operator. Canny operator is based on the second derivative. The results of the Canny (Hu 2013) operator are more close to intuitionistic consequences, so it has been widely applied in image processing. The feature in the edge detection process with the Canny edge detection operator is non maximum suppress (NMS) and the double thresholds judgment. The three performance criteria (Sonka et al. 2008) are as follows:

Signal-to-noise (SNR) :

$$\text{SNR} = \frac{\left| \int_{-w}^{+w} G(-x)h(x)\,dx \right|}{\sigma \sqrt{\int_{-W}^{+W} h^2(x)\,dx}} \tag{1}$$

Good localization performa:

$$L = \frac{\left| \int_{-W}^{+W} G'(-x)h'(x)\,dx \right|}{\sigma \sqrt{\int_{-W}^{+W} h'^2(x)\,dx}} \tag{2}$$

Only one response to a single edge:

$$D_{zca}(f') = \pi \left\{ \frac{\left| \int_{-\infty}^{+\infty} h'^2(x)\,dx \right|}{\int_{-W}^{+W} h''(x)\,dx} \right\} \qquad (3)$$

$$I = \begin{bmatrix} I_{11} & I_{12} & \cdots & I_{1q} \\ I_{21} & I_{22} & \cdots & I_{2q} \\ \cdots & \cdots & \cdots & \cdots \\ I_{p1} & I_{p2} & \cdots & I_{pq} \end{bmatrix} \qquad (4)$$

Each sub image matrix I_{kl} $(k=1,2,\cdots,p;\ l=1,2,\cdots,q)$. matrix with m_1 rowand n_1 column, that is $m_1 \times n_1$ $(m_1 = m/p, n_1 = m/q)$. The experiment found that the number of sub image is according to the whole image gray information, less than the peaks of gray histogram is the best. As shown in the experimental results and analysis.

In the above three criteria, if improve detection capabilities, and that the positioning accuracy will drop. In turn, the improvement to the edge position detection accuracy, the failure rate will increase. Canny operator is used to add a linear filter, that product optimization of the SNR and position accuracy at the step edges and additive Gaussian noise conditions, but this filter cannot satisfy the third criteria when there are noises. In this respect, the Canny operator with the optimizing analytical condition (OAC) method to make step edge has a unique response results in the case of equations (1) and (2). This filter is more complex, but we can use the first derivative model of Gaussian function to approximate. After the edge filtering, select two thresholds, and then determine the edge points. Double thresholds are used to detect and connect the edge points. The false edge will be detected with high and low thresholds for the image which has been processed by the non-maximum suppression. At the same time, there will be three conditions: First, the gray values is greater than the high threshold, it must be the true edge point; Second, it's less than the lower threshold, it must not be the true edge point. Third, the gray value of one pixel is between the high threshold and the low one, then, according to the gray values of its adjacent edge points, if there exists one greater than the high threshold, it means that this pixel is the true edge point, if not, it is the false edge point. Meanwhile, traditional Canny operator edge detection uses the same thresholds to process the whole image. That will lead to missing edges, and make the noise to be an edge point. This paper presents the image block method, and use of local gray value statistics set up multiple thresholds.

3 IMAGE PARTITION

This paper utilizes the uniform image partition method. The aim is to obtain local gray level statistics. Compared with the non uneven partitioned, uniform partitioned process is simple, easy to implement, the computational complexity is not high, the algorithm execution speed. The uniform image partition method will put one $m \times n$ image matrix I that divided into $p \times q$ sub image matrix, that is

4 MULTI-THRESHOLD OTSU ALGORITHM IN IMAGE EDGE DETECTION

4.1 Image thresholding segmentation method based on grayscale histogram

OTSU (maximum variance between clusters) is a kind of adaptive threshold value method. OTSU algorithm divided the image into the object and the background. Between them, the greater classes square error, the greater difference. When the object is divided into background or parts background, it will become smaller variances between two parts. So when the class variances reach the maximum, the probability of two part errors is placed at the minimum. This paper is equally divided image into partitions according to characteristics of gray histogram with two peaks or multi peaks. Then combine the gray statistical characteristic and OTSU threshold algorithm to calculate the thresholds. In this paper, the method of image segment combined with multi-threshold Canny operator has been introduced to improve edge detection efficiency, and the new algorithm can reduce the inflection of the noise, avoid losing detail information and improve the accuracy of edge detection.

4.2 The determination of the threshold for Canny operator

The Canny operator first carried on the filtering to image with the two-dimensional Gauss function. For each pixel with a non-zero gradient strength, two adjacent pixels are searched along its gradient direction. This process is called non-maximum suppression (Shanmuga & Kumar 2013). There are only two circumstances for the image when, after the non-maximum suppression: Simple image (image histogram has two peaks) and Complex image (image histogram has multi peaks). For simple image, the

456

image is divided into the object and the background. Assuming the two classes separated by a threshold T. Let N be the number of the image pixels. N_1 and N_2 is the object and background pixels, w_1 and w_2 are the probabilities and u_1, u_2 are their average values. The total average gray value is u. The functions have been shown as follows:

$$w_1 = \frac{N_1}{N}$$

$$w_2 = \frac{N_2}{N}$$

$$w_1 + w_2 = 1$$

$$u = w_1 \times u_1 + w_2 \times u_2$$

$$g = w_1 \times (u_1 - u)^2 + w_2 \times (u_2 - u)^2$$

Class square error has been defined as:

$$g = w_1 w_2 (u_1 - u_2)^2 \tag{5}$$

Use the traversing method to make g maximum under threshold T. T is the optimal threshold, corresponds to the bottom of gray value between neighbor peaks. Make it as a Canny high threshold T_h, then the equation $T_h = 2 \times T_l$ for low threshold T_l (Liu et al. 2011).

For Complex image, will appear more "valley". We can first calculate the threshold between two adjacent peaks. This situation can be obtained multiple thresholds. Suppose there are N peaks, and get $N-1$ thresholds $(T_1, T_2, \cdots, T_{N-1})$, thresholds $T_i (1 \le i \le N-1)$ divided the histogram into two parts, again using the method that the histogram has two peaks, then calculates the classes square error of two parts histogram which were divided by T_i. Finally, select the maximum value among the $N-1$ classes square errors, then let it be the high threshold for Canny operator.

This method can improve the ability of Canny edge detection and also can accurately select the edge detection thresholds based on the gray distribution of the image. At the same time, according to the method is proposed in this paper, processing from the overall to local, in different gray level distribution of sub images by the Otsu algorithm to choose more appropriate Canny operator threshold, make the results more continuous and contain less false edges.

5 EXPERIMENT RESULTS AND ANALYSIS

This paper deal with the original image is the Lenna (256×256) with random noise. The image is divided into four, nine and sixteen partitions and then use Otsu algorithm to select thresholds for each partition according to its histogram Figure 2. The processing results have been shown in Figure 1 (1)–(10).

(1) Original Lenna image

(2) Preprocessing image

(3) Edge detection result by traditional Canny method

(4) Image with 4 partitions

(5) Edge detection result for 4 partitions image

(6) Image with 9 partitions

(7) Edge detection result for 9 partitions image

(8) Image with 16 partitions

(9) Edge detection result for 16 partitions image

Figure 1. Experiment Results by proposed algorithm.

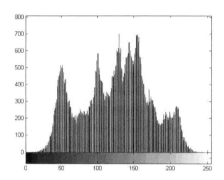

Figure 2. Histogram of the image.

Figure 1(1) is the original image with random noise. Use Gauss filter which its average value is 0 and the variance is 1 as an image filter to get rid of the noise and the preprocessing image in the Figure 1(2) can be obtained. Figure 1(3) is the result after the operation of traditional Canny operator edge detection. Figue 1(4) and 1(5), Figure 1(6) and Figure 1(7), Figure 1(8) and Figure 1(9) are the output results using partition algorithm with different segmentation numbers such as 4, 9, and 16 respectively. From Figue 1, it can be seen that when a whole image having been divided into some blocks, it can obtain distinct edge information and details compared to the traditional Canny operator for the continuous edges specially.

Experiment result shows that the number of sub-block should be selected according to the number of peaks of overall image histograms which can be identified from Figure 2. If the number of sub-block is less than or equal to the number of peak tops in the histogram, the edge detection effect can reach to the best. For example, the number of the histogram peaks is seven from Figure 2, so the 9-blocks diagram has been selected, it can be found that the best detection effect had been shown in Figure 1(6). From Figures 1 and 2, it can be concluded that using Canny operator and a kind partition method of multi-threshold Otsu algorithm to block the image can get the better detection effects on either reducing false detection ratio for edge extraction or simplifying operation steps.

6 CONCLUSIONS

In this paper, the basic principle of the edge detection algorithm based on Canny operator has been analyzed. On that basis, the new algorithm which combined with the principle of image partition refines the gray scale distribution, and by using OTSU multi-threshold method, the threshold value has been calculated and the best one has been chosen. From the experimental result, it can prove that the image edges having been highlighted and at the same time the false detection radio has been declined significantly with the new algorithm.

REFERENCES

He J., L. Hou & W. Zhang (2013). A Kind of Fast Image Edge Detection Algorithm Based on Dynamic Threshold Value. *J. Sensors & Transducers,* 23 (7): 179–183.

Deng C. X., G. B. Wang & X. R. Yang (2013). Image edge detection algorithm based on improved canny operator.*C.//Wavelet Analysis and Pattern Recognition (ICWAPR), International Conference on. IEEE.* 168–172.

Hu Z. P. (2013). Research of Curvature HOG object detection algorithm. *J. Journal of Signal Processing,* 29(11):1470–1475.

Sonka M., V. Hlavac & R. Boyle (2008). Image processing, analysis, and machine vision .*M. Toronto*: Thomson.

Shanmuga P. & A. Kumar (2013). Boundary detection of objects in digital images using bit-planes and threshold modified Canny method .*M.//Mining Intelligence and Knowledge Exploration. Springer International Publishing,* 2013: 192–200.

Liu Q. S., J. F. Dong & G. H. Liu (2011).Using the Canny edge detector and mathematical morphology operators to detect vegetation patches .*J. 2011 3rd International Conference on Digital Image Processing-ICDIP,* Chengdu, China, April, 2011. 15–17.

Electronic Engineering and Information Science – Wang (Ed.)
© 2015 Taylor & Francis Group, London, ISBN: 978-1-138-02772-5

An implicit calibration method for image distortion

S.C. Yu, P.J. Wang, S. Liu & H.Y. Xiao
The higher educational key laboratory for Measuring & Control Technology and Instrumentations of Heilongjiang Province, Harbin University of Science and Technology, China

ABSTRACT: Image distortion correction is a difficulty in the field of computer vision. The accurate distortion correction model is a fifth-order polynomial that it consists of the radial distortion coefficient ($k1$, $k2$) and tangential distortion coefficient (p_1, p_2), and it is difficult in obtaining its iterative solution. Therefore, we put forward a kind of implicit image distortion method, it is called eight parameters. We can use the correction factor λn ($1 \leq n \leq 8$) with no any the radial distortion coefficientplace of radial distortion coefficient ($k1$, $k2$) and tangential distortion coefficient (p_1, p_2), and build an implicit distortion correction model, then we use the least square method to realize the image distortion correction. The experimental results show that this method is not only simple, but also accurate.

KEYWORDS: computer vision; implicit correction; the least square method; distortion correction.

1 INTRODUCTION

Image distortion correction is a key technology in the field of computer (Wu & Sun 2004). Because of the influence of lens manufacturing precision, the image of the computer vision system would appear different degrees of distortion, the distortion can be divide into radial distortion and tangential distortion (Quan & Yu 2000). If we don't need too high precision, and because of the influence of tangential distortion is far less than the radial distortion, so We can only consider the radial distortion (Tsai 1987, Zhang 1999, Jia 2005). The influence of tangential distortion is often ignored for another reason, a correction model with radial distortion is a fifth-order polynomial, and it is difficulty in obtaining its iterative solution (Wei & Ma 1993).

So we put forward a kind of implicit image distortion method. The method is that we can use an implicit distortion correction model to take the place of the traditional fifth-order polynomial correction model, in the same way, we use the correction factor ($1 \leq n \leq 8$) with no any the radial distortion coefficientplace of radial distortion coefficient ($k1$, $k2$) and tangential dthe modeltion coefficient (P_1, P_2).then the implicit parameter of model could be achieved by the least square method.

2 EXPLICIT CORRECTION OF IMAGE DISTORTION

Assuming that (u', v') is the ideal coordinates of pixels, (u, v) is the image pixel coordinate with radial distortion and tangential distortion. The following two relationships:

$$u = k_u (k_2 u'^5 + 2k_2 u'^3 v'^2 + k_2 u' v'^4 + k_1 u'^3$$
$$+ k_1 u' v'^2 + 3p_2 u'^2 + 2p_1 u' v' + p_2 v'^2 + u') + u_0$$

$$v = k_v (k_2 u'^4 v' + 2k_2 u'^2 v'^3 + k_2 v'^5 + k_1 u'^2 v'$$
$$+ k_1 v'^3 + p_1 u'^2 + 2p_2 u' v' + 3p_1 v'^2 + v') + v_0$$

$$(1)$$

In the formula, k_u is the lan i kv is the vertical pixel unit of image, obtained by camera calibration. k_1 and k_2 are the radial distortion coefficient; p_1 and p_2 are the tangential distortion coefficient.

Formula (1) is a classic one of the explicit correction model, it is a fifth-order polynomial, including k_1, k_2, p_1 and p_2. It isn't difficult that we use the calibration data from the formula (1) to obtain k_1, k_2, p_1 and p_2, but even if we know what is the value of k_1, k_2, $p1$ and p_2, k_u a distortion of the pixel coordinates (u, v) restore the true pixel coordinates (u', v') is a very difficult work in the whole image plane. And the method of literature reference is also very little at the present. Wei[6] had put forward a method to solve this problem, finally the method didn't get the promotion because of poor precision.

Therefore, the researchers consider correcting the image distortion by the implicit parameter model with no any physical meaning.

3 IMPLICIT CORRECTION OF IMAGE DISTORTION

Implicit correction, that is, the building of distortion correction model doesn't use the distortion factor of radial distortion and tangential distortion, but will use any correction factor with no any physical meaning to realize the purpose of the simplified correction algorithm. Wei and Ma [6] proposed the implicit calibration method is one of the representative.

3.1 Implicit correction methods of wei and ma

The literature proposed implicit calibration model is as follows:

$$u' = \frac{\sum_{0 \le j+k \le N} \lambda_{jk}^{(1)} u^j v^k}{\sum_{0 \le j+k \le N} \lambda_{jk}^{(3)} u^j v^k}$$

$$v' = \frac{\sum_{0 \le j+k \le N} \lambda_{jk}^{(2)} u^j v^k}{\sum_{0 \le j+k \le N} \lambda_{jk}^{(3)} u^j v^k}$$

(2)

In the formula, $N = 3$, $\{\lambda_{jk}^{(n)}\}$ is the implicit calibration factor with no any physical meaning. After camera calibration, we can use the internal reference of video camera to get $\{\lambda_{jk}^{(n)}\}$, then a distortion of the pixel coordinates (u, v) restore the true pixel coordinates (u', v') becomes a simple work. The reason is that we already know (u, v) and $\{\lambda_{jk}^{(n)}\}$.

It can be seen from the formula (2), the literature could only use a cubic polynomial distortion correction model to get the value of $\{\lambda_{jk}^{(n)}\}$, so its precision is not high.

Therefore, we would try to use the fifth-order polynomial model of formula (1) in literature [6] to improve the correction accuracy (in the formula (2), $N = 5$). But now, there are as many as sixty-three parameters in $\{\lambda_{jk}^{(n)}\}$, using the results of the calibration are difficult to calculate the complete $\{\lambda_{jk}^{(n)}\}$. In order to solve this problem, we put forward a kind of implicit 8 parameter implicit calibration model.

3.2 A model of 8 parameter implicit calibration

This new implicit calibration model is as follows:

$$\begin{bmatrix} u' \\ v' \end{bmatrix} = \begin{bmatrix} \dfrac{u + u(\lambda_1 r^2 + \lambda_2 r^4) + 2\lambda_3 uv + \lambda_4(r^2 + 2u^2)}{H} \\ \dfrac{v + v(\lambda_1 r^2 + \lambda_2 r^4) + \lambda_3(r^2 + 2u^2) + 2\lambda_4 uv}{H} \end{bmatrix}$$

(3)

In the formula, $H = (\lambda_5 r^2 + \lambda_6 u + \lambda_7 v + \lambda_8)r^2 + 1$, $r^2 = u^2 + v^2$. $\lambda_n (1 \le n \le 8)$ doesn't have any physical meaning in this model, but λ_1, λ_2, λ_3, λ_4 and k_1, k_2, p_1, p_2 approximate into corresponding relation , and we can achieve radial and tangential distortion of image correction at the same time.

It isn't hard to see, this model is compared with model in literature [6], the number of implicit correction factor is only eight, the difficulty is reduced greatly if we use the least square method to work out.

After 8 parameters implicit calibration model was established, you can use this model to correct the distortion of the image.

3.3 Method with 8 parameters implicit calibration

Assumption, we extract n points from the camera, and measure that the image coordinate values are $(u_i, v_i)(1 \le i \le N)$. After calibration, we could use the back calculation of interior and exterior parameters to calculate the value of image coordinates is $(u'_i, v'_i)(1 \le i \le N)$. We are given three vector U_i, V_i, T as follows:

$$U_i = \begin{bmatrix} -u_i r_i^2 \\ -u_i r_i^4 \\ -2u_i v_i \\ -(r_i^2 + 2u_i^2) \\ u'_i r_i^4 \\ u'_i u_i r_i^2 \\ u'_i v_i r_i^2 \\ u'_i r_i^2 \end{bmatrix} \quad V_i = \begin{bmatrix} -v_i r_i^2 \\ -v_i r_i^4 \\ -(r_i^2 + 2v_i^2) \\ -2u_i v_i \\ v'_i r_i^4 \\ v'_i u_i r_i^2 \\ v'_i v_i r_i^2 \\ v'_i r_i^2 \end{bmatrix}$$

(4)

$$T = \begin{bmatrix} U_1, V_1, \ldots, U_i, V_i, \ldots, U_N, V_N \end{bmatrix}^T$$

(5)

In the formula (4), $r_i^2 = u_i^2 + v_i^2$.

Assumption, the vector P is made up of 8 correction factor vector, the following formula:

$$P = \begin{bmatrix} \lambda_1, \lambda_2, \lambda_3, \lambda_4, \lambda_5, \lambda_6, \lambda_7, \lambda_8 \end{bmatrix}^T$$

(6)

Assumption, the vector e as follows:

$$e = \begin{bmatrix} u_1 - u'_1, v_1 - v'_1, \ldots u_N - u'_N, v_N - v'_N \end{bmatrix}^T$$

(7)

According to the formula (3), the vector e, T, P should satisfy the following relations:

$$e = TP$$

(8)

Finally, we use the least square method to get the vector P:

$$P = (T^T T)^{-1} T^T e$$

(9)

In other words, we know that the vector P also know the eight implicit calibration $\lambda_n (1 \le n \le 8)$ of formula (3).

Then we use model of formula (3) to achieve it, namely the distortion correction is that a distortion of the pixel coordinates (u, v) restore the true pixel coordinates (u', v') in the whole image.

4 EXPERIMENT AND ANALYSIS

a) Distorted image.

b) Corrected image.

Figure 1. Experimental results.

In order to verify the validity of the algorithm in this paper, we select 20 images of literature (8) to process camera calibration and image distortion correction experiment.

The main body of camera calibration experiments is a Zhang Zhengyou calibration method, image distortion correction experiments adopt 8 parameter implicit calibration method. All twenty images for the correction, in order to compare the difference between the front and back distortion correction more intuitively, so we give the following images, the figure 1 is image before correcting distortion, the figure 1(b) is image after correcting distortion.

The figure 2(a) shows the influence of radial distortion, and the influence of tangential distortion is shown in figure 2(b).

We can see from figure 4 and figure 5, the influence of radial distortion in the image edges reach

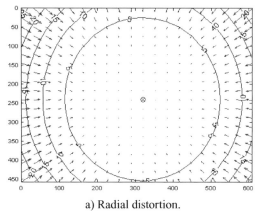

a) Radial distortion.

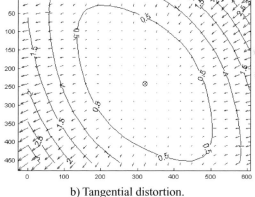

b) Tangential distortion.

Figure 2. Radial distortion and tangential distortion.

twenty-five pixels, but the tangential distortion most affect three pixels, so the influence of radial distortion is bigger on the PC of Pentium 1.7G and in the Matlab environment, the whole process of the correcting distortion takes 0.5 seconds, and after correction, the errors of horizontal and vertical coordinates are 0.11and 0.09 pixel unit.

5 CONCLUSION

In this paper, we analyze the formation of image distortion, in order to correct the image of the radial and tangential distortion at the same time, this paper proposes a new method of 8 parameters implicit calibration. The method is compared with the method of Wei and Ma, solving process is greatly simplified. The experimental results show that the proposed method performs fast and has high precision, can effectively eliminate the influence of the radial and tangential distortion image.

ACKNOWLEDGMENT

This study was supported by Heilongjiang Province ordinary university young academic backbone support plan with grant number (1253G027).

REFERENCES

Wu W. Q. & Z. X. Sun (2004). Overview of camera calibration methods for machine vision. *Application Research of Computers, 2*: 4–6.

Quan T. H. & Q. F. Yu (2000). High-accuracy calibration and correction of camera system. *Acta Automatica Sinica, 26(6)*: 748~755.

Tsai, R.Y. (1987). A versatile camera calibration technique for high-accuracy 3D machine vision metrology using off-the-shelf TV cameras and lenses. *IEEE Journal of Robotics and Automation RA ,3(2)*:323~344.

Zhang, Z.Y. (1999). Flexible camera calibration by viewing a plane from unknown orientations. *Proceedings of IEEE International Conference on Computer Vision, 1*: 666~673.

Jia H. T. & Y. C. Zhu (2005). A technology of image distortion rectification for camera. *Journal of Electronic Measurement and Instrument, 19(3)*:46–49.

Wei, G.Q. & S.D. Ma (1993). A complete two-plane camera calibration method and experimental comparisons. *4th International Conference on Computer Vision,* 439–446.

Electronic Engineering and Information Science – Wang (Ed.)
© *2015 Taylor & Francis Group, London, ISBN: 978-1-138-02772-5*

Roundness detection method based on improved Hough transform

S. C. Yu, D. Zhao, L. L. Chu & C. Liu
The higher educational key laboratory for Measuring & Control Technology and Instrumentations of Heilongjiang Province, Harbin University of Science and Technology, China

ABSTRACT: This research mainly focuses on the circle detection algorithm based on the Hough transform. By using traditional Hough transform, circle detection had high time complexity and low precision. So we used the prior knowledge of filter rod image to improve the speed of circle detection based on the Hough transform, and combined it with the regional gray center method to make the positioning accuracy which is used to improve edge detection to reach sub pixel level. Experimental results show that the proposed method has a fast speed and a high precision.

KEYWORDS: roundness detection, Hough transform, time complexity, gray centroid method.

1 INTRODUCTION

At present, the domestic cigarette manufacturers pay more attention to the quality of the production of cigarettes. Quality of cigarette rod can be tested efficiently and real-time by using machine vision technology. Roundness error detection, largely reflects the quality of cigarette production. (Huang 2001, Rosin 1999)

Machine vision technology, which combines optics, computer science and technology and digital image processing technology is an online non-contact measurement method. The principle of main machine vision measuring roundness error is through CCD (Charge Coupled Device) camera to take images of the measured object, and use digital image processing technology for processing, then extract the image features, analysis and process it, get the rounding error of measured object (Liu & He 2007), further evaluate the roundness error of the measured object. Walsh detected the appearance, quality of a paper cup with machine vision technology, and achieved the paper cup online non-contact measurement and classification (Walsh & Raftery 2012). Sharma used the machine vision technology to measure inner diameter roundness of the optical fiber, realized fast and non-contact measurement of optical fiber diameter roundness (Sharma 2010). This method has advantages of non-contact, high efficiency, high measuring precision, it becomes the main online measuring method in the non-contact measurement field.

In this paper, on the basis of the existing methods, a kind of roundness error detection method was proposed based on improving Hough transform.

2 HOUGH TRANSFORM THEORY

The principle of the Hough transform is that using the dual relations between a line and a point, and transform the line or curve in image space transform into a point in the parameter space, and use this point to indicate the line or curve in image space. The detection problem of the object given in the image becomes to obtain the peak detection problem in the parameter space, the whole image detection change into local image feature detection.

Known circular standard equation:

$$(x-a)^2 + (y-b)^2 = r^2 \qquad (1)$$

Where, (a,b) is center coordinates, r is radius of a circle.

A set of target round edge points in the image is $\{(x_i, y_i) | i \in 1, 2, 3, \dots n\}$, the edge points are mapped to the parameter space equation:

$$(a-x)^2 + (b-y)^2 = r^2 \qquad (2)$$

The equation is a three-dimensional conical surface equation, and (a,b,r) is (A, B, R) corresponding coordinates in the parameter space coordinate system. For image space $x - y$ coordinate plane, aleatoric and a determined point on the circle switch type (2) to $A - B - R$ as the parameter space coordinates. They all correspond to a three-dimensional vertical cone in parameter space, a set of target point on a circle map to the parameter space is formed in the conical surface cluster. A point on the circumference of image space mapped into the parameter space and become erect conical surface shown in Figure 1a. Due to the radius of a circle in the image space is fixed, a point on the circumference mapped into the parameter space and become three dimensional vertical conical surface cluster shown in Figure 1b.

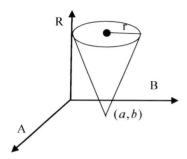

a) Points in image space

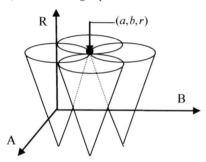

b) Circumference point set in image

Figure 1.　Mapping the image space to the parameter space.

The image goes through edge detection and get all the edge point to $\{(x_i, y_i) | i \in 1, 2, 3, ...n\}$, defining an accumulative array in Hough space is

$$H(x_s, y_s, r) = \sum_{i=1}^{n} h(x_i, y_i, x_s, y_s, r) \qquad (3)$$

Among them, $H(x_s, y_s, r)$ is a number which curve in parameter space pass determining the edge point, parameter (x_s, y_s, r) is the number of times, which points in parameter space was passed, $h(x_i, y_i, x_s, y_s, r)$ meet the following conditions:

$$h(x_i, y_i, x_s, y_s, r) = \begin{cases} 1 & g(x_i, y_i, x_s, y_s, r) = 0 \\ 0 & g(x_i, y_i, x_s, y_s, r) \neq 0 \end{cases} \qquad (4)$$

Where, $g(x_i, y_i, x_s, y_s, r) = (x_i - x_s)^2 + (y_i - y_s)^2 - r^2$ is to meet (x_s, y_s, r) as a function of the conditions which can determine the parametric equation of the circle. Only when the circle parameters satisfy the determining conditions function and the parameter space plus 1 accordingly.

Because $H(x_s, y_s, r)$ in the parameter space is composed of three parameters, $H(x_s, y_s, r)$ in parameter space is actually a three dimensional array, through each boundary point in image space transform in the parameter space and then make a vote, finally get the most vote parameter sets, which are composed of target corresponding parameters of the circle.

Because three-dimensional in the parameter space array $H(x_s, y_s, r)$ based on parameters as the subscript, according to its maximum value of the subscript can determine center coordinates and radius of the measured target circle, that is: if

$$H(x_{center}, y_{center}, R) = \max(\cup H(x_s, y_s, r)) \qquad (5)$$

Where, center coordinates are (x_{center}, y_{cemter}), the radius of the circle is R. The maximum value of three-dimensional array $H(x, y, r)$ in the parameter space indicates that the circle, obtained by it can pass most of edge points, and the coordinates of corresponding parameters express equation of circle best.

3　IMPROVED HOUGH TRANSFORM ALGORITHM

Hough transform need to take a multiple of voting program operation for all boundary points in image space, it determines the scope of the circle parameter difficultly and there still exist a lot of redundant points, when making the Hough transform to detect circle, we map and vote for redundant points, thus caused it to measured the speed of image processing is very slow, the number of the measured image size and image edge points plays a key role on the detection speed, the amount of calculation increases with the image boundary points increase. This article mainly is improved in the following two aspects: reduce the number of boundary points and shrink the image; Hough transform combine with regional gray centroid method to complete the sub pixel edge location.

The sub pixel edge detection method is mainly two aspects as follows: fast detection speed and high precision. Among this, the gray center method is a fast and effective edge sub pixel edge detection algorithm. Calculation formula of gray center method as shown in type (2-10).

$$\begin{cases} x_c = \dfrac{\displaystyle\sum_{(i,j)\in S} if(i,j)}{\displaystyle\sum_{(i,j)\in S} f(i,j)} \\[4mm] y_c = \dfrac{\displaystyle\sum_{(i,j)\in S} jf(i,j)}{\displaystyle\sum_{(i,j)\in S} f(i,j)} \end{cases} \qquad (6)$$

464

For all boundary points of Hough transform, we can do regional extension, type (6) is used to calculate the gray center coordinate of each region namely get sub pixel coordinates of the edge.

Detailed algorithm steps as follows:

Step 1: using a filter rod of a priori knowledge to obtain a filter rod center position, radius and boundary region.

Step 2: using median filter to process the filter rod image.

Step 3: selecting the center of a circle, the boundary area and the radius of fast Hough transform.

Step 4: using local gray center method to determine the extended region of the edge point coordinate with the Hough transform, and then have sub pixel edge detection, finally obtain the sub pixel center coordinate of the filter rod image edge.

4 EXPERIMENT OF ROUNDNESS DETECTION

This chapter mainly to prove the improved Hough transform algorithm in this essay, respectively for single clear images of filter rod, single fuzzy images of filter rod to detect the experiment with improved Hough transform, and compare the experiment which detects filter rod image with traditional Hough transform. It mainly divides into: typical experiment of filter rod images and the experiment compared with traditional Hough transform.

In this article, through the Matlab software write program to achieve filter rod circle algorithm with improved Hough transform, first set the center of filter rod, radius and the regional scope of boundary, use the algorithm which combined with improved Hough transform and regional gray center method to detect the edge of the clear filter rod image, and accurate positioning to sub pixel edge. In this article, the algorithm experiment of the Hough transform is improved for clear images of filter rod and fuzzy images of filter rod respectively. In this paper, through an improved Hough transform algorithm, we can get detection results of a clear filter rod image and fuzzy filter rod image as follows, Hough transform the detection result of the clear filter rod image as shown in Figure 2.

a) Hough detection

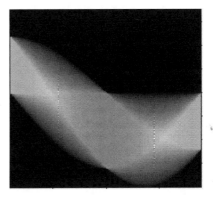

b) The space of Hough.

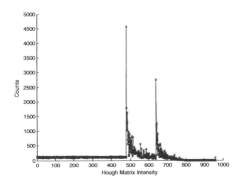

c) Hough spatial histogram.

d) Local gray center method.

Figure 2.   Algorithm testing clear image filter rods.

From Figure 2a, we can see accurate positioning of Hough transform algorithm for filter rod edge, the circle, which is determined by the Hough transform, can be approximately close to the real circle boundary of the filter rod image, it has laid a good foundation that combined with regional gray center method to obtain sub pixel edge points. Figure 2b is the result that filter rod image map to Hough space. Figure 2c shows the corresponding cumulative value of each accumulator in Hough space each accumulator, you can see a peak, subscript with peak corresponding to the accumulator array is the circle's center coordinates and radius. In Figure 2d, we can see clear filter rod image by the Hough transform to determine pixel edge and expand the area, combine with regional gray center method for the effect of sub pixel edge positioning, and has reached the sub pixel precision.

5 CONCLUSION

We improved circle algorithm on the basis of the traditional Hough transform, and combine with regional gray center method in order to edge location precision achieve sub pixel level. Through the experiment of typical filter rod image and experiment which use the traditional Hough transform to detect circle, we can prove that the improved Hough transform on the speed and precision is much better than the traditional Hough circle detection method.

ACKNOWLEDGMENT

This study was supported by Heilongjiang Province ordinary university young academic backbone support plan with grant number(1253G027).

REFERENCES

Huang J. P. (2001). A new strategy for circularity problem. *Precisionengineering*, 301–308.
Rosin P. L. (1999). Further Five-point Fit Ellipse Fitting. *Graphical Models and Image Processing, 61 (8).* 245–259.
Liu P. & J.X He. (2007). Inversion of the Radon transform on the product Lag-uerrehyper group by using generalized wavelets. *International journal of co-mputer mathematics, 84 (3).* 287–292.
Walsh D. & A.E Raftery. (2012). Accurate and efficient curve detection in ima-ges. *Pattern Recognition, 35 (7).* 1421–1429.
Sharma R., K. Rajagopal. & S. Anand. (2010). . A genetic algorithm based approach forrobust evaluation of form tolerances[J]. Journal of Manufacturi-ng Systems, , 19(1) : 46–57.

Electronic Engineering and Information Science – Wang (Ed.)
© *2015 Taylor & Francis Group, London, ISBN: 978-1-138-02772-5*

Roundness evaluation method based on least squares

S.C. Yu, J.L. Wang, L. Tian & Z.W. Pan
The Higher Educational Key Laboratory for Measuring & Control Technology and Instrumentations of Heilongjiang Province, Harbin University of Science and Technology, China

ABSTRACT: Roundness evaluation is an important means of judging the accuracy of roundness testing results, which has important significance for roundness testing quality. From the aspects of precision and algorithm complexity, the least squares fitting method is taken to fit the discrete edge point. The least square circle method is used in roundness evaluation of rods, and simulation and practical experiment are also carried out to test the proposed method. The experimental results show that the evaluation method based on the least squares can obtain a reasonable roundness evaluation.

KEYWORDS: Roundness evaluation; Least square; Roundness test; Algorithm complexity.

1 INTRODUCTION

Roundness error evaluation is a key link in the roundness measurement. There are four kinds of methods for roundness error evaluation standards stipulated by the state: (Huang 2001, Rosin 1999): The Maximum Inscribed Circle method (MIC), the Least Square Circle method (LSC), the Minimum Zone Circle method (MZC), and the minimum circumscribed circle method (MCC). The core of the roundness error evaluation method is to determine the ideal reference circle, Different ideal circle position corresponding to the different method of roundness error evaluation. In the four evaluation methods, Minimum zone circles consistent with the definition of roundness error evaluation method, but the rest of the evaluation method of roundness error evaluation of the roundness error circle and minimum area deviation general will not be huge, so the rest of the evaluation method on the workpiece when meet the requirements of its production precision, also can be used (Liu & He 2007). Scholars at home and abroad propose many new methods about roundness evaluation method. Using genetic algorithm, particle swarm optimization algorithm, neural network algorithm, etc., can be effective, high precision of roundness error assessment (Walsh & Raftery 2012), but this algorithm requires multiple iterative calculation, some algorithm parameters to be set by men, and does not guarantee results for optimal results. Using the least squares method on the uncertainty relation between the parameter is analyzed, and prove that the roundness error of uncertainty mainly depends on the measurement points and their distribution and data fitting algorithm. Due to the least square method have characteristics of specific mathematical, analytical (Sharma 2010), a small amount of calculation and

data on the elliptical deformation evaluation precision higher. According to the research object of this article of distribution of feature points and accuracy requirements, choose the least square method to evaluate the roundness of cigarette filter rods.

2 ROUNDNESS EVALUATION METHOD BASED ON LEAST SQUARES

In this paper, the edge points are used in completing the least squares fitting. By using te least square fitting, discrete edge points can be fitted to a continuous contour curve. When the curve of the standard circle as the least square circle, the so-called least square circle, or find a point in the actual contour, make the outline on each point to have centered on this circular radial distance minimum sum of squares, the point in outline is the center of the least square circle. Setting a cigarette filter rod edge points on the contour is (x_i, y_i), $i \in (1, 2, \cdots, n)$, the center coordinates of a fitting circle are (A, B), radius is R.

The equation of fitting circle is

$$(x - A)^2 + (y - B)^2 = R^2 \qquad (1)$$

It can be rewritten as the general equation of the circle

$$x^2 + y^2 + ax + by + c = 0 \qquad (2)$$

Among them

$$\begin{cases} a = -2A \\ b = -2B \\ c = A^2 + B^2 - R^2 \end{cases} \qquad (3)$$

Type (2) x and y is the known quantity, a, b and c are the desires of the unknown quantity, in the circular curve fitting process, When the tested edge points of the outline n=3, according to the type (2) can be listed three equations, Obtained the value of a, b, c is only a set of solutions, prayer circle of the three points. But when measured contour edge points n>3, Measured contour point (x_i, y_i) can't both satisfy the type (2), there is residual d_i between the actual contour by measuring point (x_i, y_i) and type (2). Represented as:

$$d_i = x_i^2 + y_i^2 + ax_i + by_i + c \qquad (4)$$

Due to the residual d_i is negative or positive, according to the characteristics of random error, its positive and negative error will be offset, According to the norm of least squares approximation, to the sum of the squares of the error to a minimum, these become extreme value problems of the requested amount as a, b, c of multivariate functions. According to the extreme value of multivariate function condition calculated a, b, c, to calculate the least squares fitting circle coordinates (A, B) and radius R. The least square value of a, b, c should be satisfied multivariate function $F(a,b,c)$ minimum.

$$F(a,b,c) = \sum_{i=1}^{n} d_i^2 = \sum_{i=1}^{n} (x_i^2 + y_i^2 + ax_i + by_i + c)^2 = MIN \quad (5)$$

When make extreme value for multivariate function $F(a,b,c)$. Meet the conditions

$$\begin{cases} \dfrac{\partial F(a,b,c)}{\partial a} = 0 \\[2mm] \dfrac{\partial F(a,b,c)}{\partial b} = 0 \\[2mm] \dfrac{\partial F(a,b,c)}{\partial c} = 0 \end{cases} \qquad (6)$$

According to the type (4-6)

$$\begin{cases} \displaystyle\sum_{i=1}^{n} (x_i^2 + y_i^2 + ax_i + by_i + c) \cdot x_i = 0 \\[3mm] \displaystyle\sum_{i=1}^{n} (x_i^2 + y_i^2 + ax_i + by_i + c) \cdot y_i = 0 \\[3mm] \displaystyle\sum_{i=1}^{n} (x_i^2 + y_i^2 + ax_i + by_i + c) = 0 \end{cases} \qquad (7)$$

Further reduction

$$\begin{cases} a\displaystyle\sum_{i=1}^{n} x_i^2 + b\sum_{i=1}^{n} x_i y_i + c\sum_{i=1}^{n} x_i + \sum_{i=1}^{n}(x_i^3 + x_i y_i^2) = 0 \\[3mm] a\displaystyle\sum_{i=1}^{n} x_i y_i + b\sum_{i=1}^{n} y_i^2 + c\sum_{i=1}^{n} y_i + \sum_{i=1}^{n}(x_i^2 y_i + y_i^3) = 0 \\[3mm] a\displaystyle\sum_{i=1}^{n} x_i + b\sum_{i=1}^{n} y_i + c \cdot n + \sum_{i=1}^{n}(x_i^2 + y_i^2) = 0 \end{cases} \quad (8)$$

Will type (4 to 8) for the corresponding parameters

$$\begin{cases} \displaystyle\sum_{i=1}^{n} x_i^2 = w_{20}, \ \sum_{i=1}^{n} x_i = w_{10}, \ \sum_{i=1}^{n} x_i y_i = w_{11} \\[3mm] \displaystyle\sum_{i=1}^{n} y_i^2 = w_{02}, \ \sum_{i=1}^{n} y_i = w_{01}, \ n = w_{00} \\[3mm] \displaystyle\sum_{i=1}^{n}(x_i^3 + x_i y_i^2) = w_{32}, \ \sum_{i=1}^{n}(x_i^2 y_i + y_i^3) = w_{23} \end{cases} \quad (9)$$

Which type (4 to 8) rewrite for

$$\begin{cases} w_{20}a + w_{11}b + w_{10}c + w_{32} = 0 \\ w_{11}a + w_{02}b + w_{01}c + w_{23} = 0 \\ w_{10}a + w_{01}b + w_{00}c + w_{20} + w_{02} = 0 \end{cases} \qquad (10)$$

By type (4-10) can be calculated a, b, c

$$\begin{cases} a = \dfrac{-w_{32}w_{01}^2 + w_{10}w_{01}w_{23} + (w_{11}w_{01} - w_{10}w_{02})(w_{20} + w_{02}) + w_{00}(w_{32}w_{02} - w_{11}w_{23})}{w_{02}w_{10}^2 - 2w_{11}w_{10}w_{01} + w_{00}w_{11}^2 + w_{20}w_{01}^2 - w_{00}w_{20}w_{02}} \\[4mm] b = \dfrac{-w_{10}w_{32}w_{01} + w_{10}^2 w_{23} + (w_{11}w_{10} - w_{20}w_{01})(w_{20} + w_{02}) - w_{00}(w_{11}w_{32} - w_{20}w_{23})}{w_{02}w_{10}^2 - 2w_{11}w_{10}w_{01} + w_{00}w_{11}^2 + w_{20}w_{01}^2 - w_{00}w_{20}w_{02}} \\[4mm] c = \dfrac{-w_{10}w_{32}w_{02} + w_{11}w_{32}w_{01} + w_{23}(w_{11}w_{10} - w_{20}w_{01}) + (w_{20}w_{02} - w_{11}^2)(w_{20} + w_{02})}{w_{02}w_{10}^2 - 2w_{11}w_{10}w_{01} + w_{00}w_{11}^2 + w_{20}w_{01}^2 - w_{00}w_{20}w_{02}} \end{cases} \quad (11)$$

The precision of center and radius can reach sub-pixel accuracy, can make filter rods roundness evaluation accuracy higher. Putting the circle's center as the center of the least square circle evaluated fitter rod roundness error. The maximum distance of the measured filter rod each edge point at the center of the circle minus the minimum distance, is the roundness error of filter rod.

3 EXPERIMENTAL RESULTS AND ANALYSIS

In order to verify the effectiveness and feasibility of the algorithm, Through the Matlab software implement this textual algorithm. First, utilizing textual algorithm to simulation image for roundness evaluation, and add noise to the simulation image, the textual algorithm for the anti-noise ability of clear filter rods image and roundness evaluation accuracy. Deal with the simulation image for blur, texting the anti-fuzzy ability to blur filter rod and roundness evaluation accuracy of textual algorithm.

First, using the Matlab software develop a standard circle, in order to consistent with the gray distribution of filter rod, the reverse phase as shown in figure 1(a), as simulation image of fitter rod image, the size of the whole picture is 640×480 a pixel. Add white Gaussian noise to simulation image, mean value is o, variance σ^2 is 0.20, as is shown in figure 1(b). In this paper, the improved Hough algorithm of filter rod clear simulated image of the calculation results to roundness error as shown in Figures 1 and 2:

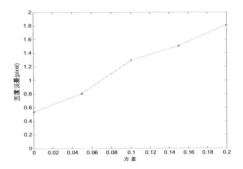

Figure 2. Simulation clear filter rod roundness error.

It can be seen from figure 2, the greater noise variance when add to clear fitter rod simulation image, the greater t roundness error correspondingly. When the noise variance is less than 0.10, the growing rate of the roundness error is very large, when the variance of noise from 0.10 to 0.20, the growing rate of the roundness error is smaller., when the Gaussian noise variance is within 0.15, the biggest roundness error is 1.5051 pixels (0.026 mm). Hough algorithm for noise has very good restraining effect.

4 CONCLUSIONS

Mainly has carried on the introduction and the model analysis to the classic roundness evaluation algorithm, study of the concrete practice process of the least squares. Through the simulation experiment and the actual experiment to fitter rod image, and analyze the experiment results, verify the validity of the algorithm in this paper. Experiments prove that the algorithm in this paper can detect fitter rod roundness effectively and high precision.

ACKNOWLEDGMENT

This study was supported by Heilongjiang Province ordinary university young academic backbone support plan with grant number (1253G027).

REFERENCES

Huang J. P. (2001). A new strategy for circularity problem. *Precisionengineering*, 301–308.
Rosin P. L. Further Five-point Fit Ellipse Fitting. (1999). *Graphical Models and Image Processing, 61 (8)*. 245–259.
Liu P. & J.X He. (2007). Inversion of the Radon transform on the product Lag-uerrehyper group by using generalized wavelets. *International journal of co-mputer mathematics, 84 (3)*. 287–292.
Walsh D. & A.E Raftery. (2012). Accurate and efficient curve detection in ima-ges. *Pattern Recognition, 35 (7)*. 1421–1429.
Sharma R., K. Rajagopal. & S. Anand. (2010). A genetic algorithm based approach forrobust evaluation of form tolerances. *Journal of Manufacturi-ng Systems. 19(1)*. 46–57.

a) no noise

b) gauss nosie(mean=0,variance=0.20)

Figure 1. Clear filter rod simulation.

Electronic Engineering and Information Science – Wang (Ed.)
© *2015 Taylor & Francis Group, London, ISBN: 978-1-138-02772-5*

Image encryption algorithm based on chaos theory and Arnold transformation research

W.C. Bi & J.H. Liu
School of Computer Science and Technology, Harbin University of Science and Technology, Harbin, China

ABSTRACT: With the increase of image resolution, traditional Arnold scrambling algorithm repeatedly scrambling will be more time-consuming, and it applies only to a square image, there are big limitations. This paper is based on the situation proposed a new method of image encryption of many areas feedback. Divide the square image area, through one-dimensional Logistic chaotic equation to generate a pseudo random sequence number control the exchange of pixel area. Experimental data show that, this algorithm solved the non square image encryption, and only a scrambling, makes image into an identification of encrypted image.

KEYWORDS: Image encryption, Arnold, Logistic, Multi-regional feedback.

1 INTRODUCTION

With the Internet and multimedia technology development and popularization security issues on images and other multimedia data are increasingly a cause for concern. Because of the digital image data having a large amount of data redundancy engage, relevance and other characteristics, using the traditional encryption methods to encrypt the digital image will make this slowly and inefficiently. Therefore, a new image encryption technology is particularly important in recent years (Zhu 2011).

Image scrambling is a commonly used digital image encryption method, it makes a digital image into a chaotic image, and makes it is non-intuitive to get real information (Liu & Sheridan 2013). Image scrambling consists of the position scrambling and grayscale scrambling. Position scrambling changes the position of each pixel in the image and grayscale scrambling changes each pixel gray value. As a common scrambling method, Arnold scrambling image processes several times to get the encrypted image. But images are not always square, and with increasing image resolution, many times scrambling will be efficiency. In this paper, an improved Arnold scrambling algorithm combined with one-dimensional chaotic equation is proposed. It can be scrambling for a square image and can get an encrypted image by an encryption (Ye 2010).

2 RELATED WORK

2.1 *Arnold scrambling algorithm principle*

Arnold transform is a kind of the chaotic mapping from torus to itself, which is also called cat face

mapping (Arnol'd & Avez 1968). Because Arnold transform has better dynamic characteristics of chaos, it gets the favor of a lot of image information security researchers, and it is applied in the field of image encryption widely. 2 d Arnold mapping equations are defined as shown in formula (1):

$$\begin{bmatrix} X_{n+1} \\ Y_{n+1} \end{bmatrix} = \begin{bmatrix} 1 & 1 \\ 1 & 1 \end{bmatrix} \begin{bmatrix} X_n \\ Y_n \end{bmatrix} \bmod 1 = M \begin{bmatrix} X_n \\ Y_n \end{bmatrix} \bmod 1 \qquad (1)$$

Because the Arnold transform has defects of fixed transformation matrix coefficient, related research expand the Arnold mapping to the generalized Arnold mapping, and then let the image pixel coordinates and gray as the initial value of general cat map, mapping parameters and the number of iterations as the key, it can be for image encryption (Ye & Wong 2012). The generalized Arnold mapping as shown in formula (2):

$$\begin{bmatrix} X_{n+1} \\ Y_{n+1} \end{bmatrix} = \begin{bmatrix} a & b \\ c & d \end{bmatrix} \begin{bmatrix} X_n \\ Y_n \end{bmatrix} \bmod N = M \begin{bmatrix} X_n \\ Y_n \end{bmatrix} \bmod N \qquad (2)$$

2.2 *Logistic chaotic mapping*

Logistic mapping is a kind of typical one-dimensional chaotic systems (May 1976), defined as shown in formula (3):

$$x_{k+1} = u \cdot x_k (1 - x_k) \quad k = 1, 2, 3, \cdots \qquad (3)$$

Because it generates a sequence fast, and the mapping has good chaos characteristics, widely applied in the field of image encryption. This paper uses a chaotic sequence to encrypt for multiple regions mutual feedback.

3 IMAGE SCRAMBLING ALGORITHM

3.1 Improved image scrambling algorithm based on Arnold transform

Suppose scrambling images as I, the length and width of I is the X and Y, respectively, divided I into several equal small square. a is the side length of square, so the calculation method of the side length a is:, so we can divide I into $n = (X \cdot Y)/a^2$ squares by taking the greatest common divisor of X and Y, then combined scrambling the n squares.

Given the Logistic map, initial parameters generate chaos sequence which length is $t + L \cdot L$, and give the front t data. Then handling the generated sequence $\{x_i | i = t + 1, t + 2, \cdots, t + L \cdot L\}$ as follows. To ensure sequence x_i^* can traverse to the n squares.

Supposed i and j are the and square, $i = 1,2,3,\cdots,n$, the pixels of the i replace the pixels of the j, $(mod(x + y, N) + 1, mod(x + 2y, N) + 1) = j(x, y)$.

Experimental results show that the efficiency of the algorithm of scrambling is higher compared with the traditional Arnold scrambling algorithm, replacement one time, we can see the effect of scrambling significantly better than an Arnold traditional algorithm. The concrete is shown in Figure 1.

3.2 Image encryption algorithm

Firstly, using the above scrambling algorithm on image I_0 to scrambling pixel to get the scrambling image I_1, then constructing chaotic sequence with Logistic system, due to the system can enter into chaos through a series cycle, so we get the sequence D that length is $(L + M \cdot N)$ after selecting initial value , μ、 x_0, which M and N respectively is the image's long and wide, $D = \{x_i | i = 1,2,3,1, L + M \cdot N\}$, giving up the front L set of values, and using after set of values, to improve the new sequence $D' = \{x_i | i = L + 1, L + 2,1, L + M \cdot N\}$ as shown in formula (4):

$$q = (q \cdot 10^6) \bmod 256 \tag{4}$$

with sequence exclusive or image pixel values and encryption image gray value, we can produce the final encrypted image. As shown in Figure 2.

Figure 1. (a. original image, b. Arnold encryption, c. improved algorithm).

Figure 2. (a. original image, b. encrypted image, c. restored image).

4 SECURITY ANALYSIS OF THE ALGORITHM

4.1 Key space

On premise of data accuracy is, Logistic map and Lorenz system have four parameters μ_1, μ_2, x_1, x_2 as the key, is key space, brute force attack will fail for orders of magnitude key space.

4.2 Gray histogram

For image encryption, an attacker can statistics characteristic of the image through the distribution histogram of image pixels to obtain important information. Therefore, scrambling of pixel values is important for image encryption, the method makes the encrypted pixel histogram to have an obvious change than before through the Lorenz system generated sequence to deal with pixel values. Cipher text image statistical characteristics are completely different from clear, clear image statistical characteristic has spread to the cipher uniform distribution histogram of the image, it can resist the attacker to restore the image by statistical characteristics. Thus, this algorithm has good ability to attack resistance statistical analysis. Experiment result is shown in Figure 3, Figure 4.

4.3 Correlation of adjacent pixels

The correlation of adjacent pixels reflects the relevance of the image location adjacent pixels, one of the goals of image encryption is to reduce the correlation of adjacent pixels to resist attacks based on statistics (Zhang & Liu 2011). Adjacent pixels includes the correlation of horizontal pixel, vertical pixels and diagonal pixels. Use the following formula to calculate the correlation of adjacent pixels.

$$\rho_{xy} = \frac{\text{cov}(x,y)}{\sqrt{D(x) \cdot D(y)}} \quad (5)$$

$$E(x) = \frac{1}{N}\sum_{i=1}^{N} x_i, \quad D(x) = \frac{1}{N}\sum_{i=1}^{N}(x - E(x))^2 \quad (6)$$

$$\text{cov}(x,y) = \frac{1}{N}\sum_{i=1}^{N}(x - E(x))(y - E(y)) \quad (7)$$

Among, x and y are gray values of adjacent pixel respectively, $E(x)$ is the expectations of x, $D(x)$ is the variance of x, $cov(x)$ is the covariance of x and y, N is the number of pixels. p_{xy} is correlation coefficient of x and y. The image encryption algorithm can resist attacks should make the encrypted correlation of adjacent pixels of image to converge to 0, makes original clear image statistics be hidden in cipher text images.

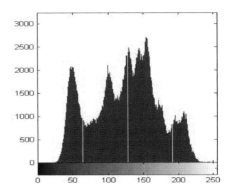

Figure 3. Gray histogram before encryption.

This paper randomly selected 1000 pixels to calculate from the original image and the encrypted image, respectively from the vertical, horizontal and diagonal directions respectively calculate the correlation of pixels, between $(x+y)$, $(x,y+1)$, $(x+1,y)$, $(x+1)(y+1)$. It is to calculate the method of averaging 10 times. The result of p_{xy} shown in the following Table 1.

Table 1. Margin settings for A4 size paper and letter size paper.

	Original	Encrypted
Vertical	0.978	0.007
Horizontal	0.984	0.026
Diagonal	0.967	0.023

The table shows that the adjacent pixels correlation coefficient of original image tends to be 1 and the adjacent pixels of the correlation coefficient to converge to 0. Namely the statistical information of the original image after encryption is full of hidden.

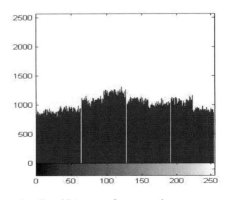

Figure 4. Gray histogram after encryption.

473

5 CONCLUSIONS

This paper studies Arnold scrambling algorithm and found its own limitations. Based on the above research, proposed an improved Arnold scrambling algorithm combined with the one-dimensional chaotic equation. In this paper, first scrambling pixels, then scrambling each pixel gray value by Logistic sequence. The experimental results show that the encryption algorithm is safe and using the improved encryption algorithm is efficient.

ACKNOWLEDGMENT

A project supported by the Natural Science Foundation of Heilongjiang Province, China (Grant No. F201304).

REFERENCES

Zhu Z., W. Zhang, K Wong et. Al (2011). A chaos-based symmetric image encryption scheme using a bit-level permutation. *J. Information Sciences. 181(6),* 1171–1186.

Liu S. & J. T. Sheridan (2013). Optical encryption by combining image scrambling techniques in fractional Fourier domains. *J. Optics Communications. 287,* 73–80.

Ye G. (2010). Image scrambling encryption algorithm of pixel bit based on chaos map. *J. Pattern Recognition Letters. 31(5),* 347–354.

Arnol'd V. I. & A. Avez (1968). Ergodic problems of classical mechanics. *M. Benjamin.*

Ye G. & K. W. Wong (2012). An efficient chaotic image encryption algorithm based on a generalized Arnold map. *J. Nonlinear dynamics. 69(4),* 2079–2087.

May R. M. (1976). Simple mathematical models with very complicated dynamics. *J. Nature. 261(5560),* 459–467.

Zhang G. & Q. Liu (2011). A novel image encryption method based on total shuffling scheme. *J. Optics Communications. 284(12),* 2775–2780.

Electronic Engineering and Information Science – Wang (Ed.)
© 2015 Taylor & Francis Group, London, ISBN: 978-1-138-02772-5

Implementation of fast unwrapping algorithm for high resolution omnidirectional images based on FPGA

J.H. Han & S.T. Liang
Computer Science and Technology, Harbin University of Science and Technology, Harbin, China

ABSTRACT: The omnidirectional images after unwrapped, can be easily identified by the naked eye. The image resolution of the omnidirectional image acquisition module collected is 2048*2048, it costs a larger system resource, not suitable for implementation in general embedded system. In this paper, according to the method of symmetric reuse and processing the image of inner area and outer area respectively, a fast cylindrical unwrapping algorithm based on the FPGA is designed. With the software Matlab, the algorithm is verified. After development and simulation through the Quartus II 9.1 and the Modelsim 6.5b, the improved algorithm for high resolution panorama is achieved on the DE2 hardware development platform of Altera.

KEYWORDS: omnidirectional; Fast unwrapping; FPGA.

1 INTRODUCTION

The omnidirectional image processing system is made up of a camera and a surface mirror, it can obtain a bigger perspective image. The omnidirectional image has a wide range of applications in many fields. But the images we collected through the omnidirectional image processing system are distorted and not suitable for the naked eye. The computer platform is used to realize the unwrapping, but it has poor real-time performance and need a lot of resources. So it is not adapted to the occasion that required equipment volume and real-time performance. In this paper, according to the theory of cylindrical unwrapping algorithms, a fast unwrapping algorithm based on the symmetric reuse partition inside and outside of the ring is designed. After simulation through the Modelsim, the algorithm was achieved on the FPGA platform.

2 OMNIDIRECTIONAL IMAGE PROCESSING PLATFORM

The platform mainly includes high resolution image acquisition module, SDRAM memory module, image unwrap module and image display module. The omnidirectional vision sensor consists of mirror lens and high resolution camera. This paper adopts the Cyclone of Altera as the hardware platform. The image acquisition module uses the ADV7190 chip to collect image signal, then input it to the SDRAM, the image unwrap module will unwrap the image to Generate rectangular images of the human eye observation. The image display module will output the image that unwrapped to the ADV71 7 chip, then through the VGA monitor to display. This platform system structure is shown in figure 1.

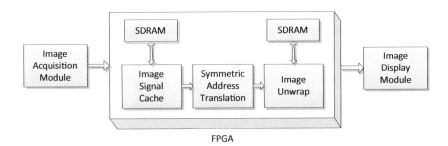

Figure 1. Platform system structure.

3 TRADITIONAL LOOK-UP TABLE METHOD

Look-up table method is a relatively practical method in the embedded devices, the method adopted the principle of optical path back and calculated the relationship of its image coordinates to rectangular image coordinates, save in the look-up table. Every time we are just according to corresponding points of the pixels in the lookup table to unwrap quickly and accurately. The traditional look-up table method diagram is shown in Figure 2.

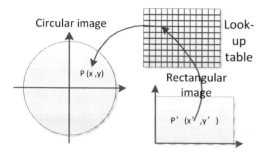

Figure 2. Look-up table method.

The look-up table method can avoid complex triangle shipment and has a faster speed and high precision, but due to the lookup table and large amount of image data it takes up higher speed memory space and the SDRAM chip has low efficiency when reading and writing a single data. It is leading to a single pixel expansion speed slower, the whole picture expansion speed slower. Under the conditions of hardware is in control and need a large space, the look-up table method is not suitable.

4 SYMMETRIC REUSE PARTITION METHOD

This method is mainly using the idea of symmetry reuse, calculated a quarter of the circular image and find out the relationship of the four parts. In figure , with the circular center as the origin of the image, rectangular coordinate system XOY is established. Use the X axis and Y axis to divide the circular images into four symmetrical fan area P0, P1, P2, P . Then, with the rectangular image lower left corner as the origin of coordinates, another rectangular coordinate system XOY is established, average rectangle image divided into P0, P1, P2, P four parts. As is shown in figure, after finishing the division the quadrant area P0, P1, P2, P respectively are expanded to the rectangular area. Only need to use the light path tracking method to calculate the relation of pixel areas' transmission, then use the symmetry principle to copy pixels can unwrap the remaining three areas of the image.

Then analyse the calculate area, on the same radius of circle, the pixels on the division is uniformity. With a different radius of the circle the pixels on the division is nonuniform, in other words, the smaller the radius the accumulation of the pixel accumulation is greater. To unwrap on the inner area is easy to distort, so in the inner area using bilinear interpolation processing to improve the degree of revivification. The improved algorithm diagram is shown in figure 4.

The algorithm assumes that the current coordinates of the pixel is $(i+u, j+v)$, $0 < u < 1$ and $0 < v < 1$, then calculate the four pixels around the current pixel, the formula is:

$$f(i+1, j+1) =$$
$$(1-u)(1-v) \times f(i,j) + (1-u)v \times f(i, j+1) \qquad (1)$$
$$+u(1-v) \times f(i+1, j) + uv \times f(i+1, j+1)$$

For the convenience of calculation, assume $u = v = 0.5$, the formula can be simplified to:

$$f(i', j') = \frac{1}{4} \begin{bmatrix} f(i,j) + f(i, j+1) + \\ f(i+1, j) + f(i+1, j+1) \end{bmatrix} \qquad (2)$$

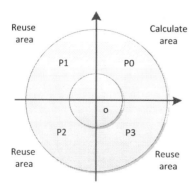

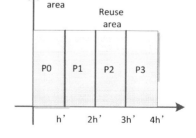

Figure 3. Symmetric reuse principle diagram.

In the process of the implementation of the algorithm, deal with the P0 area first, then using the symmetrical reuse on other three areas separately, can get the complete omnidirectional image. This algorithm not only solve the problem that the inside circular area is easy to distortion, but also calculate a quarter area of the image, greatly improved the speed of unwrapping the omnidirectional image.

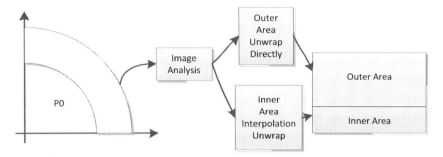

Figure 4. The improved algorithm diagram.

5 ALGORITHM SIMULATION

In order to verify the feasibility and efficiency of the proposed algorithm, use the Matlab to do the simulation. The original image was collected by image acquisition module, as shown in figure 5. Then write a function according to the algorithm in the process, it's important to pay attention of the image coordinate is integer point, not a decimal. So deal with the original image coordinates after calculation of the Integer arithmetic. And after the proper parameter adjustment, to begin the simulation, after the simulation, the unwrapped image is as shown in figure 6.

Then the other algorithm also does the simulation the purpose is to compare the symmetric reuse partition algorithm with the other algorithms, table 1 is the unwrap time for the same omnidirectional image

Figure 6. The improved algorithm diagram.

compare the traditional algorithm with the symmetric reuse partition algorithm.

Table 1. The algorithm running time (ms).

	Speed Comparison		
	Traditional Algorithm	Improved Algorithm	Speed Ratio
1	1063.5	536.2	1.98
2	1189.7	564.6	2.10
3	1107.3	545.8	2.02
4	1651.4	786.5	2.11
5	1662.1	773.3	2.15
6	1664.8	781.6	2.13

It can be seen from table 1, the new algorithm proposed in this paper faster about twice the average speed than the traditional method.

6 FPGA IMPLEMENTATION OF THE ALGORITHM

When begin to unwrap the omnidirectional images, in order to achieve the effect of rapid unwrapping, using the Verilog language which is parallel and efficiently

Figure 5. The original image.

for programming. Verllog is a hardware description language, its advantage is high execution efficiency, it's defect is impossible to deal with complex mathematical operations, such as the trigonometric function and division. According to the pixel coordinates, through address transform processing the results will be solidified into RAM, then to deal with the data in RAM to get the image pixel address of the unwrapped image. When using the SDRAM controller, adopted the "ping-pong operation" strategy. Call the macro module in FPGA to generate a ROM, then use the sine table and the cosine table to initialize the ROM. This system uses the 7200 cosine table, the precision is 16 and it's enough to meet the requirements. Then the table is generated by VisualC++, and multiplied by 1024, because of the verilog can only handle integer, moves to the right of the result of its handling last 10, the results can be a real value.

Use Verilog language to program each module, then on the Modelsim simulation software platform to write the TestBench simulation module, can get the simulation diagram shown in figure 7.

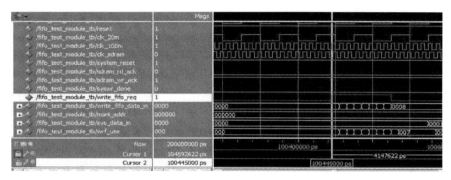

Figure 7. Simulation diagram.

The CCD camera can at a speed of 15 frames per second to collect images, so the display will need to show the image at 60 frames per second. When a frame 2048*2048 resolution image was written into SRAM, then need to read 4 frames 1024*768 resolution images from SRAM. This calculation, 6.6ms*4+8.9ms=4.9ms less than the camera's frame period 65.7 ms, so the 2048*2048 resolution omnidirectional images can be completed and satisfy unwrapped.

7 CONCLUSIONS

This article for the unwrapping of omnidirectional images presented a method of using symmetric strategy, the main idea is using the coordinates of the coordinate space in four coordinate domain symmetry and using the bilinear interpolation method to unwrap the inner area. Experiments show that, compared with the traditional methods, the new algorithm proposed in this paper not only has decreased 4 times the space of the look-up table, but also the velocity of

unwrapping is higher. Using the method of partition the inner area and the outer area, make the quality of the image improved.

REFERENCES

Nayar S.K. (1997). Catadioptric Omni-dirctional Camera. *Interational Conference on Computer Vision and Pattern Recognition, IEEE Computer Society,* 482–488.

Gaspar J., N. Winters & J. Santos-Victor (2000). Vision-based navigation and environmental representations with an omnidirectional camera. *Robotics and Automation, IEEE Transactions on, 16(6)*. 890–898.

COUTURE V. & M. S. LANGER (2010). Capturing Non-Periodic Omnistereo Motion. *IEEE Workshop on Omnidirectional Vision(OMNIVIS 2010)*. 3–4.

STURZL W. & M. V SRINIASAN (2010). Omnidirectional Imaging System with Constant Elevational Gain and Single Viewpoing. *IEEE Workshop on Omnidirectional Vision(OMNIVIS 2010)*. 1–7.

Lu H. M., H. Zhang & Z. Q. Zheng (2010). A Real-Time Local Visual Feature for Omnidirectional Vision Based on FAST and CS-LBP. *IEEE Workshop on Omnidirectional Vision (OMNIVIS 2010)*.

Electronic Engineering and Information Science – Wang (Ed.)
© 2015 Taylor & Francis Group, London, ISBN: 978-1-138-02772-5

Design of dual field-of-view zoom optical system with high resolution

Z.H. Luan, L. Zhao & X.M. Wang
School of Applied Science, Harbin University of Science and Technology, China

ABSTRACT: According to the zoom theory of optical system, a dual Field-Of-View (FOV) optical system with 8 zoom ratio is designed. The system uses a camera with high resolution of four megapixels. It has a relative aperture of f/4 and can realize 185.28mm/23.16mm step zoom. The results show the system can meet the demand of practical application. This work is valuable for the design of dual FOV zoom optical system with high resolution.

KEYWORDS: Dual field-of-view; Zoom optical system; Optical design; High resolution.

1 INTRODUCTION

As the developing of zoom lens optical design theory and the perfect of manufacturing process technology, zoom optical system has been widely used in many fields, such as space exploration (Zhou et al. 2007), military surveillance (Wang et al. 2010), scientific research (Chen & Wang 2003) and security monitoring (Luo et al. 2013). Compared with the fixed focal length optical system, dual Field-Of-View (FOV) zoom optical system can search the target in the Wide Field-Of-View (WFOV) and locate the target in the Narrow Field-Of-View (NFOV). Compare with the continuous zoom optical system, dual FOV zoom optical system has the advantages of simple opto-mechanical structure, small size, light weight and low cost. Therefore, considering the performance and cost, dual FOV zoom optical system is still the best choice in many practical projects.

In this paper, a dual FOV zoom optical system with 8 zoom ratio is designed based on a high resolution camera with 2352×1728 pixels and 7.4×7.4 μm pixel size. The WFOV and NFOV is respectively 41.18°×30.86° and 5.38°×3.95°. The F-number is 4. The results show that the optical system has a good image quality and easily applied to engineering.

2 PRINCIPLE OF ZOOM OPTICAL SYSTEM

Zoom optical system which has a varied focal length within a certain range and a stable image surface position during the zooming process can maintain good image quality. According to the different ways of zoom, zoom system can be divided into optical compensation, mechanical compensation, two zooming groups moving and multiple groups moving structure. Compare with optical compensation structure, mechanical compensation structure can effectively prevent the image surface drift and realize the continuous zoom. Compare with two zooming groups moving and multiple groups moving structure, mechanical compensation structure is more simple in the mechanical design, machining and optical system alignment. Therefore, it is widely used in the design of zoom optical system (Bai et al. 2009).

Mechanical compensation zoom optical system is composed by front fixed group, zooming group, compensating group and back fixed group . In order to maintain the image plane stability, zooming group and compensating group must satisfy the following relations:

$$f_3'(\frac{1}{\beta_3^*} + \beta_3^* - \frac{1}{\beta_3} - \beta_3) + f_2'(\frac{1}{\beta_2^*} + \beta_2^* - \frac{1}{\beta_2} - \beta_2) = 0 \quad (1)$$

Where f_2', f_3' are the focal lengths of zooming, compensating group. β_2, β_3 are the magnification of zooming, compensating group in the WFOV. β_2^*, β_3^* are the magnification of zooming, compensating group in the NFOV. According to the above formula, the initial structure parameters can be obtained by Gauss optical calculation.

3 DESIGN OF OPTICAL SYSTEM

3.1 *Performance indicators*

Considering the actual application needs, the selected detector index is 2352×1728 resolution, 7.4×7.4 µm pixel size and 31fps frame rate. It has not only high resolution, but its frame rate is enough to meet the use requirement of the human eye. Meanwhile, considering the demand of practical application and the difficulty of the design, the final design of the optical system has the parameters below:

1　Working band: visible band;
2　Detector: 2352×1728 focal plane detector, 7.4 × 7.4 µm pixel size, 31fps frame rate;
3　Focal length: Short focal length: 23.16mm; Long focal length: 185.28mm;
4　Zoom ratio: 8;
5　Field-of-view: WFOV: 41.18°×30.86°; NFOV: 5.38°×3.95°;
6　F-number: 4;
7　Total length: no longer than 350mm.

3.2 *Design result*

According to the index requirements, the optical system is designed by using Zemax optical design software. The layout of the optical system in the WFOV and NFOV is shown in Figure 1.

In the system, the front fixed group adopts common doublet-single structure. Zooming group adopts two negative lens structures. By increasing the refractive index difference, the astigmatism can be corrected and the distortion in the WFOV can be reduced. Two positive lens with high refractive index are adopted in compensating group. By decreasing the refractive index difference, the spherical aberration and coma in the NFOV can be corrected. Back fixed group adopts two groups of doublet lens to correct the residual aberration. The total system has 10 lenses which surface type are all spherical, and the total length is 342.28mm. The structure is compact and all lenses in the system are easy to fabricate. Thus, this design has high practical value.

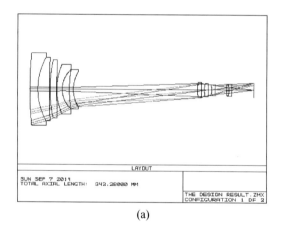

(a)

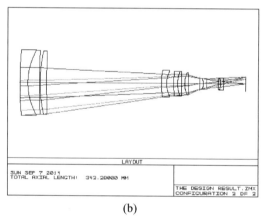

(b)

Figure 1.　Layout of optical system in the WFOV (a) and in the NFOV (b).

3.3 *Analysis of imaging quality*

The modulation transfer function (MTF), RMS radius and distortion are important indicators to evaluate imaging quality of the optical system. In this system, the pixel size of the detector is 7.4 µm, so the Nyquist frequency is $1/(2×7.4×10^{-3})$ =68lp/mm. Figure 2 shows the MTF curves in the WFOV and NFOV. It can be seen that the MTF value of each field is better than 0.4 at the Nyquist frequency. For the existing detectors, as long as MTF is achieved

0.2, the detector can get a clear image. Therefore, this system is fully able to meet the needs of practical application. Figure 3 shows the spot diagrams of the optical system in the WFOV and NFOV. It is clear that the maximum RMS radius is in a pixel size. Figure 4 shows the distortion of the optical system in the WFOV and NFOV. It can be clearly observed that the maximum distortion of an optical system is less than 4%. In conclusion, the above key indexes can meet the demand for practical application.

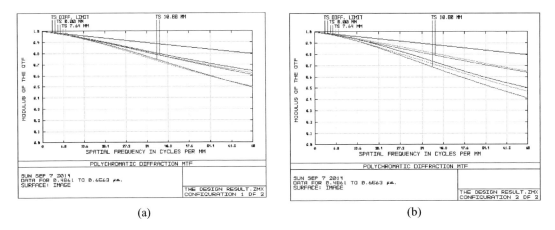

(a)   (b)

Figure 2.   MTF curves of optical system in the WFOV (a) and in the NFOV (b).

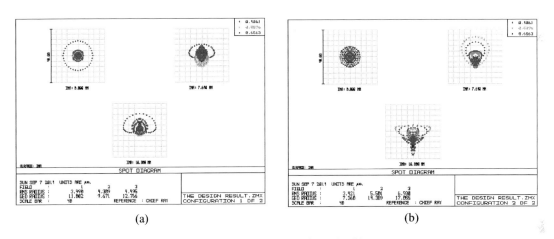

(a)   (b)

Figure 3.   Spot diagrams of optical system in the WFOV (a) and in the NFOV (b).

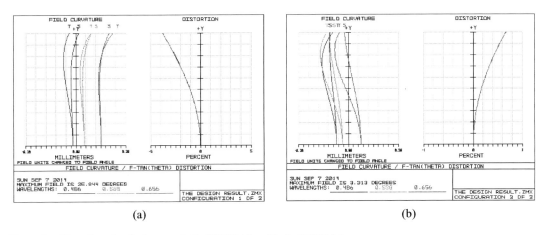

(a)   (b)

Figure 4.   Distortion of optical system in the WFOV (a) and in the NFOV (b).

481

4 CONCLUSIONS

The theory and design method of zoom optical system are discussed in detail, and a 8 zoom ratio dual FOV optical system with high resolution of 2352×1728 is designed based on the method of mechanical compensation. Through the analysis of imaging performance, the optical system can realize WFOV of 41.18°×30.86° and NFOV of 5.38°×3.95° with a relative aperture of f/4. The results show the key indexes can meet the demand for practical application. Moreover, the system structure is compact and all lense surface types are spherical which is easy to fabricate. Therefore, this work provides a significant reference for the design of dual FOV zoom optical system with high resolution.

REFERENCES

Zhou Y. P., Shu R., Tao K. Y., & Guo S. (2007). Study of photoelectric detecting and identifying of space target. *Optical Technique 33*, 68–76.

Wang A. K., Xu H. B., Yang J. F., Li T. & Run P. (2010). Application of visible video zoom system in the television-guided system. *Journal of Projectiles, Rockets, Missiles and Guidance 30*, 41–43.

Chen X. N. & Wang S. H. (2003). Focus real time test of focus-changed lens. *Journal of XI'AN University of Science and Technology 23*, 168–174.

Luo C. H., Zhang Q. Y. & Li Y. H. (2013). Design of 6mm-60mm megapixel zoom security lens. *Journal of Applied Optics 34*, 209–214.

Bai Y., Yang J. F. & Xue B. (2009). Design of uncooled LWIR thermal imager Switch-Zoom Optical System. *Infrared Technology 31*, 156–159.

Electronic Engineering and Information Science – Wang (Ed.)
© 2015 Taylor & Francis Group, London, ISBN: 978-1-138-02772-5

Latest development and analysis of image fusion

L.L. Wang
School of Computer Science and Technology, Instrument Science and Technology Postdoctoral Workstation,
Harbin University of Science and Technology, Harbin, China

Y. Shen & D.Y. Chen
School of Computer Science and Technology, Harbin University of Science and Technology, Harbin, China

X.Y. Yu
Instrument Science and Technology Postdoctoral Workstation, Harbin University of Science and Technology, Harbin,
China

ABSTRACT: Image fusion refers to the multi channel to collect on the same target image data after image processing and computer technology, favorable information extraction to maximize their respective channels, and finally integrated into the image of high quality, to improve the spatial resolution and spectral resolution on the utilization rate of image information and improve interpretation accuracy and reliability and improve the original image to monitoring computer. This paper introduces the development process of image fusion, and its development at home and abroad. And the image fusion levels were analyzed for each layer in detail and do a comparison. Detailed introduces the often used in image fusion, the method of comparative analysis. Finally, the research directions on image fusion are predicted.

KEYWORDS: Image fusion; Image fusion level; Fusion method.

1 INTRODUCTION

Nowadays, with the rapid development of computer technology, electronic information science, the sensor technology is in the rapid development accordingly. The information obtained from multi-sensor system is increasing rapidly, which presents diversity and complexity, and information processing method of the past cannot be able to satisfy this new situation, and new methods and technologies must be studied to solve the new problems (Chen 2010).

Information fusion is a new technology developed to meet the demands. Multi sensor information fusion refers to the information handle comprehensively multi-level and multi-aspect from multiple sensors, and the information obtained is more useful, richer, more accurate and more reliable. Image fusion technology is an important branch of multi-sensor information fusion, and recently twenty years, it has aroused widespread concern and research worldwide (Hu et al. 2008).

2 DEVELOPMENT STATUS AT HOME AND ABROAD

Early in the last century, 70's, the concept of information fusion was proposed in America, and the image fusion is an important aspect of information fusion From the concept of image fusion is proposed to now, with the continuous, in-depth development of the research on the theory of the fusion method, there was a great progress, and many new methods are continually put forward. In many kinds of fusion methods, in the early years there are the weighted average method, the algorithm based on the pixel gray, principal component analysis method, high-pass filter and so on. Until the mid 80's, the emergence of an image fusion method based on the decomposition of the Pyramid was put forward (Li et al. 2011), including Laplasse Pyramid, gradient Pyramid, ratio low pass Pyramid and so on. After 90's, with the widely application of wavelet theory, wavelet transform technology provides a new method for image fusion, and the image fusion technology is used widely (Wang & Wang 2014).

3 IMAGE FUSION LEVELS

3.1 Pixel-level image fusion

The principle of pixel-level image fusion is, the image signals from each sensor will be synthesized and analyzed directly in strict registration conditions. Pixel-level image fusion is performed directly on the original data layer, and the main task is to fusion and process for the measurement results of multi-sensor target and background elements. The fusion accuracy is higher, which supplies richer, more accurate and more reliable fusion images than other level fusion, and it is good for further analysis and processing of images.

3.2 Feature-level image fusion

The principle of feature-level image fusion is making a comprehension for feature information such as edge data, shape data and so on. The feature information is obtained after the pre-procession and feature extraction from images (Sun 2012). The feature level fusion is an information fusion performed at the transition stage, and it not only retains the important information in sufficient quantities, but also compresses the information, which is suitable for real-time processing.

3.3 Decision-level image fusion

The principle of decision level image fusion is to make the best decision according to certain standards and decision reliability. The decision level fusion is the highest level of information fusion, after the completion of the target extraction and classification of each sensor, fusion system will process fusion according to certain criteria and decision reliability (Chu et al. 2009).

There are advantages and disadvantages of three level image fusions, and in actual application, the method will be selected according to the characteristic and the specific requirements. The comparison of three fusion levels is shown in Table 1.

Table 1. Comparison of the three fusion levels.

Fusion levels	Pixel level	Feature level	Decision level
Information loss	small	medium	large
Real time	bad	medium	good
Precision	high	medium	low
Fault tolerance	bad	medium	optimal
Anti-interference	bad	medium	optimal
Work load	large	medium	small
Level of fusion	low	medium	high

4 COMMONLY USED IMAGE FUSION METHODS

4.1 Image fusion based on spatial domain

Weighted fusion: The principle of the weighted fusion algorithm is weighing and calculating for images. Although there are advantages such as simple in principle and high speed of this algorithm, because of high noise-signal ratio of fusion image and the poor image contrast, the edge fog often occurs (Peng 2012).

Principal component analysis: the principle of image fusion based on principal component analysis is re-combination of the related data of the original image to not relevant data, and a group principal data will be selected to represent most effective information of the image. If the difference between the original image content information is larger, the principal component analysis of image fusion can get better results.

The artificial neural network method: there are advantages such as parallel, nonlinear and distribution of knowledge storage, fault tolerance and adaptive of artificial neural network, which is a new type of reasoning method of bionics. The excellent characteristics of the neural network are distributed information storage and concurrent processing. Neural network algorithm has been used in digital image fusion.

4.2 Image fusion based on transforming domain

The principle of an image fusion method based on transforming domain: at first the related transform should be processed in order to obtain the coefficient indicated on images. After processing of appropriate algorithms, the coefficient indicated on images will be obtained, and after inverse transformation of coefficient expression, the fusion image will be reconstructed. In this kind of fusion method, multi-scale decomposition technique is often obtained, as shown in Figure 1.

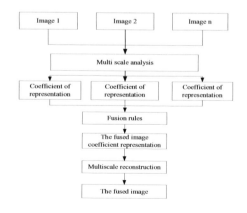

Figure 1. Structure of image fusion based on multi-scale decomposition.

Multi-scale decomposition technique can be divided into two categories: image fusion method based on the transformation of pyramid and image fusion method based on wavelet transform.

4.3 *Image fusion method based on the transformation of the pyramid*

The basic steps of a fusion method based on the pyramid: firstly the original images are pyramid decomposited separately to build a model of pyramid of each image. Decomposition layer between each image is independent of each other, so different process can be done in different image decomposition layers. In different image decomposition layer, different rules are adopted to process, then a new image of the pyramid will be obtained. The inverse transform will be executed on the new image of the pyramid. The pyramid decomposition fusion algorithm has many advantages, and there is no loss of the detail information and important information for image are highlighted. In image mosaic application, the traces of mosaic images can be weakened, so ideal image results can be obtained. But pyramid decomposition of each data layer is not independent; there is some relevance, so it is a redundant decomposition algorithm. The amount of data pyramidal decomposition is larger than the original image data, this vast amount of data decomposition has restricted its application in real-time system. At the same time, this redundancy on the decomposition of the original image data of similar circumstances, a better fusion effect, once the fusion image difference is very big, the algorithm is easy to produce unstable, making the fusion image after reconstructing lost high frequency information, edge information by fuzzy. Another disadvantage of the pyramid decomposition is, it does not have the direction, the image structure information can't be extracted completely, this also makes the fusion result is not satisfactory.

4.4 *Image fusion method based on wavelet transform*

Wavelet is a rapidly developing branch of mathematics field in recent years, and its essential idea is a stretching and translating method. In the processing of information previously, the contradiction between time domain and frequency domain information is not dealing well. Until the emergence of wavelet analysis, the technology is a good solution to this problem. In wavelet analysis, time-scale analysis method is adopted, which is from different from the traditional time-frequency analysis. In wavelet analysis, variable window is adopted. The large window is selected to process low frequency signals and smaller window is selected to process high frequency signals.

5 SUMMARY

Image fusion is a research topic with practical significance, and there are research findings in the field of image fusion. But in general, the research of image fusion technology is not yet mature; there are many problems to be solved. First of all, a comprehensive theory plays an important role in the development of technology, and the image fusion technology is the lack of such theory guidance, overall a unified theoretical framework. Secondly, in some cases, the need for real-time processing of image, then the amount of time and memory of fusion algorithm is needed to take into account, how to get stable, reliable and practical fusional images is also a hot research.

ACKNOWLEDGMENT

This work is supported by Department of Education Project in Heilongjiang province (12511097), a project of the Department of Science and Technology in Heilongjiang province (QC2012C059) and Postdoctoral Science Foundation Project in Heilongjiang province (LBH-Z11109).

REFERENCES

Chen Q. (2010). Digital Image Fusion Technique. *High Technology. 59*, 333–338.

Hu G., Z. Liu & X. P. Xu (2008). Research and Development of Image Fusion at Level. *Application Research of Computers, 25*, 650–655.

Li Y., G. Song & S. Yang (2011). Multi-sensor Image Fusion by NSCT-PCNN Transformation. *International Conference on Computer Science and Automation Engineering, 4*, 638–642.

Wang B. & C. Wang (2014). The research of remote sensing image fusion technology. *International Conference on Advances in Materials Science and Information Technologies in Industry, 513*, 3237–3240.

Li H. Y. (2008). The Research of Image Fusion Technology based on Wavelet Transform. *Application Research of Computers, 23*, 104–110.

Sun Y. (2012). Multi Sensor Image Fusion Algorithm based on Multi Resolution Analysis. Harbin: *Doctoral Dissertation of Harbin Engineering University.*

Chu B. B., L. L. Pang & D. N. Qi (2009). Review on the Technology of Image Fusion based on Multiscale Analysis. *Avionics Technology, 40*, 29–33.

Peng H. (2012). Infrared and Visible Image Fusion Algorithm Research. *Zhejiang University master thesis, 18*–25.

Electronic Engineering and Information Science – Wang (Ed.)
© 2015 Taylor & Francis Group, London, ISBN: 978-1-138-02772-5

Based on multi-band communication methods chaos phase space region detection method

C.X. Liu, J. Yi, L.M. An & W.J. Dai
Heilongjiang University. College of Physics Science and Technology, China

D.H. Li
Harbin Engineering University. College of Information and Communication Engineering, China

ABSTRACT: Non-ideal characteristics of the actual channel make chaos synchronization difficult to achieve, and since the existing chaotic communication research is still limited under laboratory conditions, it is difficult to promote the actual communications. The proposed method based on the detection of the phase space area ary chaotic communication modulation, division of phase space of Jerk oscillator, given area boundary identification method, establishes a mapping table chaotic sequence with digital information. The chaotic sequences representing different digital information sent to the sending end; At the receiving end use phase space area detection method judgment recovery m-ary digital information, this program is simple and Information occult high interference resistance. Don't need restore the chaotic carrier, effectively overcome the chaos synchronization constraints.

KEYWORDS: Jerk system; phase space area detection; ary chaotic modulation.

1 INTRODUCTION

Chaotic signals have broadband, like noise-related characteristics, and is very sensitive to the initial state of the features that can provide a lot of spreading sequence, which has broad application prospects in spread spectrum communications (Yu et al. 2004). In order to improve the information transfer rate, typically using M-ary spread spectrum communication scheme. Had it proposed a variety of M-ary chaotic spread spectrum communication scheme, changes such as the digital information would have translated into changes in system parameters confidential communications (Yalcin 2002, Huang 2005). However, by using of M-ary modulation parameters chaotic spread spectrum communications, by limiting the range of parameters, As M increases, the bit error rate is increased dramatically, the performance is not satisfactory. Work at the sending end of this paper, the design of a chaotic sequence of digital map information, and will represent the chaotic sequence of different digital information sent to the sending end. At the receiving end the use of phase space area detection method judgment recovery M-ary digital information, this program is simple and Information occult high interference resistance.

2 A REGION OF PHASE SPACE DIVISION JERK VIBRATOR

2.1 *Multi-scroll Jerk oscillator realization*

Study the following Jerk system (Liu 2007):

$$\begin{cases} \dot{x} = y \\ \dot{y} = z \\ \dot{z} = -x - y - az + f(x) \end{cases} \quad (1)$$

Where, $f(x) = \text{sgn}(x) + \text{sgn}(x+2) + \text{sgn}(x-2)$, when, System can produce 4-scroll chaotic attractors. Delayed feedback method using 4-scroll chaotic Jerk stability control. Using the non-linear characteristics of the system itself, the system control variables, In the Jerk control system The right side of the third equation requires the formula (1) is increased as a control input $u(t) = k[z(t-T) - z(t)]$, thereby obtaining a chaotic system controlled by formula (2) shown below.

$$\begin{cases} \dot{x} = y \\ \dot{y} = z \\ \dot{z} = -x - y - az + f(x) + u(t) \end{cases} \quad (2)$$

The gain is set to 0.5, delay time within the scope of [2.4s,2.8s], and available tracks clear, well-defined annular attractor (Liu 2012). 4 XY scroll chaotic signals generated after applying a delayed feedback phase diagram shown in Fig.1.

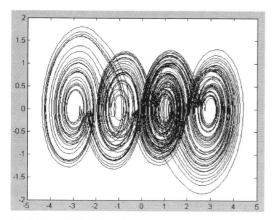

(a) 4 Scroll XY phase diagram

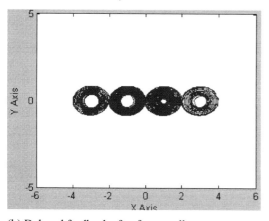

(b) Delayed feedback after four scroll

Figure 1. The 4 scroll XY phase diagram.

2.2 *Phase space zoning*

The literature gives a method to determine the parameters of multi-scroll signal phase space. Jerk system, realizes the independent control of multi-scroll signal. On the basis of the phase space can be divided into regions. The divided region of phase space is divided mainly to develop the rule of the two-dimensional phase space area and complete zoning, given the boundaries of each sub-region corresponding expressions.

Research group in 2009 for the first time proposes the use of domain division of Duffen. Jinz phase trajectory morphology were detected, and the detection

method of weak signal. In this work, the phase space area is divided into four rectangular regions. Regional division first determines the regional space for each scroll, each region of the range of variables can be determined, In any region of the phase space zone as an example. The corresponding left, right, upper, lower boundary respectively, if a moment Jerk vibrator in the region, the corresponding variables must satisfy the following relationship:

$$\begin{cases} x_u^l < x < x_r^r \\ y_{i_b}^b < y < y_{ui}^u \end{cases} \quad (3)$$

as formula (3) shows, if a certain time t, Jerk oscillator phase trajectory zone area in its corresponding region of chaotic signal x must be greater than and less than the value of the left edge of the right margin, chaotic signal Y is greater than the zone boundary and less than the upper boundary value.

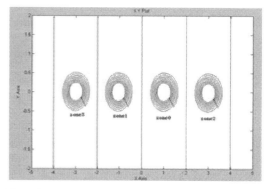

Figure 2. The corresponding phase space diagram of chaotic sequences.

3 M BINARY CHAOTIC MODULATION COMMUNICATION SCHEME

3.1 *The sending end*

The structure of the sending end is shown in Fig.3. Jerk chaotic signal generator selects different parameters and initial conditions for L iteration to generate M segment length of chaotic sequence of L (each section of chaotic sequences corresponding different scroll chaotic attractor in phase space). The digital information about M system according to the corresponding Table 1 shows the mapping to the chaotic sequence segments corresponding (mapping relation can be adjusted according to needs, and then increase the system security), and then sent to the transmitter chaotic sequence. Mapping method is used in this paper: firstly, the binary digital information through the serial parallel conversion to a N binary

488

information and then through coding module get M band digital information S, in which $N = \log_2 M$, $b_i (i = 1, 2, ..., N-1)$. Then get the chaotic sequence corresponding to the S according to the mapping relationship, M chaotic sequence represents the M band information symbol sequence S, the digital information conversion sequences for different phase space regions, into the channel.

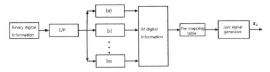

Figure 3. The M binary chaotic modulation system structure diagram.

3.2 *The receiving end*

The receiver structure is shown in fig.4. After the transmitted signal transmitted through a channel and the receiver receives the signal, we can detect the received signal by using the phase space region detection method. In the case of M=4, determine the scroll exist in phase space, a region with a scroll, is based on the mapping table that correspond to multiple information, digital information can be restored in M band. For example, region 1 has scroll, but in other regions, there are no scroll, the frame corresponding to the binary information is "1", namely the digital information is sent "01".

Figure 4. M band receiver structure of chaotic communication system diagram.

3.3 *Construction division chaotic oscillator multi – domain demodulation*

The proposing of domain segmentation method to find out a good way for the chaotic oscillator multiband demodulation, that is, using the domain segmentation to detect phase trajectory of regional space appears, and then demodulate the M-ary symbol information.

The principle of segmentation chaotic oscillator multiband demodulation is described below:

Building domain segmentation. Segmentation device can realize the detection region of phase space regions, that is able to distinguish between each sub region boundary.

When phase trajectory of the chaos baseband signal appears in a certain sub regions, accordance with the modulation multi-level information to the inverse

mapping phase track area mapping can restore M-ary symbol primitive.

The regional distribution of the chaotic attractor is shown in Fig.2, in the XY phase space, corresponding to the X axis, Y axis coordinate in a certain range, the judgment signal in phase space which area, restore the M band information using the mapping relation. The XY region of phase space is divided into 4 parts, in the chaotic signal detection module, corresponding to each region of the judgment is as follows:

If $0 < x < 2$, then scroll appeared in the region, multi band information for the "0";

If $-2 < x < 0$, then scroll appeared in the region, multi band information for the "1";

If $2 < x < 4$, then scroll appeared in the region, multi band information for the "2";

If $-4 < x < -2$, then scroll appeared in the region, multi band information for the "3".

4 FOUR BAND CHAOTIC COMMUNICATION SIMULATION SYSTEM

In order to make the digital signal transmit in a communication channel, to modulate the carrier wave of digital baseband signal. This paper selects the parameter mapping the baseband signal is mapped to the Jerk oscillator signal, after that conduct the AM carrier modulation. The simulation system structure shown in Fig.5. X and Y signals are chaos modulated signals. Signal transmission using additive white Gauss noise(AWGN channel) .

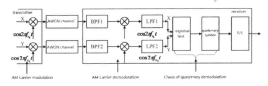

Figure 5. Chaos four hexadecimal AM communication system structure diagram.

The simulation analysis with modulation and demodulation system four hexadecimal are performed, and the error rate curve as shown in Fig.6. For comparison, graph draws the non-coherent OOK, COOK, DCSK (k=2) theoretical bit error rate curve at the same time. Can be seen: when the SNR is lower than 13dB, the chaotic four ary modulation system BER performance is better than the non coherent OOK, COOK, DCSK (k=2) system, has obvious advantages. But it should be pointed out that, performance of the M-ary chaotic communication system than the OOK system improved, is to increase 8 times the cost of spectrum requirements. However, because of the difference between modulation and demodulation system with the conventional communication system, the information have hiding effect. Obviously, further improvement of the carrier modulation method will help to reduce the spectrum requirements of the system, improve the utilization of spectrum. Compared with the existing chaotic communication system, without the need for a chaotic carrier recovery, thus effectively overcomes the synchronization control, more convenient communication system is set up in practice, and the transmission performance is better than the existing COOK, DCSK (k=2), chaos shift keying system.

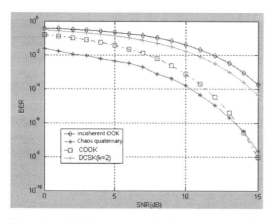

Figure 6. 4 AM chaos communication system transmission characteristic curve.

5 CONCLUSION

Using the visual Simulink model simulation method based on phase space region detection chaos M-ary modulation and demodulation of the simulation; system structure diagram gives the baseband and RF communication simulation and realization method, bit error rate curve plotted four band chaotic modulation communication system, and non coherent OOK, COOK, DCSK (k=2) system are compared. The results showed that the multi-band communication system based on chaos phase space region detection of anti-noise, transmission error rate in this paper is feasible.

REFERENCES

Yu S. M., Q. H. Lin & S. S. Qiu (2004). Multi-level Digital Chaos Shift Keying Based on Multi-scroll Chaotic Systems. *Journal of Image and Graphics, 9(12)*. 1473–1479.

Yalcin M. E., J. A. K. Suykens & J. Vandewalle (2002). Families of scroll grid attractors. *Int J Bifurcation and Chaos,12(1)*. 23–41.

Huang R. S. & H. Huang (2005). Chaos and its applications (second edition) . *Hubei: wuhan university press*. 41–76.

Liu M. H., W. Luo & D. L. Wu. (2007). Circuit implementation of Jerk system. *Microcomputer Information*. *23(7–2)*. 301–302.

Liu C. X., J. Yi, X. C. Xi et al. (2012). Research on the multi-scroll chaos generation based on Jerk model. *Procedia Engineering . (29)*. 957–961.

Electronic Engineering and Information Science – Wang (Ed.)
© 2015 Taylor & Francis Group, London, ISBN: 978-1-138-02772-5

A seamless switch method of audio files in the automatic broadcasting system

Y. Guo
School of Electronic and Information Engineering, Xihua University, Chengdu, China

Y. Li
School of Automation Engineering, UESTC, Chengdu, China

S.Y. Xiao
Auditing Department, SCU, Chengdu, China

ABSTRACT: The stability and continuity are very important for a Radio Digital Audio Workstation. The output of a workstation may be discontinuous because of an audio file error, network and database connections error, broadcasting system software or hardware error, and the virus infections, etc. This paper presents a seamless switch method of the broadcasting audio files, which can ensure continued broadcasting of the file while the program file error, network and database error and the other error statement. In addition, a realization of this method is given by DirectShow technology. This technology has been applied for Chinese national invention patents, which patent No is 201110144970.5.

KEYWORDS: Seamless Switch; DirectShow; Buffer; Broadcast.

1 INTRODUCTION

To achieve digital audio workstation security broadcast, its stability, reliability, security aspects has high requirements. To ensure 7×24 hours broadcasting is the basic requirement. In the whole process of broadcasting system, there are some problems can make the broadcasting discontinuous, such as network failure, network server intermittent cannot return data, disk error led to the audio file error.

At this stage, to solve the problems mentioned above, broadcasting system is directly switched to backup broadcast file playback. But sometimes a short pause, or backup file broadcast a file on the same server appears on this stage, when the network disk error (Cai 2008), will be unable to return data, it will directly broadcast the error, even not to mention to seamlessly switch. Therefore, this solution cannot realize seamless switch on real significance, which is broadcast from the point of view, are not able to achieve seamless switch, the effect of smoothing.

In view of this situation, we adopted the design of multiple file servers, more than one file server backup file repeat broadcast, and adding buffer mechanism when reading the file data, so as to realize seamless handoff in the broadcast effect. Because we made redundant time, enough to buffer audio data, therefore, that broadcast file error time and broadcast file

path switch and time required less time redundancy. Broadcast from the point of view, seamless handover broadcast file, strengthen the error-tolerant of the broadcast system, guaranteeing the workstation can run steadily and reliably.

2 THE SEAMLESS SWITCH PRINCIPLE AND METHOD

2.1 *The seamless switch principle*

First, the seamless switch principle of the audio file is given as follows:

1 Multiple file server design, determining a file server for the current file server, the other is the backup file server. The files of the current file server backup (Lee 2010) in all other backup servers provide physical support for multiple paths to read the file.
2 The data buffer design, data buffer has enough buffered play data, when the broadcast file error, path switch ensures the length of the data buffer play time is greater than that found the error and the time of path switch.

From the above two points, we can draw the conclusion, principle of broadcast file seamless switch is the first multipath buffer to read file data into the data buffer and then broadcast the data read from the data buffer zone, when an error is discovered and switch, because the data buffer to buffer data enough, the broadcast data will not be interrupted supply, so as to ensure the continuity of the broadcast, the broadcast aspect can achieve a seamless handover broadcast file.

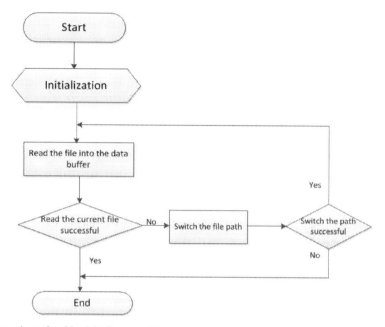

Figure 1.   The flow chart of multipath buffer to read file.

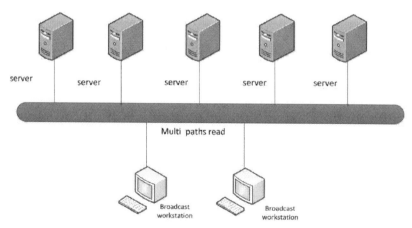

Figure 2.   Multipath file read schematic diagram.

2.2   *The seamless switch method*

The multipath file reader module and the data buffer mechanism are the key technology of this method, which will be described as follows.

The multipath file reader module and the data buffer mechanism are inseparable and complementary. Flow chart of multipath buffer file was as shown in Figure 1.

When start reading the file, an initializing operation mainly initializes multiple files read paths and sets the current path, in addition to initialize the buffer operation, set the buffer memory for the spatial distribution of the initial size. After the initialization operation, according to the current file paths read and write files into the data buffer, if read the current file successfully, then the end of the file is read; if the current file read failure or unable to return data reading process, switching file backup path switching path, if successful, will restore to the current file to read file data read. To the buffer, if all path file read failure, switching path failure, directly to the end of the file read, above is the whole process of reading multipath file.

The principle of multipath read: in order to guarantee the broadcast system broadcast process, the data source effectively and continuously, need to be able to read the file path. All of the broadcast audio files stored in a file server and save the storage path (Xu 2004), when the network error or the server cannot return data or file error, it can immediately identify the error and switch to a valid data source next path diagram, as shown in Figure 2.

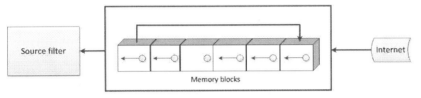

Figure 3. Buffer queue structure diagram.

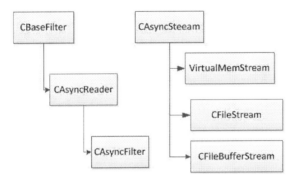

Figure 4. Source Filter inheritance graph.

The principle of data buffer mechanism: Our data buffer actually is a circular queue. A queue is a kind of first in, first out (first in, first out, abbreviated as FIFO) of the linear table data structure, it only allows to insert one end of the table, and the other end to remove elements, in the queue, allowing the insertion of the tail end is called the rear, allowing the deletion end become the head of the line (front) (Yan 1997). Source Filter to read data from the network from the buffer queue, write data is completed, returned to the buffer at the end of the queue, waiting to write data, through the operation of front pointers and the rear pointers, so reciprocating cycle using buffer queue data. The buffer queue structure diagram is shown in Figure 3.

3 DESIGN AND IMPLEMENTATION OF FILES SEAMLESS SWITCH TECHNOLOGY BASED ON DIRECTSHOW

3.1 Design and implementation of data buffer

The main function of the data buffer is writing the file data into data buffer for broadcast use, according to the capacity of the buffer and the data buffer filling strategy before or during the digital audio workstation system broadcasts. Even if a network interruption, failure or the file itself data error occur, it also won't make a play stop immediately. It provides time for security to the backup file path for switching, so as to ensure the continuous non-stop broadcasting system.

493

According to the principle of the data buffer mechanism, the data buffer is designed as a circular queue. When broadcasts in the process, it reads audio data sequentially from the circular queue and then completes the block of memory to read data, back to the circulating at the end of the queue, waiting for the next filling so that read and fill cycling.

3.2 Design and implementation of file multipath reader module

The multipath reader module for a file, is designed for broadcast system data source interrupted which may be caused by network failure or audio file error. In order to achieve the reader for multipath, it must sure that one audio file may be stored in several paths which may be on a computer or on different computers or in a disk matrix, and all the path need to be saved in database. When an error occurs, the play position of a file will be recorded and the source file will be switched into a valid one, after that read new file data from the same position, which can make the data source continuous and effective.

The above method can be achieved by designing a class CFileStream. It is not described in detail here.

3.3 Design and implementation of multipath to read source filter

We can realize this Seamless Switch method in DirectShow Filter, which is simply described as follows.

Overall demand for multipath to read Source Filter can read and buffering audio file data from the network source when the network failure disconnected or error audio file data, the ability to switch to an alternate path, continue to read the file data and the file buffer from the point of interruption.

As software development, in order to better realize the function, good scalability and robustness, the development of the filter also need the process of the design. Filter design mainly includes the following two aspects: the choice of an appropriate superclass and structural design. The hierarchy of class as shown in Figure 4.

4 SUMMARY

After the functional test of this method, we can find that this seamless switch method of audio files can make the broadcasting more coherence. Seamless switch technology can satisfy the digital audio workstations' broadcast safety requirements on audio file, while the current broadcast file is missing, or the file server network was interrupted, or the current audio file on the file server error.

REFERENCES

Cai H. G. (2008) Research and implementation of Windows pure software dual hot backup system based on the dissertation. *Nankai University.*

Lee S. (2010). Data backup and recovery scheme for CATV machine room. *Chinese cable TV*, 2010.3, 23–25.

Xu H. (2004). Nonlinear editing network system. *Gansu science and technology*, 2004.6, 33–34.

Yan Y. M. (1997). Data structure (C language version). *Tsinghua University press.*

Electronic Engineering and Information Science – Wang (Ed.)
© 2015 Taylor & Francis Group, London, ISBN: 978-1-138-02772-5

An adaptive audio mixing weight quantization algorithm

J.Y. Fan, Z.L. Zhang, Y. Fu & D.D. Hui
The Higher Educational Key Laboratory for Measuring & Control Technology and Instrumentations of Heilongjiang Province, Harbin University of Science and Technology, Harbin, China

ABSTRACT: Modern mix algorithm has a problem, which is in solving the remix overflow to lead to a reduction, distortion and long computing time of output of the audio mix. According to the G.711 specification, this paper proposed an adaptive mixing weight quantization algorithm. The algorithm of weight has nothing to do with time and input, and it can be more convenient and more efficient processing of mixing. Experiments showed that the adaptive mixing weight quantization algorithm is fairly simple, and the mixed output results are improved obviously, it can be able to satisfy the need of modern multimedia interactive video.

KEYWORDS: Multimedia conferencing; Adaptive audio mixing; Quantification; Audio processing.

1 INTRODUCTION

With video conferencing in business, medical treatment and education in a wide range of applications, modern audio video conferencing proposes new demands of voice audio, first of all, participants can hear each speaker' voice in a moment at the same time. Secondly, participants can learn the tone of the speaker's reaction and the attitude (Wang 2013). Currently, voice, audio processing mechanisms are that each speaker has a separate voice input device in the meeting, audio frequency transmitted to the device terminals through the network for mixing process and then play. The receiving end of equipment starts using multiple threads to receive audio data and processing, network bandwidth occupied very large. If the network jitters, audio effects are being played poorly after mixing, and it needs a higher quality final output device to solve this problem. If you play the mixing after output by PC, you cannot meet the requirements of the modern video conference. For this purpose, ITU-T (International Telecommunication Union-Telecommunications) (Toga 1999) proposed a centralized conference mode, in MCU (Multipoint Control Unit) it mixes the voice audio of the speaker, and then transfer the audio which after processing to the participants, and let participants receive and listen. It reduces the occupation of the terminal device and network bandwidth requirement).

Existing algorithms such as the average mixing algorithm (Cai 2006), the clamp algorithm (Yu et al. 2010), and adaptive weighted algorithm (Wang & Liao 2007) and others have obvious defects. The volume exponentially decreases with the inputs way

of audio frequency increasing in the average mixing algorithm. The output voice distorted seriously and introduced unnecessary noise in clamping algorithm. The adaptive weighted algorithm is not the usual linear superposition operation, and the calculation requires a longer time, filtering out small volume of audio input. Modern multimedia video conferencing introducing a large number of voice input, these algorithms are difficult to mixing. This paper proposes a new algorithm based on the G.711 standard, its weight does not change with time and audio input, and mixing volume after output has no fluctuated phenomenon. Operation satisfies the linear superposition. It solves the defects of the above algorithm to meet the multimedia video conferencing needs.

2 MIXING ALGORITHM

2.1 The basic mixing algorithm

Sound is a pressure wave due to the vibration of the object and the surrounding air pressure, it can be "detected" and "felt". Suppose y is sound, A is sound amplitude, ω is the angular velocity of sound vibrations, x is sound vibrations time, pure tone function model as follows

$$y = A \sin \omega x \qquad (1)$$

The functional model of voice which is heard in reality is as follows

$$y = \sin x + \frac{1}{2}\sin 2x + \frac{1}{3}\sin 3x + \frac{1}{4}\sin 4x + \cdots \qquad (2)$$

The sound is still a smooth, continuous waveform signal through sampling and quantization after being converted to an electrical signal. The quantized frequency and amplitude of the speech signal correspond to the sound frequency and volume, and the voice signal superimposed is equivalent to superposition of sound waves in the air.

According to Reference (Zou & Liu 2011), Now build the mixing model:1, Suppose that there are N-channel participate in mixing, When mixing started, first remove the audio data of a local road, this can only hear the other (is an integer of greater than or equal 1) way voice; 2, Suppose at a certain time t, the audio data of the output channel $i(i = 1,2,3,...N)$ is $output_i(t)$, a range is $[-2^{P-1}, 2^{P-1}-1]$, where P is the quantization accuracy; 3, Suppose that the $j(j = 1,2,3,...,M$, and $M=N)$ output data $Output_j(t)$, $output_i(t)$ where in $Output_j(t)$ is in addition to the other outer channel of the mix output $M-1$, $Output_{M+1}(t)$ is a mix of all audio output channel M, so

$$\begin{cases} Output_j(t) = \sum_{j=1, j\neq i}^{M} output_i(t) \\ Output_{M+1}(t) = \sum_{i=1}^{M} output_i(t) \end{cases} \quad (1)$$

Can be known by the above formula, the values of range of $output_j(t)$ is larger than $[-2^{p-1}, 2^{p-1}-1]$, the audio data overflow, joined the noise. Where M is greater, the higher frequency of overflow occurrence, by experiments, at $M=4$, since the noise is too much added by the data overflow, the mixing voice is unrecognizable, the important issue of mixing algorithm is how to deal with overflow after mixing sample values.

2.2 Adaptive weighted algorithm

Currently known, there are many algorithms to deal with the problem that the sample value overflow after mixing, but basically developed from according to Eq.1 added weight $W_{ii}(t)$ and $W_{iM+1}(t)$, different algorithms reflects in the different methods to get the value of $W_{ii}(t)$ and $W_{iM+1}(t)$. Bring weight into (1) could get mixed expressions, as follows

$$\begin{cases} Output_j(t) = \sum_{i=1, i\neq j}^{M} W_{i,j}(t) * output_i(t) \\ Output_{M+1}(t) = \sum_{i=1}^{M} W_{i,M+1}(t) * output_i(t) \end{cases} \quad (2)$$

The adaptive weighted algorithm considers its characteristics of mixing multi-channel audio signals,

so their proportions as weight to determine their output synthesized in proportion, weight $W_{ii}(t)$ and $W_{iM+1}(t)$ as follows

$$\begin{cases} W_{i,j}(t) = |output_i(t)| \Big/ \sum_{i=1, i\neq j}^{M} output_i(t) \\ W_{i,M+1}(t) = |output_i(t)| \Big/ \sum_{i=1}^{M} output_i(t) \end{cases} \quad (3)$$

The algorithm eliminates the problem of mixing overflow, and to ensure that the waveform remains substantially unchanged after mixing, without introducing large noise. But this method does not follow a linear superposition relationship. The small proportion of small voice way, after taking this method mixing it becomes smaller, and the decay amplitude large, making the original small voice hardly hear.

3 ADAPTIVE WEIGHT QUANTIZATION ALGORITHM

The average mixing algorithm, clamping algorithm and adaptive weighted algorithm are introduced weight into (3), changing to get. Weight changes over time, leading to the size of the output audio volume of mixing changing in different degrees over time, voice quality degradation, and there is no way to eliminate this defect.

3.1 Algorithm design ideas

The establishment of the model: In G.711 standard based on the low intensity signal has higher risk than the high intensity signal in the voice signal and the fact of A-law companding using segmented quantitative rules. An adaptive weight quantization algorithm based on the same idea, using segmented quantify the rules for quantifying the sampling values after linear superposition to ensure that no overflow occurs. Low-intensity signal use greater weight to ensure readability and certainly quantify of the signal, but the high intensity signal uses a smaller weight to ensure the readability and the obtained corresponding quantization percentage.

3.2 Quantized region segmentation

According to the actual situation that M with the $output_i(t)$ are a function of the time-dependent, if the weights are not associated with time, there must be independent of the weight with the M and $Output_i(t)$. By Eq.1 shows that, no matter what values M and $Output_i(t)$ are, $Output_i(t) \in (-\infty, \infty)$ meets requirements, and $|Output_i(t)|\in[0,\infty)$. Then interval length $[0, +\infty)$ is evenly divided by 2^{P-1}, it can get a lot of areas, Such as: $[0,2^{P-1}]$, ... ,$((n-1)2^{P-1}, n2^{P-1}]$, $(n2^{P-1}, (n+1)2^{P-1}]$.... If the signal $|Output_i(t)|$ is located within the interval $((n-1)2^{P-1}, n2^{P-1}]$, the signal strength of the signal is the n-level, as shown in Figure 1.

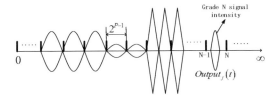

Figure 1. n-level signal strength.

3.3 Determining n interval of quantization factor

Introduction of basic quantitative factors $\alpha(\alpha>1)$, and according to the following rules: quantization factor $(\alpha-1)/\alpha$ is within the interval 0; Quantization factor $(1-(\alpha-1)/\alpha)*[(\alpha-1)/\alpha]$ is within the interval 1; Quantization factor $(1-(\alpha-1)/\alpha)*(1-(\alpha-1)/\alpha)*[(\alpha-1)/\alpha]$ is within the interval 2. By mathematical induction, quantization factor can be obtained $[(\alpha-1)/\alpha](1/\alpha)n$ within the interval n.

As discussed above, in the n-level intensity of the signal the output signal as formula 6, which sgn(x) is the sign function, mod is modulo function.

$$Output'(t) = \text{sgn}\left(Output_j(t)\right)\sum_{i=0}^{n_j-1}\frac{\alpha-1}{\alpha}\left(\frac{1}{\alpha}\right)^i 2^{P-1}$$

$$+\frac{\alpha-1}{\alpha}\left(\frac{1}{\alpha}\right)^{n_j}\left(Output_j(t)\right)\bmod\left(2^{P-1}\right) \quad (4)$$

Where $n_j=Output_i(t)/2^{P-1}, j=1,2,3,...M+1....$

For formula Eq.4, it can be proved that the value within a specified range $((n-1)2^{P-1}, n2^{P-1})$, and will not overflow.

$$\left|Output'(t)\right| = \left|\text{sgn}(Output_j(t))\sum_{i=0}^{n_j-1}\frac{\alpha-1}{\alpha}\left(\frac{1}{\alpha}\right)^i 2^{P-1} + \frac{\alpha-1}{\alpha}\left(\frac{1}{\alpha}\right)^{n_j}\left(Output_j(t)\right)\bmod\left(2^{P-1}\right)\right|$$

$$< \sum_{i=0}^{n_j-1}\frac{\alpha-1}{\alpha}\left(\frac{1}{\alpha}\right)^i 2^{P-1}+\frac{\alpha-1}{\alpha}\left(\frac{1}{\alpha}\right)^{n_j} 2^{P-1}$$

$$= \frac{\alpha-1}{\alpha}2^{P-1}\sum_{i=0}^{n_j}\left(\frac{1}{\alpha}\right)^i < \frac{\alpha-1}{\alpha}2^{P-1}\sum_{i=0}^{\infty}\left(\frac{1}{\alpha}\right)^i$$

$$= 2^{P-1}$$

The algorithm in practical applications, first the values question of basic quantify factors should be considered. To facilitate the operation, so $\alpha=2^\beta(\beta=1,2,3...)$. According to the characteristics of the algorithm, the value α too small will result in a greater contraction waveform overall distortion, too much will lead to serious distortion of the high-intensity signal, so $\alpha=8$ or 16 is better. The following describes in $\alpha=8$, for example. According to G.711 series standard, $P=16$. Quantization factor is determining the operational complexity, influencing of the time to run the program. So that $i=1,2,3,4,5,6$ corresponding to the accumulation result of a quantization factor is given in Table 2.

Table 1. Quantitative factors accumulated result.

i	1	2	3	4	5	6
$[(\partial-1)/\partial]$ $(1/\partial)^i 2^{P-1}$	3584	448	56	7	0.87	0.11
$\Sigma[(\partial-1)/\partial]$ $(1/\partial)^i 2^{P-1}$	28672	32256	32704	32760	32761	32761

As shown in Table 2, when $I \geq 5$, the cumulative result of quantified factor has changed little, effecting little on the mixing, and the effect can be ignored. We can save time by using software programming algorithms to achieve mixing algorithm and reduce delays caused by the algorithm.

4 EXPERIMENTAL RESULTS AND ANALYSIS

The speech signal in the process of sampling and quantization, limits of the hardware and software of signed short type. Commonly sound card is generally 16 bit and the quantitative accuracy is 16 bit. In the Windows operating system, the type of a sound card buffer is as the same as the signed short type of C language, the range is [−32768, 32767]. After multi-channel sound superimposing, the amplitude may be outside the scope of the sound card can accept, leading to the audio stream output after poor results. A plurality of continuous smooth wave after wave superposition is smooth, after wave superposition overflow place produces noise. Usually use 32 bit to represent a linear superposition of the latest data, after the audio mix, once again using the adaptive algorithm to decrease its amplitude, the amplitude distribution in the 16 bit range. This method is realized simple, operated fast, and it is consistent with the multi-channel audio and mixing requirements.

Usually in centralized meeting, terminal and MCU establish a unicast connection, send audio stream to MCU in real time, and then receive the audio stream from MCU. The H.323 standard specifies the system of the standard protocol family, speech codec can be used G.711, G.722, G.723.1, G.728 and G.729. Among them G.711 encoded data attributes are shown in Table 1.

Table 2. G.711 encoded data attribute in the A system.

Codec	Sampling rate(Hz)	Quantization (bites)	Target	Channel bit rate (kbps)
G.711	8000	16	64/65	Mono

We use the 3 way voice which has collected as the sound source mixed test, check the effect of the algorithm, the voice is the composition of two male voice and a female voice.

Experimental data: The configuration of the PC machine that was used in the test: Windows7 ultimate system, Intel(R) Core(TM) i5–3210M CPU @2.50GHz, Memory installed(RAM):2.0GB; The test environment: Visual Studio 2012; Audio processing format is as follows.

WaveFormat waveFormat = new WaveFormat();// Create code format examples

waveFormat.FormatTag = WaveFormatTag.Pcm; // Format type

waveFormat.Channels = 1; // Number of channels

waveFormat.SamplesPerSecond=8000;//Sampling rate

waveFormat.BitsPerSample = 16; // File sample size

waveFormat.BlockAlign = (short)(channels * (bits PerSample / (short)8)); // Block alignment

waveFormat.AverageBytesPerSecond = waveFormat. BlockAlign * samplesPerSecond; // Required average data transfer rate

The above algorithm, we intercept the mixing waveform at the same time (a total of 8000 samples), and select the output waveform of the mean value, clamping algorithm, adaptive weighted algorithm and adaptive algorithm for the experiment. The male a voice signal is shown in Figure 2. The male b voice signal is shown in Figure 3. The female voice signal is shown in Figure 4. Different mixing algorithm outputs are shown from Figure 5–9. The red dotted line represents the maximum and minimum mixing wave, the solid red box represents mixing waveform anomaly, the solid blue box represents an enlarged section.

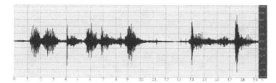

Figure 2. Male voice a.

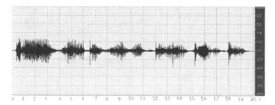

Figure 3. Male voice b.

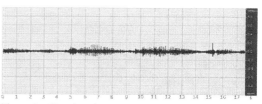

Figure 4. Female voice.

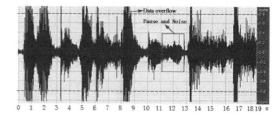

Figure 5. The basic mixing algorithm.

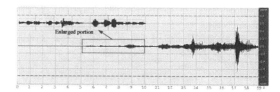

Figure 6. The average mixing algorithm.

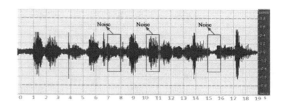

Figure 7. The clamp algorithm algorithm.

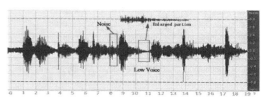

Figure 8. Adaptive weighted algorithm.

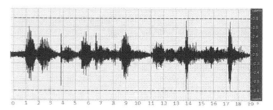

Figure 9. Adaptive audio mixing weight quantization algorithm.

Seen from time-domain waveform after mixing, basic mixing algorithm results shown in Figure 5, we can see that the data overflow occurs, resulting in waveform distortion, noise and pause phenomenon in horizontal axis 11 to 13. Figure 6 shows the average mixing algorithm results, the wave has not changed much, but the amplitude is reduced three times, namely the volume is much smaller compared to the other three algorithms. Figure 7 shows the clamp algorithm results. There is no data overflow, and the amplitude has been in maximum, like a knife cut, the horizontal axis 7–8, 10–11, 15–16 noise obviously. The sound is not smooth. Figure 8 is an adaptive weight quantization algorithm results. There is no data overflow, the amplitude decreases, the horizontal axis of 8 to 9 appear noise, 10 to 11 sounds low, the results show that the first female voice use this algorithm to output sound through the device is smaller than other three algorithms. Figure 9 is an adaptive algorithm right weight, there was no data overflow, small amplitude decreases, sounds smooth, able to clearly identify. The noise is much smaller compared to the other clamping algorithm and adaptive weight algorithm, without blasting sound.

5 CONCLUSIONS

All these suggest that this paper proposed and applied adaptive weight algorithm, to solve the problem that mixing overflow, the noise, blasting sound, the output volume small after mixing. It is possible to complete the multi-point real-time playback and also maintain details features of various voice frequencies, and has a good degree of continuous and subjective comfort in the hearing, with a wide range of applications to meet the needs of multimedia conferencing.

REFERENCES

Wang W. (2013). The design and implementation of H.232 video system based on IP Network. *Shandong University.*

Toga J. (1999). ITU-T standardization activities for interactive multimedia communications on packet-based networks: H.323 and related recommendations. *Computer Networks.*

Cai B. Q. (2006). Research on audio mixing technology in video conference. *Modern Electronics Technique. 235,* 85–87.

Yu L. & L. Y. Guo & H. Z. Tan (2010). A real-time synchronous audio Mixing and transmitting scheme in multimedia conference system. *Acta Scientiarum Naturalium Universitatis Sunyatseni. 529,* 31–36.

Wang W. L. & J. X. Liao (2007). A novel fast real-time audio mixing algotithm in multimedia conference. *Journal of Electronics & Information Technology. 690,* 29–33.

Zou L. Z. & L. Liu (2011). An Improved audio mixing algorithm for video conferencing system and its implementation. *Communications Technology. 120,* 44–46.

Electronic Engineering and Information Science – Wang (Ed.)
© 2015 Taylor & Francis Group, London, ISBN: 978-1-138-02772-5

Comparison between Hilbert-Huang transform and local mean decomposition

J.H. Cai & Q.Y. Chen
Institute of Information, Hunan University of Arts and Science, Changde, Hunan, China

ABSTRACT: Hilbert Huang Transform (HHT) and Local Mean Decomposition (LMD) are two adaptive signal processing methods. Firstly, the principle of two kinds of methods was presented. And then taking the simulated signal as example, the comparative study was carried on through the decomposition and reconstruction, instantaneous amplitude and frequency, time-frequency spectrum and marginal spectrum. Results showed that HHT and LMD decomposition have an adaptive features and the decomposed component can completely reconstruct the original signal. This feature provides conditions for de-noising. The time-frequency spectrums, coming from HHT and LMD, all have strong ability to express time-frequency characterization in details, which can be used to extract the time-frequency features of signal. The compared results indicated that LMD is superior to HHT in some respects, such as in boundary effect, iterative times and computational efficiency. And LMD avoids the phenomenon of false frequency, which often exists in HHT.

KEYWORDS: Hilbert Huang transform; local mean decomposition; decomposition and reconstruction; time-frequency analysis.

1 INSTRUCTION

In recent years, in order to break the limit that the analyzed signal must be stationary in Fourier method, scholars have put forward some analysis methods for non-stationary signals. Wigner-Ville distribution, Wavelet transform, HHT, local mean decomposition and Teager operator were proposed and widely used in non-stationary signal procession (Barbarossa1992). Hilbert-Huang transform and local mean decomposition both have some same characteristics such as direct, posterior and adaptive, because the two transforms are a decomposition based on the data itself. So they have good adaptability, which is the reason that they got the fast development in recent years (Cheng 2004). This paper achieved a comparative study of the two methods from the instantaneous amplitude and frequency, decomposition and reconstruction, time- frequency spectrum and marginal spectrum and attempted to provide some valuable reference for the selection of methods.

2 HILBERT-HUANG TRANSFORM

Signal can be decomposed into a set of intrinsic mode functions (IMFs) by the empirical mode decomposition (EMD) which is also called Huang Transform. EMD is a process of cyclic decomposition. The complete decomposition process can be found in literature (Cheng 2006). Using EMD method, the original data $s(t)$ can be expressed as a sum of the IMFs and a residue:

$$s(t) = \sum_{i=1}^{n} c_i(t) + r_n(t) \tag{1}$$

For an time series $x(t)$, its Hilbert transform $y(t)$ can be presented as

$$y(t) = \frac{1}{\pi} p \int \frac{x(t')}{t-t'} dt' \tag{2}$$

Each IMF of $s(t)$ is applied the Hilbert transform, then $H(\omega, t)$, the time-frequency spectrum, can be obtained and written as (Cheng 2009):

$$H(\omega, t) = \mathrm{Re} \sum_{j=1}^{n} a_j(t) e^{i \int \omega_j \, dt} \tag{3}$$

The function $H(\omega, t)$ is defined as Hilbert spectrum to reveal time-frequency distribution of original signal. Then, the Hilbert marginal spectrum is presented as:

$$h(\omega) = \int_0^T H(\omega, t) dt \tag{4}$$

T is the length of time series.

3 LOCAL MEAN DECOMPOSITION METHOD

Local mean decomposition(LMD) is a new adaptive method of signal decomposition proposed by Smith in 2005. It can decompose a complex non-stationary signal into several *PF* components. The specific procession can be found in literature(Smith 2005), here no longer tired. Product function is a multiplication of envelope signal $a(t)$ and pure frequency modulation signal $s(t)$.

$$PF_1(t) = a_1(t)s_1(t) \qquad (5)$$

After $PF_1(t)$ is separated from original signal $x(t)$, the above steps are repeated. At the end, $x(t)$can be decomposed into k *PF* components and u_k.

$$x(t) = \sum_{p=1}^{k} PF_p(t) + u_k(t) \qquad (6)$$

The instantaneous frequency and envelope of each product function are calculated, then they are expressed together in the form of the demodulated signal, called time-frequency distribution $DS(\omega, t)$. The scale represents the envelope variation. Finally, a demodulated signal spectrum can be calculated analogous to the Fourier spectrum, called as margin spectrum.

$$DS(\omega) = \int_0^T DS(\omega, t)dt \qquad (7)$$

T is the length of dataset and $DS(\omega, t)$ is the time-frequency distribution of demodulated signal.

4 PHOTOGRAPHS AND FIGURES

4.1 *Decomposition and reconstruction*

N LMD uses moving average to solve the extremum curves, which is done by using the three order spline interpolation in the EMD. The final result of the EMD is a series of IMFs, while LMD will decompose a signal into a series of PFs. In order to compare LMD and EMD intuitively, this paper processed a same simulated signal by the two methods respectively (without pretreatment at endpoints for two methods). The simulated signal is as following:

$$s(t) = (1 + 0.5\cos(10\pi t))\cos(160\pi t + 2\cos(12\pi t))$$
$$+2\sin(40\pi t) \qquad (8)$$

The simulated signal consists of an AM-FM signal and a sine signal. The center frequency of AM-FM signal is 80Hz, and the frequency of sinusoidal component is 20Hz. According to the decomposition principle of the two methods, after decomposition, the first-order component should be AM-FM signal (80Hz) and the two order component should be the 20Hz sinusoidal signal. The higher order component should be the decomposed residuals.

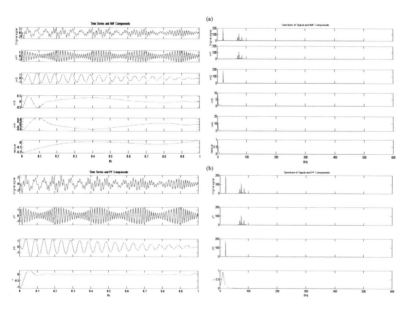

Figure 1. EMD and LMD decomposition of signal and their corresponding power spectrum of each order component: (a) from EMD ,(b) from LMD.

Fig.1(a) shows the EMD decomposition of simulated signal and its corresponding power spectrum. The termination condition of iterative decomposition was that the absolute value of average envelope function was less than 10^{-3}. In the graph, signal was decomposed into 5 layers adaptively. Fig.1(b) presented the LMD decomposition results of simulated signal and the corresponding power spectrum of *PF* component. The set condition for termination of iteration was $1 - \Delta \leq a_{1n} \leq 1 + \Delta$, where $\Delta = 10^{-3}$. In Fig.1(b), signal is decomposed into 3 layers adaptively. Obviously, LMD and EMD both were adaptive decomposition method from high frequency to low frequency.

Fig. 2 was the contrast plot between the reconstructed signal using the decomposed component and the original signal. Reconstruction is a procession to add up the decomposed component to restore the original signal. Fig. 2 was for EMD method. Fig. 3 shows the result of LMD. From the diagram, it can be seen that the error of reconstruction compared with the original signal, the amplitude is difference 15 orders of magnitude. The error is for computer calculation error, and can be negligible. Visible both EMD and LMD can reconstruct the original using the decomposed component, almost no energy loss. These characteristics of the two methods also provide a new way for noise suppression.

4.2 *Instantaneous amplitude and instantaneous frequency*

The top plot of Fig. 3 (a) shows the instantaneous amplitude a1(t) and instantaneous frequency f1(t) of the first order IMF component of the simulated signal s(t) with EMD. The bottom plot of Fig. 3 (a) expressed the same comment of the second order IMF component. From Fig. 3 (a), it can be seen that due to the severe endpoint effect of EMD and the edge effect of Hilbert transform, there is a certain distortion in the calculated instantaneous amplitude and instantaneous frequency. Fig. 3(b) shows the instantaneous amplitude and instantaneous frequency of the first and the second order PF component of the simulated signal s(t) with LMD. In Fig. 3(b), the instantaneous parameters of each PF component, obtained by LMD method, reflects the real information of original signal. It can be seen that the first PF component from

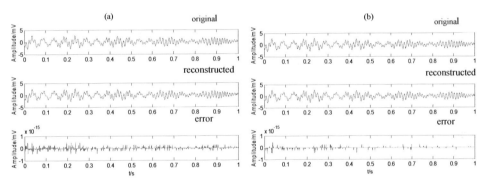

Figure 2. Reconstruction of signal with EMD or LMD method (a) The reconstructed signal with EMD (b) The reconstructed signal with LMD.

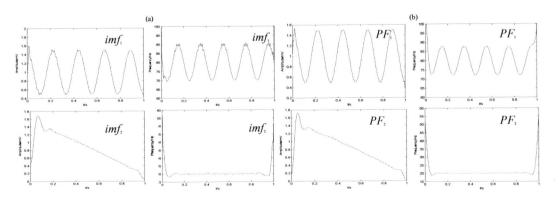

Figure 3. Instantaneous amplitude (left) and Instantaneous frequency (right) of the 1*th* and the 2*th* order component: (a) from EMD, (b) from LMD.

LMD method corresponds the 80Hz Am-Fm components existed in the original signal. The second component can exactly correspond to the 20 Hz sinusoidal component of original signal. Their amplitude also has a strict corresponding relation with signal. LMD decomposition precisely separate the components exited in the signal. The less number of iterations, the pollution degree of endpoint effect to the whole data will be the lighter. This result revealed that without pretreatment to endpoint the endpoint effect of EMD is more serious than that of LMD. LMD method is better than the method of EMD in reducing the number of iterations and restrain endpoint.

4.3 Time-frequency spectrum and marginal spectrum

By formula (3) and (7), the instantaneous parameter information of each component was assembled to obtain the time-frequency spectrum. Further, the two methods both can integral along the time axis through the time-frequency spectrum to obtain the marginal spectrum.

Fig. 4 shows the three-dimensional time- frequency spectrum and marginal spectrum. In order to observe, the X-Y planar graph of 3D graph is presented (actually, it is the two-dimensional brightness spectrum. The color represents the intensity of energy). From the left Figure of Fig.4 (a), it can be seen that the dynamic change characteristics of the energy distribution of 2 components is clear in time and frequency domain (sinusoidal components is as banded, FM components is as toothed). The right figure of Fig.4 (a) shows the marginal spectrum which reflects the accumulated energy of whole length of time. Fig.4 (b) exhibits the time-frequency spectrum and marginal spectrum obtained with LMD method. Similarly to HHT method, the spectrum can also distinguish the subtle changes of energy with time and frequency accurately. Ribbon and toothed graphics reflects the actual variation characteristics of frequency and amplitude modulation component and sinusoidal signal component. The bottom graph is corresponding to the marginal spectrum.

From Fig.4 (a) and Fig.4 (b), it can be seen that the time frequency spectrum of two kinds of method both have very high energy resolution in time domain and frequency domain, and a strong ability of mutation detection and orientation. But in Fig.4, three-dimensional from the HHT, the influence of endpoint effect is greater in the three-dimensional time-frequency spectrum based on LMD method. And some false frequency appears in the low-frequency stage. These findings can be found more clearly in the marginal spectrum. Because of the endpoint effects, Hilbert-Huang spectrum has a certain distortion. In comparison, spectrum based on LMD method has very strong ability to describe the time frequency characteristic. The false frequency, which appears in the low frequency section, is restrained in the marginal spectrum of LMD method. These are echoing the conclusions got in above section.

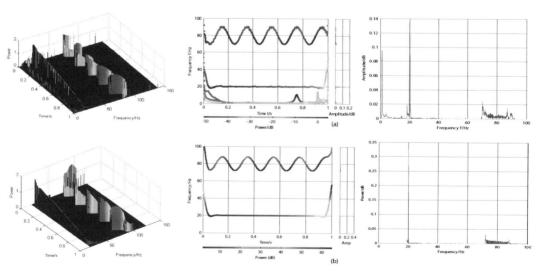

Figure 4. Three dimensional time-frequency spectrum of signal (left), two-dimensional brightness spectrum (middle) and marginal spectrum (right) (a) from HHT method (b) from LMD method.

5 CONCLUSIONS

1 EMD and LMD decomposition are an adaptive decomposition procedure from the high layers to low layers. The decomposed component can reconstruct the original signal completely, which provides conditions for signal de-noising.

2 Time-frequency spectrum characteristics of two kinds of methods described the facts that they have very high resolution for the time domain and frequency domain and can be used to extract the time-frequency feature of signals. Local mean decomposition compared with Huang transform, LMD method is better than the Huang transform in the suppression of the endpoint effect, reducing the number of iterations and the computation efficiency. The energy is more focused in LMD time-frequency spectrum than that in Hilbert spectrum, and the false frequency phenomena is avoided by using LMD method.

3 It is worth pointing out that LMD method has some deficiencies. Such as when the smoothing times is more, signals occur in advance or lag. The optimal smoothing step can not be determined and LMD has no fast algorithm.

ACKNOWLEDEGMENT

This project is supported by the National Natural Science Foundation of China (Project No: 41304098), Hunan Provincial Natural Science Foundation of China (Project No: 12JJ4034), and Young Scientific Research Fund of Hunan Provincial Education Department (Project No: 13B076). We wish to thank all those who contributed to our works and express our appreciation for the financial support.

REFERENCES

Barbarossa S. & A. Zanalda (1992). A combined Wigner-Ville and Hough transform for cross terms suppression and optimal detection and parameter estimation, *IEEE International Conference on Acoustics, Speech and Signal Processing*, 5:173–176.

Cheng, J.S. & Y. Yang (2004). Energy operator demodulating approach based on EMD and its application in mechanical fault diagnosis. *Chinese Journal of Mechanical Engineering*, 40(8): 115–118.

Cheng, J.S. & D.J. Yu (2006). Research on the intrinsic mode function (IMF) criterion in EMD method. *Mechanical Systems and Signal Processing*, 20(4):817–824.

Cheng, J.S. & K. Zhang (2009). Comparison between the methods of local mean decomposition and empirical mode decomposition, *Journal of vibration and shock*, 5(28):13–17.

Smith, J. S(2005). The local mean decomposition and its application to EEG perception data. *Journal of the Royal Society Interface*, 2(5): 443–454.

Electronic Engineering and Information Science – Wang (Ed.)
© 2015 Taylor & Francis Group, London, ISBN: 978-1-138-02772-5

A method for generating distributed network traffic with accurate speed

D.W. Hao, F. Cao & X. Zhou
Wuhan Mechanic Technology College, Wuhan, China

ABSTRACT: This paper proposes a new method of generating flexible and high-speed network traffic. Imported in the power gene, the network traffic is continually adjusted by using the reactive mechanism, until errors meet the user's initialization. By combining ON/OFF model, Distributed framework generates the network traffic with self-similarity and adjustable speed. The network traffic generated in this way is precise and controllable, and takes on self-similarity characteristic. Thus it can reflect the features of the real network traffic.

KEYWORDS: Network Traffic Generator, self-similarity, ON/OFF model, Pareto distribut.

1 INTRODUCTION

Network modeling and simulation is an important means to study network performance as well as testing and experimenting in a web-based environment. Now traffic generation techniques are faced with challenges in terms of performance, accuracy and flexibility. General-purpose processors and software based implementation can easily simulate network traffic at a lower flow rate. In the environment of Gbps rate, requirement for rate is difficult to achieve with software approach because of the limitation of general-purpose processor architecture. Currently professional testing equipments of ASIC technology has become increasingly mature. The better performance is suitable for network equipment benchmark testing which has more fixed testing purposes; the downside is that such devices are difficult to achieve, higher in prices, so it is not easy to popularize.

2 DESIGN AND IMPLEMENTATION OF A DISTRIBUTED NETWORK TRAFFIC GENERATING SYSTEM

2.1 Application environment

The method was applied to simulating network traffic in a network environment simulation. Deployment structure of the system is shown in Figure 1.

2.2 Software components

DINTSS system uses a distributed architecture, consisting of a control seat and a number of agent components. The control seat is used to design traffic

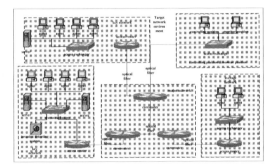

Figure 1. Deployment structure of the system.

simulation tasks and assign those tasks to each agent. With capability plug-in interface on them, agents can dispatch capability plug-ins according to the stimulating requirements from the control end, performing appropriate tasks through controlling these plug-ins, and giving feedback to the control seat; capability plug-ins are those encapsulating atomic capabilities, providing unified plug-in interface to the agents to facilitate the calling and management of agents. They are used to output data packets to network link under the control agents to generate traffic.

The core and critical part is the agents. The agents are the implementing entities of the traffic simulation, containing 12 kinds of traffic simulation plug-ins such as simulation of application service traffic, background traffic and database application, and the capacity to precisely calibrate the transmitting rate on the base of feedback mechanism.

3 TRAFFIC GENERATION METHODS

3.1 Methods for accurate calibrating send rate based on feedback mechanism

3.1.1 The use of high-precision timers

Precise timers are required in order to generate accurate time interval of data packet transmission to provide high-precision timing to simulation, while the timing functions of Windows can be only used on the ms level, unable to meet the needs of the simulation process. Therefore, using computer hardware to design precise timing operation can achieve microsecond. The high-precision timer can be achieved with API function, namely:

Query Performance Frequency () ;/ / get the hardware frequency counter of CPU;

Query Performance Counter () ;/ / get the current counter value of the hardware.

API function requires the computer's hardware to support accurate timer. Precise timer implementation is as follow:

Step 1: Firstly using Query Performance Frequency () function to obtain the high precision frequency counter operation f, the unit is time / s.

Step 2: At both ends of the code separately use timed Query Performance Counter (), to get the numerical value of the high-precision counter pre, now, differences between the numerical value are converted into time interval through f, , thus interval = (pre-now) / f.

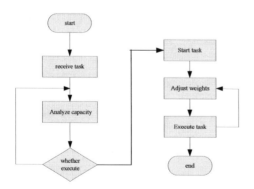

Figure 2.   Rate control flow chart.

3.1.2 Feedback mechanism

The task performing process is shown by Figure 2.

Firstly tasks are divided into even parts by control seat in accordance with the number of agents and then equally distributed to the agents. When receiving the tasks, the agents will generate traffic in maximum and calculate the largest dispatch capacity, then send back

to the control seat which will judge whether the task can be completed. If the task is beyond the ability of the agents, it will be canceled and redistributed after adding more agents or decrease the amount of total flow. If the task can be achieved, agents adjust the flow size in accordance with their respective working amounts to precisely control the rate. The overall process is that the agent adjusts the weights to reduce the transmission rate so that matching the mission requirements according to their ability limitation and the feedback transmission rate .The process is gradually approaching the infinite loop process, therefore, it shall be limited. When the frequency is adjusted to two times / s, the scale of weight change is 5%, and the size deviation of the flow is within the range set by the user, the flow rate shall meet the needs. Stop rate adjustment.

3.2 Sending method for self-similar data flow

3.2.1 Use of ON/OFF

A key problem of ON/OFF model is the determination of the periods of ON and Off. For the ON period x submit to Pareto distribution, inverse function is used to model generate Eq. 8:

$$x = k / U^{1/\alpha} , \tag{1}$$

Among which: $U \in (0,1]$, submit to even distribution, α is heavy-tailed parameter, k is On period or the minimum value of OFF period. The time value of ON period and the OFF period can be get through calculating of continuously-generated even random formula (12). What's different is that the method for determining k of the OFF period and the ON period is not the same. For the ON period, the minimum value is just the consuming time for transmitting a data packet. While for the OFF period, the minimum value is determined by the load proportion of each source, i.e. the ratio between traffic and bandwidth L_i. Suppose the load proportion of each source is the same, L_i can be expressed as Eq.9:

$$L_i = M_{on} / (M_{on} + M_{off}) \tag{2}$$

Wherein M_{on} and M_{off} are respectively the average ON and OFF period.

When $\alpha > 1$, the mathematical expectation of Pareto distribution is Eq. 10

$$E(x) = \alpha k / (\alpha - 1) \tag{3}$$

However, the uniformly distributed random number generated by the computer in the range of $(0, 1)$

is bounded. Set the minimum value generated within (0, 1) m, the maximum value generated suitable for Pareto distribution is Eq. 11 :

$$X_{max} = k / m^{1/\alpha} \qquad (4)$$

3.2.2 Generation of random number

There are various methods to generate uniformly distributed pseudo-random number in the range of (0,1], such as linear congruential method, square method, chaos method, feedback shift register (RNG), etc. One of the most commonly used is linear congruential method, which adopted the following linear recurrence relation with the remainder to produce the number of columns [3] as Eq. 15:

$$\left. \begin{array}{l} x_{n+1} = (ax_n + c)(\mathrm{mod}\, m) \\ \varepsilon_n = x_n / m \end{array} \right\}, \qquad (5)$$

Among which: a, c, x_0 are positive integer, x_0 is seed. Modulus m and multiplier a needs to be carefully selected, so that the period of the generated pseudo-random numbers is as long as possible. ε_n is the random number on (0,1]. C + + language provides a library function rand () to produce random integer. Rand () function generates the required parameters, and then congruential method generates initial random numbers, then these random numbers get improved.

The advantages over linear calculation method are less calculation and fast speed, while the disadvantage is the random number sequences generated in successive calls are related. Here the use of shuffling process [4] to destroy this serial correlation, which is as follows:

Suppose v_1, v_2, K, v_n are n random numbers generated by the linear congruential method. Now randomly take a positive integer $j(1 \le j \le n)$. v_j is the required random number, and then it is replaced by another regenerated random number. After the replacement, , casually take a random number meeting the next requirements in v_1, v_2, K, v_n. Repeat like this. The process is shown in Figure 3.

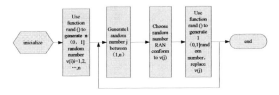

Figure 3. Chaos flow chart.

By using the above (0,1] method for generating uniformly distributed random number, random number sequence subject to the distribution can be generated. Calculations show that with the increase of random number sequences, random numbers within (0, 1] distribute more and more evenly.

3.2.3 Analysis of generating traffic

Setting: the flow size is 10Mbps, the error is 5%, = 1.1, = 1.2, = 10ms, calculating result is k_{OFF} = 28ms, duration is 600s.

FTP, P2P downloads and online video mixing application traffic collected with Kelai network analysis system is shown in Figure 4. The collected network traffic simulation is shown in Figure 5, showing the number of packet sending per unit time, and the sampling interval is 1s .

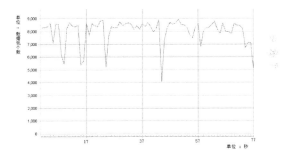

Figure 4. The real traffic figure of network mixed application.

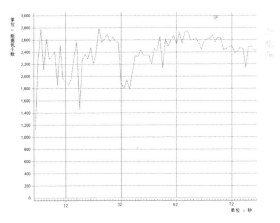

Figure 5. Simulated traffic figure.

Figure 4 and 5 shows that the network traffic generated based on the method is in line with the self-similarity of network traffic, and can simulate real network traffic. Traffic transmission rate is controlled at about 10Mbps, error within the range set by the user.

4 CONCLUSION

The author introduces the method for accurate correction of transmission rate based on feedback mechanism, self-similar network traffic generation method on ON / OFF model, and a distributed network traffic generator system designed and implemented on the basis of the above. The system can generate traffic of high rate with agent proxy, and can generate network traffic with self-similar characteristics, various combinations precisely controlled speed according to the simulated mission distributed by the control seat. The next step will be adding microscopic protocol model of real network, so that it has a microscopic characteristics of network traffic to use real-time interaction between the agents for adjusting traffic size.

REFERENCES

He J. (2004). Self-Similarity Analysis Of Network Traffic and Research of Its Application. *Hunan: Hunan University.*

Chen Y. F., Y. B. Dong & D. M. Lu (2006). Simulating Methods Of TCP Traffic Aggregation Facing Large-Scale Network. *Communications, 27 (2).* 100–106.

Zhang S. M. & Y. Li. (2006). Methods of computer-generated random numbers. *Mathematical Bulletin, 45 (3).* 44–45.

Sun X. Y. (2006). Generation Method Of Positive Overall Statistics Computer Random Number. *Dalian University Paper, 27 (6).* 4–7.

Electronic Engineering and Information Science – Wang (Ed.)
© 2015 Taylor & Francis Group, London, ISBN: 978-1-138-02772-5

Study of propagation and transformation of pulse-front tilting Gaussian pulse by transferring matrix

P.J. Sun, L.Y. Zhang, J.J. Huang & H.Y. Jin
Department of Optoelectric Information Science and Engineering, Harbin University of Science and Technology, Harbin, China

ABSTRACT: Spatial-temporal transferring matrix is employed to study the propagation of Gaussian pulses with pulse-front tilting through free space and the transformation passing angular disperse elements. The analytic expressions of output pulses are obtained and the output pulses are still Gaussian line shape which has the similar form as the input pulses. The relationships of temporal factor and tilting factor of output pulses with propagation parameter and element parameter are confirmed. In addition, for the propagation in free space, a numerical analysis is executed. The results show that the pulse width increases with the propagation length in absence of disperse media and the tilting angle decreases with the propagation length.

KEYWORDS: Pulse front tilting; Ultrashort pulses; Transferring matrix; Angular dispersion.

1 INSTRODUCTION

With the fast development of femtosecond pulse technology, ultra-short pulses have attracted a great deal of attention (Kane & Squier 1995, Fork et al. 1987). In practical applications, the precise pulse parameters are wanted, e.g., in the scheme of compressing or broadening pulses, the chirp of pulses introduced by material dispersion is needed. Another example is an achromatic phase matching device by which intensity and quality of second harmonic pulses can be greatly increased. In this device, it is necessary to precisely control pulse front tilting introduced by angular elements to reach pulse front matching and the computation of pulse front tilting is important. So Many researchers devoted themselves to study transformation and propagation of pulses. Qi & Lv (2006), Ji & Lv (2001) & Qi & Lv (2007) studied propagation and transformation of pulses in some special case. Bor et al. (1993) & Akturk *et al.* (2004) obtained the relation of angular dispersion and pulse front tilting when pulses pass by angular elements. But in this case, width and chirp of pulse will be changed, too. In this work, by ray-pulse matrix proposed by Kostebauder (1990), transformation of pulse front tilting pulses passed by free space and angular elements is studied.

2 PROPAGATION AND TRANSFORMATION OF PULSES WITH PULSE FRONT TILTING IN FREE SPACE

A pulse with pulse front tilting can be described as:

$$E_1(x,t) = E_0 \sqrt{\frac{Q_1 P_1}{q_1 p_1}} \exp\left[-\frac{k_0 x^2}{2q_1} - \frac{(t-\varepsilon_1 x)^2}{p_1}\right] \quad (1)$$

Where $Q_1 = iZ_r$, $Z_r = k_0 w_0^2/2$ is Rayleigh length, w_0 is the waist radius, k_0 is wave constant at center wavelength; $q_1 = Q_1 + z_1$, z_1 is the coordination corresponding to the waist; $p_1 = P_1 + 2i\gamma$ is a temporal parameter which governs pulse width, $P_1 = \tau_0^2$, γ is the chirp of pulses; ε_1 is the pulse front tilting parameter which decided pulse tilting. Then the pulse width can be expressed as $\Delta\tau_1 = \tau_0 (2\ln)^{1/2} (4\gamma^2/\tau_0^4+1)^{1/2}$. The tilting angle is $\tan\psi_1 = -u_g[\text{Re}(\varepsilon_1)+\text{Im}(\varepsilon_1)\text{Im}(p_1)/ \text{Re}(p_1)]$, where u_g is the group velocity.

When pulses travel in free space, the transferring matrix is

$$T_L = \begin{pmatrix} 1 & L & 0 & 0 \\ 0 & 1 & 0 & 0 \\ 0 & 0 & 1 & 0 \\ 0 & 0 & 0 & 1 \end{pmatrix} \quad (2)$$

where L is the transmitting distance. By Collins integration, the output pulses can be obtained,

$$E_{2L} = E_0 \sqrt{\frac{Q_1 P_1}{q_{2L} P_{2L}}} \exp[-\frac{ik_0 x^2}{2q_{2L}} - \frac{(\tau - \varepsilon_{2L} x)^2}{P_{2L}}] \quad (3)$$

Where $q_{2L} = q_1 + L$, $P_{2L} = p_1 - 2q_1 \varepsilon_1^2 L/(k_0 q_{2L})$, $\varepsilon_{2L} = \varepsilon_1 q_1/q_2$. From Eq. (3), the temporal parameter is added a chirp term $-2q_1 \varepsilon_1^2 L/(k_0 q_{2L})$. It indicates that the pulse width varies with the transmitting distance and so as to pulse front tilting. To simplify, we assume that the incident Gaussian pulses have no chirp at the waist. The temporal parameter of output pulses can be expressed as

$$p_{2L} = \tau_0^2 + B \frac{l^2}{l^2 + 1} - iB \frac{l}{l^2 + 1} \quad (4)$$

Where $l = L/Z_r$ is normalized transmitting distance, $B = 2\varepsilon_1^2 Z_r/k_0$. From Eq. (4), in the real part of temporal parameter, $Bl2/(l2+1)$ is always greater than zero, which induces the pulse width broaden and there is a minimum of pulse width at the waist. In this case, the tilting angle can be expressed as

$$\tan \psi_{2L} = -\frac{u_g \varepsilon_1 \tau_0^2}{(B + \tau_0^2)l^2 + \tau_0^2} \quad (5)$$

Eq. (5) shows the tilting angle decrease with transmitting distance and at the waist the tilting angle has a maximum. As an example, we consider that the pulses travel in the free space which have a Rayleigh length of 4 m with center wavelength of 800 nm, pulse width of 100 fs and tilting angle of 0.01 degree. Variation of tilting angle and pulse width with transmitting distance is indicated in Fig. 1 in which ψ_{2L}/ψ_1 and $\Delta\tau_{2L}/\Delta\tau_1$ are normalized tilting angles and pulses width, respectively.

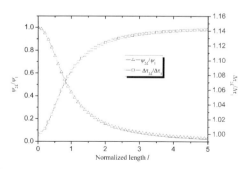

Figure 1. Pulse front tilting and pulse width vary with the length of propagation.

3 SPROPAGATION AND TRANSFORMATION OF PULSES WITH PULSE FRONT TILTING VIA DISPERSIVE ELEMENTS

As an example of dispersive element, we consider a pulse with pulse front tilting is via a grating. The transferring matrix of a grating is

$$T_G = \begin{pmatrix} m & 0 & 0 & 0 \\ 0 & 1/m & 0 & 2\pi\beta_G \\ 2\pi m\beta_G/\lambda_0 & 0 & 1 & 0 \\ 0 & 0 & 0 & 1 \end{pmatrix} \quad (6)$$

where $m = -\sin\phi/\sin\psi$, ψ is incident ray to grating surface angle, φ is reflected ray to grating surface angle. $\beta_G = \partial\theta/\partial\omega$ presents the angular dispersion of a grating. According to Kostebauder (1990), the output pulses can be expressed as

$$E_{2G} = \frac{E_0}{\sqrt{m}} \exp[-\frac{ik_0 x^2}{2q_{2G}} - \frac{(\tau - \varepsilon_{2G} x)^2}{P_1}] \quad (7)$$

where $q_{2G} = m^2 q_1$, $\varepsilon_{2G} = \varepsilon_1/m + 2\pi\beta_G/\lambda_0$. Obviously, the output pulse has the same pulse width as the incident pulse. Moreover, the tilting angle is changed due to the angular dispersion.

In the scheme of pulse broadening or compression, prism pair or grating pair is always employed. Next, we consider the transformation of a pulse with pulse front tilting via a grating pair. The distance of grating pair is set as L_1. Then the transferring matrix of grating pair is

$$T_{gp} = \begin{pmatrix} 1 & L_1/m & 0 & 2\pi L_1\beta_G \\ 0 & 1 & 0 & 0 \\ 0 & -k_0 L_1\beta_G/m & 1 & -2\pi k_0 L_1\beta_G^2 \\ 0 & 0 & 0 & 1 \end{pmatrix} \quad (8)$$

where $2\pi L_1\beta_G$ presents the spatial chirp, which indicates each frequency component exists from grating pair is separated and has the same direct due to zero angular dispersion. $-2\pi k_0 L_1\beta_G^2$ demonstrates group delay dispersion induced by a grating pair, which indicates negative group velocity dispersion can be introduced by grating pair. The output pulse can be expressed as

$$E_2 = E_0 \exp[-\frac{ik_0 x^2}{2q_{2p}} - \frac{(\tau - \varepsilon_{2p} x)^2}{P_{2p}}] \quad (9)$$

Where

$$q_{2p} = q_1 + L_1 / m^2 \, ,$$

$$p_{2p} = p_1 - 2iL_1 q_1 (\varepsilon_1 + k_0 m \beta_G)^2 / k_0 m^2 q_{2p} \, ,$$

$$\varepsilon_{2p} = q_1 \varepsilon_1 / q_{2p} - k_0 L_1 \beta_G / q_{2p} \, .$$

It can be seen that pulse width and pulse front tilting vary with the distance L_1 and the parameter of grating.

4 CONCLUSIONS

Based on ray-pulse matrix, the analytic solutions of pulses with pulse front tilting propagating in free space and angular elements are achieved. According to the analytic solutions, the relationship between pulse front tilting and the parameter of elements, and to pulse width. For free propagation, pulse width of a Gaussian pulse increases with transmitting distance and the tilting angle decrease with transmitting distance. Via a grating, pulse width is not changed and pulse front tilting is proportional to the angular dispersion of the elements.

ACKNOWLEDGMENT

The authors thank the Education Department of Heilongjiang province in China for the fund under Grant No. F201312.

REFERENCES

Akturk, S., X. Gu, E. Zeek, & R. Trebino (2004). Pulse-front tilt caused by spatial and temporal chirp. *Opt. Express* 12, 4399–4410.

Baum, P., S. Lochbrunner, & E. Riedle (2004). Tunable sub-10-fs ultraviolet pulses generated by achromatic frequency doubling. *Opt. Lett.* 29, 1686–1688.

Bor, Z., B. Racz, G. Szabo, & et al. (1993). Femtosecond pulse front tilt casued by angular dispersion. *Opt. Eng.* 32, 2501–2503.

Fork, R. L., C. H. Brito Cruz, P. C. Becker, & C. V. Shank (1987). Compression of optical pulses to six femtoseconds by using cubic phase compensation. *Optics Letters* 12, 483–485.

Huang, J. J., L. Y. Zhang, W. C. Zhang, & et al. (2013). Theory of second harmonic generation of ultrashort pulses for collinear achromatic phase matching. *J. Opt. Soc. Am. B* 30, 431–438.

Ji, X. L. & B. D. Lv (2001). Effect of the spherically aberrated lens on beam quality. *Chin. J. Lasers* 28, 347–350.

Kane, S. & J. Squier (1995). Grating compensation of third-order material dispersion in the normal dispersion regime: Sub-100-fs chirped-pulse amplification using a fiber stretcher and grating-pair compressor. *IEEE Journal of Quantum Electronics* 31, 2052–2057.

Kostenbauder, A. G. (1990). Ray-pulse matrices: a rational treatment for dispersive optical system. *IEEE Journal of Quantum Electronics* 26, 1148–1156.

Qi, H. Z. & B. D. Lv (2006). Propagation of ultrashort chirped pulsed Gaussian beams in free space. *Chin. J. Lasers* 33, 499–503.

Qi, H. Z. & B. D. Lv (2007). Temporal and spectral properties of a chirped Gaussian pulses in dispersive media. *Acta Photonica Sinica* 36, 1409–1413.

Electronic Engineering and Information Science – Wang (Ed.)
© 2015 Taylor & Francis Group, London, ISBN: 978-1-138-02772-5

The detection of QPSK-DS communication based on delay-multiplication

Z.Q. Dong

School of Electrics & Information Engineering, Xuchang University, Xuchang, Henan, China

ABSTRACT: The article introduces an efficient QPSK-DS communication detection method, which advises to achieve the detection by doing delay-multiply and filtering operation on the received signal directly. Simulation results show that the proposed method is both effective and robust, and it can meet the low-SNR need of non-cooperative communication detection. Especially, the implementation of the proposed method does not need any a priori information, it can do analyzing and processing directly on the radio (or also intermediate) frequency signals.

KEYWORDS: QPSK-DS; Delay-multiplication; Parameter estimation; Frequency-band processing.

1 INTRODUCTION

For possessing outstanding performance, spread spectrum communication has been widely used in secret communication and commercial communication for decades. And correspondingly, the detection of spread spectrum communication has become an important research content for communication monitoring and electronic warfare. C. Bouder et al. (2003) did some intensive research on the non-cooperative baseband detection of BPSK-DS communication, and proposed a detection method which is based on the analysis of the second-order moment of the auto-correlation of the received signal; J.D. Vlok et al. (2012) did some research on the detection of direct sequence spread spectrum (DSSS) communication, and a semi-blind approach was proposed; Sha Z.C. et al. (2013) discussed the baseband detection of the multi-rate DSSS communication according to the principle of the second-order moment of autocorrelation; The paper, which takes the blind detection of QPSK-DS communication as the research topic, proposes a simple algorithm for QPSK-DS communication detection.

2 THE DETECTION PRINCIPLE

2.1 *The expressing of the signal and the information extraction of the chip rate*

Commonly, QPSK-modulated direct sequence spread spectrum signal may be looked as the combination of two orthogonal BPSK-modulated DSSS signals, it may be represented as follows:

$$x(t) = d_1(t)c_1(t)\cos(2\pi f_0 t + \varphi_0) + d_Q(t)c_Q(t)\sin(2\pi f_0 t + \varphi_0) \tag{1}$$

Where $d_1(t) = \sum_{k=-\infty}^{\infty} d_{1,k} p(t - kT_s)$, $d_Q(t) = \sum_{k=-\infty}^{\infty} d_{Q,k} p(t - kT_s)$,

$d_{1,k}, d_{Q,k} \in \{-1, +1\}$, $d_{1,k}$ and $d_{Q,k}$ are the information symbol being borne by the in-phase branch and the quadrature branch respectively.

$p(t) = \begin{cases} 1 & 0 \le t < T_s \\ 0 & \text{for other } t \end{cases}$, it is the waveform of

the information symbol, T_s is the symbol period;

$$c_1(t) = \sum_{i=-\infty}^{\infty} c_{1,i} q(t - iT_c), \qquad c_Q(t) = \sum_{i=-\infty}^{\infty} c_{Q,i} q(t - iT_c),$$

$c_{1,i}$ and $c_{Q,i}$ are the spread spectrum sequence modulated on the in-phase and the quadrature branch respectively; And $c_{1,i} \in \{-1, +1\}$, it is the periodic repetition of the pseudo noise code sequence C_1 ($C_1 = [c_{1,0}, c_{1,1}, \cdots, c_{1,N-1}]$); $c_{Q,i} \in \{-1, +1\}$, it is the periodic repetition of the pseudo noise code sequence C_Q ($C_Q = [c_{Q,0}, c_{Q,1}, \cdots, c_{Q,N-1}]$), N is the length of the pseudo noise code sequence. $q(t) = \begin{cases} 1 & 0 \le t < T_c \\ 0 & \text{for other } t \end{cases}$,

and it is the chip waveform; T_c is the chip period,

$T_s = NT_c$, f_0 is the carrier frequency, φ_0 is the initial phase.

Do delay-multiplication on the signal shown in formula (1), and then low-pass filtering is implemented, there will be

$$
\begin{aligned}
x(t) \cdot x(t - t_d)\big|_{LP} = &\frac{1}{2}[d_1(t)c_1(t)d_1(t - t_d)c_1(t - t_d) \\
&+ d_Q(t)c_Q(t)d_Q(t - t_d)c_Q(t - t_d)]\cos(2\pi f_0 t_d) - \\
&\frac{1}{2}d_1(t)c_1(t)d_Q(t - t_d)c_Q(t - t_d)\sin(2\pi f_0 t_d) \\
&+ \frac{1}{2}d_Q(t)c_Q(t)d_1(t - t_d)c_1(t - t_d)]\sin(2\pi f_0 t_d)
\end{aligned}
\tag{2}
$$

Where t_d is the time-delay.

For the first term of formula (2), $d_1(t)c_1(t) \cdot d_1(t - t_d)c_1(t - t_d)$ is the periodic repetition of the bit cross of C'_I ($C'_I = [d_{I,k} \cdot d_{I,k-1} \cdot c_{I,0} \cdot c_{I,N-1}, c_{I,1} \cdot c_{I,0}, \cdots, c_{I,N-1} \cdot c_{I,N-2}]$) and an all 1 sequence, $d_Q(t)c_Q(t) \cdot d_Q(t - t_d)c_Q(t - t_d)$ is the periodic repetition of the bit cross of C'_Q ($C'_Q = [d_{Q,k} \cdot d_{Q,k-1} \cdot c_{Q,0} \cdot c_{Q,N-1}, c_{Q,1} \cdot c_{Q,0}, \cdots, c_{Q,N-1} \cdot c_{Q,N-2}]$) and an all 1 sequence, and the chip duration of C'_I and C'_Q are t_d, the duration of the 1 in the all 1 sequence is $T_c - t_d$, so the frequency spectrum corresponding to this term is discrete spectrum, and the main contributor of the discrete spectrum is the all 1 sequence (whose repetition rate is $f_{chip} = 1/T_c$); For the second term of formula (2), $d_1(t)c_1(t) \cdot d_Q(t - t_d)c_Q(t - t_d)$ is the periodic repetition of the bit cross of C'_{I-Q} ($C'_{I-Q} = [d_{I,k} \cdot d_{Q,k-1} \cdot c_{I,0} \cdot c_{Q,N-1}, d_{I,k} \cdot d_{Q,k} \cdot c_{I,1} \cdot c_{Q,0}, \cdots, d_{I,k} \cdot d_{Q,k} \cdot c_{I,N-1} \cdot c_{Q,N-2}]$) and C_{I-Q} ($C_{I-Q} = [d_{I,k} \cdot d_{Q,k} \cdot c_{I,0} \cdot c_{Q,0}, d_{I,k} \cdot d_{Q,k} \cdot c_{I,1} \cdot c_{Q,1}, \cdots, d_{I,k} \cdot d_{Q,k} \cdot c_{I,N-1} \cdot c_{Q,N-1}]$), the corresponding frequency spectrum is continuous spectrum; And for the third term of formula (2), $d_Q(t)c_Q(t) \cdot d_1(t - t_d)c_1(t - t_d)$ is the periodic repetition of the bit cross of C''_{Q-I} ($C''_{Q-I} = [d_{I,k-1} \cdot d_{Q,k} \cdot c_{I,N-1} \cdot c_{Q,0}, d_{I,k} \cdot d_{Q,k} \cdot c_{I,0} \cdot c_{Q,1}, \cdots, d_{I,k} \cdot d_{Q,k} \cdot c_{I,N-2} \cdot c_{Q,N-1}]$) and C_{I-Q}, the corresponding frequency spectrum is continuous spectrum too; Thus, it is easy to see that the frequency spectrum of the delay-multiply signal of QPSK-DS communication contains discrete spectrum, which can reflect the value-taking of the chip rate of the QPSK-DS system; In the detection of QPSK-DS communication, it is promising to achieve the detection and the chip rate estimation by checking the discrete spectrum.

2.2 The information extraction of the symbol period

In the above analysis, the periodicity of T_s may also be easily observed, and it is conducted mainly by the periodic repetition of C'_I & C'_Q (Corresponding to the first term of formula (2)); Moreover, the second term of formula (2) is a kind of baseband spread spectrum signal whose spread spectrum sequence is the bit cross of C'_{I-Q} and C_{I-Q}, and it is a signal BPSK-DS alike; The third term of formula (2) is also a signal BPSK-DS alike, which is a baseband spread spectrum signal corresponding to the periodic repetition of the bit cross of C''_{Q-I} and C_{I-Q}. For the T_s-periodicity hidden in BPSK-DS communication and the T_s-periodicity conducted by the periodic repetition of C'_I & C'_Q, it is considerable to achieve the detection of QPSK-DS communication and the T_s estimation by doing correlation analysis on the signal shown in formula (2) (C. Bouder et al. 2003, Sha Z.C. et al. 2013, P. ZHAO 2011).

2.3 The information extraction of the carrier frequency

In the above discussion, all of the analysis is done with the low pass signal shown in formula (2), if we do high pass filtering on the delay-multiplication signal of formula (1), there will be

$$
\begin{aligned}
x(t) \cdot x(t - t_d)\big|_{HP} = &\frac{1}{2}[d_1(t)c_1(t)d_1(t - t_d)c_1(t - t_d) - d_Q(t) \\
&c_Q(t)d_Q(t - t_d)c_Q(t - t_d)]\cos(4\pi f_0 t - 2\pi f_0 t_d + 2\varphi_0) \\
&+ \frac{1}{2}[d_1(t)c_1(t)d_Q(t - t_d)c_Q(t - t_d) + d_Q(t)c_Q(t) \\
&d_1(t - t_d)c_1(t - t_d)]\sin(4\pi f_0 t - 2\pi f_0 t_d + 2\varphi_0)
\end{aligned}
\tag{3}
$$

For formula (3), the first term of the equation right consist of two parts, the one is $d_1(t)c_1(t)$ $d_1(t - t_d)c_1(t - t_d)\cos(4\pi f_0 t - 2\pi f_0 t_d + 2\varphi_0)$, which is the carrier modulation of the periodic repetition of the bit cross of C'_I and an all 1 sequence, and the other, $d_Q(t)c_Q(t)d_Q(t - t_d)c_Q(t - t_d)$ $\cos(4\pi f_0 t - 2\pi f_0 t_d + 2\varphi_0)$, is the carrier modulation of the periodic repetition of the bit cross of C'_Q and an all 1 sequence; For there are subtraction operation between these two parts, the first term, in fact, is the carrier modulation of the periodic repetition of the bit cross of $C_{I'-Q'}$ ($C_{I'-Q'} = (C'_I - C'_Q) \in \{0, \pm 2\}$) and an all 0 sequence; Especially, the chip duration of $C_{I'-Q'}$ is t_d (it corresponds to a kind of BPSK-modulated spread spectrum signal), the duration of the 0 of the all 0 sequence is $T_c - t_d$ in which no signal is present (for the first term). The second term of formula (3) consist of two parts too, the former is the carrier modulation of the periodic repetition of the bit cross of $C'_{I \cdot Q}$ and $C_{I \cdot Q}$, the latter is the carrier modulation of the periodic repetition of the bit cross of $C''_{Q \cdot I}$ and $C_{I \cdot Q}$, the synthesis of these two parts is the carrier modulation of the periodic repetition of the bit cross of C''_{I+Q} ($C''_{I+Q} = (C'_{I \cdot Q} + C''_{Q \cdot I}) \in \{0, \pm 2\}$) and $C^2_{I \cdot Q}$ ($C^2_{I \cdot Q} = (2 \cdot C_{I \cdot Q}) \in \{\pm 2\}$), it is a BPSK-modulated spread spectrum signal, and the chip duration of C''_{I+Q} is t_d, the duration of the chip of $C^2_{I \cdot Q}$ is $T_c - t_d$. Thus it can be seen that the signal shown in formula (3) is the time domain crossing of a QPSK signal and a BPSK signal, and the QPSK signal is present when $iT_c \le t < iT_c + t_d$, $i \in Z$, the BPSK signal is present while $(iT_c + t_d) \le t < (i+1)T_c, i \in Z$; With the statement, the corresponding signal may be expressed as below.

$$x(t) \cdot x(t-t_d)\big|_{HP} = \frac{1}{2}[d_1(t)c_1(t)d_1(t-t_d)c_1(t-t_d) - d_Q(t)$$
$$c_Q(t)d_Q(t-t_d)c_Q(t-t_d)]\cos(4\pi f_0 t - 2\pi f_0 t_d + 2\varphi_0)$$
$$+\frac{1}{2}[d_1(t)c_1(t)d_Q(t-t_d)c_Q(t-t_d) + d_Q(t)c_Q(t)$$
$$d_1(t-t_d)c_1(t-t_d)]\sin(4\pi f_0 t - 2\pi f_0 t_d + 2\varphi_0) \qquad (4\text{-}a)$$
$$iT_c \le t < iT_c + t_d, i \in Z$$

$$x(t) \cdot x(t-t_d)\big|_{HP} = \frac{1}{2}[d_1(t)c_1(t)d_Q(t-t_d)c_Q(t-t_d) + d_Q(t)$$
$$c_Q(t)d_1(t-t_d)c_1(t-t_d)]\sin(4\pi f_0 t - 2\pi f_0 t_d + 2\varphi_0) \qquad (4\text{-}b)$$
$$iT_c + t_d \le t < (i+1)T_c, i \in Z$$

For formula (4), when t_d is much less than T_c, the main signal component is BPSK signal (such as $t_d/T_c < 1/5$). For BPSK signal, the detection may be achieved by prefilter-delay-and-multiply processing or square operation (J.F Kuehls et al. 1990). Because delay-and-multiply processing will conduct parasitic cross interference between the QPSK signal component and the BPSK signal component, and square operation can make full use of the BPSK signal, here do square processing on the signal shown in formula (4). For the useful signal component corresponding to formula (4-b), here will get

$$\left[x(t) \cdot x(t-t_d)\big|_{HP}\right]^2 = \sin^2(4\pi f_0 t - 2\pi f_0 t_d + 2\varphi_0)$$
$$= \frac{1}{2}[1 - \cos(8\pi f_0 t - 4\pi f_0 t_d + 4\varphi_0)], \qquad (5)$$
$$iT_c + t_d \le t < (i+1)T_c, i \in Z$$

For the signal shown in formula (5), discrete spectrum line will present at 4 times of the carrier frequency in the frequency domain. In the detection of QPSK-DS communication, the estimation of the carrier frequency can be achieved based on this characteristic.

3 COMPUTER SIMULATION

To validate the theoretical analysis, computer simulations are carried out by using Matlab. In the simulations, the carrier frequency of the test communication signal is 10MHz, the chip rate is 1Mchip/s, the pseudo noise codes used in the in-phase and the quadrature branch are the m-sequences whose feedback coefficients are 103 and 147 respectively. Fig.1 shows the simulation results which are gotten when the SNR is -5dB.

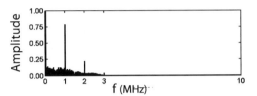

(a) Detection result of the chip rate

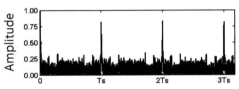

(b) Detection result of the symbol period

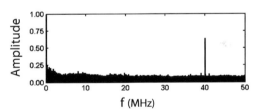

(c) Detection result of the carrier frequency

Figure 1. The processing results of the given QPSK-DS communicati on signal.

With fig.1, it is clear that the detection of QPSK-DS communication can be achieved by doing certain processing based on delay-multiply. Specifically, fig.1(a) exhibits the processing results of the delay-multiply, low-pass filtering and FFT; With fig.1(a), it is easily to see that there does exist discrete spectrum, and the spectrum lines locate at the chip rate and its times by which the detection of QPSK-DS communication and the estimation of the chip rate can be achieved; Fig.1(b) exhibits the results of the delay-multiply, low-pass filtering and coherent processing, by examining it, some correlation peaks can be easily found, and

517

the peaks are gotten when the correlation-time-delay is equivalent to T_s or its times; In non-cooperative detection, the estimation of T_s can be achieved by computing and searching the correlation peaks; Fig.1(c) exhibits the results of the delay-multiply, high-pass filtering and spectral analysis, in which the distinctive discrete spectrum line announces the presence of the communication, by the discrete spectrum line we can achieve the estimation of f_0.

By the way, because the delay-multiply processing is the base of the whole detection, t_d is an important parameter, in order to fully understand the performance of the detection algorithm, some simulations are also done on the given communication signal with diverse t_d. And the results show that the value-taking of t_d has significant influence on the detection performance. For the detection of the chip rate, the detection algorithm gets the best effect when t_d is equal to $T_c/2$; For the detection of the symbol period, the detection algorithm gets the best effect when t_d is equal to T_c; For the detection about the carrier frequency, the detection performance is good when $T_c/8 < t_d < 3T_c/8$.

4 CONCLUSIONS

The QPSK-DS communication detection algorithm proposed in this paper is easily implemented; with it, we can achieve the estimation of multi parameter. Especially, the corresponding detection processing can be done on the frequency band signal directly, which can manage to avoid the problem of the orthogonal carrier extraction and the phase tracking; And the application is significant for the non-cooperative detection of QPSK-DS communication.

REFERENCES

C. Bouder, S. Azou and G. Burel (2003). Blind estimation of a direct sequence spread spectrum transmission parameters in a non-cooperative context. *Traitement du Signal* 20 (4), 1–23.

J.D. Vlok, J.C. Olivier (2012). Non-cooperative detection of weak spread-spectrum signals in additive white Gaussian noise. *IET Communications* 6(16), 2513–2524.

Z. C. SHA, H. B. WU and X. T. REN (2013). Second-order moment of autocorrelation for signal detection in non-cooperative. DS/SS communication. *Systems Engineering and Electronics (In Chinese)* 35(8), 79–83.

P. ZHAO (2011). Autocorrelation Pitch Period Detection of Voice Signal. *Journal of Shanghai University of Electric Power (In Chinese)*27(3), 297–300.

J.F Kuehls, E. Geraniotis (1990). Presence detectionof binary-phase-shift-keyed and direct-sequence spread-spectrum signals using a prefilter-delay-and-multiply device. *IEEE Journal on Selected Areas in Communications* 8(5), 915–9331.

Electronic Engineering and Information Science – Wang (Ed.)
© 2015 Taylor & Francis Group, London, ISBN: 978-1-138-02772-5

Android-oriented mobile sharing system for multimedia projection

L.Y. Shen, S.X. Zhu & Z.C. Song
School of Computer Science and Technology, Harbin University of Science and Technology, Harbin, China

ABSTRACT: Traditional projection with cable connection brings great inconvenience for multiple users to share one projector. In view of the widely application of WiFi technology and Android mobile terminals, this paper designs and implements a mobile projection system for multiple Android mobile users. In this sharing system, mobile terminals capture, compress and send their screens to the projection gateway by running the client software via WiFi connection. Projection gateway acted by an ordinary computer decompresses, displays and projection screens by running the server software. Projection gateway supports multi-point connection and allows multiple mobile users to share one projector. The test results show that this projection sharing system implements sharing a projector for multiple Android mobile users well. It has the advantages of low cost and good usability, and it will have a good application prospect.

KEYWORDS: Multimedia; projection sharing; wireless communication; multithread program.

1 INTRODUCTION

When using the traditional projector, a VGA cable is required to connect the user's computer and the projector. Because of limited connection, a projector can be only used in a small area, and cannot be moved freely. So it is difficult to share projection for multiple users because of the lack of flexibility. Users can switch projection by changing the VGA cable connection, but it will be tedious, time consuming, and easy to interrupt the meeting coherence. If a user is far away the projector, a long VGA cable will be required. If files are copied to the computer connected with projector by using flash disks, some hidden virus can easily infect the target computer. Therefore, the projector with cable connection needs a more effective way to implement projection sharing among multiple users.

Wireless communication technology with the inherent advantages of mobility and flexibility solves the trouble introduced by cable connection and makes human life more freely and unrestrained. Especially, with greater coverage and higher transmission rate, WiFi is used more and more widely. Places such as school, meeting place, hotel, airport and cafe, all provide WiFi access point. At the same time, the Android intelligent mobile terminals with low price become more and more popular. Moreover, WiFi has become one very important part of these terminals. Therefore, with the help of WiFi wireless transmission methods and low-cost Android mobile devices, the use of the projector can be broadened greatly and help to resolve the problem of projection sharing.

So, this paper proposes a low-cost Android-oriented mobile multimedia projection sharing system based on WiFi. It will have great significance for promoting the development of the multimedia projection technology.

2 BACKGROUND

Many projection sharing schemes have been proposed until now. These schemes can be divided into three categories, which are wireless projector, wireless projection gateway, and software aided projection.

With the boom of WiFi technology, electronic equipment manufacturers, such as Epson, Toshiba and BenQ (macho 2004), have launched wireless projectors. Wireless projectors get rid of the wired connection and make projection more convenient to be shared. These kinds of projectors are mainly based on WiFi technology, can realize remote multi-point projection, and implement sharing a projector among multiple users easily. Wireless projectors use special software on personal computers to compress images to be displayed and send them to the wireless projector based on wireless protocol. Through magnification by the microcomputer embedded in wireless projector, the images are projected onto a screen. However, the price of wireless projector equipment is expensive. More importantly, if all traditional projectors with cable connection are replaced by wireless projectors, the cost will be too high because the use of traditional projector is particularly widespread.

Wireless projection gateway is an embedded design of wireless gateway. It is connected with traditional projectors with cable connection, and can send the images to the computers in the same local network with the projector based on wireless protocol. Several references (Wu & Long 2013, Xu & Zhu 2008 and Zhang 2004) have designed and implemented several wireless projection gateways. To use wireless projection gateway, the traditional wired projector needn't be replaced. Compared with wireless projector, the cost is reduced to a certain extent. However, like wireless projector, this scheme also requires the client computer to install software to implement the capturing, compression and wireless transmission of screens. And because of introducing special hardware equipment, the cost is relatively higher. Even the most simple wireless projection gateway, its price is about 1300¥. So in addition to large organizations and companies, ordinary units will not buy wireless projection gateway.

Software-aided projection is a whole software resolution to implement projection sharing based on traditional projectors. Both Qin 2012 and Zhang & La 2014 propose software schemes to implement projection sharing among multiple personal computers running Windows operating system. The computer connected to projector via VGA cable acts projection gateway, which runs the server software. Other computers, the users, send their screens to the projection gateway based wireless protocol. Projection gateway and users are in the same local network via a wireless router. Though this scheme resolves projection sharing among multiple users and needn't any additional hardware, it only works on personal computers, lacks flexibility and mobility. In 2012, Epson launched a new software product for Apple mobile devices to implement projection sharing (Engadget 2012). Later, Epson launched another version for Android mobile devices. But these schemes only work well based on Epson's wireless projectors and can't work with traditional projectors with cable connection.

Aimed at the fact that Android mobile terminals with low price and WiFi are widely used, it is necessary to design an effective wireless multimedia projection sharing system with low-cost and high performance. So this paper designs and implements a new Android-oriented wireless projection sharing system, which is easy to share one projector for multiple android mobile users in low cost.

3 MOBILE PROJECTION SHARING PROTOTYPE

The projection sharing system in this paper is a low-cost software implementation based on client/server (C/S) architecture. Ordinary computer acts projection gateway, which is connected to the projector via VGA cable and runs the server software. Android mobile devices act users, which run the client software. Both the projection gateway and Android mobile users connect the same network through a wireless router. When the system works, mobile users initiate a connection request to the projection gateway. If the connection is successful, the client software running on mobile device captures, compresses the screens of devices. Then it sends screens to the projection gateway by using UDP protocol via WiFi connection. The prototype of the mobile projection sharing system is shown in Figure 1.

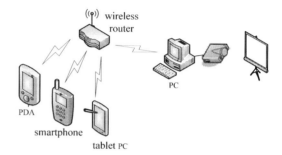

Figure 1. The prototype for mobile projection sharing system.

Our system uses C/S architecture, not only because it can make full use of the resources on both ends of the hardware environment, but also it can help to realize projection sharing more effectively and cheaply. At the same time, in order to improve the efficiency of the system, both the client and the server adopt multithread technology. Because projection gateway running the server software needn't too much emphasis on resources, almost any ordinary PC can act projection gateway. Moreover, the system doesn't have to consider any security issues. Therefore, our system can implement projection sharing cheaply, safely and effectively.

3.1 Detailed design of the mobile projection sharing system

The whole functional diagram of the mobile projection sharing system is shown in Figure 2. The client, which runs on Android mobile terminals, composes of four functions. They are screen scraping, screen compression, screen transmission and link management. The server, which runs on the projection gateway, is also composed of four modules: screen displaying, screen decompression, screen receiving and multiuser shared management.

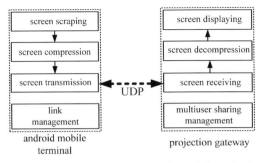

android mobile terminal | projection gateway

Figure 2. The functional diagram for mobile projection sharing system.

3.2 *The client of mobile projection sharing system*

The screens of Android mobile users need to output to the projector when projecting. So the client need scrap the screens of Android mobile devices continuously. Because the size of screen data can affect the real-time performance of the system, the screens also need to be compressed, and then the client transmits them to the server. At the same time, the client introduces a module of link management to help to share the projector among multiple mobile users. The following is the detailed description of the client.

Screen scraping. On Android mobile devices, screen data are always stored in the buffer on frame-buffer device. So screen scraping module can collect screen information through reading data from this buffer. Because screen information always is image format, this module creates a bitmap for one screen. At the same time, in order to display screen normally or magnify the screen on the server side when projection, this module also collects the screen size information for Android mobile devices. In order to improve the real-time performance of projection, screen data scraped are stored in the main memory of mobile devices.

Screen compression. The screen data which are stored in bitmap is uncompressed and they have a large amount of data. For example, if the screen resolution of a mobile terminal is 960*640 and each pixel is 4B, the size of one bitmap will be about 2.35MB. If this bitmap is transmitted directly, the real-time performance of the projection sharing system will be reduced a lot. And if the frequency of screen scraping is high, the wireless network will be more difficult to meet the requirements of such a huge amount of data transmission and the real-time performance is reduced further. Therefore, in order to maintain the continuity of projection, our scheme introduces screen compression module in which two kinds of compression method are adopted in order to fully reduce the amount of information to be transmitted. First, a compression algorithm based color depth of the bitmap which is provided by the Android system

is used to compress screen information. In this way, the size of screen information is reduced by reducing the resolution of the screen. Certainly, our system provides some options about compression ratio for users to choose the appropriate compression ratio according to demand. Then, the zip compressing algorithm which is lossless is used to compress screen further. After two-level compression, the screens to be transmitted are greatly reduced. So the real-time performance of the projection can be ensured. At the same time, compressed screens are stored in a ring sending queue which can guarantee that the screen scraped first is transmitted to the server firstly.

Screen transmission. Because projector displays what the mobile users display in real time, the

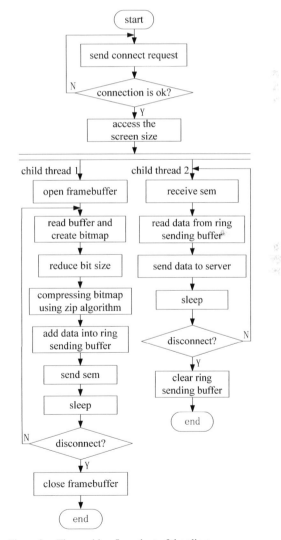

Figure 3. The working flow chart of the client.

projection screen must keep synchronized with the mobile screen. Otherwise, the effect of projection will be reduced greatly. If we transmit screen information using the TCP protocol, the sorting, reorganization, replacement of data may be delayed a lot with the pass of time, which eventually can lead to non synchronous between the projector and users. Therefore, our system uses UDP protocol to transmit screen data, which can avoid check operations on the server. In this way, time delays can be effectively reduced and the real-time of performance can be improved.

Link management. To improve the operability of the mobile users, our system introduces a link management function to aid users to control the connection with the server. Users can set the compression ratio, initiate a connection or disconnection request to the server in this module.

In order to improve the working efficiency of the client, this paper adopts multithread technology. There are three threads in the client, one main thread and two child threads. The main thread is used to connect and disconnect the link with the server. Child thread 1 is used to scrap, compress and store screens. Child thread 2 is used to send screens to the server. Semaphore mechanism is also used to synchronize two child threads. As long as the ring sending queue is not empty, child thread 2 can send screen to projection gateway. In this way, the client can make full use of the resources of the mobile terminals. The working flow chart of the client is shown in Figure 3.

3.3 The server of mobile projection sharing system

The server receives the screens via wireless network continuously, and then decompresses and displays them. So the screens of Android mobile users are projected onto a screen. The following is the detailed description of the server.

Screen receiving. This module is corresponding to the screen transmission module of the client. It is used to receive screen information sent from the clients based on UDP protocol via wireless connection.

Screen decompression. Because the screen data received are in compressed format, the server need decompress them before displaying. This module will only use an unzip algorithm to decompress the screen, and then stores them in a ring receiving queue which can ensure the order of displaying.

Screen displaying. This module is used to display the screens of Android users stored in the ring sending queue. When displaying, the screens of users are ensured to align with the center of the projection gateway screen. At the same time, users with small screen can choose to amplify the projection. The amplification coefficient is equal to the minimum of the ratio of height and width between projection gateway and terminal. For example, the terminal screen

size is 480*320 and projection gateway screen size is 1440*900, then the projection will be amplified 2 times, not 3 times.

Multiuser sharing management. When multiple mobile users share one projector, the multiuser sharing management module will work. To ensure users can connect successfully, the next user connects the projection gateway after the last one disconnects. In order to support multiple users share one project, this module will listen the link requests from the clients continuously. Once listening a connect request, the server will prepare for receiving screen.

In order to improve the efficiency of projection, the server also adopts multithread and semaphore mechanism (counting semaphore). The main thread is responsible for listening for the connections from clients. Child thread 1 implements the receiving and unzipping of screen information from users. Child thread 2 is used to display the screen information. When child thread 1 completes adding into the ring receiving queue, the value of the semaphore is added by 1. When the semaphore is not empty, child thread 2 can read from screen from the ring sending queue to display it. Thus, the screens of Android mobile users are projected onto the screen continuously. The working flow chart of the server is shown in Figure 4.

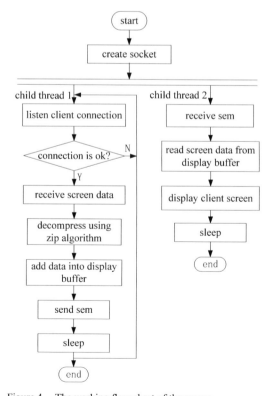

Figure 4.   The working flow chart of the server.

4 SYSTEM TEST

Our test environment is deployed according to Figure 1. An ordinary computer which connects to the projector with VGA cable acts projection gateway. JDK and the server software are installed on it. Multiple Android mobile phones on which the client software is installed act users. Projection gateway and mobile phones are connected to the same network through a wireless router.

Figure 5 shows that the screen of one Android mobile phone is displayed on the projection gateway. And Figure 5b shows that the screen of mobile device is magnified. Here, the frequency of screen scraping is one screen per 0.1s. The results show our implementation of the mobile projection sharing system works well.

(a)

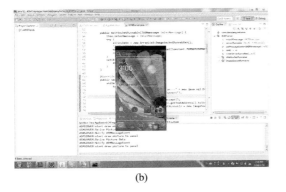

(b)

Figure 5. The display results on projection gateway.

Figure 6 shows the projection results using different compression ratios. Among three pictures in Figure 6, the compression ratio of Figure 6a is 10%, Figure 6b is 50% and Figure 6c is 90%. We can also see from Figure 6 that the projection effect with high compression ratio is acceptable.

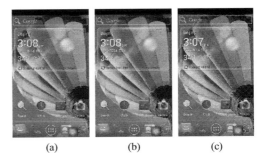

(a)   (b)   (c)

Figure 6. Projection results with different compression ratio.

5 CONCLUSIONS

In this paper, a new Android-oriented mobile sharing projection system is designed and implemented. The screens of mobile users are scraped, compressed and sent to a projection gateway which is acted by an ordinary computer, and then the gateway decompresses and projects them onto the screen. This sharing projection is implemented by software only, did not add any new hardware. System testing results show it can work well. Because of low cost and good flexibility, this Android-oriented mobile projection sharing system will have a good application prospect with the wide spread of Android terminals.

REFERENCES

Macho. 2004. *http://www.yesky.com/483/1824483.shtml.*
Wu, Z.P. & Long .H. 2013. The Research and Design of Wireless Presentation Gateway based on Connected Directly. *Computer Engineering and Design*,34: 2223–2227.
Xu D.Q. & Zhu G.X. 2008. the Design of Wireless Projector Gateway, *Computer Engineering and Design*,29:1163–1165.
Zhang Y. 2004. Wireless projection Gateway: D-Link DPG-2000W wireless projection gateway. *Personal Computer*, 08: 58.
Wang B.L. 2013. the Research and Design of Embedded Wireless Projection Gateway, *Nanjing University of Science and Technology*, Master's Thesis.
Qin K.W. 2012. Using the WiFi Direct Design and Realization of Audio and Video Sharing System Based on the Android Platform, *Software*, 07:71–73.
Zhang B. & La G.Q. 2014. Wireless Projection System Design and Implementation based on Network Control, *Information Security and Technology*, 01:60–64.
Engadget,2012.*http://digi.tech.qq.com/a/20111010/000094. htm.*

Electronic Engineering and Information Science – Wang (Ed.)
© *2015 Taylor & Francis Group, London, ISBN: 978-1-138-02772-5*

A circuit module design and realization of optical fiber communication

J.L. Bai, Y.L. He, X. Liu, J.W. Zhang, Q. Zhang & Y. Wang
Department of Electric Science and Technology, Harbin University of Science and Technology, Harbin, China

ABSTRACT: The basic optical communication system is designed completely and further researched by this paper. Although this system only contains the transmission module, the optical fiber and the receiving module, it can achieve a complete optical fiber communication. The important parts of the circuit structure have made a detailed introduction and instruction. The module is carried out on some performance tests and analysis, and the problems in the process of transmission have been improved.

KEYWORDS: Information network; Optical fiber communication system; Optical fiber.

1   INTRODUCTION

As the rapid development of the society, the requirement of a communication system is higher than before. There is huge potential of optical fiber communication thanks to its features as little amount of wastage, big bandwidth, small size, light weight as well as its convenience in transportation and installation. During the several decades of research and development, optical cables, devices and systems have been updating the species and improving the functions, the fact which makes the optical communication the most important in information network (Jiang & Huang 2004). Now, the optical network with a huge capacity of transmission has been used all around the world, no matter above or beneath the land, the fact which provides a wide and reliable stage for rapidly increased information to transmit (Huang 2000).

Optical fiber communication really started from 1970, although there are only nearly 30 years from it started, the optical fiber communication technology, both optical fiber manufacturing technology and manufacturing technology of photoelectric devices, as well as the quality of optical fiber communication system have attained remarkable achievements, it has become the main means of transport of the modern communication (Ashihara et al. 2003). The attenuation of the fiber from the beginning of the 20 dB/km, now it's low of 0.14 dB/km, it is very near to the theory of quartz optical fiber attenuation limit of 0.1 dB/km, the bandwidth of the fiber also developed, from the beginning of the 10 MHz·km to the above 1000 GHz·km present. Optical fiber communication is a big development space. The current optical fiber communication

application level according to the analysis is only around 1 to 2% of their ability. Optical fiber communication technology is not stagnant, but to a higher level (Dai et al. 2013).

2   SYSTEM DESIGN AND DEBUGGING

There are two kinds of the optical delivery system of the optical fiber communication, respectively: analog signal and digital signal. Digital optical transmission system is composed of digital signal source, a digital interface circuit, a digital driving circuit and optical transmission module. The digital signal source for waveform generator can produce triangle wave, sine wave and square wave signals. The square wave is chosen to transmit digital signal. This system uses a 64 KHz square wave signal, the frequency in this system is suitable frequency range, on this frequency work less distortion of the signals produced by this system.

Simulation of the optical delivery system is composed of analog signal source, analog interface circuit, analog driving circuit and optical transmission module. The analog signal waveform generator has to produce the sine wave signal. Analog interface circuit is the 1-level of emitter follower, used to transform the input of high impedance to the low impedance as output, to adapt to the request of the simulated driving circuit. The simulation driving circuit is a current amplifier, is used to control the voltage signal into a current control signal to drive the semiconductor light-emitting diodes. Optical delivery module converts the injection current intensity by change intensity of light, and it can realize the intensity of the optical carrier modulation.

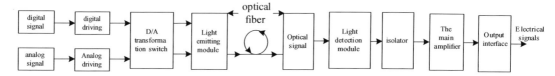

Figure 1. Structure frame of optical fiber communication.

The light receiving module mainly completes the conversion between optical signal and electrical signal, small signal detection, transmit and recovery information in digital or analog signal, and other functions. Optical receiver front-end consists of optical detector and preamplifier, holds the coupling into the optical detector light signal is converted to light current, prevent complete, current signal translate into a voltage signal at the same time, convenient for further processing. Figure 1 is the structural frame of optical fiber communication. Optical delivery and receiving system circuit is shown in Figure 2.

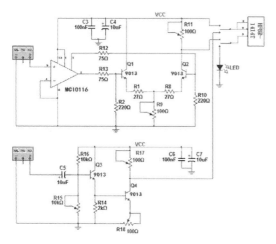

(a) Optical delivery system circuit principle diagram

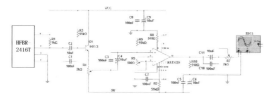

(b) Optical receiving system circuit principle diagram

Figure 2. Optical delivery and receiving system circuit principle diagram.

Because of this system includes optical transmission, the simulation software cannot be used for simulation. Figure 3 is the result of 1 KHz sine wave signal and 1 KHz square wave single through the virtual breadboard system circuit.

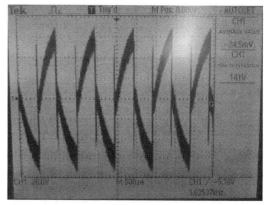

(a) 1 KHZ square wave transmission waveform figure

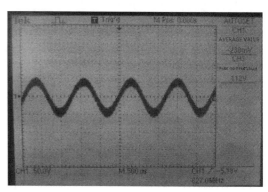

(b) 1 KHZ sine wave transmission waveform figure

Figure 3. The result of two kinds of wave propagation.

From the figure of the waveform, square wave has some distortion and spike; increase the frequency of the input signal appropriately, when the signal frequency reaches about 60 KHz, the distortion of a square wave signal is very small. Finally, transmission of digital signal is 64 KHz square wave, Analog signal as the 1 KHz positive wave. As shown in Figure 4 is the waveform which changed the transmission frequency.

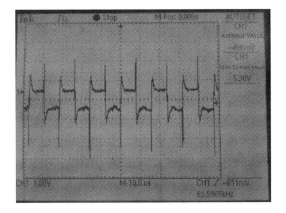

Figure 4.   The distortion of 64 KHz square wave.

There are still certain spikes of the signal in the system after transferred, and then we should add some filter circuit for the system. So we should increase a capacitor and a resistor in the circuit of the system. By the verification, we got a signal graph after filtering, as shown in Figure 5.

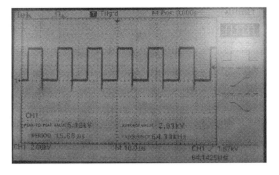

Figure 5.   After filtering of the square wave.

3   SUMMARY

The basic optical communication system is designed completely, and the optical fiber communication system mainly includes the light receiving module and optical delivery module, and the important parts of the circuit structure have made a detailed introduction and instruction. The characteristics of the system are simple in structure, easy to operate, precision need to be further strengthened. Through to the system transmission performance test and the analysis the test results, the basic transmission waveform figure is consistent with the theory.

ACKNOWLEDGEMENT

The work supported by the National Natural Science Foundation of China (Grant No. 61201075), and the Science and Technology Research Foundation of Heilongjiang Education Bureau of China (12521110), and by the China Postdoctoral Science Foundation (2012M511507), and by Science Funds for the Young Innovative Talents of HUST (No.201302), and by Frontier Research Foundation of State Key Laboratory Breeding Base of Dielectrics Engineering (DE2012B05).

REFERENCES

Jiang Y.& S. H. Huang (2004). *Journal of Yangtze University, Natural Science Edition*, Vol.1, 17.
Huang C. F. (2000). Telecommunication Engineering, 1.
Ashihara S., T. Shimura & K. Kuroda (2003). *J. Opt. Soc. Am. B, Vol. 20.* 853.
Dai J., J. H. Zeng & D. Gachet (2013). *Opt. Express, Vol. 21.* 30453.

Electronic Engineering and Information Science – Wang (Ed.)
© 2015 Taylor & Francis Group, London, ISBN: 978-1-138-02772-5

An improved acquisition method for sine-BOC(14,2) modulated signal

H. Zhang, J.L. Liu, H.B. Wu & Y.H. Hu
National Time Service Center, Chinese Academy of Sciences, Xi'an, China
University of Chinese Academy of Sciences, Beijing, China

ABSTRACT: The paper proposed an improved non-standard local replica subcarrier to make the Cross-Correlation Function (CCF) of received signal and local replica code has the single-peak, aiming at the ambiguity in acquisition and tracking of the high-order Binary Offset Carrier (BOC) modulated signals, which is caused by the multi-peaks of the auto-correlation function. And the amplitude of restructured CCF increases 3dB based on a geometric folding algorithm, meanwhile the performance of acquisition detection probability improved. The results of MATLAB simulation show the acquisition method proposed in the paper can cancel the ambiguity in actuation of sine-BOC (14, 2) signal.

KEYWORDS: Acquisition, BOC (14, 2), Non-standard local subcarrier, Geometric folding.

1   INTRODUCTION

With the development of the Global Navigation Satellite System (GNSS), the BOC modulation has been adopted in the new GNSSs (Betz 1999) Compared with the traditional PSK modulated signals, the BOC-Modulated signals can offer the opportunity for the spectrum sharing (Betz 2002), reducing the multipath interference, achieving higher precise measurement capability. However, the main disadvantage of BOC-modulated signals is that its ACF has a profile with more than one peak, which is a great drawback for receiver synchronization.

Nowadays, the method for cancelling the side peak mainly High Rate Correlation (HRC) (McGraw 1999), Sub-carrier Cancellation, Pseudo Correlation Function (PCF) (Zheng et al. 2010) and Auto-correlation Side Peak Cancellation Technique (ASPeCT) (Olivier et al. 2007). The HRC technique results in an increased robustness against multipath. But the implementation of HRC required wider bandwidth of font-end filter and higher sampling rate, which increasing the receiver's power consumption. The PCF technique that leads to the power of correlator output loss, and causes the omission for acquisition. Representatively, the ASPeCT only works well for sine-BOC (n, n) signal.

Aiming at the high-order modulated signals such as BOC (14, 2) and BOC (15, 2.5), the paper uses an improved non-standard local sub-carrier waves to generate two groups of local replica code, correlated with received signals, and obtained unambiguous ACF based on mathematical operation and Geometric Folding Algorithm. This method can cancel the side peaks of high-order BOC modulated signals effectively.

2   BOC(14,2)-MODULATED SIGNAL

The BOC-modulated signals that use a square-wave as a subcarrier based on BPSK modulated signals. Generally speaking, we defined it as (Anthony 2003) *BOC (m, n)* or *BOC (f_{sc}, f_c)*, where the parameter m denotes the ratio between the subcarrier frequency f_{sc} and the basic reference frequency f_0, and n stands for the ratio between the spread code rate and f_0, here $f_0 = 1.023$MHz. Besides, the ratio N=2m/n denotes the order of the BOC-modulated signals. As the N becomes large, the number of the peaks in the BOC ACF, which is equal to 2N-1, increased larger. The Figure 1 shows the normalized sine-BOC (14, 2) ACF along with the BPSK (1) ACF painted in star line. The numbers of peaks in sine-BOC (14, 2) are equal to 27, which are larger than BPSK (1). And the amplitude of the main peak and the two largest side peaks are almost equivalent, the ratio is 0.92, which will result in the false acquisition.

The received BOC signal can be modeled as

$$S_{BOC\_IF}(t) = d(t) \times C(t) \times S_{SC}(t) \times \sin(2\pi f_{IF} \cdot t) \quad (1)$$

Where *d(t)* represents the navigation data, *C(t)* is the PRN spreading code, f_{IF} is the carrier frequency (in Hz), S_{SC} represents the square-wave subcarrier which referred to sine and cosine respectively, in this article we use the sine square-wave.

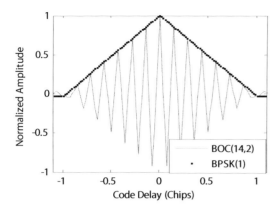

Figure 1. The ACF of BOC(14,2).

3 ACQUISITION METHOD BASED ON NON-STANDARD LOCAL SUBCARRIER

3.1 Reference replica code generation

The general theory assumed that a standard receiver generating a local replica that has the same modulation as the incoming signal. In this article, the local replica code generated by the receiver is to use the non-standard subcarrier that is different from that of the modulation of the received signal, aiming at changing the shape of the correlation function. The reference subcarrier modeled as (2):

$$
f(t) = \begin{cases} N/2, 0 \leq t \leq 2T_S \\ N/2, T_C - 2T_S \leq t \leq T_C \\ 0, others \end{cases} \quad (2)
$$

Where N represents the order of BOC-modulation; T_S represents the period of the half chips of subcarrier; T_C is the period of the PRN code chips.

The local replica non-standard code can be modeled as:

$$
S_{ref1} = \sum_{i=-\infty}^{\infty} C(t - 2T_S) \cdot f(t - i * T_C) \quad (2)
$$

$$
S_{ref1} = \sum_{i=-\infty}^{\infty} (-1) \cdot C(t + 2T_S) \cdot f(t - i * T_C) \quad (3)
$$

The local replica non-standard code could be generated in Figure 2. Essentially, the Local Replica Code1 is generated by the correlation of the early local replica PRN code with the reference replica subcarrier. And the Local Replica Code2 is generated by the correlation of the late local replica PRN code with the reference replica subcarrier.

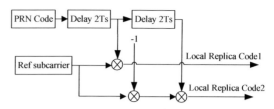

Figure 2. Block diagram of local replica non-standard code.

3.2 Restructure correlation function

The CCF of the received signal and the local replica non-standard code is defined as:

$$
R_{s,S_{ref}}(\tau) = \frac{1}{T} \int_0^T S_{BOC-IF}(t) \cdot S_{ref1}(t - \tau) dt \quad (4)
$$

Here, τ is the PRN code propagation delay in chips; T is the coherent integration time used in the correlation process.

During the period of one code chip, the CCF could be express as:

$$
R_{S\_S_{ref1}} = \begin{cases} 1 - \dfrac{|\tau - T_S|}{T_S}, |\tau - T_S| \leq T_S \\[2mm] 1 - \dfrac{|\tau + T_S|}{T_S}, |\tau + T_S| \leq T_S \\[2mm] -(1 - \dfrac{|\tau - T_C - T_S|}{T_S}), |\tau - T_C - T_S| \leq T_S \\[2mm] -(1 - \dfrac{|\tau - T_C + T_S|}{T_S}), |\tau - T_C + T_S| \leq T_S \end{cases} \quad (4)
$$

$$
R_{S\_S_{ref2}} = \begin{cases} 1 - \dfrac{|\tau - T_S|}{T_S}, |\tau - T_S| \leq T_S \\[2mm] 1 - \dfrac{|\tau + T_S|}{T_S}, |\tau + T_S| \leq T_S \\[2mm] -(1 - \dfrac{|\tau + T_C - T_S|}{T_S}), |\tau + T_C - T_S| \leq T_S \\[2mm] -(1 - \dfrac{|\tau + T_C + T_S|}{T_S}), |\tau + T_C + T_S| \leq T_S \end{cases} \quad (5)
$$

As shown in Figure 3, the $R_{S-Sref1}$ and $R_{S-Sref2}$ had the same shape in the origin, and had the reverse waves around $\tau = \pm 1$ chips. Exploit the Eq.6 to cancel the waves that around the $\tau = \pm 1$ chips, and keep the

530

waves in the origin, the results are plotted with the blue line in the Figure 4 (b).

$$R_{Combined\_corr} = \left| R_{S\_S_{ref1}} \right| + \left| R_{S\_S_{ref2}} \right| - \left| R_{S\_S_{ref1}} - R_{S\_S_{ref2}} \right| \quad (6)$$

The geometric folding algorithm is illustrated in Figure 4a, which is well documented in (Yang et al. 2009). The waves painted with a red line in Figure 4b which is using the geometric folding algorithm based on the waves painted with the blue line. As it can be easily seen, only one peak in the restructured CCF, which cancel the ambiguity of acquisition and tracking.

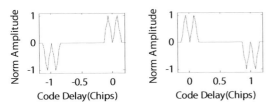

Figure 3. The CCF of local replica code and received signal.

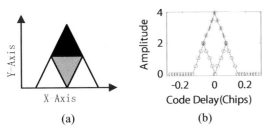

(a)　　　　(b)

Figure 4.　a) Geometric folding b) The restructured CCF.

3.3 *Acquisition method*

Consequently, the block of acquisition method is depicted in Figure 5, which based on the previous correlation. The function of the Local Ref Code module is to generate the two groups of local replica non-standard code. The Restructured module is to restructure the correlation function based on the geometric folding algorithm.

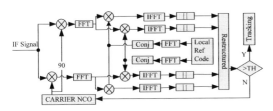

Figure 5. The block diagram of acquisition based on improved algorithm.

4 SIMULATION RESULTS

To validate and test the performance of the acquisition method, we generated the sine-BOC (14, 2) signals using MATLAB tools. The simulation parameters selected as follows: the IF is 30MHz, sampling frequency f_s is 114MHz to meet the Nyquist sampling theorem, the initial PRN code phase is 446.5 chips, the input SNR is -18dB.

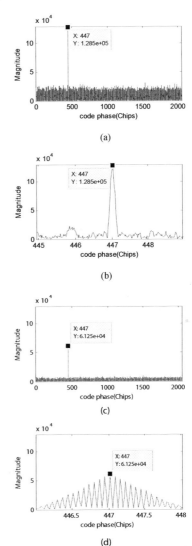

Figure 6.　a) Acquisition result based on proposed method b) Zoomed-in of figure 6. c) Acquisition result based on traditional method d) Zoomed-in of figure 6c.

The simulation results of the proposed acquisition method in article and traditional acquisition method were displayed in Figure 6a and Figure 6c. Besides, Figure 6b is the zoomed-in figure of Figure 6a and Figure 6d is the zoomed-in of Figure 6c. Compared Figure 6a with Figure 6c, the amplitude of the proposed method is 3dB larger than traditional method.

As shown in Figure 6b, 6d, the acquired PN code phase of the proposed method is obvious, whereas the acquired code phase of the traditional method is ambiguous.

Based on the Monte-Carlo simulation and the M=200, figure 7 shows the relationship of the detect probability and the input signal SNR on the condition of the false alarm probability Pf=1e-3. It is concluded that detect probability almost approaching one when the input SNR greater than -15dB.

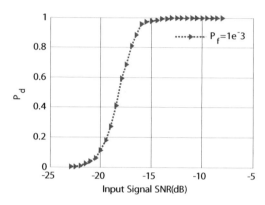

Figure 7. Relationship of Pd and Input SNR, Pf=1e-3.

5 CONCLUSIONS

Aiming at change the shape of the correlation function of the BOC-modulated signals, an improved acquisition method has been proposed in this article.

The simulation results illustrated the method work well. What's more, the algorithm proposed are also valid for the cosine-BOC(14,2), sine-BOC(15, 2.5) and cosine-BOC (15, 2.5) and other high-order modulated signals, only need to modify the reference subcarrier waveform slightly. But it has a problem that the method proposed increasing the power of the noise. Future possibilities to strengthen this technique trying to reduce the power of the noise, and design the tracking techniques based on the non-standard local replica code.

REFERENCES

Betz J. W. (1999). The Offset Carrier Modulation for GPS Modernization. *Proceedings of the ION 1999 National Technical Meeting*, San Diego, California, USA, January.

Betz J. W. (2002). Binary Offset Carrier Modulations for Radio navigation. *Journal of the Institute of Navigation*, Vol. 48, p. 227.

McGraw G.A. & M.S.Braasch (1999). GNSS Multipath Mitigation Using Gated and High Resolution Correlator Concepts, *Proceedings of the 1999 National Technical Meeting of The Institute of Navigation, San Diego*, CA, January 1999, pp. 333–342.

Zheng Y., X. W. Cui & M. Q. Lu: Pseudo-correlation-function-based unambiguous tracking technique for sine-BOC signals. *Aerospace and Electronic System, 2010, 46(4)* pp. 1782–1796.

Olivier J., C. Macabiau, E. Cannon & G. Lachapelle. (2007). Unambiguous Sine-BOC(n,n) Acquisition /Tracking Technique for Navigation Applications, *IEEE Trans. vol. 43–1*, 150–162.

Anthony R. P. (2003). BOC Modulation Waveforms. *ION GPS/GNSS, 2003*: 1044–1057, Portland.

Yang L., C. S. Pan, Y. M. Bo & Y. X. Feng (2009). A New Method for Acquisition of BOC Modulation Signal Based on Refactoring Theory to Cross-Correlation. *Journal of Astronautics, Vol. 30.*

Electronic Engineering and Information Science – Wang (Ed.)
© *2015 Taylor & Francis Group, London, ISBN: 978-1-138-02772-5*

The SAR sidelobe suppression method based on the parallel transceiver using orthogonal chaotic frequency modulation signal

L. Zhang, Z. Yu & J.W. Li
School of Electronics and Information Engineering, Beihang University, China

ABSTRACT: Since the sidelobe of the range compressed LFM signal is engendered high in SAR system, this paper proposes a novel sidelobe suppression method within SAR imaging procedure based on transforming SAR work mechanism. Chaotic sequence is introduced to the transmitter to generate N chaotic modulation signals orthogonal to each other. In the receiver, echo signal is respective imaging after orthogonal separation; the final result is the average of the imaging results. Compared with LFM signal, the signal sidelobe of the final result is effectively degraded because chaotic modulation signals have orthogonality -like performance and pin-like ambiguity function. At the same, radar use this method can owns good performance on anti-interference and anti-interception due to the noise-like spectrum.

KEYWORDS: SAR (Synthetic Aperture Radar); chaotic FM (Frequency Modulation); LFM (Linear Frequency Modulation); Bernoulli map.

1   INTRODUCTION

Synthetic Aperture Radar (SAR), which operates in all day and all weather, is a kind of positive microwave imaging radar and it can obtain high resolution radar image similar to optical image even with very low visibility. Linear Frequency Modulation (LFM) signal is transmitted and the radar image of the illuminated area can be obtained using the matched filter technique in range and azimuth directions of the echo signal. The impulse response function of SAR system is a sinc function for both range and azimuth directions and the sidelobe is relatively high due to the limited and effective area in 2 dimensional frequency domain (Cumming,Wong 2005). The sidelobe of strong signal could submerge or interrupt the mainlobe of weak signal around, which can lose the targets. Therefore, sidelobe suppression is important to improving SAR images. The typical method of sidelobe suppression is to add a window (Nuttall,1981) in the frequency domain, which can broaden the mainlobe and reduce the resolution. Apodization filtering method (Zhang, Su.2008) and neural network method (Kwan, Lee. 1993) can greatly reduce sidelobe without changing the system resolution. But the image quality of the former method is not stable and the computational load of the latter method is huge due to its iteration operation. This paper proposes a novel SAR method in which orthogonal chaotic frequency modulation signal is transmitted and received. This new system suppresses sidelobe effectively and owns a good performance on the anti-interference and the anti-interception.

2   SIGNAL ANALYSIS

2.1   Definition of chaotic FM signal

A general expression of baseband chaotic FM radar signal is expressed as follows (Flores, Solis. 2002)

$$S(t) = A_0 e^{j2\pi K\phi(t)} \qquad (1)$$

where A_0 denotes the amplitude of the signal, K represents its modulation index.

$$\phi(t) = \int_0^t \psi(\tau)d\tau \qquad (2)$$

expresses the phase function, $\psi(\tau)$ is the chaotic modulating instantaneous frequency. The energy spread over a band $K\psi_{min} \leq f \leq K\psi_{max}$, Where ψ_{min} and ψ_{max} are the minimum and maximum values of $\psi(\tau)$.

The discrete form of chaotic FM signal is expressed as

$$S(n\Delta t) = A\exp[j2\pi K\phi(n\Delta t)] = A\exp[j2\pi K\sum_{i=0}^{n}\psi(i)\Delta t] \quad (3)$$

where Δt is the sampling time, $\Delta t = 1/f_s$, and f_s is the sampling frequency. When $\psi(i) \in [-0.5, 0.5]$, i=0,1,2,..., choose

$$f_s = 2K\varphi_{max} = K \qquad (4)$$

satisfied the Nyquist criterion, and according to $S(n) \in S(n/K)$, chaotic FM radar signal is shown as

$$S(n) = A\exp[j2\pi K \sum_{i=0}^{n} \psi(i)] \qquad (5)$$

2.2 Analysis of the Bernoulli chaotic FM signal

It is proved chaotic FM signal $S(n)$ is a stationary stochastic process. The performance of $S(n)$ can be analyzed by its statistical characteristic such as spectrum, ambiguity function and correlation. Several kinds of chaotic series were generated using a typical chaotic map. We analyze the signals modulated by these kinds of series and finally select n-way Bernoulli map as the expression of modulating function $\psi(i)$ according to the experimental and comparison. It is written as

$$y_{n+1} = f(y_n) = r \cdot y_n \pmod 1 \qquad (6)$$

where r is integer and satisfies $r \geq 2$. It is demonstrated by experiments that the orthogonal properties are ideal when $r = 127$. In subsequent simulation process, we use $r = 127$. According to the transformation formula $(y_n{-}0.5) \to \psi_n$, we can gain $\psi_n \in [0.5, 0.5]$.

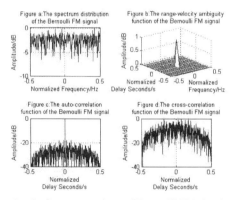

Figure 1.   Performance analyses of Bernoulli FM signal.

Figure 1 is the performance analyses of the Bernoulli FM signal. As shown in the Figure a, the spectrum distribution of the Bernoulli FM signal is relatively flat over the entire frequency band, this noise-like property has a good advantage to resist interference and interception in the transmitting procedure. We can see in the Figure b, The sidelobes of range-velocity ambiguity function are both low, it means this signal can easily obtain remarkable resolution and is

suitable as a radar signal. According to Figue c, the auto-correlation function of this radar signal with signifies narrow main lobe and low sidelobe in is similar to the impulse function, and it signifies this signal is an ideal radar signal. Figure d is the cross-correlation function of the signal.Good orthogonality is helpful to realize waveform diversity to overcommodulating interception and obtain stable imaging quality.

3   IMAGE PROCESS AND SIMULATION

3.1   Image process

Consider the schematic in Figure 2, where the basic architecture of the SAR parallel transceiver system using Bernoulli FM signal is shown. There are three steps: Bernoulli sequence is introduced to the transmitter to generate N chaotic FM signals orthogonal to each other; the echo signal is firstly orthogonally separated and then compressed to acquire the imaging result, respectively; the final imaging result can be obtained from an average of those results acquired in the last step.

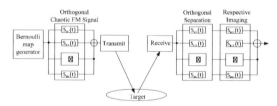

Figure 2.   Basic architecture of the SAR parallel transceiver system using Bernoulli FM signal.

3.2   Simulation

The simulation is carried out to verify the correctness of the proposed method. An FM signal is chosen to compare with chaotic FM signal. The system parameters are given in the following. The satellite altitude is 500km, the relative velocity of the satellite and ground is 7600m/s, the radar is operated at C band, the wavelength is 0.03m, the pulse repetition frequency is 3000, the Bandwidth is 90MHz, the pulse width is 1μs, the antenna azimuth dimension is 5.4m, the antenna range dimension is 1m, the antenna visual angle is 35deg, the synthetic aperture time is 1.44s.

Based on the Range-Doppler(R-D) algorithm in side-looking mode of spaceborne SAR system, we utilize the result of radar echo signal after range compression to verify the sidelobe suppression ability.

Figure 3 is the results comparison. As shown in the Figure a and figure b, Peak to Sidelobe ratio (PSR) of LFM signal is 13.4dB, PSR of single chaotic FM signal is about 45dB, which is significantly better than the former. However, one can note that the PSR of single chaotic FM signal is random. Through the mean of range compress results of multiple orthogonal signals, the final result is stable and the sidelobe is about 55dB, lower than that of single signal (as seen in Figure c).

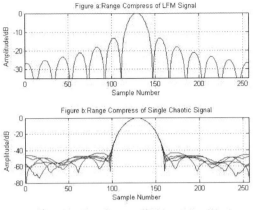

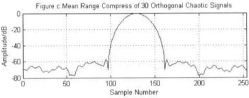

Figure 3.   Range Compress of signals.

4   CONCLUSION

The method of sidelobe suppression in SAR imaging has been discussed. N-way Bernoulli map chaotic sequence is chosen to generate an FM signal. In contrast to traditional LFM signal, our method effectively decreases sidelobe and guarantees the stability of radar system. The simulation results demonstrate the effectivity of this method. Furthermore, according to the needs of the system and the hardware equipment to obtain the optimal number of orthogonal chaotic FM signals is a further discussion.

REFERENCES

Cumming, I. G., F H.Wong (2005);Digital Processing of Synthetic Aperture Radar Data: Algorithms and Implementation. *Norwood: Artech House, Inc.*, 12–17.
Nuttall, A. H. (1981) Some windows with very good sidelobe behavior. *IEEE T A coust, Speech, Signal Proc, 29*, 84–91.
Zhang, X. Y . , W. M. Su. (2008) Application of Apodization Filtering Sidelobe Suppression Rechnique to Synthetic Aperture Radar. *Joournal of Electronics and Information Technology.*.30(4), 902–905.
Kwan, H. K., C. K. Lee. (1993);A neural network approach to pulse radar detection. *IEEE Transactions on Aerospace and Electronic Systems. 02*, 9–21.
Flores, B. C., E. A. Solis. (2002) Chaotic signals for wide band radar. *The International Society for Optical Engineering.* 4727, 100–106.

Electronic Engineering and Information Science – Wang (Ed.)
© *2015 Taylor & Francis Group, London, ISBN: 978-1-138-02772-5*

Wavelet transform application on transient component analysis for AP1000 nuclear turbo-generator

B.J. Ge, W. Guo & Y.P. Gao

College of Electrical and Electronic Engineering, Harbin University of Science and Technology, Harbin, China

ABSTRACT: This paper is the research of key technology in transient process for the world's first AP1000 third generation 1250 MVA nuclear half-speed (4-pole) turbo-generator using wavelet transform. Due to the limitations of Fourier transform, which can not carry out frequency domain analysis and time domain analysis at the same time, a method for transient component analysis based on wavelet transform is proposed in this paper. This work extracts the fundamental wave of the transient current, and gets rid of the high harmonic components. Therefore, the transient parameter values during the fault time are obtained according to the transformed current waveforms. And then, the transient component calculation results of AP1000 turbo-generator, in three-phase fault and internal phase fault conditions, are also got. The conclusions of this paper can provide theory reference for the related research of large-capacity generators.

KEYWORDS: Wavelet transform; transient process; AP1000 nuclear turbo-generator; three-phase fault; internal phase fault.

1 INTRODUCTION

Wavelet analysis is an analytical method which can represent the local features of signal in time domain as well as in frequency domain. Therefore, wavelet transform is also known as the microscope of signal analysis (Wang 2007, Adewole 2012). Also because of this characteristic, wavelet transform overcomes Fourier transform's shortcoming, but it can not completely replace Fourier transform (Sedlazeck 2009, Hao 2010). Hence, Meshing the wavelet transform and Fourier transform will get good results.

Nowadays, the third generation nuclear power technology AP1000, which is leading the trend of nuclear power development all over the world, is already on the road to realization in China. The world's first AP1000 turbo-generator came off the assembly line by Harbin Power Plant Equipment Corporation in October 2012. However, there are lots of research to do for AP1000 generator (Ge et al. 2013). Hence, the wavelet transform application in transient component analysis for AP1000 Nuclear Turbo-generator is deeply discussed in this study.

2 APPLICATION OF WAVELET TRANSFORM IN TRANSIENT CURRENT ANALYSIS

This paper adopts AP1000 turbo-generator as the model, to carry out analysis and calculation of transient components based on wavelet transform. Considering the actual situation of the generator system, this section makes the assumptions that, the initial state is under no-load operation and three-phase fault occurs suddenly. Hence, in this condition, the main parameters and the interrelation of them have been analyzed, as shown in Fig. 1.

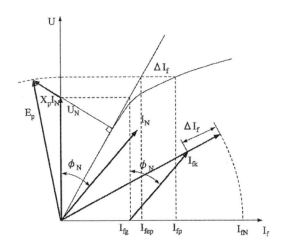

Figure 1. Main parameters under no-load operation.

In quest of transient negative-sequence component's variety law, the changes of magnetic field in the generator system during the three-phase fault have been investigated in this section through the energy analysis. By the calculation of AP1000 turbo-generator, air-gap magnetic flux density can be presented in Fig.2. As well, the magnetic flux density distribution can be demonstrated in Fig.3.

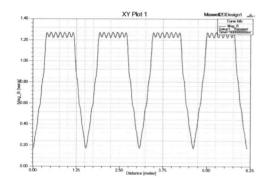

a. in no-load steady-state condition

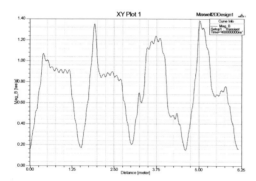

b. in three-phase fault

Figure 2. Air-gap magnetic flux density.

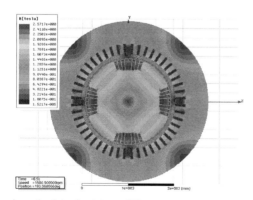

a. in no-load steady-state condition

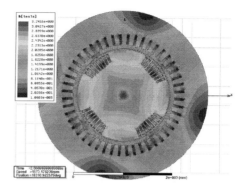

b. in three-phase fault

Figure 3.   Magnetic flux density distribution.

From the results above, the pre-and post-fault changes of electromagnetic parameters are obvious by comparison. According to the magnetic flux density features before and after the fault moment, rotor heating is produced. Because the three-phase short circuit is symmetrical fault, the energy problems found in generator are among the worst it had been caused. Therefore, the no-load three-phase short-circuit current of phase A is taken as an example to illustrate the transient current waveform treatment process based on discrete wavelet transform.

Hence, during the fault time, the three-phase current of phase A is expressed as:

$$I_a = -[(I_m{}''-I_m{}')\,e^{-t/Td''} + (I_m{}'-I_m)\,e^{-t/Td'} + I_m]$$
$$cos(\omega t + \alpha_0) + I_m{}''cos\alpha_0 e^{-t/Td} \qquad (1)$$

When the epoch angle $\alpha_0 = 90°$, the current waveforms are shown in Fig.4.

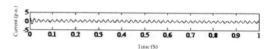

a. Transient component

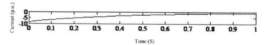

b. Non-periodic component

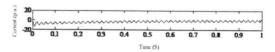

c. Current of phase A

Figure 4.   Current waveform of phase A.

The transient current waveforms of the reconstruction and wavelet details by wavelet transform are shown in Fig.5.

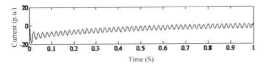

a. Reconstructed waveform of phase A

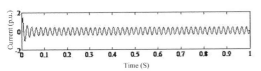

b. Wavelet detail

Figure 5. The transformed waveform of phase A.

From Fig.5 above, according to the reconstructed waveform of phase A and wavelet detail, it can be seen clearly that, the transformed current of phase A contains non-periodic component of primary current only, and wavelet detail has the periodic component of short-circuit current and harmonic components. Therefore, the transformed current of phase A contains steady current and non periodic component and the required parameters can be calculated at this time.

3 TRANSIENT COMPONENT CALCULATION

The armature winding internal fault is one of the common short circuits for generator system unites. Hence, this paper deals with 6 types of branch A2 and branch B1 short circuits in various fault positions, that are A21-B15, A21-B18, A23-B15, A23-B18, A24-B15 and A24-B18 short circuit. The transient process data results are got in Table 1.

Table 1. Parameters in different fault location.

Fault location	Phase	T_a	I''_m	$\overline{T_a}$
A21B15	A	1.01	1.146	
	B	1.01	0.43	1.00
	C	0.99	1.18	
A21B18	A	1.00	1.16	
	B	0.86	0.10	0.98
	C	1.09	1.12	
A23B15	A	1.02	0.99	
	B	1.04	0.59	1.02
	C	1.02	1.25	
A23B18	A	1.05	1.07	
	B	1.03	0.38	1.02
	C	0.99	1.22	
A24B15	A	1.03	0.88	
	B	1.02	0.69	1.00
	C	0.96	1.28	
A24B18	A	1.06	1.00	
	B	1.05	0.52	1.04
	C	1.01	1.27	

4 CONCLUSIONS

This paper presents the short-circuit transient current analysis method based on wavelet transform. The current which is obtained by finite element simulation can be calculated according to the theoretical formula after using wavelet transform. Then the transient parameters of internal fault are obtained according to the transformed current waveform, which are vital for deep research of large-capacity generators.

ACKNOWLEDGMENTS

This work was supported by the National Major Scientific and Technological Special Project of China (No. 2009ZX06004-013-04-01, Sub-project No. 2012BAF03B01-X).
    The authors appreciate the financial support from the National Major Scientific and Technological Special Project of the Chinese Government. And Harbin Electric Machinery Company Limited, Harbin, China, is also acknowledged for the technical support for this work.

REFERENCES

Wang H.Y. & G.J. Wang (2007). Application of Wavelet Transform and Fuzzy Theory for Turbo-Generator Fault Mode Classification. *The Eighth International Conference on Electronic Measurement and Instruments.* 442–445.
Adewole A.C. & R. Tzoneva (2012). Fault Detection and Classification in a Distribution Network Integrated with Distributed Generators. *IEEE Power Africa Conference and Exhibition.*

Sedlazeck, K.Richter & C.Strack (2009). Type Testing a 2000 MW Turbogenerator. *Electric Machines and Drives Conference*. 465–470.

Hao L.L., Y.G. Sun & A.M. Qiu (2010). Analysis on the Negative Sequence Impedance Directional Protection for Stator Internal Fault of Turbo Generator. *Electrical Machines and Systems*. 1421–1424.

Ge B.J., W. Guo & D.H. Zhang (2013). J. Harmonic Analysis of AP1000 Large-capacity Turbo-generator Based on BP Neural Network. *International Journal of Control and Automation*. 6, 163–175.

Electronic Engineering and Information Science – Wang (Ed.)
© 2015 Taylor & Francis Group, London, ISBN: 978-1-138-02772-5

A hardware implementation based on FPGA for BLDC fuzzy controller design

J.B. Wen & C.W. Ma
College of Electrical Engineering, Harbin University of Science and Technology, Harbin, China

ABSTRACT: Brushless DC motors have been widely researched in the field of BLDC control. Traditionally controllers are mostly based on DSP or MCU. They have varieties of problems, slow running speed, poor anti-interference ability. This paper proposes a fuzzy PI control system of BLDC based on FPGA. The system utilizes FPGA as core chip, it will be composed all the hardwares. And it produces possible results of FPGA, fast running speed, strong anti-interference ability, the ability of expanding. The system includes current and speed closed loops. The speed loop utilizes the fuzzy PI. It may improve defects of the traditional PI, such as slow response speed and big overshoot. The experiment shows that the system has advantages of fast response speed, small overshoot, high steady accuracy, stable operation.

KEYWORDS: BLDC; FPGA; speed control; fuzzy pi; anti-interference ability.

1 INTRODUCTION

Brushless DC motors with simple structure and good mechanical properties have been widely used in the fields of electric vehicles and aerospace. Traditionally controllers are based on MCU or DSP. Its software leads to poor anti-interference ability, bad flexibility and slow speed (Xia et al. 2005, Wen et al. 2012, Dejan et al. 2006).

With continuous development of EDA technology, the FPGA is more and more applied to the field of industrial control (Horvat et al. 2014). Recently, many experts proposed motor control schemes using FPGA. Many scholars (Soh et al. 2007) have successfully applied FPGA to the field of motor control, but the response speed and steady-state precision are insufficient. For these, this paper propose control system for BLDC based on FPGA.

In practical application, the traditional PID is difficult to achieve the expected control effect. The fuzzy adaptive PID algorithm can improve the problem. For the problem, many research works have been conducted, but they are mostly based on the software, they will appear the problem of slow speed. In this article, modules of the fuzzy adaptive PI are constructed all by hardware. Thus, it fundamentally solves the poor anti-interference ability and slow speed of traditionally controllers.

2 THE STRUCTURE OF THE CONTROL SYSTEM AND THE DESIGN OF FPGA

Phase commutation times are determined according to hall signals of the rotor. At the same time, the speed of the motor as feedback to speed regulator can be calculated according to the hall position signal. The output of speed regulator is seen as the given value of the current regulator, the current signal after the AD converter is being seen as the feedback of current regulator, the output of the current regulator is seen as PWM modulation signal to drive the motor. Figure 1 refers to the control system, the dashed frame is the functional module of the FPGA. All modules in an FPGA are constructed by Verilog HDL language, realize the full digital hardware of the controller, thus greatly improve the speed and control accuracy. It also fundamentally solves the poor anti-interference ability of software.

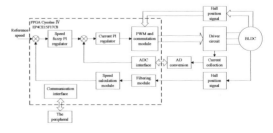

Figure 1. The diagram of control system for BLDC.

3 DESIGN OF THE CORE ALGORITHM

3.1 *Structure of the algorithm*

Traditionally controllers use the classical PID algorithm, it has the advantages of simple and easy to implement. But due to the characteristics of the BLDC motor, traditional PID is often difficult to achieve the expected effect. For this problem, this system adopts the fuzzy adaptive PI algorithm. Digital expression of the traditional PI:

$$u(k) = k_p e(k) + k_i \sum_{j=0}^{k} e(j) \qquad (1)$$

In which: u(k) refers to the output of the controller, e(k) refers to the error, k_p refers to the proportion coefficient, k_i refers to the integral coefficient, k refers to the sampling sequence.

The fuzzy algorithm is utilizing fuzzy logic to regulate k_p and k_i on-line based on error and the rate of error change. Figure 2 refers to the diagram of fuzzy PI.

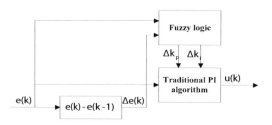

Figure 2. The diagram of fuzzy PI.

3.2 *Design of the fuzzy PI controller*

In this fuzzy system, input values are speed error E and error change EC, output values are the change of k_p Δk_p and the change of k_i Δk_i. The fuzzy set is [−6,6], quantitative level is {−6,−5,−4,−3,−2,−1, 0,1,2,3,4,5,6}. Language values are NB, NM, NS, ZO, PS, PM and PB.

The fuzzy control rule is significant for fuzzy controller. According to field test and expert experience, fuzzy PI controller fuzzy rules are based on following points:

1 When error E is larger, to eliminate error, k_p is big, k_i take a smaller value or zero. When error E is small, reduce the k_p, k_i take small value. When E is very small, to eliminate static error, k_p continue to decrease, k_i can be slightly bigger
2 When E and EC are the same sign, to reduce the error, take small k_i. When E and EC are opposite signs, take small k_p or zero to speed up the control.
3 The bigger EC, the smaller k_p and the bigger k_i.

According to the above, use the Matlab fuzzy logic toolbox, select appropriate membership functions and algorithm the output of k_p and k_i can be concluded, they are in figure 3 and figure 4. The surface is smooth, similar to continuous.

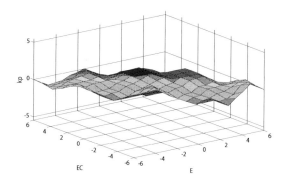

Figure 3. Output spatial surface of Δk_p.

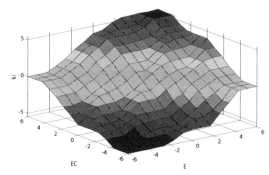

Figure 4. Output spatial surface of Δk_i.

The final k_p and k_i obtained from the following type:

$$\begin{cases} k_p = k_{p0} + \Delta k_p \\ k_i = k_{i0} + \Delta k_i \end{cases} \qquad (2)$$

In which: k_{p0} and k_{i0} refer to proportional coefficient and integral coefficient obtained by the Z-N method.

4 THE EXPERIMENT

Finally, we construct a control system for BLDC based on FPGA using traditional PI and a fuzzy PI algorithm respectively. Motor parameters: rated power is 60W, rated voltage is 24V, rated speed is 3000 RPM, rated torque is 0.18N*m. Figure 5 refers to experiment platform of control system for BLDC based on FPGA.

Figure 5. Experiment platform of control system for BLDC based on FPGA.

This article utilizes the Signaltap II sample speed data of the motor in the Quartus II environment, and synthetize speed curve using MATLAB.

Figure 6 refers to the speed curve of traditional PI and fuzzy PI with load impact when reference speed is 2000rpm. At 0.63s, load step from 0N*m to 0.1N*m, and back to 0N*m at 1.19s. Finally, the traditional PI back to steady state at 0.81s and 1.35s, while the fuzzy PI at 0.77s and 1.29s.

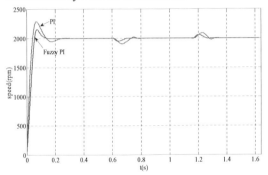

Figure 6. Speed curve when load impact.

Figure 7 refers to the speed curve of traditional PI and fuzzy PI when reference speed stepping. At

1s, reference speed step from 2000rpm to 1500rpm, and back to 2000rpm at 1.96s. Finally, the traditional PI back to steady state at 1.48s and 2.16s, while the fuzzy PI at 1.36s and 2.08s.

5 CONCLUSION

In view of the problems of the traditional motor controller, this article proposes all-hardware, fuzzy control system of BLDC based on FPGA. The experiment shows that the system has a good dynamic and static performance in wide speed range, speed range can reach 1~3000 RPM. Compared with the traditional controller, this system has advantages of high speed, strong robustness, high control precision, and fundamentally solve the poor anti-interference ability of the software controller.

REFERENCES

Xia C.L. & Z.J. Li & R. Yang & W.Y. Qi & J. Xiu (2005). Control System of Brushless DC Motor Based on Active-Disturbance Rejection Controller. *Proceedings of the Csee*, 02:85–89.

Wen J.B. & C. Zhao & Y. Zhang (2012). Design of Control System for Brushless DC Motor Used in Progressive Cavity Pump. *Journal of Harbin University of Science and Technology*, 02:80–83.

Dejan K.. (2006). FPGA Based BLDC Motor Current Control with Spectral Analysis. Power Electronics and Motion Control Conference. *EPE-PEMC 2006. 12th International*, pp.1217,1222, Aug. 30 2006-Sept. 1 2006.

Horvat R. & K. Jezernik & M.Curkovic (2014). An Event-Driven Approach to the Current Control of a BLDC Motor Using FPGA. *Industrial Electronics, IEEE Transactions on, vol.61, no.7*, pp.3719,3726, July 2014.

Soh, C.S. & C. Bi & K. K.Teo (2007). FPGA Implementation of Quasi-BLDC Drive. Power Electronics and Drive Systems, 2007. *PEDS '07. 7th International Conference on*, pp.883,888, 27–30 Nov. 2007.

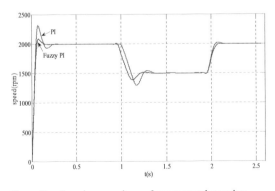

Figure 7. Speed curve when reference speed stepping.

Electronic Engineering and Information Science – Wang (Ed.)
© 2015 Taylor & Francis Group, London, ISBN: 978-1-138-02772-5

The research of code relocation about U-boot transplant based on embedded systems

H. Ai, X. Zhao & J.L. Wang

School of Automation, Harbin University of Science and Technology, Harbin, China

ABSTRACT: The Bootloader guide and the environment parameter setting are necessary for the start of the embedded system. On the basis of in-depth studies of U-boot code relocation under different startup environments and SDRAM memory allocation, this paper presented a migration scheme which has a reference significance to optimize the code execution efficiency .So as to improve the stability and the efficiency of embedded system.

KEYWORDS: U-boot, Relocation, SDRAM, Embedded system.

1 INTRODUCTION

Bootloader will run before the system kernel boot. It can complete the initialization of hardware devices, establish the mapping of the memory space, make the startup parameters and the hardware conditions of the embedded system are in a very proper status (Henkel J, et al, 2003). In general embedded system, there is no BIOS firmware like Windows system, so system startup needs Bootloader loading (Lu Wei, Pan Lian, 2010). In order to improve the executed speed and executed efficiency of code and ensure the system kernel can stable operation at a high speed, the code of Bootloader can be executed after being copied to SDRAM. And redistribute the memory of SDRAM.

2 ANALYSIS OF U-BOOT START

U-boot not only can support the processor of ARM, but also can support other processors such as MIPS, x86, PowerPC, NIOS, XScale etc. The start of U-boot is divided into two stages (Republic of Korea, 2008). The first stage is the codes depending on the architecture, they are generally programmed by assembly language. But the codes of second stage are programmed by C language (Yang Suqiu, 2013). C language can realize more complex functions and have better portability and readability.

The code of the first stage is usually put in the file of arch/arm/CPU/xxxx/start. S, and it is programmed by assembly language. Firstly, define an entrance, because an executable image file must have only one entry point. The entry is usually in address 0x00000000 of the ROM. Secondly, set up the exception vector of the system, the speed of CPU, The clock frequency and the interrupt control register. By calling low levels to initialize memory U-boot copies the code from ROM to SDRAM and initializes the stack. Finally, system runs these codes in the SDRAM, that can be completed by the instruct ldr pc. It will run the program by calling the function start armboot.

The second stage is about C code. arch/arm/lib/board.c. is not only the main function of the startup code in C language, but also the main function of U-boot. This function calls a series of initialization function, initializes Flash Devices, displays the configuration of Flash. It also initializes system memory, environment variables and related network equipment. Finally, enter the command cycle and accept a command from the serial port.

3 CODE RELOCATION

3.1 *Code relocation from Norflash.*

The efficiency of conducting read operation on Norflash is very high, but it is too slow to conduct write operation. In order to improve the efficiency of code execution code relocation is always needed in embedded system. The code of relocation is always in arch/ arm/ cpu/ xxx/ start.s. In the embedded system with both Norflash and Nandflash, the first thing is to identify the model of the system memory. When starting the system from Nandflash Norflash data can't be written, but Nandflash date can. Program of judging the type of chips is as shown in figure 1.

```
static int panduan(void)
{
        volatile int *p = (volatile int *)0;
        int val;
        val = *p;
        *p = 0x12345678;
        if (*p == 0x12345678)
        {
        *p = val;
        return 0;
        }
        else
        {
                return 1;
        }
}
```

Figure 1.   Program of judging the type of chips.

The address of Norflash is 0x00000000, Before the code is executed we need to copy the code from the Norflash to SDRAM. But simply copy the code does not guarantee that the code can run in SDRAM, because the program has a starting address at running time. That is the base address of all the symbols. So the address of this symbol is addressed + symbol which is offset of this code. When building a link, there is -Ttext $(TEXT_BASE) in LDFLAGS, means that the text section positioning base address is set TEXT BASE. Generated results such as the address of function, symbol's corresponding address all see TEXT BASE as the base address. So a particular address in SDRAM needs to be seen as the base address, then copy the code to SDRAM. After doing that the program can be executed in the SDRAM. In include/ configs/ xxxxx.h, the value of CONFIG_ SYS_TEXT_BASE has been defined. The code copy program is as shown in figure 2.

```
void copy_code_to_sdram(unsigned char *src, unsigned char *dest, unsigned int len)
{
        int i = 0;
        if (panduan())
        {
                while (i < len)
                {
                        dest[i] = src[i];
                        i++;
                }
        }
        else
        {
                nand_read_ll((unsigned int)src, dest, len);
        }
}
```

Figure 2.   Code copy program.

The U-boot code was relocated to a specified location by copying program code and will be executed in the SDRAM.

3.2   Code relocation from Nandflash.

In the embedded system Nandflash is similar to the PC hard drive which be Used to save the required kernel when the system is running. But the interface of Nandflash is different to Norflash. It contains only a few I/O pins and can be accessed sequentially, can't be accessed randomly. These characteristics determine that Nandflash is not suitable for running the program, so the code should be relocated into SDRAM. S3C2440, for example, The code will be executed from address 0x000000 after the system is powered on or reset, But when the embedded system is being powered on from the Nandflash, the code does not exist in the address. But is stored in the Nandflash. There is a memory called stepping stone buffer in S3C2440.

When the system starts from Nandflash, the first 4KB codes in the system will be copied into stepping stone automatically. So the first 4KB codes must contain the codes to relocate and the code will be executed in the stepping stone. Because the interface of Nandflash is different to the memory, it can't be accessed like memory. The read and write operations can be performed by the Nandflash controller of S3C2440. Firstly, select the Nandflash chip. Secondly, send out a command of reading. Write the address that will be read into the register. There are five address cycles for big page Nandflash, three address cycles for small page Nandflash. In order to be convenient to copy code into SDRAM, data should be read from relative registers to memory buffer after confirming the address to read. The program of sending address and reading data is as shown in figure 3.

```
static void nand_addr(unsigned int addr)
{       unsigned int col  = addr % 2048;
        unsigned int page = addr / 2048;
        volatile int i;
        NFADDR = col & 0xff;
        for (i = 0; i < 10; i++);
        NFADDR = (col >> 8) & 0xff;
        for (i = 0; i < 10; i++);

        NFADDR = page & 0xff;
        for (i = 0; i < 10; i++);
        NFADDR = (page >> 8) & 0xff;
        for (i = 0; i < 10; i++);
        NFADDR = (page >> 16) & 0xff;
        for (i = 0; i < 10; i++);
}
static unsigned char nand_data(void)
{       return NFDATA;
}
for (; (col < 2048) && (i < len); col++)
{
buf[i] = nand_data();
i++;
addr++;
}
```

Figure 3.   Program of sending address and reading data.

Finally, copy codes from memory buffer to SDRAM. the method is the same as the way to copy codes from Norflash to SDRAM.

4   MEMORY ALLOCATION OF SDRAM

The code of U-boot can be executed in the SDRAM after be relocated, but it can't be found in the SDRAM when U- boot begins to perform in SDRAM. It can be found from the link script u-boot.lds that the program executed from address 0x00000000. LMA was not set in the link script lds, but VMA has been set. VMA was set by the function LDFLAGS of config.mk in the top-level directory. The value of TEXT_BASE is 0x33f80000 that has been defined in /board/ smdk2410/ config.mk. Link code is as shown figure 4.

```
LDFLAGS += -Bstatic -T $(obj)u-boot.lds $(PLATFORM_LDFLAGS)
ifneq ($(TEXT_BASE),)
LDFLAGS += -Ttext $(TEXT_BASE)
Endif
```

Figure 4.   Link code.

The link address of code is 0x33f80000 that can be known by checking the file of U-boot.map. And this is the address in the SDRAM that codes has been copid to. Of course, this value also can be changed. Because of the capacity of SDRAM is 64M, So the address of SDRAM is from 0x30000000 to 0x40000000. A few of storage space should be left to store the code of U-boot at the high address in the SDRAM. The size of U-boot,

usually, is less than 512K, so take address 0 x33f80000 can meet the demand at the most time. If the size of U-boot is bigger than 512K, you should adjust the address to set aside more storage space. If the storage space can't match the size of the U-boot, the code won't be performed fully and the code may be truncated. The behavior of the U-boot is shown in Figure 5.

```
U-Boot 2012.04.01 (Aug 13 2014 - 18:04:25)

CPUID: 32440001
FCLK:       400 MHz
HCLK:       100 MHz
PCLK:        50 MHz
DRAM:   64 MiB
WARNING: caches not enabled
```

Figure 5.   Behavior of the U-boot.

The reason which the code of U-boot can be executed correctly after powering on is that the code executed at the first time has no relation to address. In other words, this image file can be run on any memory address. For the codes which have no relation with address the addressing is based on the value of PC. The worth of plus or minus an offset is the address to run the program.

The code will be stored in a storage space with a specified address after the code was relocated into SDRAM. Other addresses will also be distributed and reused by U-boot in the SDRAM. The pointer is 0x33f80000 after relocation. U-boot will leave some storage space according to diminishing gradually.

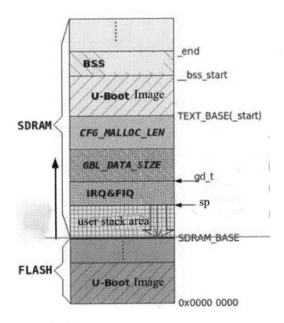

Figure 6.   Memory map.

547

First of all, set aside a memory for malloc. The size of the memory is CFG_MALLOC_LEN. This value is already defined in the macro. Then a certain space should be distributed for global data, the size of it is CFG_GBL_DATA_SIZE. Thirdly, set aside 12 bytes memory to store the anomalies. Finally, the next address is for stack. After the system kernel was boot, the system will refer environment parameters has been set in special place by U-boot. If the distribution of space cannot meet the basic parameters of storage the system can't be started normally. The memory map of SDRAM is as shown in figure 6.

5 SUMMARY

U-boot transplantation code relocation is a very complex and important task The relocated code can be executed with higher efficiency and faster speed. Code relocation played an important role like cornerstone in starting the embedded system.

REFERENCES

Henkel J, Hu, X.B., Sharon, Shuvra S.Bhat-Tacharyya (2003). Taking on the embedded system design challenge. *IEEE Computer, 5(4),* 35–37.

Lu Wei, Pan Lian (2010), The transplantation of Uboot on S3C2440, *Micro computer and application, 24,* 4–5.

Republic of Korea (2008): Samsung, Samsung Electronics. *S3C2440A 32-BIT CMOS Micro-controller User's Manual.*

Yang Suqiu (2013), Uboot transplantation in Nand Flash start mode, *Software Guide, 3,* 35–36.

Electronic Engineering and Information Science – Wang (Ed.)
© 2015 Taylor & Francis Group, London, ISBN: 978-1-138-02772-5

Research on the control strategy of regenerative braking for HEV

X.C. Tian, M.L. Zhou, Y. Zhang & L.L. Wu
College of Electrical & Electronic Engineering, Harbin University of Science and Technology, Harbin, China

ABSTRACT: Taking the regenerative braking control model and the aspect of dynamics into account, a new model has been established that the dynamic braking share changes with the load because of the deficiencies for the original share strategy of braking force in HEV simulation software ADVISOR. Simulation is done with the use of ADVISOR in HAFEI automobile. The simulation analysis shows that the effect of new control strategies of recovery braking energy is better than the old one, and it has the less emission of harmful gas and higher electrical efficiency. Meanwhile, this strategy accords with the requirement of the braking force distribution. The model effectively expands the simulation range of ADVISOR, give convenience to the HEV research.

KEYWORDS: Hybrid electric; ADVISOR; Regenerative braking; Simulation.

1   INTRODUCTION

Regenerative braking is a kind of universal technique for the hybrid electric vehicle breaking energy recovery (Carter & Zaher 2012). One of the distinguished features for regenerative braking is the remarkable ability to recover braking energy (Kelouwani 2012). This thesis constructs a braking force distribution simulation model during regenerative braking based on the vehicle simulation software ADVISOR, and performed the simulation experiment. The results showed that the method which proposed can improve the efficiency of energy recovery and fuel economy in the process of driving. Meanwhile the method also reduces the harmful gas emissions (Hegazy 2013).

2   THE SIMULATION PROCE ADVISOR SOFTWARE

ADVISOR is widely used in the simulation and analysis of hybrid power systems, it adopted a unique hybrid simulation method that simulates backward firstly and forward secondly. With ADVISOR also achieved better simulation results and higher accuracy. The simulation model is shown in Figure 1.

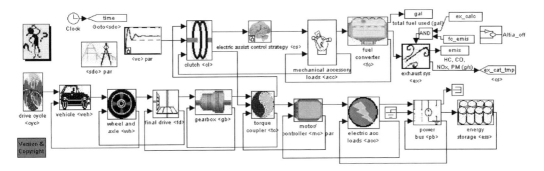

Figure 1.   Vehicle simulation model of ADVISOR3 Regenerative braking control strategy in ADVISOR software.

3   REGENERATIVE BRAKING CONTROL STRATEGY IN ADVISOR SOFTWARE

Braking model of ADVISOR is to determine the changes in each braking force from the view of kinematics, that is, each brake share of braking force changes with the speed of the car, the default of regenerative braking strategies has significant limitations, as can be seen in the figure, the higher the speed, the greater share of electric brake (Borhan 2012); the lower the speed, the smaller share of electric braking. This method can not make full use of the regenerative braking capability of the motor, and not give full play to the potential of the braking system, and do not conform to the actual working condition (Opila 2012).

4 STRAGETY AND BRAKE MODULES ON NEW BRAKING POWER DISTRIBUTION

Based on the original model for the establishment of a brake power distribution model according to load variation, the basic loading which affects the strengthen coming from the body and frame are following categories:

(1) Symmetrical vertical load: when the front and back wheels are going through the same height salients or pits, assuming the car structure on the left and right sides is symmetrical, the bearing capacity system will be subjected to the vehicle longitudinal axis relative to vertical force F_{zs}, that is

$$F_{ZS} = k_{ZS} \sum F_i \qquad (1)$$

presented by the formula (1), k_{ZS} is a vertical dynamic load coefficient symmetrically. (2) The asymmetric vertical load: the front and back wheel which comes from the flat road onto convex parts, it makes an earthing point of the left and right wheel appear height difference named h_n. In this case the load is separated in

symmetric vertical force with loads(bending condition), and due to the projection surface, then produce only symmetry in vehicle longitudinal axis vertical load named F_{ZN}, and formed torque called T_X, which the body around X axis;

$$F_{ZN} = k_{ZN} \sum F_i \qquad (2)$$

$$T_X = k_{ZN} (F_{1L} - F_{1R}) \frac{B_1}{2} \qquad (3)$$

in the formula (3), F_{1L}, F_{1R} are, respectively, the force on the left, right front wheel; B_1 is front track; presented by the formula (2), k_{ZN} is a vertical dynamic load coefficient asymmetrically.

5 ANALYSIS FOR SIMULATION RESULTS

To distinguish between the reformative control strategy and the original one, a prototype named Hafei HF3A has been chosen for the experiment. The simulation road condition is the new European driving cycle CYC_UDDS. Analysis shows that it modified the original Advisor module, and establish a fresh regenerative braking controlling module, and the backward path braking module are being modified in Figure 2, the forward path braking module are being modified in Figure 3.

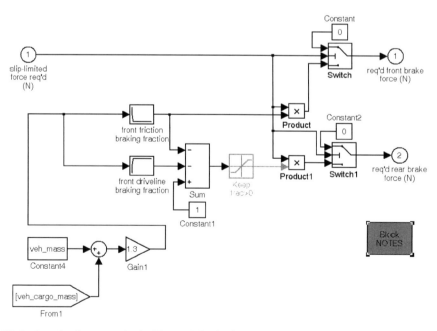

Figure 2.   The backward path regenerative braking module after being modified.

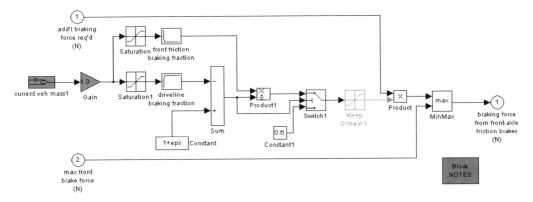

Figure 3.　The forward path regenerative braking module after being modified

The simulation results suggest that in this paper, the control strategy has obvious advantage in recycling braking energy, the battery energy reservation has arisen in the whole cycle, the emissions have improved, and motor effectiveness is considerably advanced, referring to Figure. 4.

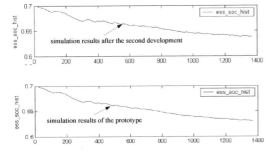

Figure 4.　The change of SOC.

The application shows its superiority as well, and expands the simulation function of ADVISOR, to use it to research the simulation of electric vehicle regenerative braking system provides a convenient.

HF3A braking power distribution curve overtops the adhesion coefficient meanwhile, the braking power is close to the mode of ideal I curves to distribute, and higher than the minimum regulatory requirements of the distribution curve, so meet the distribution requirements which braking power regulations have.

6 CONCLUSIONS

Based on regenerative braking simulation module rooting in software ADVISOR, and from the view of dynamics, the simulation model has been set up

which applies electric vehicle regenerative braking system and load-distribution based braking share and braking force. The simulation results suggest that the well-established model is correct, the actual.

ACKNOWLEDGMENTS

This work was supported by the Heilongjiang Natural Science Foundation (E201302); Heilongjiang Training Programs of Innovation and Entrepreneurship for Undergraduates (No:201410214015).

REFERENCES

Carter R., A. Cruden, P. J. Hall, & A. S. Zaher. (2012). An improved lead acidbattery pack model for use in power simulations of electric vehicles. *IEEE Trans. Energy Convers., 27(1)*:21–28 .

Kelouwani S. (2012). Two-Layer EnergyManagementArchitecture for a Fuel Cell HEVUsing RoadTripInformation. *J.VehicularTechnology,61(9)*:3851–3864.

Hegazy. (2013). PSO algorithm-based optimal power flow control of fuel cell/supercapacitor and fue cell/battery hybrid electric vehicles. *The International Journal for Computation and Mathematics in Electrical and Electronic Engineering,* 86–107.

Borhan H. (2012). Stefano Di.MPC-Based Energy Management of a Power-Split Hybrid Electric Vehicle .*J.Control Systems Technology,20(3):* 593–603.

Opila D. (2012). An Energy Management Controller to Optimally Trade Off Fuel Economy and Drivability for Hybrid Vehicles. *J.Control Systems Technology,20(6)*:1490–1505.

Electronic Engineering and Information Science – Wang (Ed.)
© 2015 Taylor & Francis Group, London, ISBN: 978-1-138-02772-5

The research and design of control system technology for circulating fluidized bed boiler

Y.C. Yang
Department of Electronic Information Engineering, Handan Polytechnic College, Handan, China

ABSTRACT: In the thermodynamic system extension project of Zonghe chemical fertilizer plants, FF Fieldbus technology was used in the automatic control system to finish the following automation functions: data acquisition, analog control, sequence control, emergency protection and other thermal automation functions, and the control system was applied in the project successfully. The system is steady and reliable for several years, it supplied the experience for application and research of Fieldbus technology in control field for Circulating Fluidized Bed Boiler.

KEYWORDS: Fieldbus; Circulating-Fluidized-Bed Boile; PLC.

1 INTRODUCTION

Fieldbus technology is the derivative products of the development of computer, automation control, and communication technologies at this stage, As the new technology in industry automation field ,it is becoming a hot spot in the world. Fieldbus technology appeared about only twenty years, but it leads the irreversible development direction, because of its digital feature that can transfer control function and device management to scene. Fieldbus technology is adopted widely not only can reduce sharply the costs of the system, improve appreciably the control quality, but also can enrich the control information to us and improved integration, openness, distributivity and interoperability of the control system. So it has great practical significance to apply Fieldbus technology actively in automation control system.

FF Fieldbus is one of the eight fieldbuses conforming to IEC 61158 standard (AI 2013), it has been supported by over 100 automation equipment manufacturer, it was early introduced into China and applied widely, The Fieldbus control system is based on Fieldbus technology, it has characteristics of digitization, distributivity and open, it carries on the advantage of PLC and DCS control system (Wang 2013, Gao 2012). but it is more advanced and more reliable than them.

2 SYSTEM HARDWARE CONSTITUTES

The thermodynamic system consists of a 75t/h Circulating Fluidized Bed boiler with assisting machines, two 80T rotary film deaerators, three high pressure feeding pumps, two temperature and pressure reducers with branch cylinder, water, steam pipeline, etc. The thermal monitor and automatic control project were designed to guarantee the boiler running safely and economically and reduce the labor strength of the operator. The fieldbus control system was adopted considering the wide equipment layout area, much locale interference source, interface with cleaning shop control system, etc.

2.1 Structure of system hardware

The structure of the whole automatic control system is shown in Fig 1. The system is slow-speed H1 fieldbus with two operator stations. The local devices connect separately with two bus wires (#1 bus, #2 bus), and the PLC data acquisition system with fieldbus interface connect with the #2 bus.

2.2 Control system characteristic and device structure

The operator stations acting as the man-machine interface devices of the whole control system adopt P I I I series high-grade microcomputer and is deployed with large screen. One of them with the function of the engineer station undertakes the function of system configuration, configuration download, local device setting, testing and fault diagnosis, etc. Two operator stations sets separately the Fieldbus PCI card into the ISA slot in the microcomputer to support the high-speed communication between the Fieldbus and application processes of the PC machine and information sharing bridge for the buses. Two operator stations and PCI cards in it are redundantly configured for reliability of the whole control system. Considering the modernization management needs of the whole factory, Interface also be reserved between the two operator stations and the MIS system.

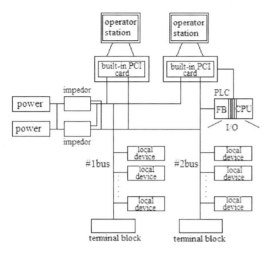

Figure 1.   System structure of automatic controlling.

The whole system is set two buses, the main and important parameter concerning economic running in the process are imported one of the two buses to be monitored, and less important parameters in process system are lead in data acquisition system consisted of the PLC. So the bus devices are separately connected into two buses. Broadly the PLC is a bus device, but PLC connects with #2 bus through the Fieldbus interface module in it, and some parameters are lead in the bus system as auxiliary parameters taking part in control function. The structure of devices connecting with two buses is shown in Table 1.

In the design, the devices connecting with two buses should be uniformly distributed to improve the overall performance of the bus, and each bus should have less than 12 devices. Moreover, devices on the same loop should connect with the same bus wire to improve the real time of the system. In addition, both of the bus have the power supply modules (PS302) and impedor (PSI302P-2) which reserves for each other to provide power for the bus.

The data acquisition system constituted by PLC adopts the programmable controller LC700 of SMAR Company with different module, including: CPU module CPU-700-C3, analog inputting (4~20mA)

Table 1.   Device connecting with bus.

Bus	Parameter names	Device Names	Device Model	Num
#1 bus	main water flow, main vapour flow , Drum level A, Drum level B, 1# Deaerator level, 2# Deaerator leve	DP transmitter	LD302D	6
	1# Deaerator pressure, 2# Deaerator pressure	Pressure transmitter	LD302M	2
	main water adjusting signal A, main water adjusting signal B, 1# deaerator pressure adjusting, 1# deaerator level adjusting, 2# deaerator pressure adjusting, 2# deaerator level adjusting	Fieldbus-currents converter	FI302	2
#2 bus	Temperature-reduction water flow	DP transmitter	LD302D	1
	1# temperature and pressure reducer pressure, 2# temperature and pressure reducer pressure, main vapor pressure, first time air pressure, second time air pressure	Pressure transmitter	LD302M	5
	main vapur temperature, 1# temperature and pressure reducer temperatrue, 2# temperature and pressure reducer temperature, coal inputing flow (3 points), bed temperature(14 points), furnace temperature	PLC and bus interface	FB700	2
	Temperature-reduction water adjusting signal A, Temperature-reduction water adjusting signal B, 1# temperature and pressure reducer temperature adjusting, 1# temperature and pressure pressure adjusting, 2# temperature and pressure reducer temperature adjusting, 2# temperature and pressure reducer pressure adjusting, 1# coal feeder control, 2# coal feeder control, 3# coal feeder control, first time ventilation door control, second time ventilation door control, fan door control	Fieldbus-currents converter	FI302	4

module M401-R-10, temperature inputting module M402-6, digital inputting module M001-2, digital outputting module M123-1 and power module PS-AC-0-1. There are 160 total I/O points, 140 points among them are used and others for system expansion. The CPU module connects with operator station by cable to realize data communication.

3 SYSTEM SOFTWARE CONSTITUTION AND REALIZATION

3.1 *Software constitution*

Running application software SYSCON (one hundred and twenty eight function modules) based on WINDOWS NT configures the fieldbus system, and through running configuration software CONF700 configures the PLC data acquisition system. Moreover, we adopt powerful industrial software AIMAX-WIN to finish the human - computer interface design.

3.2 *Control strategy realization*

#1 bus adjusts the temperature and pressure of two temperature pressure reducer and temperature of main vapor,boiler drum level based on three parameters;#2 bus adjusts pressure of two deaerators and level of them, coal feeder rotating speed, burning control and pressure of the furnace. Two buses and its devices are connected by configuration software to realize automatic adjusting control and interlock function through transferring and connecting different types of function modules of local devices.

The strategy of drum level adjusting is realized as shown in Fig 2. Cascade loop consists of level and water flow, and the level provides set value for closed loop control of water flow. The Feedforward adjusting system consists of vapor flow and water flow. Water flow changes with the vapor flow variation, not varying with the level variety. PID block adjusting the drum level has a long time constant to avoid bubble caused by the violent change of vapor flow to form momentary-false level, but the accumulated error of level formed by a feedforward adjusting system can be modified by PID adjusting loop.

4 SYSTEM CHARACTERISTIC

Programmable controller LC700 as local bus device undertakes most data acquisition and interlock protection works of the system. It connects with #2 bus through its function module FB700. LC-700 CPU connects H1 bus data into ladder logic to it through calling function blocks in FB700, on the other hand other bus devices connecting with H1 bus can use the

data of PLC-700 through calling function blocks too, such the whole control system realizes coordinating control between fieldbus and Programmable controller. Moreover system interlock protection realized by LC-700 can improve real-time of the whole system, and high-speed logic process of programmable controller is shown. The application of LC700 not only economizes lots of local devices to improve the economy of the system, but also improves performance of the whole control system.

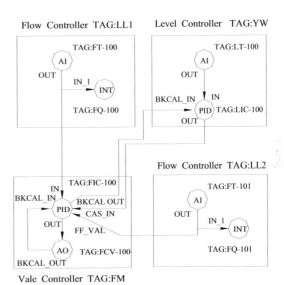

Figure 2. Automatic adjusting strategy of drum level.

5 CONCLUSION

The control system has been applied in Zonghe fertilizer plant successfully and kept reliable and stable for years, the requirement of process system is satisfied with its control precision of automatic adjusting loop,emergency protection reliability and real-time sequential control . At the same time, the system is easily maintained due to sufficient lower layer information of the Fieldbus devices (Wang 2011), moreover the advantage of the Fieldbus control system is shown through decreasing the connecting cables to reduce the fault rate caused by cables.

REFERENCES

Ai, H.(2013). The research for Boiler Control System Based on Fieldbus Profibus-DP.*Manufacturing Automation*, May 2013, 35(9):61–63.

Wang, F.Y.(2013).Machanical and Electrical Control System Analysis Based on PLC and Fieldbus. *Mechanical Engineering & Automation*, April 2013,177(2):208–209.

Gao,T.T.(2012).Research and Implementation of Integration of DCS and Fieldbus.Qingdao: *Qingdao University of Science & Technology.*

Wang, Z.H.(2011). The Application of Profibus-DP Fieldbus System in the Blast Furnace at old-area of Handan Steel CorporationJ.*China New Technologies and Products,*February 2011, 3: 45–48.

Electronic Engineering and Information Science – Wang (Ed.)
© 2015 Taylor & Francis Group, London, ISBN: 978-1-138-02772-5

An input delay approach for uncertain nonlinear time-delay system

L.Y. Fan

Department of Mathematics, Harbin University of Science and Technology, Harbin, China

ABSTRACT: In this paper, we discuss the guaranteed cost robust sampled-data control problem for nonlinear time-varying delay systems with parameter uncertainties, which appear in the state and control input matrices and are time-varying norm-bounded. By using an input delay approach, the system was mapped into a continuous one with time-delay. The objective of designing is to find a robust guaranteed cost sampled-data control law to ensure the closed-loop system asymptotic stability and a prescribed performance index, wich in this paper is less than a given bound for all admissible uncertainties. The theorems are obtained to provide sufficient conditions on the existence of guaranteed cost robust sampled-data control law by introducing Lyapunov stability theory, in terms of the Linear Matrix Inequalities (LMIs).

KEYWORDS: Input delay approach; Time-varying delay; Sampled-data control.

1 INTRODUCTION

Because digital technique usually is applied into modern control systems, many attentions are focused on the sampled-data control theory. As known, there are three main approaches used for the sampled-data control of linear uncertain systems in the forms of linear matrix inequalities (LMIs). One of them is the input delay approach, firstly proposed by Fridman, Shaked and Suplin, where the sampled-data system was transformed into a continuous-time one with an input delay (Gao et al. 2009). The second approach studied the representation of the sampled-data system as impulsive model. The impulsive model approach was used for sampled-data stabilization of linear uncertain systems in the case of bounded sampling intervals. The last one is lifting technique, which is not applied to a class of system with uncertain sample or uncertain matrices (Liu et al. 2010).

Over the past decades, discussions on the robust stability and stabilization problem are rich in control system literature. As well as known, time-delay and parameter uncertainties often occur in many dynamic systems and are frequently a cause of instability and performance degradation (Gao et al. 2008). The main task is to design a controller which is not only stable but also guarantee an adequate level of performance. One design approach to this problem is called guaranteed, cost control approach, firstly presented by Chang and Peng. This approach is better to provide an upper bound to a given cost function, therefore, the system performance degradation incurred by the uncertainties is guaranteed to be less than this bound (Jiang

et al. 2008). Based on this advantage, lots of significant results have been obtained for the discrete-time case and the continuous-time case. In the recent years, these approaches have been extended to uncertain nonlinear systems. Although these approaches have proven to be efficient in solving various control problems, the guaranteed cost sampled-data control problem for uncertain nonlinear systems have no much work to been done yet (Fredman 2010). This is the objective of this paper.

In the present paper, we investigated the design of a guaranteed cost robust sampled-data controller for a class of uncertain time-varying delay nonlinear system. By applying the input delay approach, sufficient conditions are derived in terms of LMIs, which guarantee the asymptotically stable and performance level less than a certain bound for all admissible uncertainties. At the same time, we also proposed the design of optimal guaranteed cost sampling control law. The feasibility and effectiveness of the proposed technique are examined through simulation example (Naghshtabriza 2008).

2 FORMULATION OF THE PROBLEM

We consider the following state-space representation for a class of uncertain nonlinear system.

$$
\begin{cases}
\dot{x}(t) = (A + \Delta A)x(t) + (A_d + \Delta A_d)x(t - \tau(t)) \\
\qquad + (B + \Delta B)u(t) + \mathbf{f}(t, x(t)) \\
x(t) = \varphi(t) \quad t \in [0, h]
\end{cases}
\tag{1}
$$

where $x(t) \in R^n$ is the state vector assumed to be Lebesgue measurable, $u(t) \in R^m$ is the control input and $f: R^+ \times R^n \to R^n$ is uncertain nonlinear term. A, B, A_d are known real constant matrices of appropriate dimensions, ΔA, ΔB, ΔA_d are real-value unknown matrices representing parameter uncertainties.

In the paper, it is assumed that the admissible parameter uncertainties are the following form

$$[\Delta A \quad \Delta B \quad \Delta A_d] = MF(t)[N_1 \quad N_2 \quad N_3],$$

$$F^\mathrm{T}(t)F(t) \le I \; \forall t$$

where M, N_1, N_2 and N_3 are real constant matrices and $F(t) \in R^{i \times j}$ is an unknown time varying matrix function. We assumed that all the elements of $F(t)$ are Lebesgue measurable.

Assumption 1 The uncertain nonlinear term f satisfies $f^\mathrm{T} f \le x^\mathrm{T} Hx$, where H is known positive definite matrix.

Assumption 2 $t_{k+1} - t_k \le h = \; \forall k \ge 0$.

We define the following continuous quadratic cost function

$$J = \int_0^\infty \left[x^\mathrm{T}(t)Qx(t) + u^\mathrm{T}(t)Ru(t) \right] \tag{2}$$

where $Q \in R^{n \times n}$, $R \in R^{m \times m}$ are known positive definite and symmetric matrices.

The following digital control law is represented as a delayed control

$$u(t) = u_d(t_k) = u_d(t - \tau(t))$$

$$\tau(t) = t - t_k, \quad t_k \le t < t_{k+1} \tag{3}$$

3 MAIN RESULTS

In this section, we employed an input delay approach to solve the problem formulated in section1.

Theorem 2.1 Assuming that there exist constants $\varepsilon > 0$, $\alpha > 0$, $W \in R^{m \times n}$, and positive definite and symmetric matrices $X, S \in R^{n \times n}$ such that the following matrix inequality holds:

$$\begin{bmatrix} \Omega & A_d X + BW & M & XN_1^\mathrm{T} & 0 \\ * & 0 & 0 & W^\mathrm{T}N_2^\mathrm{T} + XN_3^\mathrm{T} & W^\mathrm{T} \\ * & * & -\varepsilon I & 0 & 0 \\ * & * & * & -\varepsilon^{-1}I & 0 \\ * & * & * & * & -Z \end{bmatrix} < 0 \tag{4}$$

Then

(1) $u^*(t) = WX^{-1}x(t - \tau(t))$ is a guaranteed cost robust sampled-data controller of system (1);

(2) the cost function of system (2) as follows:

$$J^* = \varphi^\mathrm{T}(0)P\varphi^\mathrm{T}(0) + \int_{-h}^0 x^\mathrm{T}(r)Sx(r)\mathrm{d}r$$

The following theorem extended this optimization problem.

Theorem 2.2 We consider the uncertain system (1) under sampled measurements with cost function (2), if the following optimization problem

$$\min_{\varepsilon, \beta, W, X, D, \Phi} \beta + \mathrm{tr}(\Phi)$$

s.t. (1) inequality (4) hold;

$$\begin{bmatrix} -\beta & \varphi^\mathrm{T}(0) \\ \varphi(0) & -X \end{bmatrix} < 0, \quad \begin{bmatrix} -\Phi & N^T \\ N & -D \end{bmatrix} < 0$$

(where $\mathrm{tr}(\Phi)$ represents the trace of Φ, $D = S^{-1}$ and N satisfies $\int_{-h}^0 x^\mathrm{T}(r)x(r)\mathrm{d}r = NN^\mathrm{T}$). has a feasible solution ε, β, X, W, D, Φ. Then the controller u^* is an optimal state-feedback guaranteed cost sampled-data controller, which ensures the minimization of the guaranteed cost.

4 NUMERICAL EXAMPLE

We consider the following uncertain nonlinear system (1), and the parameters as followed:

$$A = \begin{bmatrix} -0.8 & -0.1 & -0.3 & 0.1 \\ 0.6 & -0.5 & 0.2 & 1 \\ 0.1 & 0.3 & 0.2 & -0.2 \\ 0.1 & 0 & 0 & 0.1 \end{bmatrix}, B = \begin{bmatrix} 0 \\ 1.1 \\ 0.8 \\ 0.4 \end{bmatrix}$$

$$A_d = \begin{bmatrix} -0.3 & -0.2 & 0.4 & 0.3 \\ -0.2 & 0.3 & 0.5 & -0.2 \\ 0 & -0.1 & 0 & 0 \\ 0.3 & 0 & -0.1 & 0.1 \end{bmatrix}, x(0) = \begin{bmatrix} 0.6796 \\ 2 \\ 0 \\ 0 \end{bmatrix}$$

$$Q = \begin{bmatrix} 1 & 0 & 0 & 0 \\ 0 & 1 & 0 & 0 \\ 0 & 0 & 1 & 0 \\ 0 & 0 & 0 & 1 \end{bmatrix}, H = \begin{bmatrix} 0.5 & 0 & 0 & 0 \\ 0 & 0.1 & 0 & 0 \\ 0 & 0 & 0.3 & 0 \\ 0 & 0 & 0 & 0.1 \end{bmatrix}$$

$$M = \begin{bmatrix} -0.6 \\ 0.2 \\ 0.2 \\ 0.1 \end{bmatrix}, E_1 = [0.8 \quad 0.2 \quad 0.3 \quad 0]$$

$$E_2 = 0.11, E_3 = [0.2 \quad 0 \quad 0.8 \quad 0.2], R = 1$$

$$h = 2.7, \tau(t) = 2.5$$

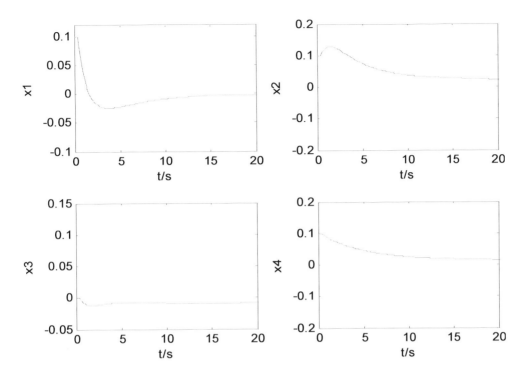

Figure 1.   State response curve of nonlinear time-delay perturbed system.

By applying mincx solver of the Matlab LMI Control Toolbox to solve the LMIs (4), we derived a set of feasible solutions as follows (other associated matrices are omitted here):

$$u^* = [-0.0376 \quad -0.0771 \quad -0.4312 \quad -0.0130]$$
$$^*x(t - \tau(t))$$

$$u^*_{opt} = \begin{bmatrix} 1.3195 & -0.0771 & -0.4312 & -0.0130 \end{bmatrix}$$
$$* x(t - \tau(t))$$

$$J^* = 27.0314 , \quad J^*_{opt} = 16.1219$$

The state responses are shown in Fig.1. Analyzing the above simulation results, it is concluded that the performance level of the optimal guaranteed cost sampled-data controller is obviously smaller than that of general guaranteed cost sampled-data controller. Under the action of the optimal controller, the system (1) is asymptotically stable.

5   CONCLUSIONS

Based on the Lyapunov stability theory, an input delay approach has been employed to design a guaranteed cost robust sampled-data controller, which ensures not only the asymptotic stability but also a bound of quadratic performance level for the closed-loop system with all admissible parametric uncertainties. Sufficient conditions have been obtained in terms of LMIs. Upper bound is derived by solving these LMIs. A convex optimization problem has been introduced to select the problem of optimal guaranteed cost sampled-data controller.

ACKNOWLEDGMENT

This work was supported by science and technology studies foundation of the Heilongjiang educational committee of 2014(12541161). The author also gratefully acknowledges the helpful comments and suggestions of the reviewers, which have improved the presentation.

REFERENCES

Gao H.J., J.L. Wu & P. Shi. (2009). Robust sampled-data H_Y control stochastic sampling. *Automatica*, *45(7)*,1729–1736.
Liu K., V. Suplin & E. Fridman. (2010). A novel discontinuous Lyapunov functional approach to networked-based

stabilization. *Proc. of the 19th international Symposium on mathematical Theory of Networks and Systems.*

Gao H., T.Chen & J. Lam (2008). Network-based H_γ output tracking control. *IEEE Transactions on Automatic Control, 53(3),* 655–667,.

Jiang X., Q. L. Han, S. Liu, & A. Xue (2008). A new H_γ stabilization criterion for network control systems. *IEEE Transactions on Automatic Control, 53(4),* 1025–1032.

Fridman E. (2010). A Refined Input Delay Approach to Sampled-Data Control. *Automatica, 46(2),* 421–427.

Naghshtabriza P., J. Hespanha & A. Teel (2008). Exponential stability of impulsive systems with application to uncertain sampled-data systems. *Systems&Control Letters, 5(57),* 378–385.

Electronic Engineering and Information Science – Wang (Ed.)
© 2015 Taylor & Francis Group, London, ISBN: 978-1-138-02772-5

The P-Q decoupling control and simulation of the DFIG wind power generation system

M. Fu, K.K. Gai, Y.M. Lin & Y. Chen

College of Electrical & Electronic Engineering, Harbin University of Science and Technology, China

ABSTRACT: In this paper, considering the influence on the operation of the DFIG wind power generator during the grid voltage fluctuation but not at fault. Using vector control strategy, the paper has built the DFIG wind power generator mathematical model of the two-phase synchronous rotating coordinate system, and presented based on stator flux oriented vector control strategy. The results show that the control strategy can be eliminated the influence when the grid voltage fluctuation, and implement the active power and reactive power decoupling control effectively.

KEYWORDS: Doubly-fed wind power generation system, Power grid voltage fluctuation, P-Q decoupling control, Vector control.

1 INTRODUCTION

For the active and reactive power decoupling problem of doubly-fed generator system, there is a large number of literature on it. The literature (Liu et al. 2010), respectively, using sliding mode control, the resistance to disturbance control, fuzzy control and double pulse width transformer, vector control, etc., method, has strong robustness but fluctuations. Only under the condition of considering the power grid voltage stability, the results have limitations (Zhao et al. 2006).

This paper adopts vector control strategy based on stator flux oriented, considering the influence of doubly-fed generator and its control strategy two factors that the power grid voltage fluctuation and the wind speed change on (Bao et al. 2010). The simulation verifies the accuracy of the decoupling control strategy which has a good engineering application value by MATLAB/simulink.

2 THE P-Q DECOUPLING MODEL OF DFIG WIND GENERATION SYSTEM

Doubly-fed generator is a high order, nonlinear and strong coupling multivariable system. The system structure as shown in the figure below:

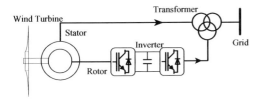

Figure 1. The structure of DFIG wind power system.

2.1 *The mathematical model of DFIG*

In synchronous rotating d-q coordinate system, all adopts the motor convention in stator and rotor. The mathematical model of DFIG is shown as below:

The stator and rotor voltage equations:

$$\begin{cases} u_{ds} = r_s i_{ds} - \omega_1 \psi_{qs} + D\psi_{ds} \\ u_{qs} = r_s i_{qs} + \omega_1 \psi_{ds} + D\psi_{qs} \\ u_{dr} = r_r i_{dr} - (\omega_1 - \omega_r)\psi_{qr} + D\psi_{dr} \\ u_{qr} = r_r i_{qr} + (\omega_1 - \omega_r)\psi_{dr} + D\psi_{qr} \end{cases} \tag{1}$$

The stator and rotor flux linkage equations:

$$\begin{cases} \psi_{ds} = L_s i_{ds} + L_m i_{dr} \\ \psi_{qs} = L_s i_{qs} + L_m i_{qr} \\ \psi_{dr} = L_m i_{ds} + L_r i_{dr} \\ \psi_{qr} = L_m i_{qs} + L_r i_{qr} \end{cases} \tag{2}$$

The stator active power and reactive power equations:

$$\begin{cases} P = u_{ds} i_{ds} + u_{qs} i_{qs} \\ Q = u_{qs} i_{ds} - u_{ds} i_{qs} \end{cases} \tag{3}$$

In the above equations: u_{ds}, u_{qs}, u_{dr}, u_{qr}— the stator and rotor voltage respectively d and q axis components; i_{ds}, i_{qs}, i_{dr}, i_{qr}— the stator and rotor current respectively d and q axis components; ψ_{ds}, ψ_{qs}, ψ_{dr}, ψ_{qr}—the stator and rotor flux respectively d and q axis components; r_s, r_r—the resistance of the stator and rotor respectively; L_s, L_r, L_m—the self inductance and mutual inductance of the stator and rotor respectively; ω_1, ω_r—the synchronous speed and actual speed respectively; D—the differential operator.

In the condition that the system frequency is a power frequency or fluctuating ±0.5 (50±0.5 HZ), as the stator winding resistance is much smaller than its reactance, the resistance of the stator winding is negligible. Chose d axis in the stator flux vector, namely the phase and the speed of d axis and the ψ_s is the same. The coordinate system of stator flux linkage oriented as shown in Figure 2.

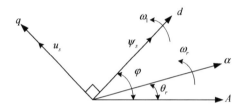

Figure 2. The stator flux oriented coordinate system.

Selecting the coordinate system: $\psi_{ds} = \psi_s$. It is then plugged into the Eq.1, Eq.2 :

$$\begin{cases} u_{ds} = 0 \\ u_{qs} = \omega_1 \psi_s \\ i_{ds} = -i_{dr} L_m/L_s + \psi_s/L_s \\ i_{qs} = -i_{qr} L_m/L_s \end{cases} \quad (4)$$

The next step to get the stator power equation:

$$\begin{cases} P = -u_s i_{qr} L_m/L_s \\ Q = u_s(-i_{dr} L_m/L_s + \psi_s/L_s) \end{cases} \quad (5)$$

Thus, in the stator flux vector control strategy, the active power of the stator has a relationship with the stator q axis component, and the reactive power has a relationship with the stator d axis component. Respectively, adjusting i_{qr} and i_{dr} can realize the stator active and reactive power decoupling control.

The rotor voltage equations can be written as follows:

$$\begin{cases} u_{dr} = u'_{dr} + \Delta u_{dr} \\ u_{qr} = u'_{qr} + \Delta u_{qr} \end{cases} \quad (6)$$

In the above equations:

$$u'_{dr} = (r_r + bp)i_{dr}, \; u'_{qr} = (r_r + bp)i_{qr},$$

$$\Delta u_{dr} = -b(\omega_1 - \omega_r)i_{qr}, \; \Delta u_{qr} = a(\omega_1 - \omega_2)\Psi_s + b(\omega_1 - \omega_r)i_{dr},$$

$$a = L_m/L_s, \; b = L_r(1 - L_m^2/L_s L_r).$$

2.2 The design of the flux observer

Based on the control strategy, the flux linkage observer is one of the necessary parts, which tests the phase and amplitude of the stator flux (Cartwright et al. 2004). According to the analysis above, the stator voltage vector and the stator flux vector on the phase difference 90° and amplitude in multiples ω_1 of the synchronous speed after ignoring the stator resistance. Flux observer is designed in a simplified method. The stator flux linkage approximation for: $\psi_s = u_s/\omega_1$. K/P coordinate transformation is transforming the rectangular coordinate system into the polar coordinate system; K/P transform expressions are shown as following:

$$\begin{cases} U_s = \sqrt{U_{\alpha s}^2 + U_{\beta s}^2} \\ \theta_1 = \arctan U_{\beta s}/U_{\alpha s} \end{cases} \quad (7)$$

The principle diagram of the flux observer is shown in Figure 3:

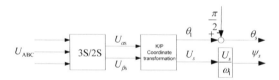

Figure 3. The structure of flux observer.

2.3 Vector control strategy based on the stator flux oriented

In the condition that the network side converter control bus voltage constant, the bus voltage keeps 450V. The rotor side of DFIG system adopts double loop closed loop PI control structure, which outer loop is a power control loop and the inner is a current control loop. Active power control is mainly based on maximum wind power tracking control strategy to achieve. The paper adopts the given maximum wind power tracking strategy (Hu et al. 2005). The reactive power is adjusted by power grid needs. The power reference value, respectively, compared with the reference value is concluded after the rotor the d-q axis current handled by PI controller. The d-q axis rotor voltage decoupling components handled by PI controller, which obtained by comparing the reference value with actual rotor current. Added it to the rotor voltage compensation, and get the d-q axis of the rotor voltage reference value. The diagram of control system is shown in Figure 4.

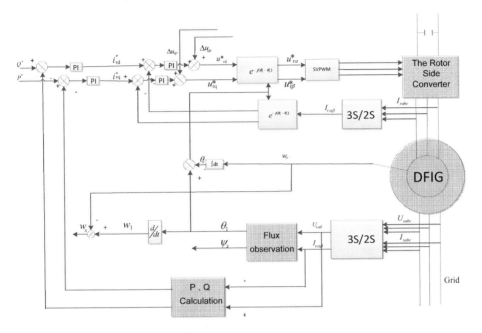

Figure 4. The stator flux oriented vector control.

3 THE P-Q DECOUPLING CONTROL SYSTEM SIMULATION

Using MATLAB/simulink as a platform, build a doubly-fed wind power system model. Decoupling the active and reactive power with the vector control strategy, and verify the grid voltage fluctuation on the influence of active and reactive power decoupling control. System parameters are shown as follows.

Table 1. The parameter design of double fed induction generator.

Parameter name	Rated voltage	Stator and rotor self inductance (L_s, L_r)	Stator and rotor mutual inductance (L_m)	Moment of inertia (J)
Value	380 V	0.071 H	0.069 H	0.089 $kg \cdot m^2$

Parameter name	Rotor resistance (r_r)	Stator resistance (r_s)	Logarithmic (n_p)	DC bus voltage
Value	0.816 Ω	0.435 Ω	2	450 V

3.1 The simulation under the grid voltage stability

The type of change is step. Sets the wind speed from the beginning of 2s change from 5 m/s to 5.5 m/s. At the time of 5 s, the second change is from 5.5 m/s to 6.5 m/s and the last is from 6.5 m/s to 6 m/s at 8s, then maintain a constant. Setting the PI controller parameters $k_p = 2$, $k_i = 0.5$. The waveform is as follows (a) - (c).

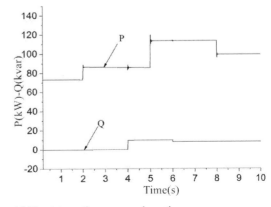

(a) The stator active power and reactive power

563

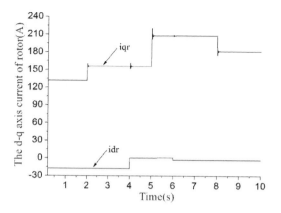

(b) The d - q axis current of rotor

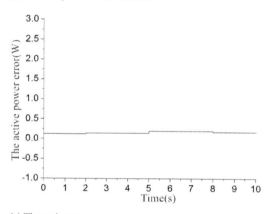

(c) The active power error

Figure 5.   The waveform under the grid constant voltage.

Figure 6(a), with the increase of wind speed, active power increases, the wind speed decreases, and active power reduced. It shows the DFIG following, the output power of the wind turbine well, and verifying the accuracy of the maximum wind power tracking control strategy based on the power given. At the same time, in the 2s, 5s, 8s, the active power changes but reactive power remains the same. In the 4s, 6s, reactive power change, active power remains the same. It verifies the strategy based on stator flux oriented can effectively realize the active and reactive power decoupling. Figure 6(b), the active current component i_{dr} and reactive i_{qr} also realize the decoupling, therefore, respectively adjusting i_{qr} and i_{dr} can realize the active and reactive power decoupling. Figure 6(c), the picture shows the error value of the reference power and the actual power value is almost zero, which can be seen that the maximum wind power tracking is realized.

3.2   The simulation under the grid voltage fluctuation

Under the condition of power grid frequency constant, in the 0-1.5 s, the grid voltage keeps constant, which is 380V. In the 1.5-5 s the voltage starts to drop. The type of variable is a ramp and the rate of change is -5v/s which drops to 361V at 5s. The parameters of PI controllers are set $k_p = 5$, $k_i = 0.2$.

The results are shown as(a)-(c):

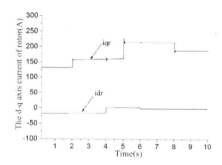

(a) The d - q axis current of rotor

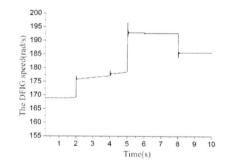

(b) DFIG speed (angular velocity rad/s)

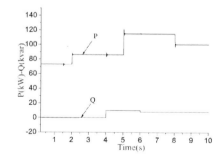

(c) The stator active power and reactive power

Figure 6.   The waveform under the grid voltage fluctuation.

When the grid voltage fluctuates, Figure 6(a), the active component of rotor current increases, and the reactive components unchanged. Figure 6(b), the speed increases. Figure 6(c), the active and reactive power realizes decoupling controlling, and it remains the same when the power grid voltage changed. The experimental results show that when the grid voltage decreases, the rotor speed increases by increasing the excitation current, to achieve the maximum wind power tracking.

Therefore, when the grid voltage fluctuates, the control strategy based on the stator flux oriented makes the DFIG correspond the changed wind speed rapidly and adjusts the rotor excitation system, increases speed and meets the requirements of the stator voltage fluctuation. By using PI controller, it has a short response time and small overshoot, and realizes the active power and reactive power decoupling control.

4   CONCLUSION

This paper presents a mathematical model of DFIG and builds a vector control system based on stator flux oriented, by the means of correcting parameters, finally realizes the active, reactive power decoupling control in situations of the power grid voltage constant and fluctuation. The experimental results show that the control strategy can effectively realize decoupling in both of the two cases and have the advantages of fast response, small overshoot, which has a certain theoretical and practical significance of researching the variable speed constant frequency doubly-fed wind power generation system.

REFERENCES

Liu Y.T., J.H. Yang & J.F. Xie (2010). Sliding Mode Power Decoupled Control of Active and Reactive Power in Doubly Fed Wind Power Generator.*EMCA, 37(4),* 39–43.

Zhao D.L., J.D. Guo & H.H. Xue (2006). The Study and Realization on The Decouping Control of Active and Reactive Power of A Variable-Speed Constant-Frequency Doubly-Fed Induction Generator. *Acta Energiae Solaris Sinica,27(2),* 174–179.

Bao W., Z.D. Yin & Z.H. Ren (2010). Study on AC-Excited Control System of VSCF Doubly-Fed Wind Power Generation System. *ELECTRIC DRIVE , 40(1),* 27–32.

Cartwright P. et al. (2004) Coordinated Voltage Control Strategy for a Doubly fed Induction Generator (DFIG) based Wind Fame. *IEEE Proc Gener Transm. Distrib., 15(14),* 495–502.

Hu J.B., Y.K. He & Q.H. Liu. (2005). Optimized Active Power Reference Based Maximum Wind Energy Tracking Control Strategy.*Automation of Electric Power Systems,29 (24),* 32–38.

Electronic Engineering and Information Science – Wang (Ed.)
© 2015 Taylor & Francis Group, London, ISBN: 978-1-138-02772-5

The fault diagnostic simulation and research of wind generator electrical signal in different conditions

X.J. Shi, J.C. Zhang, J.K. Zhang, H. Du & J.S. Si
School of Mechanical and Power Engineering, Harbin University of Science and Technology, Harbin, China

ABSTRACT: In order to research the mechanical fault induced mechanism of electrical signals of wind generator stator in different conditions, a simulation model of doubly-fed wind generator has been built with the environment of SIMULINK. Simulating the dynamic response of the gear fault of the wind generator transmission system in the generator stator current signal in the condition of off-grid and grid. According to the demodulation principles of Hilbert transform amplitude and frequency, researching the fault feature extractions under different conditions. The results of simulation and analysis show that the amplitude modulation characteristic information of the current signal can be detected in the conditions of grid and off-grid. In the condition of off-grid, the grid voltage has no impact on the generator output. The fault frequency modulation characteristics can also be detected in the electrical signal. It shows the condition of off-grid is more favorable to the extraction and analysis of the fault signal.

KEYWORDS: Wind generators; Stator electrical signal; Gear fault; Fault diagnostic; SIMULINK.

1 INTRODUCTION

At present, the proportion of wind power in the power system is constantly increasing, which has become the third largest energy after thermal power and hydropower (Shen and Li 2013). However, the cost of maintenance is higher and higher with the growing of wind turbine monomer generating capacity. Statistically, the maintenance costs accounted for 25% ~ 30% of the total cost (Chen et al. 2011). Although the most wind field is equipped with a complete Supervisory Control And Data Acquisition (SCADA) monitoring system, most wind field does not contain the vibration test project, because the cost of conventional vibration diagnosis sensor is high and the problems of installation and maintenance are exist. In view of this situation, a variety of indirect methods which are used for fault diagnosis of wind turbine research are gradually increasing in recent years. For example, methods of use existing SCADA data for indirect diagnostic, modeling methods and electrical parameters, etc. (Chen et al. 2011) Among them, diagnostic method which fully uses a wind turbine electrical parameters (such as voltage, current, etc.) stands out, because it has low cost, easy to maintain, and the electrical parameters transducer can be centrally installed at the substation. This method does not need install vibration and other physical sensors in the equipment, only use voltage or current converter, so it is also known as non-sensor diagnostic method.

Diagnostic method without sensors is mainly used for fault diagnosis of the motor, it is certainly applied in motor braking, eccentric and motor bearings or drive system fault diagnosis. The most commonly used types of wind turbines, such as cage-type constant speed generator, Doubly Fed Induction Generator (DFIG) belong to the induction motor. According this, diagnostic methods without sensor also can be applied to wind turbines. At present, there are many reports about this method at home and abroad, for example Yang and Gong et al. (2010) proposed wavelet filter method to realize wind turbine bearing diagnosis. Amirat et al. (2010) adopted stator current amplitude demodulation method and empirical mode decomposition method for the doubly-fed induction generator bearing fault diagnosis, and achieved good results. The results of these studies show that the diagnostic method without sensor is feasible in the wind turbine fault diagnosis.

However, there is a big difference between the working status of generators and motors. The load of the motor is easy to control and the motor current is directly related to the load, but has little to do with the grid; the generator load is related to the total load on the net, but the load on the net can't be controlled by individual generators, therefore, it needs more in-depth and detailed theoretical researches to transplant diagnostic methods without sensor into the generator system, especially the wind turbine system.

In this paper, take the gear fault diagnosis of wind turbine transmission chain as objects, establish the simulation model in theory. The paper mainly simulates the environment of grid and off-grid, take computer simulation and analysis under different working condition. We also explore the affect process of diagnosis feature extraction results under the changeable working condition.

2 MATHEMATICAL MODEL OF DFIG GENERATOR

According to the AC motor theory, the dynamic characteristics of the asynchronous generator in the d–q rotating coordinate system can be described by the following equation (Wang, Wu and Zhang 2009, Wang et al. 2011)

The voltage equation:

$$\begin{cases} u_{sd} = -p\Psi_{sd} + \omega_s\Psi_{sd} - R_s i_{sd} \\ u_{sq} = -p\Psi_{sq} + \omega_s\Psi_{sq} - R_s i_{sq} \\ u_{rd} = p\Psi_{rd} + \omega_s\Psi_{rd} + R_r i_{rd} \\ u_{rq} = p\Psi_{rq} + \omega_s\Psi_{rq} + R_r i_{rq} \end{cases} \quad (1)$$

where u_{sq}, u_{sd}, i_{sq} and i_{sd} are the d–q coordinate components of the stator voltage and current vector, u_{rq}, u_{rd}, i_{rq} and i_{rd} are the d–q coordinate components of the rotor voltage and current vectors, R_s, R_r are the resistance of the stator and rotor, Ψ_{sq}, Ψ_{rq}, Ψ_{sd}, Ψ_{rd} are the d–q coordinate components of the rotor ,stator flux vector, $\omega_s = \omega_1 - \omega_r$, ω_s is d–q coordinate system relative to the angular velocity of the rotor, Including ω_1, ω_s for the synchronous speed and rotor angular velocity, p is differential operator, $p = d/dt$.

Flux-linkage equation:

$$\begin{cases} \Psi_{sd} = L_s i_{sd} - L_m i_{rd} \\ \Psi_{sq} = L_s i_{qd} - L_m i_{rd} \\ \Psi_{rd} = L_r i_{rd} - L_m i_{sd} \\ \Psi_{rq} = L_r i_{rq} - L_m i_{sq} \end{cases} \quad (2)$$

where L_s, L_r, L_m are the self inductance and mutual inductance of the stator and rotor.

According to the definition of induction motor electromagnetic torque, we can get the asynchronous generator electromagnetic torque T_e equation at d–q coordinate system:

$$T_e = \frac{3}{2} p_n (i_{sq}\Psi_{sd} - i_{sd}\Psi_{sq}) \quad (3)$$

Equation of motion of the induction generator:

$$T_L = T_e + J\frac{1}{p_n}\frac{d\omega_m}{dt} \quad (4)$$

where T_L is drag torque from wind turbines, ω_m is rotating mechanical angular velocity of wind turbines, J is the rotational inertia of the system.

3 THE SIMULATION MODEL OF DOUBLE-FED WIND GENERATOR

This thesis takes DFIG as research object, call a variety of electrical simulation components provided by SIMULINK, establish the simulation model which is shown in Figure 1. The simulation model includes a generator module, the module generator rotor excitation frequency, power and load modules, electrical parameter measurement module and fault signal simulation module. Doubly-fed induction wind generator model is replaced by the wound rotor asynchronous motor module, which is set as generator operation mode. Stator winding of the generator is connected to three-phase resistive load, or connected with power grids through a three-phase circuit breaker, Simulate operated conditions of off-grid and grid. Supply the unidirectional excitation power to the rotor of the generator by unidirectional Pulse Width Modulation (PWM) power converter module, which only simulate synchronous or sub-synchronous operation mode. The generator mechanical input takes the speed regulator type and the reference coordinate system take synchronization reference coordinate system, the initial states are zero.

The basic parameters of the simulation system: Rated power of the generator is 3 kW, Rated voltage is 220VAC, Rated speed is 1400 rpm, number of pole pairs $p_n = 2$, the stator resistance, inductance $R_s = 1.9188\,\Omega$, $L_s = 0.24122$ H, converted to the stator side rotor resistance, inductance $R_r = 2.5712\Omega$, $L_r = 0.24122$ H, stator and rotor mutual inductance $L_m = 0.234$H.

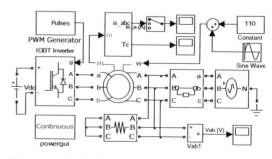

Figure 1. Doubly-fed induction generator and control system simulation model.

568

When steady state simulation runs, the rotor excitation frequency $f_2 = 15$ Hz, rotor speed $\omega_r = 109.9557$ rad/s, the frequency of the generator stator f_1:

$$f_1 = p_n f_r + f_2 = p_n \frac{\omega_r}{2\pi} + f_2$$
$$= 2\frac{109.9557}{2\pi} + 15 \approx 50 \quad \text{Hz} \tag{5}$$

Where *fear* is the rotational frequency of generator shaft, p_n is the number of pole pairs.

4 THE SIMULATION RESULTS ANALYSIS OF NORMAL WIND TURBINE

4.1 *The simulation results of normal operation under the grid environment*

If the three-phase circuit breaker is closed in Figure 1 and the generator stator winding is connected to the grid, we can simulate the grid environment. When the side of generator rotor is supplied with 15Hz frequency excitation current by PWM inverter, the doubly-fed induction generator speed is stable, aiming at the steady state of the stator current and voltage signal, we can calculate the amplitude spectrum, amplitude envelope spectrum and instantaneous frequency demodulation spectrum by the Hilbert demodulation principle (Shi 2009). As shown in Figure 2, (a) is a current signal; (b) is a voltage signal (the same applies below). We can conclude that the majority frequency of the stator current and the voltage signal is 50 Hz, there is no other relevant characteristic frequency except an unknown 9.74 Hz ingredient in low frequency of the amplitude demodulation spectrum.

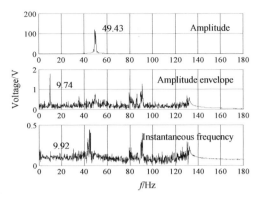

(b) Stator voltage signal

Figure 2. In the condition of grid, the normal steady generator stator current and voltage spectrum (f_2=15 Hz, ω_r =109.9557 rad/s, power frequency 220VAC, 50Hz).

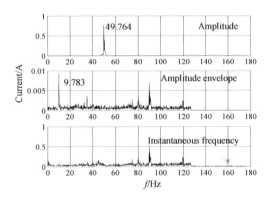

(a) Stator current signal

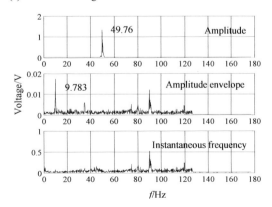

(b) Stator voltage signal

Figure 3. In the condition of off-grid, the normal steady generator stator current and voltage spectrum (f_2=15 Hz, ω_r =109.9557 rad/s).

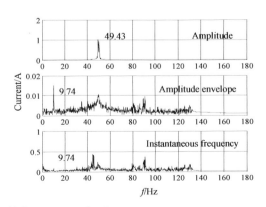

(a) Stator current signal

4.2 The simulation results of the off-grid under normal operation environment

Disconnect the three-phase circuit breaker from the system in Figure 1, we can simulate the off-grid environment. If the other simulation conditions are the same as grid, the steady signal amplitude spectrum, amplitude envelope spectrum and instantaneous frequency spectral demodulation spectrum of the generator stator current and voltage respectively show in Figure 3. We can see that it is similar to Figure 2, there are no significant frequency components, except an unknown frequency component of 9.783 Hz in the low frequency band of the amplitude envelope spectrum and instantaneous frequency demodulation spectrum.

5 SIMULATION RESULTS ANALYSIS OF THE SIMULATION OF GEAR BROKEN TOOTH

5.1 The simulation results of the simulated fault at the running state under the grid environment

It is generally considered that the broken teeth part mesh would produce instantaneous fluctuation of torque and speed of the drive shaft, and passed to the generator rotor through transmission mechanism, causing the stator current modulation phenomenon. Input a changing speed signal at the speed input terminal of the motor model, we can simulate the fault. When simulating the generator input speed as 109.9557+sin(31.4t) rad/s, that is the frequency of the speed is 5 Hz, the fluctuations of amplitude in (1), the rest of the parameters are the same as 3.1. At the moment, the generator stator current, the steady state voltage amplitude spectrum, the amplitude envelope spectrum and instantaneous frequency demodulation spectrum show as in Figure 4. In the current amplitude envelope spectrum of Figure 4 (a), we can clearly distinguish the modulated component 4.82 Hz due to velocity fluctuations, which is considered as input fault frequency of 5 Hz simulated gear, In Figure 4 (a), we can also find the modulation frequency component in the instantaneous frequency demodulation spectrum; however, in the voltage signal of Figure 4 (b), the frequency component is not recognized, which influenced by the grid.

5.2 The simulation results of the simulated fault at the running state under the off-grid environment

The other simulation conditions are the same as grid, we disconnect the three-phase circuit breaker in the simulation model from the network, simulating the time-domain signal at steady state, amplitude spectrum, amplitude envelope spectrum and instantaneous frequency demodulation spectrum of stator voltage and current under the condition of off-grid.

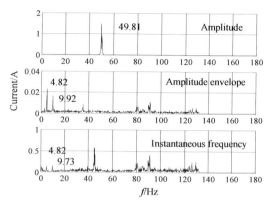

(a) Stator current signal

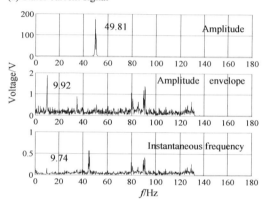

(b) Stator voltage signal

Figure 4. In the condition of grid, the simulation of steady generator stator current and voltage spectrum under the environment of failure (f_2=15 Hz, ω_r =109.9557 rad/s, power frequency 220VAC, 50Hz).

They are shown in Figure 5. Compared with Figure4, the amplitude modulation phenomenon of the stator current and voltage is not obvious, but the frequency modulation phenomenon is very significant. From the Figure5, we can clearly distinguish 4.849 Hz component due to the speed fluctuations in the current and voltage instantaneous frequency demodulation spectrum, which indicate that the speed fluctuation due to broken tooth failure causes the fluctuation of generated power frequency of off-grid condition and its fluctuation frequency equals the failure frequency.

6 CONCLUSION

Through the fault simulation analysis of a double-fed wind generator, we can conclude that the electrical generator stator can reflect mechanical fault information about transmission equipment accurately. There are the following rules: (1) In the grid environment, the fault information is mainly reflected in the stator current signal, the voltage signal is not sensitive; in the off-grid environment, the fault information of the generator stator current and voltage are more sensitive. The above phenomenon indicates that the off-grid environment is more conducive to fault signal extraction and analysis. (2) In the grid environment, the fault information is mainly reflected in the amplitude envelope spectrum, which indicates that the fault information and the main power supply are mainly amplitude modulation relationship; In the off-grid environment, the fault information is mainly reflected in frequency demodulation signal, which indicates that the fault information mainly occurs in the form of frequency modulation, the fluctuation phenomenon of generator power frequency occurs at the same time. Therefore, diagnostic method without sensor can be applied to both the grid and off-grid environment, it can be effective only by choosing suitable demodulation methods according to the circumstance.

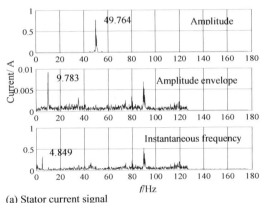

(a) Stator current signal

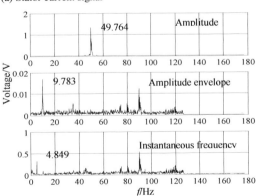

(b) Stator voltage signal

Figure 5. In the condition of off-grid, the simulation of steady generator stator current and voltage spectrum under the environment of failure (f_2=15 Hz, ω_r =109.9557 +sin(31.4t) rad/s)

ACKNOWLEDGMENT

This article is funded by the national natural science foundation of China (51275136).

REFERENCES

Y.M. Shen, and F. Li. 2013. A survey of diagnosis methods for wind power system. *Control Engineering of China*, 20(3),789–793.
X.F. Chen, J.M. Li, & H. Cheng, et al. 2011. Reseasch and application of condition monitoring and fault diagnosis technology in wind turbines. *Journal of Mechanical Engineering*, 47(9), 45–52.
W. X. Yang, P. J. Tavner & C.J. Crabtree, et al. 2010. Cost-effective condition monitoring for wind turbines. *IEEE Transaction on Industrial Electronics*, 57 (1), 263~271.
X. Gong, W. Qiao, and W. Zhou. 2010. Incipient Bearing Fault Detection via Wind Generator Stator Current and Wavelet Filter, in: Proceedings -IECON 2010, 36th Annual Conference of the IEEE Industrial Electronics Society, 2615–2620.
Y. Amirat, V. Choqueuse & M.E.H. Benbouzid, et al. 2010. Bearing fault detection in DFIG based wind turbines using the first intrinsic mode function. 19th International Conference on Electrical Machines-ICEM 2010, 1–6.
X.W. Wang, Z.Y. Zhang, & Y.L.Wu. 2009. Modeling and simulation of variable speed constant frequency double fed wind generator. *Computer Simulation*, 26(10), 294–306.
R.L. Wang, D. Xie & X.T. Wang, et, al. 2011. Modeling and simulation of DFIG for torsional vibration. *Electrical Automation*, 33(1), 51–54.
X.J. Shi, H. Guo, J.P. Shao. 2009. Torsional vibration test of mechanical system using sensorless detection method. *Journal of Vibration, Measurement & Diagnosis*, 29(3), 352–355.

Study of ambient temperature and relative humidity compensation method for methane gas sensor of thermal conduction

Y. Hu, X.B. Ding & Y.Yu

Higher Educational Key Laboratory for Measuring & Control Technology and Instrumentation of Heilongjiang Province, Harbin University of Science and Technology, Harbin, China

ABSTRACT: In order to improve the sensitivity of the thermal conductivity methane sensor, this paper proposes a method for measuring the methane gas of low density by combining constant temperature detection and digital compensates. Verification was carried out by thermodynamic analysis and the experimental results. Experiments show that the effects of environmental humidity on the output of a thermal conductivity sensor without compensation can up to 4.76% at 25°C, while after compensation, it can lower to 0.14%. The result shows that this method can reduce the effects of environmental humidity and temperature on measurement accuracy, and find an effective and available method by using a single thermal conduction sensor to measure the methane density with full range.

KEYWORDS: thermal conduction gas sensor; full range; temperature compensation; constant temperature detection.

1   INTRODUCTION

As gas explosion has been a serious threat to a coal mine, safety, production of coal mine becomes the focus of China's energy industry. In addition, as the span of gas concentration in high gassy mine, effectively monitoring is needed in the process of recovery and utilization. Therefore, full range for mine's methane gas detection technology has great significance to ensure safety production in the coal industry, reduction of accidents, loss of life and property, energy utilization and environmental protection. Enterprises and research institutions domestic and abroad adopt different methods and principles to develop a variety of gas monitored products. Currently, coal mine gas alarm device has a low measurement accuracy and poor stability, and it cannot be used to simultaneously measure the full scale methane gas by using single-sensitive component. For existing problems of methane detection, the method of the thermal conduction gas sensor based on constant temperature is presented in this article. The aim is to achieve the full range for mine's methane gas detection through using a single-sensitive component.

2   THERMODYNAMIC ANALYSIS

The thermal conduction gas sensors are typically used in high concentrations of gas measurement with anti-poisoning ability and high stability. In order to complete the detection method and the applied research of the full scale methane based on thermal conductivity, a theoretical analysis of the mechanism of thermal conductivity gas sensor is needed to carry out. According to the basic theory of heat transfer, the static characteristic equation of sensor, that is a heat balance equation, is established.

Set the operating current of the sensor is I, resistance is r. The sensor heat is converted only from the heating power of the working current, that is $Q_{in}=I^2r$. Known from the theory of heat transfer, the heat transfer has three basic ways: heat transfer, thermal radiation and thermal convection. In the production process of the sensor, in order to reduce the axial thermal conductivity of the resistance wire, length to diameter ratio of the resistance wire is usually between $2\times10^3\sim3\times10^3$. Thereby, the axial heat conduction influence of sensor can be ignored. When thermal conduction gas sensor is working, the sensitive body gets effective protection. So, thermal convection loss caused by airflow can also be ignored.

The ideal operating temperature of thermal conductivity sensor is 400°C~500°C. In this paper, the constant temperature detection method is used, therefore, the heat radiation loss is constant and much less than the conduction heat loss. The heat radiation loss can be ignored. Q_{out} is the heat of thermal conduction loss, it can be expressed as:

$$Q_{out} = \lambda S(T - T_0) \tag{1}$$

When $Q_{in}=Q_{out}$ meets thermal equilibrium condition

$$I^2 r = \lambda S(T - T_0) \qquad (2)$$

Suppose that tested gas is a single gas (methane gas), and background gas is air. Volumetric concentrations of tested methane gas and air in the gas mixture are C_m and C_c, thermal conductivities of test methane gas and air are λ_m and λ_c, thermal conductivity of gas mixture is λ_x. As $C_m + C_c = 100\%$, thermal coefficient of gas mixture can be achieved

$$\lambda_x = \lambda_c C_c + \lambda_m C_m = \lambda_c (1 - C_m) + \lambda_m C_m \qquad (3)$$

Figure 1 is the schematic diagram of thermal conductivity sensor based on constant temperature detection.

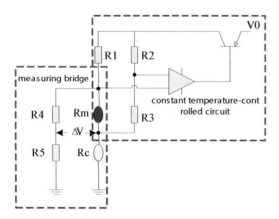

Figure 1. Schematic diagram of thermal conductivity sensor.

According to Figure 1, Heat-balance condition of the reference component and sensitive component is followed:

$$\begin{cases} I^2 R_c = \lambda_c S \Delta t_1 \\ I^2 R_m = \lambda_x S \Delta t_2 \end{cases} \qquad (4)$$

The formula above: λ_c is the thermal conductivity of air; λ_x is the thermal conductivity of gas mixture; R_c is the resistance of the reference component; R_m is the resistance of the sensitive component; Δt_1 is the temperature difference between the reference component and the environment (°C); Δt_2 is the temperature difference between the sensor component and the environment (°C).

Relationships between temperature and resistances of the reference component and sensitive component are followed:

$$\begin{cases} R_c = R_0(1 + \alpha t + \alpha \Delta t_1) \\ R_m = R_0(1 + \alpha t + \alpha \Delta t_2) \end{cases} \qquad (5)$$

In the formula above: R_0 is the resistance of the reference component when its temperature is $0(°C)$; t is environmental temperature (°C); α is the temperature coefficient of platinum wire resistance inside sensors.

Voltages of the reference component and sensitive component can be got:

$$\begin{cases} V_m = IR_m = IR_0(1 + \alpha t + \alpha \Delta t_1) \\ V_c = IR_c = IR_0(1 + \alpha t + \alpha \Delta t_2) \end{cases} \qquad (6)$$

When the measuring bridge satisfies the condition $R_4 = R_5$, the output ΔV is:

$$\Delta V = \frac{V_c - V_m}{2} \qquad (7)$$

According to equation (3), equation (4), equation (5), equation (6) and equation (7), it can be derived:

$$\frac{\Delta V}{V_c} = \frac{\alpha \Delta t_2}{2(1 + \alpha t)} \cdot \frac{\lambda_m - \lambda_c}{\lambda_c} C_m \qquad (8)$$

The expression of methane concentration is:

$$C_m = \frac{2(1 + \alpha t)}{\alpha \Delta t_2} \cdot \frac{\lambda_c}{\lambda_m - \lambda_c} \cdot \frac{\Delta V}{V_c} \qquad (9)$$

We can see from equation (9), the methane concentration is in direct ratio with the output of the bridge, and in inverse ratio with the voltage of the reference component. It also has a direct relation with the temperature coefficient of platinum wire resistance, environmental temperature and the temperature of the sensitive component.

According to the theory analysis above, the concentration of measured gas has relations with the temperature of the sensitive component. Conventional constant-current detection method supplies power for the sensor by using a constant-current source, when the methane concentration or ambient temperature changes, the sensor cannot guarantee a constant temperature, so the temperature of the sensitive component is unknown or unpredictable.

According to equation (9), if using the constant temperature control circuit to realize thermostatic control of the sensitive component, the sensor has a high linear output in a wide concentration range after data processing and achieves ambient temperature compensation. The key to achieve constant temperature detection is how to ensure that the sensor is always working at predetermined temperature when environmental temperature and the methane concentration changes. The variable current source technology is used in this paper to achieve constant temperature detection. By setting the resistivity ratio of R_1, R_2, R_3 and R_m, thermostatic control circuit makes the sensitive component stay in a desired operating temperature. The thermostatic control circuit shown in Figure 1, the circuit changes the heating power by adjusting the working current in order to keep a constant temperature of the sensitive component.

3 THE IMPACT OF TEMPERATURE AND RELATIVE HUMIDITY

Each gas has its own specific thermal conductivity. When the difference between the thermal conductivity of the two gases are large, the heat capacity of the mixed gas will change due to containing different elements. According to the heat capacity difference of the mixed gas, composition analysis of the gas can be achieved. Since water vapor has different thermal conductivities at different temperature and relative humidity. When the thermal conductivity of water vapor and the thermal conductivity of air and methane gas have large differences, the measurement accuracy will be affected. When measuring the methane gas of low density, humidity becomes an important factor that affects measurement accuracy.

In order to solve the influence of water vapor on the measurement, the variation of thermal conductivity of air, methane and dry saturated water vapor should be understood. Assuming the sensitive component works at the absolute temperature T, regulations of thermal conductivity of air and methane changes can be obtained, based on the data from *Matheson gas data sheet*:

$$\begin{cases} \lambda_k = -0.003 + 6.2 \times 10^{-5} \times T + 7.0 \times 10^{-8} \times T^2 \\ \lambda_{CH_4} = -0.009 + 1.4 \times 10^{-4} \times T + 3.3 \times 10^{-8} \times T^2 \end{cases} \quad (10)$$

According to the record in the *Handbook of chemical property data (inorganic volume)*, the quadratic fitting equation of the thermal conductivity of dry saturated water vapor varying with temperature can be obtained as follows:

$$\lambda_{H_2O} = 0.28 - 0.0012 \times T + 0.000001 \times T^2 \quad (11)$$

Using equation (10) and equation (11), we can get the curve of change rules of the thermal conductivity of air, methane gas and dry saturated water vapor. Figure 2 shows it.

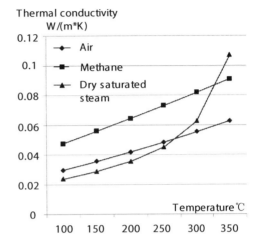

Figure 2. The change rules of thermal conductivity under different temperature.

From Figure 2, the changes of the thermal conductivity of water vapor affected by temperature are obvious. At about 250°C, the thermal conductivity of the water vapor is close to the air's. At about 350°C, the thermal conductivity of the water vapor is close to the methane's. The thermal conductivity of the water vapor increased significantly with the rise of temperature. Conclusions can be obtained: 1. When sensitive elements work at about 250°C, the effects of water vapor on the measurement accuracy are smaller; 2. The effects of water vapor are increasing with the rise of the operating temperature of the sensitive element. In order to eliminate the effects of water vapor on sensitive components in the high-temperature state, this paper presents a digital compensation method to solve this problem.

The digital compensation method is to calculate the volume fraction of water vapor in the air based on relative humidity. Due to the constant temperature detection, it can regulate the working current to maintain constant temperature of the sensor. By equation (11), we can know that the current thermal conductivity of water vapor is constant, in this way, water vapor can be seen as a gas known concentration and thermal conductivity, and the original mixture can be regarded as a mixed gas of water vapor, air and methane gas.

Assuming water vapor's thermal conductivity coefficient is λ_s, volume fraction is C_s. Because of $C_m + C_c + C_s = 100\%$, so equation (3) can be rewritten as:

$$\lambda_x = \lambda_c C_c + \lambda_m C_m + \lambda_s C_s \qquad (12)$$

The expression of the concentration of methane gas can be deduced:

$$C_m = \frac{2(1+\alpha t)}{\alpha \Delta t_2} \cdot \frac{\lambda_c}{\lambda_m - \lambda_c} \cdot \frac{\Delta V}{V_c} - \frac{\lambda_c - \lambda_s}{\lambda_c - \lambda_m} \cdot C_s \qquad (13)$$

In this paper, digital temperature and humidity sensor $HDT21$ is used for the temperature and humidity measurement, the measured humidity is relative humidity. Assuming this is the percentage of vapor pressure P_1 in air and the saturated vapor pressure P_2 at the same temperature:

$$RH(\%) = \frac{P_1}{P_2} \times 100\% \qquad (14)$$

Table 1. Relationships between ambient temperature and the saturated vapor pressure.

Ambient temperature[°C]	0	5	10	15
Saturated vapor pressure[KPa]	0.611	0.872	1.228	1.705
Ambient temperature[°C]	20	25	30	35
Saturated vapor pressure[KPa]	2.338	3.169	4.245	5.626

In Table 1, the saturated vapor pressure in corresponding temperature can be found, and then the vapor pressure in air can be calculated,

$$C_s = \frac{V_1}{V_3} = \frac{P_1}{P_3} \qquad (15)$$

V_3 is the current air pressure. P_3 is the current atmospheric pressure, here takes the standard atmospheric pressure (101KPa) as P_3. The volume fraction of water vapor C_s can be got. So $HDT21$ can be used for the measurement of the methane gas of low density, the aim of which is to eliminate the effects of humidity on the measurement accuracy.

4 APPLICATION AND RESULTS ANALYSIS

According to Table 1, the saturated vapor pressure of water is 3.169KPa at 25°C. Based on equation (14) and equation (15), the concentration of water vapor in air with 30% relative humidity is 0.924% while in air with 80% relative humidity is 2.512%.

Thermal conductivity sensor worked under the constant 450°C through the constant temperature control

circuit. When ambient temperature is 25°C, the sensitivity experiment of thermal conductivity sensor is implemented in the environment which methane concentration is 0%, 5% and 20%. Experimental data is as shown in Table 2.

Table 2. Experimental data of thermal conductivity gas sensor's sensitivity. (Ambient temperature 25°C).

Test gas	Output Voltage [mV]	Changes in output voltage	1%Methane relative sensitivity
Dry air	410.2	0	0
5%Methane gas	608.7	198.5	39.7
20%Methane gas	1210.9	800.7	40.1

According to the data in Table 2, every time the methane relative sensitivity of thermal conductivity gas sensor changes 1%, the output voltage changes 39.9 mV.

Ambient temperature at 25°C, while the sensitivity experiment of thermal conductivity sensor is implemented in the environment in which relative humidity is 0%, 30% and 80%. Experimental data is as shown in Table 3.

According to the data in Table 3, the output voltage changes 74.6mV when the relative humidity changes from 0 to 30% in air, and the output voltage changes 116.7mV when the relative humidity changes from 0 to 80% in air. It can be seen that changes in output voltage are about 78.4mV which are caused by the 1% volume concentration of water vapor. The

Table 3. Experimental data of thermal conductivity gas sensor's humidity drift (Ambient temperature 25°C).

Test gas	dry air	30% Relative humidity of the air	80% Relative humidity of the air
The concentration of water vapor volume[%]	0	0.93	2.51
Output Voltage [mV]	410.2	484.8	601.5
Changes in output voltage[mV]	0	74.6	116.7
Uncompensated measurement value(methane equivalent)[%]	0.05	1.89	4.76
Compensated measurement value(methane equivalent)[%]	0.05	0.10	0.14

experimental results indicate that the effects of 1% volume concentration of water vapor is equivalent to 1.94% volume concentration of methane on the output voltage. In other words, the compensation coefficient of humidity is 1.94 in the measurement condition above.

The analyses above and the measured results are processed to obtain the uncompensated measured value and the compensated measured value, two columns of data are shown in Table 3. The value after the compensation is reduced to less than 0.2% of methane equivalent. It can be seen that the thermal conductivity sensor based on constant temperature detection effectively reduces the influences of environmental humidity on the measurement accuracy after digital compensation.

5 CONCLUSIONS

On the basis of analyzing the thermal conductivity sensor's characteristics, this paper puts forward a method of compensating environmental humidity by using a digital sensor which can eliminate the influences of water vapor on this sensor. This method effectively improves detection accuracy and solves the key technical problems about the measurement of the full-scale methane gas by using a single thermal conductivity sensor.

ACKNOWLEDGMENT

This project is supported by the National Natural Science Foundation of China (Grant No.61179023).

REFERENCES

Wang H.Y., J. Cao & C.G. An (2009). Applicable research of thermal conductivity sensor for gases based on MEMS. *Chinese Journal of Sensors and Actuators.* 7, 1050–1054.
Kliche K., S.Billat, F.Hedrich, C.Ziegler&R.Zengerle (2011). Sensor for gas analysis based on thermal conductivity, specific heat capacity and thermal diffusivity. *IEEE 24th International Conference on Micro Electro Mechanical Systems.* 1189–1192.
Du B.X, J.R. Chen&J.Yi (2010). Working principle and improvement of testing method of thermal conductivity sensor. *Chemical Engineering & Equipment.* 2, 64–66.
Rastrello F., P.Placidi&A.Scorzoni (2011). Measurements, FEM simulation and spice modeling of a thermal conductivity detector. *IEEE Instrumentation and Measurement Technology Conference.* 651–655.
Jin Y.Z.&P.G. Jiang (2014). Groundbreaking research in thermal bridge sensor technology. *Analytical Instrumentation.* 72–75.
Sun J.H., D.F. Cui&X. Chen (2011). Design, modeling, micro fabrication and characterization of novel micro thermal conductivity detector. *Sensors and Actuators.* 936–941.
McGraw-Hill (2003). Matheson gas data book. *Chemical Industry Press Publications.* Beijing.

Electronic Engineering and Information Science – Wang (Ed.)
© *2015 Taylor & Francis Group, London, ISBN: 978-1-138-02772-5*

Study of the reciprocity gap method for Maxwell's equations

J.Y. Sun & J.T. Zhang
College of Measurement-Control Technology and Communications Engineering, Harbin University of Science and Technology, Harbin, Heilongjiang, China

ABSTRACT: As a qualitative method in inverse scattering theory, the reciprocity gap functional method (RGM) could be used in Multiple transmitter and multiple receiving (MIMO) radar applications since RGM is designed to process multistatic data. Some fundamental study of the RGM is presented in this paper. The RGM method needs to solve ill-posed linear integral equations. Analytical results for a circular domain are derived, which give hints on what to expect when a regularized scheme is employed. The ill-posedness of the inverse problem is illustrated by numerical examples. The singular values for special cased are given. The results are valuable in future study since they provide guidance of the RGM for more general cases.

KEYWORDS: MIMO Radar, Inverse Scattering, Reciprocity Gap, Maxwell's Equations.

1 INTRODUCTION

Multiple Input and Multiple Output (MIMO) systems have important applications in wireless communication systems, such as WiFi and WiMax systems. It is due to the fact that it has been shown that the capacity of a MIMO system is proportional to the minimum of the transmitting and receiving antennas in Teletar (1995), Foschini & Gans (1998), which is due to the fading characteristics of a wireless channel. In contrast, although MIMO radar concept has been recently attracted much attention in both signal processing and radar communities, see for example in Rabideau&Parker (2003), Bliss&Forsythe (2003), Robey et al. (2004), Fishler et al. (2006), its fundamental limit in our opinion is not clear yet (for multiple transmit antennas). This is because in wireless communications, fading is due to the superposition of reflected multi-paths from various reflectors and we are NOT interested in reflectors but in transmitting signals. While in radar applications, we are NOT interested in transmitting signals, but interested in reflectors and thus the MIMO advantage (spatial diversity) of a MIMO (multiple transmit antennas) system in combating fading in wireless systems is not clear in radar systems although it has been claimed in some MIMO radar literature under various assumptions. Penetrating obstacle such as walls using electromagnetic waves offers a powerful tool for both military and civilian applications including through-wall target detection and rescue. The application of MIMO radar can significantly increase the signal-to- clutter ratio (SCR) and image resolution. The mathematical theory of MIMO radar is critical to decide the geometry for multiple radar transmitters and receivers in order to obtain clear images.

Recently, a new direct method, called the reciprocity gap functional method (RGM) in inverse scattering was proposed in Colton&Haddar (2005). The RGM can be applied to processing multistatic data and has the possibility to be designed to process MIMO radar data. The method is very simple. One only needs to solve ill-posed linear integral equations and no prior knowledge is required. A numerical example for buried targets is presented in Colton&Haddar (2005) to show the performance of the method. The RGM is used to reconstruct surface impedance and conductivity (e.g., see Di Cristo & Sun (2006), Di Cristo & Sun (2007), Cakoni et al. (2011), Aramini et al. (2010), Griesmaier (2009), Harigaa et al. (2010), inverse mixed impedance problems in elasticity Athanasiadis et al. (2010), and the index of refraction.

2 THE RECIPROCITY GAP METHOD

2.1 *The forward and inverse problem*

The scattering of a point source incident field by an obstacle is studied at first on the inverse problem. For simplicity, we only present the study in two dimensions. Then we illustrate the RGM and introduce the regularization techniques which are used to solve the ill-posed linear integral equations.

We consider the case of a perfectly conducting obstacle D in a homogeneous background in $\mathbb{R}^2$, which is contained in the interior of another bounded domain Ω. We assume that electromagnetic wave propagates in an isotropic medium. Wes et the electric permittivity $\varepsilon = \varepsilon_r \varepsilon_0 > 0$, magnetic permeability $\mu = \mu_r \mu_0 > 0$, electric conductivity $\sigma = \sigma(x)$ where $x = (x, y)$, and the electromagnetic fields Υ and T are time-harmonic. Let ω be the frequency and $\Upsilon = Ee^{-i\omega t}$, and $T = He^{-i\omega t}$. Then E and H satisfy the following equation

$$\nabla \times E - i\mu\omega H = 0 \tag{1}$$

$$\nabla \times H + (i\varepsilon\omega - \sigma)E = 0 \tag{2}$$

The equation is called the time-harmonic Maxwell equation. Eliminating H, the vector Helmholtz equation is obtained

$$\Delta E + (\varepsilon\mu\omega^2 + \sigma\mu\omega i)E = 0 \tag{3}$$

Let the wave number k be given by $k^2 = \varepsilon_0 \mu_0 \omega^2$ and the refractive index $n = n(x)$ be given by

$$n(x) := \frac{\mu_r}{\varepsilon_0}\left(\varepsilon + i\frac{\sigma(x)}{\omega}\right) \tag{4}$$

2.2 The reciprocity gap functional

Denote by $H(\Omega)$ the set $\{v \in H^1(\Omega): \Delta v + k^2 n(x)v = 0 \text{ in } \Omega\}$ and by U the set of solutions to Eq. 4- Eq.7 for all $x_0 \in \partial B$ $v \in H(\Omega)$ and $u \in U$ The reciprocity gap functional is defined by

$$R(u, v) = \int_{\partial\Omega}\left(u\frac{\partial v}{\partial\mu} - v\frac{\partial u}{\partial\mu}\right)ds \tag{5}$$

It is known that $H(\Omega)$ contains Herglotz wave functions given by

$$v_{g,k_1}(x) = \int_S e^{ik_1\hat{d}\cdot x}g(\hat{d})ds(\hat{d}) \tag{6}$$

where $g \in L^2(S)$, with $S = \{\hat{d} \in R^2 : |\hat{d}| = 1\}$ an $k_1^2 = k^2 n_0$ for some constant n_0. Herglotz wave functions are dense in $H^1(D)$. Let $\Phi_{k_1,z} = \frac{i}{4}H_0^{(1)}(k_1|x - z|)$ For $z \in Y$, a sampling domain inside Ω and outside D, a solution $g \in L^2(S^1)$ needs to satisfy

$$R(u, v_g) = R(u, \Phi_{k_1,z}) \quad \text{for all } u \in U \tag{7}$$

where v_g is a Herglotz wave function defined by Eq.8.

In the same way, one can formulate the RGM for inverse scattering by an inhomogeneous medium.

We first recall some results in [7], which justify that the obstacle D can be reconstructed using the solution v of the ill-posed linear integral equation Eq.5.

We look at the case when k^2 is not a Dirichlet eigenvalue for D. If $z \in D$, there exists a sequence $\{g_n\}$, such that

$$\lim_{n\to\infty} R(u, v_{g_n}) = R(u, \Phi_z) \quad \text{for all } u \in U$$

Moreover, v_{g_n} converges in $H^1(D)$ and $v_{g_n} \to \Phi_z$ in $H^{1/2}(\partial D)$.

If $z \in \Omega \setminus D$ then for every sequence v_{g_n}, such that $\lim_{n\to\infty} R(u, v_{g_n}) = R(u, \Phi_z)$ for all $u \in U$

Furthermore,

$$\lim_{n\to\infty}\left\|v_{g_n}\right\|_{H_1(D)} = \infty$$

Now equation Eq.5 can be written in the following equation

$$Ag(\cdot, z) = \phi(\cdot, z) \tag{8}$$

where $A : L^2(S) \to L^2(C)$ is the integral operator and $K : S \times C \in C$ is defined by

$$K(\hat{d}, x_0) = R(u(., x_0), v(., \hat{d}))$$

with $v(x, \hat{d}) = \exp(ik\hat{d}\cdot x)$ and where

$$\varphi(x_0, z) = R(u(\cdot, x_0), \Phi_z)$$

The RGM is based on the fact that $g(\cdot, z)$ being the solution to Eq.8 where A is replaced by a regularized operator $\lim_{n\to\infty}\left\|v_{g_n}\right\|_{L^2(S)}$ is relatively large for $z \in \Omega \setminus D$ and relatively small for $z \in D$.

Tikhonov regularization has been used to solve the ill-posed integral equations. One needs to solve

$$(\alpha + A^*A)g_\alpha(\cdot, z) = A^*\varphi(\cdot, z)$$

In the above equation, α is the regularization parameter and A^* denotes the adjoint of A. If A is a compact operator with singular system $(\sigma_n; v_n, u_n)$, we have

$$g_\alpha = \sum_{n=1}^{\infty}\frac{\sigma_n}{\sigma_n^2 + \alpha}<\varphi, u_n>v_n$$

It is critical to study the singular values of A in the implementation of the RGM.

3 SIMULATIONS

Due to the above result, the sequence of the norm of the regularized solution is bounded when is inside and unbounded otherwise. This can be seen by plotting

580

the eigenvalues of the reciprocity gap operator and eigenvalues of the expansion of the right hand side of Eq.5.

Next two examples are provided to show the effectiveness of the RGM to reconstruct targets. The first example is a flower shape target. The reconstruction is shown in Fig.1. The second example is a kite. The reconstruction is shown in Fig. 2. We see that the reconstruction given a clear boundary of the scattering targets.

We consider the singular values of four different wavenumbers. In Fig. 3, we plot the singular values of for the flower target for wavenumber. We see the behavior of the singular values are similar to the analytic result we derived above. The behavior for the kite case is also consistent with our analysis. Since the plot is similar to the flower case, we did not show it here.

4 CONCLUSIONS

In this paper, we study some fundamentals of the RGM. For a special case, we show the singular values of the ill-posed integral equation in the RGM. Some analytical results in the case of a circular scatterer is presented by the use of the RGM. The singular values of the reciprocity gap operator are also derived analytically. The results in this paper provide useful study of the RGM and is helpful for future studies in this direction.

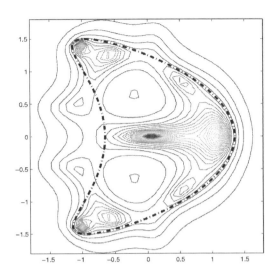

Figure 2. The reconstruction of the kite for $k = 3$. The dashed line is the exact boundary of the flower.

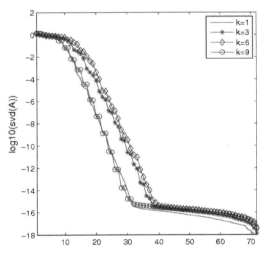

Figure 3. Singular values of A for $k = 1, 3, 6, 9$ (flower).

REFERENCES

Teletar,E.(1995). Capacity of multi-antena Gaussian channels, *AT&T Bell Labs, Tech. Rep.*

Foschini,G. J.&M. J.Gans(1998). On limits of wireless communications in a fading environment when using multiple antennas,*J.Wireless Pers. Commun.*6:311–335.

Rabideau,D. L.&P. Parker(2003). Ubiquitous MIMO multifunction digital array radar.*Proc. Asilomar Conf. Signals, Systems, and Computers*. Pacific Grove, CA.

Bliss,D.W.&K.W.Forsythe(2003), Multiple-input multiple-output (MIMO) radar and imaging. *Proc. Asilomar Conf. Signals, Systems, and Computers.*Pacific Grove, CA.

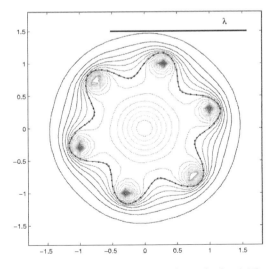

Figure 1. The reconstruction of the flower for $k = 3$. The dashed line is the exact boundary of the flower.

Robey, F., S. Coutts, D. Weikle, J. McHarg& K. Cuomo(2004).MIMO radar theory and experi- mental results. *Proc. Asilomar Conf. Signals, Systems, and Computers*.Pacific Grove, CA.

Fishler E., A. Haimovich, R. Blum, D. Chizhik, L. Cimini, & R. Valenzuela(2006). Spatial diversity in radars - models and detection performance.*J.IEEE Trans. Signal Process*.54: 823–838.

Colton,C.&H.Haddar(2005).An application of the reciprocity gap functional to inverse scattering theory. *J.Inverse Problems* 21:83–398.

Di Cristo, M. & J. Sun(2006). An inverse scattering problem for a partially coated buried obstacle. *J.Inverse Problems* 22:2331–2350.

Di Cristo, M. & J. Sun(2007). The determination of the support and surface conductivity of a par- tially coated buried object. *J.Inverse Problems* 23: 1161–1179.

Athanasiadis,C.E., D. Natroshvili, V. Sevroglou &I.G. Stratis(2010).An application of the reciprocity gap functional to inverse mixed impedance problems in elasticity.*J.Inverse Problems* 26 (8):085011.

Cakoni, F., D. Colton & P. Monk(2011). The Linear Sampling Method in Inverse Electromagnetic Scattering.*CBMS-NSF Regional Conference Series in Applied Mathematics* 80. SIAM, Philadelphia.

Aramini,R., G. Caviglia, A. Massa & M. Piana(2010). The linear sampling method and energy conservation. *J.Inverse Problems* 26(5):055004.

Griesmaier,R.(2009).Reciprocity gap MUSIC imaging for an inverse scattering problem in two-layered media. *J.Inverse Probl. Imaging* 3(3):389403.

Harigaa, N. T., A. Ben Abdaa, R. Bouhlilac & J. de Dreuzy(2010).Definition and interests of reciprocity and reciprocity gap principles for groundwater flow prob- lems.*J.Advances in Water Resources*,33(8): 899–904.

Electronic Engineering and Information Science – Wang (Ed.)
© *2015 Taylor & Francis Group, London, ISBN: 978-1-138-02772-5*

A survey on automatic pavement surface cracking detection systems

X.M. Sun, Z.Y. Xiao, Y.F. Cai, T.J. Zang, L.J. Qi, L. Bao & L. Huang
The higher Educational Key Laboratory for Measuring & Control Technology and Instrumentations of Heilongjiang Province, Harbin University of Science and Technology, Harbin, China

ABSTRACT: The detection of pavement surface cracking is crucial for maintenance of pavements. Traditional methods cannot fulfill the requirements of the large quantity of pavements nowadays. With the developments of optical, computer , sensor technologies, the instruments for detection of pavement cracking have been investigated all over the world. The state-of-art automatic pavement cracking detection systems are briefed in this survey. Based on the analysis of the existing 2D and 3D automatic pavement crack detection technologies, we make comparisons for various pavement cracking detection systems, point out the pros and cons of each detection method, and predict the future directions of automatic pavement crack detection technologies.

KEYWORDS: Pavement surface cracking; Cracking detection; Automatic detection system.

1 INTRODUCTION

Road traffic plays an important role in the national economy, while due to the shortage such as high risk and personal subjective factors, traditional manual detection has been unable to fulfill the demand for road development (Gao 2003, Li 2007). In the late 1980s, USA, Japan and France started to research the 2D pavement cracking detection equipment(NCHRP 2004). However, it is difficult to distinguish the real cracks from road oil, shadows and tyre tracks(Chen 1999). Moreover, 3D information especially the cracking depth is difficult to obtain from the 2 D images. Application of laser scanning, radar and stereo vision technology makes it possible to detect the 3D cracking information. Since the 1990s, a number of 3D pavement surface cracking detection systems have appeared, such as the G.I.E system(Bursanescu et al.1997), the Phoenix Scientific system, the 3D stereo vision pavement detection system proposed by Kelvin et al (Wang 2002). In this paper, we summary the existing domestic and foreign automatic pavement cracking detection systems and data acquisition methods, and discuss the future direction of pavement cracking detection technology.

2 OVERVIEW OF FOREIGN AUTOMATIC PAVEMENT SURFACE CRACKING DETECTION TECHNOLOGY

In the late 1980s, several developed countries started to research the automatic pavement detection equipment(Liu 2007), which mainly includes two directions, one is establishing the 3D cracking model and the other is analyzing the 2D cracking information(Wang 2002).We will respectively introduce the 2D pavement surface cracking detection systems and the 3D ones in the following part.

2.1 *2D pavement cracking detection system*

As is shown in Figure 1, the 2D detection system mainly captures the pavement images with lighting system and use cameras to record the information. According to the different data output mode, the detection systems can be classified as the analog and digital data output system.

Most initial study is about researching the analog data output system, including the 35mm photographed and video method, which means to store the information by changing the chemical, mechanical, and magnetic status of the film. The detection of vehicle Roadrecon 70 studied by PASCO in the US and PAVUE system in Sweden are both based on the analog mode (Lin,1994). As the digital image technology became mature, the analog method was rarely used in recent years. The digital image method can be classified as area scanning and line scanning(NCHRP 2004).The area scanning method takes photos with 1/2 or the entire width of the road, such as ARAN 4900C digital pavement data acquisition vehicle. The line scanning method is used when we need continuous images with high resolution. By gathering a series of transverse lines with the whole road width, then we connect all these lines to form a continuous image, the RoadCrack system developed by Australian in CSIRO Laboratory belongs to this method. Due to the characteristic of line scanning, the shadow of the detection vehicle inevitably makes a continuous shadow in the image of line scanning, special lighting system can avoid this problem. According to the above study, the analysis of 2D pavement surface cracking detection technology is shown in Table 1:

Table 1. Analysis of 2d pavement surface cracking detection technology.

Classification	Realization Method	Problem
Analog Method	35mm photographed & video method	Uneasy to convert to digital form and store on PC
Digital Method	Area Array CCD	Easy to produce distorted and overlapped images
	Line Array CCD	Images contain shadows of the vehicle

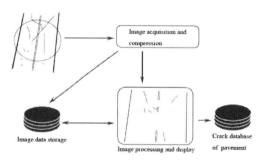

Figure 1. Procedure of 2D pavement cracking detection system.

2.2 3D pavement surface cracking detection system

Because the depth is difficult to get with 2D system, 3D pavement cracking detection technology is becoming the new direction. Bursanescu (Bursanescu et al.1997) introduced 3D pavement laser vision system developed by G.I.E, as is shown in Figure 2, the core technology is the BIRIS laser sensor. The BIRIS laser sensor emits three lasers to the detected surface simultaneously, the CCD capture the image of the laser reflected by the pavement surface to analyze the crack 3D information. There are 6 BIRIS laser sensors in this system, when vehicles are in high speed, sensors scan the longitudinal and transverse information of the surface continuously to get the three-dimensional vision images. Then we can obtain the damage information such as cracks and other distress by analyzing images.

Laurent et al (Laurent et al. 1997) proposed the laser cracking detection system (LCDS), which uses laser sensor, as is shown in Figure 3, the sensor is made up of high-power laser light source and detection system. The scanning oscillating mirror has two surfaces, one is used to scan the laser point on the pavement, and the other sends the information

into the optics acquisition system. The LCD system launches two sensors which are two meters apart to each other at the back of the detection vehicle, covering the 4-meter-wide road. The line scanning location detection instrument sends the transverse profile into the computer, and the computer extracts the cracks through the cracking detection algorithm.

The Phoenix Scientific in California uses laser and radar testing technology to detect the 3D pavement information. By measuring the reflecting time from the laser to the reflector, the system establishes the 3D model moving longitudinally along the driving direction of vehicles. However, the system can only provide the location and severity of the cracks, not more details (Mei 2001).

Kelvin et al. applied the stereo vision technology into establishing the 3D pavement model, as is shown in Figure 4. The basic principle of the stereo imaging technology is to capture the same image from two different angles, to form a stereoscopic image. The second generation digital highway data vehicle (DHDV) achieves obtaining 3D data through this theory, which uses two high-definition cameras with 2048 pixels to take photos of 2-meter-wild pavement, and obtain the geometric data of the pavement and provide evidence for detecting and distinguishing the cracks automatically.

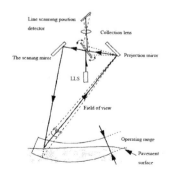

Figure 2. Schematic diagram of BIRIS sensor.

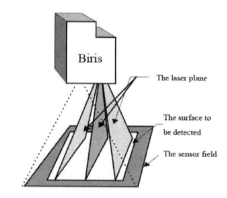

Figure 3. Laser scanning theory of LCDS system.

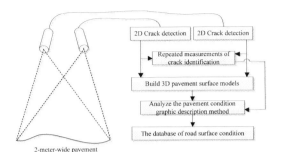

2-meter-wide pavement

Figure 4. Pavement cracking stereo vision detection diagram.

According to above research, the analysis of 3D pavement cracking detection system is shown as that in Table 2. The detection systems can be classified in laser, radar and stereo image. Due to the shortwave characteristic of laser, laser sensor can only analyze visible light, but the data it obtains is clearer than radar does, and the stereo image technology performs poor on data precision, so laser detection will be the tendency in 3D pavement cracking detection area.

Table 2. Analysis of 3D pavement surface cracking detection technology.

Classification	Characteristics	Representative Products
3D Laser	Rapid, High-Precision	3D Laser Vision System by G.I.E,LCDS
3D Radar	Rapid, Less Information	Pavement Detection System by Phoenix Scientific

3 THE DOMESTIC DEVELOPMENT SITUATION

Since the late 1980s, road traffic has developed rapidly in our nation; the total mileage and high-level mileage have both increased remarkably. Therefore, developing the automatic highway damage detection system with our own property rights is imminent (Zhang 2007). Xi'an Research Institute of Applied Optics developed asphalt pavement damage detector, which made up of video surveillance and data processing. Detection vehicle with video system drive at the speed of 15km/h, and sends the recorder with the crack information. Zhenmin Tang et al. in Nanjing University of Science and Technology developed an intelligent pavement detection vehicle JG-1 with the speed 70km/h, which is made up of data acquisition and data processing, the vehicle detects the pavement cracks by image processing. ZOYON and Hubei

HELI Special Automobile Manufacturing Co.,Ltd developed the intelligent pavement detection vehicle ZOYON–RTM together, which has the lighting system to avoid shadows. Harbin Institute of Science and Technology developed multifunctional pavement detection vehicle HIT Guochang with independent intellectual property, which uses high strength light flash illumination system to achieve acquiring the images of pavement damage continuously.

Currently, most domestic pavement cracking detection systems use 2D technology to obtain images, which is difficult to distinguish real damage from road oil, shadow and tire tracks. Therefore, 3D pavement detection system will be the main direction of domestic research.

4 CONCLUSION

Pavement cracking detection systems can be divided into two kinds, 3D and 2D detection system. The 2D system is difficult to distinguish the real cracks from road oil, shadows and tire tracks. The 3D detection system can avoid the above problem, but it has low precision and low acquisition speed. Therefore, to develop automatic 3D pavement cracking detection system with high precision, high speed and high data fusion will be the main direction in the future.

ACKNOWLEGEMENT

University Students' Innovation and Entrepreneurship Training Project of harbin university of science and technology(number: 201310214040)

REFERENCES

Gao, J.Z. (2003) . Research on Automatic Pavement Distress Detection Based on Image Analysis. *NanJing: NanJing University of science and technology,* 85–87.
Li, Q.Q. & X.L. Liu. (2007). An algorithm to image-based pavement cracks geometry features extraction. *Sciencepapers Online2(7),* 517–522.
NCHRP synthesis 334. (2004). Automated pavement Distress Collection Techniques. *TRB 2004 Annual Meeting, WashingTon, D.C.,* 11–14.
Cheng, H.D., J. R.Chen, , C Glazier,., & Y.G.Hu, (1999). Novel approach to pavement cracking detection based on fuzzy se theory . *Comput.Civ.Eng.13(4),* 270–280.
Bursanescu L& F.Blasi (1997). Automated Pavement Distress Data Collection and Analysis: a 3-D Approach. *Proceedings of the international conference on recent advances in 3-D digital imaging and modeling,* 311–7.
Wang, G. (2002). Automated Pavement Distress Survey: A Review and A New Direction. *In 2002 Pavement Evaluation Conference, Virginia,* 21–25.

Liu, W.Y. (2007). Development summary of automatic detection system for highway pavement. *Journal of China & Foreign Highway* 27(2), 30–33.

Lin, Y.S.(1994). Study on the line scan camera used in Pavement Distress Image. *Taiwan: National Central University*: 6–7.

Laurent J, M,Talbot ,M.Doucent (1997). Road surface inspection using laser scanner adapted for the high precision measurements of large flat surfaces. *Proceeding of the international conference on recent advances in 3-D digital imaging and modeling*, 303.

Mei, X.Y. (2001). Automatic detection of pavement surface crack depth on florida roadways. *Tampa: Univ. of South Florida*,12–15.

Zhang, Y. (2007). Design and Implement of High Level Road Infrastructure Image Collection and Management System. *WuHan: Wuhan University of Technology*, 6–7.

Electronic Engineering and Information Science – Wang (Ed.)
© 2015 Taylor & Francis Group, London, ISBN: 978-1-138-02772-5

Robust online multi-object identifying and tracking

J. Su, G.S. Yin & Z.Y. Luo
Department College of Computer Science and Technology, Harbin Engineering University, China

ABSTRACT: In order to realize multiple target identifying and tracking in complex scenes, a robust online identifying and tracking method is proposed, which is implemented on a novel Bayesian tracking model by using the ability of managing multi-modal distributions. The association between tracked targets and detections need not to be explicitly computed in this model. First, accurately identify multi-targets whose number is completely unknown by using Mumford-shah model. Secondly track targets using a joint multi-filter. The proposed algorithm can be proved to be robust to erroneous, distorted and missing detections. Its superior performance can be proved compared with the formal work and its meaning has been seen in unmanned video surveillance.

KEYWORDS: Multiple target tracking; Bayesian model; Particle filtering; Confidence.

1   INTRODUCTION

To realize the robust visual tracking in a variety of complex environments, researchers proposed many methods. Research on Real-time Multi-target Tracking Algorithm Based on MSPF (Wang, Fang & Cong 2012) predicts the extent possible by the use of MSPF for each vehicle in the next frame, uses different detection strategies for simple or multiple targets to avoid a global search and improve the tracking speed; by constructing the importance density function based on the latest observations, the algorithm can achieve an accurate and robust tracking in the part of the block and cross-vehicle. Paper (Mei, Ling, Wu, Blasch & Bai 2011) proposes an efficient BPR-L1 tracker which using a two stage sample probability scheme. This tracker has the advantage of minimum error bound, low computational expensive and occlusion detection. Recently, tracking-by-detection methods have shown impressive performance improvement thanks to the development of object detectors (Dalal & Triggs 2005; Wu & Nevatia 2007), that provide reliable detections even in crowded scenes. The tracking-by-detection methods generally build long trajectories of objects by associating detections provided by detectors. They can be roughly categorized into batch and online methods. Paper (Seung-Hwan Bae & Kuk-Jin Yoon 2014), propose a robust online multi-object tracking method which provide the inspiration for us.

This paper proposed a visual tracking method of multiple moving targets based on improving Mumford-shah model. The key work is to identify multi-objects in a dynamic confusion scene by accurately extracting the edge-features of every target and properly determining and amending edge-feature set of every target in complex environments. Based on this work, the accuracy of tracking, multi-objects will be improved.

2   IMPROVED MUMFORD-SHAH MODEL

Mumford-Shah model (Guo & Luo, 2007) can be improved supposes multiple targets are discrete independent entities initially. The areas of targets can be expressed as a number of areas without overlap regions. Through combining local edge information with the overall uniform regional information, we can locate the boundary curve and accurately identify multi-targets in complex environments while reduce blur edges, noise. During the process of tracking targets template information is used to realize quickly identifying and tracking targets.

There are a few hypotheses in our method show as follows

1   Suppose the image sequence include multiple targets and initial regions of targets are not overlapped for each other. Background includes non-connectivity regions.
2   Suppose a curve of C composed by several client regions separates the image into $m+n$ regions.
3   Suppose target regions fulfill $\phi > 0$. There are m target regions recorded as $C_{i,inside}(i=1,2,\cdots,m)$ and $c_{i,inside}$ is the characteristic value of $C_{i,inside}$.
4   Suppose background regions fulfill $\phi < 0$. There are n background regions recorded as $C_{j,outside}(j=1,2,\cdots,n)$ and $C_{j,outside}$ is the characteristic value of $C_{j,outside}$.

5 $c_{i,inside}$ and $C_{j,outside}$ are brightness values corresponding to their regions. The improved M-S model is shown as formula 1

$$E(C, c_{i,inside}, c_{j,outside}) = \sum_{i=1}^{m} \int_{c_{i,inside}} \alpha_i \mid u - c_{i,inside} \mid^2 dx\, dy$$
$$+ \sum_{j=1}^{n} \int_{c_{j,outside}} \beta_j \mid u - c_{j,outside} \mid^2 dx\, dy \quad (1)$$
$$+ u \cdot length(C) + v \cdot area(C_{inside})$$

Region marks are used to mark up target regions and background regions using a template that is corresponding to the image.

The initial values of M and N is zero. Scan the pixels of the whole image and do as the following rules

1 If $\phi(x, y) > 0$, $M + = 1$ and region is marked with M. Implement region's growth and mark the region with M.
2 If $\phi(x, y) = 0$, mark the region with 0.
3 If $\phi(x, y) < 0$, $N - = 1$ and region is marked with N. Implement region's growth and mark the region with N.

Repeat above steps until finishing scanning. Once model finishing iterative calculation, do region marking.

3 MULTIPLE TARGETS RECOGNITION

3.1 Identifying multiple targets

Before tracking targets, we must judge whether there are targets and the number of the targets. Background subtraction can completely split motion information and inter-frame difference method can detect differential image. First set the global threshold of frame image difference to detect motion in the scene area. Then complete accurate detection for foreground target using the improved M-S model and marked template that are shown as above. Build target marked template base, which is used to match moving targets.

First, we carry out initial detection for moving targets by using the method of background subtraction. Secondly, define moving targets feature and build the accurate templates for moving rigid targets using improved M-S model. Predict the area of moving targets in the next frame by using the joint particle filter model. By matching the templates, we can identify the number of targets and the moving information on the targets'. If the number of the moving targets has changed, repeat above steps and identify the newborn targets and disappeared targets. Build new templates for the newborn targets and renew the template base. If the number of targets is decreased, there are two situations. On the one hand, there may be some disappeared targets. In this condition, we can find the disappeared targets by matching templates. On the other hand, if the shape of some targets changed, there may be shielded targets and we need to judge shielded targets. During tracking, to capture the appearance variations of the target, the template in the template base must be updated and if none of the template is similar to the tracking result new tracking result must be added to the template base.

3.2 Judge shielded targets

If targets are partly shielded, acquiring edge-feature of targets becomes difficult. When the target number in real scene is constant and the poses of targets will not do very little change and when targets are partly shielded, there are two possible states in marked template.

1 The number of targets in marked template is constant and there are small changes in the contour of some targets.
2 The number of targets in marked template is reduced and contour of some target is changed.

The first state shows the target whose contour changed is not shielded by other targets. We can recover its contour using Hough transformation and expansion method and then extract its edge-features.

The second state shows there are at least two targets shielding for each other. In this case, we need to break up targets using the following method.

1 Compare current marked template with the initial marked template and identify the targets being shielded for each other using edge features.
2 Mark unshielded part of the target in template according to the initial marked template.
3 Recover the contour of each target using Hough transformation and expansion method.
4 Separately re-mark the shielded part of targets using the method of region marks in template.
5 According the re-marked template, we define edge-features set of every target using the method of defining edge-feature vector.

4 EXPERIMENTS

In these two sequences, the target undergoes illumination change, noise and partly shielding. Figure 1(a) was the results of sequences 1. Figure 1(b) shows the results of sequences 2.

Experiments are carried out assuming that multi-targets are slowly moving. Fail tracking is shown in figure 2(a) and tracking success rates are shown in Figure 2(b).

(a)Results of slowly moving multi-targets undergoing noise

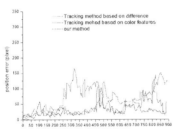

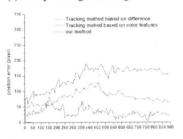

(b)Tracking multiple targets undergoing worse illumination

Figure 1.　Results of track multi-targets.

(a) slowly moving multi-targets

(b) under worse illumination

Figure 2.　Position error while track targets.

5　CONCLUSIONS

A multiple target identifying and tracking in complex scenes has been proposed for surveillance and counting applications. Because the key work for improving success rates of tracking is to accurately identify multi-objects in dynamic confusion scene, our work is focused on accurately extracting the edge-features of every target and properly determining and amending edge-feature set of every target in complex environments. A limitation of this algorithm is that the algorithm cannot judge and track accurately when the shape of tracking targets changes frequently.

ACKNOWLEDGMENTS

This research is supported by the Foundation of Heilongjiang Educational Committee (12521115).

PREFERENCES

Wang X.H., Fang L.L. & Cong Z.H. (2012). Research on Real-time Multi-target Tracking Algorithm Based on MSPF. *J. ACTA AUTOMATICA SINICA. 1(38)*, 139–144.

Mei, X. ,H. Ling, Y. Wu, E. Blasch & L. Bai(2011). Minimum errorbounded efficient *l* 1 tracker with occlusion detection. *in Proc. IEEE conference of Computer Visual Pattern Recognition*, 1257–1264.

Dalal ,N. & B. Triggs(2005). Histograms of oriented gradients for human detection. *In Proc. 2005 Conference on Computer Vision and Pattern Recognition*, 886–893.

Wu, B. & R. Nevatia (2007). Detection and tracking of multiple, partially occluded humans by bayesian combination of edge let based part detectors. *J. International Journal of Computer Vision. 75(2)*, 247–266.

Seung-Hwan Bae & Kuk-Jin Yoon(2014). Robust Online Multi-Object Tracking based on Tracklet Confidence and Online Discriminative Appearance Learning. *In Proc. 2014 Conference on Computer Vision and Pattern Recognition.*, 1–8.

Guo, Y .Y. & X. N. Luo, (2007). Multi-Targets extracted based on single level set. *J.Chinese Journal of Computer. 1(30)*, 120–128.

Electronic Engineering and Information Science – Wang (Ed.)
© 2015 Taylor & Francis Group, London, ISBN: 978-1-138-02772-5

Ankle stress monitoring device based on WIFI

Y.L. Wang, X.X. Duan & J. Yue
Department of Electrical Engineering, Cangzhou Vocational Technical College, Cangzhou, Hebei, China

ABSTRACT: Ankle stress monitoring device based on WIFI can calculate the patients' ankle current stress value by single chip microcomputer to ensure that patients do rehabilitation training in the safe range. Medical personnel can real-time monitor and record the patients' rehabilitation process through WIFI. The ankle stress monitoring device provides reference data for medical personnel instructing the patients to control the rehabilitation training intensity. And it can ensure the rehabilitation effects, shorten the rehabilitation period and reduce the probability of secondary injury.

KEYWORDS: Ankle; Rehabilitation; WIFI.

1 INTRODUCTION

The rehabilitation Medicine study shows that suitable bone stress stimulation can accelerate tissue healing in functional rehabilitation training. And it can help recovering the ligament proprioception, accelerating the recovery rate of patients (Li, 2008). But the ankle rehabilitation, stress is difficult to measure directly. And so the rehabilitative effects cannot be guaranteed because clinicians guide the patients with an ankle injury to do rehabilitation training by experience. The patients control ankle rehabilitation training intensity of subjective feeling in rehabilitation training. The intensity of rehabilitation may be too high to lead the secondary injury of the ankle. An ankle stress monitoring device based on WIFI is designed. And it is used in the lower limb rehabilitation equipment to realize the stress quantification control in ankle rehabilitation process.

2 GETTING STARTED

There are three stress points to support body weight in each foot when people stand. The stress points are the heel nodules, the first and fifth metatarsal. The human foot is a physiological system which is made up of bones, joints, muscles, ligaments and other components. The bones bear pressure, the contracture of muscle produces power, the tendon and the muscle fix and connect the joints. Ankle has three revolute pairs and one prismatic pair. It is difficult to measure the change of the distance and the angle between the three axes of revolution. The movement of the ankle and the simulation results show that the distance of the three axes of revolution is very short. And so

ankle can be approximated to spherical pair in mechanism (Pan, 2007, Yu, 2006). According to the spiral theory three revolute pairs can be equivalent to one spherical pair, so the ankle model can be equivalent to PRRR series branch. Ankle's deformation in the vertical direction is slightly in the rehabilitation process. And so PRRR series branch can be simplified as a RRR series branch (Duan, 2009).

Patients with ankle injury move slowly in the rehabilitation process. The sick foot bears part weight. And so the three stress points of the foot bear stress at the same time. The series branch can be regarded as an open kinematic chain which is composed of a series of links if the bottom of the foot is regarded as a moving platform. The two adjacent links are linked by revolute pairs. The force and torque, which the moving platform bears can be calculated if the stresses of the moving platform's three stress points can be measured. The ankle angle can be measured by angle sensors. The stress of the moving platform can be measured by force sensors.

3 ANKLE STRESS MONITORING DEVICE DESIGN

The schematic diagram of the ankle rehabilitation device with the ankle stress monitoring device is shown in figure 1. The force plate in which four force sensors were installed is fixed at the bottom of the ankle rehabilitation device's bracket. And so the ankle rehabilitation force can be measured (Duan, 2010, Duan, 2011). The bracket is made of low temperature thermoplastic material. For fixing the patient's calf and ankle, the bracket's size and shape are made according to the patients' ankle parameters. Four resistance strain sensors are used to detect the force. Two force sensors are installed under the touchdown point of heel nodules symmetrically. One force sensor is installed under the touchdown point of the first metatarsal. And the other force sensor is installed under the touchdown point of the fifth metatarsal.

The control circuit system hardware structure is shown in figure 2. When patients use the rehabilitation device to carry on the rehabilitation the four force sensor output voltage signals corresponding to the force. These voltage signals will be converted to 24-bits digital signals by A/D converter TM7711 and these digital signals will be sent to the single-chip microcomputer after digital filtering. The single-chip microcomputer will calculate the current force of ankle and compare with the rehabilitation force limit which is set by clinicians. If the current force of ankle achieves or exceeds the rehabilitation force limit the single-chip microcomputer will give an alarm by light and voice. The medical personnel can real-time monitor and record the patients' rehabilitation process through WIFI.

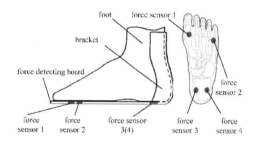

Figure 1. Schematic diagram of the ankle rehabilitation device.

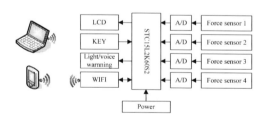

Figure 2. Structure of system hardware.

STC15L2K60S2 produced by STC MCU Limited is selected for the MCU of ankle stress monitoring device. As a new generation of MCS51, the STC15 series MCU has four features which are high speed, high stability, low power consumption and strong anti-disturbance. In addition, it has an internal reset which has 8 level optional threshold voltage of reset, four programmable I/O ports, three 16-bit reloadable

timer/counter, 10-bits A/D converter, 2048 bytes static random access memory, 1024 bytes data EEPROM, 60K bytes read only memory, two asynchronous serial ports, interrupt of low voltage detecting and a watch dog timer. It can be applied to portable instruments which take lithium batteries as power supply to meet the needs of force signal acquisition, processing and data communication.

As a single channel simulation front-end, A/D converter TM7711 can accept the low level input signal from the sensor directly and then become serial number output. It uses sigma-delta conversion technology to achieve the 24-bits without missing code performance. TM7711 has an internal clock oscillator and low noise amplifier, a second-wire communication interface and a full-differential analog input with a difference benchmark input. Its nonlinearity is 0.003% and its rate of data output is 40Hz. It can work on a single DC 3.3v power and reduce the synchronization interference suppression of 50 Hz or 60 Hz power. The expansion circuit schematic diagram of TM7711 is shown in figure 3.

The WIFI module selects HLK-RM04 module. This module is a kind of embedded UART-ETH-WIFI module and has a built-in TCP/IP protocol stack and a universal serial interface. The module can realize the conversion between the three interfaces which are user serial port, Ethernet and WIFI. And so the traditional serial port devices whose configuration needn't be changed can transfer own data through internet by HLK-RM04 module. The expansion circuit schematic diagram of the HLK-RM04 module is shown in figure 4.

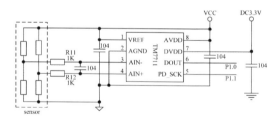

Figure 3. Expansion circuit schematic diagram of TM7711.

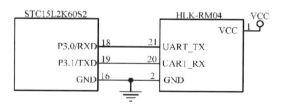

Figure 4. Expansion circuit schematic diagram of HLK-RM04 module.

4 DATA ANALYSIS

Human foot bottom has many degrees of freedom. The force of each force sensor changes with the change of the foot bottom's shape even if ankle doesn't rotate. The four force sensors acquire the support force which the moving platform bears. The rehabilitation force keeps constant when the rehabilitation weight keeps constant. So the ankle stress can be calculated on the basis of the rehabilitation force.

The ankle stress monitoring device is installed to the rehabilitation device. The rehabilitation device is used by Qinhuangdao orthopedic hospitals. Forty eligible elderly patients with fracture are divided into two groups evenly and randomly. One is experiment group, the other is a control group. The patients in the experiment group use the rehabilitation device with the ankle stress monitoring device. The patients in the control group use the rehabilitation device without the ankle stress monitoring device. The bone healing time in the experimental group is 12 to 16 weeks, the average bone healing time is 13.7 weeks. The bone healing time in the control group is 14 to 20 weeks, the average bone healing time is 16.6 weeks (Wang, 2011).

5 SUMMARY

The ankle stress monitoring device based on WIFI is used in medical rehabilitation equipment. And so the ankle rehabilitation force and torque are measured indirectly. It is convenient for doctors to guide patients. It is helpful for patients to control bone stress effectively. The rehabilitation effects of patients with ankle injury are improved. The rehabilitation period of patients with ankle injury is shortened.

REFERENCES

D. Li (2008). Trajectory planning and control of "bsms" ankle rehabilatian robot. *D. Yanshan University*.16–19.

W. Pan(2007). Foot joint rehabilitation robot system design and development. *D. Yanshan University*.17–20.

H.B. YU(2006). Design of Parallel Robot for Ankle Rehabilitation. *D. Yanshan University*.18–21.

X.X. Duan(2009). Research and simulation of four-dimensional force sensor Based on 3-SPS/PS institution. *D. Yanshan University*.50–55.

X.X. Duan, Y.L. Wang & Q. Wang(2010). Design of the measuring circuit for ankle rehabilitation force. *J. Electronic Measurement Technology. 33,* 8–10.

X.X. Duan, Y.L. Wang & Q. Wang(2011). Study on ankle rehabilitation force measuring device. *J. Journal of Hebei University of Science and Technology. 32,* 460–464.

Q. Wang, X.X. Duan & J.T. Yan (2011). Application of pressure-protective brace in the rehabilitation training for the elder patients after surgical operation of femoral intertrochanteric fracture. *J. Clinical Medicine of China.27,* 19–21.

Electronic Engineering and Information Science – Wang (Ed.)
© 2015 Taylor & Francis Group, London, ISBN: 978-1-138-02772-5

Thing-of-Internet based breeding and sprouting measurement and monitor system to facilitate STEM course education

G.S. Xi, R.C. Jia, Z.F. Liu, Z.M. Luo & J.F. Zhang
The higher educational key laboratory for Measuring & Control Technology and Instrumentations of Heilongjiang Province, Harbin University of Science and Technology, Harbin, Heilongjiang, China

ABSTRACT: The system is applied to production of pregermination of rice seed. Depending on theory of agriculture of rice seed pregermination, the technique of rice sprout is made. The monitor and control system is designed for realized the manufacturing technique. MCGSE software is used to design the monitoring interface for supervising the temperature. Master Controller is designed as the master of apparatuses. By the thing of Internet connection with Master Controller, the host computer can control the apparatuses in the field.

KEYWORDS: Thing of Internet, Temperature, MCGSE.

1 INTRODUCTION

Sangiovanni Vincentelli develops greenhouse control system that can complete plant irrigation and handling, spraying pesticides and pest controlling (2010). Zuowei H has been developed the computer networks and remote sensing technology to various productions (2011). The Intelligent monitoring system of agricultural technology is a new technology. The thing about the Internet based breeding and sprouting measurement is highly practical.

The agriculture intelligent monitoring system is used for rice production links, bud and seedling for rice cultivation. Because of the lack of clear production processes and parameters, this method is easy to build on the seeds of death. It can bring to the farmers' economic losses (Hashimoto & Murase & Morimoto 2012).

2 HARDWARE DESIGN OF SYSTEM

The system can send or receive data between the serial interface and network which does not have any analysis, and can be compatible with the user's original software platform. By the thing-of- Internet to network, the PC can control all the system in the same network. This system collects the data through the DS18B20. The diagram of the measuring system is shown in Figure 1.

STC15F2K60S2 has the characteristics of low cost, high reliability and low power consumption. It contains the central processing unit (CPU), program memory, data memory, timer, I/O port, high-speed

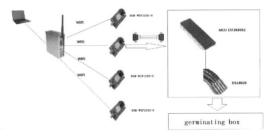

Figure 1. The measuring system based on WIFI.

A/D conversion, watchdog, UART high-speed COM, high-speed synchronous serial port, high precision clock and reset.

USR-WIFI232-G directly connects to the device which have COM and can access the wireless WIFI network very conveniently to control the device. The module has the advantages that support many network protocols and the intelligent network function. Serial transparent transmission mode the WIFI module can directly connect to COM to reduce the complexity of the using.

3 THE DESIGN OF SOFTEARE

The development of 15F2K60S2 is KeiluVision4 for C51 and the C language is applied to implement the software programming. The instruction makes DS18B20 completing the data collection, and store data in the memory of DS18B20. The flow chart of DS18B20 is shown in Figure 2.

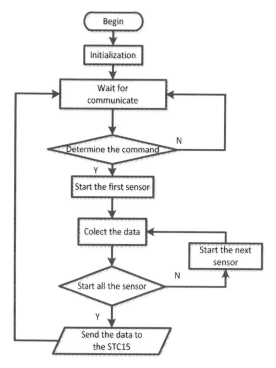

Figure 2. The flow chart of DS18B20.

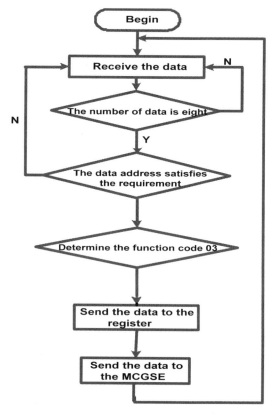

Figure 3. The flow chart of MODBUS TCP.

The system uses the Modbus Tcp protocol to communicate with the PC. In this program, the main function code is used in 03. Its function is sending data to the registers. The flowchart of Modbus Tcp is shown in Figure 3.

The monitoring interface for supervising is MCGSE. The configuration software gives an alarm when the temperature exceeds boundary. By the real time measurement, MCGSE can draw a real-time curve to prepare for subsequent analysis. The configuration interface is shown in Figure 4.

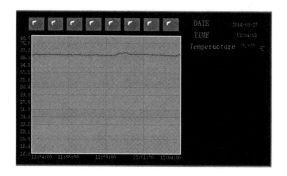

Figure 4. The configuration interface of MCGSE.

4 SENSOR CALIBRATION

The performance of the system is tested with ESPEC of temperature and humidity testing box PVS-3KP. Diagram of calibration is shown in Figure 5.

When perform temperature calibration test, every of 5 °C sets up a measuring point (humidity value sets to the value of 75% relative humidity). The test results as shown above, the error range of the device is 0.24 degrees. After identification, the sensor can meet the actual needs.

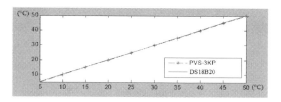

Figure 5. The diagram of calibration.

5 CONCLUSIONS

DS18B20 collects temperature to obtain the change of temperature and humidity characteristics of germinating box. It is the acquisition time of 2000 minutes. The data of temperature are shown in Figure 6.

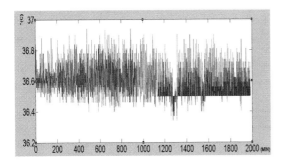

Figure 6. The data of temperature.

The best temperature control between 36 °C and 37 °C. The system sets temperature at 36.5±0.3 °C which meets the requirements of agriculture germination. As the system is portable, lecturers can take it to the classroom, facilitating STEM (Science Technology Engineering and Mathematics) course education. Feedback from students reflects the effectiveness and efficacy of this pedagogy and detailed analysis will be discussed in future work.

ACKNOWLEDGEMENTS

This work was supported by Natural Science Foundation of Heilongjiang Province (Grant No. F201421), Harbin Scientific Innovation Project for Elite Young Researcher (Grant No. 2013RFQXJ003), Scientific Research and Talent Project of Education Department of Heilongjiang Province (Grant No.12541109, 12541140), Higher Education Reform Program of Harbin University of Science and Technology (Grant No. A201300007), 2013 Bilingual Teaching Project of Heilongjiang Affiliated Universities, and Higher Education Reform Program of Heilongjiang Province (Grant No.J

REFERENCES

Sangiovanni Vincentelli A&Di Natale M (2010). Embedded system design for automotive applications. *IEEE Computer. 40(10)*, 42–51.

Zuowei H., Ming Z., Ximei Z (2011).Design and realization of embedded vehicle terminal based on WIFI. *Computer Measurement & Control.* 2205–2208.

Hashimoto Y, Murase H&Morimoto T (2012). Intelligent systems for agriculture. *Control Systems, IEEE. 21(5),* 71–85.

Electronic Engineering and Information Science – Wang (Ed.)
© 2015 Taylor & Francis Group, London, ISBN: 978-1-138-02772-5

Temperature and humidity wireless monitoring system design using undergraduate innovation training program

L. Wang, Z.F. Liu, Z.M. Luo, T. Su & R.C. Jia
The higher educational key laboratory for Measuring & Control Technology and Instrumentations of Heilongjiang Province, Harbin University of Science and Technology, Harbin, Heilongjiang, China

ABSTRACT: A wireless monitoring system of temperature and humidity is designed in this article, which is based on nRF24L01+ RF chip and VC++ display interface. The range of experimental error was analyzed. AM2301 sensors are applied in this system to realize the data acquisition of temperature and humidity in the microcontroller. In order to avoid the inconvenience of wired communications, short-range RF communication technology was introduced. WLAN is constructed to realize the communication and control of the computer and microcontroller by nRF24L01+. VC++ is used to develop GUI (Graphic User Interface) to complete real-time data display and storage of historical data. With the help of standard temperature and humidity calibration instrument, the temperature and relative humidity error of the system were obtained. The whole system can realize remote monitoring and automatic data storage.

KEYWORDS: AM2301; nRF24L01+RF; VC++; Monitoring System.

1 INTRODUCTION

With the development and popularity of radio frequency technique, the high reliability acquisition of wireless temperature and humidity has become more convenient (Jiang & Bei 2010, Xinrong 2011).

On the network organization, nRF905 has the advantages of the practical hardware interface and higher power, and is easy to expand and extend for allowing the network topology (XU & ZHANG 2012, Zhiyuan & Yingju 2012).

The temperature and humidity control system is designed by CC2430 mainly used in indoors and is more and more intelligent, including computer control technology, sensors and data acquisition in terms of technology applications (Bo 2011).

Based on the analysis above, this paper adopts the short distance radio frequency technology based on nRF24l01+ chips and Visual C++ which contains a serial control module. This system is lower cost, operated easily, transmission stability and widely used such as home or office room monitoring, small or medium greenhouse etc.

2 HARDWARE DESIGN OF MEASUREMENT SYSTEM

The entire design incorporates a master station and substation wireless communication mode. The substation is mainly applied to the field for detecting the temperature and humidity and is responsible for data wirelessly transmitted to the master station; the Master station is responsible for receiving the data from every substation and processing relative data, at the same time, it can achieve curve display and store historical data. The schematic of system solutions is shown in Fig 1.

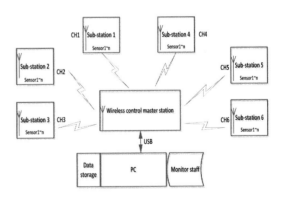

Figure 1. Schematic of system solutions.

In the hardware design of the system, the radio frequency chip nRF24L01+ is applied to wireless data transmission partly for the system. MCU is regarded as the main control chip. We use the AM2301 sensor as the instrument of temperature and humility detection.

The AM2301's advantages of supporting one bus serial data interface, make the system integration easier to achieve, and the operation more convenient. Its shape is ultra-small, lower-power. And it transfers data reliably over 20 meters.

NRF24L01+ chip works in the 2.4GHz band, operated at 0~2Mb/s, built-in hardware CRC (Cyclic Redundancy Check) and multi-point communication address control, and integrated frequency synthesizer, crystal oscillator and modem.

IAP15F2K61S2 and STC15W204S were the controller chips. Radio frequency chip is directly connected with general I/O port in the acquisition system via the SPI bus interface. The Functional connection between the master station and slave station is shown in Fig 2.

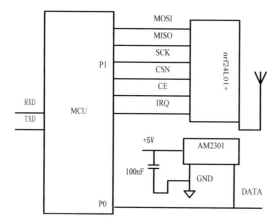

Figure 2. A functional connection between master station and slave station.

3 SYSTEM SOFTWARE DESIGN

This part of the software design and development environment is Keil uVision4 for C51 with the C language programming.

The temperature and humidity acquisition node, according to the characteristics of a single bus timing requirements of AM2301 sensors, software programming achieve acquisition and transmission of temperature and humidity. The sensor acquisition and wireless transmission are shown in Fig 3.

The master's interface design selects Visual C++ develop graphical visualization window, the operating environment is Visual Studio 2010 development environment.

4 SENSOR CALIBRATION

The performance of the system is tested with ESPEC of temperature and humidity testing box PVS-3KP. Diagram of calibration is shown in Fig 4.

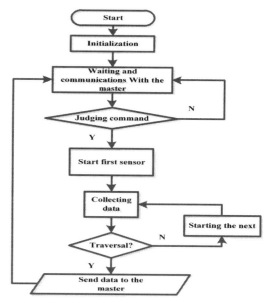

Figure 3. Temperature and humidity acquisition & wireless transmission.

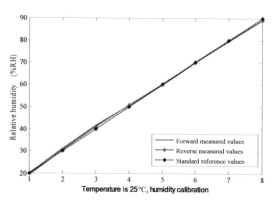

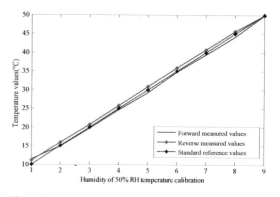

Figure 4. The diagram of calibration.

Actual results show that: the temperature range at 10°C~50°C, the error does not exceed ±0.2°C, relative humidity measurement range is 10 to 90%RH, the error does not exceed ±2%RH (at 25°C ambient temperature measurement), and they meet the requirements of practical application.

5 CONCLUSIONS

In this article, temperature and humidity wireless monitoring system is described consisting of STC microcontroller, nRF24L01+, AM2301 temperature and humidity sensors and Visual Studio 2010. The temperature error of the system is ±0. 2°C, relative humidity error at ±2% RH, the measured data show that the system measures the temperature and humidity changes, and the real-time data storage to facilitate further analysis. Through this innovative training program, several second and third year undergraduate students worked with graduate students under the guidance of their supervisor. This project provided an opportunity for them to learn from each other and cooperate in a team work. Until now, three out of five participants have passed national graduate entrance exam and the rest received company offers. In addition, this team won the silver medal in the electronic competition in Northeast China Region.

ACKNOWLEDGEMENTS

This work was supported by Natural Science Foundation of Heilongjiang Province (Grant No. F201421), Harbin Scientific Innovation Project for Elite Young Researcher (Grant No. 2013RFQXJ003), Scientific Research and Talent Project of Education Department of Heilongjiang Province (Grant No.12541109, 12541140), Higher Education Reform Program of Harbin University of Science and Technology (Grant No. A201300007), 2013 Bilingual Teaching Project of Heilongjiang Affiliated Universities, and Higher Education Reform Program of Heilongjiang Province (Grant No.JG2013010302).

REFERENCES

Jiang X.&B. Jiang (2010). Design for wireless temperature and humidity monitoring system of the intelligent greenhouse. *2010 2nd International Conference on Computer Engineering and Technology.3*, 59–63.

Zhang X.R (2011). Research of temperature and humidity monitoring system based on WSN and fuzzy control. *2011 International Conference on Electronics and Optoelectronics (ICEOE 2011).4*, 300–303.

Xu L. & H.W. Zhang (2012). A design of wireless temperature and humidity monitoring system. *2010 IEEE.* 13–16.

Gao Z.Y.& Y.J. Jia (2012). A design of temperature and humidity remote monitoring system based on wireless sensor network technology. *2012 International Conference on Control Engineering and Communication Technology.* 896–899.

Chang B(2011), July. Design of Indoor Temperature and Humidity Monitoring System Based on CC2430 and Fuzzy-PID. *2011 Cross Strait Quad-Regional Radio Science and Wireless Technology Conference.* 26-30:980–983.

Electronic Engineering and Information Science – Wang (Ed.)
© 2015 Taylor & Francis Group, London, ISBN: 978-1-138-02772-5

The design of wind-field meteorological data monitoring system

Y. Yu, X.B. Ding, Y. Hu & C.Y. Li
Higher Educational Key Laboratory for Measuring & Control Technology and Instrumentation of Heilongjiang Province, Harbin University of Science and Technology, Harbin, Heilongjiang, China

ABSTRACT: The system uses MSP430 microcontroller with low power consumption to acquire temperature, humidity, wind speed, wind direction and other meteorological data of wind fields, and use the ZigBee wireless communication technology for data transmission, and have the data real-time display and memory mass storage by SD card. The wind field meteorological data acquisition instrument of this paper designing has advantages of low power consumption, good network expansibility, large storage and easy operation etc, which can be used in all kinds of measuring wind project in the wind fields, and can provide detailed and accurate project decision data to support for wind fields project decision such as wind energy resources evaluation, the site selection of the wind fields, the wind generator model selection and installation, the reliability research of the wind generator, power transmission and transformation system design ect.

KEYWORDS: Wind field; Meteorological data; MSP430; ZigBee; SD card.

1   INTRODUCTION

The primary basis of wind field approving and initiating economy of running is the analysis of the previous research on wind farm meteorological date. In the past, meteorological data of Chinese wind farms adopted directly the local meteorological department data. It is difficult to meet the special requirements of the wind data for construction of the wind farm. In addition, most of wind farm distribute in remote areas of the traffic inconvenience, poor environment, so data acquisition is extremely inconvenient. In the absence of the project we hope can with less investment to obtain the wind data in recent years.

2   GETTING STARTED

The system is mainly to achieve monitoring wind speed, wind direction, temperature and humidity data of wind field . On the whole, it is composed of data acquisition, ZigBee wireless transmission, data processing. The block diagram of overall system is shown in Figure 1.

The sensor signal process into MCU A/D channel after conditioned by conditioning circuit. After MCU A/D conversion processing, the signal is sent out by MCU through short distance wireless radio frequency chip in this way completed the data acquisition process. After receiving data through short distance wireless RF chip, all data will be stored in the SD card, so completes the process of data.

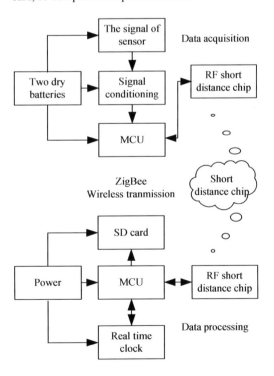

Figure 1.   The block diagram of overall system.

3 LAYOUT OF TEXT

The system adopts TI MSP430F2618 MCU microcontroller. MSP430F2618 has ultra-low power consumption and up to 16MIPS frequency, besides storage space is 116KB+256KB Flash Memory, 8KB RAM, with 4 universal serial communication interface(USCI) module, supports USART and I2C mode.

3.1 The design of data acquisition module

The temperature sensor is the DS18B20 digital temperature sensor. Humidity sensor is the capacitive humidity sensor. Wind speed sensor adopts a wind vane driving a wheel rotation, which rotates a circle, so photoelectric tube receives the light emitted by the light emitting diode in a state of conduction. The signal processed by the signal conditioning circuit and MSP430F2618, measured the wheel speed value that is the value of wind speed. The wind direction sensor adopts bracket driving a rotating 360 degrees rotation sliding rheostat. When the wind blows to the baffle, the bracket rotates with the wind, so sliding rheostat rotates. When the rotation of the bracket stops, the value of sliding rheostat is the value of wind direction. Data acquisition module is powered by two dry batteries. The data acquisition module is shown in Figure 2.

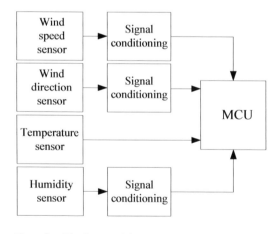

Figure 2.   The data acquisition module.

3.2 The design of ZigBee wireless module

ZigBee is a kind of wireless communication technical standard with short distance, low power consumption, low complexity.

ZigBee supports three network topology structure including star, tree and mesh. It defines three types of equipment, including the Coordinator, Router and End Device. The Coordinator is used for starting up and configuration of network. It is responsible for normal network and maintaining communication with other equipment in the same network. A ZigBee network can only have one coordinator. The Router is a kind of support supporting related equipment, is able to forward information to other devices. The End Device can perform related functions, and can use the ZigBee network communicating with other equipment. In this system, the data acquisition module is the end device, data processing module is the coordinator.

Short distance wireless RF chip adopts CC2520 of TI. CC2520 is the second generation of ZigBee chips, has higher transmitting power and receiver sensitivity, and low power consumption.

3.3 The design of data processing module

The system needs to store the collected data, we use the SD card. The SD card can be read and written in SD and SPI mode, but we use the SPI mode. In order to facilitate the PC read date from SD card, we establish the FAT16 file system in the SD card.

In order to demarcate time of every meteorological data received, we use the DS3231 clock chip to acquire the data processing time. The communication between MSP430F2618 and DS3231 is I²C mode.

In addition, the keys are used for setting the time interval settings of data acquisition; liquid crystal displays the current time data. Data processing module is powered by the power supply. The data processing module diagram is shown in Figure 3.

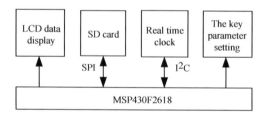

Figure 3.   The data processing module diagram.

4 SOFTWARE DESIGN

MCU program draws on the ZigBee protocol stack of TI company(Z-Stack). Z-Stack is constructed by the idea of the operating system, using event round robin. When no events happened, the system enters a low-power mode. When an event occurs, the system is woke up and begins to handle the event, then the system enters a low power mode. If several events are occurred at the same time, determining the priorities, then processing the event one by one. This software framework can greatly reduce the power consumption of the system.

4.1 *The Design of data acquisition program*

The data acquisition program flow chart is shown in Figure 4. The data acquisition module timing gather wind speed, wind direction, temperature and humidity, sends the data of wind field to the data processing module, and receives the set command of parameters from data processing module to set the acquisition time interval.

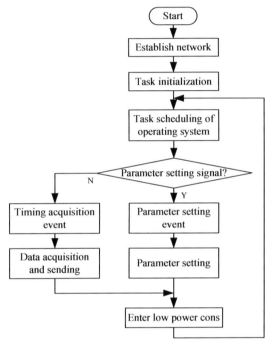

Figure 4. The data acquisition program flow chart.

4.2 *The design of data processing module program*

Data processing module timing receives the date from the acquisition module, and stores the data into SD card, besides can send commands of setting the parameters to data acquisition module to set the data acquisition time interval parameter. The program flow chart is shown in Figure 5.

5 CONCLUSION

The design of wind-field meteorological data monitoring system is feasible. In this paper, ZigBee network of the system adopts point to point communication. If ZigBee structures as star or mesh network, it will be able to measure meteorological data in multiple point of different positions, and integrate in the cluster monitoring system in wind field.

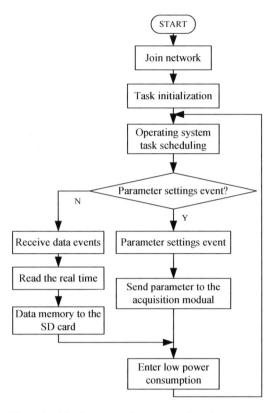

Figure 5. The data processing program flow chart.

REFERENCES

Lan Liu, Na He, Mingyan Lu. (2009). The Wireless Building Monitoring Systems Based on MSP430F2618 and CC2520. *Journal of University of South China(Secience and Technology)* . vol4, 30–32.

Fucheng Cao, Guangming Zhang. (2011).Design of Anemometric System Based on ZigBee in Wind Farm. *Renewable Energy*. vol5, 13–17.

Junbin Li, Yongzhong Hu. (2011). Design of ZigBee Network Based on CC2530. *Electronic Design Engineering*. Vol16, 31–36.

Wang G.F, Y.S. Zhao&Y.S. Fan (2013). Research on error compensation algorithm for wind speed and direction measurement. *Chinese Journal of Scientific Instrument*. 4,11–15.

Li R.X&H.B. Yu (2007). Study of realizing technology on ZigBee wireless communication protocol. *Computer Engineering and Applications*. 43, 143–145.

Cheng Q.M., Y.M. Cheng, M.M. Wang, Y.F. Wang (2010). Development of wind speed measuring technologies in wind power generation. *Process Automation Instrumentation*. 31,156–160.

Electronic Engineering and Information Science – Wang (Ed.)
© *2015 Taylor & Francis Group, London, ISBN: 978-1-138-02772-5*

Research on the impacts of working-wavelength on the precision of the non-source temperature calibration method based on power function

K. Sun, J. Yu, H. Sun & W. Li

The Higher Educational Key Laboratory for Measuring & Control Technology and Instrumentations of Heilongjiang Province, Harbin University of Science and Technology, Harbin, China

ABSTRACT: In the field of multi-spectral temperature measurement, the high-temperature blackbody furnace is applied to calibrate the Multi-Spectral Pyrometer (MSP). The material of the blackbody furnace is graphite and its melting point is 3000°C. As a result, the MSP is not able to be calibrated above 3000°C (non-source temperature) and the measurement range of MSP is limited to lower than 3000°C. To solve the problem above, the non-source temperature calibration method based on curve similarity principle (NCCSP) is proposed. The core idea of the NCCSP is that apply the calibration data below 3000°C to predict the calibration data above 3000°C. The working-wavelength number of MSP is from a few to hundreds, under which the NCCSP is more accurate needs to be further researched. Therefore, the impacts on the NCCSP accuracy of the working-wavelength selection are researched to get the working-wavelength selection basis.

KEYWORDS: MSP; NCCSP; Working-wavelength.

1 INTRODUCTION

Calibration technique is a key technology in the field of multi-spectral temperature measurement, in a sense, to study it even more important than the study of MSP instrument itself (Sun 2011, Sun 2013, Sun 2012). At present, the MSP used in high-temperature measurement has already had high resolution and high signal to noise ratio. However, the non-source temperature calibration falls far behind the development of the MSP and the existing methods which are the one-point calibration method and the warming filter calibration method have already seriously hindered the precision and application range of the pyrometer (Coppa 1988, He 2003). In order to break through the limitation of calibration of non-source temperature, the NCCSP has been put forward in a previous paper (He, 1999). Due to the number of the MSP work-wavelength ranging from a few to a few hundred, the work-wavelength under which the application of the NCCSP has higher accuracy needs to be researched. In this article, the impacts on the NCCSP accuracy of the wavelength are researched to get the work-wavelength selection regular.

2 PRINCIPLE OF THE NCCSP

The actual temperature T of blackbody furnace has the following relationship with output voltage U when the working wavelength is λ:

$$T = a + bU^c \tag{1}$$

In function (1), a, b and c are the unknown value, use the derivative least squares method (Sun 2011), let

$$Q_0 = \sum_{i=1}^{n} w_i \left[(T_i - a - bU_i^c) \right]^2 \tag{2}$$

$$Q_1 = \sum_{i=1}^{n-1} w_i' \left[(T_i' - c\frac{\overline{T}_i}{U_i} + ca\frac{1}{U_i}) \right]^2 \tag{3}$$

Then, the model parameters a, b and c can be obtained from the following functions.

$$\frac{\partial Q_0}{\partial b} = -2\sum_{i=1}^{n} w_i \left[(T_i - a - bU_i^c) \right] \times U_i^c = 0 \tag{4}$$

$$\frac{\partial Q_1}{\partial a} = -2c\sum_{i=1}^{n-1} w_i' \left[(T_i' - c\frac{\overline{T}_i}{U_i} + ca\frac{1}{U_i}) \right] \times \frac{1}{U_i} = 0 \tag{5}$$

$$\frac{\partial Q_1}{\partial c} = -2\sum_{i=1}^{n-1} w_i' \left[(T_i' - c\frac{\overline{T}_i}{U_i} + ca\frac{1}{U_i}) \right] \times \left(\frac{\overline{T}_i}{U_i} - a\frac{1}{U_i} \right) = 0 \tag{6}$$

$$c = \frac{S_{tt}S_{xx'} - S_{tx}S_{tx'}}{S_{tt}S_{xx} - S_{tx}^2} \tag{7}$$

607

$$a = \frac{S_{tx}S_{xx'} - S_{xx}S_{tx'}}{S_{tt}S_{xx'} - S_{tx}S_{tx'}} \tag{8}$$

$$b = \frac{S_{xm} - aS_{1m}}{S_{2m}} \tag{9}$$

$$S_{tt} = \frac{1}{n-1}\sum_{i=1}^{n-1}\frac{w_i'}{\overline{T}_i^2} \tag{10}$$

$$S_{xx} = \frac{1}{n-1}\sum_{i=1}^{n-1}\frac{w_i'\overline{U}_i^2}{\overline{T}_i^2} \tag{11}$$

$$S_{tx} = \frac{1}{n-1}\sum_{i=1}^{n-1}\frac{w_i'\overline{U}_i}{\overline{T}_i^2} \tag{12}$$

$$S_{tx'} = \frac{1}{n-1}\sum_{i=1}^{n-1}\frac{w_i'U_i}{\overline{T}_i} \tag{13}$$

$$S_{xx'} = \frac{1}{n-1}\sum_{i=1}^{n-1}\frac{w_i'\overline{U}_iU_i}{\overline{T}_i} \tag{14}$$

$$S_{xc} = \frac{1}{n}\sum_{i=1}^{n}w_i\overline{U}_iT_i^c \tag{15}$$

$$S_{1c} = \frac{1}{n}\sum_{i=1}^{n}w_iT_i^c \tag{16}$$

$$S_{2c} = \frac{1}{n}\sum_{i=1}^{n}w_iT_i^{2c} \tag{17}$$

$$S_{xc-1} = \frac{1}{n}\sum_{i=1}^{n}w_i\overline{U}_iT^{c-1} \tag{18}$$

$$S_{1c-1} = \frac{1}{n}\sum_{i=1}^{n}w_iT_i^{c-1} \tag{19}$$

$$S_{2c-1} = \frac{1}{n}\sum_{i=1}^{n}w_iT_i^{2c-1} \tag{20}$$

$$U_i' = \frac{U_{i+1} - U_i}{T_{i+1} - T_i}, \qquad i = 1, 2, \cdots, n-1 \tag{21}$$

$$\overline{T}_i = \eta T_{i+1} + (1-\eta)T_i, \qquad i = 1, 2, \cdots, n-1 \tag{22}$$

$$\overline{U}_i = \eta U_{i+1} + (1-\eta)U_i, \qquad i = 1, 2, \cdots, n-1 \tag{23}$$

Where w_i and w_i' are the corresponding weight of known temperature T_i. The rules of value-taking are 1) the weight value near extrapolation range is large; 2) the weight value of actual measurement is large. However, in the process of 100°C-interval calibration of the known temperature range, every calibration point is subjected to random and systemic errors, and the relationship of which is hard to predict. Therefore, in order to reduce the interference of the error of

one point to the whole curve, let $w_i = w_i'$, and use the known temperature point near the extrapolation range. In function (22) and (23), parameter ε is set as 1/2 in actual practice.

3 THE IMPACTS ON THE NCCSP ACCURACY OF THE WORK-WAVELENGTH SELECTION

Theoretically, in high temperature measurement, the commonly used spectral range of MSP is from 0.4μm to 1.1μm in which blackbody temperature-radiation curves under different work-wavelengths is used to verify the method. The specific method is as follows: the Planck function calculation values of five 100°C-interval points from 2600°C to 3000°C are taken as the calibration value of source temperature range. The Planck calculation value at temperature point over 3000°C is taken as the output of MSP. Then, through NCCSP, the corresponding temperature value is obtained. Table 1 to table 3 shows the results when λ=0.4μm, λ=0.6μm, and λ=0.8μm.

Table 1. The results and errors of theoretical extrapolation when λ is 0.4 μm.

ER	EV	E	ER	EV	E
°C	°C	%	°C	°C	%
100	3102.363	0.0762	1000	3964.947	−0.8763
200	3202.065	0.0645	1100	4055.068	−1.0959
300	3300.097	0.0302	1200	4139.157	−1.3353
400	3399.056	−0.0278	1300	4231.475	−1.5936
500	3496.182	−0.1099	1400	4317.739	−1.8696
600	3592.206	−0.2164	1500	4402.703	−2.1621
700	3687.163	−0.3469	1600	4486.368	−2.4702
800	3780.965	−0.5009	1700	4568.736	−2.7928
900	3873.570	−0.6777	1800	4649.814	−3.1289

Extrapolation range – ER; Extrapolation value – EV; Error – E

Table 2. The results and errors of theoretical extrapolation when λ is 0.6 μm.

ER	EV	E	ER	EV	E
°C	°C	%	°C	°C	%
100	3100.714	0.0230	1000	3962.028	−0.9494
200	3199.996	−0.0001	1100	4052.362	−1.1619
300	3298.578	−0.0431	1200	4141.498	−1.3929
400	3396.355	−0.1072	1300	4229.421	−1.6413
500	3493.238	−0.1932	1400	4316.123	−1.9063
600	3589.151	−0.3013	1500	4401.601	−2.1866
700	3684.033	−0.4315	1600	4485.856	−2.4814
800	3777.832	−0.5834	1700	4568.891	−2.7896
900	3870.507	−0.7562	1800	4650.712	−3.1102

Extrapolation range – ER; Extrapolation value – EV; Error – E

608

Table 3. The results and errors of theoretical extrapolation when λ is 0.8 μm.

ER	EV	E	ER	EV	E
°C	°C	%	°C	°C	%
100	3100.176	0.0005	1000	3963.484	−0.9129
200	3199.374	−0.0196	1100	4054.010	−1.1097
300	3297.934	−0.0626	1200	4144.423	−1.3233
400	3395.758	−0.1248	1300	4233.238	−1.5526
500	3492.762	−0.2068	1400	4320.942	−1.7968
600	3588.880	−0.3089	1500	4407.531	−2.0549
700	3684.053	−0.4310	1600	4493.008	−2.3259
800	3778.234	−0.5728	1700	4577.376	−2.6090
900	3871.388	−0.7336	1800	4660.642	−2.9033

Extrapolation range – ER; Extrapolation value – EV; Error – E

The model parameters a, b, and c under working-wavelengths of 0.4μm, 0.5μm, 0.6μm, 0.7μm, 0.8μm, 0.9μm, 1.0μm and 1.1μm and the non-source temperature extrapolation ranges when the calibration precisions are greater than 3‰, 1% and 3% are shown in table 4.

Table 4. The theoretical extrapolation ranges and model parameters.

WW				ERT 3‰	ERT 1%	ERT 3%
μm	MPa	MPb	MPc	°C	°C	°C
0.4	1481.85	193.12	0.167	600	1000	1700
0.5	1498.33	85.79	0.210	600	1000	1700
0.6	1506.05	40.16	0.252	600	1000	1700
0.7	1508.06	19.75	0.294	500	1000	1700
0.8	1505.48	10.22	0.335	600	1000	1800
0.9	1498.99	5.57	0.374	600	1100	1900
1.0	1489.22	3.20	0.411	600	1100	1900
1.1	1476.75	1.93	0.446	600	1100	2000

Working-wavelength – WW; Model parameters – MP;
Extrapolation range when precisions are greater than – ERT;

As can be seen from the table 4, when the calibration precisions are greater than 3%, the minimum extrapolation range is 1700°C. Take 17 100-interval points from 3100°C to 4700°C to calculate the variances and means of the errors under different working-wavelengths. The results are shown in table 5.

Table 5. The means and variances of theoretical extrapolation errors.

WW		
μm	M	V
0.4	−0.00936	0.00318
0.5	−0.00984	0.00326
0.6	−0.00994	0.00324
0.7	−0.00980	0.00318
0.8	−0.00949	0.00306
0.9	−0.00909	0.00292
1.0	−0.00862	0.00276
1.1	−0.00814	0.00261

Working-wavelength – WW; Means – M; Varinances – V

4 CONCLUSION

The non-source temperature calibration method based on curve similarity principle has been put forward in the previous paper. In this paper, the NCCSP is further researched. The working-wavelength selection effects on the NCCSP precision is researched when the wavelength range is 0.4μm to 1.1μm and then get the working-wavelength selection method. That is when the working-wavelength is from 0.4μm to 0.6μm, the wavelength selection is arbitrary. When working-wavelength is from 0.6μm to 1.1μm, the wavelength selection is to select large wavelengths as possible. The regular above provides a theoretical basis for working-wavelength selection of the NCCSP.

ACKNOWLEDGMENTS

This research is supported by Technology Research Project of Heilongjiang Province Education Department (12541104). The constructive comments from the reviewers are of this paper gratefully acknowledged which have helped the author to improve the paper.

REFERENCES

Sun, K. & X.G. Sun,(2011). Development of Multi-spectral Thermometer for Explosion Flame True Temperature Measurement. *Spectrosc. Spectral. Anal. 31(3)*, 849–853.
Sun, K. & X.G. Sun,. (2013). Research on the Anti-Random Error Abilities of the Two Non-source Temperature Calibration Methods of Multi-spectral Pyrometer. *Spectrosc. Spectral. Anal. 33(6)*,1723–1727.

Sun, K. (2012). The non-source temperature calibration method of multi-spectral pyrometer .*In 2012 IMCCC*, Harbin, 15–18.

P. Coppa, G. Ruffino & A. Spena. (1988). Pyrometer Wavelength Function: Its Determination and Error Analysis. *High Temperature - High Pressure. 20*, 479–490.

He, J. (1999). A New Calibration Method of Narrow-band Radiation Pyrometer. *Chin. J. Sci. Instru.20(3),* 268–270.

Sun, K. & X.G..Sun, (2011). Development of Multi-spectral Thermometer for Explosion Flame True Temperature Measurement. *Spectrosc. Spectral. Anal. 31(3),* 849–853.

Electronic Engineering and Information Science – Wang (Ed.)
© 2015 Taylor & Francis Group, London, ISBN: 978-1-138-02772-5

Internet-based of a wireless monitoring system for temperature regulating device design

J.F. Zhang, Z.M. Luo, Z.F. Liu & S. Zhao
The higher educational key laboratory for Measuring & Control Technology and Instrumentations of Heilongjiang Province, Harbin University of Science and Technology, Harbin, Heilongjiang, China

G.Y. Huang
Heilongjiang Provincial Institute of Measurement & Verification, Harbin, Heilongjiang, China

ABSTRACT: A wireless monitoring system for the temperature regulating device is designed in this article, which is based on wireless sensor network and the Android operating system. To collect the temperature comprehensively and accurately, the signal acquisition module is formed by four digital temperature sensors: DS18B20, which are placed in four key points. Wireless network based on IEEE 802.11 standard is used for achieving real time data display because of the generality and reliability of the standard. IAP15F2K60S2 is the control chip to read the temperature from digital temperature sensors and send the data for display and storage. The monitor module consists of a mobile device with the Android operating system and a touch screen. The historical data was saved as txt format in the mobile device for data analysis and was displayed on the touch screen periodically. To guarantee the accuracy of the system data, the sensors in the system were tested with a standard temperature calibration instrument.

KEYWORDS: Wireless Network; DS18B20; Android; Touch Screen.

1   INTRODUCTION

Temperature regulate device has been widely used in many fields as a testing equipment, such as breeding in agriculture, electronic equipment testing and microorganism culture in medical science. Based on past experiments, the monitoring system Fuzzy-PID control on Lab VIEW has been used in the control of the constant temperature box, which can make the system has better dynamic characteristics and has good ideal steady-state quality (Xinhua 2008). Pt100 has been used as the temperature sensor in a constant temperature box, and the control chip on the box is the STC12C5620AD MCU (Ma 2009). It is possible to use Internet of Things to combine in a particular operational entity all the bits of the world around us (Bari 2013). The operational facilities of the surmised construction of the Internet of things come out of a cellular automaton infrastructure of the physical world, on top of which develops the all-embracing mechanism of the Holographic Universe (Berkovich & Al-Shargi 2010).

By applying the technology of the Internet of Things in a temperature regulate device, the monitoring system can accompany the control and monitor of the temperature regulate device with a mobile device far away from the device.

2   HARDWARE DESIGN OF MONITORING SYSTEM

The system contains a real-time signal acquisition module, wireless transmission module, control module and monitor module. There are many nodes in the system, which are distinguished by their unique IP address. The working process relies on digital temperature sensor DS18B20 to collect real time temperature and the microcomputer reads the temperature data through the I/O interface. And then send the data through serial port one to the touch screen with the MODBUS Agreement and through serial port two to the master IP address with the UDP protocol. The touch screen can get the real time temperature by unpacking the MODBUS Agreement, and display the data with a curve, besides, the data was saved in the memory of the touch screen. At the same time, the mobile device with the Android operating system can receive the temperature data of any node by unpacking the UDP protocol after visiting the IP address of the node. Schematic of the system is shown in Figure 1.

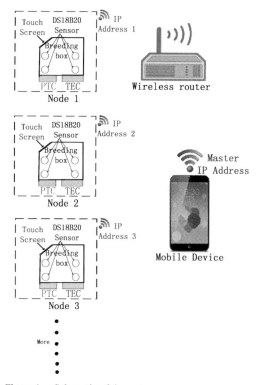

Figure 1. Schematic of the system.

The signal acquisition module is formed of four DS18B20 sensors, which has a temperature measurement range from −10°C to 85°C with an accuracy of ±0.5°C. Besides, the DS18B20 communicates over a 1-Wire bus that by definition requires only one data line for communication with a central microprocessor. After a lot of experiments, we find four key points which can represent the temperature in the environment to place the sensors.

Wireless transmission relies a module which can transmit signals through serial port and wireless network. With a wide range of baud rate from 1200 to 500000 bps and support multiple security authentication mechanisms, it is possible to accompany high speed and security transmit. The WI-FI module also support most of the mainstream protocol, such as the User Datagram Protocol, which is used in the monitor system. We allot each WI-FI module a unique IP address, which is used for the mobile device to read the signal of every node.

IAP15F2K60S2 microcomputer is the core of the control module. By controlling the positive temperate coefficient and the thermo-electric cooler with the relay according to real-time temperature, the microcomputer can keep the temperature in a constant range. A pump is used in the control module for circulating water in the breeding box to keep the temperature in the box even.

With a touch screen on the breeding box and a mobile device with the Android operating system, we can monitor the temperature environment of the breeding box somewhere. Both of the interactive modes can show the curve of the temperature, besides, the temperature was saved in a curve format and txt format, which is easy to import in a computer and it is possible to do any analysis with the data later.

3 SOFTWARE DESIGN OF MONITORING SYSTEM

This part of the software design and development environment contains Keil uVision4 for C51 with the C language programming, Eclipse for Android with the JAVA programming language and MCGS with the graphic language. The software schematic of the system is shown in Figure 2.

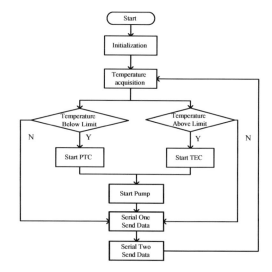

Figure 2. Software schematic of the system.

The Android Package for the monitor system is developed by Eclipse with JAVA programming language. After unpacking the UDP protocol, the Android client can get the data of any nodes by visiting the unique IP address of the node. To reduce the burden of the microcomputer and improve the arithmetic speed, the system analyzes the data procession and modifies the error data with the CPU of the mobile device. The data after the modification was displayed on the mobile device with the chart engine AChartEngine and saved on the SD card every 20 seconds.

4 SENSOR ACCURACY TEST

The sensor is tested by ESPEC of temperature and humidity testing box PVS-3KP, which has a temperature accuracy of 0.1°C. From 10°C to 50°C every 5°C is a test point. We test each point with the DS18B20 sensors for 2 hours, and get the average temperature of each point. The result of accuracy test is shown in Figure 3.

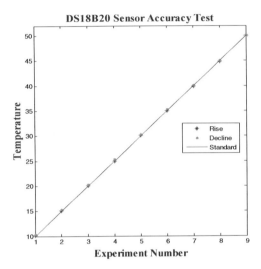

Figure 3. The result of sensor accuracy test.

Actual results show that: the temperature range at 10°C~50°C, the error does not exceed ±0.2°C, they meet the requirements of practical application.

5 THE SYSTEM IN ACTION

To test the monitoring system, the system was applied to the germination phase of the rice breeding process, which lasts 24 hours. In this phase, the temperature should be controlled in a constant range from 30°C to 32°C. After 20 times of the experiment, the temperature can be adjusted in the range of 30.5°C to 31.6°C. After analyzing the experimental data with MATLAB software, the mean temperature of each experiment is shown in Tab. 1.

6 CONCLUSIONS

A description of a monitoring system that contains IAP microcomputer, DS18B20 sensor, WI-FI module, Android client, the positive temperate coefficient and the thermo-electric cooler was presented. After

Table 1. Mean temperature of each experiment.

Experiment	1	2	3	4	5
Mean	30.79	30.79	30.82	30.80	30.80
Experiment	6	7	8	9	10
Mean	30.79	30.80	30.81	30.82	30.79
Experiment	11	12	13	14	15
Mean	30.81	30.78	30.78	30.78	30.78
Experiment	16	17	18	19	20
Mean	30.74	30.78	30.78	30.78	30.77

test the accuracy of the sensor, we can draw a conclusion that the error does not exceed ±0.2°C. After 20 days of the experiment, by control the positive temperate coefficient and the thermo-electric cooler to increase and decrease the temperature, the constant temperature range can be controlled in ±0. 3°C.

ACKNOWLEDGEMENTS

This work was supported by Natural Science Foundation of Heilongjiang Province (Grant No. F201421), Harbin Scientific Innovation Project for Elite Young Researcher (Grant No. 2013RFQXJ003), Scientific Research and Talent Project of Education Department of Heilongjiang Province (Grant No.12541109, 12541140), Higher Education Reform Program of Harbin University of Science and Technology (Grant No. A201300007), 2013 Bilingual Teaching Project of Heilongjiang Affiliated Universities, and Higher Education Reform Program of Heilongjiang Province (Grant No.JG2013010302).

REFERENCES

Jiang X.H.(2008). Designing a Temperature Measurement and Control System for Constant Temperature Reciprocator Platelet Preservation Box Based on Lab VIEW. *Fourth International Conference on Natural Computation*.48–50.

Ma.J.W. (2009). Intelligent Control and Implementation of a New Type Constant Temperature and Humidity Box. *Proceedings of the 29th Chinese Control Conference*.4263–4264.

Bari, N. Mani, G. Berkovich, S.(2013). Internet of Things as a Methodological Concept. *Computing for Geospatial Research and Application*.48–55.

S. Berkovich, H. Al-Shargi. (2010). Constructive Approach to Fundamental Science. *University Publishers, San Diego*.50–53.

Electronic Engineering and Information Science – Wang (Ed.)
© 2015 Taylor & Francis Group, London, ISBN: 978-1-138-02772-5

MRC automatic measurement system of imaging device by FPGA

T.Z. Liu, W.J. Li, P.S. Zhao, D.Y. Yang & C. Wang
School of Electrical and Electronic Engineering, Harbin University of Science & Technology, Harbin, China

ABSTRACT: The evaluation method of photoelectric imaging device has been a research priority topic. MRC is one of the important parameters which can evaluate the imaging device. The traditional MRC measurement principle based on the subjective judgment of the human eye leads to the results which largely depend on the person and it makes the evaluation uncertainty. This paper presents a method of automatic MRC measurement that does not rely on the human eye. FPGA can achieve the judgment of the sharpness of the image by extracting feature vectors of image. In this paper, the entire experimental apparatus is designed and the corresponding experiment is done. The experimental results show that the method of MRC measurement is feasible and very innovative.

KEYWORDS: FPGA; MRC; objective measurement; photoelectric; imaging.

1   INTRODUCTION

Along with the development of electronic technology and optical imaging system, optical imaging technique has become more and more popular. In order to get better and better image quality of optical imaging devices, it needs a comprehensive evaluation system of the optical imaging device to identify the advantages and disadvantages of the imaging system.

Image quality evaluation of visible light imaging system has two parameters: modulation transfer function (MTF) and minimum resolvable contrast (MRC) (Zhou et al. 2002). To any of the visible light imaging systems as long as the two parameters are measured, the performance of the overall system can almost be determined.

MRC is a new kind of subjective evaluation parameter which can qualitatively describe the contrast of the optical imaging system (Li et al. 2006). It considers many factors, just as sensitivity, noise, the target spatial frequency and human visual characteristics of the system threshold contrast. Compared with the MTF which can only illustrate spatial frequency, MRC can be more comprehensive to reflect the limit of performance of the optical imaging system. It can be used to estimate the sight of the optical imaging system.

2   TRADITIONAL MEASUREMENT PRINCIPLE OF MRC

The measurement principle of MRC (Jin et al. 2000) is: the test patterns with different spatial frequencies or different sizes are placed in the background, with the contrast of the test patterns changing, the technician observes the test patterns through visible light imaging system. When the observer can just distinguish the test patterns, the contrast of the test patterns is defined as the distinguished contrast in the minimum spatial frequency or the size of the space in the imaging system. The contrast of the test patterns is defined as the ratio of the target luminance and the background brightness.

Because of the human factor of uncertainty, the standard definition of judgment of the human eye has directly impact on the experimental results, so the computer is used instead of the human eye to achieve the purpose of discerning clarity and realize the identification of automation.

3   PROPOSAL OF MRC OBJECTIVE MEASUREMENT TECHNIQUE

3.1   *Process of objective measurement*

With the improvement of computer technology, computer instead of the human eye has become possible. The automatic measurement method of objective MRC is that FPGA is used instead of the brain which uptakes video images to identify and find out the clear picture (Xiong 2004). However, the threshold value affects the conversion processing result which is from a computer for image processing (binaryzation) to some extent, it adds and subtracts the same amount of the same magnitude on the basis of standard threshold. The new four threshold values are made on the standard upper and lower and it calculates MRC values based on 5 threshold values respectively, thus the average value of the last five thresholds values shall be MRC value of the imaging system in this spatial frequency.

$$MRC = \frac{MRC_A + MRC_B + MRC_C + MRC_D + MRC_E}{5} \quad (1)$$

The specific measurement process is shown in Figure 1.

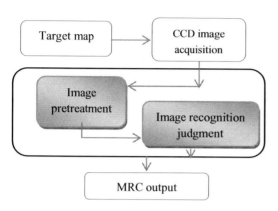

Figure 1.   MRC measurement technique.

3.2   *CCD image acquisition*

CCD has many advantages, for example, self-scanning, high-resolution, high sensitivity, small volume, accurate pixel location, less consumption, long service life and convenient signal processing. It is easy to connect with computer interface. With the rapid development of nearly 30 years, CCD has become very important high-tech in the field of photoelectric imaging and testing. CCD camera is used instead of the device of video input to achieve the purpose of image acquisition.

3.3   *Image pretreatment*

Because FPGA is different from human brain, the recognition of images is realized by identifying data volume. The picture is turned into a kind of simple data type that can be handled by FPGA. And the data is simple in order to ensure its computational speed. Because of the following selection of feature vector, images are needed to handle by gray-scale and binaryzation (Batlle et al. 2006).

3.4   *Selection of eigenvectors*

Eigenvectors based on the relationship between the target and background selection are shown in Formula (2) and (3).

The difference of gray value L:

$$L = \left| L_{max} - L_{min} \right| \quad (2)$$

where L_{max} is the maximum gray value on the background, and L_{min} is the minimum gray value on the background.

The ratio of length to width S:

$$S = \frac{A}{B} \quad (3)$$

where A is the stripe length, and B is the stripe width.

3.5   *Feature extraction*

To extract image feature, the composition of the image must be understood. There are mainly two kinds of commonly used color model, one is the RGB model, and the other is YCbCr model. In the mode of YCbCr, Y is the brightness, Cb is chroma and Cr represents color value. If the image information is displayed in YCbCr mode, the Y value is the gray value, so it is easy to get gray difference. If Y is set as a threshold value, it will be white when the value is greater than this threshold, or it is black, then the image can be obtained. According to the synchronization signal, the length to width ratio can be obtained from the ratio of number of pixel (Yan 2005).

4   DESIGN OF AUTOMATIC MEASUREMENT SYSTEM

4.1   *Hardware design*

Hardware design mainly divided into the design of measurement device and the design of detection device.

The first part, the design of measurement system, the task of measurement device is generating an uniform light before and after target to produce different contrast. A tungsten lamp is regarded as light source of device, and it enters the target integrating sphere and background integrating sphere through the beam splitters respectively. The effect of integrating sphere makes the light into space uniform, so as to achieve the purpose of foreground and background illumination uniformity. Target on the two integral balls joint is used to provide different spatial resolution.

The second part, the design of detection device is shown in Figure 2. The CCD camera is used instead of video capture source. The camera acquisition is analog quantity, so the RCA interface is used. A special video A / D chip is added to convert the video signal to the digital analog. The digital quantity is converted into FPGA. Through processing image data, FPGA outputs the results when it meets characteristic value, at the same time, FPGA also controls the step motor to measure the target space into standard rate of change. In order to verify the accuracy of identification, an LCD screen is added.

Through enabling flag, when it meets the conditions of the feature vector, the signal which is collected by CCD camera can be observed synchronously.

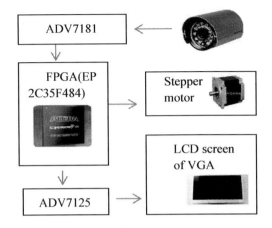

Figure 2.   Design of the detection system.

4.2 *Software design*

EP2C35F484 of Altera is used as the FPGA chip. Quartus ll 8.0 is selected as simulation software of Altera. When the system is powered on, at first the internal register of ADV7181B is configured by I²C model. Because the camera output is an analog video signal in PAL mode, ADV7181B is needed to configure in analog video signal input of PAL mode, and the signal needs to be converted to a digital video signal in CCIR656 format. ADV7181B inputs luminance signal of real-time digital video image obtained by the conversion. Chrominance signal (TD_DAT) and the line-field synchronization signal (TD_HS / VS) are sent into FPGA, then the video signal is converted to YCbCr signal, and the YCbCr signal is converted to RGB signal, at last the signal is transformed to LCD screen through the VGA chip ADV7125.The I²C model is shown in Figure 3:

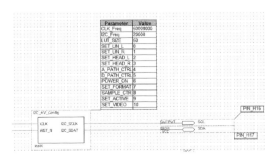

Figure 3.   I²C model.

Video data processing and display model and the image gray value of the two modules are shown in Figure 4 and Figure 5:

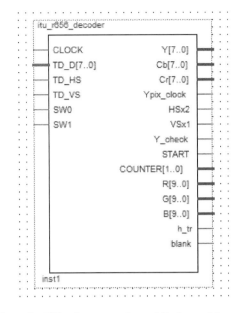

Figure 4.   Video data processing and display model.

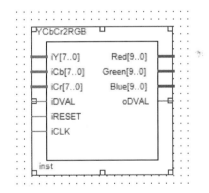

Figure 5.   Image gray value of the two modules.

It shows a large TV_TO_VGA model in Figure 6. The TV_TO_VGA model can be seen in Figure 7. It is a RTL structure diagram of the whole system model. The input of YCbCr2RGB model can be adjusted in the gray transform. The specific method is that the input value of IY[7:0] is kept in constant, and the input values of ICb[7:0] and ICr[7:0] are turned to (8 'h80). When the two values change, it needs to change the values of IY[7:0] to ((Y[7:0]>95)? 8 'hff:8' to hoo).

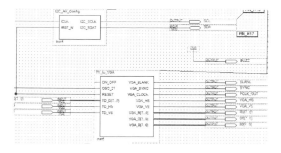

Figure 6. TV_TO_VGA model.

Figure 7. TV_TO_VGA RTL model structure diagram.

5 RESULTS OF EXPERIMENT

Experiment luminance is measured by a luminance meter to achieve; the accuracy of the apparatus is shown in Table 1:

Table 1. Device luminance test.

No.	Device display value (cd/m²)	Luminance meter display value (cd/m²)	Absolute difference (cd/m²)
1	10.5	10.4	0.1
2	14.3	14.5	-0.2
3	17.0	17.1	-0.1
4	20.2	20.3	-0.1
5	24.5	24.4	0.1
6	28.2	28.1	0.1
7	32.3	32.2	0.1
8	39.6	39.4	0.2
9	43.0	42.9	0.1
10	47.3	47.2	0.1

Table 2 shows the results of the curve means measured at different resolutions. Figure 8 shows the actual

effect, left for the target color, characteristic values that match the aspect ratio of the intermediate binary state; the right is consistent with gray-scale difference of gray scale.

Table 2. Actual measurement data.

No.	Spatial frequency (c/mm)	Background luminance display value (cd/m²)	Target brightness display value (cd/m²)	Contrast
1	0.500	38.2	38.4	1.006
2	0.625	38.2	38.6	1.013
3	0.833	38.2	40.4	1.060
4	1.250	38.2	41.6	1.088

Figure 8. Detection device measured map.

6 CONCLUSIONS

This paper proposes a method of MRC automatic measurement based on computer automatic identification through the principle of MRC objective measurement. In addition, the MRC automatic measurement system of imaging device by FPGA is designed. The purpose of accurate identification can be achieved by converting the output image signal, thus the MRC of image system can be measured. The method solves the problem of current automatic measurement values of MRC by computer and takes the length to width ratio and difference of gray value as the characteristic vector innovatively. Furthermore, it ensures the accuracy of detection.

ACKNOWLEDGEMENTS

This work was supported by the Heilongjiang Natural Science Foundation (F201125).

REFERENCES

Batlle, J., Marti, J. & Ridao, P. 2006. A New FPGA/DSP-Based Parallel Architecture for Real-Time Image Processing. *Real-Time Imaging* 8: 345–347.

Jin, W. Q., Gao, Z.Y., Su, X. G. & Guo H. 2000. On the Problem of Matching Between Optoelectronics Imaging System and Human Vision. *Infrared Technology* 22: 40–42.

Li, W. J., Zhang, Y., Dai, J. M. & Chen, Y. H. 2006. Study on the Measurement Techniques of MRC in Visible Imaging System. *Acta Metrologica Sinica* 27(1): 32–34.

Xiong, X. H. 2004. Digital image quality assessment. *Science of Surveying and Mapping* 29(1): 68–70.

Yan, M. 2005. Application of FPGA in digital image processing. *Electronic Technology* 1: 74–75.

Zhou, Y., Jin, W. Q., Gao, Z.Y. & Liu, G. R. 2002. Minimum Resolvable Contrast (MRC) Study for CCD Low-light-level Imaging System. *Proc. SPIE* 4952: 591–597.

Electronic Engineering and Information Science – Wang (Ed.)
© 2015 Taylor & Francis Group, London, ISBN: 978-1-138-02772-5

Active power filter based on dq detection arithmetic and one-cycle control

Q. Yan, W.J. Li, T.Q. Wu, J. Feng & Y. Sheng
School of Electrical and Electronic Engineering, Harbin University of Science &Technology, Harbin, China

ABSTRACT: At present, the problem of harmonic pollution is increasingly serious. Active power filter is one of the most effective means to reduce harmonic pollution in power grid. In consideration of good real-time of dq detection arithmetic and excellent dynamic response of One-Cycle Control (OCC), the active power filter is designed with them. The system of active power filter is modelled on PSCAD/EMTDC software, and its compensation performance is simulated. Before and after compensation, the waveforms of source current and voltage are extracted and analyzed. The feasibility of the system is further validated. The experimental results show that active power filter based on dq detection arithmetic and OCC can effectively compensate for the harmonic component, and make it reduce markedly.

KEYWORDS: active power filter; dq detection arithmetic; one-cycle control; harmonic pollution.

1   INTRODUCTION

With the development of modern industry, electronic devices inject large amounts of harmonics to the power grid, and the power quality is increasingly decreased (Qiao et al. 2008). Active power filter (APF), as a kind of equipment for harmonic suppression, has been applied widely in electric power system (Hu et al. 2004).

Nowadays, the instantaneous space vector method based on instantaneous reactive power theory is a detection method of APF, which is the most widely used. It mainly includes the p-q method (Haroen et al. 2005), ip-iq method introduced by literature (Chen et al. 2009), and the adaptive detection method described by literature (Xu et al. 2013). The p-q method is only suitable for the detection of grid voltage under no harmonic distortion situation; the ip-iq method is not only effective to the grid voltage without distortion, also with distortion, but this detection circuit is too complex. Similarly, the adaptive detection method can accurately detect the harmonic current in power grid voltage distortion, while with a slow dynamic response. This paper introduces a harmonic current detection method based on dq coordinate transformation. It can accurately detect the harmonic current whether the power distorts or not.

One-cycle control (OCC) is a new nonlinear control method. Compared with the traditional control, it has the characteristics of quick response, constant switching frequency, strong robustness and simple control circuit etc.

In this paper, the parallel APF system is modelled on PSCAD/EMTDC software. The simulation verifies the superiority of system on the compensation performance for different loads.

2   THE DQ DETECTION ARITHMETIC

The dq detection arithmetic is the use of park transforming from the three-phase current in the a-b-c coordinate system into the d-q-0 coordinate system. The load current includes harmonic component of positive sequence, negative sequence and zero sequence. Park transform is applied to the three components. The zero sequence component value becomes zero.

When the three-phase load current contains each harmonic component, the expression is:

$$\begin{cases} i_a = i_{1n} \sin(n\omega t + \varphi_{1n}) + i_{2n} \sin(n\omega t + \varphi_{2n}) \\ i_b = i_{1n} \sin(n\omega t + \varphi_{1n} - \dfrac{2\pi}{3}) + i_{2n} \sin(n\omega t + \varphi_{2n} - \dfrac{2\pi}{3}) \\ i_c = i_{1n} \sin(n\omega t + \varphi_{1n} + \dfrac{2\pi}{3}) + i_{2n} \sin(n\omega t + \varphi_{2n} + \dfrac{2\pi}{3}) \end{cases} \quad (1)$$

Among them, i_{1n} is the effective value of each positive sequence current. φ_{1n} is the initial phase angle of each positive sequence current. i_{2n} is the effective value of each negative sequence current. φ_{2n} is the initial phase angle of each negative sequence current.

ω is the fundamental frequency. After park transform is performed on the last formula. It turns into:

$$\begin{bmatrix} i_d \\ i_q \end{bmatrix} = C_{dq} \begin{bmatrix} i_a \\ i_b \\ i_c \end{bmatrix} = \begin{bmatrix} i_{1n}\sin[(n-1)\omega t + \varphi_{1n}] \\ -i_{1n}\cos[(n-1)\omega t + \varphi_{1n}] \end{bmatrix}$$

$$\begin{bmatrix} i_{2n}\sin[(n+1)\omega t + \varphi_{2n}] \\ -i_{2n}\cos[(n+1)\omega t + \varphi_{2n}] \end{bmatrix} \quad (2)$$

From the analysis of Formula 2, there exists a fundamental frequency difference between the original signal and the signal after dq transform. Among them, the negative sequence component after transform adds one than the number of fundamental wave, and the positive sequence component transform reduces one. At the same time, the fundamental wave is converted into a DC signal.

The dq detection method is shown in Figure 1. Among them, the input currents of the three-phase loads are i_a, i_b and i_c. After dq transform the three-phase load currents become the d axis component i_d and q axis component i_q, including DC component $\bar{i}_d$, $\bar{i}_q$ and AC component $\tilde{i}_d$, $\tilde{i}_q$. The DC component of d axis $\bar{i}_d$ and q axis $\bar{i}_q$ can be obtained by the low pass filter (LPF). After inverse dq transform matrix, the fundamental component of detected currents i_{af}, i_{bf}, i_{cf} can be obtained by $\bar{i}_d$, $\bar{i}_q$. Among them, the difference between the load current and the fundamental current component is the need for harmonic current detection.

According to the theoretical analysis above, it can be seen that the advantage of dq detection method is regardless of instantaneous reactive power. As a result, voltage detection is omitted in the calculation process of harmonic current. It also directly eliminates the interference of positive, negative and zero sequence components of harmonic in the calculation of harmonic current.

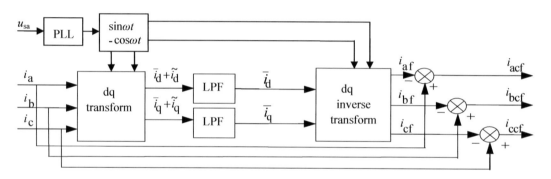

Figure 1. Principle diagram of the dq arithmetic.

3 REALIZATION OF ONE-CYCLE CONTROL

3.1 OCC theory derivation

One-cycle control is applicable to control switching circuit. The basic idea is: in each clock cycle, the average value of switch variable is the same or proportional to control reference quantity.

The control target of using one-cycle control method in active power filter is to make source current follow the change of source voltage, hence it can simultaneously compensate harmonic component and reactive power component. But one point needs to be emphasized; APF requires a higher capacity in this case. And there is a prerequisite for the derivation of control equations, which assumes the source voltage is balanced without distortion. If there is distortion in source voltage, the current which tracks voltage is influenced. In this paper, the method of one -cycle

control and dq detection method are used in three-phase APF in order to compensate harmonic component of load flexibly and in real time.

Load current and compensation current generated by APF come together to form three-phase power current. The load current can be regarded as the sum of fundamental component and harmonic component, where

$$\begin{cases} i_{sa} = i_{Lfa} + i_{Lha} + i_{Ca} \\ i_{sb} = i_{Lfb} + i_{Lhb} + i_{Cb} \\ i_{sc} = i_{Lfc} + i_{Lhc} + i_{Cc} \end{cases} \quad (3)$$

The goal of active power filter is to minimize harmonic component, so that the remaining only contains fundamental component. Hence the fundamental component of load current that the source current tracks is regarded as the control target; whose difference magnified h times is control reference quantity of OCC. The control equation is expressed in Formula 4:

$$
\begin{cases}
\dfrac{1}{T_s}\displaystyle\int_0^{dT_s} i_{sa}\,dt = d_{an}\bar{i}_{sa} = u_{refa} = h(i_{Lfa} - i_{sa}) \\[2ex]
\dfrac{1}{T_s}\displaystyle\int_0^{dT_s} i_{sb}\,dt = d_{bn}\bar{i}_{sb} = u_{refb} = h(i_{Lfb} - i_{sb}) \\[2ex]
\dfrac{1}{T_s}\displaystyle\int_0^{dT_s} i_{sc}\,dt = d_{cn}\bar{i}_{sc} = u_{refc} = h(i_{Lfc} - i_{sc})
\end{cases} \quad (4)
$$

In the Formula 4, d_{an}, d_{bn}, d_{cn} are the duty ratio. T_s is a switching cycle, also the time constant of integrator. Formula 4 shows that the value of h becomes more and more large, and then the power current value is closer to the reference value; in consequence control objective is completed.

3.2 OCC control process

Figure 2 is the structure diagram of one-cycle controller. Each controller includes the reset integrator, comparator and RS flip-flop. At the same time, two absolute value circuits are joined to maintain the ability that the reset integrator keeps constant frequency integration. CLK is a 5 kHz frequency clock pulse signal. Through the driving circuit, output signal can be used to control the corresponding main circuit switch.

OCC control process can be seen from the Figure 2. For a phase, when the PI regulator output value u_{refa} is greater than zero, the reference current is greater than the power current of a phase, the output of Q_{ap} remains as "0", and the output of Q_{an} remains as "1". When output value u_{refa} u_{refa} of the PI regulator is lower than zero, the state of Q_{ap} depends upon the output Q_a of flip-flop. After each clock cycle begins, the clock pulse follows, and integrator begins working whose function is to do integral for the power current from zero. When the integrator output value $|u_{int\,a}|$ equals with the $|u_{refa}|$, the output value of comparator is rapidly turned over, and the output value $|u_{int\,a}|$ makes corresponding change. The integrator resets when the next cycle comes, and then continues executing the same action as the last cycle (Jin et al. 2006).

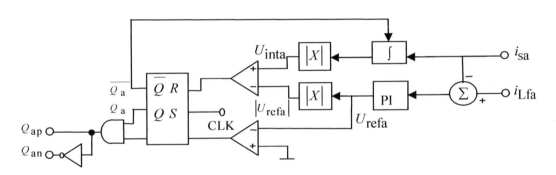

Figure 2. Controller structure of APF OCC.

4 EXPERIMENT RESULTS AND ANALYSIS

4.1 Waveforms analysis

In order to further verify the correctness and feasibility of one-cycle control of APF in this paper, a simulation system of APF based on dq detection and one-cycle control is built in the PSCAD/EMTDC simulation software. APF system based on dq detection and one-cycle control is simulated. The source voltage is 220V. The frequency is 50Hz. The capacitance value is 1000µF, whose voltage is 1200V. Three-phase uncontrolled rectifier bridge with RL

load is used to be the load, among them, $R=10\Omega$, $L=2mH$.

Taking a phase for example, the waveforms of load current, compensation current and source current after compensation are shown in Figure 3. Meanwhile, in order to further verify the dynamic performance of APF, sudden load is joined in the model at 0.230s. And the results of simulation waveforms are shown in Figure 4.

It can be seen from Figure 4, when sudden load is joined, APF can also detect the change of load current timely. In about 1/4 power cycle, APF can enter into stable compensating state.

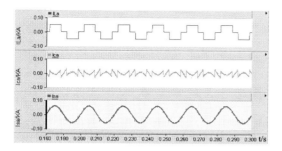

Figure 3. Current waveforms of steady state.

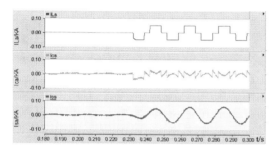

Figure 4. Current waveforms of sudden load.

4.2 THD analysis

The total harmonic distortion (THD) is a fundamental index to measure the compensation performance of APF. THD and the spectral analysis of a phase source current waveform before and after compensation are shown in Figure 5 and Figure 6.

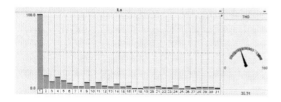

Figure 5. Spectral analysis and THD before compensation.

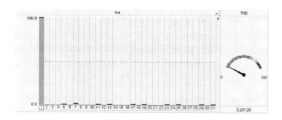

Figure 6. Spectral analysis and THD after compensation.

It can be seen from Figure 5 and Figure 6, the distortion of a phase source current is serious before compensation. THD reaches 35.31%, but the THD has dropped from 35.31% to 3.25% after compensation.

5 CONCLUSIONS

In this paper, from the point of harmonic pollution in AC power network, dq detection arithmetic and one-cycle control are selected for the system, after comparing the kinds of detection method and control strategy of the APF. Parallel three-phase APF system is found, and the systematic theoretical analysis is completed. Simulation model is built, and the compensation performance of the system is simulated on the PSCAD/EMTDC software. It can be seen from the simulation results that APF has a good filtering effect, whether with the sudden load or RL load. Such results also show that the system is correct and executable.

REFERENCES

Qiao, C. M., Jin, T. T. & Smedley, K. M. 2008. One-cycle control of three-phase active power filter with vector operation. *IEEE Transactions on Power Electronics* 51(2): 455–463.

Hu, Z. B., Zhang, B. & Deng, W. H. 2004. Feasibility study on one cycle control for PWM switched converters. *IEEE 35th Annual Power Electronics Specialists Conference Aachen Germany* 5: 3359–3365.

Haroen, Y. & Riyadi, S. 2005. Analysis of Instantaneous Representative Active Power equality based Control Method for Three Phase Shunt Active Power Filter. *IEEE Transactions on Power Electronics* 1(2): 542–547.

Chen, Q.G. & Zhao, C.M. 2009. Simulation of Algorithm and Parameters for Shunt Active Power Filter in Three-phase and Four-wire System. *Electric Machines and Control* 13(1): 20–24.

Xu, H.,Wang, Q.J.,Qi, X. & Li, G.L. 2013. Study and Implementation of Adaptive Harmonic Detecting Algorithm. *Electric Drive* 43(5): 51–54.

Jin, T. T. & Smedley, K. M. 2006. Operation of One-Cycle Controlled Three-Phase Active Power Filter With Unbalanced Source and Load. *IEEE Transactions on Power Electronics* 21(5): 1403–1412.

Electronic Engineering and Information Science – Wang (Ed.)
© 2015 Taylor & Francis Group, London, ISBN: 978-1-138-02772-5

Reactive power measurement meter design based on FPGA

J.H. Han & G.L. Zhang

Harbin University of Science and Technology, College of computer science and technology, Harbin, China

ABSTRACT: Recent years, due to the use of a large number of nonlinear loads in Grid system, the imbalance of the reactive-power of the Grid system becomes increasingly serious. The emergence of power quality problems has a direct effect on the safe operation of the power grid. The realization of the reactive-power measurement is the key technology for the compensation of reactive-power. How fast and accurately to finish the reactive power measurement, has been a hot research area in the field of electrical measurement. An FPGA with its powerful parallel computing ability and flexible programmability, gradually replaced the traditional Processor, and Programmable System on Chip has become a hot spot in the field of electronic design. In this paper, we use SOPC technology to design a measuring instrument of reactive-power which via Hilbert transforms method. The experimental result shows that the method to measure instrument of reactive-power, has high accuracy, real-time strong.

KEYWORDS: Reactive power, SOPC Technology, Measuring instrument.

1 INTRODUCTION

Reactive power measurement in power system measurement is an important technical indicator. It often requires real-time reactive power, accuracy and effectiveness. According to the definition of reactive power by C. Budeanu proposed (Zheng. X. P 2006), there are lots of methods for reactive power measurement. Right now, there are several major measurement methods such as : fast Fourier transform (FFT) based method (Qiu.H.F & H.Zhou 2007), the instantaneous detection method (Xue.H &R.G.Yang 2002), the phase shift method (Chen.G.T. & J.K.Wu & H. L. Zhang 2007), as well as a digital filter method based on Hilbert (Zu.Y.X. & H. Pang & D.X. Li. 2003, Pang. H & Z. J. Wang & J.Y. Chen. 2006, Ansari Rashid. 1987) and based on Hilbert transform and so on. A higher accuracy and wide using can be obtained by using the Hilbert transform. This article used the SOPC technology, which make use of Hilbert transform to solve the reactive power in the former FFT computation and a windowing function to eliminate the spectral leakage to improve the accuracy of the system. With the help of FPGA powerful parallel computing power, it improved the real-time of the measurement system.

2 THE OVERALL DESIGN OF THE SYSTEM

The system is mainly used in the design Reactive power measurement of the power grid. In order to achieve the accuracy requirement, the system is divided into a signal acquisition module, management module, communication and storage module and communication module. The overall system block diagram is shown in figure 1.

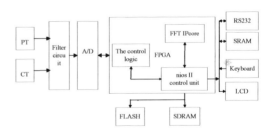

Figure 1. Total system design.

3 THE HARDWARE DESIGN

The sensor unit is often an important factor for the accuracy of the system. The precision level of a system depends on the accuracy the sensor can achieve. This design uses DVDI-001 small precision AC transformer, making strong voltage signal into weak signal.

3.1 The design of signal conditioning circuit

Conditioning circuit is preprocessing of A/D signal acquisition. In order to provide a safe input signal, we can through the step-down processing transform the signal into −5V~+5V. The Pin is input, and Out as output. When the SW is connected to the upper port, the output voltage is the input voltage of 9/100. When the SW connects the lower port, the output voltage of 1/100 as the primary voltage. This design uses the upper port, to assure proper input voltage A/D.

3.2 The design of signal A/D acquisition module

This design uses 16 bit A/D converter AD73360, which is 6 channel and 16 bit processor. AD73360 chip using serial interface mode, which make it simple to realize the connection to the external circuit. We design an AD73360 controller in an FPGA chip, to read and write control. The AD sampling is set to 6400 Hz. The master clock frequency in the AD acquisition chip is set to 13.1072MHz, Which obtain from the external frequency using the PLL in PFGA.

3.3 The design of data storage module

In this design, we need a very high demand on the operational speed, so the data storage module is very important. This design uses the IS61LV51216 made by the ISSI, which has the 512Mbit memory storage capacity and the data width is 16 bits, working in the 3.3V supply voltage.

3.4 The design of communication module

Nios II system can be integrated two forms of serial kernel, and one of them is JTAG_UART. Data interact through the JTAG communication port and the PC machine, generally using for program debugging. Another is what we usually say that the UART, and its data communicate with peripheral by RS-232. Using 9600 baud rate, data bits are set to 8. The parity bit is set to be none, and do not send the check bit.

4 THE SOFTWARE DESIGN

We build software platform in the SOPC, and configure the Nios II kernel into the fast type. We can adjust the cost of the system, according to the product requirement, and quickly bring products to market. We can avoid the System generation brings the loss due to FPGA programmable flexibility.

To achieve accurate measurement results, the system is divided into three functional modules, respectively: The signal acquisition module, the data processing module, the data storage module.

4.1 The signal acquisition subprogram design

The signal acquisition program mainly controls the A / D converter for sampling an electrical signal, storing and transmitting it to the data processing module. According to the sequence diagram of AD73360 chip,

govern the sampling chip with reading and writing, for the sake of making the signal acquisition module continuous acquiring of signals in accordance with a certain frequency. Finally, the data collected from the three analog channels are deposited into the size of the 2-port 1024 * 16bits. We can use the LPM module to implement.

In the design, we select the 1024 consecutive points in the sampling point, and pass them to the preprocessing module. The more points we used in a single cycle sampling, the higher accuracy of the measurement results. So we need require sufficient sampling frequency for the purpose of improving the measurement accuracy of the results. In ADC control module, we choose a 12-bit address register, and put the RAM written address in the high eight bits of the register. Through this operation, we can achieve four consecutive repeat sampling data written to the same address in a RAM. That is, only the fourth sample data is valid. In the pre-treatment sample waveform, we store the sample data in RAM according to the order. When the wave Reduce, the reduced data are provided. As shown in the waveform, with this approach, the written speed of RAM memory is four times shorter than the RAM memory of data preprocessing.

4.2 The design of data pretreatment

The frequency of the grid signal is usually unstable, which produces certain fluctuation and unable to realize synchronous sampling. At the time of spectrum analysis, the non synchronous sampling often leads to signal spectrum leakage among every harmonic, and the fast Fourier transform causes barrier effect, and the solution of the two problems is the key to improve the harmonic parameter measurement accuracy.

We usually adopt choose an appropriate window function method for restraining spectrum leakage, and the method of processing fence effect is usually resolved through window function interpolation. In practical application, we can select the appropriate window function to pretreatment according to the sampling period in the grid for getting the precision of meeting the demand of harmonic parameters. In this design, Hamming window function is used as the pretreatment of the window function.

The window function is obtained from the MATLAB. The hex file is generated, and a zero is added into the LPM_ROM memory module in turn after getting 1023 window sequence. The sampling data and window sequence in LPM_ROM are read at the same time, the two groups of data were sent to an ALPFP_MULT multiplier for corresponding multiplication. The FIFO memory is used to store the first 512 data of the multiplier output. The output of the FIFO data and multiplier date correspond to the floating-point addition, when the multiplier outputting 513th data. Figure 2 is the schematic diagram of the pretreatment.

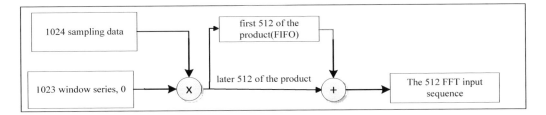

Figure 2. Schematic diagram of preprocessing.

4.3 The design of FFT module

FFT module is a key part of the Hilbert transform. FFT Mega Core functions provide a multi-stage pipeline technology, which greatly improved the system's processing speed. In our design, we call the IP core to implement the transformation of FFT, which can improve operational performance.

There are four kinds of I/O structure in FFT Mega Core function, including Streaming, Variable Streaming, Burst, Buffered Burt. We adopt a Variable Streaming structure in our paper. The structure of the input and output data are continuous, data accuracy can reach 32. We selected the default configuration for four output engine structure, designing the handle points to 512, choosing natural sequence data as I/O sequence, and using floating-point data representation as our data type.

4.4 Hilbert transform

The ideal Hilbert transform is amplitude is 1, the negative frequency part 90 degrees phase shift.

$$H(\omega) = \begin{cases} -j & \omega > 0 \\ j & \omega < 0 \end{cases} \tag{1}$$

In type 2, matrix H is the transfer matrix of Hilbert transform, and we achieve the Hilbert transform through the product of the matrix H and u(n).

$$H = Lm(\frac{1}{N} \begin{bmatrix} 1 & 2 & \cdots & 2 \\ 1 & 2W_N^{-m} & \cdots & 2W_N^{-mh} \\ \vdots & \vdots & & \vdots \\ 1 & 2W_N^{-(N-1)m} & \cdots & 2W_N^{-(N-1)mh} \end{bmatrix} 1 \times$$

$$\begin{bmatrix} 1 & 1 & \cdots & 1 \\ 1 & 2W_N^{-m} & \cdots & 2W_N^{-mh} \\ \vdots & \vdots & & \vdots \\ 1 & 2W_N^{-(N-1)m} & \cdots & 2W_N^{-(N-1)mh} \end{bmatrix}) \tag{2}$$

4.5 The design of data storage module

Every time after power off will produce the error in the measurement results. In order to ensure the accuracy of measurement, when the general power of each has changed 0.01kwh, the power data will be stored in a memory. Because every time when the power is on the memory of the data will change, an external SRAM can read the initial SRAM value when it's about power to rewrite the SRAM after each parameter in the program changes. So that it can be read into memory as the initial value, after the system re-power.

5 THE SIMULATION RESULTS AND ANALYSIS

When the power grid frequency is 50Hz, we obtained signal time domain waveform through the grid signal acquisition. Then we obtained the signal frequency domain waveform through the FFT transform after window function, as shown in figure 3.

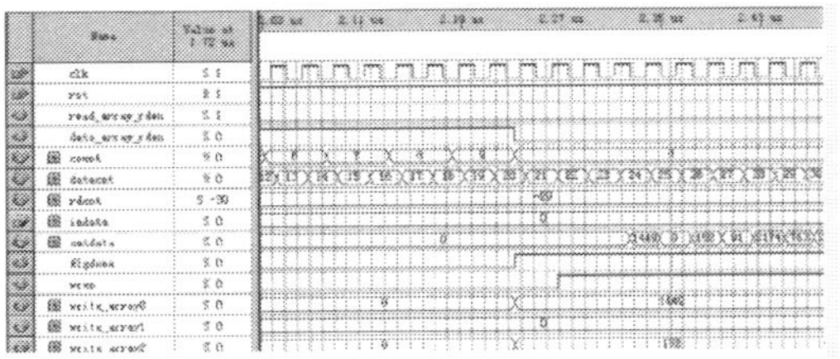

Figure 3. The signal acquisition waveform.

We select the 7 times the effective value of non sinusoidal signals of harmonic voltage, current, and phase harmonic difference as an experimental model. We can calculate the accurate value of 0.8722 according to the definition of reactive power, as shown in Table 1. We select the sampling frequency is 64KHZ, the truncated 4 cycles, for each cycle sampling 64 points, a total of 256 sampling points.

627

Table 1. Voltage, current signal model.

Harmonic components	U/V	I/A	Φ/°
The fundamental	10.00	10.00	45
The second harmonic	0.20	0.30	−60
The Third harmonic	0.50	0.05	45
The fourth harmonic	0.10	0.30	−30
The fifth harmonic	0.30	0.50	60
The sixth harmonic	0.10	0.10	−15
The seventh harmonic	0.05	0.07	30

When the grid frequency respectively 49.5, 49.7, 50, 50.3, 50.5Hz, we get the results shown in table 2, after FFT algorithm processing of added by Hamming window and interpolation.

Table 2. The calculation results and the relative error.

Frequency/Hz	power	calculated value	relative error
49.5	Reactive power	0.8765	−0.4791
49.7	Reactive power	0.8734	−0.1259
50.0	Reactive power	0.8722	0
50.3	Reactive power	0.8714	0.1031
50.5	Reactive power	0.8685	0.4376

The experimental results show that we have relative error is within 0.5, reached the expected requirements.

6 CONCLUSION

In recent years, FPGA has a rapid development in the field of military industry and the digital communication, because of its ability to strong. We used FPGA for the measurement of the reactive power of the power grid, reducing the measurement error and improving the accuracy.

REFERENCES

Zheng.X.P.(2006)The definition and arithmetic for the reactive power.*Electrical Measurement& Instrumentation*, 43(6): 1–4.
Qiu.H.F & H.Zhou. (2007)Summary on reactive power measurement in power system, *Electrical Measurement & Instrumentation*, 44 (1) : 5–9.
Xue.H &R.G.Yang. (2002)A novel detection theory of instaneous reactive and harmonic current. *Proceedings of the CSU EPSA*, 14(2): 8–11.
Chen.G.T. & J.K.Wu & H.L.Zhang. (2007) Phase shift algorithms for reactive electric power and energy in power systems. *Journal of Electric Power*, 22(4): 428–431.
Zu.Y.X. & H.Pang & D.X. Li. (2003)A method of reactive power measurement based on Hilbert digital filtering. *Automation of Electric Power Systems*, 27(16): 50–52, 70.
Pang.H & Z.J.Wang & J.Y. Chen. (2006) Method of reactive power measurement based on two pair s of Hilbert digital filters. *Automation o f Electric Power Systems*, 30(18): 45–48.
Ansari Rashid. (1987) IIR discrete-time Hilbert transformer. *IEEE Tran son Acoustics, Speech, and Signal Processing*, 35(8): 1116–1119.

Electronic Engineering and Information Science – Wang (Ed.)
© 2015 Taylor & Francis Group, London, ISBN: 978-1-138-02772-5

System of human basic parameters signal acquisition

L.F. Zhang & W. Jiang
Institute of physics and Electronic Information Engineering of Qinghai Nationalities University, Qinghai, Xining, China

ABSTRACT: Human pulse information reflects health status of human internal organs. The design ideas of this paper was based on research projects of national commission of ethnic affairs from the Qinghai Nationalities University, which called extraction of both characteristic parameters and measurement of Qinghai-Tibet Plateau people's human pulse. This paper mainly introduces system design and the pulse measurement method of pulse measurement by a piezoelectric sensor which mentioned by G. Q. MIE&Z. X. FANG (2006). In consideration of the practicality and portability of the instrument in the measuring process, using the LabVIEW to develop the human pulse wave and body temperature virtual measurement instrument.

KEYWORDS: Pulse measurement; Temperature gathering; Data acquisition; Sensor.

1   INTRODUCTION

With the application of large-scale integrated circuits, large amounts of data acquisition and statistical analysis become more and more simple, the collection of human body basic physiological parameters can be realized in the school laboratory. Firstly, obtain the corresponding signal respectively by hardware circuit, then sent to the virtual instrument for real-time display. Virtual instrument is powerful, realistic appearance and flexible interface, so it can display various forms of parameters (R.WANG et al 2007, M.Kostic 1998), also can process the data acquisition. As a programming language, the completed virtual instrument can be generated as an application, dynamic link library or installation package, it can still run divorced from the LabVIEW environment.

2   DESIGN OF TEMPERATURE, PULSE SIGNAL TEST PANEL

The acquisition of temperature mainly uses the single type digital temperature sensor DS18B20 to test the temperature of the human body, the human body pulse waveform acquisition using piezoelectric pulse sensor HK-2000B, the output is an analog smaller signal, add the signal in magnify circuit and filter circuit for processing, then sent the waveform which has been processed through the data acquisition card after to a virtual instrument and display. Considering the temperature and the pulse waveform display state can be adjusted to make it easier to observe, this VI select buttons, knobs, thermometer as the input control used to set the alarm temperature and adjust the amplitude and frequency of a pulse signal. Using waveform graph and square led as the output indicator controls to observe the collected signal waveform and judge whether the temperature has reached alarming system that set up in advance.

3   IMPLEMENTATION OF UNIT CONTROL CIRCUIT FUNCTION IN LABVIEW

3.1   *Temperature display and adjustment unit*

In X.B. HUANG (2008), it is shown that, the temperature obtained by the temperature sensor DS18B20 in real-time. Input a simulated temperature as test instrument developing in LabVIEW, using the comparison function to set an upper limit alarm temperature value is 37 ℃, then Comparing the temperature, which received by DS18B20 to the alarm value automatically in the case structure so that to judge whether the limit is reached, if the upper limit temperature exceeded fever alarm indicator light. On the other hand, alarm lamps always extinguish. Temperature input in the test instrument can be displayed in floating-point and thermometer two forms on display panel.

3.2   *Pulse and reshape waveform display unit*

The pulse waveform signal is mainly transmitted by means of human pulse waveform acquisition circuit. The core device is the piezoelectric Sensor of the human pulse (HK-2000B), the human pulse signal is small signal, the amplitude is relatively small, it is prone to distortion when adjust the waveform parameters, the human eye has limitation in observation. In order to observe easily, set pulse triggering level is 0 in the interior when the waveform displaying, shaping the input signal to the corresponding pulse waveform used to display the approximate heartbeats number.

3.3 Pulse amplitude, frequency adjustment unit

The oscilloscope is a typical test and display instrument in the laboratory teaching, for the convenience of observing, measure and record the graphics, how to adjust the amplitude and scanning frequency of waveform reasonable are the necessary steps in the experiment. The pulse waveform test using knob controls for adjusting amplitude and frequency for the duration of the whole design. In order to grasp the waveform variety law, the regulation is mainly expressed in the manual way (G.Q. MIE& H.Z.SUN 2008), the optimum amplitude display range of shaped waveform was set up so that can observe and test conveniently.

3.4 Heart rate test unit

Heart rate tests mainly abstracts a part of pulse signal from human pulse waveform which collected by the sensor HK-2000B in trigger way, select the appropriate parameters in the first waveform in a series output signal as a criterion applied to every subsequent signal in signal array, as long as the parameter meets the requirements the counter will add 1. Finally, the total number of signal statistics Characterized the heart rate, of course, this method may be remains certain deviation in the process of measurement due

to the randomness of human pulse(Y.ZHANG et al 2005,Pestel G et al2009).Therefore,Set 30s, 60s, 90s, 120s four measurement time, at last according to the measurement results under different conditions to judge the accuracy . The data obtained in the process of measurement will be displayed on the virtual instrument front panel in digital form.

4 BODY TEMPERATURE, PULSE TESTER

This VI has the characteristics of concise appearance and simple operation. Pulse signal waveform has an ideal display after modulated by the amplifier circuit and filter circuit. The front panel of human basic parameters signals acquisition system as shown in figure.1, and the design block diagram as shown in figure.2.

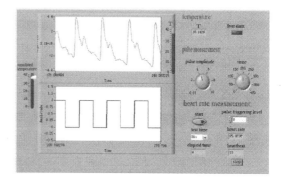

Figure 1. The front panel of human basic parameters signal acquisition system.

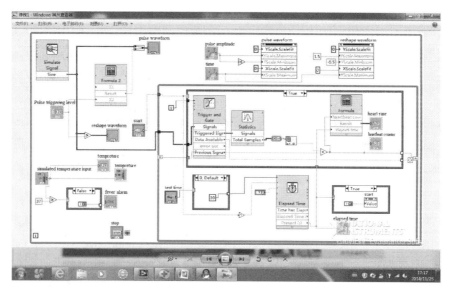

Figure 2. Block diagram of human basic parameters signal acquisition system.

5 CONCLUSIONS

This system has been finished at school laboratory. Mainly introduces the whole design process, adding the pulse and temperature signal into the tester and achieve a good result. We should take account of the different characteristics of pulse signal in the future for improving accuracy of the instrument, adding more parameters acquisition function to enhance its practicability.

REFERENCES

R.WANG et al(2007). The application of LabVIEW in data acquisition .*J. Mechanics,*113–117.

X.B. HUANG(2008). Temperature monitoring system for based on AT89S52 MCU and DS18B20.*J. Micro Computer Information*, 237–240.

Y.ZHANG et al (2005). The ECG signal acquisition system based on LabVIEW. *J. Chinese medical equipment,* 93–97.

G.Q. MIE. & Z.X. FANG (2006). Measurement and analysis of human pulse.*J. Shanghai Biomedical Engineering*, 133–138.

Pestel G et al (2009).Automatic algorithm for monitoring systolic pressure variation and difference in pulse pressure .*J. Anesthesia and Analgesia,*342–345.

G.Q. MIE.& H.Z.SUN(2008). The design of the pulse testing instrument based on LabVIEW. *J. Biomedical Engineering and Clinical Medicine*, 224–228.

M.Kostic(1998). Data Acquisition and Control for an Innovative Apparatus Using LabVIEW .*J.Virtual Instrument.*64–68.

Electronic Engineering and Information Science – Wang (Ed.)
© *2015 Taylor & Francis Group, London, ISBN: 978-1-138-02772-5*

Study of stereo vision measurement methods based on coding structured light

Y.J. Qiao, L.L. Wan, H.R. Wang & M.J. Zheng
Harbin University of Science and Technology, Harbin, Heilongjiang Province, China

ABSTRACT: A binocular stereo vision measurement technology based on coding structured light method has been proposed, which will project gray-code fringe patterns and phase-shifting projection gratings onto the surface of mechanical parts. Three-dimensional coordinates of points can be obtained through gray-code and phase decoding. In stereo vision measurement, non-sinusoidal phase-shifting gratings influence the precision markedly. A grating phase self-calibration method is presented to adjust the raster pattern and achieve the best grating fringe.

1 INTRODUCTION

For measuring the large machining pieces or structures, stereo vision measurement technology as a non-contact measurement method, for its advantages, fast measurement, real-time and high accuracy, has been extensively studied (Zhang, G.J. 2005). Firstly, gray-code patterns and phase-shifting gratings are generated by a computer, and moreover grating phase self-calibration is necessary (Koninck, T.P. & Van, G.L. 2006). Secondly, a set of stripe encoded patterns is projected onto the measured object by a digital projector and their images are taken by two CCD cameras from different angles. The image processing prepares for gray-code and phase decoding. Finally, according to constraint of the epipolar line, gray-code value and phase value, depth information through matching corresponding points is got. The stereo matching process is used to obtain complete measurement information (Chen, M.Y. & Guo, H.W. 2010). So the vision measurement method not only improves the precision, but also reduces the measurement time.

According to the different coding methods, the structured light coding technologies are divided into time coding, spatial coding and direct coding. Spatial coding is usually used to measure a moving object, direct coding has the disadvantage of low resolution and is more sensitive to noise, and among them time coding comparatively meets the measurement requirements (Frosion, I. & Borghese, N.A. 2008). Time coding method contains binary code and gray-code which it has higher resolution and higher anti-interruption ability, compared with binary code (Hartly, R. & Zisserman, A. 2000). On the other hand, black and white, gray-code stripes cannot be too narrow, otherwise it would bring heavy difficulty on boundary extraction, even can't be decoded (Zhang, L. & Curless, B. 2002). So phase-shifting makes up for the weaknesses. To work on the two coding methods, combining with both advantages in the use of industrial measuring system, it can supply high precision and wide measuring range in the industry produces, especially for the large machining pieces or structures that need to be observed in short time.

During the measurement, because the measurement data will be affected by the accuracy of sinusoidal projection grating, a self-calibration method for a vision measurement system based on stereo vision technology and gray-code is proposed (Pratibha, G. 2007). It allows automatic adjustment of raster pattern to improve the performance of grating projection, making the sinusoidal projection grating self-adaptive.

2 GRAY-CODE CODING AND DECODING

Based on stereo vision technology, gray-code, coding is adopted by dividing the measurement range into sections and accurately identifying each section to determine the identification unit and get the decoding sequence. By using white-black stripes on DPL as object, CCD as component of image grabbing and a computer to do image processing work, N pieces of Time sequence coding images with white-black stripes are got and converted into binary images only contains two gray levels 0 and 1, and finally the gray value for each gray-code value is converted into the binary and decimal value, which is shown in figure 1 and figure 2.

It is following for gray-code, coding and decoding that the images are divided into 2^N regions and each region has the same gray-code value, in order to achieve the coarse matching by corresponding to the same gray value. However, each pixel point can not be clearly defined. To match each pixel point, the phase-shifting method is introduced to subdivide each region.

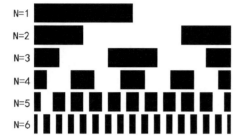

N=1
N=2
N=3
N=4
N=5
N=6

Figure 1.   Gray-code coding pattern.

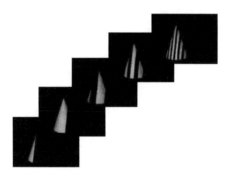

Figure 2.   Time coding sequence.

3   PHASE UNWRAPPING

After the different phase movements, phase-shifting method is realized by the obtained phase-shifting fringe patterns. This paper adopts four-step phase-shifting measurement method. A series of gratings as shown in figure 3 and figure 4 are projected on the surface of the measured object, and the distorted grating images are collected into the computer by the CCD cameras, then the phase collapsed in the range (−π, π) will be obtained. Through moving the sinusoidal grating for 4 times, each moving distance for 1/4 period, the moved phase of fringe pattern is π/2, which generates a light-intensity function $I_k(x, y)$ (k=1, 2, 3, 4). In the same way, four phase patterns and four corresponding phase formulas would be obtained by three equidistance movements.

Assuming the projection of the light intensity subjects to standard sine distribution, the light intensity distribution functions of the deformation stripes obtained by CCD cameras are given below.

$$I_1(x, y) = I'(x, y) + I''(x, y)\cos[\theta(x, y)], \qquad (1)$$

$$I_2(x, y) = I'(x, y) + I''(x, y)\cos[\theta(x, y) + \pi / 2], \quad (2)$$

$$I_3(x, y) = I'(x, y) + I''(x, y)\cos[\theta(x, y) + \pi], \qquad (3)$$

$$I_4(x, y) = I'(x, y) + I''(x, y)\cos[\theta(x, y) + 3\pi / 2], \quad (4)$$

Where $I_k(x, y)$ is the gray value of Picture k, $I'(x, y)$ is the background value of fringe light density, $I''(x, y)$ is the intensity modulation, and $\theta(x, y)$ is an unknown phase field to be solved.

Figure 3.   Sinusoidal grating.

Figure 4.   Amplified sinusoidal grating.

The following equation is derived,

$$\varphi(x, y) = \arctan[\frac{I_4(x, y) - I_2(x, y)}{I_3(x, y) - I_1(x, y)}] \qquad (5)$$

where is the basic formula for four-step phase shift method. The $\theta(x, y)$ is called a phase principal value.

The value range of $\theta(x, y)$ locates in the range $(-\pi, \pi)$ is solved by arc tangent function characteristic and the positive and negative of numerator and denominator in the formula, which needs to be phase unwrapping. When the gray-code period equals to the phase period, a difference of one phase period is equal to one gray-code value. Taking the zero phase point as the first pixel in the row direction of the picture, the phase absolute value of each pixel could be successfully obtained.

$$\psi(x, y) = G(x, y) + \varphi(x, y) \qquad (6)$$

where $\theta(x, y)$ is the phase principal value, $G(x, y)$ is the gray-code value of the pixel point, and $\psi(x, y)$ is the unwrapping absolute phase value.

Of course, the surface modulation and noise, gray-code value range can not completely correspond to the phase interval, so it needs to do period calibration processing in actual measurement.

4 SELF-CALIBRATION OF THE MEASUREMENT SYSTEM

When a DLP projector is used to project the grating, the projected pixel unit is directly set as a standard sine distribution to project sinusoidal gratings. In fact, such grating strips do not meet the standard sinusoidal distribution, which is an important factor affecting the measurement.

I^s is the raster mode, the gray distribution of projector pixel unit, I^p is the intensity distribution of projected grating strips in space, and I is the intensity distribution of fringe image. A set of sinusoidal grating fringe I^s is projected into space by the projector, forming a fringe projection I^p. The fringe image I is generated by the cameras. θ^s and θ^p correspond to the phase field of I^s and I^p respectively. Here self-calibration phase is realized by establishing the mapping relationship of θ^s and θ^p.

The following is the specific solution: the grating image $I_i (i=1,2,3 \ldots T)$ is put into the following formula. I', I'' and θ^s are solved by the least square method.

$$I_i = I' + I'' \cos(\theta^s + i \frac{2\pi}{T}) , i=1,2,3 \ldots T. \qquad (7)$$

When getting the value of I', I'' and θ^s, the phase of each pixel is known,

$$\theta_i^p(x, y) = \arccos \frac{I_i(x, y) - I'(x, y)}{I''(x, y)} , i=1,2,3 \ldots T, \qquad (8)$$

$$\theta_i^s(x, y) = \theta^s(x, y) + i \frac{2\pi}{T} , i=1,2,3 \ldots T. \qquad (9)$$

Thus, the mapping relationship is used to correct sine mode I^s, making the sinusoidal of I corrected.

5 MEASUREMENT SYSTEM AND RESULTS

The vision system vision measurement system consisted of two CCD cameras, a DLP projector and a computer, which is shown in figure 5. The projector was used to project coding stripes. The computer was used to generate coding stripes and obtain three-dimensional information with the fringe images collected by CCD cameras.

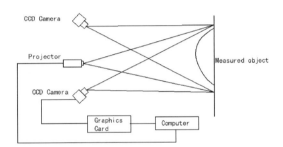

Figure 5. Measurement system.

First calibrate parameters of cameras with a calibration plate as shown in figure 6 and figure 7. Then control gray-code and phase-shifting structured light patterns with a computer and control two cameras to respectively capture image sequence, after self-calibration vision system. Finally, process these image sequences and output the three-dimensional cloud point data of the measured surface with MATLAB.

Figure 6. A calibration plate collected by left camera.

Figure 7.   A calibration plate collected by left camera.

In the experiments, this method is applied to measure a cone plaster model shown as figure 8, whose range image of three-dimensional reconstruction is shown as figure 9. The smooth surface of reconstruction model is helpful to realize the complete three-dimensional measurement.

Figure 8.   Cone plaster model.

Figure 9.   Range image.

6   CONCLUSION

With the rapid development of science and technology and industrial production, many applications are implemented in a three dimensional environment, so collecting three-dimensional scene information reliably and studying the three-dimensional structure of objects are an important measurement means. Therefore, the fast and accurate measurement of the shape of three dimensional objects becomes a critical issue. Based on the previous research, using binocular stereo vision technology, the structured light of gray-code and phase-shifting is used to encode the measured object. For the phase of grating projection, self-calibration vision measurement system is achieved to greatly improve the measurement accuracy. It has been well validated in the experimental system.

REFERENCES

Zhang, G.J. (2005). *Machine Vision*, Beijing: Science Press.
Da, F.P. 2011. *Three-dimensional Precision Measurement of Projector Grating*, Beijing: Science Press.
Koninck, T.P. & Van, G.L. (2006). Real-time range acquisition by adaptive structured light. *IEEE Transation on Pattern Analysis and Machine Intelligence.* 28(3):432–445.
Chen, M.Y. & Guo, H.W. (2010). Algorithom immune to tilt phase-shifting error for phase-shifting interometers. *Applied Optics.* 39:3894–3898.
Frosion, I. & Borghese, N.A. (2008). Real-time accurate circle fitting with occlusion. *Pattern Recognition.* 41(7):1041–1055.
Hartly, R. & Zisserman, A. (2000). *Multiple View Geometry in Computer Vision*, London: Cambridge University Press.
Zhang, L. & Curless, B. (2002). Rapid shape acquisition using color structured light and multi-pass dynamic programming. *3D Data Processing Visualization and Transmission.* 2002:24–36.
Pratibha, G. (2007). *Gray Code Composite Pattern Structured Light Illumination*, Kentucky: University of Kentucky.

Electronic Engineering and Information Science – Wang (Ed.)
© *2015 Taylor & Francis Group, London, ISBN: 978-1-138-02772-5*

Research on splicing method in three-dimensional measurement of hydraulic turbine blades surface and simulation

J.H. Yuan, H.J. Yuan & B. Niu
Harbin University of Science & Technology, Harbin, China

ABSTRACT: The splicing of three-dimensional segmentation measurement of the hydraulic turbine blade is achieved in this paper. The comparatively large size of hydraulic turbine blades that could not be taken to the complete picture, multiple images must be obtained based on the segmentation measurement technology. Installing landmark in public areas of the adjacent segment of the hydraulic turbine blade, splicing images with the principle of least squares method, the entire blade three-dimensional surface information is obtained. According to the principle of the structure light measurement, the three-dimensional measurement system is designed, as well as its parameters are defined. The mathematical model of the three-dimensional measurement is set up. The hydraulic turbine blade is simulated using 3ds max, meanwhile hydraulic turbine blade model is built. Designing three-dimensional measurement simulation system, the turbine blades were measured and the three-dimensional coordinates of the hydraulic turbine blade surface. The reconstruction of the three-dimensional Hydraulic turbine blade is given according to the measured data.

KEYWORDS: Splicing; three-dimensional measurement; hydraulic turbine blade; landmark.

1 INTRODUCTION

Hydraulic turbine blades are a core part of large hydro-power equipment and measurement of the blade has a crucial impact on the overall operational efficiency and service. As the demand improved constantly for the output power and efficiency, the specifications and precision demand for the hydraulic turbine blades are increasingly higher than before, the profile sizes of the hydraulic turbine blade are increasingly larger, we are often unable to access to get all the data once when in the process of three-dimensional measurement of the hydraulic turbine blades. Based on the size of the target structure in this paper, acquiring multiple images by the way of the segmentations shooting, three-dimensional parameters are extracted from the hydraulic turbine blades and splicing to form the overall three-dimensional data and to reconstruct the hydraulic turbine blade surface information.

2 THE PRINCIPLE OF THREE-DIMENSIONAL MEASUREMENT OF HYDRAULIC TURBINE BLADES

A common means of measurement are used for the three-dimensional structure of an optical measurement method (Hanin et al 2005 and Yuan 2006) of the hydraulic turbine blades in this paper. Figure 1 shows the principle of the 3D measuring system.

The principle is two known angles and one side can obtain the other side. The coding structure light which through the projector at a certain angle can project the hydraulic turbine blades and filmed by camera, projectors and projected points on the blade and camera form a triangle, known the distance between projectors and the camera, the angle between projectors and the projection point and the angle between the camera and the projection point.

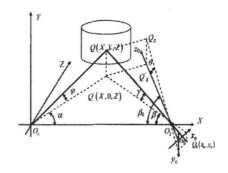

Figure 1. The theory of measuring on color-encoded.

O_1 is the lens' center of the DLP; O_2 is the lens' center of the camera. $O_1O_2=B$; Q is a spot projected to the object surface by DLP. Spot Q is on the coordinate system of object $XYZO_1$, Q_0 is the image of spot Q captured by the camera, and it is on the image surface $X_0O_0Y_0$ of the coordinate system of image $X_0Y_0\,Z_0O_0$. $X_0O_0Z_0$ is on the same surface with XO_1Z. Q', Q_z and Q'_z are images of spot Q in the surface XO_1Z, $Y_0O_0Z_0$ and the axis O_0Z_0. Based on the triangulation principle, the 3D coordinates of the points on the object surface can be calculated. According to the formula (1) to (3) coordinate values of the projection point X, Y, Z can be get. In where, f is the focus of the camera's lens, β_0 is the angle between the optic axis O_0Z_0 and O_1O_2. f, B, β_0 are constants. x_0, y_0 are determined by the pixel of CCD in the camera. Angle α is the projection angle.

$$Z = \frac{B}{\cot\alpha + \dfrac{x_0 + f\cot\beta_0}{f - x_0\cot\beta_0}} \qquad (1)$$

$$Y = \frac{Z \cdot y_0}{f\sin\beta_0 - x_0\cos\beta_0} \qquad (2)$$

$$X = Z\cot\alpha \qquad (3)$$

3 LANDMARKS THREE-DIMENSIONAL STITCHING TECHNIQUES AND METHODS

When measured on the surface of the hydraulic turbine blades, due to the larger blade, it needs to be segmentation measurement for the blades model that can get the complete shape information of the hydraulic turbine blades (LIU et al. 2009). Then spliced the image measurement data together, which to be dealt with separately throughout the image processing techniques that is named three- dimensional shape splicing technology. Projecting the landmarks to the blade surface, the characteristics of these landmarks are easy to be identified and matched. This is landmarks stitching technology which is getting the result, according to the coordinate transformation matrix as well as achieving the stitching.

3.1 Image stitching method

Firstly, establishing the turbine blade model, measuring blade model should be divided into two sections, it needs to be measured in both the field of view. The captured image in the two fields of view must have overlapping part. There are two projectors in this measuring system, the first projector projects the color code stripe pattern to the blade, and the second projector projects the multiple landmarks to the blade. After the camera obtains images in the first field of view, the first projector and the camera must be moved synchronously for getting images in

the second field of view. The coordinates of the blade surface can be measured using these two images of the blade with the color code stripe pattern obtained by the camera in the two fields of view. Using two images of the blade with landmarks in the two fields of view, after images identifying and images registration, three-dimensional coordinates of the landmarks could be obtained. Getting the coordinate transformation matrix between them according to the variation model of the blade, image stitching is realized.

3.2 Three-dimensional coordinate transformation

Three-dimensional coordinate transformation is through the promotion of two-dimensional coordinate transformation form, a 3×3 matrix can be expressed in a two-dimensional coordinate transformation converting. In the two-dimensional space, setting a coordinates(x,y), h is a constant which is not equal to 0, then the homogeneous coordinates of the point can be expressed as (x',y',h), and $x'=hx$, $y'=hy$. For simplicity, assuming that h is 1, then the homogeneous coordinates of the point is $(x',y',1)$. It's a two-dimensional linear transformation form usually is written formula (4).

$$\begin{bmatrix} x' & y' & 1 \end{bmatrix} = \begin{bmatrix} x & y & 1 \end{bmatrix} \begin{bmatrix} a_1 & a_2 & a_3 \\ b_1 & b_2 & b_3 \\ c_1 & c_2 & c_3 \end{bmatrix} \qquad (4)$$

The general form of this transformation for the formula (5), the former conversion coordinates of the point is expressed as Q, the changed coordinates are expressed as Q', The transformation matrix is expressed as T, so different transformation follow the different transformation matrix.

$$Q' = QT \qquad (5)$$

Similarly, three-dimensional coordinate transformation of points in three-dimensional coordinates can be used a homogeneous transformation matrix of 4×4 represents. Assuming that $(x,\ y,\ z)$ are the coordinates of point in three-dimensional space, $(x\ y\ z\ 1)$ for its homogeneous coordinates, its three-dimensional linear transformation is (6). Formula (7) is the three-dimensional coordinate transformation matrix.

$$(x\ \ y\ \ z\ \ 1) = (x,\ \ y,\ \ z)\begin{bmatrix} S_{11} & S_{12} & S_{13} & S_{14} \\ S_{21} & S_{22} & S_{23} & S_{24} \\ S_{31} & S_{32} & S_{33} & S_{34} \\ S_{41} & S_{42} & S_{43} & S_{44} \end{bmatrix} \qquad (6)$$

$$S_{3D} = \begin{bmatrix} S_{11} & S_{12} & S_{13} & S_{14} \\ S_{21} & S_{22} & S_{23} & S_{24} \\ S_{31} & S_{32} & S_{33} & S_{34} \\ S_{41} & S_{42} & S_{43} & S_{44} \end{bmatrix} \qquad (7)$$

In the three-dimensional coordinate transformation matrix S_{3D}, dividing S_{3D} into four sub-matrix according to the different functions $\begin{bmatrix} s_{11} & s_{12} & s_{13} \\ s_{21} & s_{22} & s_{23} \\ s_{31} & s_{32} & s_{33} \end{bmatrix}$,

$\begin{bmatrix} s_{14} & s_{24} & s_{34} \end{bmatrix}$, $\begin{bmatrix} s_{41} & s_{42} & s_{43} \end{bmatrix}$ and $\begin{bmatrix} s_{44} \end{bmatrix}$,

$\begin{bmatrix} s_{11} & s_{12} & s_{13} \\ s_{21} & s_{22} & s_{23} \\ s_{31} & s_{32} & s_{33} \end{bmatrix}$ sub-matrix represents the object

rotation, scaling, shearing and other conversion function; $\begin{bmatrix} s_{14} & s_{24} & s_{34} \end{bmatrix}$ is in the charge of the object projection transformation; $\begin{bmatrix} s_{41} & s_{42} & s_{43} \end{bmatrix}$ represents the translation transformation function; $\begin{bmatrix} s_{44} \end{bmatrix}$ represents the conversion function of overall proportion. When measuring Segmentation in this paper, only pan the blade model in the two fields of the view, so just considering the sub-matrix $\begin{bmatrix} s_{41} & s_{42} & s_{43} \end{bmatrix}$.

Assuming that a pixel of the blade model image data in a visual field coordinate $A = (x_1 \ y_1 \ z_1 \ 1)$, its data coordinate in field 2 is $B = (x_2 \ y_2 \ z_2 \ 1)$, then the relationship between these two coordinates can be expressed by the formula (8) the same point coordinates transformation in different field can be made through the formula.

$$AS_{3D} = B \qquad (8)$$

3.3 *Stitching algorithm steps*

The data splicing procedure is divided to the following steps based on least-squares method. The first step, the pixel coordinates of the landmark are extracted by processing the blade images which are in two fields of the view. The two corresponding landmarks are registration using landmark recognition technology. If the right number of landmark point is more than three pairs after registration, then to splice. The matrix of their coordinates is written P and Q. The second step, according to the formula $S_{3D} = P^+Q$ to calculate the coordinate transformation matrix S_{3D}, P^+ represents P generalized inverse, $P^+ = P^T (PP^T)^{-1}$. The third step, the landmark pixel

coordinates in different fields of view are written into matrix P and Q, and into the formula to achieve splicing.

According to the splicing theory, the more right landmarks are registration, the less error and the finer stitching.

4 SETTING LANDMARKS

Setting the first view field, simulation projector is replaced by the target spotlight. The spotlight is placed in the origin of the world coordinate system O_1, the angle between the optical axis of the projector and the x-axis is 60°, and the opening angle of the projector is 60°. Simulation camera placed at position away x axis 900mm from the projector, the angle between its optical axis and the x-axis is 60°, and the opening angle is 60°. Due to the larger blade model, according to the actual situation, the blade model is placed in the center of the (800, 900, 24.57) which get the optimal effect.

The projector 2 is placed in the x-axis, coordinates (800,0,0) for projecting landmarks in simulation system.

Designing the second view Field, the relative position of the projector 2 and blade model should remain unchanged. The blade model and projector 2 are panned 560mm to the left, x-axis coordinates of both the blade model and the projector 2 changed to 240.

After the binarization processing of the color coding pattern of two view fields, decoding the color-coded pattern, three dimensional coordinate value of the point of the blade surface can be measured by decoding, and the three-dimensional shape can be reconstructed.

5 DETECTION TECHNOLOGY OF THE CIRCULAR LANDMARK

Hough transformation is one of the basic geometric methods which identified from the image processing (Zhang et al. 2006), it is an edge connection technology which based on the image global segmentation result (Kimme et al 2007). An improved Hough transform circle detection algorithm(Liu et al. 2009) is adopted in this paper.

Processing the object image which contains the landmarks circle in the first field of view, after using the Hough transform circle detection, the tested results are shown in figure 2.

Similarly, the same processing is done with a captured landmark circular image in the second field of view. The tested results are shown in figure 3.

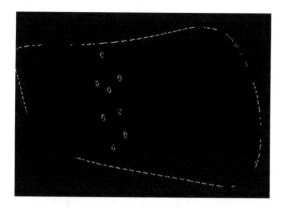

Figure 2: Circle detection in the first field 1 of view.

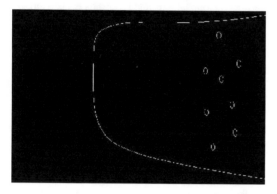

Figure 3.   Circle detection in the second field of view.

Accordance with the registration principle of the landmarks circle, after the landmarks circle registration, the coordinate transformation matrix is calculated under both fields of view according to the least squares method.

$$
S_{3D} = \begin{bmatrix}
0.5389 & 0.2230 & 1.7057 & 0.0007 \\
0.0798 & 0.4508 & -0.6264 & -0.0002 \\
0.1700 & -0.2361 & -0.4752 & -0.0002 \\
-430.0805 & 190.4295 & 928.9358 & 1.0000
\end{bmatrix}
$$

Reconstruction figure of splicing result is shown in Figure 4.

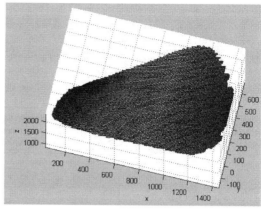

Figure 4.   Reconstruction figure of splicing result.

6   CONCLUSION

Experimental results show that the sectional measurement splicing method which used in this paper is simple and feasible, easy to operate. However, the error still exists when using the Hough circle detection technology. In order to get greater accuracy in future measurement, the Hough center of the detection accuracy needs to be further improved.

REFERENCES

C. Kimme, D. Balland, J. Sklansky(2007). Finding circles by an array of accumulators. *J. Commun, ACM. 15.* 11–15.

Hanin S, Kiyoshi T(2005). 3D profile measurement using color multi-line stripe pattern with one shot scanning. *J. Integrated computer-aided engineering. 12*, 333–341.

Huijuan Yuan(2006). Combination of color stripe encoding three-dimensional measurement techniques.

LIU Jian-wei, LIANG Jin, LIANG Xin-he(2009). A Novel Rapid Measurement Approach for Blade of Large Water Turbine. *J. Opto-Electronic Engineering. 36.* 50–55.

Liangjiang Liu, Yaonan Wang(2009). A circle detection method based on Hough transform. *J. Microcomputer Information. 5.* 274–276.

Zhang Xiao, Peng Wei(2006). Circular object detection based on Hough transform. *J. Sensors and Microsystems. 25.* 62–64.

Electronic Engineering and Information Science – Wang (Ed.)
© 2015 Taylor & Francis Group, London, ISBN: 978-1-138-02772-5

Design and implementation of granary temperature and humidity detection system

Z.M. Su, J.J. Sun, E.J. Zhang & H.J. Yuan

The Higher Educational Key Laboratory for Measuring & Control Technology and Instrumentations of Heilongjiang Province; Harbin University of Science & Technology, Harbin, China

ABSTRACT: For measuring granary temperature and humidity, this paper developed a set of data acquisition and wireless transmission device which is cost-effective, convenient and reliable. In this system with the functions of wireless transmission, PC monitoring, threshold alarm and so on, STC series single-chip microcomputer is used as main control chip, DHT11 digital sensor is used as the temperature and humidity sensor, and the cost-effective nRF905 chip is selected for realized wireless transmission.

KEYWORDS: Detection; digital temperature and humidity sensor; radio frequency; wireless transmission; single-chip microcomputer.

1   INTRODUCTION

Food is essential to human survival material condition, is a necessary precondition for the country's economic development process, so the granary management science is crucial, and the most important problem in the granary management is to monitor the temperature and humidity changes in the grain heap. Therefore, using the modular engineering ideas, a set of the detection system of granary temperature and humidity is researched and designed, and the system is cost-effective, user-convenient and reliable.

The design can realize the granary of temperature and humidity data acquisition, and display of temperature and humidity data at the scene, and when the temperature and humidity exceeds the upper and lower limits of the system Settings, buzzer alarm processing, and the relay to control the corresponding equipment, at the same time, temperature and humidity data through the wireless transmission module transmitted to the PC system.

2   SYSTEM SCHEME

The idea of modular design is adopted in system design. The system is divided into three modules: temperature and humidity data acquisition module, an intermediate control module, a PC interface module. Figure 1 shows the block diagram of the measuring system.

Temperature and humidity acquisition module is composed of mainly temperature and humidity

Figure 1.   The block diagram of the measuring system.

sensor, single-chip microcomputer and nRF905 wireless communication of the sender. The sensor obtains real-time temperature and humidity data, and sends them to the single-chip microcomputer. The processed data by the single-chip microcomputer are sent to the intermediate control module through the wireless communication (Huang Haizhen 2011).

The intermediate control module is composed of the nRF905 wireless transmission receiver, single-chip microcomputer and voice alarm control circuit. The nRF905 wireless communication receiver receives the temperature and humidity data sent by the acquisition module. The data after the single-chip microcomputer processing are sent to the PC to realize human-computer interaction through a serial communication. When the temperature and humidity transfinite, voice control circuit alarm (Li Qiang 2013).

PC interface module, through a serial port connected to the intermediate control module, receiving temperature and humidity data, is mainly responsible for providing the human-computer interaction interface, real-time monitoring the temperature and humidity of distal granary (Ma Hanting et al 2013).

3 SYSTEM HARDWARE

3.1 *The temperature and humidity acquisition module*

This module includes a single chip microcomputer, temperature and humidity sensor, power circuit, display circuit, alarm circuit, keyboard circuit, control circuit and wireless (emission) communication circuit, its main function is to monitor the temperature and humidity of the patch, and the real-time data of temperature and humidity transmission through nRF905 wireless transceiver module to the intermediate control module (Xue Mindi 2010). The basic design, block diagram is shown in Figure 2.

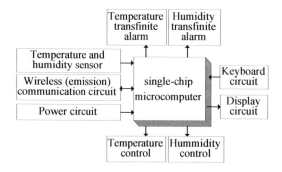

Figure 2. The basic block diagram of temperature and humidity acquisition module.

3.2 *Intermediate control module*

The intermediate control module is the link between temperature and humidity data acquisition module and PC, the stability of its hardware and software determines the stability of the whole system (Yu Chengbo et al 2010), the basic block diagram is shown in Figure 3.

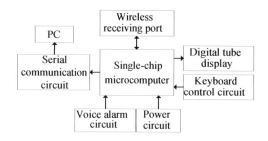

Figure 3. The basic block diagram of the intermediate control module.

4 SYSTEM SOFTWARE

4.1 *Temperature and humidity acquisition module programming*

This part of the main program first initializes the system, mainly includes: LCD12864 liquid crystal display initialization, nRF905 wireless transmission terminal initialization, timer initialization and power-on initialization, etc. Then the single-chip microcomputer will send a request to the temperature and humidity sensor DHT11 for busy state detection, when DHT11 returns false, the single-chip microcomputer reads temperature and humidity data of the DHT11, and through LCD12864 display. At this time, the single-chip microcomputer will determine the nRF905 working state, send instruction to the nRF905 through the program simulated SPI interface, and send the temperature and humidity data to the intermediate control module via nRF905 when nRF905 in a non-working state. The flow chart is shown in Figure 4.

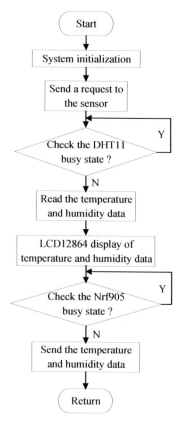

Figure 4. The main program flowchart of acquisition module.

4.2 *Intermediate control module programming*

Before receiving data, first initializing the system and setting up the nRF905 is in Standby mode, at the same time receiving the address written while writing the configuration register. After that, by setting CE = 1, TXEN = 0, putting the nRF905 in reception mode. In reception mode the nRF905 will automatically receive air carrier, at this time, the single-chip microcomputer will keep testing DR port, and when DR = 1, suggests that the validity of data reception is complete. Single-chip microcomputer sets CE to 0, nRF905 into Standby mode, while CSN port is set to 0, waiting for the next instruction. It reads buffer command by sending and receiving data to read out valid information received. The flow chart is shown in Figure 5.

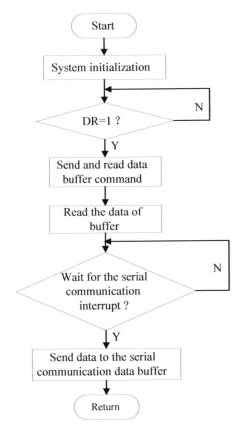

Figure 5. The flowchart of intermediate control receives data.

4.3 *PC interaction module programming*

The part use LabVIEW 8.6 programming to realize man-machine interface. Using the LabVIEW programming PC, because it provides users with an independent standard I/O low-level function can be convenient to call. Not only can save a lot of development time and reduce the cost of hardware, also for future system upgrades, leaving a large space, but also makes the system has good portability.

PC programming steps are as follows: (1) The serial initialization configuration parameters, such as baud rate, data bits and parity bits, etc. (2) Using the read and write serial port function of communication module VISA to read the serial data, after the first will determine the data header, if true, then enter the condition structure of the program; (3) Close the serial port.

In addition, you can adjust the temperature and humidity gauge value by adding a comparison function, thereby controlling the normal range of the temperature and humidity, and display transfinite alarm.

5 SYSTEM TESTING AND RELIABILITY ANALYSIS

There is no doubt that it is an essential step to debug function in the late of system design. In the process of debugging, meeting a lot of problems: (1) Lack of engineering document management thoughts, the confused file storage brings a lot of inconvenience to late modification and reference; (2) Package components go by feeling without actual measurement; (3) Without considering the actual element size, element layout is not reasonable; (4) Wiring width setting unreasonably; (5) It appeared chip pin connection errors when designed PCB, and solve the error through the jumper; (6) The voice of the board buzzer of acquisition part is too low, it is because the single-chip microcomputer IO port to drive the buzzer insufficiently and needs transistor drive; (7) Because of the unreasonable LCD display position, it leads to a large number of external connection and the occupied space; (8) It increases development costs because of the component placement loose and footprint; (9) The middle receiving part of serial communication appears errors partly , because DB9 component package pins do not correspond the actual element to the order, leading us not to communicate with the PC and the single-chip microcomputer programming procedures. After repeated testing, we can achieve basic system design functions.

Finally, through the testing experiment, we compare the measured temperature value the standard temperature value in the line graph, is shown in Figure 6. Similarly, comparing the measured humidity value with the standard value of humidity in the line graph, is shown in Figure 7.

Experiments show that the system temperature measurement range between 0~50°, and the accuracy of the temperature measurement result is good, the measurement error does not exceed ± 5% RH, basically meeting the system requirements.

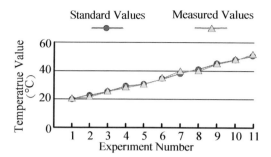

Figure 6. Temperature measured value and the standard value comparison.

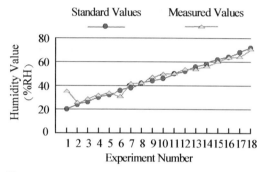

Figure 7. Humidity measured value and the standard value comparison.

6 CONCLUSION

The wireless temperature and humidity acquisition system can collect temperature and humidity data ranging from the receive port ranges within 100m easily and reliably. Along with the system upgrading ceaselessly, the system with data acquisition speed faster, stronger stability and higher reliability will be more widely used in various other areas.

REFERENCES

Huang Haizhen(2011). Design of laboratory temperature and humidity monitoring system. *D. Master degree thesis of Tianjin University*. 1–5.

Li Qiang(2013). Pigpen temperature and humidity wireless monitoring system based on AVR. *J. Industrial applications and exchange. 32*. 78–81.

Ma Hanting, Cheng Jingxia, Sun Teng(2013). Design of FPGA-based multi-point remote temperature and humidity monitoring. *J. Information and communication. 9*. 68–69.

Xue Mindi(2010). Temperature and humidity wireless measurement system with low power consumption based on nRF905. *J. Modern electronic technology. 33*. 135–138.

Yu Chengbo, Zhang Yimeng, Zhang Jin(2010). Grain depot environmental monitoring system based on ZigBee. *J. Journal of Chongqing University of Technology (Natural Science). 24*. 51–54.

Electronic Engineering and Information Science – Wang (Ed.)
© 2015 Taylor & Francis Group, London, ISBN: 978-1-138-02772-5

Research on calibration device of relay protection tester

J.H. Yuan, B. Niu & Y. Wang
Harbin University of Science and Technology, Harbin, China

ABSTRACT: The relay tester calibration equipment is studied and developed in this paper. A relay tester's calibration method is proposed according to the principle of relay tester. The hardware and software components of relay tester calibration equipment are designed. The calibration equipment is not only powerful function, high accuracy, but also replacing the previous multiple instruments and equipment combinations for testing, shorten the inspection cycle. So it has greater significance for safe and stable operation of power systems, with higher value of practical application and promotion.

KEYWORDS: Power grid security; calibration equipment; relay protection; power measurement.

1 INTRODUCTION

With the rapid development of our country's electric power industry, microprocessor based relay protection device has become mature and widely used (Guo et al. 2011). Advanced technology such as computer, automatic test of new relay protection tester are constantly emerging (Zhang et al. 2005). Application of traditional techniques and equipment for verification of their performance indicators has been difficult to achieve the comprehensive, convenient and efficient requirements (Yuan et al. 2007 & Li et al. 2008). In foreign countries, a similar test, calibration is realized by using the "Automatic Test System"(Lai et al.2005). But investment in equipment of reaching up to hundreds of thousands dollars is difficult to be accepted by domestic users.

According to the problems of domestic relay protection tester calibration, in accordance with the relevant national standards "Microprocessor-based Relay Protection Calibration Device Technical Conditions"(DL/ T624-1997) and core technology of "Relay Protection Experiment Instrument Calibration Device" patent, a new type of intelligent equipment-relay protection experiment instrument calibration device is studied to meet domestic main performance, indicators of relay protection experiment instrument calibration requirements.

2 THE WORKING PRINCIPLE AND HARDWARE DESIGN

Relay protection experiment instrument calibration device is mainly composed of signal transformation, A/D conversion, signal recognition, microprocessors and man-machine interface module. Device principle block diagram is shown in Figure 1.

After sampling, a rectifier, a voltage comparator compares the input signal is converted to MCU recognizable signal transition, the timer starts the MCU, and then carried out through the analog switch contacts, power actions to achieve verification operation time. Circuit design is shown in Figure 2.

A rectifier and a comparator constitute a signal shaping circuit, the input signal by the partial pressure, rectification, filtering, clipping, dual high-speed voltage comparator amplifies shaping, via opt coupler isolated output, ensuring MCU receives a recognizable signal. Since the gate operation generated more change from low to high, the required inverter inverting the signal to ensure proper MCU interrupt.

Above devices are the high-speed response devices, the comparator insensitive domain (Amir et al. 2006) have also been through the same phase amplifier circuit improve its response time in the nanosecond; The design uses MCU internal timer completion time measurement, an LCD display timing process, therefore, the system clock stability and program execution time delay is the main factor affecting the operating time measurement accuracy. Using 24MHz active crystal, ensure that the clock signal is stable, while reducing the delay time, to meet the measurement requirements to provide adequate hardware.

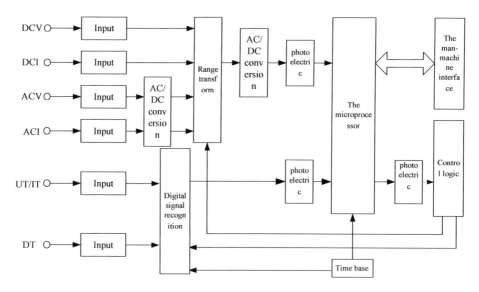

Figure 1.　Calibration device design of overall structure block diagram.

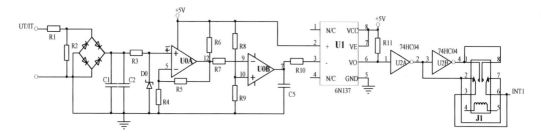

Figure 2.　Operation time measuring circuit.

3　SOFTWARE DESIGN

Using the C language compiler software system completed at Keil uVision2 integrated development environment. Due to the complexity of the control of the entire system, the program is large, modular programming ideas, so that the software structure is clear, easy to debug, modify. Programs include data acquisition, processing, LCD display, self, self-calibration and other modules, the main program calls the function module enables calibration device operation, as shown in Figure 3.

After the user selects the power measurement functions, perform the A/D initialization program. The falling edge of the BUSY signal width measurements, BUSY signal INT0 interrupt is triggered by T0, get A/D conversion result. Program flow is shown in Figure 4.

After the user selects pass (off) electrical operation time measurement function, execute the function module initialization, opening timing T2, the capture function and INT1 interrupt.

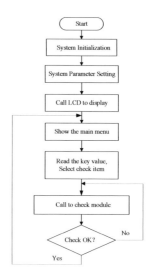

Figure 3.　The main flow chart figure.

4 DATA PROCESSING

The data processing according to certain algorithms on the raw data obtained by the data acquisition module computing, processing, converted to the corresponding value of a physical quantity.

Power measurement calculated as:

$$U = k \cdot (x - 10001) - a \qquad (2)$$

$$I = U / r \qquad (3)$$

Where x is the number of pulse measurement; k for A / D conversion slope calibration variables; a is an A / D zero calibration variables; r is the current sampling resistor.

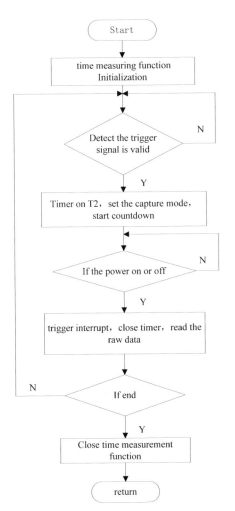

Figure 4.   A / D conversion process flow chart.

Pass (off) electrical measurements to calculate the operating time

$$t = (65536 - n) \cdot y + m - c \qquad (4)$$

Where n is the initial value of T2 is set; y is the number of overflows T2; m is the difference between the current value of T2 is set to the initial value; c is the time measurement calibration variables.

Run the self-calibration module, the pro- gram automatically calculates the formula (2), the formula (4) in a k, a, c equivalent and stored in the EEPROM in the system, the calibration apparatus to achieve a self-calibration.

5 CONCLUSION

A relay tester calibration device similar to a standard resistor, its main function is protection tester primary measurement calibration. This study calibration device realizes the relay tester, AC and DC voltage, current, and fast and accurate test pass, power operation time, by measuring the value of the corresponding parameter relay tester than right, found errors exist, and then on the relay tester debugging calibration work to make it accurate.

The relay tester calibration device features strong, accurate, able to replace the conventional use multiple instruments, equipment calibration mode combinations, not only supplement the domestic relay tester device validation studies are few, and can indirectly reduce testing cycle relay device testing, normalization, normalization of economic development to contribute to the social order of life, provide a reliable guarantee for the stable and secure society and people's lives and property.

REFERENCES

Amir M. Sodagar, Khalil Najafi(2006). Extremely Wide-Range Supply Independent CMOS Voltage References for Telemetry-Powering Applications. *J. Analog Integrated Circuits and Signal Processing.* 46, 253–261.

DL/T624-1997(1997). Technical conditions with microcomputer relay protection test device. *the People's Republic of China power industry.* Beijing: China power press.

Guo A, Guo J(2011). Power system relay protection technology present situation and the development. *J. Opencast Mining Technology.* 06, 66–68+71.

Li J(2008). The development of relay protection tester test analysis device. *D. LAN Zhou University of Technology.* Master.

Lai G, Xiao M, Xia R(2005). Automatic test system development present situation abroad were reviewed. *J. Journal of Detection & Control. 03*, 26–30.

Yuan R, Zhao J, Ding H(2007). Type of microcomputer relay protection tester testing technology research. *J. North China Electric Power. 06*, 30–33.

Zhang F(2005). The present situation of the type of microcomputer relay protection tester and development. *J. electric power construction. 07*, 56–58.

Electronic Engineering and Information Science – Wang (Ed.)
© 2015 Taylor & Francis Group, London, ISBN: 978-1-138-02772-5

Research on performance degradation of SnO_2 gas sensor based on accelerated test

J.H. Yuan, Z.M. Su, B. Niu & Z. Zhou
Harbin University of Science & Technology, Harbin, China

ABSTRACT: Performance degradation of the SnO_2 gas sensor is studied in this paper. Accelerate degradation model is established in voltage as the accelerated stress. Constant stress accelerated degradation testing is designed and executed. The life of the SnO_2 gas sensor under the normal stress level is obtained.

KEYWORDS: Performance degradation, Gas sensor, Accelerated test.

1 INTRODUCTION

The SnO_2 gas sensor has features of high sensitivity, fast response time, fast recovery time and long-term stability. Because of its excellent performance, low cost, they are widely used in daily life and industrial production. However, due to the serious consequences of gas leakage misstatement, omission, so people's requirement for its performance, life, reliability is also growing. How to quickly and accurately obtain the reliability indices of SnO_2 gas sensor become an urgent problem to be solved. The reliability of SnO_2 gas sensor is low, traditional test methods typically require more test samples and a longer test period, in order to quickly assess their reliability and shorten the product development cycle and reduce the development costs, need to adopt the Accelerated Testing (Accelerated Testing, AT) technology.

Accelerated Degradation Testing (ADT) and Accelerated Life Testing (ALT) are two important parts of the Accelerated Testing technology (Chen et al. 2007). In this paper, the Accelerated Degradation Testing (ADT) without changing the failure mechanism of product by improving the stress level to speed up product performance degradation, life, performance in- formation obtained through reasonable extrapolation of statistical methods to get the reliability or accelerated testing method of life estimation under the condition of normal stress. In ADT, the "failure" is generally defined as a performance parameter degenerated to below a given project indicators (i.e., failure threshold). The data of product function parameter followers time degradation is called degradation data. In addition to the failure data of the product, product performance degradation data is also an important data in reliability analysis. NAIR has pointed out that degradation data for reliability assessment is a rich source of information (NAIR et al. 1988).

ADT overcomes ALT only record failure time of products, product failure and specific process of failure are not recorded. The ADT through product degradation data processing solves the problem very well, and solves the problem does not apply to the current traditional reliability theory and engineering applications. ADT is a strong complement to ALT, and is an important development direction of the acceleration testing.

This paper studies the SnO_2 gas sensor by using the voltage as the Accelerated Stress, establish the Accelerate Degradation Model firstly, and then conduct the CSADT. Finally, obtain each reliability index of the SnO_2 gas sensor under the normal stress level by statistical analysis the reliability test data.

2 SETTING ACCELERATED DEGRADATION TRAJECTORY MODEL

The ADT under the constant stress, it is assumed that there are n_k samples under the stress level S_k, measuring performance degradation in $t_{i1}, t_{i2}, ... (i=1,2...n_k)$ time and measuring the performance degradation data $y_{i1k}, y_{i2k}, ... y_{ijk}$ (j=1,2...,m_i) m_i times, and y_{ij} is the performance degradation data of the sample i at the moment j. The $y_k(t_{ij})=y_{ijk}$. Does not require the number of testing degradation data is same under the different stress level, the time of testing degradation data and the number of samples can also differ (Wang et al. 2013) .

Set the actual degradation path of the product for $D(t)$ $t>0$, actually the value of $D(t)$ can be monitored in the time $t_{i1}, t_{i2}, ... (i=1,2...n_k)$, the sample i at the moment t_{ij} under the stress level S_k is monitored as:

$$y_{ijk} = D(t_{ij}, \beta_{1i}, \beta_{2i}, ..., \beta_{pi}) + \varepsilon_{ijk} , i=1,2...n; j=1,2...$$
$$m_i; k=1,2...r \qquad (1)$$

$D(t_{ij}, \beta_{1i}, \beta_{2i}, ..., \beta_{pi})$ is the actual trajectory of the sample i at the moment $t_{ij}, \beta_{1i}, \beta_{2i}, ..., \beta_{pi}$ are parameters vector which associated with accelerated stress, $\beta_{ijk} \sim N(0, \beta_e)$ are measurement error which obey the zero mean Gaussian distribution.

The accelerated trajectory of each of the SnO_2 gas sensors under each stress can be obtained through Accelerated degradation paths which created by the first step. Then, according to the failure thresholds calculated for pseudo life for each of the SnO_2 gas sensors. When the amount of degradation product reaches a critical level (defined as the failure threshold), assuming that a failure at this time, so the product fault time can be defined as the actual degradation path $D(t)$ of the product reached the time of critical degradation level, due to the failure time is not the actual life, defined as pseudo failure life.

Assume that pseudo longevity obey the Lognormal Distribution, by a degradation trajectory (1) the pseudo longevity

$$T = e^{\mu + \frac{\sigma^2}{2}} \tag{2}$$

Taking into account the failure of the product data for small sample (n <50), so K–S testing is applied. $F_0(t)$ is the distribution function of a distribution. $F(t)$ is the Empirical Distribution Function of samples.

The null hypothesis: $F(t)=F_0(t)$

The alternative hypothesis: $F(t) \neq F_0(t)$

$F_0(t)$ can be regarded as theoretical distribution, the sample observations from small to large in order of priority $t(1), t(2), ..., t_n$, value $F_0(t_i), (i=1...n)$ of the distribution function can be calculated after substituting.

According to the experience of the sample distribution can approximately calculate the corresponding cumulative frequency $F_n[t_{(i)}]$, it is calculated using an approximate median rank formula for

$$F_{n(t_i)} = \frac{i - 0.3}{n + 0.4} \tag{3}$$

Firstly, according to the characteristics of the life test, selecting the appropriate formula to calculate each failure data corresponding cumulative failure probability of $F_0(t_i)$, then $F_n(t_i)$ is calculated based on the formula chosen distribution function, and then calculating the difference between $F_0(t_i)$ and $F_n(t_i)$.

$$D_n = \sup_{-\infty < t_i < \infty} |F_n(t_i) - F_0(t_i)| \tag{4}$$

Given the significant level α, and according to the

$$P[D_n > D_{n,\alpha}] = \alpha \tag{5}$$

A determination is made. If meet $D_n \leq D_{n,\alpha}$, which can be inferred that the above has made confidence $(1-\alpha) \times 100\%$, whereby the type of primary distribution determined is correct.

According to the selection of stress, establishing the relationships between voltage stress and the characteristics of SnO_2 gas sensor life, and then calculating the life of the SnO_2 gas sensors under the normal stress. Before deciding accelerate the degradation model, first make the following assumptions: under the accelerated stress level S_k, pseudo life of SnO_2 gas sensors are obey the lognormal distribution $LN(\mu_k, \sigma_k^2)$, $k=1, 2, ..., r$.

In the accelerated degradation test, the voltage as the accelerated stress, so choose the inverse power law model(Mao et al.2003).

$$\xi = Av^{-C} \tag{6}$$

In the formula, ξ is a life characteristics, such as medium life, average life and characteristic life, etc, A is a positive constant. C is a positive constant which related to the activation, v is the stress, often taking the voltage.

In order to calculate conveniently, the inverse power law model taking the logarithm transformation

$$\ln \xi = a + b \ln v \tag{7}$$

In the formula, $A=\ln a$, $b=-c$. They are undetermined parameters, moreover, $v > 0$.

For the parameters of the inverse power law model, the amount of expression after the substitution can also be estimated using the least squares method.

3 THE DESIGN OF ADT OF SnO_2 GAS SENSOR

In the process of the plan formulation of ADT, often need to determine the number of the product sample, the types of test stress, the stress way, the stress level, the test time and the measured interval, etc.

3.1 The number of the sample

Due to the ADT is a kind of forecast product reliability testing technique based on the physical failure model and combining with the method of mathematical statistics extrapolation, so the more sample quantity the better failure data we have. But in fact, it is difficult to meet the demand of such a large number of samples, because the large sample number means more experimental cost and testing time. According to related provisions in the literature (GB2689.1-1981), decide to use 8 samples in each test temperature in this test, a total of 24.

3.2 The selection of the acceleration stress

In the process of reliability test will produce a performance degradation data, which contains a lot of life information, there is a certain relationship between failure and performance parameters which is continuous degradation, the degradation of product performance can lead to failure, therefore, based on reliability analysis of performance degradation has become a hot research in recent years (ZEHUA et al. 2005). At the same time, developing characteristics of modern electronic equipment led to some new reliability problems, such as complex working conditions (temperature, power supply)and the complex task environment. Electric power electronic devices in the current, voltage or power under the action of stress, the stronger the stress, the faster the failure rate, the shorter the device lifetime(YANG et al.1982). Therefore, there is a very significant research on accelerated testing of electrical stress. In the actual using process, the voltage has a significant impact on the performance of SnO_2 gas sensors, so the voltage stress is chosen as accelerated stress of the test.

Constant Stress Accelerated Degradation Test (CSADT) has many examples. This is because the accelerated life test method has the simple operation and more successful data processing method. Although it is not the shortest time required for the test, it still exponentially shorter time than the average life test, so it is often actually used. Therefore, this test could take constant stress as the stress applied method.

The CSADT stress level selection principle is as follows: the stress level is less than the product work limits ensuring that the product degradation failure mechanism is same with the mechanism under the normal stress; The initial stress level can approximate to the normal stress level to improve the credibility of the extrapolation results; The number of the stress level are between 3–5.

Based on the above principle, we select the $U_1=5.15V$, $U_2=5.30V$ and $U_3=5.45V$ as the stress.

3.3 Selecting parameters of performance degradation and the failure criterion

The SnO_2 gas sensor made of sensitive element fixed in the cavity of the plastic or stainless steel cavity. Miniature Al_2O_3 ceramic tube, SnO_2 sensitive layer, measuring electrode and heater are made of the sensitive element. The heater provides the necessary working conditions for the gas sensor. The main performance parameters of the SnO_2 gas sensors are the sensitivity, the response time, the recovery time and the resistance R_0 under the clean air, etc. Among them, the change of the R_0 value is the main failure model, so we choose R_0 as the performance degradation

parameter. Informed by preliminary test, when the R_0 is change more than 40%, the SnO_2 gas sensor can't work normally, so the threshold is the degradation value more than 40%.

3.4 The test equipment

A homemade closed gas sensor box, a power supply and a resistance measurement device are used.

3.5 The test steps

From the qualified products of the same production batch, selecting 24 SnO_2 gas sensors as the sample randomly and measuring the R_0 of each test samples after preheat 48 hours. The R_0 is the initial values of the corresponding sample. The test method is in accordance with the literature(GBT 15653-1995).

24 for test samples are divided into three groups, followed by constant voltage test 5.15V, 5.30V and 5.45V. At regular intervals and recovery after 90 minutes under no applied voltage measurement value R_0 under the clean air, repeating the above steps until the completion of the all voltage level test.

3.6 The example analysis

Through the analysis of the accelerated degradation data of various samples, obtained each sample degradation path, and then according to the failure threshold, we can get the pseudo failure life in 5.15V, 5.30V and 5.45V, based on the pseudo failure life time, assume that the pseudo failure data of the sensor obey the logarithmic normal distribution, based on the reliability assessment method of the logarithmic normal distribution, can get the logarithmic normal distribution parameter estimation of the product pseudo failure life under different stress level. The estimated results are shown in table 1.

Table 1 shows: the shape parameters s of the Logarithmic Normal distribution approximate the same under the 5.15V, 5.30V and 5.45V, which means that the degradation (failure) mechanism of the samples has not changed in the process of ADT, thus accelerate the model equation can be used to analysis.

Table 1. The parameter estimation under each voltage stress level.

voltage	s	m
V	$k\Omega$	$k\Omega$
5.15	0.339	7.115
5.30	0.406	7.103
5.45	0.413	7.034

So we can use the data from the table 1 to calculate the acceleration equation parameters estimation: a=9.588, b=−1.498. So acceleration equation is

$$\ln \xi = 9.588 - 1.498 \ln v \tag{8}$$

Then we can get the failure life distribution parameters under the 5V, σ_0=0.386KΩ, μ_0=7.177 KΩ.

So, when acceleration stress is of voltage 5V, reliability of the sensor under given time is

$$R(t) = 1 - \Phi\left(\frac{\ln t - \mu_0}{\sigma_0}\right) = 1 - \Phi\left(\frac{\ln t - 7.177}{0.386}\right) \tag{9}$$

When $R(t)$=0.9, by calculating, the life of the sensor SnO_2 is t=2145h.

4 CONCLUSIONS

The CSADT of SnO_2 gas sensor is designed in this paper. The acceleration model under the voltage stress is established, and the distribution type of the life is determined. The life of SnO_2 gas sensor under the reliability of 0.9 is obtained using the accelerated degradation data.

REFERENCES

Chen Aimin, Chen Xun, Zhang Chunhua(2007). A Comprehensive Review of Accelerated Degradation testing. *J. ACTAARMAMENTARII. 28*, 1002–1007.

GB2689.1–1981, the general for constant stress life test and accelerated life test method.

GBT 15653–1995, metal oxide semiconductor gas sensitive element total specification.

MaoS.(2003).The acceleration model of accelerated life test. *J. Quality and Reliability. 02*, 15–17.

NAIR V N(1988). Discussion of estimation of reliability in field performance studies by JDJF lawless .*J. Techno metrics. 30*, 179–383.

Wang Y, Li X(2013). Accelerated test research based on the degradation trajectory of lithiumion battery. *J. Equipment Environmental Engineering. 02*, 30–33.

YANG Jiajian, GUAN Chengxun(1982). Physical basis of failure. *M. Science Press Publishing*, Beijing.

ZEHUA C, SHURONG Z(2005). Lifetime distribution based degradation analysis. *J. IEEE Transactions on Reliability. 54*, 3–10.

Electronic Engineering and Information Science – Wang (Ed.)
© *2015 Taylor & Francis Group, London, ISBN: 978-1-138-02772-5*

The 3D surface measurement and simulation for turbine blade surface based on color encoded structural light

J.H. Yuan, H.J. Yuan & B. Niu
Harbin University of Science & Technology, Harbin, China

ABSTRACT: This paper proposes a color-encoded structural light approach for the three-dimensional measurement of turbine blade surface. It adopts RGB color mode design coding scheme, by using red, green, blue, yellow, magenta, cyan six kinds of color as basic colors for color-coded colored stripes. Each three stripes are grouped together, while the color combinations for each group are different. The coding pattern based on these color stripes is projected to the blade, and a CCD camera then captures the image modulated by the blade. We then decode this image to obtain the three-dimensional coordinates of the measured point on the blade surface. According to the structure light measurement principle, the three-dimensional measurement system and determine the parameters of the measurement system are designed, and the mathematical model of three-dimensional measurement is also established. This paper proposes a model for the turbine blade through the simulation by using 3ds max software. We design a three-dimensional measurement system for the simulation, and obtain the three-dimensional coordinates of the blade surface by measuring the turbine blade. Finally, the three-dimensional blade reconstructed from the measured data is established.

KEYWORDS: color stripes encoded, three-dimensional surface, hydraulic turbine blade.

1 INTRODUCTION

Currently, China's hydropower business is booming, and it has widely served in all areas of the national economy of life. During the turbine operation, due to the influence of various complex underwater environments, turbine blades are eroded, appeared pitting, and even part of the organization is peeled. These easily lead to resonate and other adverse reactions, loss of the turbine's efficiency, resulting in the insufficient power output. Thus, the measurement and maintenance for hydro key equipment have become the focus of research. Since turbine blades are composed of very complex free-form surfaces, it is very difficult to be measured. Also note that due to its long-term in the complex underwater and its complex external environment, the intensity of maintenance personnel labor is intensive. Therefore, many hydropower stations use welding robots to repair eroded blades. The research on the 3D surface measurement can provide a theoretical basis for robot welding. There exist various different methodologies for measuring the turbine blade. This paper proposes a color-coded structural light measurement technique based on a machine vision technology, for measuring the turbine blade. There are two steps: in the first step, various types of two-dimensional light information are obtained through observation system, in the second step, the

three-dimensional structural information contained in the two-dimensional light is reconstructed by the program.

2 MEASUREMENT PRINCIPLE

Structural light measuring method (Yuan. 2006) is the more popular and highly used means of measuring at present. Point structural light method is the simplest and most basic method (Yu et al. 1997), however, it suffers from some disadvantages, which will be discussed in details. First, point-by-point scanning must be performed in order to get the shape information on the entire blade surface. Second, the three-dimensional coordinate of each point is determined by the above formula, and thus the calculation burden increases. Last but not the least, it results in sharp increased time in image acquisition time and processing capacity and long time of image processing time, which in turn seriously affect the measuring efficiency. These drawbacks can be overcome by linear structural light measurement method. In such a method, since the point becomes the line, the information for all the points on this line can be obtained by scanning the blade surface. Compared to the point structural light method, the measuring efficiency has been greatly improved. However, the line light source must constantly change the projection angle so that the surface of the entire blade within the field of view can be scanned and many images can be taken. Through this, the three-dimensional coordinates of all points are obtained. Therefore, the measurement efficiency still needs to be improved.

According to the idea of transforming the point in the line, the line can be further projected into a two-dimensional pattern on the blade. By encoding the pattern in a certain way, it becomes coding pattern method (Tajima et al. 1990, Chen. et al 2003, Chen et al. 2005). Coding structural light measurement is one of the most reliable three-dimensional measurement techniques.

3 ENCODING AND DECODING PRINCIPLE

Compared with the geometric feature, the color feature is easier to identify. Therefore, using the color images for image recognition processing color images results in a very good result. Since three colors, red, green, and blue are easy to be identified during the image processing, RGB color mode coding has been used. This color stripe coding method uses the change of the channels of red (R), green (G), blue (B) and the superposition on each other to get a variety of colors. RGB color mode uses the RGB model to assign a value of intensity 0-255 to RGB components for each pixel in the image. For example, for red color, R component has a value 255, while both G and B components have value 0. In the experiment, we only assign two values, 0 or 255 of each color in order to maximize the difference between colors, which in turn yields the best recognition performance. This also improves the ability of anti-interference. Coding patterns are generated by 6 colors: red, green, blue, and their combinations, cyan, magenta and yellow. Based on the above analysis, the values of R,G,B components for each color is obtained, as shown in Table 1. Respective colors R, G, B values.

Table 1. The values of RGB for different colors.

Color	Red	Green	Blue	Yellow	Magenta	Cyan
R	255	0	0	255	255	0
G	0	255	0	255	255	255
B	0	0	255	255	0	255

In the process, each color represents as a stripe. We then combine 6 different color stripes in the way such that three adjacent color stripes form a group, the order of all colors among all the stripes cannot be repeated, for example, Such as cyan, green, blue; green, blue, magenta; blue, magenta, red; magenta, red, yellow; red, yellow, red; yellow, red, red; red, red, green; red, green, cyan; green, cyan, yellow, the width of each stripe is preferably in the 2 pixels. Since the

interference among the color stripes may occur due to the light intensity, we deliberately a black stripe across all the color stripes. The width of the black stripe is set to be 4 pixels. There are in total 6^3-(6-1) =211groups of eligible combinations. Thus, the number of color stripes being used is 213. Figure 1 shows the resulting coding pattern for combinations of the color stripes.

Figure 1. The coding pattern of the color stripes.

There are various combinations of the color stripes. Different initial color choices lead to different coding combinations of the color stripes. After many tests, we find that the only way to get the longest period of the code combinations is to set red, red, red as the initial color stripe, and to order all other colors are in ordered.

4 BUILDING TURBINE BLADE MODEL

We first model the turbine blade in 3ds max software. As shown in Figure 2, according to the actual data of the turbine blade, we establish a rectangular area in the Front view.

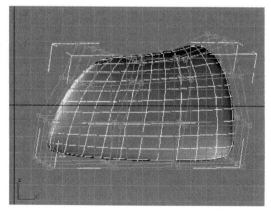

Figure 2. Model of the turbine blade.

The number of segments of the length, width, and height are properly set. This is to make the rectangular subdivided so that each part can be edited. We then enter the edit panel, select "FFD $4 \times 4 \times 4$" editor on the Edit command panel, and use rotation, translation, and other functions to edit the control point, so that the model of the blade is in a preliminary with shape. Finally, we choose "MeshSmooth" editor on the Edit command panel to make the surface of the blade model smooth. We next open the Material Editor to edit the material of the blade surface. Since turbine blades are generally made of stainless steel, we choose the metal pattern in the material editor so that the blade model has a metallic sheen. In the "specular highlights" adjustment panel, we set the high light level to 238, and the gloss to 23, close the self-luminous option and set the opacity to 100%. Figure 3 is a rendered graph of the blade model being attached to materials.

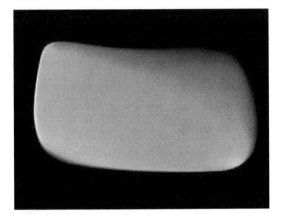

Figure 3.   Rendered graph of turbine blade.

5   SETTING MEASURING SYSTEM

The projector should be set up according to the principle of structural light projector, as shown in Figure 4. The triangle in the figure is composed of the projection lamp, the blade model and the camera, i.e., the center O_1 of projection lamp, the camera lens O_2 and the projected point Q on the blade by spotlight together constitute the triangle. $O_1O_2=B$; Point Q is located in the object coordinate $XYZO_1$, and the image Point Q_0 formed on image sensor is located in the image coordinate $x_0o_0y_0$ with coordinates (x_0,y_0), thus $x_0o_0z_0$ and XO_1Z are in the same plane. The object point Q is located on XO_1Z plane, $y_0o_0z_0$ plane and its projection on o_0z_0 axis are Q', Q_Z and Q'_Z, respectively. β_0 is the angle between the camera Optical axis and X axis. The projection angle α can be computed according to of color decoding.

According to Figure 4 can be learned formulas (1)–(3)

$$Z = \frac{B}{\cot\alpha + \dfrac{x_0 + f\cot\beta_0}{f - x_0\cot\beta_0}} \qquad (1)$$

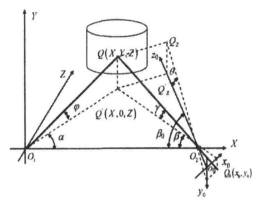

Figure 4.   Point principle structural light method.

$$Y = \frac{Z \cdot y_0}{f\sin\beta_0 - x_0\cos\beta_0} \qquad (2)$$

$$X = Z\cot\alpha \qquad (3)$$

In the above equations, f, B, β_0 are all predictable parameters. For example, the focal length of the camera lens is given by f, the angle between the optical axis of the camera and the X axis is predetermined angle β_0, and the distance between the center of the spotlight and the center of the camera lens is denoted by B. The coordinates (x_0, y_0) of the image point Q_0 of the point Q is known, the projection angle a is also a known variable which can be obtained according to the color decoding. Thus, by plugging these known variables into equations (1)–(3), we obtain the three-dimensional coordinate values X, Y, Z of the point Q. The deep image of the blade in the projection field of view can be obtained by changing the projection angle along the directions of φ, a. The above three formulas for computing the values X, Y, Z is the basic formulas of structural light method, and is also the basis for the three-dimensional measurements of the structural light.

We then set up the projector accordingly, with the simulation projector replaced by target spotlight, as shown in Figure 4. Let us place the spotlight in the origin O_1 in the world coordinate, i.e., the center of the projection system has coordinates (0,0,0). Also note that the angle between the optical axis and the x-axis is 60°, i.e., $\alpha_0 = 60°$, and the projection angle is 60°, i.e., $\alpha_1 = 30°$. We next set up the spotlight according to the parameters determined before. The light intensity coefficient is 20cd. The color of the light projected by the spotlight is white, since only white light can least affect the projected color-coded patterns; the starting and the ending point of the near attenuation of target spotlight are set to 100mm and

900mm, respectively; the starting and the ending point of the far attenuation of target spotlight are set to 2000mm and 5000mm, respectively.

The measuring system is set according to the principle of structural light. The simulation projector replaced by target spotlight which projects the color-coded pattern, being placed at the point on the x axis with coordinates (0,0,0), as shown in Figure 5 a). The simulation video camera is placed at the point on x axis whose distance to the projector is 900mm, as shown in Figure 5 b). The angle between the optical axis of the projector and the x-axis is 60 °, and also the angle between the optical axis of the camera and the x-axis is 60 °.

Since the blade model is large, it requires two steps for measuring the blade model. We need to measure in two fields of view separately, then splice the images in order to synthesise the entire blade.

6 SIMULATION RESULTS

Figure 6 shows the images obtained in two fields of view. The three-dimensional coordinates of turbine blade surface are calculated by decoding and splicing images obtained in two fields of view. The results reconstructed are shown in Figure 7.

7 CONCLUSIONS

Experimental results show that the proposed three-dimensional color-coded structural light measurement method for measuring the turbine blade is feasible, is easy to operate, and process data faster compared to the point structural light and line structural measurement methods.

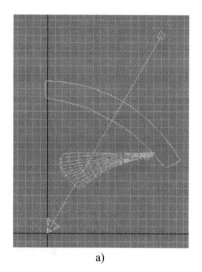

a)

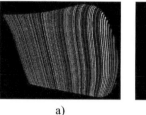 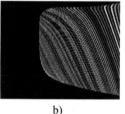

a)   b)

a) Pattern shot in the first field of view
b) Pattern shot in the second field of view

Figure 6.   Two images of the coding pattern projection.

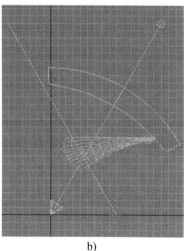

b)

a): The set-up of the projector
b): The set-up of the video camera
Figure 5.   The set-up of the projector and the video camera.

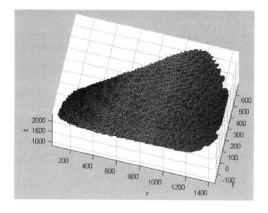

Figure 7.   Reconstruction of the turbine blade.

REFERENCES

L. C. Chen, S. H. Tsai, K. C. Fan(2005). A New Three dimensional Profilometer for Surface Profile Measurement Using Digital Fringe Projection and Phase Shifting. *J. Key Engineering Materials. 295.* 471–476.

J. Tajima, M. Iwakawa(1990). 3D Data Acquisition by Rainbow Range Finder. *C. International Conference on Pattern Recognition.* 309–313.

S. Y. Chen, Y. F. Li(2003). Self-recalibration of a Colour encoded Light System for Automated Three- dimensional Measurements. *J. Measurement Science & Technology. 14,* 33–40.

Yuan Huijuan(2006). Combination of color stripe encoding three-dimensional measurement techniques. *D. Harbin University of Science and Technology.*

Yu X, Zhang J, Wu L, et al.(1997). The progress of the structured light 3d vision technology. *J. The space measuring technology. 5,* 43–48.

Electronic Engineering and Information Science – Wang (Ed.)
© 2015 Taylor & Francis Group, London, ISBN: 978-1-138-02772-5

RTR-1000 rapid triaxial rock testing system and its application

Q. Hu & Y. Mei
Southwest Petroleum University, Chengdu, Sichuan, China

L. Zhang
Changqing General Drilling Company, Xi'an, Shanxi, China

ABSTRACT: A new experimental system for rock mechanics under high temperature and high pressure, RTR-1000 rapid triaxial rock testing system, is introduced in this paper and its operational procedure, reliability, accuracy and characteristics are analyzed. It can be used to study strength property and deformation behavior of rock, concrete and other materials at different temperature and different pressure by all kinds of tests, such as triaxial compression test, uniaxial compression test, cyclic loading-unloading test, triaxial shear test, and creep test, etc. Through the tests, it is easy to obtain variety of characteristics parameters of test specimen, such as pore volume compressibility, Poisson's ratio, modulus of elasticity, etc. This testing system is one of the essential experimental systems in the field of modern rock mechanics testing and research owing to its simple and safe structure, easy and convenient operation, high test precision, and wide application.

KEYWORDS: Rock mechanics; Testing system; Triaxial test; Uniaxial compression; Stress.

1 INTRODUCTION

Triaxial rock testing system is an important tool for studying rock mechanics, and it is widely used in the research field of oil and gas exploration, geotechnical engineering, construction engineering, and geological hazard, etc (Liu, X., Q. Zhang & Z. Yue 2013). At present, the domestic triaxial rock testing systems are divided into two types: conventional triaxial testing machine and true triaxial testing machine, and according to different research purposes, there were developed some new testing machines with added special functions, such as creep, static and dynamic state, different temperature, infiltration, and other recombination (Tang, H. & S. Li 2004, Sun. X., M. He & C. Liu 2005, Zhang. K., Z. Yin & Z. Xu 2003, Chen. A., J. Gu, J. Shen, Z. Ming & X. Zhang 2004). Internationally, the most famous electro-hydraulic servo rock testing machines are produced by GCTS (Geotechnical Consulting and Testing Systems) Corporation, MTS (Mechanical Testing and Sensing Solutions) Systems Corporation, Instron Corporation and Shimadzu Corporation (Huang. R., D. Xu, X. Fu, X. Yu, Y. Huang & Y. Liu 2008, Wang. B., J. Zhu & A. Wu 2010, Zhu. J., Y. Jiang & L. Wang 2010, Zhao. Y., Z. Wan, Y. Zhang, F. Qu, G. Xie, X. Wei & W. Ma 2008). In addition, France TOP INDUSTRIE Corporation, Germany Schenk Corporation and Switzerland W+B Corporation also produce rock testing machines (Martin, C.D. &

R. Christiansson 2009, Xu. J., S. Peng & G. Yin 2010, Wu. A. 2010). Now, most existing triaxial rock testing systems are equipped with all-digital servo feedback technology and advanced technology in different countries, such as EDC digital measure and control instrument of Germany DOLI Corporation, servo valve of the U.S. MOOG Corporation, servo motor of Japanese Panasonic Corporation, mute oil pump of Japanese NACHI Corporation and other key components and parts (Lv. J., Y. Cui, J. Liu & H. Lu 2010, Zhang. Y. 2004, Chen. J. 2008). So, there are remarkable improvements in its system functionality, performance, stability, etc.

RTR-1000 rapid triaxial rock testing system developed by GCTS Corporation in U.S. is the typical representative of modern rock testing machines (Gao. C., J. Xu, P. He & J. Liu 2005). The whole experimentation is under computer automatic control, and it can finish various mechanics tests by simulating the formation under different pressure and temperature conditions (Liu. Z., X. Li, G. Zhao, Q. Li & W. Wang 2010). Then, the testing results can provide accurate basic data for petroleum exploration and development research, such as rock fragmentation, well stability, formation fracture pressure prediction, production casing load calculation and design, oil and gas production and making development plans, hydraulic and acid fracturing in oil-wells, and geostress prediction, etc (Li. Y., Z. Ma, Y. He, B. Wang & X. Liang 2006). Therefore, it can be applied not only in the industry areas, such as oil and gas drilling, coalbed methane exploitation and environmental geology, etc., but also in civil engineering, mining industry and nuclear waste storage.

2 RTR-1000 TESTING SYSTEM

2.1 System features

RTR-1000 testing system is an international advanced new product developed by GCTS Corporation in U.S. It is one of the most advanced rock mechanics testing systems at home and abroad, and it is equipped with mute oil pump, electromotor, and closed-loop servo control. Experimented with this system, after entering the specimen size, installation location parameters, and the testing condition parameters into the computer, the whole experimentation is under computer automatic control, the testing data will be collected and saved by the computer automatically, and graphical curves will be generated automatically in computer. Also, the testing data and graphical curves will be saved in computer for later analysis. The system interface can realize the all-digital control of the various dynamic load tests. This system supports function generator, dynamic mode control transformation, test programming, multi-mode data collection, data processing, the generating reports of data, and other functions. For the rock, concrete and other materials, it can finish such basic performance tests as uniaxial loading, triaxial compression, destruction test under high temperature and high pressure, porosity seepage test, and tensile strength testing under room temperature, etc. At the same time, the system can realize the real-time monitoring and multichannel data acquisition and processing in the whole experimentation.

2.2 Components of testing system

RTR-1000 testing system is composed of the following main components, and its system principle diagram is shown in Figure 1.

1 Main frame. It has four-poster loading frame surrounded by protective glass and the oil pump upper installed. So, the main frame makes the testing system appearance beautiful, enhances loading stiffness, and increases testing security.
2 Control cabinet. The whole control system includes all-digital servo controller, various sensors (such as force sensors, displacement sensors, and strain sensors), and servo valves, etc. The Controller displays and controls (compares with the setting parameters) the amplified signal from various sensors, and then adjusts the opening of the servo valve to meet the setting value.

Simultaneously it sends the data to the computer. Finally, data display, data process, and curves drawing are finished in the computer. Therefore, it has the advantages of high precision, full protection and high reliability.

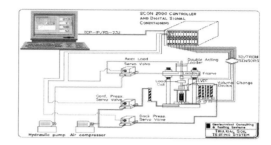

Figure 1. RTR-1000 testing system principle diagram.

3 Hydraulic system. As we all know, there are many advantages in hydraulic transmission system, such as smooth transmission, low weight, small size, large carrying capacity, stepless speed regulation, overload protection and easy automation, etc. In this testing system, the advanced mute oil pump is equipped, which makes the system running smooth, low noise, small vibration, fast response, and stable operation under high-pressure condition.

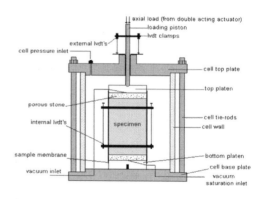

Figure 2. Structure diagram of triaxial cell.

4 Triaxial cell. The triaxial cell can be automatically lifted and slipped by hydraulic control system. There are no bolts and other fasteners in this cell for simplifying testing operations, improving the stress conditions of the test specimen, and raising the testing efficiency and accuracy. Its structure diagram is shown in Fig. 2. The whole cell can be automatically sealed, lifted, locked. And the test specimen can be installed easily, safely and quickly in this cell. Generally, the diameters of test specimen are 1 inch, 1½ inches, or 2 inches.

5 Experimental measurement and Data acquisition system. All the testing parameters are measured by experimental measurement system. After test specimen installation and parameters input (installation location, axial force, and confining pressure, etc.), the whole testing procedure, including loading method, loading speed, various data acquisition and data storage, etc., is automatically controlled by computer. So, this system has more advantages in terms of good stability, high precision, high degree of automation, increasing reliability, and avoiding personal error, etc.

2.3 *Main test indicators*

The main technical indicators of RTR-1000 testing system are as following.

Confining pressure: 140MPa; Temperature: 150°C; Pore pressure: 140MPa; Maximum axial load: 1000KN; Axial deformation range: ±2.5mm; Radial deformation range: ±2.5mm; Pressure testing accuracy: 0.01MPa; Liquid volume testing accuracy: 0.01cc; Deformation testing accuracy: 0.001mm; Diameter of test specimen: 1 ", 1½", 2 "; Dynamic frequency: 0-10Hz.

3 FUNCTIONS AND APPLICATIONS

Many types of rock mechanics experiments can be conducted with RTR-1000 testing system, and this system can be widely used in the research field of oil and gas exploration, geotechnical engineering, construction engineering, and geological hazard, etc.

Only in the field of rock mechanics, there are 16 types of rock experiments through this testing system. They are ① uniaxial compression / tension / creep testing / relaxation test / fracture toughness test; ② triaxial compression test / creep / relaxation test; ③tensile strength test; ④ rock triaxial test under high temperature and high pressure; ⑤ rock triaxial / uniaxial dynamic test; ⑥ triaxial creep experiments; ⑦ permeability test; ⑧ indirect tensile test (Brazilian test); ⑨ fracture toughness test; ⑩ triaxial ultrasonic wave velocity (transverse wave, longitudinal wave) test; ⑪ acoustic emission test (KAISE effect); ⑫ triaxial acoustic emission test; ⑬ hydraulic fracturing test; ⑭ CVA radial velocity anisotropy test; ⑮ DSA differential strain analysis; ⑯ rock true triaxial test (54.7*54.7*108mm cube sample). In addition, according to the special requirements, some experiments can be extended. Through the tests, variety of rock mechanics parameters can be achieved, for example, elastic modulus, Poisson's ratio, compressive strength, bulk modulus, shear modulus, and stress sensitivity, etc.

Also, the testing operation is easy and simple. After making the test specimen according to testing requirement, the test specimen and various sensors should be correctly installed and connected firstly. Next, all the testing parameters and conditions should be inputted into the computer, such as the parameters and conditions of rock specimen size, loading, pressure, and temperature, etc. Finally, the whole experimental process is controlled by the computer. Therefore, the testing operation is very easy and simple, the testing data acquisition is real-time and convenient, and the testing results are accurate and reliable with RTR-1000 rapid triaxial rock testing system.

4 PARTIAL EXPERIMENTAL RESULTS OF RTR-1000 TESTING SYSTEM

For the length of this paper and representative testing, the partial experimental results of uniaxial compression testing and triaxial rock mechanics testing are shown in Figure 3-4.

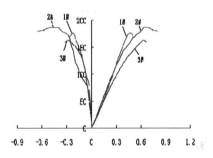

Figure 3. Stress-strain curves of uniaxial compression testing.

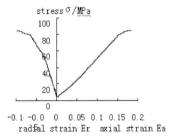

Figure 4. Stress-strain curve (confining pressure: 20MPa).

Through uniaxial compression testing, such rock mechanics parameters as uniaxial compressive strength, elasticity modulus, and Poisson's ratio, etc. can be easily achieved, and various data and curves can also be outputted by computer. All the data and curves can be extracted and analyzed according to your requirements later.

Through triaxial compression testing, not only the stress-strain curves under different confining pressure and loading condition, but also the test specimen deformation curves during the whole loading procedure can be achieved and displayed. Various rock mechanics parameters can be achieved, for example, elastic modulus, Poisson's ratio, compressive strength, bulk modulus, shear modulus, fracture toughness, cohesive force, and internal frictional angle, etc. through the tests. So, the rock mechanics characteristics can be studied accurately and comprehensively. Similarly, all the data and curves can be extracted and analyzed according to your requirements later.

5 CONCLUSIONS

After the studies of RTR-1000 testing system structure, feature, function, application, testing operation and result, the main conclusions can be drawn as following.

Firstly, as the advanced and practical testing equipment, RTR-1000 rapid triaxial rock testing system is an important tool for studying rock mechanics, and it is widely used in the research field of oil and gas exploration, geotechnical engineering, construction engineering, and geological hazard, etc.

Secondly, there are more than 16 types of experiments with RTR-1000 rapid triaxial rock testing system, such as uniaxial compression testing, triaxial testing, creep testing, tensile strength test, etc. under different temperature, pressure, loading conditions. In addition, according to the special requirements, the part content of some experiment can be extended.

Thirdly, various rock mechanics parameters can be achieved, for example, elastic modulus, Poisson's ratio, compressive strength, bulk modulus, shear modulus, fracture toughness, cohesive force, internal frictional angle, and stress sensitivity, etc. through the variety of tests.

Finally, the advantages of this system include beautiful appearance, testing security, simple operation, perfect function, high degree of automation, high precision, and reliability. The testing data and curves are collected, processed and displayed timely, and all the data and curves can be extracted and analyzed according to your requirements later. It is very useful for our research.

REFERENCES

Liu. X., Q. Zhang & Z. Yue (2013). Current Situation and Development Trends of Rock Triaxial Testing Machines. *Rock and Soil Mechanics*. 34, 600–607.

Tang. H. & S. Li (2004). MTS815 Full-digitally Servo-controlled Rock Mechanics Testing Machine and its Application. *Mining Research and Development*. 24, 28–31.

Sun. X. & M. He & C. Liu (2005). Development of Nonlinear Triaxial Mechanical Experiment System for Soft Rock Specimen. *Chinese J. of Rock Mechanics and Engineering*. 26, 2870–2874.

Zhang. K., Z. Yin & Z. Xu (2003). Development and Application of True Triaxial Mechanical Experiment in China. *Geotechnical Engineering Technique*. 5, 289–293.

Chen. A., J. Gu, J. Shen, Z. Ming & X. Zhang (2004). Development and Application of Multifunctional Apparatus for Geotechnical Engineering Model Tests. *Chinese J. of Rock Mechanics and Engineering*. 23, 372–378.

Huang. R., D. Xu, X. Fu, X. Yu, Y. Huang & Y. Liu (2008). Development and Research of High Pressure Permeability Testing System for Rocks. *Chinese J. of Rock Mechanics and Engineering*. 27, 1181–1992.

Wang. B., J. Zhu & A. Wu (2010). Some Improvements of Deformation Measurement Techniques on MTS815 System. *J. of Yangtze River Scientific Research Institute*. 27, 94–98.

Zhu. J., Y. Jiang & L. Wang (2010). Advance of Laboratory Test Technique in Rock Mechanics. *Chinese J. of Solid Mechanics*. 31, 209–215.

Zhao. Y., Z. Wan, Y. Zhang, F. Qu, G. Xie, X. Wei & W. Ma (2008). Research and Development of 20MN Servo-Controlled Rock Triaxial Testing System with High Temperature and High Pressure. *Chinese J. of Rock Mechanics and Engineering*. 27, 1–8.

Martin, C.D. & R. Christiansson (2009). Estimating the Potential for Spalling around a Deep Nuclear Waste Repository in Crystalline Rock. *International J. of Rock Mechanics & Mining Sciences*. 46, 219–228.

Xu. J., S. Peng & G. Yin (2010). Development and Application of Triaxial Servo-Controlled Seepage Equipment for Thermo-fluid-solid Coupling of Coal Containing Methane. *Chinese Mechanics and Engineering*. 29, 907–914.

Wu. A. (2010). A General Review to Developments of Rock Mechanical Test Technology and Its Applications. *Report on Advance in Rock Mechanics and Rock Engineering in 2009–2010*.

Lv. J., Y. Cui, J. Liu & H. Lu (2010). Analysis of Status Quo of Rock Strength Theories. *J. of North China University of Technology*. 22, 73–78.

Zhang. Y. (2004). *Rock Mechanics*. Beijing: China Architecture & Building Press Publications.

Chen. J. (2008). Triaxial Testing Research of Rock Deformation and Acoustic-Emission Characteristic. *J. of Wuhan University of Technology*. 30, 94–96.

Gao. C., J. Xu, P. He & J. Liu (2005). Study on Mechanical Properties of Marble under Loading and Unloading Conditions. *Chinese J. of Rock Mechanic and Engineering*. 24, 457–461.

Liu. Z., X. Li, G. Zhao, Q. Li & W. Wang (2010). Three-dimensional energy dissipation laws and reasonable matches between backfill and rock. *Chinese J. of Rock Mechanic and Engineering*. 29, 344–348.

Li. Y., Z. Ma, Y. He, B. Wang & X. Liang (2006). Experimental Research on Permeability of Rocks of Coal Bearing Strata. *J. of Experimental Mechanics*. 21, 129–134.

Electronic Engineering and Information Science – Wang (Ed.)
© 2015 Taylor & Francis Group, London, ISBN: 978-1-138-02772-5

Signal sorting a new combinatorial jamming strategy countering phased array radar

J. Han, M.H. He, M.Y. Feng & L.R. Guo
Air Force Early Warning Academy, Wuhan, China

ABSTRACT: In order to interfere with the new system phased array radar effectively with sophisticated anti-jamming technology and to improve the utilization rate of jamming resources, the paper advances a new combinatorial jamming strategy. In this paper, it starts from synthetically analyzing the anti-jamming technology characteristics of phased array radar, respectively studies on the theory of azimuth saturated jamming, which is able to counter sidelobe cancellation (SLC), and noise convolution jamming, which can obtain pulse compression gain. On this basis, fuses the two jamming techniques together combined with the actual jammers and advances a combinatorial jamming strategy. Finally analyzes and validates the effectiveness of the strategy as well as its specific application methods by simulation experiment.

KEYWORDS: phased array radar, combinatorial jamming, strategy.

1 INTRODUCTION

At present, new system phased array radar with anti-jamming technology such as sidelobe cancellation (SLC), sidelobe blanking (SLB) and pulse compression (PC) has been widely used in early warning surveillance, air-defense countermissile and other fields. It will be too hard for a jamming strategy just aiming at single anti-jamming technique to get expected interference effect with the synthetical application of various anti-jamming techniques. However, due to the urgent demand for information operations, to effectively interfere and suppress the phased array radar has become increasingly necessary. So, an integrated anti-jamming program is critically needed to fight against the whole anti-jamming system of phased array radar, which can be used to strengthen the accuracy of interference effect and to increase the utilization ratio of jamming resources.

Recently, smart noise ECM has drawn tremendous attention among all sorts of radar jamming techniques. It can be interpreted in reference (Schleher 1999) as one that can gain pulse compression gain and have little possibility of being influenced by anti-jamming technology such as sidelobe cancellation (SLC) and sidelobe blanking (SLB). However, this interpretation isn't perfectly correct, in view of the misunderstandings of smart noise ECM, Professor Qiu Jie in Navy Aviation Engineering Academy has probed into the essence of smart jamming, explaining that smart jamming can still be affected by sidelobe blanking (SLB). Besides, reference (Bai et al. 2009) makes a study on open-loop sidelobe cancellation, explaining that

it possesses the capability of cancelling pulse jamming under specific conditions because of the high convergence speed of open-loop sidelobe cancellation, which can prove from one aspect that open-loop sidelobe cancellation can resist smart noise ECM. Therefore, smart noise ECM can still be influenced by anti-jamming technology such as sidelobe cancellation (SLC) and sidelobe blanking (SLB), whose most typical feature is to obtain pulse compression gain.

In order to fight against sidelobe anti-jamming technology, reference has advanced azimuth saturated jamming technique (Hu, Jin & Li 2003), explaining that sidelobe cancellation (SLB) won't take effect when the number of interference sources is greater than the dimensionality of sidelobe cancellation. Nevertheless, available distributed jamming signals are often radio-frequency noise signals so that they can't obtain pulse compression gain after being processed by pulse compression (PC) technology, which lowers the utilization rate of jamming resources, hence higher power of radio-frequency jamming signals is indispensable to achieve expected results.

As the above analyzed, it may have no effect to interfere with the new system phased array radar with synthetical anti-jamming measures such as sidelobe cancellation (SLC), sidelobe blanking (SLB) and pulse compression (PC) merely by employing smart noise ECM or azimuth saturated jamming. But if we can fuse the above two jamming techniques together to produce a combinatorial jamming strategy possessing both advantages of the two jamming techniques, then the problem that phased array radar is hard to effectively interfere will be solved.

2 GETTING STARTED

With the rapid development of modern digital signal processing technology, new system phased array radar usually have wonderful ECCM capabilities by synthetically using various anti-jamming measures, whose signal processing circuit is shown in Fig.1.

If noise jamming signals can enter the radar receiver from sidelobe, the jamming signals will firstly be suppressed by sidelobe cancellation (SLC), sidelobe blanking (SLB) and other sidelobe anti-jamming techniques, whose interference cancellation ratio is usually 20~30dB (LI et al. 2012). Then, owing to the noncoherence between noise jamming and radar echo signals, pulse compression gain has no way to be obtained, leading jamming-to-signal ratio to be further reduced. Although it won't produce tremendous effects on noise jamming signals by MTI processing, suppression of sidelobe cancellation (SLC) and pulse compression (PC) will directly result in low power on the input terminal of radar constant false alarm rate detection, which is not enough to affect the target detection on phased array radar.

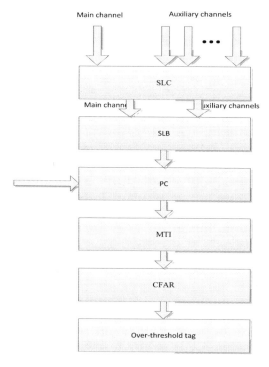

Figure 1. Signal processing circuit of phased array radar.

If convolution jamming signals can enter the radar receiver from sidelobe, then because of the high convergence speed of open-loop sidelobe cancellation, they can converge within the lasting time of convolution jamming and will also be affected by sidelobe cancellation (SLC), sidelobe blanking (SLB) and other anti-jamming techniques, causing strong suppression of jamming power. Consequently, wonderful interference effect cannot be obtained despite some pulse compression gain.

Therefore, as for new system phased array radar with multiple anti-jamming measures, the main reasons for constraining sidelobe interference effect are sidelobe cancellation (SLC), sidelobe blanking (SLB) and other anti-jamming techniques as well as the suppression of noise jamming by pulse compression (PC) processing. But since phased array radar has flexible beams, it can't be guaranteed that all jamming signals can enter the radar receiver from the main lobe. In fact, sidelobe jamming is still indispensable in many cases.

3 AZIMUTH SATURATED JAMMING TECHNIQUE TO FIGHT AGAINST SIDELOBE CANCELLATION (SLC)

Azimuth saturated jamming technique is capable of fighting against sidelobe cancellation (SLC), whose number of distributed jammers is greater than the dimensionality of sidelobe cancellation, making the jamming signals from sidelobe can't be largely cancelled and its basic principle is as follows:

Assume that the combined directional pattern after sidelobe cancellation is estimated to form a zero point in the jamming direction θ, including M auxiliary antennas.

$$G(\theta) + w_1 e^{j\varphi_1(\theta)} + w_2 e^{j\varphi_2(\theta)} + \cdots + w_M e^{j\varphi_M(\theta)} = 0 \quad (1)$$

$G(\theta)$ is expressed as the directional pattern function of the main antenna in Equation (1), w_n is the sidelobe cancellation coefficient of the N_{th} auxiliary antenna and the calculation method of w can be referred to the Reference (Wang X.G. 2011), phi(theta) is the phase difference of the N_{th} auxiliary omnidirectional antenna relative to the main antenna in the jamming direction. When the jamming direction $\boldsymbol{\theta} = [\theta_1, \theta_1, \cdots, \theta_N]$, Equation (1) is transformed into:

$$\begin{cases} G(\theta_1) + w_1 e^{j\varphi_1(\theta_1)} + w_2 e^{j\varphi_2(\theta_1)} + \cdots + w_M e^{j\varphi_M(\theta_1)} = 0 \\ G(\theta_2) + w_1 e^{j\varphi_1(\theta_2)} + w_2 e^{j\varphi_2(\theta_2)} + \cdots + w_M e^{j\varphi_M(\theta_2)} = 0 \\ \qquad\qquad\qquad \cdots \\ G(\theta_N) + w_1 e^{j\varphi_1(\theta_N)} + w_2 e^{j\varphi_2(\theta_N)} + \cdots + w_M e^{j\varphi_M(\theta_N)} = 0 \end{cases} \quad (2)$$

When $N > M$, the linear equation sets as Equation (2) have no solution, that is, when the number of jammers is greater than the dimensionality of sidelobe cancellation, weighted vector w doesn't exist to form zero point in all jamming directions, thus causing sidelobe cancellation (SLC) processing to be invalid.

3.1 Convolution jamming to fight against pulse compression (PC)

At present, using convolution modulation to produce smart noise jamming is a common method that can interfere with frequency-agile pulse compression radar without the need for frequency measurement, using pulse compression gain can lower the requirement for jamming power as well. Its basic principle is:

Suppose the radar transmitter signal is chirp signal $s(t)$, then the response function of the target is:

$$h(t) = \sigma\delta(t - t_R) \tag{3}$$

Where σ is the reflex cross sectional area of the target, t_R is the time delay of the target echo, so the target echo is:

$$s_r(t) = s(t) \otimes h(t) \tag{4}$$

Pulse compression coefficient through the radar signal processor is $s^*(-t)$, then the expression after matched filtering is:

$$s_{pc}(t) = F^{-1}(S_{pc}(f)) = F^{-1}(|S(f)|^2) \otimes h(t) \tag{5}$$

Where $F^{-1}(|S(f)|^2)$ is called point expansion function of chirp signals. As is shown in Equation (5), when a random function convolutes with a chirp signal, its pulse compression output signal is the convolution of the function and a point expansion function, that's to say it can obtain radar pulse compression gain. If the target response function $h(t)$ is transformed into a jamming modulation signal, then the jamming signal can also obtain the pulse compression gain. Jamming power gain of convolution jamming after pulse compression processing is:

$$K_d = \frac{J_{out}}{J_{in}} = \frac{T + T_n}{T + 1/B} > 1 \tag{6}$$

Where T is the length of the jamming signal involving in convolution, T_n is the length of the chirp signal involving in convolution and B is the bandwidth of the chirp signal. As for common radio frequency noise, pulse compression gain is 1 compared to its jamming power, so convolution jamming has a higher energy utilization rate than common radio frequency noise.

Besides, since the chirp signal has the characteristic of strong coupling in time and frequency, delay time between the jamming false target and the actual target is:

$$\Delta t = -\frac{\Delta f}{k} \tag{7}$$

Where Δf is the frequency shift of the signal, k is modulation slope, the jamming signal can precede the target echo signal by tuning the frequency shift $\Delta f(\Delta f < B)$, thus getting wonderful suppression effect.

3.2 The combinatorial jamming strategy

Combining azimuth saturated jamming with convolution jamming, the combinatorial jamming strategy is advanced to fight against the reality of low sensitivity in distributed jamming receivers that it can't totally receive sidelobe radar signals. The strategy can be expressed as: (1) Employ distributed jamming whose number of jammers is greater than the dimensionality of sidelobe cancellation; (2) Employ noise convolution jamming as soon as the jammer receives radar signals or employ radio frequency noise jamming if not; (3) Do proper frequency shift modulation in noise convolution jamming to make sure the jamming signals can cover the target signals in time domain.

As is seen from the expression of combinatorial jamming strategy, the strategy is capable of fighting against sidelobe cancellation like saturation distributed jamming, and the convolution jamming part can obtain radar pulse compression gain at the same time. Therefore, it's a new jamming strategy capable of fighting against sidelobe cancellation, sidelobe blanking and other anti-jamming technology simultaneously.

4 SIMULATION

4.1 Simulation conditions

The linear array antenna is employed in main antenna of radar with 30 elements, and the ratio of element spacing to the wavelength of transmitter signal is 0.5. There are 3 auxiliary omnidirectional antennas, respectively situated on the center and two flanks of the linear array, whose gain is greater than the first sidelobe gain of the main antenna slightly. The mainlobe width of the antenna $\theta_o = 5°$, transmitter chirp signal bandwidth of radar $B = 4\text{MHz}$, pulse width $\tau = 40\mu s$, incidence angle of the target echoes $\theta = 0°$, sampling frequency $F_s = 8\text{MHz}$, signal-to-noise ratio of the radar echoes $SNR = 10\text{dB}$, constant false alarm rate $P_d = 10^{-6}$. CA-CFAR detection pattern is employed, including 32 reference units and 1 protection unit. The signal processing circuit of radar is shown in Fig.1. Assume that only when the interference incidence angle $|\theta| \leq 6°$ can the interference receiver receive transmitter signals from sidelobe of radar and noise convolution jamming is employed. Fig. 2 has clearly shown the constant false alarm rate detection results without any jamming signals, from which we know that the amplitude of echo signals exceeds the constant false alarm rate detection threshold at the 320_{th} and the 321_{th} sampling point, hence echo signals being successfully found.

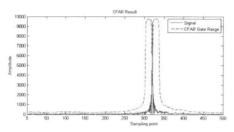

Figure 2. Target echo figure without jamming.

4.2 Analysis against saturated distributed jamming

A number of the distributed jamming stations $N = 4$, incidence angle $\theta = [5°, 8°, -10°, 10°]$ in turn, jamming-to-signal ratio of a single jammer to the input terminal of radar receiver $JNR = 25dB$, the interference effect when respectively adopting combinatorial and saturated distributed jamming strategy is shown in Fig. 3.

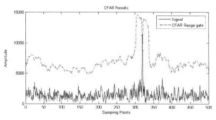

(a) Interference effect figure by saturated distributed jamming

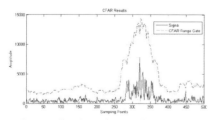

(b) Interference effect figure by combinatorial jamming

Figure 3. Contrast Figure of Interference Effect After CFAR Detection.

As is shown in Fig.3, under the same circumstances, although saturated distributed jamming can't be influenced by sidelobe cancellation processing, jamming signals are unable to suppress the target echoes successfully like combinatorial jamming in Fig.3 (b) owing to the lack of pulse compression gain, and target echo signals can still be found through constant false alarm detection.

In order to analyze the superiority of combinatorial over saturated distributed jamming strategy more visually, Monte Carlo simulation experiment is carried on. Suppose that the number of distributed jamming stations $N = 4$, jamming incidence angle $\theta A[5°, 8°, -10°, 10°]$ in turn, jamming-to-signal ratio of a single jammer to the input terminal of radar receiver $JNR = 5:1:50dB$. Conduct Monte Carlo experiment for 500 times at each jamming-to-signal ratio and calculate the target detection probability, then the experiment results can be seen in Fig.4.

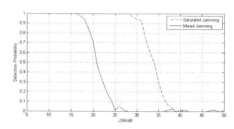

Figure 4. Contrast Figure of Interference Detection Probability.

From Fig.4 we can see, on same condition of jammer quantity and jamming incidence angles, if target detection probability is cut down to 0.1, jamming-to-signal ratio of each jammer needs to be 24.2dB by combinatorial jamming, while the ratio has to be 37.1dB by saturated distributed jamming, so 12.9dB jamming power is saved for a single jammer by using combinatorial jamming strategy and the utilization rate of jamming power is improved. From the view of interference effect, target detection probability can be affected when jamming-to-signal ratio exceeds 16.3dB by combinatorial jamming strategy, while the ratio has to be 27.6dB by saturated distributed jamming, greater than that of combinatorial jamming strategy up to 11.3dB, so combinatorial jamming strategy can also contributed to lowering the requirement for jamming power.

4.3 Application analysis of combinatorial jamming strateg

1 Power Distribution of Convolution and Radio Frequency Noise Jamming

Adopt combinatorial jamming strategy, jamming incidence angle $\theta A[5°,8°,10°,-10°]$ in turn, jamming-to-signal ratio of convolution jamming is respectively 20 dB and 70 dB, jamming-to-signal ratio of radio frequency $JSR = 10:1:70\text{dB}$. Conduct Monte Carlo experiment for 500 times at each jamming-to-signal ratio and calculate the target detection probability, then the experiment results can be seen in Fig.5.

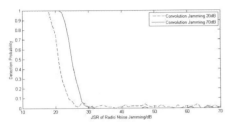

Figure 5.   Interference effect of combinatorial jamming.

From Fig.5 we can see, under given conditions of jamming-to-signal ratio in convolution jamming, different radio frequency noise power can influence the interference effect. When jamming-to-signal ratio of convolution jamming is 20dB , if the ratio of radio frequency noise jamming is also 20dB, then the detection probability will be 0.64, which is unable to get an excellent suppression effect, but when the jamming-to-signal ratio of radio frequency noise jamming is added up to 24.8dB, then the detection probability can be cut down to 0.1.

From contrast analysis of the two curves, it isn't difficult to know that employing 20dB power of convolution jamming is better than that of 70dB in interference effect when jamming-to-signal ratio of radio frequency noise is lower than 30dB, which results from big difference of power between convolution jamming power and radio frequency noise jamming so that distributed jamming joint force can't be well produced.

Therefore, through the description of the above two points, when making power options in combinatorial jamming strategy, power of convolution jamming and radio frequency noise jamming doesn't need to be perfectly equivalent, the combination of higher convolution jamming power and lower radio frequency noise jamming power may not acquire excellent interference effect, but under given convolution jamming power, proper increased power of radio frequency jamming can improve it.

2   Distribution of Convolution and Radio Frequency Noise Jamming in Quantity

Jamming incidence angle $\theta A[5°,8°,9°,-10°]$ in turn, convolution jamming and radio frequency noise jamming have equivalent jamming power, given other

constant conditions, calculate the target detection probability through 500 times of Monte Carlo experiments. Necessary jamming-to-signal ratio under a different number of convolution jammers is given in table 1 when target detection probability of radar is cut down to 0.1.

Table 1. Relationship between the number of convolution jamming and the interference effect.

Num.	Convolution total	Noise total	Jamming-to-signal ratio /dB
1	0	4	37.1
2	1	3	24.2
3	2	2	22.8
4	3	1	21.3
5	4	0	18.9

From table 1 we can clearly find, with the increasing number of convolution jammers, under given incidence angle, necessary jamming-to-signal ratio to make the radar detection probability cut down to 0.1 is reduced gradually, difference of jamming-to-signal ratio between the two jamming techniques is 18.2dB. So the utilization rate of jamming power can be improved with the increasing number of convolution jamming when jammer total is fixed.

Therefore, to acquire excellent interference effect, it's suggested to employ convolution jamming when jamming receiver can receive the radar signals or employ radio frequency noise jamming if not.

5   CONCLUSION

The paper has advanced a combinatorial strategy that fuse saturation distributed jamming technique and convolution jamming technique together, proving the effectiveness of the combinatorial strategy by theoretical analysis and validating its superiority over common saturation distributed jamming technique as well as the application method by simulation experiment.

The strategy possesses the characteristics of saturation distributed jamming and convolution jamming simultaneously, capable of effectively fighting against the open-loop sidelobe cancellation system with high convergence speed and obtaining some pulse compression gain, thus improving the utilization rate of jamming power. Besides, the new strategy has taken the actual point into consideration that the jamming receiver has rather low sensitivity, that is, the jamming receiver begins to release convolution jamming soon after receiving radar signals, which is a valid jamming method with strong practical application values to fight against new system phased array radar.

REFERENCES

Schleher D.C. (1999). Electronic warfare in the information age. Norwood MA: Artech House.

Bai W.X., W. Zhang & M. Miao (2009). Study on the countermeasure technology against side-lobe interference. *Systems Engineering and Electronics (01)*, 86–90.

Hu S.L., J.W. Jin & X.M. Li (2003). Study of the Multi-direction Saturated Jamming against the Sidelobe Cancellation of Radar. *Radar & Ecm.(3)*, 45–49.

Li S., Y.Z. Li, G.Y. Zhang & Y.K. Guo (2012). A Study on Twinkle Jamming Project Against the Adaptive Sidelobe Canceling System. *Modern Radar.34(2)*, 51–54.

Wang X.G.(2011) research for Radar jamming sidelobe cancelling machine[D]. *University of Electronic Science and Technology of China.*

Electronic Engineering and Information Science – Wang (Ed.)
© *2015 Taylor & Francis Group, London, ISBN: 978-1-138-02772-5*

An cruise vehicle trajectory planning scheme based on the artificial potential field

L.F. Pan, X.X. Liu, Y. Guo & Y.X. Li
Xi'an Research Institute of Hi-Tech Hongqing Town, Xi'an, Shaanxi, P.R China

ABSTRACT: The Artificial Potential Field (APF) method was introduced to solve the cruise vehicle path planning problem. Using a height adjust function to keep the cruise flight altitude, two schemes were proposed to design the lateral maneuver trajectory. The virtual repulsive force sources strategy was a basic APF planning method, but sometimes the simulation results were not the global optimal; the virtual attractive force source strategy was an additional APF guide method that needs experts' experiences which could guide the vehicle flight towards the direction that would be safer as we suggested. Simulation results show it is an easy and fast approach that's based the APF method the cruise vehicle would generate an optimal or nearly optimal path in the complex battlefield.

KEYWORDS: APF-Artificial Potential Field; cruise vehicle; lateral maneuver; trajectory planning.

1 GENERAL INSTRUCTIONS

The cruise vehicles must avoid getting into no fly zones because of political reasons or intercept threats. Generally, these no fly zones can be described as circular areas with given center and radius. Usually the cruise path is designed before launching, but due to various emergency situations the planned trajectory in advance should be modified or re-planned.

With the simpler structure the Artificial Potential Field (APF) method is widely employed in real-time obstacle avoidance. The basic idea of the APF theory is that the movement of the robot in the surroundings can be considered as a movement in the suppositional artificial force field with obstacles and the goal. The goal attracts the robot with attractive force and the obstacles produce repulsive forces. Both attractive force and repulsive forces combine the resultant force, which can control the robot movement and confirm the robot position (KABAMBA P.T. et al. 2006).

In this paper the traditional APF theory is introduced into the research of cruise vehicle maneuvering penetration, where the target point and guidance points are attractive poles, and no fly zones are repulsive poles. Through the descent gradient direction of the Artificial Potential Field a collision free path to the target would be found. Simulation results indicate that the APF method can realize optimal path planning in complex environments.

2 THE ARTIFICIAL POTENTIAL FIELD

Here the vehicle is supposed as a mass point, If $\vec{r}$ is the current position vector of the vehicle, $\vec{r}_{goal}$ is the goal position vector, and $\vec{r}_{threat}^i$ is i^{th} threat center position vector, then the attractive potential function of the goal position is $U_{att}(\vec{r}, \vec{r}_{goal})$, the repulsive potential function of the ith threat is $U_{rep}^i(\vec{r}, \vec{r}_{threat}^i)$, and the attractive force and repulsive force separately is $\mathbf{F}_{att}(\vec{r}, \vec{r}_{goal})$ and $\mathbf{F}_{rep}(\vec{r}, \vec{r}_{threat}^i)$. So the resultant potential function $U(\vec{r})$ and the resultant force $\mathbf{F}(\vec{r})$ are as follows,

$$U(\vec{r}) = U_{att}(\vec{r}, \vec{r}_{goal}) + \sum U_{rep}^i(\vec{r}, \vec{r}_{threat}^i) \qquad (1)$$

$$\mathbf{F}(\vec{r}) = \mathbf{F}_{att}(\vec{r}, \vec{r}_{goal}) + \sum \mathbf{F}_{rep}^i(\vec{r}, \vec{r}_{threat}^i) \qquad (2)$$

Where,

$$U_{att}(\vec{r}, \vec{r}_{goal}) = \frac{1}{2} k_{att} \left| \vec{r} - \vec{r}_{goal} \right|^2 \qquad (3)$$

$$U_{rep}^i(\vec{r}, \vec{r}_{threat}^i) = \begin{cases} \frac{1}{2} k_{rep} \left(\frac{1}{\rho^i} - \frac{1}{\rho_0^i} \right)^2 & \rho^i \le \rho_0^i \\ 0 & \rho^i > \rho_0^i \end{cases} \qquad (4)$$

$$\mathbf{F}_{att}(\vec{r}, \vec{r}_{goal}) = -k_{att} \left| \vec{r} - \vec{r}_{goal} \right| \qquad (5)$$

$$\mathbf{F}_{rep}(\mathbf{X}) = \begin{cases} -k_{rep}\left(\dfrac{1}{\rho^i} - \dfrac{1}{\rho^i_0}\right)\dfrac{1}{\left(\vec{r}-\vec{r}^i\right)^2} & \rho^i \le \rho^i_0 \\ \\ 0 & \rho^i > \rho^i_0 \end{cases} \quad (6)$$

$$\rho^i = \left\| \vec{r} - \vec{r}^i \right\| \quad (7)$$

Here, ρ^i is the Euclidean distance from the vector $\vec{r}$ to $\vec{r}^i$ of the ith obstacle, ρ^i_0 is the influence radius of the ith threat, k_{att} and k_{rep} are proportional gains of the corresponding function respectively.

Though the APF method is easily used for a vehicle to reach the goal and avoid threats, the disadvantage is the vehicle often gets stuck in a local minimum of the potential. The conditions that dead ends appear are various due to different threat configurations. Lots of approaches have been proposed to get out from dead ends for a robot, such as heuristic method or global recovery. In this paper, we use the method of references (Liu C.Q. et al. 2000, Park M.G. & M.C. LEE 2003, Park M.G. & M.C. Lee 2003).

3   THE HEIGHT CONTROL FUNCTION

The cruise aircraft flies in the three-dimensional space, generally the repulsion from the ground is far greater than the repulsion from the sky (the repulsive force from above is zero most of the time). So the potential force function to hold the cruise altitude should to be modified. As we know, the flight height of the vehicle was lower, the probability to be found by enemy detectors or destroyed by ground defense systems is smaller, but flying too low will tend to increase the crash probability of collision with the ground (Ender St. & J. Kluza 2007).

Therefore, to keep a certain flight height greater or equal to the allowed minimum cruising altitude H_{min}, a height adjustment function was introduced in the paper. The height adjustment function is defined as the attractive potential field U_{alt}, the size of which is proportional to the difference between the actual flight altitude h and the allowed minimum cruising altitude H_{min}.

$$U_{alt} = \begin{cases} 0.5\lambda(h - H_{min})^2 & h > H_{min} \\ 0 & h \le H_{min} \end{cases} \quad (8)$$

Where, λ is a proportional gain. And the height control virtual force F_{alt} is computed as follows:

$$F_{alt} = \begin{cases} -\lambda(h - H_{min})^2 & h > H_{min} \\ 0 & h \le H_{min} \end{cases} \quad (9)$$

4   LATERAL MANEUVERING SCHEME

4.1   *Virtual repulsive force sources strategy*

The influence regions like air defenses and antimissile systems, the early warning radar and other regions that should be avoided usually can be simplified as circles, the center point of which is O, and the radius of which is R in Fig.1 below.

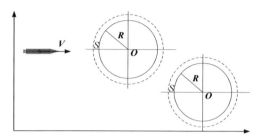

Figure 1.   Virtual repulsive force regions.

Considering the certain flight speed V and the minimum turning radius D of the cruise vehicle, we extend a distance S to the radius R of the virtual repulse force region. The vehicle starts evasive action at the outside edges of the area which are dashed lines. To avoid the threat the cruise vehicle starts to turn at the point A, and ends at the point B in Fig.2.

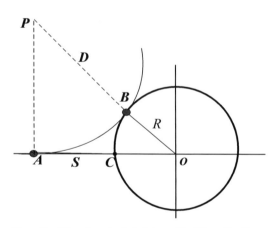

Figure 2.   The min extended distance D of the radius R.

According the Pythagorean Theorem the min extended distance S is

$$S = \sqrt{(D+R)^2 - D^2} - R \quad (10)$$

Supposing the cruise vehicle's velocity in the horizontal plane is a constant value $V = 1000$ m/s; the starting position of the cruise flight is (0,0) m, and the target position is (1000000, 0) my; the maximum available lateral overload is 8 g. The Matlab simulation result is shown in Fig.3 below.

4.2 *Virtual attractive force source strategy*

Because the APF method belongs to the local path planning approaches, the simulation result above may not be the global optimal one. Known from the analysis of the simulation result in Fig.3, if the vehicle passes by the circle C above instead of below, the trajectory would be safer as shown in Fig. 4.

To overcome the shortcoming of the APF method, the virtual attractive force source strategy was introduced to guide the vehicle towards safe navigation points designed by experts' experiences. Using the same simulation data before, we got the new trajectory in Fig.4.

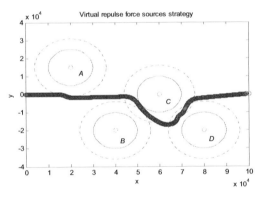

Figure 3.   Virtual repulse force sources strategy.

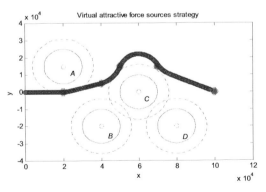

Figure 4.   Virtual attractive force sources strategy.

5   CONCLUSIONS

In this paper, the Artificial Potential Field (APF) method was extended to solve the cruise vehicle trajectory planning problem. Simulation results show it is an easy and fast approach that's based the APF method the cruise vehicle would generate an optimal or nearly optimal path in the complex battlefield.

REFERENCES

KABAMBA P.T., S.M. MEERKOV & F.H. ZEITZIII (2006). Optimal path planning for unmanned combat aerial vehicles to defeat radar tracking. *Journal of Guidance, Control, and Dynamics 29(2)*, 279–288.
Liu C.Q., H. KRISHNAN& L.S. Yong (2000). Virtual obstacle concept for local-minimum-recovery in potential-field based navigation. *Proceedings of the IEEE International Conference on Robotics & Automation*, 983–988.
Park M.G. & M.C. LEE (2003). Artificial potential field based path planning for mobile robots using a virtual obstacle concept. *Proceedings of IEEE/ASME International Conference on Advanced Intelligent Mechatronic*, 735–740.
Park M.G. & M.C. Lee (2003). Artificial Potential Field Based Path Planning for Mobile Robots Using a Virtual Obstacle Concept. *IEEE/ASME International Conference on Advanced Intelligent Mechatronics*, 735–740.
Ender St. & J. Kluza (2007). Low-complexity spacecraft guidance using artificial potential functions. *The 3rd International Conference on Recent Advances in Space Technologies,June,*14–16.

Electronic Engineering and Information Science – Wang (Ed.)
© 2015 Taylor & Francis Group, London, ISBN: 978-1-138-02772-5

Surface roughness prediction and parameter optimization of high speed milling based on the DAAGA

X.L. Sui, C. Hua, Y.Q. Chen & J.T. Chen
Harbin University of Science and Technology, Harbin, Heilongjiang, China

ABSTRACT: Considering the influence of cutting parameters on surface roughness in the process of high speed milling of titanium alloy, the surface roughness prediction model is established by the fuzzy system. On this basis, a new method (DAAGA) is proposed to fuzzy systems for training and establish the surface roughness prediction model and the milling parameter optimization based on the trained data. While this method is introduced, the excellent genetic model of Genetic Algorithm and the improvement of local optimization ability in Ant Colony Optimization are guaranteed. In addition, iterative adjustment threshold is used which makes the disadvantage of premature convergence of the genetic algorithm and the useless cross in traditional hybrid algorithm solved. Moreover, the size of the solution population increases and the search space is expanded. The experiments show, the surface roughness predict more accurate by DAAGA , and the optimization capability is greatly improved.

KEYWORDS: High Speed Milling; DAAGA; Fuzzy System; Surface Roughness; Parameter Optimization.

1 GENERAL INSTRUCTIONS

In this paper, mainly studying titanium alloy milling process, the method of predicting surface roughness which based on the fuzzy system and the milling parameter optimization is established. In addition, the use of DAAGA (Dynamic Ant Algorithm Genetic Algorithm) for fuzzy system to train is proposed in this paper. Then an accurate prediction model is established by this method and based on the model the milling parameter optimization model is established. The call time of the Genetic Algorithm and Ant Colony Optimization can be controlled dynamically by using the method. And combined with the corresponding pheromone updating method and the iterative adjustment threshold, the convergence speed, the convergence performance and the optimization capabilities are improved greatly.

2 DESIGN OF DAAGA ALGORITH

DAAGA is a new Dynamic Ant Algorithm Genetic Algorithm (Cakir M.C. 2009). Its framework is as follows: the solutions of Ant Colony Optimization is regarded as the seeds of genetic manipulation, then the initial population of Genetic Algorithm is optimized. The best fusion point evaluation strategy is used to determine whether invoking the Ant Colony Optimization (Wang G. & W.H. Zhang 2010). Within the scope of the set number of iterations, if n successive generations are satisfied:

$$\Delta f_n < \Delta f_{n-1} \tag{1}$$

Genetic operation is end. Then the pheromones is generated, the global optimal solution is updated and the Ant Colony Optimization is called. The above steps are circularly executed.

The rule of optimal individual in each generation of genetic algorithm and the global optimal solution to update the pheromone is as follows:

$$\tau_{ij}(t+1) = \rho \tau_{ij}(t) + \Delta \tau^{best}(t) \tag{2}$$

Different from traditional Ant Algorithm Genetic Algorithm, iterative thresholds is designed in DAAGA to adjust to the operation of genetic algorithm and the size of the ant population:

$$\psi = \xi \times NK_{max} \tag{3}$$

3 SURFACE ROUGHNESS PREDICTION MODEL BASED ON FUZZY SYSTEM

For the research, the axial depth of cut, the radial depth of cut and the spindle speed, feed speed are selected as inputs of fuzzy system, and the milling surface roughness are selected as the output. And the fuzzy system is trained by DAAGA.

First the orthogonal table which is based on the standard of four factors and four levels is used to arrange experiments, and 16 groups of experimental data scored will be as the training sample. Then respectively, the recursive least squares algorithm, Ant Colony Optimization and DAAGA are used to train the fuzzy system. The convergence condition of all kinds of algorithms are showed in figure 1.

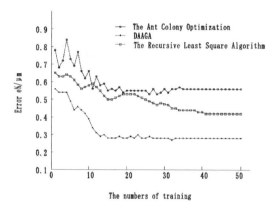

Figure 1. Training process of the three algorithms.

Comparing the final convergence effect, the DAAGA is better than Ant Colony Optimization and the recursive least squares algorithm.

The surface roughness prediction model based on the data from the fuzzy system which trained by DAAGA are established, if the regression equation is the exponential form, the experience formula of the surface roughness prediction which is obtained by training data is as follows:

$$Ra = e^{-0.201}a_p^{0.1614}a_e^{0.1041}n^{-0.7519}v_f^{0.8738} \qquad (4)$$

4   ESTABLISHMENT OF THE SURFACE ROUGHNESS OPTIMIZATION MODEL

4.1   *Optimization process of the optimal solution based on DAAGA*

DAAGA is adopted in this paper, the structure of which not only can dynamically control the call time of the Genetic Algorithm and Ant Colony Optimization, and combined with the corresponding pheromone updating method, the convergence speed and convergence of the algorithm also can be improved. The reference of iterative adjustment threshold, for Genetic Algorithm, greatly ensures the population genetic

patterns to the next generation (Zain A.M.et al. 2010, Wang G. & M. Wan 2011). At the same time, through the variants the searching space is expanded further and searching a better solution is more advantageous for next Ant Colony Optimization. For the Ant Colony Optimization, the appropriate increase of the size of the group in the late algorithm can better exert its local optimization ability and more quickly find the optimal solution . The optimized structure flow chart 2 is as follows.

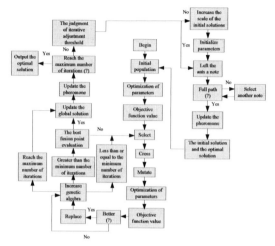

Figure 2.   Structure flow chart of searching the optimal solution based on DAAGA.

First searching the optimization from the initial parameters by using Ant Colony Optimization, forming the initial population and updating the global optimal solutions. Then the population of solutions form Ant Colony Optimization and the global optimal solutions are used as the seeds of genetic operation. In the process of genetic operation, firstly, a judgment on genetic iteration number should be made and start the best fusion point assessment depending on the results. Then the global optimal solution and the pheromone are updated. Finally a determine whether the number of iterations reach the maximum number of iterations should be made and then the judgment of iterative adjustment thresholds would be done.

5   EXPERIMENTAL VERIFICATION

5.1   *Experimental conditions*

The experiment is carried out in the UCP710 type five axis high speed machining center. Extreme pressure water-based lubrication is used and the work piece is titanium alloy plate.Cutter is carbide end milling

Table 1. Compare of the experimental parameters and optimization parameters.

	Serial Number	V_f (mm/min)	n (r/min)	a_p (mm)	a_c (mm)	R_a (μm)	F_x (N)	F_y (N)
The experiment parameters	X1-1	100	600	0.5	1	0.177	26,25	62.62
	X1-2	100	600	1	1.5	0.325	77.36	161.4
	...	...	...	...	...	...	...	...
	X1-9	150	1000	1	1	0.297	161.8	180.1
Optimization of parameters		135	830	1.78	7.42	0.335	45.7	218.8
						0.298	45.43	215.4
Experimental verification		135	830	1.78	7.42	0.354	43.74	214.1
						0.317	44.32	215.1

cutter.The parameters of Genetic Algorithm: the size of the group *MAXGEN* = 400;the largest genetic algebra *MAXGEN* = 400; choose probability *GGAP* = 0.8;crossover probability *GJIAOCHA* = 0.7;mutation probability *GBIANYI* = 0.5;

Feed direction component:

$$F_X = 4305n^{-0.069}v_f^{1.092}a_p^{0.754}a_c^{-0.224} \quad (5)$$

Cutting width direction component:

$$F_X = 4305n^{-0.069}v_f^{1.092}a_p^{0.754}a_c^{-0.224} \quad (6)$$

Residual stress

$$S = 61n^{0.383}v_f^{0.154}a_p^{0.178}a_c^{-0.235} \quad (7)$$

Tool life:

$$T = 36n^{-5.89}v_f^{-2.11}a_p^{-0.199}a_c^{-0.281} + 23.19 \quad (8)$$

The optimization results and the experimental results of high speed milling TC4 B-1are shown in Table 1. From the table, the material removal rate increased by 60.5% after the milling parameters optimized rate increased by 30.1%. And the components of the direction of the feed and the cutting width significantly decrease, which can more effectively reduce the vibration and improve the tool life.

For persuasive conclusion, the experiment about the roughness and milling force with the optimized parameters is done. The verification results show that with the optimization parameters, the cutting process is smoother, the chip is removed smoothly, and the noise of the machine is smaller. And the surface roughness, the feed direction component and the width of cutting direction component accord well with those of theoretical calculation, which shows the mentioned optimization method is reasonable and effective.

6 CONCLUSION

The fuzzy system is trained by DAAGA in this paper. The surface roughness prediction model established by the trained fuzzy system is proved more accurate. The convergence speed and convergence of the algorithm are also improved further. In addition, the milling parameter optimization model is also established based on DAAGA. Through the experiments, the surface roughness, the feed direction component and the width of cutting direction component are proved according well with those of theoretical calculation. So this optimization method is reasonable and effective and has important guiding significance for the production practice.

ACKNOWLEDGMENT

This research was sponsored by Excellent Academic Leaders Project of Harbin science and technology innovation talents of special fund (2013RFXXJ064).

REFERENCES

Cakir M.C., C. Ensarioglu & I. Demirayak (2009). Mathematical modeling of surface roughness for evaluating the effects of cutting parameters and coating material. *Journal of Materials Processing Technology* 209 (1), 102–109.
Wang G. & W.H. Zhang (2010). Prediction of diameter error of work piece in turning process using fuzzy system based on recursive least square algorithm. *Journal of Mechanical Strength 32 (6)*, 953–960.
Zain A.M., H. Haron & S. Sharif (2010). Prediction of surface roughness in the end milling machining using artificial neural network. *Expert System with application 37(2)*, 1755–1768.
Wang G. & M. Wan (2011). Modeling of milling force by using fuzzy system optimized by particle swarm algorithm. *Journal of Mechanical Engineering 47(13)*, 123–130.
Zain A.M.et al. 2010, Wang G. & M. Wan 2011.

Electronic Engineering and Information Science – Wang (Ed.)
© 2015 Taylor & Francis Group, London, ISBN: 978-1-138-02772-5

Structure design and movement mechanism analysis for underwater micro-robot

S.J. Ren & J. B. Dong
School of Electrical and Control Engineering, Heilongjiang university of Science and Technology, Harbin, Heilongjiang Province, China

ABSTRACT: In this paper, structure design of the new type underwater micro-robot is given, and permanent magnet materials embedded in the robot are analyzed and selected, driven wing geometry model of the robot is built based on the principle of swimming propulsion, finally, the movement performance of the micro-robot is analyzed. Tests show that the scheme of micro underwater robot without cable is feasible.

KEYWORDS: NdFeB; Micro-robot; Motion control.

1   INTRODUCTION

In the real project, people often have to work in some environment with liquid, such as underwater adventures, underwater rescue, the maintenance of liquid lines, etc. As a result of the limitation of the body's own conditions, in this kind of environment, human beings are often difficult to give full play. Currently, the research work on underwater robot is carried in many countries, while the research on swimming micro-robot that works in water is an important branch of the field (Jiang Y.J. et al. 2006).

2   THE STRUCTURE OF THE UNDER WATER MICRO-ROBOT

Swimming fish are bathing with high efficiency and excellent maneuverability, which have a close relationship on that they have streamlined appearance, organization structure and the swimming control mechanism itself, among which the main reason is that they have good drag reduction mechanism to overcome water resistance in the swimming process.

The micro-robot studied has replicated the fish swimming mechanism and the shape. So Figure1 shows that the robot is mainly composed of three parts, the front-end part is made of EVA resin materials and magnetic materials. Its shape imitates Carangidae division mode promoting fish, precursor has good "streamlined", which make swimming resistance decrease. The middle section is an EPE flexible material which has certain stiffness and higher vibration frequency which is advantageous for the robot to produce better movement speed. The

end is the tail fin, movement of the robot body is promoted by tail's swinging.

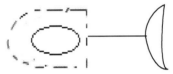

Figure 1.   Micro-robot body shape.

3   SELECTION OF PERMANENT MAGNETIC MATERIAL

For underwater micro-robot ontology, its movement mainly relies on the external magnetic field to drive the permanent magnetic material placed inside the robot head, which make it move with force in the alternating magnetic field. So the selection of permanent magnetic material in the robot head has very important significance for the robot movement.

Permanent magnetic material can be roughly divided into sintered permanent magnet and bonded permanent magnet. Sintered permanent magnet is made by magnetic powder sintering poured into a mold casting, bonded permanent magnet is molding by mixing with trace plastic or rubber in the magnetic powder (Li X.Y. et al. 2013).

On the choice of permanent magnets, there are many kinds of permanent magnetic material, commonly used can be divided into three categories: one is aluminum nickel and cobalt (Al - Ni - Co) permanent magnet materials; second is permanent magnet ferrite ($BaFe_{12}O_{19}$ $SrFe_{12}O_{19}$); Third is rare earth permanent magnet material, this class can be divided into three generation again, that is SmCo5, Sm$_2$Co$_1$ and NdFeB type. Main parameters that characterize the performance of permanent magnetic material are the residual magnetic induction intensity Br, coercive force Hcb and maximum magnetic energy product(BH) max , which higher, the better the performance of permanent magnet material (Zhao R., Y.W. Li & T.S. Zhao 2009).

Table 1. Performance parameters comparison of the several types permanent magnet materials.

parameter	unit	NdFeB	Sm$_2$Co$_1$	ferrite	Al-Ni -Co
remanence	T	1.25	1.12	0.44	1.15
magnetic fluxdensity coercive force	kA/m	915.4	533.32	222.88	127.36
Intrinsic coerc ive force	kA/m	1098.48	543.24	230.84	127.36
maximum magnetic energy product	kJ/m^3	286.56	246.76	36.32	87.56

It can be seen from table 1, the five parameters of the remanence Br, magnetic flux density coercive force of Hcb, intrinsic coercive force Hcj, maximum magnetic energy product(BH)max and reversible permeability μr, etc, NdFeB permanent magnet material performance is excellent.

4 MOVEMENT PERFORMANCE ANALYSIS OF THE MICRO SWIMMING ROBOT

Movement of micro swimming robot is realized by swimming propulsion way, so main research object of micro-robot focuses on "spinal curve". Corresponding to the shape of the micro-robot, we can simplify the tail as geometric model shown in figure 2.

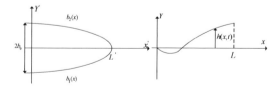

Figure 2. Plane coordinate geometry model of tail swimming of the micro robot.

Figure 2 shows the micro-robot tail's the structure size and state parameters shape in the coordinate system $O'X'Y'$ and OXZ, Among which, $O'X'Y'$ is coordinate system that the micro-robot tail placed naturally, OXZ is coordinate system in the process of swimming of the micro-robot. Micro swimming robot is swimming through a liquid in the form of wave motion, its swimming form is similar to a list of sine waves, which motion equations can be expressed as follows (Zhang Y.S., W. Liu & Z.Y. Jia 2005).

$$z = h(x,t) = \alpha(x)\sin[2\pi(\frac{x}{\lambda} - \frac{t}{T})] \quad (1)$$

$$x' = \int_0^x \sqrt{1 + (\frac{\partial h(x,t)}{\partial x})^2}\, dx \quad (2)$$

$$L = k\lambda \quad (3)$$

Formula (1) shows the wave shape of the tail of the micro swimming robot in the process of swimming;

Formula (2) shows that the relationship between the micro swimming robot any point on the tail when placed naturally in the coordinate system and swimming in the coordinate system;

Formula (3) shows the relationship of wave number and wavelength that tail formed in the process of swimming (Ren S.J., J.B. Dong & A.H. Wang 2011).

Micro-robots are able to forward swimming propulsion, is due to its "spinal curve" which drive the liquid enveloped spray backward , it is similar to the backward flow, thus forward swimming propulsion is realized by its reaction . Then the liquid quality that enveloped in "spinal curve" can expressed as :

$$M_1 = \rho\int_0^L b(x)\alpha(x)\sin[2\pi(x/\lambda - t/T)]dx \quad (4)$$

Among which ρ is liquid density, liquid relative "spine curve" speed V , at the starting moment, the enveloped liquid will be "spine curve" squeezed and pushed to reach relative wave velocity V to the micro-robot , Ground is set as reference coordinate system, set up the relative speed of micro-robot to the ground as V_B , the speed of the enveloped liquid relative to the ground as V_W , as a result of the micro-robot's resistance in water is increasing with speed increasing, so the start moment, the resistance of the robot can be neglected. Assuming that the quality of the micro-robot as M_0 ,so the application of conservation of momentum is:

$$M_1 \times V_W = M_0 \times V_B \quad (5)$$

And there is

$$V = f\lambda \quad (6)$$

Assuming that

$$V = f\lambda$$

$$Y = \frac{M_1}{M_0 + M_1} = \frac{2\rho \int_0^L b(x)\alpha(x)\sin[2\pi(x/\lambda - t/T)]dx}{2\rho \int_0^L b(x)\alpha(x)\sin[2\pi(x/\lambda - t/T)] + M_0} \quad (7)$$

So

$$V_B = Yf\lambda \quad (8)$$

5 MICRO-ROBOT MOTION TESTS WITH DIFFERENT WINDING DRIVING WAYS

At present, the commonly used method for pipe robot driven by the external magnetic field is uniform winding enameled wire outside the pipe to drive the robot body movement, enameled wire is winded into a small section of the solenoid to make it move along the pipe, which in turn drives the robot body movement. Figure 3 and Figure 4 shows that two different experiments to prove the feasibility of these two kinds of driving mode.

Figure 3. Horizontal pipeline with uniform winding mode.

Figure 4. Horizontal pipeline with mobile mode.

Pipe with two different winding ways are placed horizontal respectively, the micro-robot is put in into them, from a 100 cm distance, their speed are tested.

6 CONCLUSION

The structure of the micro-robot is designed on the basis of the fish bionics foundation. The movement mechanism of micro-robot is analyzed by fish swimming movement way and the geometric model of it is found and formula is derived, and the advancing speed of the robot is analyzed.

ACKNOWLEDGEMENTS

Fund project: Heilongjiang province education department project (12541695);
National Natural Science Foundation of China (51304075).

REFERENCES

Jiang Y.J., J.C. Li & Y.P. Yu et al. (2006). Development of Underwater Swimming Robot. *Robot 28*, 229–234.
Li X.Y., Z.B. Li & L. Mao et al (2013). Manufacturing and Wireless Driving of a Permanent Magnetic Micro-robot. *Robot 35*, 513–520.
Zhao R., Y.W. Li & T.S. Zhao (2009). Present situation of researches and trend of development of micro-robots. *Mechanical Design 26*, 1–2.
Zhang Y.S., W. Liu & Z.Y. Jia (2005). Biomimetic swimming properties of a wireless micro-robot driven by outside magnetic field. *Journal of mechanical engineering 10*, 51–56.
Ren S.J., J.B. Dong & A.H. Wang (2011). Design and motion analysis of wireless micro imitating fish swimming robot. *Journal of Heilongjiang Institute of Science & Technology 21*, 486–488.

Electronic Engineering and Information Science – Wang (Ed.)
© *2015 Taylor & Francis Group, London, ISBN: 978-1-138-02772-5*

Optimization of pressure-equalizing groove distribution of gas thrust bearing

Y.J. Qiao, M.J. Zheng, L.L. Wan & H.R. Wang
Harbin University of Science and Technology, Harbin, Heilongjiang Province, China

ABSTRACT: In order to improve the bearing capacity of annular static pressure aerostatic thrust bearing and reduce the gas consumption, this research numerically analyzes and simulates the flow field inner bearing of different pressure-equalizing groove distribution of CFD simulation software, which is based on the mathematical model of gas static pressure aerostatic thrust bearing. The research studies bearing characteristics and gas consumption of the gas static pressure aerostatic thrust bearing. It also compares the bearing capacity of supply hole sequential distributions and interlaced distribution. The conclusion is that the annular static pressure gas thrust bearing with double rows connected pressure-equalizing groove and interlaced distribution of supply hole have good bearing capacity.

KEYWORDS: bearing capacity; gas thrust bearing; pressure-equalizing groove; CFD simulation.

1 INTRODUCTION

Recently, many domestic and overseas scholars do a lot of researches of pressure-equalizing groove, which can suppress the pressure attenuation far away from the orifice and improves the bearing capacity of static pressure aerostatic thrust bearing (Du, J.J. & G.Q. Zhang 2012). For example, M. M. Kho*nsari* from Louisiana State University built a model of screw-type pressure-equalizing groove and optimized the structure parameters of groove profile in 2012. But the study of the pressure-equalizing groove distribution is rarely reported (Guo, L.B. 2007). Domestic institutes and researchers study the bearing characteristics of aerostatic thrust bearing in different parameters of perturbation method and flow continuity equation. It cannot reflect the influence of a single parameter to the bearing (Kazimierski, Z. & J. Trojnarski 1980). The CDF calculation software was used to simulate gas-lubricated aerostatic thrust bearing under different distribution of pressure-equalizing groove, meanwhile, the influence of pressure-equalizing groove distribution and bearing characteristics was analyzed.

2 THE WORKING PRINCIPLE AND THE MATEMATICAL MODEL OF THE AEROSTATIC THRUST BEARING

2.1 The working principle of the aerostatic thrust bearing.

The bearing supply high-pressure gas to the film gap forming the gas film (Khonsari, M.M. & Y.F. Qiu 2012).

The aerostatic thrust bearing uses the pressure difference of the airspace to support the external load. Just like what the Figure 1 shows.

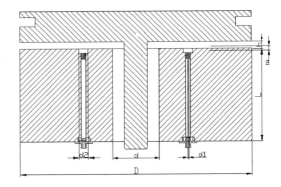

Figure 1. Single rows supply hole of aerostatic thrust bearing.

The parameters of the structure of the aerostatic thrust bearing shows in Table 1,

Table 1. Structural parameters.

outside diameter D: 200mm	bearing thickness L: 80mm	Groove depth: 0.05mm
internal diameter d: 60mm	air space diameter d2: 2mm	Groove width: 0.4mm
air space depth h: 0.1mm	Supply hole d1: 0.4mm	the number of supply hole: 12

2.2 The mathematical model for the numerical simulation calculation

In the CFD software calculates, the basic dynamic lubrication theory of gas flow is N-S equations. The basic idea of the lubrication equations numerically solving is the finite volume method. It discrete the lubricating gas film region to be some gas infinitesimal and calculates the physical quantities of the infinitesimal by dividing the flow field grid. Lubrication equations including continuity equation and momentum equation, its general expression like the Eq.1 shows.

$$\frac{\partial}{\partial t}(\rho\delta)+div(\rho u\delta)=div(\Gamma grad\delta)+S_{\delta} \qquad (1)$$

The character δ is common variable; Γ is generalized diffusion; $S\delta$ is generalized original item. Fluid Forms within the aerostatic thrust bearing are mainly formed by the gas pressure and flow rate of the formation of nonlinear coupling. It shows strong pressure and velocity coupling effects. The pressure-velocity coupling equations as shows in Eq.2,

$$\frac{\partial}{\partial t}(\rho\varphi)+div(\rho u\varphi)$$
$$=div(\Gamma grad\varphi)-\frac{\partial p}{\partial\eta}+S_{\varphi} \qquad (2)$$

According to the finite volume method, rewrite the Eq.2 to Eq.3,

$$\int_{V}\frac{\partial}{\partial t}(\rho\partial)dV+\int_{V}div(\rho\partial u)dV$$
$$=\int_{V}div(\Gamma grad\,\partial)dV+\int_{V}S_{\partial}dV \qquad (3)$$

In the Eq.3, the character V is the volume of gas film lubrication. The volume integral is converted to area integration by gauss divergence Eq.4,

$$\int_{\Delta t}\frac{\partial}{\partial t}\left(\int_{V}\rho\partial dV\right)dt+\int_{\Delta t}\int_{A}\eta\cdot(\rho\delta u)dAdt$$
$$=\int_{\Delta t}\int_{A}\eta\cdot(\Gamma grad\delta)dAdt+\int_{\Delta t}\int_{v}S_{\delta}dV \qquad (4)$$

In the Eq.4, A is gas film lubrication area; the character Δt is the time change amount. Eq.4 is the theoretical basis of CFD calculations. The CFD Software iteratively calculates the initial pressure by SOMPLE algorithm. We can get the converged solution of pressure through constantly revise the initial pressure. The bearing characteristics are express by bearing capacity and static stiffness

3 SIMULATION OF GAS AEROSTATIC THRUST BEARING IN DIFFERENT PRESSURE-EQUALIZING GROOVE DISTRIBUTION

3.1 Meshing

The simulated process of aerostatic thrust bearing actually is the process of solving the mathematical model. From the theoretical analysis, we can get the result that the more supply holes and pressure-equalizing grooves, the bigger bearing capacity and static stiffness, but gas consumption is also increasing. When the pressure-equalizing groove and supply holes increase to a certain number, the bearing capacity and static stiffness stop increasing (Long W. & G. Bao 2010). Bearing characteristics improve insignificant. In order to improve connective function of supply holes and increase bearing capacity, the four different simulated models were built. Just like figure.2 shows, single row pressure-equalizing groove, double rows pressure-equalizing grooves, double rows connected pressure-equalizing groove and interlaced distribution of supply holes, double rows connected pressure-equalizing groove and sequential distribution of supply hole. The distribution circle radius of single is 65mm (Hong X.Z. & T.L. Yun 2012). The distribution circle radius of double rows is 45mm and 67mm (Hou, Y. & Y.L. Xiong 2000).

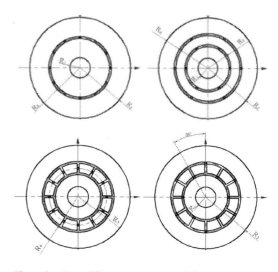

Figure 2. Four different pressure-equalizing groove.

When making flow field models, the gas supply channel is 1mm (Sun J.H. & Z.Q. Wu 2013). According to professional CAD software Solid Works established flotation flow field geometry, the geometric model import ICEM to meshing (Wu, D.Z. & J.Z.Tao 2010). The simple profile of flow field and an enlarged view of grid as shown in figure.3 and figure.4, when Film thickness is 20μm ,the grid nodes of single pressure-equalizing groove is 430904; the grid nodes of double pressure-equalizing groove is 864328, the grid nodes of sequential distribution of supply hole is 463196; the grid nodes interlaced distribution of supply holes is 923368.

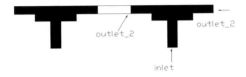

Figure 3. Sample gas fluid model.

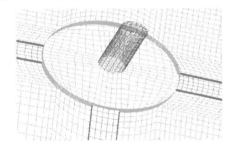

Figure 4. An enlarged view of grid.

3.2 CFD Simulation

This text use laminar calculated model for analysis those models by FLUENT. The pressure on supply hole is 0.3MP. We calculate the flow field grid model of the different film thickness and different pressure-equalizing groove under those conditions, solving the Reynolds equation of the flow field. The under-relaxation factor is 0.07. We use separation implicit solution to solve equations under above conditions. The gas film static pressure contour filling profile shows in the Figure .5 when the film thickness is 20μm.

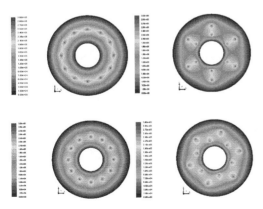

Figure 5. The gas film static pressure contour filling profile when Film thickness is 20μm.

Using MATLAB to fit the simulation result and getting the bearing characteristic curves, as shown in the figure.6 and figure.7; the static stiffness curves show in figure8 and figure.9; the curves of consumed mass flow show in figure.10 and figure.11.

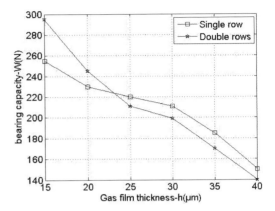

Figure 6. Fitting curves of load coefficient.

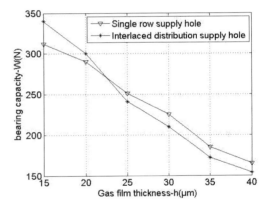

Figure 7. Fitting curves of load coefficient.

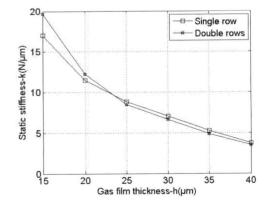

Figure 8. Fitting curve of static stiffness.

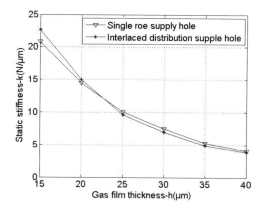

Figure 9.   Fitting curve of static stiffness.

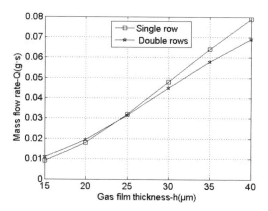

Figure 10.   Fitting curve of consumed mass flow.

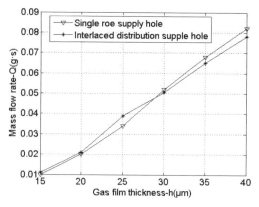

Figure 11.   Fitting curve of consumed mass flow.

4   RESULT ANALYSIS

From the bearing characteristic curves, the bearing capacity and static stiffness both decreases along with the increase of gas film thickness, but the consumed mass flow is increased. The bearing characteristics of double rows connected pressure-equalizing groove are bigger than single and double rows as a whole. From the figure.6 and figure.8, when the gas film thickness is less than 20um, the bearing characteristics of double pressure-equalizing groove are bigger than single pressure-equalizing groove. From the figure.7 and figure.9, also can conclude that when the gas film thickness is less than 27um, the bearing characteristics of the interlaced distribution of supply hole is bigger than sequential distribution of supply hole. Comparing figure10 and figure.11, analyze the reason of this phenomenon, we can know that the consumed mass flow grows linearly with the increase of gas film thickness and the pressure of flow field decrease.

Under the condition of the cheap film thickness, the inhibitory effect to pressure attenuation of double pressure-equalizing groove is greater than single pressure-equalizing groove. The coupling and throttling effect of double rows connected pressure-equalizing groove is far greater than single and double rows pressure-equalizing groove due to connecting effect of radial pressure-equalizing groove. The gas film static pressure contour filling profile shows in the figure.5 that an area of high pressure of double rows connected pressure-equalizing groove is obviously bigger than single and double row pressure-equalizing groove. The coupling effect of interlaced distribution of supply holes is greater than a sequential distribution of supply hole. So the distributions of double rows connected pressure-equalizing groove is superior to single and double row pressure-equalizing groove. An area of high pressure interlaced distribution of supply hole is superior to sequential distribution of supply hole.

5   CONCLUSION

Finally, we can get some conclusion based on simulation of aerostatic thrust bearing which have different distributions of pressure-equalizing groove. The flow field results are consistent with the actual, can it can effectively respond to the changes in the flow field within the bearing. Analyzing the curves of bearing characteristics and consumed mass flow, we can draw several conclusions:

1   The bearing capacity and static stiffness are declining along with the increase of film thickness. When the film thickness is less than 22um, the bearing capacity of double rows pressure-equalizing groove is greater than the single row pressure-equalizing

groove; but the consumed mass flow of single row pressure-equalizing groove is less.

2 The bearing characteristics of double rows connected pressure-equalizing groove is greater than single and double row pressure-equalizing groove.

3 The bearing capacity of double rows connected pressure-equalizing groove is bigger than the double circumferential rows pressure-equalizing groove due to the connecting effect of radial pressure-equalizing groove. But the consumed mass flow increases slightly. The structure of double rows connected pressure-equalizing groove and interlaced distribution of supply hole have good bearing capacity.

REFERENCES

Du, J.J. & G.Q. Zhang (2012). The relationship between pressure-equalizing groove and bearing characteristics of the static gas journal bearings. *Chinese journal of mechanical engineering 48(15)*, 106–112.

Guo, L.B. (2007). The parameter design of aerostatic thrust bearing with more supply holes. *Lubrication Engineering 32(4)*, 108–111.

Kazimierski, Z. & J. Trojnarski (1980). Investigations of externally pressurized gas bearings with feeding systems. *Journal of Lubrication Technology 102(1)*, 59–64.

Khonsari, M.M. & Y.F. Qiu (2012). Thermo hydrodynamic analysis of spiral groove mechanical face seal for liquid applications. *Journal of tribology 134(3)*, 1–11.

Long W. & G. Bao (2010). Entrance effect on load capacity of orifice compensated aerostatic bearing with feeding pocket. *Chinese journal of mechanical engineering 23(4)*, 451–458.

Hong X.Z. & T.L. Yun (2012). CFD Investigation on the performance of aerostatic thrust bearing with exhaust slots used in low-vacuum condition. Proceedings of 2012 international conference on mechanical engineering and material science. Atlantis Press: Shanghai Jiaotong University: 201–204.

Hou, Y. & Y.L. Xiong (2000). The research of gas aerostatic bearing with Multi-row supply holes. *Journal of Xi'an Jiao tong University 34(11)*, 41–43.

Sun J.H. & Z.Q. Wu (2013). Flotation ring with the pressure-equalizing groove pad design. *Light Industry Machinery 31(1)*, 80–82.

Wu, D.Z. & J.Z.Tao (2010). The numerical Simulation and Experimental Research of gas aerostatic bearing by FLUENT. *Mechanical design and manufacturing* 150–151.

Electronic Engineering and Information Science – Wang (Ed.)
© 2015 Taylor & Francis Group, London, ISBN: 978-1-138-02772-5

Investigation on frequency doubling characteristics of one-dimensional BBO photonic crystals

Z.Y. Yang, J.J. Huang, L.Y. Zhang, S.Z. Pu & H.N. Wu
Department of Optoelectric Information Science and Engineering, Harbin University of Science and Technology, Harbin, China

ABSTRACT: The transfer matrix method was employed to study frequency doubling in one-dimensional nonlinear photonic crystals. A general expression of the second harmonic field was deduced by introducing the phase mismatch factor, where the operation of the matrices involved was simplified. In the case of angle detuning, the second harmonic generation was simulated and the variation of conversion efficiency with the number of layers was analyzed in a BBO nonlinear photonic crystal. It is found that due to phase modulation characteristics of Bragg reflection, the phase matching for bulk BBO crystals is no longer suitable for photonic crystals.

KEYWORDS: Photonic crystals; Frequency doubling; Transfer matrix method; Conversion efficiency; Phase matching

1 INTRODUCTION

Yabluonovitch and John put forward the concept of photonic crystals, due to its unique properties photonic crystals are concerned with the majority of scholars. Photonic crystals are periodic materials which band-gap characteristic is applied to many optical devices. Moreover, due to its local states in which mode density is higher and incident pulses travel slowly, then there are stronger nonlinear effects (Dolgova et al. 2002, Liscidinia et al. 2004, Joseph et al. 2014). Nearly 20 years, the researches of transmission characteristics (Wang et al. 2001) and dispersion characteristics (Centini et al. & Tarasishin et al. 1999) are gradually becoming mature. But there are a few researches to study phase matching characteristics of photonic crystals. Therefore, study of optical field coupling in nonlinear photonic crystals is very meaningful. Transfer matrix method is the common method of studying one dimensional photonic crystals' transmission and nonlinear process (Bethune 1989, Jeong et al. 1999, Li et al. 2007, Saleh et al. 2008). In this paper, we expand and improve the frequency doubling model, and on this account we study the characteristics of frequency doubling in BBO photonic crystals.

2 FREQUENCY DOUBLING MODEL

In this section, we consider a layered material with periodically modulated linear and nonlinear optical parameters. Plane wave with frequency ω propagate through the media. As a result of multiple interference and reflection, light field has forward wave and backward wave.

From reference (Li et al. 2007), we can obtain total transfer equations after fundamental frequency (FF) wave passed i layers can be described as:

$$\begin{pmatrix} \Omega_{2i-1}^+ \\ \Omega_{2i-1}^- \end{pmatrix} = D_1^{-1} (D_2 P_2 D_2^{-1} D_1 P_1 D_1^{-1})^{i-1} D_0 \begin{pmatrix} \Omega_0^+ \\ \Omega_0^- \end{pmatrix}$$

$$\begin{pmatrix} \Omega_{2i}^+ \\ \Omega_{2i}^- \end{pmatrix} = D_2^{-1} D_1 P_1 D_1^{-1} (D_2 P_2 D_2^{-1} D_1 P_1 D_1^{-1})^{i-1} D_0 \begin{pmatrix} \Omega_0^+ \\ \Omega_0^- \end{pmatrix} \tag{1}$$

Coupling equations of forward and backward second harmonic (SH) waves are:

$$dE_i^\pm (2\omega, z)/dz = \pm 2i\omega^2 / k_i^{(2)} c^2 P_i^\pm (2\omega, z) e^{\mp ik_i^{(2)} z} \tag{2}$$

where nonlinear polarization can be expressed as:

$$P_i^\pm (2\omega, z) = \varepsilon_0 \chi^{(2)} \left[\Omega_i^\pm (z) \right]^2 e^{\pm i2k_i^{(1)} z} \tag{3}$$

The amplitudes of FF wave $\Omega_i^+(z)$ and SH wave $E_i^\pm(z)$ are normalized $B_i^\pm(z)$ and $C_i^\pm(z)$, respectively.

Amplitude of output SH wave which output from the right side of photonic crystal is:

$$\begin{pmatrix} C_t^+ & C_t^- \end{pmatrix}^T = D_0^{-1} S^N D_0 \begin{pmatrix} C_0^+ & C_0^- \end{pmatrix}^T +$$

$$\sum_{j=1}^{N} D_0^{-1} S^N M g_1 Q_1 \begin{pmatrix} A_1 e^{i\Delta k d_1/2} & 0 \\ 0 & -A_1 e^{i\Delta k d_1/2} \end{pmatrix} \begin{pmatrix} \left(B_{2j-1}^+\right)^2 \\ \left(B_{2j-1}^-\right)^2 \end{pmatrix}$$

$$+\sum_{j=1}^{N} D_0^{-1} S^N g_2 Q_2 \begin{pmatrix} A_2 e^{i\Delta k d_2/2} & 0 \\ 0 & -A_2 e^{i\Delta k d_2/2} \end{pmatrix} \begin{pmatrix} \left(B_{2j}^+\right)^2 \\ \left(B_{2j}^-\right)^2 \end{pmatrix} \quad (4)$$

where:

$$A_i = 2i\Omega_0^+ d_i \chi_i^{(2)} \omega^2 / k_i^{(2)} c^2, \quad \Delta k = 2\omega \left(n_i^{(1)} - n_i^{(2)}\right)/c,$$

$$S^N = \psi_N S - \psi_{N-1} I, \quad \psi_N = \sin N\beta / \sin \beta,$$

$$\cos \beta = \mathrm{Re}(1/t), \quad S = g_2 Q_2 g_2^{-1} g_1 Q_1 g_1^{-1},$$

$$g_i = \begin{pmatrix} 1 & 1 \\ n_i^{(2)} & -n_i^{(2)} \end{pmatrix},$$

$$Q_i = \begin{pmatrix} \exp\left(ik_i^{(2)} d_i\right) & 0 \\ 0 & \exp\left(-ik_i^{(2)} d_i\right) \end{pmatrix} \quad (5)$$

so we can infer the SH conversion efficiency as:

$$\eta_i^\pm = \left| C_i^\pm \right|^2 \quad (6)$$

3 NUMERICAL RESULTS

BBO is an excellent frequency doubling crystal which is widely used in the visible light area. The photonic crystal is composed of air and BBO crystal. The incident FF wavelength is 800 nm with intensity is 4.5MW/cm². The photonic crystal have 30 layers, in which BBO layer's thickness is $d_1 = \lambda_0/4n_0$, another air layer is $d_0 = \lambda_0/2$, where n_0 is BBO's ordinary refractive index for 800 nm, and λ_0 is set as 700 nm. Some dispersion formula of BBO crystals comes from reference (Dmitriev et al. 2003). Figure 1 shows the variation of conversion efficiency with different layers for different incident angles of FF waves (incident angle defines as the angle of incident light and axis). Through calculations SH transmittances corresponding to Figure 1a–d are 0.87, 0.95, 0.95 and 0.86, respectively. Because there is a Bragg reflection of photonic crystal the SH amplitude deviation on both sides of the interface is very large, but the amplitude in every layer is linear change. Obviously, with the increase of the layer number conversion efficiency's overall trend is rising. The incident angle in Figure 1b is exactly the pure BBO crystal phase matching angle, which conversion efficiency is higher than (a) and (d) but less than (c). SH transmittance in Figure 1b and c are almost equal, so we can eliminate the influence of transmittance on SH conversion efficiency.

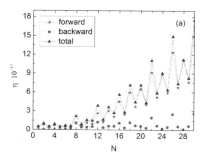

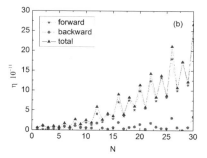

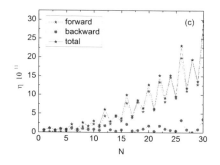

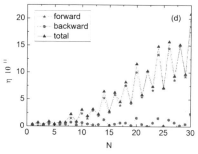

Figure 1. Conversion efficiency versus number of layers for different angles (a) 20° (b) 29.2° (c) 40° (d) 50°.

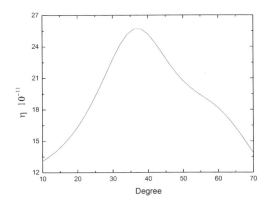

Figure 2. Variation of the conversion efficiency with angle.

Forward conversion efficiency changing with angle is shown in Figure 2. When the incident angle is about 37°, conversion efficiency is the largest, and this angle completely deviate from BBO crystal's phase matching angle (29°). The full width at half maximum is larger than the acceptable angle's width of BBO (>180°). It is multiple reflections and interference in photonic crystal increases matching bandwidth. All in all, pure BBO crystal phase matching condition does not mean that SH conversion efficiency in nonlinear photonic crystal is the largest.

4 CONCLUSIONS

In this paper, we used the transfer matrix method to calculate SH generation in one-dimensional nonlinear photonic crystal under the undepleted pump approximation. Amplitude expressions of frequency doubling wave are simplified to simulate frequency doubling in a photonic crystal composed of air and BBO, so we obtained the FF and SH field distribution. Calculations showed that with the increase of the layer numbers of photonic crystal, the overall trend of SH conversion efficiency has been on the rise. When the incident angle is about 37°, conversion efficiency is the largest, and it has a large deviation of the matchng angle of pure BBO crystal (29°). Because of the existence of Bragg reflection in photonic crystals, the phase of transmission field has been modulated,

so traditional phase matching conditions are no longer suitable.

ACKNOWLEDGMENT

The authors thank the Education Department of Heilongjiang province in China for the fund under Grant No. 12511113.

REFERENCES

Bethune, D. S. (1989). Optical harmonic generation and mixing in multilayer media: analysis using optical transfer matrix techniques. *J. Opt. Soc. Am. B, 6,* 910–916.
Centini, M., C. Sibilia , & M. Scalora et al. (1999). Dispersive properties of finite, one-dimensional photonic band gap structures: Applications to nonlinear quadratic interactions. *Phys. Rev. E, 60(4),* 4891–4898.
Dolgova, T. V., A. I. Maidykovski, & M. G. Martemyanov et al. (2002). Giant microcavity enhancement of second-harmonic generation in all-silicon photonic crystals. *Appl. Phys. Lett. 81(15),* 2725–2727.
Dmitriev,V. G. & G. G.Gurzadyan (2003). Handbook of Nonlinear Optical Crystals. Springer.
Joseph, S., M.ShahidKhan, & A. K.Hafiz (2014). Parameters for efficient growth of second harmonic field in nonlinear photonic crystals. *Phys. Lett. A, 378(18–19), 1296–1302.*
Jeong, Y. & B. Lee (1999). Matrix analysis for layered quasi-phase-matched media considering multiple reflection and pump wave depletion. *IEEE J. Quantum Electron 35,* 162–172.
Liscidinia, M. & L. Andreani (2004). Highly efficient second-harmonic generation in doubly resonant planar microcavities. *Appl. Phys. Lett., 85(11),* 1883–1885.
Li, J. J., Z. Y. Li & D. Z. Zhang (2007). Second harmonic generation in one-dimensional nonlinear photonic crystals solved by the transfer matrix method. *Phys. Rev. E, 75(5),* 056606-1-056606–7.
Saleh, M. F., L. D. Negro, & B. E. A. Saleh (2008). Second-order parametric interactions in 1-D photonic-crystal microcavity structures. *Opt. Express, 16(8),* 5261–5276.
Tarasishin, A. V., A. M. Zheltikov & S. A. Magnitskii (1999). Matched second-harmonic generation of ultrashort laser pulses in photonic crystals. *JETP Lett., 70(12),* 819–825.
Wang H., & Y. P. Li (2001). An eigen matrix method for obtaining the band structure of photonic crystals. *Acta Phys. Sin. 50(11),* 2172–2178.

Electronic Engineering and Information Science – Wang (Ed.)
© *2015 Taylor & Francis Group, London, ISBN: 978-1-138-02772-5*

Polarization and dielectric properties of ferroelectric thin films with structural transition zones

J. Zhou, W.G. Xie, C.B. Yu, L. Xu, Z.J. Li, P.N. Sun & H.G. Ye
College of Physics Science and Technology, Heilongjiang University, Harbin, China

T.Q. Lü
Center of Condensed Matter Science and Technology, Harbin Institute of Technology, Harbin, China

ABSTRACT: The polarization and the susceptibility of the ferroelectric thin film are investigated by using the modified transverse Ising model. The Structure Transition Zones (STZ) are introduced to a finite size thin film. The results demonstrate that the STZ is a vital factor that affects the mean polarization when the ratio of the parameter s_t to the lateral size is big, and the thin film can be at the paraelectric phase when the lateral size is small and the temperature is high.

KEYWORDS: ferroelectric thin film, transverse Ising model, phase transformation.

1 INTRODUCTION

Surface and size effects on ferroelectric thin films have become one of subjects of experimental (Alexe et al. 2001, Stolichnov et al. 2002, Spanier et al. 2006, Ohno et al. 2006) and theoretical (Wang et al. 2000, Jiang et al. 2005, Zhou ct al. 2008, Wang et al. 1999, Wesselinowa et al. 2004, Michael et al. 2007) studies in recent years. Experimentally, Alexe and co-workers (2001) fabricated the epitaxial Pb(Zr,Ti)O$_3$ thin films by a film patterning process. They found that the morphotropic phase boundary was sensitive to the lateral size of the Pb(Zr,Ti)O$_3$ thin film. Theoretically, transverse Ising model (TIM) has been proved to be an effective method to investigate surface and size effects in ferroelectrics. Wang et al. (2000) investigated the size effects on the polarization and the susceptibility of ferroelectric particles by considering the long-range interactions. The result indicates that the critical size of the ferroelectric particle is sensitive to the interaction range. Jiang et al. (2005) investigated the critical behavior of the phase transition of the thin film by using the transverse Ising model, and found that the critical value of the pseudo-spin interaction showed a strong dependence on the thickness of the thin film. Recently, Zhou and co-workers (2008) proposed a "multi-step" model, namely structural transition zones (STZ) were introduced to the traditional model. And they studied the dependence of the lateral STZ, the psudo-spin interaction, the lateral size and the thickness on the polarization and the critical temperature of the thin film. The motivation of this paper is to study the lateral size effect on the susceptibility and the polarization of the thin film with the STZ.

2 FORMALISM

Figure 1 shows the schematic illustration of the model of the ferroelectric thin film with STZ. S_i^x and S_i^z severally denote the lateral size and the thickness. In figure 1 (a), every pseudo-spin layer is comprised of s (i or J_{ij}) square loops, and the parameter m represents the site of the each square loop, and s_t is the sum of the square loops in the lateral STZ, namely the sum of the transition square loops. The site of the each pseudo-spin layer is numbered by k in figure 1 (b), and E and $k = L$ both represent the surface layers.

The Hamiltonian of the TIM is described as:

$$H = -\sum_i \Omega_i S_i^x - \sum_{ij} J_{ij} S_i^z S_j^z - 2\mu E \sum_i S_i^z, \qquad (1)$$

where S_i^x and S_i^z stand for the spin-1/2 operators at lattice site i, Ω_i is the tunneling frequency, J_{ij} denotes nearest-neighbor pseudo-spin interactions, μ stands for the dipole moment and E denotes the external electric field. It is assumed that $J_{ij} = J_a(m, k)$ if two sites are both in a square loop; $J_{ij} = J_b(m, k)$ if two sites lie in the diverse square loops; $J_{ij} = J_e(m, k)$ if two sites lie in the diverse pseudo-spin layers.

Because the surface of the thin film is symmetric, pseudo-spin interactions J_{ij} for the upper half-film are described as

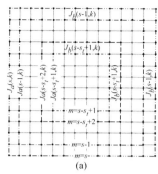 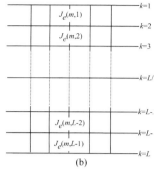

(a)   (b)

Figure 1.   Schematic illustration of the model of the ferroelectric thin film with $N \times N \times L$ size.

$$\begin{cases} J_a(m,k) = J_\infty e^{-1/(a_1 \times N \times L \times k)}, & 1 \le m \le (s-s_t) \quad 1 \le k \le L/2 \text{ or } 1 \le k \le (L+1)/2 & (2a) \\ J_a(m,k) = J_s e^{-1/(a_1 \times N \times L \times k)}, & m=s \quad 1 \le k \le L/2 \text{ or } 1 \le k \le (L+1)/2 & (2b) \\ J_a(m,k) = J_a(s,k) + (J_a(2,k) - J_a(s,k)) \times (1 - e^{-\beta_1 (m_t-1)/s_t}), \\ \quad m = s+1-m_t \quad 1 \le m_t \le s_t \quad 1 \le k \le L/2 \text{ or } 1 \le k \le (L+1)/2 & (2c) \end{cases}$$

$$\begin{cases} J_b(m,k) = J_\infty e^{-1/(a_1 \times N \times L \times k)}, & 1 \le m \le (s-s_t) \quad 1 \le k \le L/2 \text{ or } 1 \le k \le (L+1)/2 & (3a) \\ J_b(m,k) = J_a(s,k) + (J_a(2,k) - J_a(s,k)) \times (1 - e^{-\beta_2 m_t/s_t}), \\ \quad m = s-m_t \quad 1 \le m_t \le (s_t-1) \quad 1 \le k \le L/2 \text{ or } 1 \le k \le (L+1)/2 & (3b) \end{cases}$$

$$\begin{cases} J_e(m,k) = J_\infty e^{-1/(a_2 \times N \times L \times k)}, & 1 \le m \le (s-s_t) \quad 1 \le k \le L/2 \text{ or } 1 \le k \le (L+1)/2 & (4a) \\ J_e(m,k) = J_{es} e^{-1/(a_2 \times N \times L \times k)}, & m=s \quad 1 \le k \le L/2 \text{ or } 1 \le k \le (L+1)/2 & (4b) \\ J_e(m,k) = J_e(s,k) + (J_e(1,k) - J_e(s,k)) \times (1 - e^{-\beta_3 (m_t-1)/s_t}), \\ \quad m = s+1-m_t \quad 1 \le m_t \le s_t \quad 1 \le k \le L/2 \text{ or } 1 \le k \le (L+1)/2 & (4c) \end{cases}$$

where J_∞ is the pseudo-spin interaction when the lateral size is boundless, J_s and J_{es} are the pseudo-spin interactions when two pseudo-spins are in a edge loop and in the diverse layers for $L = \infty$, $\beta_1, \beta_2, \beta_3$, α_1 and α_2 denote the adjustable parameters of the interaction intensity.

Adopting the mean-field approximation, the thermal average $\langle S_i^z \rangle$ can be given by

$$\langle S_i^z \rangle = (\langle H_i^z \rangle / 2|\mathbf{H}_i|) \tanh(|\mathbf{H}_i|/2k_B T), \quad (5)$$

Where

$$\langle H_i^z \rangle = \sum_i J_{ij} \langle S_j^z \rangle + 2\mu E, \quad (6)$$

$$|\mathbf{H}_i| = \sqrt{\Omega_i^2 + (\langle H_i^z \rangle)^2}. \quad (7)$$

The mean polarization is

$$\bar{P} = \frac{1}{(N^2 \times L)} \sum_{i=1}^{N^2 \times L} P_i = \frac{2n\mu}{(N^2 \times L)} \sum_{i=1}^{N^2 \times L} \langle S_i^z \rangle. \quad (8)$$

where n stands for the number of pseudo-spins per volume.

When $T \to T_C$, the critical temperature is determined by

$$\sum_j J_{ij} \langle S_j^z \rangle - (2\Omega_i / \tanh(\Omega_i / 2k_B T)) \langle S_i^z \rangle = 0. \quad (9)$$

The mean susceptibility is written as

$$\bar{\chi} = \frac{1}{(N^2 \times L)} \sum_{i=1}^{N^2 \times L} \chi_i = \frac{1}{(N^2 \times L)} \sum_{i=1}^{N^2 \times L} \frac{\partial P_i}{\partial E}\bigg|_{E=0}. \quad (10)$$

3   NUMERICAL RESULTS AND DISCUSSIONS

The numerical results are presented in this section based on equations (8)–(10). We set $\Omega_i / J = 0.1$ and $\beta_3 = \beta_1$. T_b is the critical temperature of the bulk materials and set $t = T / T_b$.

The temperature dependence of the mean polarization with different s_t, β_1 and N is illustrated in figure 2. For convenience of discussion, we set $\beta_1 = \beta_2$. In the case of $N = 50$, the mean polarization is improved with the decrease in s_t when β_1 is well-defined, and the mean polarization is improved with the increase in β_1 when s_t is well-defined. Therefore, shrinking the lateral STZ and strengthening the pseudo-spin interactions in the lateral STZ help to improve the mean polarization of the thin film with the STZ. But the critical temperature of the thin film depends insensitively on parameters s_t and β_1. It can also be seen from figure 2 that the mean polarization is diversely affected by the lateral STZ when the lateral size is changed. The influence of lateral STZ on the mean polarization of the smaller lateral size film becomes more evident when parameters s_t and β_1 are fixed. The larger is the ratio of the parameter s_t to the lateral size N, the more obviously the lateral structural transition zones affect the mean polarization.

Figure 3 presents the dependence of the critical temperature on the lateral size for different parameters J_∞ and α_1. For convenience of discussion, we set $\alpha_1 = \alpha_2$. One can see from figure 3 that the critical temperature of the thin film decreases to zero when the lateral size reaches a critical value. And the increase of J_∞ (or α_1) comeswith an obvious

enhancement of the critical temperature and an evident decrease of the critical lateral size. This is reasonable because the stronger pseudo-spin interactions at the core of the film can improve the critical temperature and decrease the critical lateral size. Therefore, strengthening the pseudo-spin interactions at the core of the film can shrink the electric device sizes.

Figure 4 shows the mean susceptibility as a function of the lateral size when the temperature t is changed. Three curves all show a susceptibility vertex, which indicates that the lateral size effect will occur. The susceptibility vertex moves to the lager lateral size when the temperature is increased, which means that the thin film can be at the paraelectric phase when the lateral size is smaller and the temperature is higher.

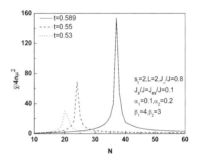

Figure 4. The dependence of the mean susceptibility on the lateral size with different t.

4 CONCLUSIONS

In conclusion, the lateral size effects on ferroelectric thin films with STZ have been investigated on the basis of the transverse Ising model. The conclusions can be obtained as the following. (1)The mean polarizations of thin films with the different lateral sizes are diversely influenced by the lateral STZ. The lateral STZ can be an important factor that affects the mean polarization when the ratio of the parameter s_t to the lateral size N is big. (2) Strengthening the pseudo-spin interaction at the core of the thin film is an effective way to downscale the critical lateral size and improve the critical temperature. (3)The susceptibility vertex moves to the larger lateral size of the thin film when the temperature is higher.

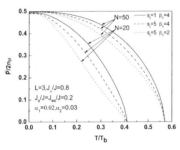

Figure 2. The mean polarization of the three-layer thin film with different s_t, β_1 and N.

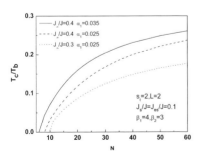

Figure 3. The dependence of the critical temperature on the lateral size with different J_∞ and α_1.

ACKNOWLEDGEMENT

This project is sponsored by Scientific Research Fund of Heilongjiang Provincial Education Department (Grant No: 11551378).

REFERENCES

Alexe M., C. Harnagea, D. Hesse & U. Gösele (2001). Polarization imprint and size effects in mesoscopic ferroelectric structures. *Appl. Phys.Lett. 79*, 242–244.

Stolichnov I., E. Colla, A. Tagantsev, S. S. N. Bharadwaja, S. Hong, N. Setter, J. S. Cross & M. Tsukada(2002). Unusual size effect on the polarization patterns in micron-size Pb(Zr,Ti)O$_3$ film capacitors. *Appl. Phys. Lett. 80*, 4804–4806.

Spanier J. E., A. M. Kolpak, J. J. Urban, I. Grinberg, L. Ouyang, W. S. Yun, A. M. Rappe & H. Park(2006). Ferroelectric phase transitions in individual single-crystalline BaTiO$_3$ nanowrires. *Nano Lett. 6* , 735–739.

Ohno T., D. Suzuki, K. Ishikawa, M. Horiuchi, T. Matsuda & H. Suzuki (2006). Size Effect for Ba(Zr$_x$Ti$_{1-x}$)O$_3$ (x=0.05) Nano-Particles. *Ferroelectrics. 337*, 25–32.

Wang C. L., Y. Xin, X. S. Wang & W. L. Zhong(2000). Size Effects of Ferroelectric Particles Described by the Transverse Ising Model. *Phys. Rev. B .62* , 11423–11427.

Jiang Q. & Y. M. Tao(2005). Critical Line between the First-Order and the Second-Order Phase Transition for Ferroelectric Thin Films Described by TIM. *Physics Letters A. 336*, 216–222.

Zhou J., T. Q. Lü, L. Cui, H. Chen & W. W Cao(2008). Phase transformation properties of finite size ferroelectric thin film with structural transition zones. *J. Appl. Phys. 104,* 124105-1-124105–7.

Wang X.G., S. H. Pan & G. Z. Yang(1999). Effects of multi-surface modification on Curie temperature of ferroelectric films. *J. Phys.: Condens. Matter. 11,* 6581–6588.

Wesselinowa J. M. & S. Trimper(2004). Thickness dependence of the dielectric function of ferroelectric thin films. *phys. stat. sol. (b). 241*, 1141–1148.

Michael Th., S.Trimper & J. M. Wesselinowa(2007). Size effects on static and dynamic properties of ferroelectric nanoparticles. *Phys.Rev. B . 76*, 94107–1-94107–7.

Electronic Engineering and Information Science – Wang (Ed.)
© 2015 Taylor & Francis Group, London, ISBN: 978-1-138-02772-5

Influence of the surface transition layer on the susceptibility distribution of the ferroelectric thin film

W.G. Xie, J. Zhou & X. Liang
College of Physics Science and Technology, Heilongjiang University, Harbin, China

X.Y. Yang & X. Zhang
College of Mathematical Science, Heilongjiang University, Harbin, China

ABSTRACT: The susceptibility distribution of the ferroelectric thin film is investigated by using the modified transverse Ising model. And the Surface Transition Layer (STL) is taken into account in the ferroelectric thin film. The influence of the temperature and the STL parameters on the susceptibility distribution is discussed. The results demonstrate that the susceptibility distribution depends strongly on the STL and the temperature.

KEYWORDS: ferroelectric thin film; transverse Ising model; susceptibility distribution.

1   INTRODUCTION

Although surface and size effects were firstly reported in 1950s, it has recently motivated renewed interest because of the rapid progress in material fabrication techniques (Glass et al. 1977, Ishikawa et al. 1988). Theoretically, the Transverse Ising Model (TIM) is an effective method that studies the surface and size effects in ferroelectrics. Based on the TIM, Wang et al. (1992, 2000) and Sy (1993) investigated the surface parameters and the size dependence of the mean polarization of ferroelectrics. Teng and co-workers (2004, 2005, 2006) applied the Green's function technique to a modified TIM to study the surface and size effects on the mean polarization of the ferroelectric thin film. Recently, Chen and co-workers (2008) put forward a "multi-step" model, i.e., the surface transition layer (STL) was introduced to the traditional model. They discussed the influence of the STL on the polarization and the critical temperature of the thin film, and thought that the STL is a vital factor that affects the phase transformation properties of the ferroelectric thin films. The purpose of this work is to investigation the susceptibility distribution of the thin film with the STL.

2   FORMALISM

The Hamiltonian is described as

$$H = -\sum_i \Omega_i S_i^x - \frac{1}{2}\sum_{ij} J_{ij} S_i^z S_j^z - 2\mu E \sum_i S_i^z, \quad (1)$$

where S_i^x and S_i^z stands for the spin-1/2 operators at lattice site i, Ω_i is the tunneling frequency, J_{ij} denotes nearest-neighbor pseudo-spin interactions, μ stands for the dipole moment and E denotes the external electric field.

Figure 1 shows the schematic of N layer ferroelectric thin film with STL. We assume that the tunneling frequency of every layer is same and the surface of the thin film is symmetrical. N is the film thickness. n_s denotes the number of the STL in the thin film. We assume that $J_{ij} = J_a(m)$ if two pseudo-spins are in a film layer; $J_{ij} = J_e(m)$ if two pseudo-spins are in the diverse film layers; $J_{ij} = J_a(m) = J_e(m) = J$ if the two pseudo-spins are not in the STL.

Because the surface of the thin film is symmetrical, the pseudo-spin interactions J_{ij} for the upper half-film can be expressed as

$$\begin{cases} J_a(m) = \alpha\left(\dfrac{2m}{N}\right)^\sigma J, & m = 1 \sim n_s, \quad n_s \le \dfrac{N}{2}, \quad (2a) \\[2ex] J_a(m) = J, & \dfrac{N}{2} \ge m > n_s, \quad (2b) \end{cases}$$

$$\begin{cases} J_e(m) = \beta\left(\dfrac{2m}{N}\right)^\sigma J, & m = 1 \sim n_s, \quad n_s \le \dfrac{N}{2}, \quad (3a) \\[2ex] J_e(m) = J, & \dfrac{N}{2} \ge m > n_s, \quad (3b) \end{cases}$$

where m labels the location of the each pseudo-spin layers. The parameter σ reflects the pseudo-spin interaction intensity of the interlayer and intra-layer. The parameters α and β are the adjustive parameters.

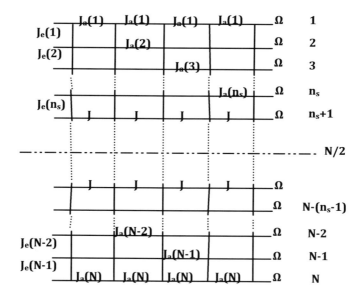

Figure 1.   Model of the ferroelectric thin film with STL.

Under the mean field approximation, the thermal average $\langle S_i^z \rangle$ can be described as

$$\langle S_i^z \rangle = \left(\langle H_i^z \rangle / 2 |\mathbf{H}_i| \right) \tanh \left(|\mathbf{H}_i| / 2 K_B T \right), \qquad (4)$$

Where

$$\langle H_i^z \rangle = \sum_i J_{ij} \langle S_j^z \rangle + 2\mu E, \qquad (5)$$

$$|\mathbf{H}_i| = \sqrt{\Omega_i^2 + \left(\langle H_i^z \rangle \right)^2}. \qquad (6)$$

The mean susceptibility is given by

$$\bar{\chi} = \frac{1}{N} \sum_{i=1}^{N} \chi_i = \frac{1}{N} \sum_{i=1}^{N} \left. \frac{\partial P_i}{\partial E} \right|_{E=0} = \frac{2n\mu}{N} \sum_{i=1}^{N} \left. \frac{\partial \langle S_i^z \rangle}{\partial E} \right|_{E=0}. \qquad (7)$$

where n stands for the number of pseudo-spins per volume.

3   NUMERICAL RESULTS AND DISCUSSIONS

In this section, we present numerical results based on equation (7). T_b is the critical temperature of the bulk materials. For simplicity, we assign $\Omega / J = 1.0$, $N = 10$ and set $t = T / T_b$.

The susceptibitliy distribution along the thickness direction of the thin film is plotted in figure 2 when the temperature t is different. It can be seen that the

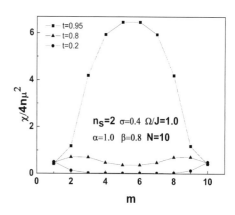

Figure 2.   The susceptibility distribution of the ten-layer thin film with different t.

susceptibility of the thin film is improved when the temperature is increased. The reason is that the ferroelectricity of the thin film is depressed with the increase of the temperature. One can also see that the shape of the susceptibility distribution varies with the change of the temperature. Compared with the middle layer, the susceptibility at the surface layer is big when the temperature is lower. The reverse case occurs when the temperature is higher. It is ascribed to the fact that the mean polarization is diversely influenced by the STL when the temperature is changed. Compared with the low temperature, the STL can obviously depress the polarization of the thin film at the high temperature. Therefore, the susceptibility of the thin film can be improved by increasing of the temperature, and the STL is a vital factor that affects the susceptibility distribution of the thin film.

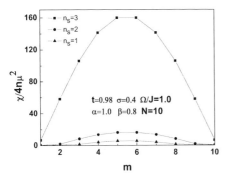

Figure 3. The susceptibility distribution of the ten-layer thin film with different n_s.

How the STL affects the susceptibility distribution is shown in figures 3–5 in details. Figure 3 shows the susceptibility distribution with different n_s. One can see that the susceptibility of each layer can be improved by thickening the STL. This is because the mean polarization is depressed with the enhancement of the parameter n_s. Thus thickening the STL can contribute to improve the susceptibility of the thin film.

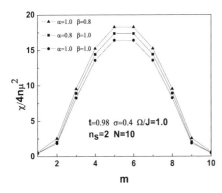

Figure 4. The susceptibility distribution of the ten-layer thin film with different α and β.

Figure 4 shows the susceptibility distribution with different α and β when $t = 0.98$. It can be seen that the susceptibility of each layer increases with the reduction in the parameter α (or β) when the parameter β (or α) is fixed. This is because that the mean polarization is depressed with the decrease of the parameters α and β. The parameter α has a more

obvious influence on the susceptibility compared with the parameter β. The reason is that the intra-layer interactions have a more obvious influence on the polarization in comparison with the interlayer interactions.

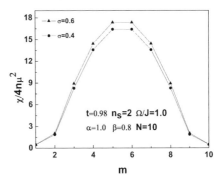

Figure 5. The susceptibility distribution of the ten-layer thin film with different σ.

Figure 5 shows the susceptibility distribution of the thin film when the parameter σ is different. It is demonstrated that the susceptibility of each layer increases with the enhancement of the parameter σ. The pseudo-spin interactions in the STL are weakened with the increase of the parameter σ, which leads to the increase of the susceptibility of the thin film. Therefore, weakening the intra-layer and inter-layer interactions in the STL can improve the susceptibility of the thin film.

4 CONCLUSIONS

In summary, the susceptibility distribution of the thin film with the STL is investigated by using the modified TIM. The conclusions can be obtained as the following. (1) The STL is an important factor that affects the susceptibility distribution of the thin film. The susceptibility of the thin film can be improved by thickening the STL and weakening the intra-layer and interlayer interactions in the STL. (2) The change of the temperature can influence the shape of the susceptibility distribution, and the susceptibility of the thin film is improved with the increase of the temperature.

ACKNOWLEDGEMENT

This project is sponsored by Scientific Research Fund of Heilongjiang Provincial Education Department (No: 12533045).

REFERENCES

Glass A. M. , K. Nassau & J. W. Shiever (1977). Evolution of ferroelectricity in ultrafine-grained $Pb_5Ge_3O_{11}$ crystallized from the glass. *J. Appl. Phys. 48*, 5213–5216.

Ishikawa K., K.Yoshikawa & N.Okada(1988) . Size effect on the ferroelectric phase transition in $PbTiO_3$ ultrafine particles. *Phys. Rev. B 37*, 5852–5855.

Wang C. L., W. L. Zhong & P. L. Zhang(1992). The Curie temperature of the ultra-thin ferroelectric films. *J. Phys.: Condens. Matter. 3* , 4743–4749.

Wang C. L., Y. Xin, X. S. Wang & W. L. Zhong(2000). Size effects of ferroelectric particles described by the transverse Ising model. *Phys. Rev. B .62* , 11423–11427.

Sy H. K. (1993). Surface modification in ferroelectric transitions. *J. Phys.: Condens. Matter. 5* ,1213–1220.

Teng B. H. & H. K. Sy (2004). Green's function investigation of transition properties of the transverse Ising model. *Phys. Rev. B .70,* 104115–1- 104115–5.

Teng B. H. & H. K. Sy(2005). Transition regions in the parameter space based on the transverse Ising model with a four-spin interaction. *Europhys. Lett. 72*, 823–829.

Teng B. H. & H. K. Sy(2006). Phase diagrams of the transverse Ising model with a four-spin interaction. *Europhys. Lett. 73*, 601–606.

Chen H., T. Q. Lü, L. Cui & W. W. Cao (2008). Dielectric properties of ferroelectric thin films with surface transition layers. *Physica A .387*, 1963–1971.

Electronic Engineering and Information Science – Wang (Ed.)
© 2015 Taylor & Francis Group, London, ISBN: 978-1-138-02772-5

Strong optical limiting of a molecular metal cluster

D.Z. Xin
Chaoyang Teachers' College, Chaoyang, Liaoning Province, PR. China

Y.C. Gao
Key Laboratory of Electronics Engineering, College of Heilongjiang Province, Heilongjiang University, Harbin, PR. China

ABSTRACT: Optical limiting response of a molecular metal cluster $[(n\text{-}Bu)_4N]_3[MoS_4Ag_3I_4]$ in DMF solution was investigated using 8-ns and 40-ps laser at 532nm wavelength. It was found that the metal cluster displays strong optical limiting even better than well reported limiter, fullerene C_{60} in toluene. And the picosecond optical limiting is much better than nanosecond optical limiting. The pulse width effect and origin of the optical limiting were discussed using a model based on effective excited state absorption.

KEYWORDS: Optical limiting; Molecular metal cluster; Excited-state absorption.

1 INTRODUCTION

Photonic devices require that the frequency and intensity of light can be controlled. This encourages people to search and study new materials for the fabrication of optical switching and limiting devices. Particularly, there has been increasing efforts made to exploit effective optical limiters to protect sensors and human eyes against laser threats. Typically, the optical nonlinear properties that can be employed to realize the purpose include two photon absorption (TPA), reverse saturable absorption (RSA), free carrier absorption, nonlinear refraction, and nonlinear scattering (Tutt et al. 1993, Perry et al. 1997).

Over the past decades, organic optical limiting materials such as metallophthalocyanines, metalloporphyrins (Perry et al. 1996, Blau et al. 1985, Li et al. 1994) and fullerenes (Tutt et al. 1992, Li et al. 1997, Sun et al. 1999) have been studied widely. Recently, metal clusters have also been found to be a new kind of optical limiting materials. Their unique inorganic heavy atom, organic ligands and versatile cluster structures make the optimization of optical limiting properties through molecular modification possible. In most studies (Xiong et al. 2000, Song et al. 2001, Chang et al. 2010, Xu et al. 2009) of metal cluster compounds, the origin of optical limiting was reported to be reverse saturable absorption (RSA). But some authors (Xia et al. 1998, Xiong et al. 2000) believed that, nonlinear scattering is responsible for the optical limiting of molecular metal cluster. They

proposed that microplasmas or microbubbles induced by the linear absorption result in the strong scattering of the incident light.

In this paper, we present the results of experimental investigations of optical limiting of a cubic metallic cluster in DMF irradiated by 8-ns and 40-ps laser at a wavelength of 532nm. Experimental results showed that the optical limiting behavior is pulse-dependent. Specifically, 40-ps optical limiting was much better than 8-ns optical limiting, which proves indirectly that optical limiting in the molecular metal cluster is not due to nonlinear scattering. We also found that, the metal cluster can provide better optical limiting than C_{60} in toluene under both nanosecond and picosecond laser pulse excitation. Experimental results were discussed using a simplified equation based on effective excited absorption.

2 EXPERIMENTS

The sample is metal cluster compound $[(n\text{-}Bu)_4N]_3[MoS_4Ag_3I_4]$. It was prepared from the reaction of AgI, $(NH_4)_2MoS_4$ and $n\text{-}Bu_4I$. As shown in Figure.1, the structure of the cluster anion $[MoS_4Ag_3I_4]^{3-}$ can be viewed as a cube in which the four metal atoms and the four non-metal atoms are statistically distributed. The linear absorption spectrum of the metal cluster DMF solution is shown in Figure 2, Two weak absorption peaks can be found around 320nm and 485nm, respectively. While at 532nm, the absorption is not strong.

In optical limiting experiments, linearly polarized 8-ns and 40-ps laser pulses at 532nm provided by a Continuum Np70 ns/ps hybrid Nd: YAG laser system was applied. The laser pulse was adjusted by an attenuator. Then a laser was separated into two beams using a splitter. One was used to monitor the incident laser pulse energy, and the other was focused onto the sample using a lens with a focal length of 308mm. The two beams were detected using two energy detectors linked to energy meter (Model 2835-C, Newport). A computer was used to record and process data. In experiment, laser pulse was chosen to be 1Hz. The linear transmission of the sample is adjusted to be 75%, and the path length of the sample is 5mm. For comparison the same experiment was also conducted in C_{60}.

3 RESULTS AND DISCUSSIONS

8-ns laser optical limiting properties are shown in Figure 3 The transmission curves in Figure. 4 make results more clearly. In Figure. 3, we can see that, the optical limiting of metallic cluster is much stronger than that of C_{60} under nanosecond laser pulse. In Figure 4, we can find that, with the increase of input fluence, the transmission of metallic cluster decreases faster than that of C_{60}. The optical limiting threshold (input fluence at which the transmission decreases to be half of the linear transmission) is 322mJ/cm^2 and 571mJ/cm^2 for the metal cluster and C_{60}, respectively. This further confirms that, the metallic cluster shows strong optical limiting than C_{60} for nanosecond laser radiation at 532nm.

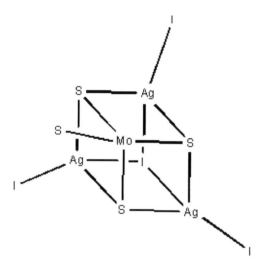

Figure 1. Structure of the cluster anion $[MoS_4Ag_3I_4]^{3-}$.

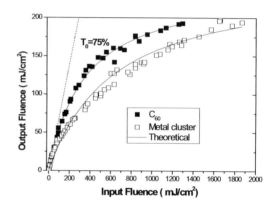

Figure 3. Comparison of optical limiting properties of metal cluster in DMF and C60 in toluene with the same linear transmittance of 75% at 532nm for nanosecond laser pulse.

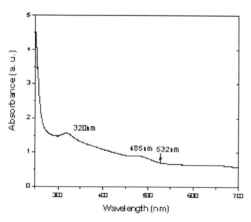

Figure 2. Linear absorption spectrum of metal cluster.

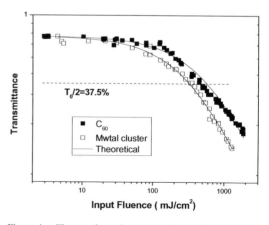

Figure 4. Fluence-dependent transmittance of metal cluster in DMF and C_{60} in toluene with the same linear transmittance of 75% at 532nm for nanosecond laser pulse.

Under 40-ps laser pulses, as shown in Figure 5, the metallic cluster in DMF also exhibits better optical limiting than C_{60}. Additionally, these results can be supported by Figure 6, where the optical limiting threshold is 98mJ/cm2 for the metallic cluster much lower than that for C_{60} with a corresponding value of 155mJ/cm2. Experimental results indicate that the metallic cluster in DMF has a stronger optical limiting capacity than C_{60} for both nanosecond and picosecond laser pulse. Furthermore, the metal cluster exhibits better optical limiting for 40-ps laser pulse than in 8-ns laser pulse.

Many investigations have proved that the optical limiting in metal clusters results mainly from reverse saturable absorption (Xiong et al. 2000, Song et al. 2001, Chang et al. 2010, Xu et al. 2009). However, some researchers (Xia et al. 1998, Xiong et al. 2000) believed that, for metal cluster nonlinear scattering is the main origin of optical limiting. In general, the optical limiting based on nonlinear scattering is pulse-width dependent, and the optical limiting is better for laser with longer pulse-width (Vivien et al. 2001). But, in our experiments, we found the metal cluster shows better optical limiting under relatively shorter laser pulses. So instead of nonlinear scattering, we believe RSA plays an important role in the optical limiting.

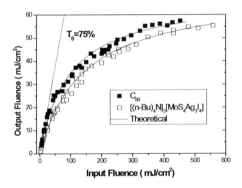

Figure 5. Comparison of optical limiting properties of metal cluster in DMF and C_{60} in toluene with the same linear transmittance of 75% at 532nm for picosecond laser pulse.

Under picosecond excitation, the RSA properties of the metal cluster can be analyzed using a three-level model. However, under nanosecond laser, because intersystem crossing can occur, RSA should be described using five–level system including two triplets and three singlets. In fact, nonlinear absorption can also be analyzed using effective excited-state absorption cross-section σ_{eff}. Thus, the output fluence as a function of input fluence can be written as follows (Sanghadasa et al. 2001):

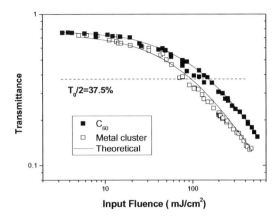

Figure 6. Fluence-dependent transmittance of metal cluster in DMF and C_{60} in toluene with the same linear transmittance of 75% at 532nm for picosecond laser pulse.

$$F_0 = F_i e^{-\alpha_0 L} \Big/ [1 + \frac{F_i \sigma_{eff}}{2hn}(1 - e^{-\alpha_0 L})] \qquad (1)$$

Where F_i is input fluence, F_0 output fluence, L the length of the cell. And α_0 is the linear absorption coefficient and can be expressed as $\alpha_0 = \sigma_0 N$, using ground-state absorption cross-section S_0 and the population density N. σ_0 can be calculated using $T_0 = e^{-\sigma_0 NL}$ after measuring the linear transmission T_0. So σ_{eff} can be used to fit the optical limiting data of the metal cluster and C_{60}. Table 1 lists the ground state absorption cross-section σ_0, effective excited-state absorption cross-section σ_{eff} and σ_{eff}/σ_0 (Ebbesen et al. 1991). The solid curves in Figures 3–6 show a good fit of theoretical value with experimental data.

Table 1. Optical parameters of [(n-Bu)$_4$N]$_3$[MoS$_4$Ag$_3$I$_4$] and C_{60} for nanosecond and picosecond laser pulses.

Sample	Pulse duration	σ_0 $10^{-18}\,cm^2$	σ_{eff} $10^{-18}\,cm^2$	$\dfrac{\sigma_{eff}}{\sigma_0}$
Metal cluster	ns	4.2	26.7	6.3
	ps		28.9	6.8
C_{60}	ns	3.5	13.1	3.7
	ps		18.5	5.2

A figure of merit for RSA can be defined as: σ_{eff}/σ_0 which can also be used as a figure of merit for optical limiting. The larger the value of σ_{eff}/σ_0 is, the better the optical limiting will be. In our experiments, the value of σ_{eff}/σ_0 is 6.3 for nanosecond laser pulses and 6.8 for picosecond laser pulses. While for C_{60} in toluene under nanosecond and picosecond laser pulses, the values are 3.7 and 5.2, respectively. So we can know that, the metal cluster is better optical limiting material than C_{60}. In fact, it is the different for different laser pulses that cause the Pulse-width-dependent optical limiting.

4 CONCLUSIONS

Optical limiting of $[(n\text{-}Bu)_4N]_3[MoS_4Ag_3I_4]$ were experimentally investigated. The investigation indicates that, under both nanosecond and picosecond laser pulses the metal cluster can show better optical limiting than C_{60}. And the optical limiting of metal cluster is pulse-duration-dependent. The results were analyzed using a model based on effective excited state absorption. The ratios of effective excited state absorption cross-section to ground state absorption cross-section were obtained. The study indicates that excited state absorption is a responsible mechanism for optical limiting in the cluster.

ACKNOWLEDGMENT

This work was supported by the National Nature Science Foundation of China(61275117, 61108018, 51372072), Heilongjiang Province Science Foundation (F201112, F201032), and High Level Team Project of Heilongjiang University (Hdtd2010–15).

REFERENCES

Blau, W. & Byrne, H. Dennis, W.M. Kelly, J.M. 1985. Reverse satuable absorption in tetraphenylporphyrins. *Opt. Comm.* 56(1):25–29.

Chang, Q. & Wang, Y. Zheng, H. Song, Y. 2010. Optical limiting properties of $W_2Ag_4S_8(PPh_3)_4$ and $Mo_2Ag_4S_8(PPh_3)_4$. *Physica B: Condensed Matter* 405(17): 3732–3734.

Ebbesen, T.W. & Tanigaki, K. Kuroshima, S. 1991. Excite-state properties of C60. *Chem. Phys. Lett.* 181(6):501–504.

Li, C. & Zhang, L. Yang, M. Wang,H. Wang, Y. 1994. Dynamic and steady-state behaviors of reverse saturable absorption in metallophthalocyanine. *Phys. Rev. A.* 49(2): 1149–1157.

Li, F. & Song, Y. Yang, K. Liu, S. Li, C.1997. Measurements of the triplet state nonlinearity of C60 in toluene using a Z-Scan technique with a nanosecond laser. *Appl. Phys. Lett.* 71(15):2073–2075.

Perry, J.W. 1997. *Nonlinear optics of organic molecules and polymers.* CRC Press: NewYork.

Perry, J.W. & Mansour, K. Lee, L.-Y. Wu, X.-L. Bedworth, P. V. Chen, C. Ng, T. D. Marder, S. R. 1996. Organic Optical Limiter with A Strong Nonlinear Absorptive Response. *Science* 273 (5281):1553–1556.

Sanghadasa, M. & Shin, I.-S. Clark, R.D. Guo, H.S. Penn, B.G. 2001. Optical limiting behavior of octa-decyloxy metallo-phthalocyanines. *J. Appl. Phys.* 90(1):31–37.

Song, Y.L. & Wang, Y.X. Zhang, Q.F. Zhou, F.X. Xin, Q. Ye, H.G. 2001. Excited state nonlinearity and optical limiting reponse of cubane-like shaped metal clusters $WSe_4(MPPh_3)_3Cl$ (M=Ag, Cu). *Mater. Lett.* 51(1): 85–87.

Sun,Y.-P. & Riggs, J.E. 1999. Organic and inorganic optical limiting materials: from fullerenes to nanoparticles. *International reviews in physical chemistry.* 18(1): 43–90.

Tutt, L.W. & Boggess,T. F. 1993. A review of optical limiting mechanisms and devices using organics, fullerenes, semiconductors and other materials. *Prog. Quantum. Electron* 17(4): 299–338.

Tutt, L.W. & Kost, A. 1992. Optical limiting performance of C_{60} and C_{70} solution. *Nature* 356(6366):225–226.

Vivien, L. & Riehl, D. Lancon, Hache, P.F. Anglaret, E. 2001. Pulse duration and wavelength effects on the optical limiting behavior of carbon nanotube suspension. *Opt. Lett.* 26 (4): 223–225.

Xia, T. & Dogariu, A. Mansour, K. Hagan,D.J. Said, A.A. Van Stryland, E. W. Shi, S. 1998. Nonlinear reponse and optical limiting in inorganic metal cluster $Mo_2Ag_4S_8(PPh_3)_4$ solutions. *J. Opt. Soc. Am. B.* 15(5): 1497–1501.

Xu, H. & Song, Y.L. Meng, X.R. Hou, H.W. Tang, M.S. Fan, Y.T. 2009. Strong optical limiting effects of two Ag(I)-bridged metal-organic polymers. *Chem. Phys.* 359: 101–110.

Xiong, Y.N. & Ji, W. Zhang, Q.F. Xin, X.Q. 2000. Optical nonlinearities of inorganic metal cluster $\mu_3\text{-}MoSe_4Ag_3(PPh_3)_3Cl$. *J. Appl. Phys.* 88(3): 1225–1229.

Xiong, Y. & Zhang, Q. Sun, X. Tan, W. Xin, X. Ji, W. 2000. Nonlinear optical response of inorganic metal cluster $MoCu_3Se_4(PPh_3)_3Cl$ solution. *Appl. Phys. A: Mater.* 70(1):85–88.

Electronic Engineering and Information Science – Wang (Ed.)
© 2015 Taylor & Francis Group, London, ISBN: 978-1-138-02772-5

Investigation on preparation and refining effect of Al-5Ti-1B-xCe master alloys

F.W. Kang, Q. Liu, Z.Z. Yu, B. Gao & H.L. Geng
School of Materials Science and Engineering, Harbin University of Science and Technology, China

Q. Zhang
Northeast Light Alloys Co., Ltd, Harbin, China

ABSTRACT: Al-5Ti-1B-xCe master alloys were prepared by melting reaction method, and its refining effect was examined with purity aluminum in this paper. The microstructural evolutions of Al-5Ti-1B-xCe master alloys were analyzed by optical microscopy, XRD and SEM. The results indicated that the $Ti_2Al_{20}Ce$ phase was achieved due to adding Ce element, and with the increase of Ce element from 1.5% to 2.0%, the $TiAl_3$ phase decreased and $Ti_2Al_{20}Ce$ phase was more and distributed uniformly in the matrix. The effect of refining pure Al by Al-5Ti-1B-2.0Ce master alloy was better than that by Al-5Ti-1B-1.5Ce for an amount of 1.0% (wt.%) refiner and held time for 30min at 750°C.

KEYWORDS: Al-5Ti-1B-xCe master alloy; refining effect; microstructure.

1 INTRODUCTION

To improve mechanical properties and workability of aluminum and its alloys, it is now an effective method to add grain refiners to aluminum melt. The function of grain refinement is to control the grain size and grain morphology during casting and extrusion for aluminum alloys, so that good mechanical properties and stable workability can be achieved. At present the purity and its alloys are often refined by the Al-Ti-B master alloys, which are very effective as the refiners of Al alloys, but they also have some problems during almost the half century application (Birol, 2006). One of them is that agglomeration and coalescence of the borides, blockage of filters in continuous refinement, defects during subsequent forming operations and poisoning by certain elements such as Zr, Cr, or V (Ding, Liu and Yu 2007, Bunn , Schumacher, Kearns and Boothroyd1999).

It is well known that rare-earth elements possess many advantages with modification, cleaning for molten alloys, and so on. In recent years, some researchers proposed a novel Al-Ti-B-RE grain refiner which combined the advantages of Al-Ti-B grain refiner and rare earth element to overcome the above-mentioned the shortcomings of Al-Ti-B grain refiner (Wang, Xu and Muhammad submitted2014, Wang, Xu, Kang, Tang and Zhang2014).

In this paper, the Al-5Ti-1B-xCe (x=1.5, 2.0) master alloys were prepared by adding Al-20Ce alloy into the made-home Al-5Ti-1B alloy by the melt reaction process. The microstructure characteristics and refining effect of Al-5Ti-1B-xCe were investigated by an Olympus optical microscope (OM), X-ray diffraction (XRD), scanning electron microscope (SEM).

2 EXPERIMENTAL PROCEDURES

The raw materials included purity aluminum (99.7%), K2TiF6, KBF4. The Ce element was added in the form of Al-20Ce intermediate alloy. The purity Al was melted in the graphite crucible by electrical furnace. The preheated K2TiF6 and KBF4 were added into the purity Al molten and stirred manually by a graphite rod. After 10min, the crucible loading the molten alloy and slag was taken out of the electric furnace to separate the slag from the melt. The different quantity Al-20Ce alloys were added into the Al molten and the crucible was then put back into the electrical furnace to keep the temperature of the melt, being held for 10min, and then poured Al molten into a steel mold with a diameter of 60mm and height of 200mm. The addition amount of Al-20Ce was 0, 1.5%, 2.0% (Wt.%), respectively. The prepared Al-5Ti-1B-xCe ingots were extruded to a diameter of 9.5mm rod by an extrusion machine. The refining effects of the rod Al-5Ti-1B-xCe master alloys were examined by purity Al.

To reveal the microstructures of Al-5Ti-1B-xCe ingots, the samples were polished and etch by the Keller's reagent and then analyzed by an Olympus GX-71–6230A optical microscope and an FEI Sirion SEM equipped with an energy dispersive system. X'Pert PRO XRD was employed to identify the phases present in the ingots.

3 RESULTS AND DISCUSSION

3.1 *Microstructure characteristics of different Ce content*

The chemical composition of the prepared Al-5Ti-1B-xCe master alloys is shown in table 1. Fig.1 shows the features of the microstructure at different Ce element contents, 0%, 1.5%, 2.0%, respectively. It is obvious that only TiAl3 phases are observed without adding Ce element in Fig.1 (a), with the increase of Ce element content, more and finer white light particles were examined by optical microscopy in Fig.1 (b) and (c), especially when the content of Ce element increases from 1.5 to 2.0%. These white particles should be composed of TiAl3 and Ti2Al20Ce.

Table 1. Chemical composition of Al-5Ti-1B-xCe master alloys (Wt.%).

Master alloy	Ti	B	Ce
Al-5Ti-1B	4.81	0.95	0
Al-5Ti-1B-1.5Ce	4.77	0.98	1.12
Al-5Ti-1B-2Ce	4.75	0.96	1.63

Fig. 2 shows the XRD pattern obtained from produced.= Al-5Ti-1B-xCe master alloys. In Fig.2 (a), due to no adding a rare earth element Ce, only TiAl3, TiB2 phases were found in addition to α-Al matrix. Fig.2 (b) and (c) shown that the Ti2Al20Ce phase was formed after adding rare earth elements Ce. Indexing of the XRD peaks had shown that with the increasing of Ce element content, the amount of Ti2Al20Ce phase increases in Fig.2 (b) and (c).

The SEM micrographs and energy spectrum analysis of Al-5Ti-1B-xCe master alloys are illustrated in Fig.3 (a) and (b). As seen from Fig.3, with the increase of Ce content, the Ti2Al20Ce particles are finer and more homogeneous distribution in the matrix. However, TiAl3 particles decreased because TiAl3 and Ce generated Ti2Al20Ce by the response equation (1) as follows:

$$TiAl_3 + Ce \rightarrow Ti_2Al_{20}Ce \qquad (1)$$

(a)0%Ce

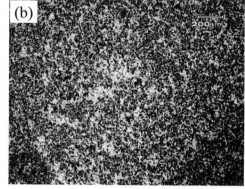

(b)1.5% Ce

(c)2.0% Ce

Figure 1. Optical micrographs of Al-5Ti-1B-xCe alloy.

3.2 *Evaluation refining effect*

Commercial-purity aluminum alloy was refined by Al-5Ti-1B-1.5Ce and Al-5Ti-1B-2.0Ce master alloys at the same testing conditions, respectively. The refining effects were evaluated by average grain size, as shown in Fig. 4 and Fig.5. The results shown that

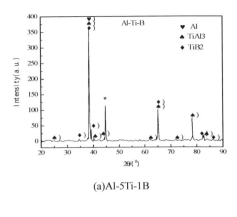

(a)Al-5Ti-1B

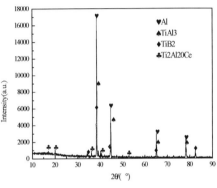

(b) Al-5Ti-1B-1.5Ce

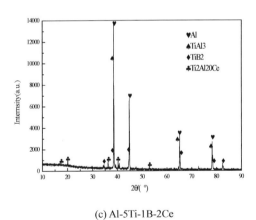

(c) Al-5Ti-1B-2Ce

Figure 2.   XRD pattern of the produced Al-5Ti-1B-xCe master alloys.

finer grain size was obtained by Al-5Ti-1B-2.0Ce master alloy at 750°C for holding 30min, amount of refiner 1.0%wt. The reason was that Al-5Ti-1B-2.0Ce master alloy possessed more Ti2Al20Ce phase than

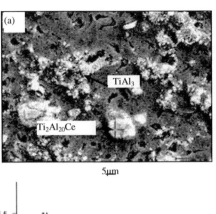

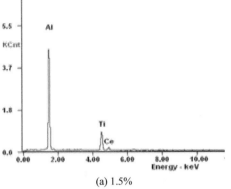

(a) 1.5%

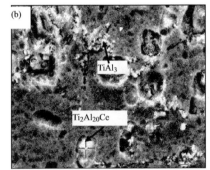

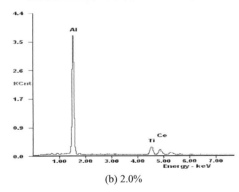

(b) 2.0%

Figure 3.   SEM images and EDS analysis of different Ce content Al-5Ti-1B-xCe alloy.

705

Al-5Ti-1B-1.5Ce master alloy, the Ti2Al20Ce phase improved refining performance.

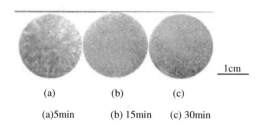

(a)   (b)   (c)

(a)5min   (b) 15min   (c) 30min

Figure 4.   Effect of different holding time on microstructures of purity Al at 750°C by 1.0%wt. Al-5Ti-1B-1.5Ce master alloy.

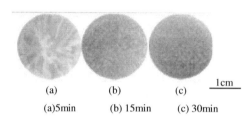

(a)   (b)   (c)

(a)5min   (b) 15min   (c) 30min

Figure 5.   Effect of different holding time on microstructures of purity Al at 750°C by 1.0%wt. Al-5Ti-1B-2.0Ce master alloy.

4   SUMMARY

Al-5Ti-1B-1.5Ce and Al-5Ti-1B-2.0Ce master alloys grain refiners were prepared by the melt reaction method. The master alloys were composed of α-Al, Al3Ti, TiB2 and Ti2Al20Ce phases. With the increase of Ce element from 1.5% to 2.0% wt., the amount of Ti2Al20Ce phase increased and TiAl3 phase decreased. The refining effect of Al-5Ti-1B-2.0Ce master alloy was superior to Al-5Ti-1B-1.5Ce at 750°C, addition of 1.0%wt. for holding 30min.

ACKNOWLEDGMENT

This research was financially supported by Undergraduate Training Programs of Innovation and Entrepreneurship of Harbin University of Science and Technology (2013).

REFERENCES

Birol, Y. (2006): submitted *Journal of Alloys and Compounds*.
Ding, H.M, X.F.Liu and L.N.Yu(2007): submitted *Journal of Materials and Science*.
Bunn, A.M., P.Schumacher, M.A.Kearns and C.B.Boothroyd(1999): submitted *Materials and Science of Technolog.y*.
Wang, X.J. ,C.Xu and A.Muhammad submitted(2014) Transactions of Nonferrous *Metals Society of China*.
Wang, H.J, J.Xu, Y.L.Kang, M.O.Tang and Z.F.Zhang(2014): submitted *Journal of Materials Engineering and Performance*.

Electronic Engineering and Information Science – Wang (Ed.)
© *2015 Taylor & Francis Group, London, ISBN: 978-1-138-02772-5*

Physics-chemistry mechanism discussion of preparing HAP with sol-gel method

J. Wei, J. Han, F.X. Li & S. Li
School of Applied Sciences, Harbin University of Science and Technology

ABSTRACT: HAP powder was prepared from $Ca(NO_3)_2 \cdot 4H_2O$ and P_2O_5 with Sol-gel method. In the first, we gave the sol-gel process. Secondly, we make use of analytical methods of FTIR and XRD to analysis the Hydroxyapatite precursor powder and the Hydroxyapatite precursor powders calcined at different temperature. At last, the chemical mechanism of sol-gel synthesizing Hydroxyapatite were investigated through analyzing the ethanol solution of P_2O_5 and $Ca(NO_3)_2 \cdot 4H_2O$.

KEYWORDS: HAP; Sol-gel; Physics-Chemistry Mechanism.

1 INTRODUCTION

Sol-gel method is a common method to prepare ultrafine ceramic powders and thin films. In this process, we use liquid chemical reagents as raw material, mix them together and let them react in the liquid phase to generate the sol system, then the products you need can be obtained by drying, dehydrating and sintering. Compared with the traditional process, the sol-gel method has many advantages: first, the synthesis temperature is low, materials which some traditional methods are hard to get or not at all can be obtained by this method; Secondly, the material particles of preparation are small and uniform, the proportion of each component is easy to control and nanocomposites can be prepared with the sole-gel method; In addition, the equipment is simple, technology is flexible and products have high purity.

Based on the various advantages of sol-gel method, people have been applying this method to prepare HAP. Layrolle (Layrolle, Lto and Tateishi 1998, Layrolle and Lebugle 1994) et al. prepared HAP with $Ca(OEt)_2$ and H_3PO_4, Cameron S.Chai (Chai, Gross, and Ben-Nissan, 1998) et al. used triethyl phosphate and $Ca(OEt)_2$ as a precursor for the preparation of HAP; Wenjian Wen(Wen and Baptista 1998) et al. prepared HAP by mixing Ethylene glycol calcium and P_2O_5 in anhydrous ethanol; Wenjian Wen prepared HAP by mixing $Ca(NO_3)_2 \cdot 4H_2O$ and P_2O_5 in anhydrous ethanol; Yunjing Song et al. improved the preparation process on the basis of the predecessors, which used lower-cost $Ca(NO_3)_2 \cdot 4H_2O$ and P_2O_5 in anhydrous ethanol as raw materials and prepared HAP powder in a relatively short period; Recently, Il-Seok

Kimet al. reported that the products of HAP powder had better performance using a similar method.

In this study, HAP precursors were prepared by mixing $Ca(NO_3)_2 \cdot 4H_2O$ and P_2O_5 in anhydrous ethanol with Sol-gel method, and HAP powders were obtained through the different heat-treated temperature. Then, HAP precursors and HAP powder samples were carried out test and analysis by FTIR and XRD, chemical synthesis mechanism in which HAP powders synthesized by the sol-gel method were discussed.

2 EXPERIMENTAL STUDIES

2.1 Selection of raw material

Select pure $Ca(NO_3)_2 \cdot 4H_2O$ (≥99.0%, Tianjin bodi Chemicals Co., GmbH), P_2O_5 (≥98.0%, Shanghai Ling Feng Chemical Reagent Co., GmbH) and anhydrous ethanol CH_3CH_2OH(≥99.7%, Sales center of Tianjin City North glassy procurement).

2.2 The preparation of nano-hydroxyapatite material

To configure the solution whose $Ca(NO_3)_2 \cdot 4H_2O$ and P_2O_5 was in the molar ration that Ca/P is 1.67, P_2O_5 was slowly added into the $Ca(NO_3)_2 \cdot 4H_2O$ under the condition of 80°C temperature bath, and kept stirring. When appeared white gel, kept stirring for 10 min, and then stayed the stabel sol in the 80°C temperature bath for 30 min. Gelation is aged and dries at 100°Cfor 24h. After grinding, hydroxyapatite precursor was getting and then hydroxyapatite powder can be synthesized under different temperature and holding time.

2.3 *Analysis and test of HAP*

Domestic X-ray diffraction (XRD) at the scanning speed of 0.03⁰/S, CuKα and Ni filter are utilized to complete XRD test for the samples (accelerating voltage is 40Kv, tube current is 50A), in which purity of crystallinity and HAP phase are analyzed. All kinds of sol-gel solution and functional group of powder calcined at different temperature are analyzed with AVATAR 370 Fourier transform infrared spectrometer (FT-IR), made by Thermo Nicolet Corporation, to analyze the chemical reaction mechanism of the sol-gel process.

3 RESULT AND ANALYSIS

In order to discuss the chemical mechanism of the preparation of HAP powders in our experiment, we get a mixture after adding an alcohol solution of P_2O_5 into an alcohol solution of CA $(NO_3)_2$ and stirring for 1h, and HAP powder samples through calcining at different heat-treated temperature (200, 500 and 600°C), then the mixture, precursor and samples are analyzed by FTIR, the precursor and samples are analyzed by XRD.

3.1 *XRD analysis*

XRD pattern of precursor powder of HAP is shown in Fig. 1, XRD pattern of HAP sample after heat-treated at different temperature for 1h are shown in Fig. 2 and Fig. 3, respectively. According to Fig. 1, HAP precursor powder is mainly composed of $CaHPO_4$, organic amorphous phase, as well as a small amount of $Ca(NO_3)_2$. Fig. 2 shows that $CaHPO_4$ and $Ca(NO_3)_2$ content in powders increase through heat-treated at 200°C, meanwhile, HAP phase appears. However, Fig. 2 shows that $CaHPO_4$ content reduces significantly, $Ca(NO_3)_2$ drop and HAP increase. From Fig. 3, we know that there is no $CaHPO_4$ in a powder sample after heat-treated at 500°C and 600°C, the sample is mainly composed of HAP and contains trace amounts of $Ca(NO_3)_2$, the content of $Ca(NO_3)_2$ decrease with increasing temperature, there is almost no $Ca(NO_3)_2$ under 600°C.

3.2 *FTIR analysis*

Infrared spectra of HAP powder after heat-treated at different temperature for 1h are shown in Fig. 4–6.

In Fig.4 and 5, acromions at 1100 cm⁻¹, 1040 cm⁻¹ (v(P-O)), 565 cm⁻¹(δ(O-P-O)) and 934 cm⁻¹(v(P-O-P)) are caused by HAP, sharp absorption peak at

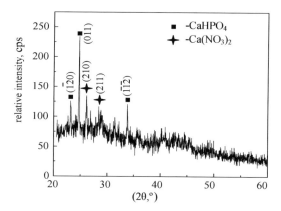

Figure 1.   XRD pattern of precursor powder of HAP.

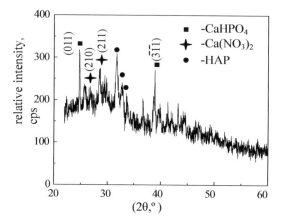

Figure 2.   XRD pattern of HAP sample after heat-treated at 200°C for 1h.

380 cm⁻¹ belongs to NO_3^-. Acromions (v(P-O-P)) of three samples(A, B, C) at 602 cm⁻¹ and 565 cm⁻¹ (δ(O-P-O)), 1040 and 1100 cm⁻¹ (v(P-O)), 959 cm⁻¹ are caused by HAP in the Fig. 6, Sharp absorption peaks in A and B at 1380 cm⁻¹ are caused by NO_3^-, there is absorption peaks caused by NO_3^- in C.

3.3 *Analysis of physical-chemical mechanism of HAP powder*

For the results of XRD and FTIR analysis, HAP powder is prepared by sol-gel method under different heat-treated temperature, and chemical rea-ctions in this process are as follows:

The reactions of P_2O_5 with ethanol medium are: $P_2O_5 + H_2O \rightarrow H_3PO_4$, $H_3PO_4 + CH_3CH_2OH \rightarrow PO_m(OH)_n(OEt)_o + H_2O$ ($1 \le m \le 3$, n,o≤ 3).

The reactions are in the mixture of P_2O_5 and $Ca(NO_3)_2$: $H_3PO_4 + Ca(NO_3)_2 \rightarrow CaHPO_4$ (the pH value of test solution is located in the range of 1~3

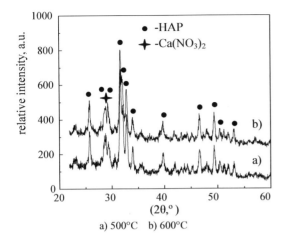

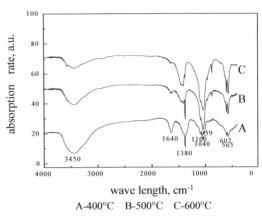

Figure 3. XRD patterns of HAP sample after heat-treated at different temperature for 1 h.

a) 500°C b) 600°C

Figure 6. Infrared spectra of HAP powder after heat-treated at different temperature for 1 h.

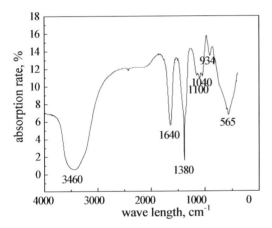

Figure 4. Infrared spectra of HAP powder after heat-treated at 200°C for 1 h.

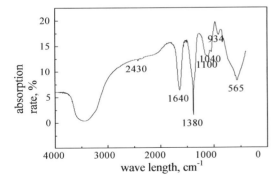

Figure 5. Infrared spectra of HAP powder after heat-treated at 300°C for 1 h.

during the tria) + HNO_3 (NO_2+H_2O) (With high viscosity, mixed solution overflows the tan acid gas), H_3PO_4 or $CaHPO_4 + Ca(NO_3)_2 \rightarrow$ ACP + others, $xCa(NO_3)_2 + yPO_m(OH)_n(OEt)_o \rightarrow xCa(NO_3)_2 \cdot y[PO_m(OH)_n(OEt)_o]$ (x>y>0).

Precursor powders which dried at 100°C for 24h and heat-treated at 200°C for 1h can be obtained by the following reaction: $xCa(NO_3)_2 \cdot y[PO_m(OH)_n(OEt)_o]$ + O_2 + $Ca(NO_3)_2$ + ACP→ HAP + $Ca(NO_3)_2$ + $CaHPO_4 + H_2O + NO_2$. This equation corresponds to the previous Fig. 1 and Fig. 2, that is, corresponds to increasing the content of $CaHPO_4$ after precursor powders undergo heat-treated at 200°C for 1h.

The reactions of powders at 500°C-600°C can be written as follows: $Ca(NO_3)_2 \rightarrow CaO + NO_2$, CaO + HAP→HAP(nonstoichiometric ratio). When temper-ature reaches 600°C, $Ca(NO_3)_2$ is not included in powder, as shown in the Fig. 3 and Fig. 6C. With the relevant data, $Ca(NO_3)_2$ is fully decomposed in the range of 500 to 600°C.

In conclusion, The chemical synthesis mechanism of preparation of HAP powder with sol-gel can be summarized as follows:

a) $P_2O_5 + H_2O \rightarrow H_3PO_4$;
b) $H_3PO_4 + CH_3CH_2OH \rightarrow PO_m(OH)_n(OEt)_o + H_2O$ ($1 \leq m \leq 3$, n, $o \leq 3$);
c) $H_3PO_4 + Ca(NO_3)_2 \rightarrow CaHPO_4 + HNO_3(NO_2+H_2O)$;
d) H_3PO_4 or $CaHPO_4 + Ca(NO_3)_2 \rightarrow$ ACP + others;
e) $xCa(NO_3)_2 + yPO_m(OH)_n(OEt)_o \rightarrow xCa(NO_3)_2 \cdot y[PO_m(OH)_n(OEt)_o](x>y>0)$;
f) $xCa(NO_3)_2 \cdot y[PO_m(OH)_n(OEt)_o] + O_2 + Ca(NO_3)_2$ + ACP → HAP + $Ca(NO_3)_2 + CaHPO_4 + H_2O + NO_2$ (heat-treated temperature≤200°C);
g) $Ca(NO_3)_2 \rightarrow CaO + NO_2$ (500°C≤heat-treated temperature≤600°C);
h) CaO + HAP → nonstoichiometric ratio HAP (heat-treated temperature 500°C).

4 CONCLUSION

1 Using $Ca(NO_3)_2$ and P_2O_5 as raw material, HAP powder is prepared by sol-gel method.
2 According to XRD analysis, chemical composition of the preparation of powders vary with heat-treated temperature, soaking time is 1h between the anneal-ing temperatures from 400°C to 600°C, we get high purity of HAP powders.
3 FTIR analysis demonstrates how the functional Group of powders after different heat-treated is changing.
4 Based on XRD and FTIR analysis, we get physical-chemical reaction mechanism and related physical-chemical reaction equations for the preparation of HAP powders under different heat-treated temperature.

ACKNOWLEDGMENTS

This research was funded by the National Natural Science Foundation of China (No. 51101046), Natural Science Foundation of Heilongjiang Province (No.A200811) and by the Educational Department Scientific Technology Program of Heilongjiang (No. 11521045).

REFERENCES

Zhen, C.Q. & J.G.Ran, (2003). *New type of inorganic materials*, Beijing: Changqiong Zhen science.
Layrolle, P., A.Lto, & T. Tateishi (1998). Sol-Gel synthesis of amorphous calcium phosphate and sintering into micropor-ous hydroxyapatite bioceramics. *J.Am. Ceram Soc.* 81(6), 1421–1428.
Layrolle, P. & A.Lebugle, (1994). Characterization and reactivi-ty of nanosized calcium phosphates prepared in anhudrous ethanol. *Chem. Mater.* 6,1996–2004.
Chai, C.S., Gross, K.A.& Ben-Nissan, B. 1998. Critical ageing of hydroxyapatite sol-gel solution. *Biomaterials.*19, 2291–2296.
Wen, W.J. & J.L.Baptista,(1998). Alkoxide route for preparing hydroxyapatite and its coating. *Biomaterials.* 19, 125–131.

Electronic Engineering and Information Science – Wang (Ed.)
© *2015 Taylor & Francis Group, London, ISBN: 978-1-138-02772-5*

Emulsion dehydration characteristics under non-uniform electric field

C.H. Song, Q.G. Chen, W. Liang, T.Y. Zheng & X.L. Wei
Key Laboratory of Engineering Dielectrics and Its Application, MOE, Harbin University of Science and Technology,
Harbin, China

ABSTRACT: In order to research emulsion dehydration characteristics under non-uniform electric field, the experimental study on the emulsion dehydration was carried out under non-uniform electric field with coaxial cylindrical electrode structure. Meanwhile, the research works on the characteristics of droplet deformation and motion behavior were also carried out by analyzing the dielectrophoresis and electrodispersion, as well as observing the migration of water droplets. The results show that the dielectrophoresis force is proportional to electric field intensity, water droplet radius and non-uniform coefficient of electric field. Within a certain range, the improvement of non-uniform coefficient can increase the dehydration rate and speed of the emulsion. However, when the non-uniform coefficient is too high, the droplet will be ruptured and the dehydration rate will decrease. The electrodispersion for water droplets is related to surface charge density, droplet size and surface tension. When the droplet is charged, its electric intensity of rupture will decrease.

KEYWORDS: Emulsion; Electric dehydration; Dielectrophoresis; Electrodispersion; Non-uniform.

1 INTRODUCTION

Dehydration of crude oil is an essential process in the petroleum industries. According to the different types of applied electric field, electric dehydration can be divided into uniform and non-uniform electric dehydration. At present, in the study of the emulsion dehydration, many works have been paid to the dehydration mechanism and motion characteristics of water droplets under uniform electric field (Chen et al. 2014, Eow & Ghadiri, 2003). However, fewer efforts have been spent studying the experiments and motion behaviors under the non-uniform coefficient of the electric field (Alinezhad et al. 2010, Wei et al. 2013, Sameer et al. 2013).

In order to research emulsion electric dehydration characteristics under non-uniform electric field, the experimental study on the electric field intensity and non-uniform coefficient was carried out under non-uniform electric field. Meanwhile, the characteristics of emulsion dehydration are obtained through analyzing the dielectrophoresis force and electrodispersion, as well as observing the motion behavior of water droplets under non-uniform electric field. The dehydration mechanisms under non-uniform field were revealed in theory, which can provide theoretical guidance for improving the efficiency of electric dehydration.

2 EXPERIMENTAL SYSTEM AND METHODS

The experimental system for emulsion dehydration is shown in Figure 1, which consists of mixing tank, lab Homogenizer, metering pump, electric dehydrator, high-voltage power supply, waste liquid tank, high-speed camera and computer.

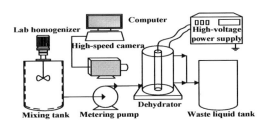

Figure 1. Emulsion dehydration test system.

The mixing tank consists of insulated containers and emulsification machine. The dehydrator is composed of a transparent cylinder with acrylic material and steel tubes. Dehydration power is provided by high-voltage power supply device with the voltage amplitude from 0 to 15 kV, and frequency 50 Hz. The experiment time is set to 30 minutes. The phenomenon of emulsion dehydration and motion of water droplets are recorded by high-speed camera, which is connected to computer. The water content of the emulsion is measured by Karl Fischer Moisture Titrator.

In the dehydration test system, steel tubes are used as concentric cylindrical dehydration electrodes, the non-uniform coefficient can be expressed as

$$f = \frac{E_{max}}{E_{av}} = \frac{r_2 - r_1}{r_1} \cdot \frac{1}{\ln(r_2 / r_1)} \quad (1)$$

Where, r_1 and r_2 are the radius of internal electrode and external electrode respectively, E_{max} and E_{av} are the maximum and average electric field intensity respectively, f is the non-uniform coefficient.

In order to obtain the stable emulsion with the water content of 10 %, industrial mineral oil and distilled water are mixed at the volume ratio of 9:1 in the mixing tank with the 0.01 volume fraction of Span-80 emulsifier added. The mixture is agitated by the lab Homogenizer with the speed of 6000 rad/min. Finally, the experiment begins after the emulsion is pumped into the electric dehydrator. Dehydration rate is calculated by $\eta = (P_0 - P_t)/P_0$, P_0 is the initial water content of emulsion, P_t is the water content of emulsion at time t.

It can be seen that concentric cylindrical electrode is a highly non-uniform electric field with the condition of $r_1/r_2 < 1/e$, slightly uneven electric field with the condition of $r_1/r_2 > 1/e$. Four types of electrodes used during the test in this paper as shown in Table 1.

Table 1. Electrode size and non-uniform coefficient.

Electrode type	Internal electrode radius mm	External electrode radius mm	Non-uniform coefficient	Electric field type
f_1	6	25	2.22	Highly non-uniform
f_2	6	20	1.94	Highly non-uniform
f_3	6	15	1.64	Slightly uneven
f_4	6	10	1.31	Slightly uneven

3 EXPERIMENTAL RESULTS AND ANALYSIS

3.1 Electric dehydration in the non-uniform electric field

This section studies the influence of electric field intensity and non-uniform coefficient of the electric field on the separation efficiency of emulsion.

Emulsion dehydration rate changes under different average electric field intensities and non-uniform coefficients are shown in Figure 2.

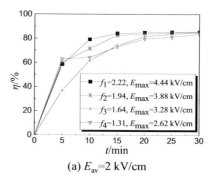

(a) E_{av}=2 kV/cm

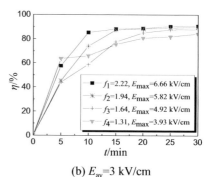

(b) E_{av}=3 kV/cm

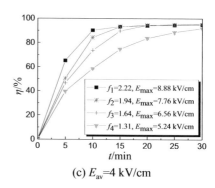

(c) E_{av}=4 kV/cm

Figure 2. Emulsion dehydration rate under different average electric field intensities and non-uniform intensities.

In order to analyze the separation efficiency of emulsion under different average electric field intensities and non-uniform coefficients, the consuming time t_{80} (when the dehydration rate reaching 80%) and finial dehydration rate P_{30} (when dehydration time reaching 30 minutes) are concluded as shown in Table 2.

From the results in Figure 2, it can be seen that the increase rate of dehydration slows down gradually in the dehydration process. The reason is that with the

Table 2. The consuming time t80 and dehydration rate P30.

Electrode type	The consuming time t_{80} min			The finial dehydration rate P_{30} %		
	2 kV/cm	3 kV/cm	4 kV/cm	2 kV/cm	3 kV/cm	4 kV/cm
f_1	10	6	6	1.465	0.844	0.497
f_2	12	10	9	1.470	0.975	0.463
f_3	18	15	10	1.476	0.985	0.371
f_4	20	18	16	1.617	1.527	1.233

decrease of droplets in number and size, the collision frequency in a specific time decreases, which causes the decrease of dehydration rate.

Considering the Table 2, it is clear that the separation efficiency and dehydration rate is proportional to electric field intensity and non-uniform coefficient of electric field. This is due to the voltage gradient and greater accumulation of electric flux in the center, which have a great effect on the separation efficiency. Nevertheless, it is noticeable that in very high electric field intensity and non-uniform coefficient, the reverse effect happens, which results in drop breakage. By comparing the change of water content, the influence of electric field intensity of dehydration rate is more obvious than the effect of non-uniform coefficient under the electrode type of higher non-uniform coefficient.

3.2 The dielectrophoresis force of water droplet in non-uniform electric field

Droplets are polarized in non-uniform electric field. There are the same quality positive and negative polarization charges at both ends of water droplets. The electric field intensities of both ends of water droplets are different, so that the resultant force of electric forces is formed. The resultant force can be expressed as

$$F_{DEP} = q.d \frac{\partial E(l)}{\partial r} = \mu.\Delta E(l) \qquad (2)$$

Where, μ is the dipole moment of water droplet, $E(l)$ is the electric field intensity at the field point l, $\nabla E(l)$ is the electric field gradient of field point l, l is the distance between droplet and the center of electrode, d is the diameter of droplet, q is the polarization charge.

Water droplet is polarized to regard as induced dipoles in non-uniform electric field, according to the theory of eccentric electric dipole, the center of the induced dipole will no longer be in the geometric center of the droplets, the center of water droplets will deviate towards the direction of electric field intensity, and produce eccentricity (Ye et al. 2004). The dipole moment of water droplets could be calculated by the model of eccentric electric dipole under the

non-uniform electric field which consists of concentric cylindrical electrodes. The force of dielectrophoresis can be expressed as

$$F_{DEP} = -64\pi\varepsilon_m R^3 K_e f^2 E_{av}^2 r_1^2 \frac{1}{l^2 (\sqrt[3]{l} + \sqrt[3]{l+d})^3} e \qquad (3)$$

Where, R is the radius of droplet, e is the direction of the maximum electric field gradient from $E(l)$ to $E(l+d)$, K_e is written as (Alinezhad et al. 2010):

$$K_e = \frac{\varepsilon_w - \varepsilon_m}{\varepsilon_w + 2\varepsilon_m}, f_0 > f_{wm}; K_e = \frac{k_w - k_m}{k_w + 2k_m}, f_0 < f_{wm} \qquad (4)$$

Where, subscripts w and m are particle and medium respectively. ε stands for the permittivity, k stands for the conductivity, f_0 stands for the frequency of the electric field. The f_{wm} refers to the critical frequency (the so-called Maxwell-Wagner) that differentiates the conductivity regime from the permittivity regime, defined as:

$$f_{wm} = \frac{k_w + 2k_m}{2\pi(\varepsilon_w + 2\varepsilon_m)} \qquad (5)$$

It can be seen that dielectrophoresis is closely related to the average electric field intensity, water droplet radius, electrode size, non-uniform coefficient of electric field and position of water droplet under non-uniform electric field of cylindrical electrode. The dielectrophoresis force increases by increasing the average electric field strength, droplet diameter, electric field non-uniformity coefficient or electrode size or decreasing the distance between the droplet and the center of the electrode. The bigger dielectrophoresis force could make the water droplet aggregate faster near the concentrated area of electric field, which becomes more facilitated to increase the collision rate of the droplets and makes the coalescence efficiency higher. Hence, the dehydration rate increases gradually with the increase of the field intensity and non-uniform coefficient in the experiment.

713

3.3 Electric dispersion of a droplet in non-uniform electric field

When droplets are polarized in the electric field, the polarization charges at both ends of water droplets will be generated, and droplets will be stretched into the shape of ellipsoid by the electric field force. Water droplets will have a tendency to be ruptured with the increase of electric field force. Within the certain electric field, the water droplets cannot be ruptured because the additional pressure caused by interfacial tension of water droplet prevents the water droplet deformation. But electric dispersion will occur when the electric field force per unit area exceeds the additional pressure (Wang et al. 2009), which can be expressed as

$$f_e = \frac{dF}{dS} > P \qquad (6)$$

Where $P = \gamma(R_{11}^{-1}+R_{12}^{-1})$ and $f_e = \sigma E$, F stands for the electric field force, S stands for the surface area, P is the additional pressure, f_e is the electric field force per unit area of the droplet, σ is the surface charge density, E is the maximum electric field intensity, R_{11} and R_{12} is the main curvature radius of water droplet, γ is the surface tension of water droplet.

The surface charge density equals the polarized charge density for the neutral water droplet under non-uniform electric field. Hence, the critical electric field intensity of droplet breakage can be expressed as

$$E_{ed} \geq \frac{\gamma}{\sigma_p}(\frac{1}{R_{11}}+\frac{1}{R_{12}}) \qquad (7)$$

Where σ_p is the surface density of polarization charge, E_{ed} is the critical electric field intensity of droplet breakage.

Water droplet is charged when it contacts with the plate. The surface charge of water droplet is redistributed which lead to the decrease of surface tension of water droplet (Wang et al. 2013). After the water droplet is charged, the critical electric field intensity of droplet breakage can be expressed as

$$E_{ed} \geq \frac{1}{\sigma_p+\sigma_e}(\gamma-\Delta\gamma)(\frac{1}{R_{11}}+\frac{1}{R_{12}}) \qquad (8)$$

Where $\sigma_e = Q/(4\pi R^2)$ and $\Delta\gamma = Q^2/(64\varepsilon_w\pi^2R^3)$, σ_e is the surface density of charge that water droplet is injected, $\Delta\gamma$ stands for the surface tension variation, Q is the charge that water droplet is injected.

By the comparison between equation (7) and equation (8), it can be seen that the critical electric field intensity of rupture will reduce after the water droplet is charged. The higher electric field intensity could generate the more charge on both sides of the droplet, which will be easier to lead to the electric dispersion in the process of dehydration. Therefore the dehydration rate of emulsion reduces in very high electric field intensity and non-uniform coefficient.

3.4 The motion behavior of droplets in non-uniform electric field

The motion behaviors of water droplet are shown in Figure 3. During the test, the electric dehydrator is filled with white oil, and the water droplets with a certain radius are injected into the dehydrator between the two electrodes by micro syringe. The motion of water droplets is recorded by high-speed camera at a speed of 100 frames per second, which is connected to the computer.

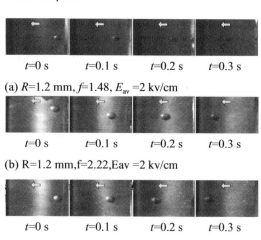

$t=0$ s $t=0.1$ s $t=0.2$ s $t=0.3$ s

(a) $R=1.2$ mm, $f=1.48$, $E_{av} =2$ kv/cm

$t=0$ s $t=0.1$ s $t=0.2$ s $t=0.3$ s

(b) R=1.2 mm, f=2.22, Eav =2 kv/cm

$t=0$ s $t=0.1$ s $t=0.2$ s $t=0.3$ s

(c) R=1.2 mm, f=2.22, Eav =3 kv/cm

$t=0$ s $t=0.1$ s $t=0.2$ s $t=0.3$ s

(d) $R=1.5$ mm, $f=2.22$, $E_{av} =3$ kv/cm

Figure 3. Water droplet motion under different electric field intensities and non-uniform intensities.

The increase of the non-uniform coefficient and the decrease of distance between droplet and electrode make the motion velocity and the accelerated velocity of water droplet faster, which are shown in Figure 3 (a)-(b). As shown in Figure 3 (b)-3(c), the motion velocity and the accelerated velocity of water droplet also increases with the average electric field intensity. As shown in Figure 3 (c)-3(d), it can be seen that water droplet with the biggest radius moves faster than that with a smaller radius.

In order to observe the phenomenon of electric dispersion of water droplets under the high electric field intensity and non-uniform coefficient, the droplet with 1.5 mm in radius is injected into industrial mineral oil subjected to the electric field E_{av} =4 kV/cm with the non-uniform coefficient of f=2.22. The motion of water droplet is shown in Figure 4.

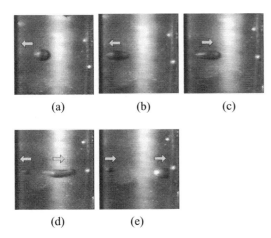

(a)   (b)   (c)

(d)   (e)

Figure 4. Water droplet electrodispersion under the non-uniform electric field.

4 CONCLUSION

The researches on the mechanism of emulsion electric dehydration under the non-uniform electric field were carried out in this paper, and the results can be summarized as follows:

a. Within a certain range, the improvement of non-uniform coefficient of the electric field can increase emulsion dehydration speed. However, the too higher non-uniform coefficient of electric field will lead to electric dispersion for the water droplets.

b. The dielectrophoresis force is proportional to electric field intensity, water droplet radius and non-uniform coefficient of electric field under non-uniform electric field with coaxial cylindrical electrode structure.

c. The electric field intensity of electrodispersion for water droplets is related to surface charge density of a water droplet, droplet size and surface tension. When the water droplet is charged, its rupture electric field intensity will decrease.

REFERENCES

Chen, Q. G., Liang, W. & Song, C. H. 2014. Effect of electric field strength on crude oil emulsion's demulsification and dehydration. *High Voltage Engineering* 40(1): 173–180.

Eow, J. S. & Ghadiri, M. 2003. Deformation and break-up of aqueous droplets in oils under high electric field strengths. *Chemical Engineering and Processing* 42(4): 259–272.

Alinezhad, K., Hosseini, M., Movagarnejad, K. & Salehi, M. 2010. Experimental and modeling approach to study separation of water in crude oil emulsion under non-uniform electrical field. *Korean Journal of Chemical Engineering* 27(1): 198–205.

Wei, W., Zhang, Y.W. & Gu, Z. L. 2013. The electrorheological droplet's deformation and mechanical beavior. *Chinese Science Bulletin* 58(3): 197–205.

Sameer, M., Rochish, M. & Thaokar. 2013. Drop motion, deformation, and cyclic motion in a non-uniform electric field in the viscous limit. *Physics of Fluids* 25(7): 072105.

Ye, Q. Z., Li, J. & Zhang, J. C. 2004. A displaced dipole model for a two-cylinder. *IEEE Transactions on Dielectrics and Electrical Insulation* 11(3): 542–550.

Wang, Z. H., Liao, Z. F., Gao, Q. J. & Wang, J. Q. 2009. Study on Oil Corona Discharge and Atomization Experiment with Electrostatic Spray. *High Voltage Engineering* 35(3): 636–640.

Wang, Z. T., Wang, J. F. & Gu, L. P. 2013. Theoretical and Experimental Investigation on Mechanism of Biodiesel Droplets Electrostatic Breakup. *High Voltage Engineering* 39(1): 135–140.

Electronic Engineering and Information Science – Wang (Ed.)
© 2015 Taylor & Francis Group, London, ISBN: 978-1-138-02772-5

Polyaniline/graphene conductive composite materials research

Z.H. Wang, B. Li, W.J. Qiao, J. Shao, D.Q. Jin & J.L. Li
Harbin University of Science and Technology, Harbin, China

ABSTRACT: Graphene is a kind of new carbon materials with excellent performances, it has a high specific surface area and abundant functional groups on the surface. Therefore the surface modification of graphene composites becomes another research focus. This paper focuses on polyaniline/graphene conductive composites, Testing with different proportion of compound of polyaniline and graphene within the functional groups by Fourier infrared spectrum(FTIR), and then use Scanning Electron Microscope (SEM) to study on graphene and their composites with surface morphology characterization. For the research on capacitive characteristics of composite materials, Respectively using cyclic Voltammetry and constant current charge and discharge to test for comparing different ratios of composite materials. Come to the conclusion: the specific surface area of the composite material utilization ratio, and the capacitance characteristics are different when the content of graphene is different. Its charging and discharging process is a standard electric double layer capacitance. The composite materials with 50% graphene reached 327F/g in 500mA/g.

KEYWORDS: Graphene Oxide; Hummers method; XRD; AFM;

1 INTRODUCTION

In 1977, Alan j. Heeger, Allen. Jennifer and Hideki Shirakawa succeeded in doping high conductivity iodine with polyacetylene, then people have great interest in the area of conducting polymer. In 2004, we have a new definition of graphene(Iijima,1991). It is a dense, wrapped in honeycomb crystal lattice of carbon atoms on the two-dimensional crystal carbon materials(Stoller, Park and Zhu 2008) and it has good electrical conductivity and large specific surface area. Polyaniline material chemistry has strong stability. It also has a good electrical conductivity and high turns capacitance characteristics. Its raw material sources, low cost, easy synthesis leads to a broad application prospect(Saliger, Fiseher and Herta, 1998), such as super capacitor(Conway,1991) electrode materials. Graphene is a research hotspot in recent years, due to the particularity of its structure, making it having excellent electrical conductivity. Polyaniline/graphene can be compatible with the advantages of the two and become another research hot spot.

2 EXPERIMENT

Conductive polyaniline was prepared by traditional solution, graphene is made by the Improved method which is obtained by improving the Hummers method. Then put them compound, to prepare of the polyaniline/graphene conductive composites with excellent properties (Du, Xiao and Meng (2008). Testing with different proportion of compounds of polyaniline and graphene within the functional groups by Fourier infrared spectrum(FTIR), and then use scanning electron microscope (SEM) to study on graphene and their composites with surface morphology characterization. Finally, making the polyaniline/graphene composites into electrode, and to study its electrochemical properties. It is the focus of the research that the electrochemical properties of the conductive composite material when the content of graphene is different.

3 RESULTS AND DISCUSSION

3.1 *Section 1*

Figure 1 shows the polyaniline, graphene and their Fourier infrared absorption spectra of the composite material. Infrared spectra produced from molecules to the incident photon energy absorption of vibration energy level transition, the infrared absorption spectrum is mainly used to identify the existence of the functional groups, and thus determine the structure of organic compounds. We can see from the graph, graphene appears C-C bond, then TO and LO phonons at 1115cm-1 and 1128cm-1department and the peak corresponding at 3431cm-1 is –OH. It is due to the graphene surface adsorption of water molecules slightly. Composite materials also appeared TO and LO phnons peak of C-C bond at 115cm-1 and 1128cm-1, but the intensity relative to the graphene is weak. The infrared absorption peak of composite samples at 1312cm-1 and 1572cm-1 correspond to the deformation of quinine ring and stretching vibration peak of the benzene ring. That is characteristic of polyaniline infrared peak. There is a strong interaction between polyaniline and graphene, and this interaction enhances the ability of electron delocalization in the polyaniline chain molecules, this ability can enhance its conductivity.

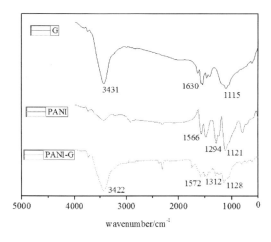

Figure 1.   FTIR chart of the polyaniline/graphene composite material.

3.2 Section 2

Figure 2 shows the scanning electron microscope (SEM) images of the polyaniline/graphene composites under different magnifications. It can be observed from the figure that there is so much deposition of polyaniline in the fold of the graphene surface. With further observation, we can see the similar fibrous form of polyaniline. It is the formation of porous structure. The porous structure is advantageous to the electrolyte ion conduction and charge and discharge rate. Moreover, graphene provides a larger surface area, and polyaniline makes composite to have great potential for use in the super capacitor. As can be seen from the fig.2(b), polyaniline is growing independently. It is unlike the growth of graphene that the growth of polyaniline/graphene conductive composites is morphic. It is not only along the graphene growth, but also completely covered the graphene. We can see that on the surface of composites showed rough polyaniline coating material from fig.2(c). This structure is conducive to the increasement of surface area of the composite electrode materials, it also has great contribution on the capacitor. In addition, you can clearly see that there is a lot of space in composite electrode materials, this will help improve the ionic conduction rate and cycle stability of the electrode.

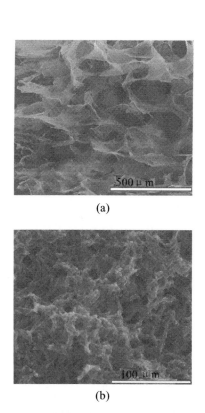

(a)

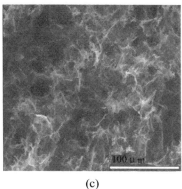

(b)

(c)

Figure 2.   SEM images of the polyaniline/graphene composites under different magnifications.
(a) graphene; (b) polyaniline; (c) polyaniline/graphene conductive composites.

3.3 Section 3

Figure 3 shows polyaniline, graphene and under the different ratio of polyaniline/graphene composites electrodes in the same scan range (0 ~ 0.8 V). The different scan rate (0.05, 0.02 and 0.005 V/s) under the cyclic voltammograms test curve. Cyclic voltammetry is the common way to study the performance of electrode capacitance. Graphene can be seen from the Figure 3 that the shape of the CV curves of the electrode is "kind of rectangular" shape. The shape of the current-voltage(CV) curves of the electrode is closer to the rectangle, but their CV curve are basic symmetric about zero current line. Especially under 0.05 V/s high scan rate, degree of rectangular curve is still good, showing that two kinds of graphene electrodes have a good performance of electric double layer capacitors, but also compatible with polyaniline is a bit high power capacity. So, the electrode materials are suitable for capacitor. It shows under the same experimental conditions, when the ratio of polyaniline content increasing, capacitance performance will improve. The background current of polyaniline/graphene composite electrodes is higher. It suggests that the relative to the graphene electrodes, polyaniline/graphene composites electrode has better capacitance properties.

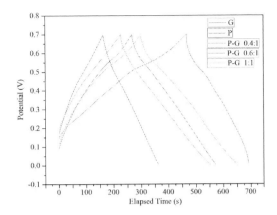

Figure 4. Polyaniline/graphene composites in the constant current charge and discharge test time and voltage of the diagram.

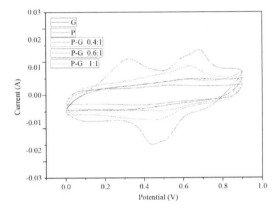

Figure 3. CV curves of composite electrode in a solution with different scanning speed.

3.4 *Section 4*

Figure 4 shows polyaniline, graphene and with different ratios of polyaniline/graphene composites in the current density of 0.1 A/g under the condition of constant current charge and discharge test time and voltage of the diagram. It can be seen from the graph curve that it has a very good symmetrical triangle shape, showing that graphene and polyaniline/graphene composites is a standard charge and discharge process of electric double layer capacitance. Here, the charge and discharge performance of graphene is still far away differ with the theoretical value(Wang ,Shi, Huang, Ma and Wang 2009). This is mainly due to the preparation of graphene still exists a certain reunion, the specific surface area of graphene cannot be fully taken advantage of. So the capacitance properties of polyaniline/graphene composites can't very well.

4 CONCLUSIONS

During the research, We learn that polyaniline/graphene composites in the process of charging and discharging is the electric double layer capacitance exist together with the constraint capacitance. The content of polyaniline has a great influence on the electrochemical properties of composite materials, with the increase of the content of polyaniline, the electrochemical performance of composite gets improved. Structure characterization results show that polyaniline and graphene get a good composition. The composite materials with 50% graphene reached 327F/g in 500mA/g. Test result shows that the composite materials have good application value in supercapacitor.

ACKNOWLEDGEMENT

This work was financially supported by the Provincial innovative experimental project of Harbin university of technology.

REFERENCES

Iijima,S.(1991)Helical microtubules of graphitic carbon,Nature, Vol 354, p. 7.
Stoller M D, S. Park & Y. Zhu(2008),etal,Graphene-based ultracapacitors,*Nano Letters, Vol* 10,p. 3498.
Saliger R U, C. Fiseher & Herta, (1998)etal,High surface area carbon aerogels for supercapacitors,*Journal of Non-Crystalline Solids. Vol* 225, p. 81.
Conway, B. E.(1991), Transition from supercapacitor to "batter" behavior in electrochemical energy storage, *Electrochemical Society, Vol* 138,p. 1539.
Du X S, M. Xiao, & Y. Z. Meng(2008), Facile synthesis of highly conductive polyaniline/graphene nanocomposites, *Eur Polym,Vol* 40,p. 1489.
Wang Y, Z. Shi, Y. Huang, Y. Ma, & C. Wang(2009), M. Y. C, Supercapacitor devices based on graphene materials, *J Phys Chem C, Vol* 113,p. 13103.

Electronic Engineering and Information Science – Wang (Ed.)
© 2015 Taylor & Francis Group, London, ISBN: 978-1-138-02772-5

Three methods for preparation of graphene

Z.H. Wang , B.Li , W.J. Qiao , J. Shao, D.Q. Jin & J.L. Li
Harbin University of Science and Technology, Harbin, China

ABSTRACT: In order to improve the conventional methods to obtain a better way, analysis on the performance of the grapheme oxide which is prepared by different methods is necessary. We choose three methods for preparing grapheme oxide. The first method is the widely used Hummers method. The second is the Improved method which is obtained by removing the Hummers method. The third is the Improved- method which is obtained by removing phosphoric acid from the Hummers method. The main difference of these methods is the procedure of oxidation. We have used the method of X-ray diffractometry (XRD), atomic force microscope (AFM) and two electrode testing method to analyze the yield, structure, form, size, functional groups, degree of oxidation of the Graphene Oxide which were prepared by three methods and focused on the degree of oxidation and particle size of the Graphene Oxide. And then the advantages and disadvantages of the three methods are discussed. After a detailed analysis of the experimental data, we got a conclusion that the grapheme oxide, which are prepared by Improved-method has a vast property.

KEYWORDS: Graphene Oxide; Hummers method; XRD; AFM.

1 INTRODUCTION

Graphene has many excellent properties on the mechanical, electrical, thermal and so on various aspects(Iijima,1991). At present there are many ways to the preparation of graphene, which mainly include the micro mechanical stripping method, chemical vapor deposition method, oxidation-reduction method. Redox process due to its simple preparation, low cost of preparation and the preparation of a very high rate become an effective way to realize the industrialization of the preparation of graphene. By redox preparation of graphene is the main steps of the preparation of graphene oxide, the stand or fall of graphene oxide performance is directly related to the size of the degree of oxidation to restore after the properties of graphene. In addition, as a result of graphene oxide containing a large amount of oxygen containing functional groups, make its than graphene has been applied in many areas are more likely to, so how to preparation of the excellent performance of graphene oxide has very important significance.

2 EXPERIMENT

We have adopted three kinds of different methods for the preparation of graphene oxide, these three methods are mainly in the preparation of graphite oxide, other experimental steps are basically the same. The proportion of each component in graphene oxide was identified by X-ray diffractometry (XRD). Graphene oxide functional groups in the number and variety were observed with FTIR. Graphene oxide distribution of surface atoms measured on an AFM.

3 RESULTS AND DISCUSSION

3.1 *Section 1*

Fig.1(a) using the Improved method, fig.1(b) using Hummers method (Novoselov, Geim and Morozov 2004), fig.1(c) using the Improved method. It can be seen from fig.1 (b) that the precipitation of graphite oxide and impurities in the solution is more. There are a lot of black particles, solution color darker. And Figure 1a and c is more pure color and more bright. Which can deduce the improved method and improved method of graphite oxidized more completely (Marcano, Kosynkin and Berlin, 2010), Hummers method of graphite oxide of graphite is not oxidation, which can be further concluded that improved method and the improved method than Hummers method with large. Then through to the final preparation of graphene oxide weighing, Improved method and Improved method of graphene oxide significantly more than Hummers method, which proves that the above speculation is correct.

Figure 1. graphene oxide prepared by the three methods. (a) Hummers method; (b) Improved method; (c) Improved- method.

3.2 *Section 2*

Fig. 2 shows three diffraction peak in the diffraction diagram of graphite. In Hunmmers method, the Improved method and Improved method of diffraction diagram, in the vicinity of 26.5° has a small diffraction peak. It shows that all three methods in the preparation of graphene oxide left part of the oxidized graphite. In addition by the three figures can be seen, three approaches to the preparation of graphene oxide are near 10° appeared an obvious diffraction peak. After accurate analysis it is found that graphene oxide of the preparation of various methods of diffraction peak corresponding to the layer spacing is respectively.

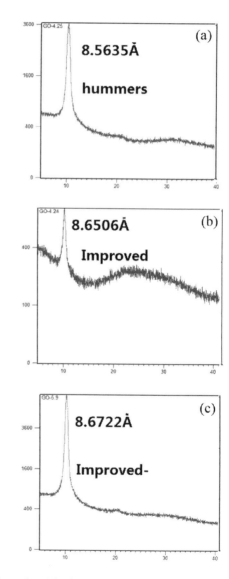

Figure 2. Adoption of different methods for preparation of graphene oxide XRD diagrams. (a) Hummers method; (b) Improved method;(c) Improved- method.

3.3 *Section 3*

As shown in Fig.3 for the preparation of graphene oxide three methods using atomic force microscope images, can be seen from the three picture, using the method of Improved preparation of graphene oxide layer using Improved method of graphene oxide particle size is big, Hummers method preparation of graphene oxide particle size of the smallest, which can estimate the Hummers on graphene layer of the degree of damage is bigger, in the process of oxidation is not easy to generate a large area of graphene oxide layer, and the Improved method and Improved method in the process of oxidation on the graphene layer destruction is lesser, but using other characterization methods and fig.3 can see Improved method and Improved- method of graphene oxidation degree is bigger than Hummers method, so there must be a mechanism that Improved-method and Improved method in guarantee a high degree of oxidation and gives the overall structure of the graphene protection, that result in a larger piece of layer(Zhang, Yang and Evans 2008).

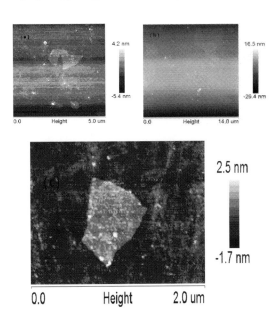

Figure 3. Images of three kinds of graphene oxide in AFM. (a) Hummers method; (b) Improved method; (c) Improved- method.

3.4 *Section 4*

Through two electrodes testing method for preparation of three kinds of methods to test the electrochemical properties of graphene oxide, and fig.4 (a) is a different scan rate, the Current – voltage(C-V) curve of the graphene oxide, in 0~0.8V voltage range,

with the increase of scan rate, the current density has increased in proportion(Tien and Teng,2010), that graphene oxide has a good capacitive performance, preparation and Improved methods of graphene oxide capacitive performance is best. Fig.4 (b) describe the capacitance change with the scanning speed, scanning speed faster, capacitance performance is poorer, Improved- same method preparation of graphene oxide properties is best.

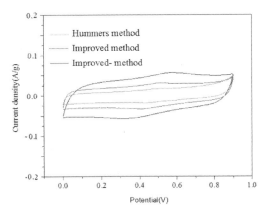

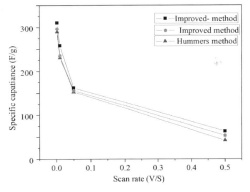

Figure 4. The change of C-V with circulation charge and discharge and Capacitance along with the change of scanning rate curve. (a) CV curve of the graphene oxide; (b) Capacitance along with the change of scanning rate curve.

4 CONCLUSIONS

By using two electrode testing method, X-ray diffraction (XRD), atomic force microscope (AFM) and the method of direct observation of three methods had been used to characterize the preparation of graphene oxide. After analyzing the result of the characterization the following conclusions were drawn.

The three methods of oxidation degree are greater than Improved methods, but the difference between the two is not big, and the oxidation of Hummers method was significantly lower than the former two, and Hummers of graphite oxide is incomplete, the black is not of graphite oxide residual. The particle size of the Graphene oxide made by Improved- is biggest, but similar to that made by Improved methods, Hummers method of graphene oxide particle size is minimal.The output of the Graphene oxide made by Improved- is largest in three ways, the second Improved method, the minimum is Hummers method. Three kinds of graphene oxide capacitor performance Improved- methods is best, Improved methods and Hummers method are poor.

ACKNOWLEDGEMENT

This work was financially supported by Provincial innovative experimental project of Harbin University of Technology.

REFERENCES

Iijima, S. (1991)Helical microtubules of graphitic carbon,- Nature Vol 354,p. 7.

Novoselov K., A.Geim,& S.Morozov (2004) etc. Electric field effect in atomically thin carbon films,Science, Vol 5696, p. 666.

Marcano, D. C., D. V.Kosynkin, & J. M.Berlin, (2010) etc., Improved Synthesis of Graphene Oxide,ACS Nano, Vol 8, p. 4806.

Zhang, X., W.Yang & D.G. Evans(2008), Layer-by-layer self-assembly of manganese oxide nanosheets/polyethylenimine multilayer films as electrodes for supercapacitors.Journal of Powe Sources. Vol 184, p. 696.

Tien, C.P, H.Teng, (2010)Polymer/graphite oxide composites as high-performance materials for electric double layer capacitors,Journal of Power Sources, Vol 184. p. 2414.

Electronic Engineering and Information Science – Wang (Ed.)
© *2015 Taylor & Francis Group, London, ISBN: 978-1-138-02772-5*

Matching phase and group velocities in second harmonic generation for one-dimensional nonlinear photonic crystals with linear dispersion

H.N. Wu, J.J. Huang, L.Y. Zhang, S.Z. Pu & Z.Y. Yang
Department of Optoelectric Information Science and Engineering, Harbin University of Science and Technology, Harbin, China

ABSTRACT: The simultaneous equations of phase matching and group velocity matching of second harmonic generation in one-dimensional nonlinear photonic crystals are developed by considering dispersion effects. The restriction of refractive index ratios of nonlinear and linear media for phase matching and group velocity matching is achieved. The analysis illuminates that a larger refractive index contrast of two materials can be compensated by increasing the dispersion of the nonlinear media. A brief comparison between results with and without dispersions is discussed at the end of this paper and it is demonstrated the differences are weak when the layer thicknesses and the mismatching of refractive index in nonlinear media are not large.

KEYWORDS: nonlinear photonic crystals, phase matching, group velocity matching, second harmonic generation.

1 INTRODUCTION

1.1 *Type area*

Recent years, nonlinear photonic crystals have attracted a great deal of attention on their characteristics of nonlinear frequency conversion in a periodic structure and optical field enhancement caused by the Bragg resonance peak (Bendickson & Dowling 1996, Chutinan & Noda 2000, Scalora 1994, Scalora et al. 1997, Vecchi et al. 2004), and particularly extensive studies have been done in improving the conversion efficiency of second harmonic generation (SHG) in one-dimensional photonic crystals (Centini et al. 1999, Kiehne et al. 1999, Li et al. 2007, Ren & Li 2012). Owing to the band-edge effect of photonic crystals, the photonic mode density is large and the electric fields can be enhanced at the band-edge (Dumeige et al. 2002), not unexpectedly, it is available to obtain SHG more efficiently. However, the increase of photonic crystal periods may lead to narrow transmission peaks unsuited for short light pulse. In addition, with the further discussing of short light pulse in the past decade, studies show that dramatic field enhancement may result in large and sensitive group delays for ultra-short pulse (Scalora et al. 1996). According to the previous studies, contrasted with enhancing the incident light intensity at the resonance peak or photonic band edge, group velocity dispersion also plays an important role in SHG for the picosecond or sub-picosecond pulses in multi-period photonic crystals, it has gained significant importance in the investigation of the phase matching (PM) and group velocity matching (GVM)

condition. Based on the idea proposed by Tarasishin et al in 2001 (Tarasishin et al. 1999, Tarasishin et al. 2001) and our pioneer work (Wu et al. 2015), in this paper, we focus on finding the restrictive condition on selecting materials with satisfying PM and GVM simultaneously and taking into account the material dispersion in nonlinear media in one-dimensional nonlinear photonic crystals and compared with the case without the dispersion as shown in Wu et al. (2015).

2 EQUATIONS OF PM AND GVM WITH DISPERSION

In this section, we consider a multi-period photonic crystal of the linear layer with thickness a and nonlinear layer with thickness b, and the non-dispersive linear media refractive index is n_a, the refractive indices of the fundamental (FF) and the second harmonic (SH) in nonlinear media are $n_{1b}(\omega_0)$ and $n_{2b}(\omega_0)$, respectively. According to Maxwell equations and Bloch' theorem, the dispersion relation of FF and SHG can be achieved as Tarasishin et al. (2001)

$$\cos[k(\omega_0)d] = \cos(k_0 n_a a)\cos[k_0 n_{1b}(\omega_0)b] \\ - N_1(\omega_0)\sin(k_0 n_a a)\sin[k_0 n_{1b}(\omega_0)b], \quad (1)$$

$$\cos[k'(2\omega_0)d] = \cos(2k_0 n_a a)\cos[2k_0 n_{2b}(\omega_0)b] \\ - N_2(\omega_0)\sin(2k_0 n_a a)\sin[2k_0 n_{2b}(\omega_0)b], \quad (2)$$

where $N_1(\omega_0) = [n_a^2(\omega_0) + n_{1b}^2(\omega_0)]/2n_a(\omega_0)n_{1b}(\omega_0)$, $N_2(\omega_0) = [n_a^2(\omega_0) + n_{2b}^2(\omega_0)]/2n_a(\omega_0)n_{2b}(\omega_0)$, $d = a + b$, $k_0 = \omega_0/c$, $k_0(\omega_0)$ and $k'_0(\omega_0)$ are the efficient wave vectors of FF and SH, c is the vacuum speed of light. For the convenience of calculation, we replace the variables in Eq. (1) and Eq. (2) with four new parameters as Wu et al. (2015), that is, $p_a = k_0 n_a a$, $p_b = k_0 n_{1b}(\omega_0) b$, $R = n_{1b}(\omega_0)/n_a$ and $\delta n = [n_{2b}(\omega_0) - n_{1b}(\omega_0)]/n_{1b}(\omega_0)$ respectively. Here, p_a and p_b are the phase variables, R and δn represent the refractive index ratios of nonlinear and linear media, and FF and SH in nonlinear layers in one-dimensional nonlinear photonic crystals. Hence, Eq. (1) and Eq. (2) are simplified as

$$\cos[k(\omega_0)d] = \cos(p_a)\cos(p_b) - \tfrac{1}{2}(R + 1/R)\sin(p_a)\sin(p_b), \quad (3)$$

$$\cos[k'(2\omega_0)d] = \cos(2p_a)\cos[2p_b(1 + \delta n)] - \tfrac{1}{2}[R(1 + \delta n) + 1/R(1 + \delta n)]. \quad (4)$$
$$\times \sin(2p_a)\sin[2p_b(1 + \delta n)]$$

As is well known, PM condition of FF and SH is expressed in the form

$$k'(2\omega_0) = 2k(\omega_0). \quad (5)$$

Meanwhile according to the relation between group velocity and electromagnetic density of modes, GVM condition can be treated as

$$\frac{\partial k(\omega)}{\partial \omega}\bigg|_{\omega = \omega_0} = \frac{\partial k'(\omega)}{\partial \omega}\bigg|_{\omega = 2\omega_0}. \quad (6)$$

To derive Eq. (3) and Eq. (4), the electromagnetic density of modes can be written as

$$
\frac{\partial k(\omega)}{\partial \omega}\bigg|_{\omega = \omega_0} = \frac{1}{\omega_0 d \sin(k(\omega_0)d)} \bigg\{ [p_b(1 + \frac{\delta_n}{R})
$$
$$
+ \tfrac{1}{2}(R + 1/R)p_a]\cos(p_a)\sin(p_b)
$$
$$
+ [p_a + \tfrac{1}{2}(R + 1/R)p_b(1 + \frac{\delta_n}{R})] \quad , (7)
$$
$$
\times \sin(p_a)\cos(p_b) + \frac{1}{2}\sin(p_a)\sin(p_b)
$$
$$
\times (\delta_n - \frac{\delta_n}{R^2})\}
$$

$$
\frac{\partial k'(\omega)}{\partial \omega}\bigg|_{\omega = 2\omega_0} = \frac{1}{\omega_0 d \sin(k'(2\omega_0)d)}((1 + \delta_n)
$$
$$
\left\{ p_b(1 + \frac{\delta_n}{R}) + \frac{1}{2}\left[R + \frac{1}{R(1 + \delta_n)^2} \right] \right.
$$
$$
\left. \times p_a \right\}\cos(2p_a)\sin[2p_b(1 + \delta_n)]
$$
$$
+ \left\{ p_a + \frac{1}{2}\left[R(1 + \delta_n)^2 + \frac{1}{R} \right] \right. \quad . (8)
$$
$$
\left. \times p_b(1 + \frac{\delta_n}{R}) \right\}\sin(2p_a)
$$
$$
\times \cos[2p_b(1 + \delta_n)])
$$
$$
+ \frac{1}{2}\sin(2p_a)\sin[2p_b(1 + \delta_n)]
$$
$$
\times [\delta_n(1 + \delta_n) - \frac{\delta_n}{R^2(1 + \delta_n)}])
$$

In Eq. (7) and Eq. (8), we simplify the dispersion relations of nonlinear media as linear approximation. From the equations mentioned above, PM condition and GVM condition can be solved with considering material dispersion, so that the refractive index restriction of photonic crystals is determined and the selection of materials in photonic crystals become more convenient in the actual operation.

3 NUMERICAL RESULTS OF REFRACTIVE INDEX RATIOS

To improve the conversion efficiency of SHG in nonlinear photonic crystals, the refractive indices and the thicknesses of nonlinear and linear layers must satisfy Eq. (5) and Eq. (6) simultaneously. By applying numerical method to solve the simultaneous equations, we have demonstrated qualitatively the relationship of refractive index ratios, this is the minimum restrictive conditions to accomplish PM and GVM in nonlinear photonic crystals and the layer thicknesses of crystals do not have to limit a single value and have selectively degenerate solutions what make the preparation of photonic crystals more realizable. With setting as $0.5 \leq R \leq 1.5$, ranges of p_a and p_b both belong to $(0, 2\pi)$, the relationship between the contrast ratio of refractive index R and mismatching quantity of refractive index in nonlinear media δn is plotted in Fig. 1.

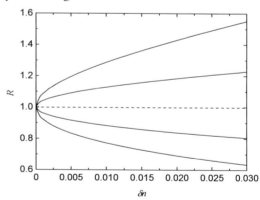

Figure 1. The minimum contrast ratio of refractive index R versus the mismatching quantity of refractive index in nonlinear media δn, and the parts of the dotted line above and below present the normal and abnormal regions respectively.

Figure 1 displays four curves that correspond to a different layer thickness which is caused by degenerate solutions as mentioned in Wu et al. (2015). The layer thickness corresponding to the two curves closed to the dotted line is smaller than that corresponding to the other two. In fact, not only the four curves in Fig. 1 but lots of other curves can be obtained from the simultaneous equations. The curves will be further away from the dotted line with the increasing layer thicknesses. It means that no matter in the normal regions or abnormal regions, when the mismatching quantity of nonlinear medium increases, the refractive index ratios always change towards the directions away from $R=1$. That is, only the high contrast ratio of refractive index of nonlinear and linear medium can compensate the larger phase mismatch and group velocity mismatch in a nonlinear medium. Consistent with the previous study ignoring dispersion, this case gets a similar result. Figure 2 shows the difference between them.

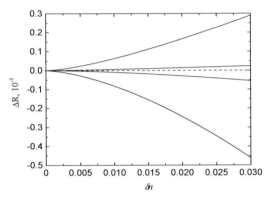

Figure 2. The absolute differences ΔR between the value of minimum refractive index ratios R with and without dispersion versus the mismatching quantity of refractive index in nonlinear media δn.

From Figure 2, we can illustrate that the dispersion terms in Eq. (7) and Eq. (8) are weak enough when the layer thickness and mismatching quantity both are small, but the difference ΔR grow rapidly with the increase of δn and the layer thickness obviously, in other words, the material dispersion should be taken into account when we select the larger layer thicknesses of photonic crystals in the practical application.

4 CONCLUSION

We propose an extensive discussion of the conditions of the PM and GVM in one-dimensional nonlinear photonic crystals, simultaneous equations are amended by taking the dispersion of pulse into account. The refractive index ratios of nonlinear and linear medium satisfied the equations of the PM and GVM have been presented and the results demonstrate the multiplicity of solutions still exists. Comparing previous studies and current results, the dispersion of pulse show less of an effect on the restriction condition of refractive index for PM and

GVM in SHG process when the layer thicknesses of photonic crystals are not too large. No matter the dispersion are neglected or not, the mismatching between FF and SH in nonlinear medium can be compensated by extending the refractive index ratios of nonlinear and linear medium. This consequence provides more information to devise and forge photonic crystals for SHG.

ACKNOWLEDGMENT

The authors thank the Natural Science Foundation of Heilongjiang province of China for a fund under Grant No. F201312.

REFERENCES

Bendickson, J. M. & J. P. Dowling (1996). Analytic expressions for the electromagnetic mode density in finite, one-dimensional, photonic band-gap structures. *Phys. Rev. E 53*, 4108–4120.

Centini, M., C. Sibilia, M. Scalora, G. D'Aguanno, M. Bertolotti, M. J. Bloemer, C. M. Bowden, & I. Nefedov (1999). Dispersive properties of finite, one-dimensional photonic band gap structures: Applications to nonlinear quadratic interactions. *Phys. Rev. E 60*, 4893–4897.

Chutinan, A. & S. Noda (2000). Waveguides and waveguide bends in two-dimensional photonic crystal slabs. *Phys. Rev. B 62*, 4489–4492.

Dumeige, Y., I. Sagnes, P. Monnier, P. Vidakovic, I. Abram, C. Mériadec & A. Levenson (2002). Phase-Matched Frequency Doubling at Photonic Band Edges Efficiency Scaling as the Fifth Power of the Length. *Phys. Rev. L 89*, 2–4.

Kiehne, G. T., A. E. Kryukov, & J. B. Ketterson (1999). A numerical study of optical second-harmonic generation in a one-dimensional photonic structure. *Appl. Phys. Lett. 75*, 1677–1678.

Li, J. J., Z. Y. Li & D. Z. Zhang (2007). Second harmonic generation in one-dimensional nonlinear photonic crystals solved by the transfer matrix method. *Phys. Rev. E 75*, 1–6.

Ren, M. L. & Z. Y. Li (2012). High conversion efficiency of second harmonic generation in a short nonlinear photonic crystal with distributed Bragg reflector mirrors. *Appl Phys A 107*, 72–75.

Scalora, M. J., P. Bowling, C. M. Bowden & M. J. Bloemer (1994). Optical Limiting and Switching of Ultrashort Pulses in Nonlinear Photonic Band Gap Materials. *Phys. Rev. L 73*, 1368–1371.

Scalora, M., M. J. Bloemer, A. S. Manka, J. P. Dowling, C. M. Bowden, R. Viswanathan & J. W. Haus (1997). Pulsed second-harmonic generation in nonlinear, one-dimensional, periodic structures. *Phys. Rev. A 56*, 3168–70.

Scalora, M., R. J. Flynn, S. B. Reinhardt & R. L. Fork (1996). Ultrashort pulse propagation at the photonic band edge: Large tunable group delay with minimal distortion and loss. *Phys. Rev. E 54*, 1079–1081.

Tarasishin, A. V., A. M. Zheltikov, & S. A. Magnitski (1999). Matched second-harmonic generation of ultrashort laser pulses in photonic crystals. *JETP Lett. 70*, 821–824.

Tarasishin, A. V., S. A. Magnitskii, & A. M. Zheltikov (2001). Matching Phase and Group Velocities in Second-Harmonic Generation in Finite One-Dimensional Photonic Band-Gap Structures. *Laser Phys. 11*, 32–38.

Vecchi, G., J. Torres, D. Coquillat, M. L. V. d'Yerville & A. M. Malvezzi (2004). Enhancement of visible second-harmonic generation in epitaxial GaN-based two-dimensional photonic crystal structures. *Appl. Phys. Lett. 84*, 1245–1247.

Wu, H. N., J. J. Huang, S. Z. Pu, L. Y. Zhang & Z. Y. Yang (2015). Restriction of the phase matching condition on refractive index modulation of 1D photonic crystals in second harmonic generation. *Laser Phys. 25*, 1–5.

Electronic Engineering and Information Science – Wang (Ed.)
© 2015 Taylor & Francis Group, London, ISBN: 978-1-138-02772-5

Surface protection technology research of Mg-Li alloy

L. Yao, L. Chen, Y.Q. Chen & X.X. Sun
Mudanjiang Normal College, Mudanjiang, China
Heilongjiang Province Key Laboratory of superhard materials, Mudanjiang, China

D. Sun
Harbin University of Science and Technology, Harbin, China

ABSTRACT: Surface protection of Mg-Li alloy has significance to extending the scope of application and prolonging service life of alloy. Decreasing the corrosion rates of Mg alloy can be achieved by purifying the component of the alloy, amending the unreasonable design, reducing the impurity in the smelting process and mitigating surface contamination. So does Mg-Li alloy. But we usually adopt the surface protection technology to form a shield layer on the surface of the alloy to isolate alloy and environment, or put alloy into anticorrosive inhibitor to prohibit corrosion process of Mg-Li alloy. At present the major methods are mainly chemical converting film, anodic oxidation, electroplating and chemical plating, sol-gel coating and vapor deposition technique.

KEYWORDS: Mg-Li alloy; surface; protection.

1 INTRODUCTION

1.1 *Chemical conversion*

After Mg-Li alloy is processed by chemical conversion, a layer of indissolvable compound membrane is formed on the surface of matrix, which has good binding power with matrix and prevents corrosive medium erosion of the matrix. The most sophisticated chemical conversion treatment is chromate conversion that the biggest disadvantage of being high toxicity and carcinogenicity of hexavalent chromium, thus people research and development chromium-free conversion coating of Mg-Li alloy. These conversions generally can be divided into two aspects: the conversion treatment of organic compound solvents and inorganic salt solution. The former includes phytic acid conversion (Grav and Luan 2002, Liu, Guo & Huang, 2006), silane derivatives conversion (Zucchi, Grassi and Frignani 2006) and so on; the latter includes stannate conversion (Anieai, Masi and Santamaria 2005), phosphate-potassium permanganate conversion (Zhao, Wu and Luo 2006, Zhao,Wu and Luo 2006), rare-earth conversion coating and phosphorization, etc.

1.2 *Phytic acid conversion*

Phytic acid ($C_6H_{18}O_{24}P_6$) also called myo-inositol hexakisphosphate, and it has a significant role in the protection treatment on the surface of metal because of its Special molecular structure and physicochemical properties . At present it is less that the research of phytic acid conversion on the Mg-Li alloy, and there are only a few of scholars having preliminary research about it. Gao et al. researched the conversion coating composition and corrosion resisting property of phytic acid conversion coating on the Mg-Li alloy. It is reported that there were tiny cracks on the coating covered with white irregular flocculent particles. Phytic acid conversion coating was the clathrate formed by metal ions of the alloy and phytic acid reaction, and a dense layer of protective film was formed on the metal surface. This improves the corrosion resistance of alloy so that the corrosion potential is 0.2V higher than the traditional chromate conversion, and reduces the cathodic current density greatly.

1.3 *Stannate conversion*

The composition, pH and reaction temperature of stannate conversion processing liquid are the important conditions of forming qualified conversion coating. Zhang et al. determined the compositions of the stannate conversion solution are: NaOH 8g/L, $Na_2SnO_3 \cdot 3H_2O$ 55g/L, $Na_4P_2O_7$ 40g/L, $NaC_2H_3O_2$ 8g/L [11]. And the conversion coating was prepared on the surface of Mg-Li alloy in this process. The research shows that stannate conversion coating is composed of approximately spherical uniform particles and relatively dense. Meanwhile the main ingredient of coating is $MgSnO_3$ and the structural property is crystalline state. The film formation time of stannate conversion coating has a significant effect on the anode polarization current of Mg-Li alloy. When the film formation is 45min, the anode polarization current has minimal value and the protection of collective Mg-Li alloy is the strongest. Meanwhile, it has good binding power with matrix and improves the corrosion resistance of Mg-Li alloy to some extent.

1.4 Rare earth conversion

The rare earth conversion process has characteristics of no toxicity, no pollution and good protective effect for Al, Zn and other metals including their alloys, which is a preferred environment-friendly chemical treatment technology instead of chromate conversion at present. Rare earth conversion coating process usually uses the impregnation method, and only Yang et al. researched it in Lanthanum nitrate solution both at home and abroad. The results show that the main composition of the rare earth conversion coating is La $(OH)_3$ and it has a protective effect for alloy.

2 EXPERIMENT CONDITION AND METHOD

2.1 Corrosion mechanism

Sol-gel coating process is a method that metal organic or inorganic compounds form an oxide or other solid compound by sol-gel and heat treatment. The forming process includes the hydrolysis of metal alkoxide, condensation, polymerization and the formation of sol, the formation of gel by further hydrolysis and polycondensation and getting the required protective film by recoating process. Sol-gel method operation is simple, low cost, wide adaptability of materials and process adaptability is strong, but it causes easily the separation of the coating and the generation of cracks because the thermal expansion coefficients are different between Sol-gel coating and the metal substrate. So the practical technology about sol-gel coating is less reported.

At the bare metal, cathodic hydrogen evolution reaction happened:

$$2H_2O + 2e \rightarrow H_2 + 2OH^- \qquad (1)$$

With the metal ion concentration in the solution, the corrosion products:

$$Mg^{2+} + 2OH^- \rightarrow Mg(OH)_2 \qquad (2)$$

$$Li^+ + OH^- \rightarrow LiOH \rightarrow Li_2O_2 \qquad (3)$$

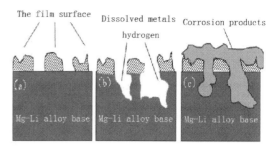

Figure 1.   Mechanisms of corrosion of Mg-Li alloy.

2.2 Vapor deposition layer

Vapor deposition technology deposits monolayer or multilayer films on the all kinds of material by the reaction between the gas phase, so that the materials obtain a variety of superior performance required. During the past 20 years, as the miniaturization of electronic components, high reliability, high integration and various advanced the development of science and technology, vapor deposition technology obtained the rapid development and wide application. Usually based on the generation of gas phase material, it can be divided into physical vapor deposition (PVD) and chemical vapor deposition (CVD). Physical vapor deposition is a method which makes plating material gasification into atoms, molecules or ionization into ion under vacuum condition to deposit to the substrate surface directly by high temperature evaporation, sputtering, ion plating and other physical methods. PVD has been widely used because of the advantages of fast deposition rate and no pollution, but it has also some defects that the adhesion strength and uniformity of the film layer are poor. So it must be handled properly before depositing, and ion implantation is an effective method of surface modification. While CVD is a process that supplying one or more compounds and elemental gas containing the component elements of the film to matrix and preparing metal or compound thin film on the matrix by the influential action of gas phase or the chemical reaction on the surface of the matrix. This method has the advantage of film-forming temperature can be far below the melting temperature of coating metal, some refractory metal surface can be obtained, and the film is not affected by the shape of the sample. Its disadvantage is that the chemical guide agent is more expensive and toxic and the processing cost of toxic by-products is high.

3 CONCLUSIONS

Adopting the surface protection technology to form a shield layer on the surface of the alloy to isolate alloy and environment, or put alloy into anti-corrosive inhibitor to prohibit corrosion process of Mg-Li alloy. At present the major methods are mainly chemical converting film, anodic oxidation, electroplating and chemical plating, sol-gel coating and vapor deposition technique.

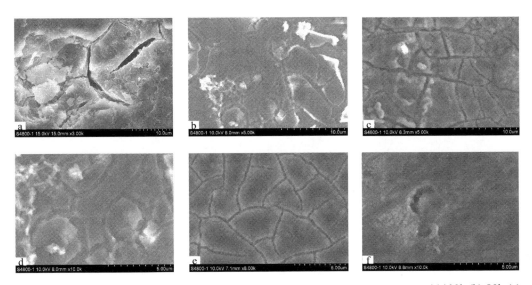

Figure 2. Surface morphology of the sample after the immersion test in 0.35wt.%NaCl solution. (a)100h (b) 80h (c) 60h (d) 50h (e) 30h (f) 50h.

ACKNOWLEDGMENT

Supported by Foundation of Heilongjiang Educational Committee (12543080), Science and technology project of Mudanjiang city (Z2013g004).

REFERENCES

Grav, J. E.& B. Luan(2002). Protective coatings on magnesium and its alloys-a critical review. J. Journal of Alloys and Compounds, 336:88–113P.

Liu, J. R., Y. N. Guo & Huang, W. D.(2006). Study on the corrosion resistance of phytic acid conversion coating for magnesium alloys. J. Surface and Coating Technology, 201(3–4);1536–1541P.

Zucchi F, V. Grassi &T. A .Frignani (2006). Influence of a saline treatment on the corrosion resistance of a WE43 magnesium alloy. J. Surface and Coatings Technology, 200(12–13):4136–4143P.

Anieai, L., R.Masi & M. Santamaria (2005). J.Corrosion Science, , 47(12):2883–2900P.

Zhao, M,S. S .Wu , J. R.Luo (2006). J.Surface and Coatings technology, 200(18–19):5407–5412P.

Kwo, Z .C. & S.S.Teng (2003). Conversion-coating treatment for magnesium alloys by a permanganate-phosphate solution. J.Materials Chemistry and Physics,2003, 80(1):191–200P.

Electronic Engineering and Information Science – Wang (Ed.)
© 2015 Taylor & Francis Group, London, ISBN: 978-1-138-02772-5

The effect of surface corrosion of Mg-Li alloy

L. Yao, L. Chen, Y.Q. Chen & Y.J. Zheng
Mudanjiang Normal College, Mudanjiang, China
Heilongjiang Province Key Laboratory of superhard materials, Mudanjiang, China

D. Sun
Harbin University of Science and Technology, Harbin, China

ABSTRACT: The density of Mg and Li are 1.74 g/cm³ and 0.53 g/cm³ respectively. With the increased content of Li, the density of Mg-Li alloy decreases and is less than the density of Al-Li alloy used as the novel aeronautical material, which generally is 1.3 g/cm³. When the content of Li exceeds 35%, the density of Mg-Li alloy is less than 1 g/cm³ so that Mg-Li alloy can float on the water. While the content of Li is larger than 5.7% in Mg alloy, the crystal structure translates from close-packed hexagonal structure to body-centered cubic structure, which would improve the processability and reduce the specific gravity of alloy. As a novel metal construction material, Mg-Li alloy has many inherent virtues. In addition to low density and high elastic modulus, it also has a strong ability to resist penetration of high-energy particles, an excellent performance of shock absorption and noise reduction, and finishing.

KEYWORDS: Mg-Li alloy; corrosion; mechanism.

1 INTRODUCTION

Mg reserve is extremely abundant element on earth, can oxide, carbide, chloride, silicate, phosphate, and other forms exist, many ore and contain a certain amount of mg in seawater. Mg content in the earth's crust can account for about 2.7%, reserves ranks the sixth in the surface layer of the earth's crust, our country is one of the richest countries in the world for mg resources (Li, 2005). The density of Mg is 1.74 g/cm³ only two-thirds of aluminum, steel ¼ (Felix Lee, Jie and zeng 2005), is one of the lightest metal. On the crystallography, mg has close six-party structure, but the mechanical properties, such as strength and hardness of pure mg is low. It cannot be directly used as structural material. By adding Mg-Li alloy elements and the formation of solid solution, its mechanical properties of alloy can be improved (Zhang, Gao and Zyxel 2005). Mg alloy with low density, high specific strength, good electromagnetic shielding performance, casting performance, radiation protection performance are good and the advantages of the alloy is it can be 100% recycled, so the development and utilization of Mg-Li alloy products is the world's development trend, and Mg-Li alloy is known as the "21st century green engineering materials" (Xu,Xu 2005,Han, and Long 2003,Yang and Hao (2004)).

2 EXPERIMENT CONDITION AND METHOD

In this experiment, the magnetron sputtering machine is made in Shenyang. The factory is adopted to prepare film. The target materials have metal Cr and Zr and their purities reach 99.99%. Ar gas whose purity reaches 99.99% as work gas and N2 whose purity reaches 99.99% as reaction gas are well mixed before importing them into the vacuum chamber. The gas flux is controlled by CIE mass flow controller, and this system also equips a precise MKS gas pressure measuring instrument to monitor the gas pressure of vacuum chamber in the process of sputtering.

2.1 Corrosion mechanism of Mg-Li alloy

Mg-Li alloy has many excellent properties, but due to the difference of corrosion resistance, the application is limited. The application of Mg-Li alloy is the key to solve the problem of corrosion of metal. Mg-Li alloy corrosion resistance is poor and is mainly caused by the negative differential effect of magnesium, so first understand before know Mg-Li alloy corrosion mechanism of pure magnesium corrosion reactions.

2.2 Corrosion mechanism of Mg

About corrosion and electrochemical mechanism of Mg in summary to roughly five kinds, namely promoting the dissolved hydrogen takes magnesium, the unit price of magnesium ion dissolves transition, surface membrane damage, loss and comprehensive theory. First four although theory can explain some experimental phenomena, but also has the unreasonable place, and in reasonable theory into a comprehensive, can get a more comprehensive magnesium corrosion theory model. As Fig. 1, the theoretical model thinks Mg surface with a layer of incomplete surface film, and its incomplete increase with the increase of polarization potential. That is to say, the surface of Mg membrane rupture or no film in the area of the will as the electrode potential is moving and get bigger, is this mask the bursting of the corrosion of Mg plays a decisive role.

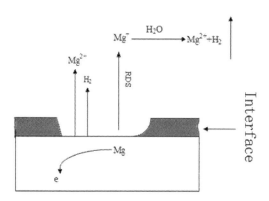

Figure 1.　The corrosion mechanism of Mg.

This theory also suggests that membrane rupture is exposed on the surface of Mg, Mg metal corrosion mainly by the unit price of Mg ion dissolving mechanism, namely, metal Mg first lose electrons and Mg ions into the unit price, then react with water to form a more stable bivalent Mg products, and release the hydrogen, so the total anode solution process can be written as:

$$Mg + \left[y/(1+y) \right] H_2O \rightarrow Mg^{2+} + \left[y/(1+y) \right] OH^-$$

$$+ \left\{ y/\left[2(1+y) \right] \right\} H_2 + \left[1 + 1/(1+y) \right] e \qquad (1)$$

In addition, the metal Mg naked, cathodic hydrogen evolution reaction may be easier than surface film covering, the total cathode reaction can be written as:

$$2H_2O + 2e \rightarrow H_2 + 2OH^- \qquad (2)$$

By the total anodic reaction and the total total cathodic reaction available Mg corrosion equation:

$$Mg + 2H_2O \rightarrow Mg^{2+} + 2OH^- + H_2 \qquad (3)$$

Per dissolve a magnesium atom can produce a hydrogen molecule.

3　CORROSION BEHAVIOR OF Mg-Li ALLOY

The main components in Mg-Li alloy chemically active Mg and Li, exposed to the air will be generated in the alloy surface layer of oxide film, the membrane is not necessarily uniform complete, some parts of the film will appear relatively thin may even membrane layer state of nudity, result in the alloy substrate in these parts of the resistance is weak, so the corrosion behavior choice in these areas.

In the part of the corrosion, the main alloy elements magnesium first, on the basis of monovalent ions and then react with water to form a more stable bivalent Mg, anode reaction mainly for:

$$Mg \rightarrow Mg^+ + e \qquad (4)$$

$$2Mg^+ + 2H_2O \rightarrow 2Mg^{2+} + 2OH^- + H_2$$
(In neutral and acidic solution) $\qquad (5)$

$$2Mg^+ + 2H^+ \rightarrow 2Mg^{2+} + H_2 \text{ (In acidic solution)} \qquad (6)$$

$$Li \rightarrow Li^+ + e \qquad (7)$$

$$Al \rightarrow Al^{3+} + 3e \qquad (8)$$

And at the bare metal, cathodic hydrogen evolution reaction happened:

$$2H_2O + 2e \rightarrow H_2 + 2OH^- \qquad (9)$$

OH⁻ with the metal ion concentration in the solution, namely the corrosion products:

$$Mg^{2+} + 2OH^- \rightarrow Mg(OH)_2 \qquad (10)$$

$$Li^+ + OH^- \rightarrow LiOH \rightarrow Li_2O_2 \text{ (Only under the alkaline condition)} \qquad (11)$$

4 CONCLUSIONS

More clearly illustrates the process of metal corrosion, Mg-Li alloy oxide film formed by Incomplete in the air, there will be a bare metal:

1　As shown in the corrosive medium, the erosion starts first at the bare area and then continues on the surface film. The aluminum alloy of the main elements of Mg and Li in the anodic oxidation and the cathodic hydrogen evolution in alloy exposed areas. In addition, there are other elements in the alloy are dissolved. Metallic elements of the anodic dissolution rate increase, at the same time because of the destruction of the surface film is more serious, the bare substrate area will also gradually increase, so the corrosion occurs rapidly.

2　To the late corrosion, because of the serious local or uneven corrosion, make Mg-Li alloy parts covered by corrosion products, after removing the corrosion products formed deep corrosion pit.

ACKNOWLEDGMENT

Supported by Nature Science Foundation of Heilongjiang Province (201341), Supported by Foundation of Heilongjiang Educational Committee (12543080), Science and Technology Project of Mudanjiang city (G2013e1233).

REFERENCES

Li,W.X. (2005). Magnesium and magnesium alloys .M. *Central south university press.*

Felix Lee, D. Jie & X.Q. zeng(2005). Research progress of high performance rare-earth magnesium alloys .J. *Materials review,* (8) : 51–53.

Zhang, Y.H. D.M. gao & zyxel(2005). Casting magnesium alloys. *Journal of application and research progress of foundry technology,* 26 (5) : 423–425.

Xu,Z.M. & X.M.Xu (2005). Surface treatment of aluminum and magnesium .M. *Shanghai: Shanghai science and technology literature publishing house.*

HanX.Y. & J.M.Long (2003). Magnesium and magnesium alloy surface treatment and application situation and development .J. Journal of light metals, 2:45–51.

YangG. & Q.T. hao (2004). Research status of magnesium lithium alloy is .J. Journal of foundry technology, 25 (1) : 19–21.

Electronic Engineering and Information Science – Wang (Ed.)
© 2015 Taylor & Francis Group, London, ISBN: 978-1-138-02772-5

The influence of N2 flow rate on the structure of CrN film

L. Yao, L. Chen, Y.Q. Chen & X.X. Sun
Mudanjiang Normal College, Mudanjiang, China
Heilongjiang Province Key Laboratory of superhard materials, Mudanjiang, China

D. Sun
Harbin University of Science and Technology, Harbin, China

ABSTRACT: In the process of sputtering, the phase compositions of thin film are, in order, Cr, Cr2N+Cr, CrN with the increased flow rate of nitrogen. The pure CrN or Cr_2N have better hardness and corrosion resistance, and the phase composition of CrN can be increased by rising the flow rate of nitrogen, which improves elasticity modulus, hardness and corrosion resistance of the thin film. But the deposition rate of would be reduced because of the excessive flow rate of nitrogen. Thus it has practical significance to acquire the single-phase CrN thin film under a high deposition rate. The samples of CrN thin film are prepared in different nitrogen flow rates. This paper discusses the influence of nitrogen flow rate on the property of CrN thin film and comparatively analyzes the property of samples prepared through different techniques (A1–A5).

KEYWORDS: CrN film; structure; flow rate.

1 INTRODUCTION

At present, magnetron sputtering has been widely adopted as a sputtering method, and the main reason is that the sputtering rate of magnetron sputtering can be an order of magnitude higher than that of other sputtering methods (Fornies, Escobar Galindo, Sanchez and Albella, 2006, Zhanga, Rapaud, Bonasso, Mercs, Donga and Coddet.2008, Lin,Wu,Zhang,Mishra and, Moore 2009). This attributes to the high ionization efficiency of electron in the magnetic field which effectively improves the current density of target and sputtering rate, while the target voltage declines significantly with increasing degree of gas ionization. Under low pressure the low-probability of sputtering atom being scattered by gas molecules is another reason. Because magnetic field increases the collision probability between electron and gas molecules effectively, the working gas pressure is also reduced so that the energy of incidence to the basal surface is increased in the process of depositing film and the possibility of film pollution is reduced to some extent(Barshilia and Rajam 2006,Nam,Jung and Han(2000)).

The basic device of magnetron sputtering improves the electrode structure on the basis of the device of DC sputtering or RF sputtering. Generally, the permanent magnet is installed in the inside of target cathode, and the magnetic orientation is perpendicular to the magnetic orientation of cathode dark space. The theory is that under the effect of electric field electrons collide with argon atoms in the process of accelerated flying to substrate. If electron has enough energy, Ar+ and electron can be ionized and glow discharge plasmas are produced. Secondary electrons produce deflection by the Lorentz force of magnetic field in the process of accelerated flying to substrate and make the circular motion in the form of cycloidal or spiral linear composite on the plane of target [9,10]. The route of these electrons is very long and they are limited in plasma region which is close to the plane of target. In this region there are a mass of Ar+ to be ionized to bombard target. With the increasing number of collisions, the electron energy gradually reduces and the electron is away from the plane of target.

2 EXPERIMENT CONDITION AND METHOD

In this experiment, the magnetron sputtering machine made in Shenyang Instrument Factory is adopted to prepare film. The target materials have metal Cr and Zr and their purities reach 99.99%. Ar gas whose purity reaches 99.99% as work gas and N_2 whose purity reaches 99.99% as reaction gas are well mixed before importing them into vacuum chamber. The gas flux is controlled by CIE mass flow controller, and this system also equips a precise MKS gas pressure measuring instrument to monitor the gas pressure of vacuum chamber in the process of sputtering.

The performance of sputtering coating is codetermined by work gas pressure, flux of N_2, deposition temperature, sputtering power, target-substrate distance and other parameters. In the experiment the fixed target-substrate distance is 30mm, as well as total pressure of sputtering chamber keeps at 0.5Pa and the sputtering time is 30min. The influence of N_2 flux on structure and performance of CrN film is emphatically considered.

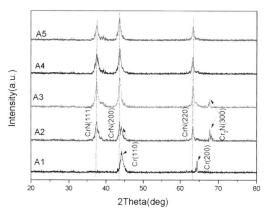

Figure 1.   XRD patterns of the films at different N_2 flux.

3   THE INFLUENCE OF N_2 FLUX ON THE STRUCTURE OF CRN FILM

Fig.1 shows the XRD patterns of CrN film deposited on the base of Mg-Li alloy at different N_2 flux. By the corresponding 2θ value of diffraction peak, we refer to related literatures to know that when 2θ values are 37.62°, 43.38° and 63.66°, it shows as the (111), (200) and (220) crystal face on the construction of face-centered cubic CrN relatively; when 2θ values are 37.61° and 67.82°, it shows as the (110) and (300) crystal face on the construction of hexagonal close-packed structure CrN relatively.

As shown in Fig. 1, the main phase of film is consist of CrN and Cr_2N. As the change of N_2 flux, the component of film will change correspondingly. When N_2 flux is 0sccm, the film is composed of Cr with body-centered cubic structure. With the increase of N_2 flux, Cr_2N and CrN will gradually occur, where preparation of the CrN film mainly depends on ionization of N_2 and evaporation ionization of Cr, meanwhile the more ionization rate the more possibility for the forming of CrN. So, the single-phase of CrN can be generated when the N_2 flux approaches 30scc. The production of Cr_2N phase is mainly due to the existence of Ar ion decreasing the combination of Cr atom with N ion in the sputtering chamber at the beginning of coating film stage, so the film contains more Cr_2N phases.

From the picture, the thickness of film being deposited has remarkable variation with the change of the N_2 flux. The deposition rate of the film shows high when the N_2 flux is lower. With the increase of the N_2 flux, the rate which is corresponding to sample A3 reaches highest value. Then, the deposition rate decreases rapidly by the reason of the phenomenon of target poisoning during the process of coating film. To explain in detail, N ion combines with the Cr atom on the target surface to form the CrN compound. Cr_2N will be generated firstly because of the lower N_2 flux to form Cr_2N. Then with the increase of the N_2 flux, the number of CrN on the surface of Cr target will be also increased, and the evaporation rate of the Cr target decreases so that the deposition rate of film declines significantly. As a result, with changing reactive gas flux, the deposition rate of film has a remarkable change, and this presents the hysteretic characteristic.

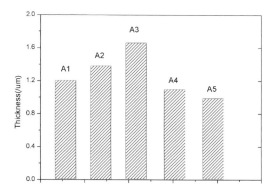

Figure 2.   The relationship between N_2 flux and films.

When the flux of reaction gas is lower, the deposition rate of film is higher. When the flux of reaction gas is increased to a certain threshold, the deposition rate of film would fall suddenly. At this moment, the adsorption rate of active gas is higher than the sputtering rate and the corresponding chemical reactions occur on the target material. The sputtering target of incident ion is not metal target but surface compound being constantly formed by sputtering. So there are two different kinds of sputtering patterns in the process of sputtering. One is the metal pattern with higher sputtering rate and the other is the compound pattern with lower sputtering rate.

Meanwhile, only when the flux of active gas flows down to lower level, the sputtering rate and deposition rate can increase to the original level. As a result, to improve sputtering efficient we hope to ensure the composition of film and control the sputtering process curve near the point E as far as possible.

4 CONCLUSIONS

Under the condition of invariable working pressure, the increase of nitrogen flux causes reducing partial pressure of argon, increasing nitrogen rate and decreasing sputtering rate of Cr target so that the growing rate of film is reduced. This increases the chance of each atomic layer being bombarded by high-energy electron and reduces particle energy arriving at the substrate. Meanwhile, the decreasing deposition rate, the formation of more defects in film, the decline of membrane density and the deterioration of quality are also attributed to the increase of nitrogen flux. The film defect is a main reason that the protective film of alloy surface loses protection; moreover the nitrogen flux effects the composition of film. So the film having standard chemical ratio presents the best performance of corrosion resistance.

ACKNOWLEDGMENT

Supported by nature science foundation of Heilongjiang province (201341), Supported by Foundation of Heilongjiang Educational Committee (12543080), Science and technology project of Mudanjiang city (G2013e1233).

REFERENCES

Fornies, E. , Escobar Galindo, R. ,Sanchez, O. & Albella, J .M.(2006). Growth of CrN_x films by DC reactive magnetron sputtering at constant N_2/Ar gas flow.J. *Surface and Coatings Technology.* 200: 6047–6053.

Zhanga,Z.G. ,O. Rapaud, N. Bonasso, D. Mercs, C. Donga &C. Coddet(2008). Control ofmicrostructures and properties of dc magnetron sputtering deposited chromium nitride films. *Vacuum.*J. 82:501–509.

Lin,J., Z.L.Wu,, X.H.Zhang,, B. Mishra, J.J. Moore, W.D(2009). Sproul. A comparative study of CrNx coatings synthesized by DC and pulsed DC magnetron sputtering .J. *Thin Solid Films*, 517(6): 1887–1894.

Barshilia,H.C. & K.S. Rajam(2006). Reactive sputtering of hard nitride coatings using asymmetric-bipolar pulsed DC generator. J.*Surface and Coatings Technology*, 201:1827–1835.

Nam, K.H. M.J. Jung & J.G. Han(2000). A study on the high rate deposition of CrNx films with controlled microstructure by magnetron sputtering. J. *Surface and Coatings Technology*, 131:222–227.

Electronic Engineering and Information Science – Wang (Ed.)
© 2015 Taylor & Francis Group, London, ISBN: 978-1-138-02772-5

Growing CrN film via magnetron sputtering on the surface of Mg-Li alloy

L. Yao, L. Chen, Y.Q. Chen & X.X. Sun
Mudanjiang Normal College, Mudanjiang, China
Heilongjiang Province Key Laboratory of superhard materials, Mudanjiang, China

D. Sun
Harbin University of Science and Technology, Harbin, China

ABSTRACT: In the process of preparing CrN thin film via magnetron sputtering, the process parameter is an extremely important influence factor which has a direct effect on components, microstructure, wear resistance and corrosion resistance of film. This paper mainly focuses on three vital parameters in the process of thin film deposition: nitrogen flow rate, substrate temperature and sputtering power. The effect of the above-mentioned process parameters on CrN film is discussed by researching the component, thickness and corrosion resistance of film in many different process parameters, thus we optimize the process parameters of preparing CrN film.

KEYWORDS: Mg-Li alloy; CrN film; surface.

1 INTRODUCTION

With advances in technology, Mg-Li alloy has attracted increasing attentions of the researchers in many areas because of its excellent performance, but it has not a large-scale application so far. The reason is mainly that the corrosion resistance and mechanical properties of alloy have not been fundamentally resolved(Chang,Wang and Chu 2006,Wang, Wu and Li 2006, Mg-Li ally2005). Only a few researches about corrosion resistance of Mg-Li alloy are reported, and at present the technologies used such as micro-arc oxidation, physical vapor deposition and ion injection have high equipment requirement and complicated process. While the traditional chemical conversion, metal coating and organic coating techniques are usually used in magnesium-rare earth, magnesium- aluminum and other alloys (Zhang, gao and zyxel 2005,Xu,and Xu 2005). So it is one of the heat pots of present research area and has important practical significance to study the corrosion behavior of Mg-Li alloy, develop the surface treatment technology of Mg-Li alloy with no pollution, environmentally friendly and low cost to improve the corrosion resistance of Mg-Li alloy and explore the corrosion and protection mechanism of Mg-Li alloy .

2 EXPERIMENT CONDITION AND METHOD

In this experiment, we use magnetron sputtering machine made in Shenyang Instrument Factory to prepare film. This device has a DC power source, a RF power and a cylindrical vacuum deposition chamber, and the cathode sputtering target is sealed by a copper plate connecting water cooling system to ensure good electrical conductivity and have effect in water cooling. A baffle is set between sputtering target and the base platform in order to prevent the target material from being polluted in the process of sputtering. Different sputtering powers such as DC or RF can be connected to the sputtering target and the substrate, but just DC sputtering power is used in this experiment. The target materials have metal Cr and Zr and their purities reach 99.99%. Ar gas whose purity reaches 99.99% as work gas and N_2 whose purity reaches 99.99% as reaction gas are well mixed before importing them into vacuum chamber. The gas flux is controlled by CIE mass flow controller, and this system also equips a precise MKS gas pressure measuring instrument to monitor the gas pressure of vacuum chamber in the process of sputtering.

2.1 *Preparation of CrN film*

In the process of the practical experiment, the composition of CrN film is controlled by adjusting the flux of Ar gas and N_2 gas and sputtering power, and the film thickness is controlled by sputtering time. The experiment process is shown in Fig.1. Before putting the sample into vacuum chamber, they need to be pre-processed, and the process flow diagram is shown in Fig1. Then this sample is put into vacuum chamber. At first we reduce the gas pressure of workroom until

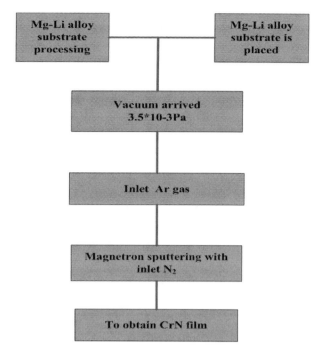

Figure 1.   Diagram for sputtering process.

it is less than 5×10^{-3}Pa by using mechanical pump and the ultimate vacuum of this device is 5×10^{-4}Pa. When the temperature needs to be increased in the preparation process, we open the baking unit for heating and make temperature control in the set temperature with temperature controller to ensure the temperature can change form room temperature to 300°C. When the temperature reaches the set temperature, work gas-Ar gas is filled into vacuum chamber. Remove the gases and impurities of surface adsorption when the gas pressure of Ar gas reaches 0.2Pa. In order to increase the adhesion performance of the film, N_2 gas is filled gradually until the flux reaches the standard flow required in experiments after sputtering Cr layer for 5min. Their fluxes have their own mass flowmeter to control respectively and the unit is sccm. The total pressure is set by the pressure controller.

2.2   The influence of process parameters on the film structure and performance

The performance of sputtering coating is codetermined by work gas pressure, flux of N_2, deposition temperature, sputtering power, target-substrate distance and other parameters [15,16]. In the experiment the fixed target-substrate distance is 30mm, as well as total pressure of sputtering chamber keeps at 0.5Pa and the sputtering time is 30min. The influence of N_2 flux on structure and performance of CrN film is emphatically considered. In the process of sputtering, the phase compositions of thin film are, in order, Cr, Cr2N+Cr, CrN with the increased flux of nitrogen. The pure CrN or Cr_2N have better hardness and corrosion resistance, and the phase composition of CrN can be increased by raising the flux of nitrogen, which improves elasticity modulus, hardness and corrosion resistance of the thin film. But the deposition rate of would be reduced because of the excessive flux of nitrogen. Thus it has practical significance to acquire the single-phase CrN thin film under a high deposition rate [5–10]. The samples of CrN thin film prepared in different nitrogen fluxs are shown in table 1. This paper comparatively analyzes the property of samples prepared through different techniques (A1–A5) and discusses the influence of nitrogen flux on the property of CrN film. The influence of gas flux, deposition temperature and sputtering power on the structure and performance of film.

Table 1. Films deposited in the different nitrogen flux.

Sample	Rate of flow (sc/cm)	Sputtering power (W)	Deposition temperature (°C)	Sputtering pressure (Pa)	Sputtering time (min)
A1	0				
A2	10				
A3	20	200W	100°C	0.5Pa	30min
A4	30				
A5	40				

3 CONCLUSIONS

1 Preparing CrN film by magnetron sputtering device, we studied the effect of different process parameters (nitrogen flow, substrate temperature, sputtering power) on microstructure of CrN film, film thickness and corrosion resistance, and obtained a set of better preparation parameters by optimizing.
2 CrN/ZrN multilayer film was deposited by using the optimized process parameters, and the difference of microstructures between monolayer and multilayer was analyzed.

We researched the effect of film on the corrosion resistance performance of Mg-Li alloy basement and discussed the corrosion mechanism of Mg-Li alloy and CrN film.

ACKNOWLEDGMENT

Supported by Foundation of Heilongjiang Educational Committee (12543080), Supported by nature science foundation of Heilongjiang province (201341), Science and technology project of Mudanjiang city (Z2013g004).

REFERENCES

Chang, T. H. J.Y.Wang & Chu C.l. (2006). Mechanical microstructures of various Mg-Li alloys. J.*Materials Letters*, 60: 3272–3276.
Wang, S. J.,G. Q. Wu & R. H. Li (2006). Microstructures and mechanical properties of 5wt. % Al2YP/Mg-Li composite. J. *Materials Letters*, 60: 1863–1865.
Li.MgW.X. & Mg-Li ally(2005). *Changsha: Central south university press*, 2005.
Zhang Y.H., d.M. gao & zyxel(2005). Casting magnesium alloys. *Journal of application and research progress of foundry technology*, 26 (5) : 423–425.
Xu, Z.M. & X.M. Xu(2005). Surface treatment of aluminum and magnesium. *Shanghai: Shanghai science and technology literature publishing house.*

Electronic Engineering and Information Science – Wang (Ed.)
© *2015 Taylor & Francis Group, London, ISBN: 978-1-138-02772-5*

Adsorption of NO and NO$_2$ on monolayer MoS$_2$: First principle study

H. Luo, Y.J. Cao, J.M. Feng, L. Lan & N. Zhang
College of Applied Science, Harbin University of Science and Technology, Harbin, China

ABSTRACT: Using first principles, we investigate the structure, electrical and magnetic properties of NO and NO$_2$ adsorbed on monolayer MoS$_2$. By changing different adsorption height, the most stable adsorption system, adsorption energy and charge transfer are obtained, and physical adsorption is obtained between the gas and the monolayer. Mulliken charge analysis shows that, NO and NO$_2$ exist as charge acceptors, NO and NO$_2$ accept 0.01e and 0.02e respectively from monolayer MoS$_2$, which will cause the change of the conductive properties of monolayer MoS$_2$. In addition, the energy band structure of the adsorption system has moved and the adsorption system has a strong magnetic moment. The analysis showed that charge accumulation and depletion of MoS$_2$ result of the motion of the energy band structure, and the magnetic moments almost concentrate in the gas molecules. Our results show that monolayer MoS$_2$ has good sensitivity for the adsorption of gas molecules, and is suitable for gas sensor materials.

KEYWORDS: First principle; Gas adsorption; MoS$_2$.

1 INTRODUCTION

Although nitric oxides have important applications in biomedicine, but nitrogen oxides, including NO and NO$_2$, are toxic gas, have properties of combustion, irritation, strong oxidation corrosiveness, will cause great harm on human health and the environment. Therefore, the detection of NO and NO$_2$ gas has an important role in agriculture, medicine and monitoring of environmental pollution. MoS$_2$ has attracted wide attention as a 2D layered material in recent years. MoS$_2$ is a transition metal chalcogenides, monolayer MoS$_2$ is composed of S-Mo-S layer, where the Mo-atom layer is sandwiched between two S-atom layers, and the interaction between layers is weak Vander Vauls (Keong Koh et al 2012). Compared to graphene with zero band gap, MoS$_2$ has a band gap of 1.8eV (Ding et al 2011, Kadantsev et al 2012), which makes MoS$_2$ have excellent electrical, optical and catalytic properties. Since Radisavljevic et al produced thin film transistor based on monolayer MoS$_2$ (Radisavljevic et al. 2011), grapheme-like MoS$_2$ obtained wide application in photoelectricity (Yin et al. 2012), catalyst (Chang et al 2013) and integrated circuit (Radisavljevic et al. 2011) etc. In recent years, theory and experiment researches show that MoS$_2$ is a very good material for sensor (Li et al. 2012, He et al. 2012, Yue et al. 2013, Zhao et al. 2014), but the reaction mechanism of gas molecules and MoS$_2$ is seldom studied. This paper focuses on the research of gas molecules NO and NO$_2$ of nitrogen oxides adsorbed on monolayer MoS$_2$, adsorption energy and

charge transfer are analyzed between gas molecules and monolayer MoS$_2$, the reasons of band structure change are also analyzed. The hope of this study can provide theoretical basis for MoS$_2$ as the sensor material.

2 CALCULATION METHOD

In this paper, the first principle is performed by using Castep software package of Material Studio 5.5 based on density functional theory (DFT). The ultra-soft pseudo-potential plane wave is used to describe the ion-electron interaction, the generalized gradient approximation (GGA) functional of Perdew-Burke-Ernzerh (PBE) is used to treat the exchange and correlation potentials. All calculations are performed with a 4x4x1 supercell of monolayer MoS$_2$ containing 16 Mo atoms and 32 S atoms. In order to avoid inter layer interactions, a vacuum layer of 16 Å is used in c direction. In geometry optimization, 340eV of plane wave cut-off energy and 2x2x1 Monkhorst-Pack (Monkhorst et al 1976) k-point sampling for the Brillouin zone are set. In energy calculation, 400eV of plane wave cut-off energy and 3x3x1 k-point sampling are set. 0.015 seperation and 3x3x1 k point sampling are set in band structure calculation and density of state calculation, respectively. In the process of optimization, all the structures are fully relaxed, where energy convergence precision is 1.0e-5eV/atom, the inter atomic forces do not exceed 0.03eV/Å. The adsorption energy of gas molecules on MoS_2 is defined as $E_{ad}=E_{gas+MoS2}-E_{MoS2}-E_{gas}$ to calculate, where $E_{gas+MoS2}$ is the total energy of optimized structure of gas molecules adsorbed on MoS_2, E_{MoS2} and E_{gas} are the total energy of the pristine MoS_2 and gas molecules respectively.

3   RESULTS AND DISCUSSION

This paper mainly studies the structure and electrical properties of NO and NO_2 gas molecules adsorbed on monolayer MoS_2. Firstly, the pristine MoS_2 supercell is geometry optimized. The optimized lattice constants is 3.17, consistent very well with the previous theoretical results (Zhao et al 2014, Ataca et al 2011). According to the previous theoretical analysis, the analysis of electronic structure is practically independent of the orientations and adsorption sites (Zhao et al 2014, Leenaerts et al 2008), therefore we select only the same position with different height of the gas molecules, then geometry optimization and energy calculations are performed. From the above calculations, we select the most stable height configuration with lowest adsorption energy to carry out theoretical analysis. The optimal adsorption configurations are shown in Figure 1. As shown in Figure 1 (a), at NO adsorption, N atom is located below O atom and the height between the center of mass of NO molecule and the top S-layer of the MoS_2 is 3.29 Å, and the optimized N and O bond length d_{N-O} is 1.19 Å. As shown in Figure 1 (b), at NO_2 adsorption, NO_2 is located on the top of S-Mo bond symmetrically and vertically, the height between the center of mass of NO_2 molecule and the top S-layer of the MoS_2 is 3.24 Å, and the optimized N and O bond length d_{N-O} is 1.23 Å, O-N-O bond angle is 130.545°, the calculation results are consistent with Zhao et al (Zhao et al 2014).

In the position of figure 1 (a) and (b), adsorption height of gas molecules adsorbed on MoS_2 is analyzed, adsorption energy versus height is shown in Figure 2. The height represents the distance between the center of mass of gas molecule and the top S-layer of the MoS_2, adsorption energy can be calculated as $E_{ad}=E_{gas+MoS2}-E_{MoS2}-E_{gas}$. It can be seen from the figure 2, when the adsorption height of NO gas molecule is 3.29Å, it has the largest adsorption energy of −30meV. When the adsorption height of NO_2 gas molecule is 3.24Å, it has the largest adsorption energy of −61meV, which is 2 times larger than the maximum adsorption energy of NO. As the adsorption height increases, the adsorption energy decreases gradually, which illustrates the adsorption properties of gas molecule on MoS_2 gradually weaken. Due to the larger adsorption height and smaller adsorption energy, the interaction between gas molecules and MoS_2 is physical adsorption.

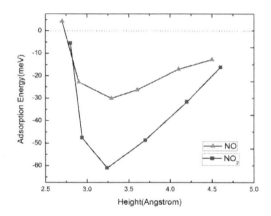

Figure 2.   Adsorption energy versus height on NO and NO_2 gas molecules.

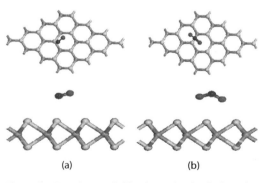

(a)   (b)

Figure 1.   Top views and side views of optimal adsorption configuration of (a)NO and (b)NO_2 gas molecules adsorbed on MoS_2. Yellow and wathet spheres represent S and Mo atoms respectively, blue and red spheres represent N and O atoms respectively.

In order to observe the influence of adsorbed gas molecules on the electrical properties of MoS_2, we analyzed the band structure and the spin density of state (DOS) of monolayer MoS_2. As shown in Figure 3, monolayer MoS_2 is a direct band gap semiconductor with band gap of 1.729eV, consistent with previous report (Ding et al. 2011, Kadantsev et al 2012, Yue et al. 2013). The conduction band energy is contributed mainly to the Mo-4d orbit, the valence band energy is determined by the Mo-4d and S-3p orbit. Spin up and spin down of DOS is completely symmetrical, the magnetic moment of the system is 0, does not show magnetism.

When the gas molecules adsorb on monolayer MoS_2 with optimal adsorption configuration, the introduction of gas molecules causes the change of band structure and DOS of monolayer MoS_2. But due

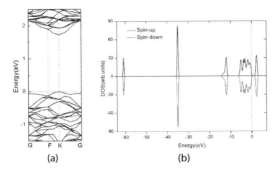

(a)   (b)

Figure 3.   (a) band structure, (b)spin DOS of monolayer MoS$_2$ respectively.

to the smaller adsorption energy, the band structure is not significantly altered when NO, NO$_2$ are adsorbed, just in the MoS$_2$ band gap, impurity states are introduced. We can see from fig 4(a), that in the band structure of NO adsorption system, two up-spin states and a down-spin state are introduced. By observing the local density of state (LDOS) of system with NO adsorbed on MoS$_2$, we can see that the impurity states are mainly contributed to the N-2p orbit, and

also contributed to O-2p orbit. The hybridizations of Mo-4d, S-3p, N-2p and O-2p orbits cause the electron transfer between MoS$_2$ and NO molecule, leading to the overall downward of band structure. From Fig 4(b), in the NO$_2$ adsorption system, the impurity state of down spin is introduced, which is contributed to the N-2p orbit from LDOS of system with NO$_2$ adsorbed on MoS$_2$. The hybridizations of Mo-4d, S-3p, N-2p, N-2s and O-2p orbits cause the electron transfer between MoS$_2$ and NO$_2$ molecule, but the band structure of NO$_2$ adsorption system almost does not move, the reason is that the electrons transfer between many obits of NO$_2$ and MoS$_2$, leading to the stable energy of the system.

NO and NO$_2$ molecules orbits contain unpaired electrons, so these two kinds of gas molecules have magnetism. Both of the adsorption systems have magnetic moment of 1μB which mainly concentrate in the NO and NO$_2$ molecules. This accounts for smaller adsorption energy and physical adsorption between MoS$_2$ and gas molecules.

In order to further analyze the interaction between gas molecules and monolayer MoS$_2$, we performed Mulliken charge analysis (as shown in table 1). We can see from system with NO molecule adsorbed on MoS$_2$, that N-2p orbit accepts 0.03e, but N-2s orbit

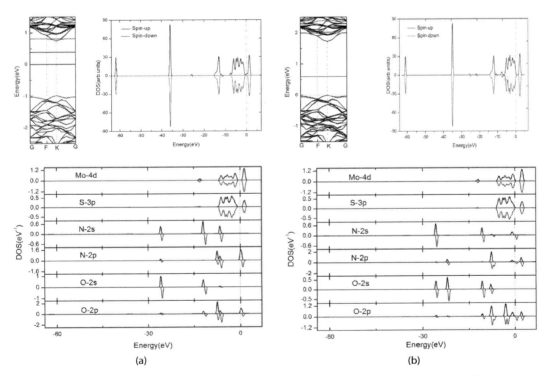

(a)   (b)

Figure 4.   Energy band structure, spin DOS and LDOS for (a) NO adsorbed on monolayer MoS$_2$ (b) NO$_2$ adsorbed on monolayer MoS$_2$.

and O-2p orbit donate 0.01e respectively, electron transfer occurs in monolayer MoS$_2$ and NO gas molecule, NO accepts 0.01e finally. Due to the electron energy of NO molecule is lower, making energy of NO adsorption system lower after electrons transfer from NO to MoS$_2$. We can see from system with NO$_2$ molecule adsorbed on MoS$_2$, that N-2s orbit accepts 0.03e and N-2p donates 0.01e, NO$_2$ accepts 0.02e finally. After NO$_2$ molecule accepts and donates electrons, the energy tends to be stable, so the band structure of adsorption system with NO$_2$ adsorbed on MoS$_2$ hardly produce downward move. Electrons NO$_2$ gas molecule gains is more than that NO gas molecule gains, which is related with adsorption energy between gas molecules and MoS$_2$. Electron transfer between gas molecules and monolayer MoS$_2$ leads to the change of the resistance of MoS$_2$, which cause change of current at applied voltage. The gas molecules get electrons from MoS$_2$, which makes carrier mobility of n-type doped MoS$_2$ semiconductor decrease, makes carrier concentration of p-type doped MoS$_2$ semiconductor increase, so that the current of transistor based on MoS$_2$ changes, then NO and NO$_2$ gas molecules are detected.

Table 1. Mulliken charge analysis. s/p indicate the number of electrons in 2s orbit and 2p orbit. s/p of NO and NO$_2$ represent 2s and 2p electrons of the pristine NO and NO$_2$ molecules, s/p of NO ad MoS$_2$ and NO$_2$ ad MoS$_2$ represent 2s and 2p electrons of NO and NO$_2$ adsorption system.

s/p	NO	NO ad MoS$_2$	NO$_2$	NO$_2$ ad MoS$_2$
N	1.79/3.06	1.78/3.09	1.38/3.18	1.41/3.17
O	1.84/4.32	1.83/4.32	1.85/4.37	1.85/4.37

4 CONCLUSION

Using first principles, the structure and electrical properties of gas molecules adsorbed on MoS$_2$ are calculated and analyzed, which provides theoretical analysis for exploring MoS$_2$ as sensor materials. Calculated results show that the monolayer MoS$_2$ is a direct band gap semiconductor with no magnetic moment. Impurity states are introduced with gas molecules adsorbed on monolayer MoS$_2$, near the Fermi level, mainly by the hybridization of Mo-4d, S-3p, N-2s and N-2p orbits, which results in the electron transfer. The band structure of NO adsorption system

has overall downward, and NO molecule obtains 0.01e from monolayer MoS$_2$, NO$_2$ molecule obtains 0.02e from monolayer MoS$_2$. The adsorption of NO and NO$_2$ gas molecules cause a magnetic moment of 1μB, which mainly concentrated in the NO and NO$_2$ molecules. Our research shows that the adsorption of NO and NO$_2$ gas molecules could change the carrier concentration of monolayer MoS$_2$ semiconductor, is a promising material for sensor application.

REFERENCES

Ataca, C., H.Sahin, , E.Akturk, et al. (2011). Mechanical and Electronic Properties of MoS$_2$ Nanoribbons and Their Defects, *The Journal of Physical Chemistry C*, 115: 3934–3941.

Chang, Y.H., F.Y.Wu, , T.Y.Chen, et al. (2013).Three-Dimensional Molybdenum Sulfide Sponges for Electrocatalytic Water Splitting. *Small*, 10:895–900.

Ding, Y., Y.L.Wang, J.Ni, et al. (2011). First principles study of structural, vibrational and electronic properties of graphene-like, MX2 (M=Mo, Nb, W, Ta; X=S, Se, Te) monolayers. *Physica B*, 406:2254–2260.

He, Q., Z. Zeng, & Z.Yin, (2012). Fabrication of flexible MoS2 thin-film transistor arrays for practical gas-sensing applications. *Small*, 8(19):2994–2999.

Kadantsev, E.S. & P. Hawrylak, (2012). Electronic structure of a single MoS$_2$ monolayer. *Solid State Communications*, 152: 909–913.

Keong Koh, E.W., C.H.Chiu, Y.K.Lim, et al. (2012). Hydrogen adsorption on and diffusion through MoS$_2$ monolayer: First-principles study. *International Journal of Hydrogen Energy*, 37:14323–14328.

Leenaerts, O., B. Partoens, & F.M.Peeters, (2008). Adsorption of H$_2$O, NH$_3$, CO, NO$_2$ and NO on graphene: a first-principles study. *Phys Rev B*, 77:125416.

Li, H., Z.Yin, , Q.He, et al. (2012). Fabrication of single and multilayer MoS$_2$ film-based field-effect transistors for sensing NO at room temperature. *Small*, 8: 63–67.

Monkhorst, H.J. & J.D.Pack, (1976). Special points for Brillouin-zone integrations. *Phys Rev B*, 13:5188–5192.

Radisavljevic, B., A.Radenovic, J. Brivio, et al.(2011). Single-layer MoS$_2$ transistors. *Nature Nanotechnology*, 6.

Radisavljevic, B., M.B.Whitwick, & A. Kis, (2011). Integrated circuits and logic operations based on single-layer MoS$_2$. *ACS Nano*, 5: 9934–9938.

Yin, Z.Y., H.Li, , H.Li, et al. (2012). Single-Layer MoS$_2$ Phototransistors. *ACS Nano*, 6:74–80.

Yue, Q., Z.Z.Shao, , S.L.Chang, et al. (2013). Adsorption of gas molecules on monolayer MoS$_2$ and effect of applied electric field. *Nanoscale Research Letters*, 8:425.

Zhao, S.J., J.M.Xue, & W.Kang, (2014). Gas Adsorption on MoS$_2$ monolayer from first-principles calculations. *Chemical Physics Letters*, 595:35–42.

Electronic Engineering and Information Science – Wang (Ed.)
© 2015 Taylor & Francis Group, London, ISBN: 978-1-138-02772-5

Effect of composite films preparation on polyimide performance

N. Zhang, J.H. Yin, J.L. Li, Z.S. Lin & S.C. Wu
School of Applied Science, Harbin University of Science and Technology, Harbin, PR China
Key laboratory of Engineering Dielectrics and Its Application, Ministry of Education, Harbin University of Science and Technology, Harbin, PR China

ABSTRACT: The design and preparation of polymers have become the hot issue in the study of material science. This paper outlines the main characteristics of nano hybrid polyimide film, simply previews the preparation of composite films by in situ polymerization method. Preparation of polyamide acid and imidization process are the main factors of affecting the performance of polyimide composite films. The results show that fully dried reagent is a prerequisite for successful experiments, carrying out experients at low temperatures and selecting appropriate imidization temperature are a guarantee of success.

KEYWORDS: polyimide; composite film; imidization; preparation.

1 INTRODUCTION

The polyimide is widely known with the excellent dielectric properties, mechanical properties, thermal stability, chemical resistance and radiation resistance for its unique structure of rigid bar. The Polyimide film acts as the substrate of composite materials, developing flourishing in the fields of aerospace, electronics, motorcycles, precision instruments and automated office machinery (Li et al. 2007, Srivastava et al. 2009, Shen et al. 2013). With the promotion for the miniaturization of electrical appliances, high voltage and variable frequency speed adjusting technology, the new requirements of thermal expansion coefficient, low dielectric constant and high resistance to corona life of composite film materials are put forward. The traditional polyimide film has been unable to meet the demand. On the one hand, the researchers from domestic and abroad to improve the performance of the Polyimide film by optimizing the process of the film and introducing advanced equipment, but this way brings about the large investment and long cycle (Sonerud et al. 2009, Tian et al. 2011, Zheng et al. 2012). On the other hand, by nano-modifying the performance of the PI film, the nano-hybrid film is obtained with a special function, this method has become the main way to improve the performance of Polyimide film for its low cost and quick effect.

2 PREPARATION OF COMPOSITE FILMS

Composite films were prepared by in situ polymerization method. The thickness is about 30 μm.

2.1 Experimental Materials

4, 4'- diamino dipheny lether (ODA); pyromellitic dianhydride (PMDA); N , N - dimethylacetamide (DMAC); nano - particle.

2.2 Sample preparation

The preparation process flow diagram of the nanocomposite film is shown in Figure 1.

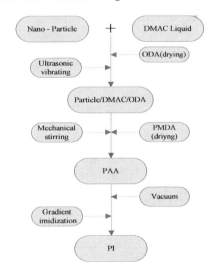

Figure 1. Preparation process flow diagram.

According to the composite proportion, nano-particle and ODA are accurately weighed and put in three neck flask, DMAC solvent of certain volume is added to dissolve the ODA completely, then ultrasonic vibration 4 h is carried on. PDMA of formula ratio is added in portions, precursor polyamide acid (PAA) slurry of compatible viscosity is obtained after mechanical agitation. Continue stirring 15 min, the polyamide acid solution of more mixing uniformity is obtained. Letting stand for 12 h, a thin film of certain thickness is paved on the clean glass plate. Then the thin film is placed in a high temperature oven and processed in imidization. After the immunization is completed, the composite film is removed from the glass plate, nanocomposite film of 30 μm is completely prepared.

3 RESULTS AND DISCUSS

3.1 *Preparation of the polyamide acid.*

Preparation of the polyamide acid is one of the keys to prepare polyimide composite film successfully, which affects the performance of the polyimide. The factors of affecting the performance of the polyimide composite film are as follows:

1 ODA and PMDA need to dry: Because reagent contains more moisture which will cause the molecular weight of polyamide acid not big enough and cause the polymer chain not long enough. Without drying, the toughness of film made is not enough and easy to break.

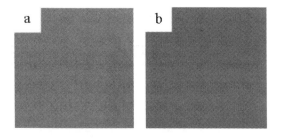

Figure 2. Films before folding experiments: (a) ODA and PMDA are drying, (b) ODA and PMDA are not drying.

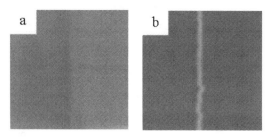

Figure 3. Films after folding experiments: (a) ODA and PMDA are drying, (b) ODA and PMDA are not drying.

Figure 2 is the photograph for the same component composite films before folding experiments, Figure 3 is the image after folding experiments. Figure 2 shows composite films surface is smooth, no significant difference. According to Figure 3(a), composite film which is made of ODA and PMDA without drying process is easy to rattle. As shown in Figure 3(b), the composite film made of ODA and PMDA with drying process shows good toughness, only a crease is on the film surface and the film does not break after folding experiments. In order to ensure that polyamide acid has higher molecular weight, ODA and PMDA and the reaction vessel should be sufficiently dried before using them. Drying reagent is the first step of the experiment, but also a very important step.

2 Fully stirred: In order to make ODA with PMDA fully polymerization, we should stir them at room temperature over six hours. If the stirring is insufficient and the degree of polymerization is not enough, it will influence the performance of composite films.

3 The temperature: The formation of polyamide acid is an exothermic reaction, reducing the temperature is beneficial to the formation of polyamide acid. Therefore, synthesis of polyamide acid is usually at a low temperature (-10 °C ~ room temperature). In addition, the degradation of polyamide acid is very sensitive to temperature, which is an endothermic reaction. The degradation process requires activation energy, increasing the temperature accelerates the degradation. On the other hand, high temperature enhances the motion ability of molecular chain, where some weak sections are easy to be exposed and degraded by other groups. Too high temperature will cause part of polyamide acid cyclization and dehydration, resulting in precipitation, and dehydration will accelerate the degradation of macromolecules. So the polyamide acid usually requires preservation of low temperature.

3.2 *The imidization process*

Only when the hydrogen bond between polyamide acid and solvent conducts decomplexation, imidization reaction can be carried out. The decomplexation of hydrogen bond shows that solvent molecules are no longer bound to polyamide acid molecules and become free solvent molecules, which in the imidization process has "plasticizer" effect. The plasticizing degree mainly reflects in the solvent diffusion to the surface and volatile time.

Figure 4 is the composite film image at initial imidization temperature 30 °C and with maintaining

12 hours at this temperature. Figure 4 indicts that embrittlement exists on the film surface, parts of the region have been burst in the heating process, and degree of imidization is low. Low initial imidization temperature produces less "plasticizer", it weakens the plasticication of the molecular chain. This not only results in reducing the mechanical properties of the material, but also decreases imidization rate, and degree of imidization make is low in the limited time.

Figure 4.　Initial imidization temperature 30 °C and maintain 12 hours.

Figure 5 is the composite film images at initialization temperature 80 °C and with maintaining different hours at this temperature. According to Figure 5(a), the composite film is prepared at the initial imidization temperature 80 °C and maintained 1 hour, all of the film has been burst during the heating process. However, as shown in Figure 5(b), film maintained 12 hours at temperature 80 °C keeps smooth.

High initial imidization temperature can make the polyamide acid and solvent decomplexation, which obtain a higher degree of imidization in very short time. The maintaining time is longer, the plastication of residual solvent moleculars on molecular chain is stronger, the rate of imidization is faster.

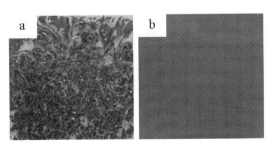

Figure 5.　Keep initialization temperature 80 °C different hours: (a)1 hour, (b)12 hours.

4　CONCLUSION

Composite film was prepared by adopting the in-situ polymerization, the factors to affect the performance of polyimide composite film are analyzed, the conclusions are as follows:

1　The ODA, PMDA and their reaction vessel need to dry fully before experiment, which can guarantee the composite film with good toughness performance. This way can reduce the influence of moisture on experiment, and the film made by this way is not easy to break.
2　In the process of experiment, ODA and PMDA need to mix fully to make polymerization reaction.
3　The experimental process and the way to save the polyamide acid are conducted in the low temperature environment.
4　In the process of experiment, the initial imidization temperature was chosen higher and maintained for a long time, which can make the degree of imidization for thin film above the average.

ACKNOWLEDGMENTS

The research work was supported by National Natu-ral Science Foundation of China (51077028, 51307046, E070502) and Heilongjiang Natural Science Foundation of China (A201006).

REFERENCES

Li H.Y., G. Liu & B. Liu (2007). Dielectric properties of polyimide/Al2O3 hybrids synthesized by in-situ polymerization. *J. Material Letters 61*, 1507–1511.
Srivastava R., S. Banerjee & D. Jehnichen (2009). In situ preparation of polyimide composites based on functionalized carbon nanotubes. *J. Macromolecular Materials and Engineering 294(2)*, 96–102.
Shen J., X.L. Li & Y. Zhang (2013). Synthesis and characterization of highly soluble and optically transparent polyimides derived from novel fluorinated pyridine containing aromatic diamine. *J. High Performance Polymers 25(3)*, 268–277.
Sonerud B., T. Bengtsson & J. Blennow (2009). Dielectric heating in insulating materials subjected to voltage waveforms with high harmonic content. *J. IEEE Transactions on Dielectrics and Electrical Insulation 16(4)*, 926–933.
Tian F.Q., Q.Q. Lei & X. Wang (2011). Effect of deep trapping states on space charge suppression in polyethylene/ZnO nanocomposite. *J. Applied Physics Letters 99(14)*, 103–106.
Zheng Z.Y., G.L. Yu & Z.J. Zhang (2012). Characterization of SiO2/polyimide/SiO2 composite films and its applications in field emission display. *J. Journal of Vacuum Science and Technology 32(3)*, 214–218.

Electronic Engineering and Information Science – Wang (Ed.)
© 2015 Taylor & Francis Group, London, ISBN: 978-1-138-02772-5

The research of EDFA signal out and total ASE

C.P. Lang, T. Shen, Y.Q. Li, H.Y. Yang, J.D. Xie & Y.C. Li
School of Applied Sciences, Harbin University of Science and Technology, Harbin, China

Y.H. Deng
School of Automation, Harbin University of Science and Technology, Harbin, China

S.Q. Li
Office of Academic Affairs, Harbin University of Science and Technology, Harbin, China

ABSTRACT: EDFA signal light output power and the total power of the spontaneous emission by numerical simulation on the basis of the spontaneous emission of erbium-doped fiber amplifier output optical signal and amplification factors, discuss and validate the conclusions in the EDFA design personnel to provide reference.

KEYWORDS: Signal out; Amplified spontaneous emission; Numerical simulation.

1   INTRODUCTION

Fiber amplifier is generally composed of the gain medium, pump and input/output coupling structure. At present, there are three kinds of fiber amplifiers mainly include EDFA, semiconductor optical amplifier and fiber Raman amplifier (Roriz 2014). According to its application in optical fiber networks. There are three main different USES optical fiber amplifier (Yanga et al. 2014). In order to improve the power of the transmitter, we use at its side. To extend the transmission distance, we relay amplifier to compensate for the optical fiber transmission loss in optical fiber transmission line. In the passage by Oasix software, the output of the optical signals in EDFA and amplified spontaneous emission in the above are simulated under different conditions (Qu et al. 2014). To draw charts and use the output data, analysis, we expound the output light signal and amplified spontaneous emission and the relationship with above situation changes by drawing charts and using the output data, analysis, and summarize combining chart finally.

2   MEASUREMENT PRINCIPLE

With high energy to pump laser erbium-doped fiber, it can make the erbium ions bound electrons from the ground state energy of a large number of excitation on high level. However, high energy level, is not stable. No radiation attenuation and erbium ions will soon experience. Doesn't the release of photons fall into metastable energy levels. Excited electrons will issue a wide spectrum of light, which called fluorescence.

When the pump is strong enough, you can make the reversal of the electron is formed between energy levels, and distribution, at this time if the wavelength of incident signal light falls on the fluorescent with time. When pumping and signal light through the erbium-doped fiber at the same time. Fluorescence with the energy will shift to the signal light. The signal light can go through the process of stimulating radiation harvest energy from ion system, thereby increasing the amplified signal. Because of the pump light and signal light in optical fiber transmission in guided wave form, So between the two can exchange energy effectively, the efficiency is very high. Spontaneous radiation is random is coherent light, can be in any direction. Only that part of the spontaneous radiation within the critical Angle is to the body and coupling into the optical area can be further stimulated radiation (Yang et al. 2014). So coupling into the beam transmission path would then be amplified spontaneous emission light. ASE is the noise of the optical amplifier and its frequency band is very wide, can occupy the whole gain bandwidth. In order to reduce the noise, it is usually equipped with optical filter in the optical amplifier.

3   THE ESTABLISHMENT OF MEASUREMENT SYSTEM

In order to efficiently complete the increasingly complex communication system simulation research, On the basis of the analysis of optical fiber communication system, simulation software is used to establish the simulation model of high speed large capacity optical fiber system, for high-speed optical fiber communication system features of laser modulation frequency bias current and the relationship between the loss and dispersion of the fiber of the simulation results, besides, it can be designed using the simulation software is flexible to meet the need of the simulation system. using the modern computer simulation of the optical fiber communication system, and can be directly set laser, optical fiber and model of testing instruments ,based on the device, the performance of the system analysis and testing.

4 RESULTS AND THE ANALYSIS OF EXPERIMENT

Experiments respectively test the fiber length, the forward pump power, back to the pump power, signal light power, temperature and structure of the output light signal of different pump and the influence of the amplified spontaneous emission, by mathematical methods smoothing curve as shown in Figures 1–6. The output optical signal under different conditions and amplified spontaneous emission spectrum line is not the same, but are generally on the rise, and the output light signal and the ultimate size of amplified spontaneous emission is roughly the same.

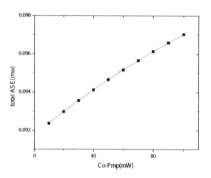

Figure 1. EDF-HE980 output optical signal forward along with the wave under the pump.

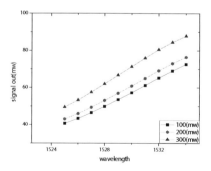

Figure 2. EDF-HE980 the output light signal in different fiber length.

Figure 1 selection for optical fiber EDF - HE980 with 10 m fiber length, the signal light intensity for 100 mW, the forward pump power of 100 mW, 200 mW, 300 mW are shown in the curve. As shown in pictures, overall change of rules is the same as EDF - HE980. However, due to a different distribution of erbium ions in the optical fiber, curve and EDF - HE980 is slightly different, when near the wavelength at 1525 nm, three curves may intersect. Output optical signal increment in wavelength arrived at near 1532 mm less gradually. Gradually increase again in wavelength near more than 1534 mm.

Figure 2 selection for optical fiber EDF - HE980 with the signal light intensity for 100 mW, the forward pump power of 100 mW, 10 m 30 m and 50m fiber length. As shown in pictures, overall change of rules is the same as EDF - HE980, However, due to a different distribution of erbium ions in the optical fiber, curve and EDF - HE980 is slightly different, When the fiber length of 10 m, wavelength is near 1532 mm, the output light signal increases gradually decreasing, while the wavelength near more than 1531, the input optical signals incremental increase gradually.

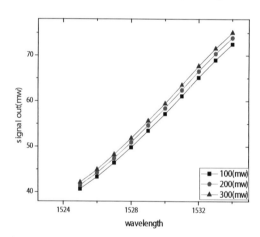

Figure 3. The curves of EDF-HG980 's output light signals varies with wavelength at different input light signals.

Figure 3 is the one chosen from experimental results which meets the requirements of optical fiber EDF-HG980 with10 m fiber length, the forward pump power for 100 mW, the signal light intensity 100 mW, 200 mW, 300 mW. However, because of different distribution of erbium ions in the optical fiber, the curve is slightly different from EDF-HE980's.With the increase of input light signal, the corresponding incremental increase amplitude than EDF - HE980, the output optical signal increment is gradually decreased when the wavelength is at about 1532 mm and increase again when the wavelength is at about 1534 mm.

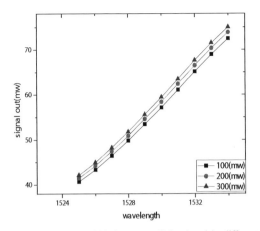

Figure 4. EDF-HE980 the output light signal in different pump power.

Figure 4 is the one chosen from experimental results which meets, the optical fiber EDF-HG980 with 10 m fiber length, the signal light intensity for 100 mW, the forward pump power of 10 mW at first and no more than 100 mW and measuring the figure every 10mv. Over all specific rules is the same as EDF- HE980.However, the slope of the course is slightly small, the output value is relatively large.

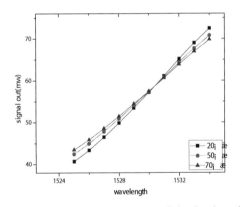

Figure 5. The EDF-HG980 's output light signals varies with temperature at different input light signals.

Figure 5 is the one chosen from experimental results which meets the requirements of optical fiber EDF-HG980 with fiber length 10 m,the forward pump power 100 mW, the signal light intensity 100 mW, the temperature of 20°, 50°, 70° are shown in the curve. With the increase of wavelength, the output optical signal has gradually increased. When the wavelength is at about 1525 nm, three curves may intersect; When the fiber length is 10 m and wavelength is about 1532 mm, the output light signal increases gradually decreasing, while the wavelength is near

more than 1531, the input optical signals incremental increase gradually. EDF-HE980' amplified spontaneous radiation range is relatively large with the change of forward pump power.

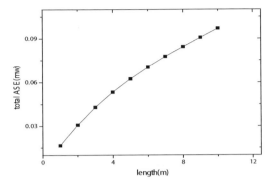

Figure 6. The curves of EDF-HG980 's amplified spontaneous radiation with fiber length.

Figure 6 is the one chose from experimental results which meets the requirements of optical fiber EDF-HG980 with 10 m fiber length, the forward pump power of 100 mW, the signal light intensity for 100 mW, Initial fiber length 1 m, the maximum of 10 m, every 1 m all measurement data are shown in the curve at a time. However, the increment amplified spontaneous radiation is gradually decreased when the length of optical fiber is from 5to 9 m.when the length of optical fiber is more than 9 m, the increment amplified spontaneous radiation is gradually increased.

5 CONCLUSION

The structure of the pump doesn't influence the amplified effect EDFA on output optical signals. The output optical signal increases with the increase of pump power, decreases with the increase of optical fiber length, increases with the increase of input optical signal, increases with the increase of temperature. The structure of the pump doesn't influence the amplified spontaneous radiation of EDFA. The amplified spontaneous radiation increases with the increase of pump power, increases with the increase of optical fiber length, increases with input optical signal, increases with the increase of temperature.

ACKNOWLEDGMENTS

This work was financially supported by the Science and Technology Research Project of Heilongjiang Province Education Bureau (No.12531134).

REFERENCES

Roriz, P., L. Carvalho, O. Frazãob, J. L. Santosb, & J. A. Simõesa (2014). From conventional sensors to fibre optic sensors for strain and force measurements in biomechanics applications: A review. *J. Biomech. 47*, 1251–1261.

Yanga, H., Z., X. G. Qiaoa, D. Luoc, K. S. Limb, W. Y. Chongb, & S. W. Harunb (2014). A review of recent developed and applications of plastic fiber optic displacement sensors. *Measurement. 48*, 333–345.

Qu, Z. L., Q. Zhao, & Y. G. Meng (2014). Im provement of sensitivity of eddy current sensors for nano-scale thickness measurement of Cu films. *NDT & E International. 61*, 53–57.

Yang, L., A. Frank, R. Wüest, B. Ülenaltin, M. Len ner, G. M. Müller, & K. Bohnert (2014). A Study on Different Types of Fiber Coils for Fiber Optic Current Sensors. *Key Engineering Materials. 605*, 283–286.

Electronic Engineering and Information Science – Wang (Ed.)
© 2015 Taylor & Francis Group, London, ISBN: 978-1-138-02772-5

Simulation of fiber optical current transformer base a novel structure

J.D. Xie, T. Shen, L.L. Liu, S. Meng, W.B. Liu, J.D. Li, C.P. Lang, Y.Q. Li & Y.C. Li
School of Applied Sciences, Harbin University of Science and Technology, Harbin, China

Y.H. Deng
School of Automation, Harbin University of Science and Technology, Harbin, China

S.Q. Li
Office of Academic Affairs, Harbin University of Science and Technology, Harbin, China

ABSTRACT: Based on the magneto-optical medium Faraday effect, it designs an all-optical fiber current sensor which based on the new film. Through the optical software Optisysterm, in the light of the structure of all-optical current sensor makes emulation. Through the software emulation, it verifies the feasibility of the design of all parts components and overall structure. Software of Origin does observation and analysis of simulation results for, respectively in the 1 T within 1/8, 1/4, 1/2 ,1, respectively, find the sine wave, square wave and square wave state filtering signal to the three kinds of modulation signal in the corresponding period history. Analyze the corresponding relationship, and the rationality of design scheme is verified.

KEYWORDS: Faraday effect; Optisysterm; Fiber Optical Current Transformer.

1 INTRODUCTION

With the new development of power system, the voltage class and the improvement of automation, The traditional electromagnetic current transformer gradually exposed its limitations (Kucuksari & Karady 2012). In recent years, with the rapid development of the optical fiber sensing technology. The fiber optical current transformer's (FOCT) research and development of globally had been Promoted (Geng et al. 2012, Wang & Quan 2012). FOCT has high linearity, sensitivity, safety operation, a high degree of isolation, small size, low cost, high ability of resistant to vibration & electric – magnetic interference, wide frequency domain response, AC - DC measurable, easy to connect to the electronic equipment, etc. It has become a strong candidate for the traditional electromagnetic current transformer. So far, researchers have proposed various FOCT implementation scheme, but still failed to get the popularization and application of large area in the power system, the main reason is the problem of temperature drift and reduce of long-term operational stability. Therefore, achieve real practical need further in-depth study. The Key components are the magneto-optic material of magneto-optic effect and the structure of the Fiber Optical Current Transformer under the interaction between optical field and magnetic field (Yu et al. 2012). They determine the sensitivity of the OCT,

the measurement range of electricity, the wastage of the system, and the temperature range in applicable environment, etc. In this paper, we will introduce the full optical fiber current sensor based on new thin film materials. It is expected to get practical application in power system.

2 OPTICAL ELECTRICITY MUTUAL INDUCTOR'S BASIC OPERATING PRINCIPLE

Optical electricity mutual inductor divides into active type and passive type, the sorts by if the sensing head needs power-up. The optical fiber type electricity mutual inductor which presented is passive optical electricity mutual inductor (Joseba et al. 2013). Its mainly operating principle is Faraday magneto-optic effect.

Faraday magneto-optic ray effect propagates along the direction of magnetic field in some medium. It's rotation angle Δφ is in direct proportion to the magnetic field intensity H and the length of light path L:

$$\Delta\varphi = VHL \qquad (1)$$

The V is the fiber optical Verdet constant. For FOCT, the electromagnetic produced by current carrying conductor around the space meet the Ampere loop rule, so the rotation angle in optical fiber is:

$$\Delta\varphi = V \oint H \, dl = NVI \qquad (2)$$

The N is the fiber sensing coil's total number of turns, and I is the electricity in optical fiber.

3 FOCT'S SIMULATION AND EXPERIMENT

3.1 Simulat FOCT basic on OPtisysterm

OptiSysterm is a new type of innovative optical communication system simulation software, which has the feature of designing, measurement and optimizing every type of optical network elements and its physical layer, suitable for long-distance communication system to LANS and MANS. There are powerful environmental simulation ability, abundant component libraries and real system's classification definition in OptiSystem. Its function can extend by additional user component libraries and complete interface, to become a series of emulational instruments widely used by the optical fiber system.

3.2 AFOCT 's theoretical model

To make the experimental structure of an AFOCT light path operate chronically and steadily, the experiment has been simplified. This light path chooses relatively simple construction of simple optical way. Figure 1 is the theory picture of single light path construction. The light sends by the light source L turns out to be parallel light after through the lens L1, and turns to linearly polarized light through polarizer P1, then enters into magneto optical medium C. Finally, it goes through analyzer P2 and lens L2 into photoelectric detector D. When the magnetic field around the magneto-optic material changes, measure the variation of Faraday rotation to measure the variation of electricity.

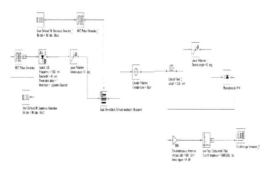

Figure 1. Simple optical way of schematic diagram of AFOCT.

This simulation system is: the first make simulation of the light source through the polarizer to become polarized light, through λ/4 wave plate and the sensing fiber, add a modulation signal to the polarizer and λ/4 wave plate to modulate, then become polarizer through polarized light. Finally, the output signals are processed and analyzed by PIN transformation. By changing the modulation signal to analog current changes in the simulation of the actual environment, compare the output signal, observe the phenomenon.

The simulation theory under the Opti System software was set up. Simulate the structure shown in figure 2. In order to concisely display, leave out some auxiliary accessories and measuring instruments. Assume the frequency of light source is 1550 nm, the spectral width is 40 nm. The output signal is Lagrange-Gaussian model. Call up ideal polarizer of OptiSystem software, which is set up to 0 degrees same with the actual polarizing angle. The ideal circularly polarized light conversion model in the Optisystem software. Assuming the phase delay is consistent with the actual data, set to 90 degrees, handle the signal using a transimpedance amplifier and low-pass bass Voss filter after the photoelectric conversion.

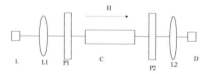

Figure 2. Simulate structure diagram.

Respectively, graphic date of sinusoidal modulation signal and filtered signal derived from the OptiSystem software corresponding to the retrieved date into 1T, 1/2T, 1/4T, 1/8T to Origin. Then, making the relation curves between sinusoidal signal respectively in 1T, 1/2T, 1/4T, 1/8T filter output signal and the corresponding periodic modulation signal.

It can be seen form figure 3 that the filter output signal of each modulation signal corresponding to is not the only one.In two points in one period,the relationship between two kinds of signals is also more complicated.

Fitting analysis of the two results of Figure 4. The relative relations curve within the 1/4T can be fitted to the curve of the samples like $y = ax^3 + bx^2 + cx$, and the curve within the 1/8T can be fitted to the curve of the samples like $y = ax^2 + bx + c$. To compare the fitting results,as shown in Figure 5.

The left curve relations conforms to this relation $y = ax^3 + bx^2 + cx$, among them, a=0.2512, b=−0.05487,c=0.18414. The standard deviation. It can be seen from the table, the right curve relations conforms to this relation in the 1/8 T of the fitting relationship :y=a*x^3+b*x^2+c*x, among them, the standard deviation also can be seen from the table.

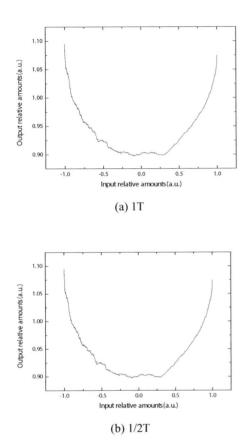

(a) 1T

(b) 1/2T

Figure 3.　Sinusoidal singal simulation analysis graphics in 1T(a) and 1/2T(b).

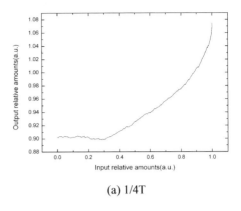

(a) 1/4T

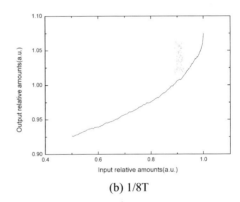

(b) 1/8T

Figure 4.　Sinusoidal singal simulation analysis graphics in 1/4T (a), 1/8T (b).

4　CONCLUSION

In this paper, using optical software Optisysterm to simulate the measurement of the relation of all fiber optical current transformer.It can be seen that The relationship between the modulation signal and the filtered signal which is not only in the corresponding within 1T, 1/2T. In a quarter cycle, it can be fitted into a curve of $y = ax^3 + bx^2 + cx$ approximately, in which a=0.02512, b=-0.05487, c=0.18414. While in 1/8T, there is a consistent one-to-one match between the curve relationship,it can be fitted into a curve of $y = ax^2 + bx + c$ approximately, in which a=−0.55606, b=−0.53008, c=0. The thesis conclusion from the research of the fiber optical current transformer is practical to provide references.

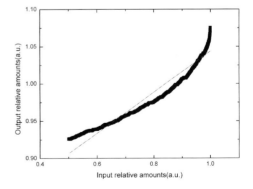

Figure 5.　Sinusoidal signal simulation result fiting analysis graphics.

ACKNOWLEDGEMENTS

This work was financially supported by the Science and Technology Research Project of Heilongjiang Province Education Bureau (No.12531134).

REFERENCES

Kucuksari, S., & G. Karady (2012). Complete Model Development for an Optical Current Transformer. *IEEE Transactions on power delivery 27*, 1755–1762.

Geng, J., J.Yang, & H. Yang (2012). Secondary-Side Signal Processing of Fiber-Optic Current Transducer Based on LTC1068, *Electronic Science and Tech. 25*, 69–71.

Wang, N., & W. Quan (2012). Application of the Fiber Optical Current Transformer in the 110kV Smart Substation, *Asia-Pacific Power and Energy Engineering Conference*. 1–4.

Yu, W., S. Li, & G. Zhang (2012). Self-healing Optical Current Transformer and Its Application in Smart Substation, *Electronic Information and Electrical Engng. 19*, 340–345.

Joseba, Z., C. Luciano, & A. Gotzon (2013). Design and Development of a Low-Cost Optical Current Sensor, *Sensor*s. 13, 13584–13595.

Electronic Engineering and Information Science – Wang (Ed.)
© 2015 Taylor & Francis Group, London, ISBN: 978-1-138-02772-5

Synthesis of yttrium iron garnet by conventional solid-state method

H.L. Dai, T. Shen, C. Hu & Y. Feng
School of Applied Science, Harbin University of Science and Technology, Harbin, China

ABSTRACT: In this article, $Y_3Fe_5O_{12}$ (YIG) powders were produced by conventional solid-state methods. The samples were prepared using the starting molar ratio Y_2O_3/Fe_2O_3=3:5. Yttria and ferrite were calcined at 1080°C for 4 hours and sintered at 1330°C for 3 hours in the air. The samples contained YIG and a small amount of associated phases, as found by X-ray diffraction (XRD). In this way we produced high purity YIG by a cheaper and a simple route.

KEYWORDS: YIG, Conventional solid-state method, XRD, $YFeO_3$.

1 INTRODUCTION

Ferromagnetic garnets after their discovery in 1956, grabs much scientific and technological interest because of their high resistivity and application in micro-wave and magneto-optical devices (Holmquist W. 1961, Zhang X. 2003, Huang M. 2002). For these applications, it is seen that rare-earth iron garnet that is synthetically produced from ferrite based materials is the most popular material being used nowadays.

RIG can be depicted by the chemical formula of $R_3Fe_5O_{12}$. YIG is the most representative compounds among the RIG (Zhang X. 2012, Popov M 2011), until today, the best material to be used for high frequency microwave applications (1-10GHz range) (Fechine P. 2008).

The first known method in producing YIG was developed in 1957 (Holmquist W. 1961), through the mixed oxides method and later known as the solid-state reaction method. Nowadays, many synthesizing techniques have been introduced such as microwave assisted conventional ceramic approaches, co-precipitation, sol-gel, combustion synthesis(Jesus FS. 2012, Pinit K. 2014, Abbas Z. 2009). It is shown that the best way to produce single phase YIG with high density would possibly require expensive technology such as CVD and PVD (Valenzuela R 1994). Nevertheless, as a cheaper cost with a simple technique, the conventional solid-state method is still an attractive route to be employed although high processing temperature is needed.

In this paper, the conventional solid-state methods were used to obtain YIG powder and the phase structure was characterized by XRD with Cu Ka radiation. At last, YIG ceramics were done at a high temperature with the calcined powder.

2 EXPERIMENT

YIG ferrite powders which having stoichiometric compositions of $Y_3Fe_5O_{12}$ were prepared by the solid state reaction. The two weighted raw materials, Fe_2O_3 (99.97%purity) and Y_2O_3 (99.999% purity) were mixed with ethanol and ground for 6 hours to form the precursor. The precursor was calcined at 1080°C for 4 hours. After investigated by X-ray diffraction, the calcined powder was pressed into disks with a diameter of 30mm and thickness of 3.5mm and then sintered at 1330°C for 3 hours.

The precursor was characterized by Thermogravinmetric Anaiysis (TGA) and Differential Scanning Calovimltry (DSC) (STD Q600 V20.9 Build 20) at a heating rate of 5°C/min in the air. The phase structure of calcined poewders were characterized by XRD with Cu Ka radiation.

3 RESULTS AND DISCUSSION

The DSC-TGA plot of the precursor is shown in Figure 1. There was an endothermic peak at about 150°C accompanied by a steady weight loss, which was caused by the elimination of the residual water in the precursor. The second exothermic peaks appear at about 1080°C with little weightlessness. Reaction equation of the precursor during the heating process was shown as follows:

$$5Fe_2O_3 + 3Y_2O_3 \rightarrow 2Y_3Fe_5O_{12} \qquad (1)$$

The second exothermic peak appeared at about 1080°C which corresponds to the crystallization of the amorphous phase.

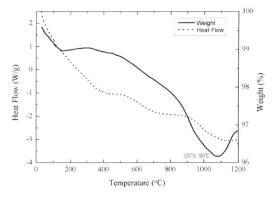

Figure 1.   DSC-TGA plots of the precursor.

According the analysis to the DSC-TGA plot, the precursors were calcined at 1080°C. Figure 2 shows the XRD patterns of the calcined powders. As it can be seen from figure 2, the single garnet phase crystalline YIG powders were formed at a high calcin temperature of 1080°C. As expected, the experimental results in conformity with the inference from the DSC-TGA result.

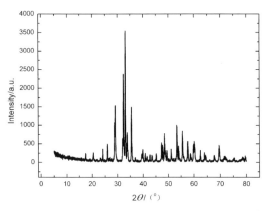

Figure 2.   XRD patterns of the samples calcined at 1080° for 4 h.

We chose 1080°C as the calcin temperature, the calcined powder added with 8 wt.% polyvinyl was sintered at 1330°C for 3 hours. Figure 3 and Figure 4 shows the intended pattern of the samples which one was not calcined and the other was sintered.

As is shown in the two pictures, the sample that was sintered transform into gray from red. It can be inferred that the density of the sample has been changed.

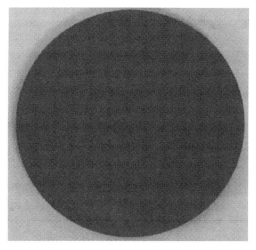

Figure 3.   Pattern of sample that was not sintered.

Figure 4.   Pattern of sample that was sintered.

4   CONCLUSION

Single phase YIG powders is synthesized through conventional solid-state mathod at 1080°C for 4h, Which was a cheaper and a simple route.

ACKNOWLEDGEMENTS

The authors thank National Natural Science Foundation of China (51307036), Natural Science Foundation of Heilongjiang Province of China (E201303) and the Education Department of Heilongjiang province of China (12531134).

REFERENCES

Holmquist, W., C. Kooi, & R. Moss (1961). Reaction kinetics of polycrystalline yttrium iron garnet. *Am. Ceram. Soc. 44*, 194–196.

Zhang, X. & S. Zhang (2003). A new Bi-substituted compounded rare-earth-iron garnet of $(TbYbBi)_3Fe_5O_{12}$ with wideband and temperature-stabilized properties. *Mater. Chem. Phys. 80*, 452–456.

Huang, M., & S. Zhang (2002). Growth and characterization of cerium-substituted yttrium iron garnet single crystal for magneto-optical applications. *Appl. Phys. A: Mater. Sci. Process 180*, 177–180.

Zhang, X., X. Han, & M. Balinskiy (2012). Compact, widely tunable half lambda YIG oscillator. *IEEE International Conference in Frequency Control Symposium 12*, 549–512.

Popov, M., I. Zavislyak, & G. Srinivasan (2011). Yttrium iron garnet magneti-cally tunable magnetodielectric resonator for milimeter and submillimeter wavelength. *International Crimean Conference in Microwave 21*, 219–221.

Fechine, P. B. A., R. S. T. Moretzsohn, R. C. S. Costa, J. Derov, J. W. Stewart, & A. J. Drehman (2008). Magnetodielectric properties of $Y_3Fe_5O_{12}$ and $Gd_3Fe_5O_{12}$ dielectric ferrite resonator antennas. *Microw. Opt. Tech. Lett. 50*, 2852–2857.

Jesus, F. S., C. A. Cortes, R. Valenzuela, S. Ammarm, & M. Bolarin (2012). Synthesis of $Y_3Fe_5O_{12}$ (YIG) assisted by high-energy ball miling. *Ceram. Int. 38*, 5257–5263.

Abbas, Z., R. Al-habashi, & K. Khalid (2009). Garnet ferrite $(Y_3Fe_5O_{12})$ nanoparticles prepared via modified conventional mixong oxides (MCMO) method. *Eur. J. Sci. Res. 36*, 154–160.

Pinit K., P. Santi, & M. Santi (2014). X-ray absorption spectroscopy study on yttrium iron garnet $(Y_3Fe_5O_{12})$ nanocrystalline powders synthesized using egg white-based sol-gel route. *Microelectronic Engineering,126*, 148–52.

Valenzuela, R (1994).Magnetic ceramics. 1st ed. *New York: Cambridge University Press.*

Miao B. F., S. Y. Huang, & D. Qu et al (2014). Physical Origins of the New Magnetoresistance in Pt/YIG. *Phys. Rev. Lett. 112*, 236601.

Gaudisson, T., U. Acevedo, & S. Nowak et al (2013). Combining Soft Chemistry and Spark Plasma Sintering to Produce Highly Dense and Finely Grained Soft Ferrimagnetic $Y_3Fe_5O_{12}$ (YIG) Ceramics. *J. Am. Ceram. Soc. 96*, 3094–3099.

Zhuang, N., W. Chen, & L. Shi et al (2013). A new technique to grow incongruent melting Ga: YIG crystals: the edge-defined film-fed growth method. J. Appl. Crystallogr. 46, 746–751.

Heinrich, B., C. Burrowes, & E. Montoya et al (2011). Spin pumping at the magnetic insulator (YIG)/normal metal (Au) interfaces. *Phys. Rev. Lett. 107*, 066604.

Electronic Engineering and Information Science – Wang (Ed.)
© 2015 Taylor & Francis Group, London, ISBN: 978-1-138-02772-5

Instructing writing with online writing labs—taking OWL as an example

J.F. Pan
School of Foreign Languages, Hainan University, Haikou, China

ABSTRACT: Online writing labs, with encyclopedic self-access writing resources, can be valuable writing instruction materials for tertiary EFL students in China today. The article takes Online Writing Lab (OWL) at Purdue University as an example, illustrating that the lab can be employed to instruct students to write in process approach and genre approach and can be used to enhance students' writing skills. Moreover, OWL is effective for teachers to prepare for writing courses. The increasing enrollment of Chinese universities makes Online writing labs supplementary and indispensable writing resources in writing instruction since the labs have learner-centered layout, detailed explanations and practice, and platforms for peers to exchange and share experience. Therefore, college English writing teachers are supposed to instruct students to use online teaching labs appropriately in different writing contexts.

KEYWORDS: OWL; CILL; web page.

1 INTRODUCTION OF ONLINE WRITING LABS

These days, the internet offers us many new possibilities for teaching and learning. One kind of online resource that we might be able to use is the online writing lab / writing resource site. Some of these online sites contain a rich variety of well-prepared and freely available materials, for example, Hong Kong Polytechnic University, Centre for Independent Language Learning (CILL) resources for writing, Paradigm Online Writing Assistant, and UniLearning. This article is to explore an Online Writing Lab (OWL) at Purdue University, which is brimming with handy writing resources and instructional materials. Through exploration and illustration with examples, it is found that online writing labs can be used to instruct tertiary Chinese students to write English essays in writing courses; moreover, they can be employed as self-instruction materials for the students to improve their writing skills and capabilities

2 PROCESS WRITING APPROACH SHOWN IN OWL

Wring consists of many processes or components, which do not have to take place in a particular linear sequence. Process writing approach includes four basic steps: (1) prewriting, (2) composing/drafting, (3) revising, and (4) editing (Tribble, 1996). Coffin and his partners enriched this approach by elaborating it with seven recursive stages in the figure of the writing process approach (Coffin et al. 2003). OWL contains and displays this writers view clearly. The web page (snapshot 1) named 'the writing process' and its subsequent subsections in the right column are to help us "with the writing process: pre-writing (invention), developing research questions and outlines, composing thesis statements, and proofreading." Moreover, the web page states, "While the writing process may be different for each person and for each particular assignment, the resources contained in this section follow the general work flow of pre-writing, organizing, and revising." We can see although identical terms are not used to describe different parts of writing, writing is considered to be a process with different elements called different ways and divided either generally or in detail.

The subsections of the web page give very comprehensive instructions in nine aspects: writing task resource list, starting the writing process, prewriting (invention), writer's block, Stasis Theory, creating a thesis statement, developing an outline, reverse outlining, proofreading. Because of the convenience of web technology, OWL resources give us an encyclopedic view in the writing process. Take Invention, one of the subsections for example. When we click it, we can see further information about tips for how to start a writing assignment, including that writing takes time, use the rhetorical elements as a guide to think through your writhing, and keep in mind the purpose of writing assignment and prewriting strategies, and we are offered very clear and constructive ideas in each part.

Figure 1. The writing process.

Figure 2. Writing task resource list.

3 GENRE WRITING APPROACH SHOWN IN OWL

"Genre refers to abstract, socially recognized ways of using language." (Hyland, 2003) Genre-based view of writing emphasizes that texts work as typical ways of accomplishing the particular social purpose in particular contexts (Paltridge & Tardy, 2006). Based on this approach, the purpose of writing can be applying for a job, introducing a new supermarket, asking for a leave, and so on. When we write in a kind of genre, at least we should have a clear picture of the purpose of our writing, how to organize our writing and what kind grammar and language we should use.

OWL houses a large amount of genres and detailed explanations of how to write each of them. Logging on the 'The Writing Process' web page (Figure 1) and clicking the first subtitle 'Writing Task Resource List', we come to a web page (Figure 2) with the third subtitle called When You're Ready to Compose Your Writing Task. Under this subtitle, there are more than twenty different genres, below which are subtitles of their different types and writing features. Each subtitle can be further clicked to find a minute view about exact styles and their organization and language strategies. The categories of the genres include abstract, academic research paper, an argument or position paper, bibliography or annotated bibliography, book report and so on. We can find that genre-based view of writing is part of process writing (since we click the same web page, the Writing Process), because the familiarity of the genre that we are going to write is part of the prewriting process.

4 AN ILLUSTRATION OF USING OWL TO ACHIEVE CONCISE WRITING

As has been mentioned, OWL can be employed to instruct students to write in process approach and genre approach. Besides, the online writing lab also provides a number of effective ways to enhance writers' writing skills. For instance, it illustrates in a detailed way how to achieve conciseness in writing, a very important aspect of writing English essays. For some Chinese students, they may write very long sentences with several clauses and complex structures, especially in academic writing, partly because they consider it as a sign of proficiency in this language. Sometimes they just write and write without spending time pruning the product, with the result of its redundancy and wordiness. While for other Chinese students, they may realize that writing should be processed in a concise and effective way, but they are not quite sure about how to achieve conciseness. The tips in the web pages can kindle their mind as to the importance and approaches to achieving conciseness in writing. Just as what is put on the web, like bad employees, words that do not accomplish enough should be fired. A range of small and ambiguous words should be deleted or replaced with strong and specific words. Since a large variety of nouns, verbs, and adjectives exist, more things have a corresponding description.

There are several web pages and sections concerning concise writing, giving very specific and detailed guidance. A web page with the headline of 'conciseness' displays a couple of strategic tips: replacing several vague words with more powerful and specific words, interrogating every word in a sentence and combining sentences. In the first example of tip one, the long verb phrase 'talked about several of the merits of' is changed into 'touted'. (What a powerful but short-spelt word!) 'Conciseness' can be found in the list of the right column of this page. Except 'conciseness', there are also three sub-sections under this topic, 'eliminating words', 'changing phrases', 'avoiding common pitfalls', which all give very detailed instructions on how to cut off superfluous words. Paramedic Method is also a feasible resort which includes seven steps to cut a wordy sentence into a concise one. The seven steps are: 1. Circle the prepositions (of, in, about, for, onto, into); 2. Draw a box around the "is" verb forms; 3. Ask, "Where's the action?" 4. Change the "action" into a simple verb; 5. Move the doer into the subject (Who's kicking whom)

766

6. Eliminate any unnecessary slow windups; 7. Eliminate any redundancies. After following the steps to rewrite some of the prolix sentences in the practice part on this web page, students can find the adjusted sentences are much easier to understand, more persuasive and reader-centered than the original ones. The method works very well to make English writing concise and effective and they will be very helpful for students to write abstracts and articles in any genre.

5 USING THE ONLINE RESOURCE FOR WRITING COURSE PREPARATION

OWL will not only facilitate writing instruction in the classroom, but also will be very helpful for teachers to prepare for writing course. For instance, if an English writing teacher is preparing for a lesson about teaching curriculum vitae and is considering designing some brainstorming questions, he or she can type CV in the search box on the website, click it and then come to a page called Writing the Curriculum Vitae. On the webpage, there are seven question-and- answer strategies about how to write an effective CV. The questions are short, but some answers are long and complicated. Among the seven questions available, the teacher can select three or so, which reflect major facts about CV. These facts can help students have a general idea about CV. Then the teacher can revise the answers in a very simple and learner-friendly way. The three questions selected and answers revised might be:

1 What is a Curriculum Vitae? The curriculum vitae is, an overview of one's life's accomplishments, most specifically those that are relevant to the academic realm.
2 How is a CV different from a resume? The most noticeable difference between most CVs and most resumes is the length. Resumes are usually limited to a page while CVs often run to three or more pages.
3 What should I include? Your CV should include your name, contact information, an overview of your education, your academic and related employment (especially teaching, editorial, or administrative experience), your research projects (including conference papers and publications), and your departmental and community service.

6 OVERALL SUITABILITY AS A SELF-ACCESS RESOURCE FOR TERTIARY EFL STUDENTS IN CHINA

The valuable resources of OWL and other online writing labs can become supplementary and indispensable writing resources for tertiary EFL students in China today. Many of these sites were created to be self-access resources for students. "Self-access" means that students are directed to refer to information on the site or to work through the explanations and activities, on their own, as and when they themselves feel they need to, independently of the teacher. For college students, English majors or non-English majors, they have already learned English for about ten years, so they have acquired certain basic knowledge about grammar and language usage, which makes the self-access learning possible. During their exploration of the English writing lab, they can refer to handy online dictionaries when they run into some new words. They can seek help from their classmates when they meet some puzzles. Furthermore, most of today's college students are digital natives, and they are very adept at computer and online surfing, so exploring and learning online is not difficult for them.

With the development of Chinese society and the increase of life standard, it is nothing new for a college student to have a personal computer. Even for those who do not have their own computers, they can make use of online writing labs, for Computer labs and web bars are available at every university. The result of the increase enrollment of Chinese universities is the large size of classrooms, which makes it impossible for teachers to pay close heed to every student, especially in the writing class. Online writing resources can make some compensation in this way. When students are interested in a writing aspect, for example, a genre, or have some problems, they can refer to the online writing labs, which have very clear and learner-centered layout, very detailed explanations and practice, and platforms for peers to exchange and share experience.

After all, many of the writing labs available online were created specifically to meet the needs of students at particular universities or in particular contexts. In using their resources, or in directing our students to use these resources, we need to assess whether the materials are appropriate for what we are trying to teach in our own contexts.

REFERENCES

Coffin, C., M.J. Curry, S.Goodman, A.Hewings, , T. M.Lillis, & J.Swann, (2003). *Teaching academic writing: A toolkit for higher education*. London: Routledge. 32–43.
Hyland, K. (2003). Genre-based pedagogies: A social response to process. *Journal of Second Language Writing 12(1)*, 17–19.
Paltridge, M. J. Reiff, & C. Tardy. (2006). Crossing the boundaries of genre studies: commentaries by experts. *Journal of Second Language Writing, 15(3)*, 234–249.
Tribble, C. (1996). *Writing Oxford* , Oxford University Press, 256–268.

Electronic Engineering and Information Science – Wang (Ed.)
© 2015 Taylor & Francis Group, London, ISBN: 978-1-138-02772-5

Study of demodulator based on DFB laser wavelength modulation of a small band FBG

M.Z. Wu, Y.L. Xiong, H. Liang, N.K. Ren & Z.Y. Ma
Department of Applied Science, Harbin University of Science and Technology, China

ABSTRACT: This paper presents a method based on the features of a distributed feedback laser (DFB) dynamic scanning to achieve the wavelength demodulation of fiber Bragg grating (FBG). Using LabVIEW software to build up a simulation platform to search peaks of the transmission spectrum of Fabry-Perot (FP) etalon and the reflection spectrum of FBG. The wavelength demodulation of FBG was carried out by combining the C language and LabVIEW. Setting up the temperature sensing system based on the DFB dynamic scanning, and then the experiments are carried out. The temperature sensitivity of the FBG is 10.6pm/°C, which is basically consistent with the results of the standard wavelength demodulator.

KEYWORDS: Distributed feedback laser; Fiber Bragg Grating; Sensor; Wavelength demodulation.

1 INTRODUCTION

Study of demodulator based on DFB laser wavelength modulation of a small band FBG is the study of small band wavelength demodulation based on the widely use of fiber grating sensors. Currently, many demodulation methods have been put forward (Song et al. 2000, Nunes et al. 2004), such as the broadband ASE light source and tunable wavelength filter are used for the FBG wavelength demodulation. Fiber Bragg grating (FBG) is a passive fiber component and develops quickly in the last decade. Which have the advantages of wavelength encoding and quasi-distributed of measurement. Certain wavelengths of light can be propagated by FBG, and the rest of the narrow band of light can also be reflected. The reflection wavelength which carries the information about the external environment, and the center wavelength will shift with the external conditions change. And the power of multi-channel synchronous scanning is only about 1μW and also easy to lose signal. Besides, the price is expensive and cannot for subsequent development (Jiang et al. 2011, Niu et al. 2012). To solve these drawbacks, the research was proposed that utilizing tuning distributed feedback laser to demodulate the FBG wavelength information.

2 PRINCIPLE

The research is based on the sensing of FBG, the DFB lasers dynamic synchronous scanning FBG and F-P etalon, and then build up the relationship between wavelength and time of both the reflection

spectrum of FBG and the standard transmission spectrum of F-P. In order to realize the demodulation of the DFB reflection spectrum, two aspects of works have been done. The programs of peak searching algorithm and the wavelength detection are included in the software design. And besides, the designs of amplifying circuit and the tuned circuit of DFB lasers are included in the hardware design. Figure 1 shows the schematic.

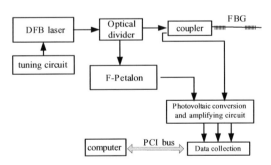

Figure 1. Schematic.

As to achieve the sensing of external information, FBG has been used. We use laser light as sources because of its good monochromaticity, directionality, coherence and high brightness, so it is suitable for sensing application. The DFB laser is used in the experiment. This is a distributed feedback laser with a Fiber Bragg grating inside. This belongs to the side-firing semiconductor lasers, which has a good monochrome, and can launch a narrow band of light. The center wavelength and the bandwidth of the laser we used in the study are 1550nm and 0.2nm respectively.

The center wavelength of the narrowband light of DFB laser moves within a certain range, and scan the FBG and F-P etalon at the same time. Through the F-P etalon we can achieve the real time detection of the center wavelength of FBG. We take a wavelength tuning circuit outside the laser. By changing the temperature, the range of wavelength scanning is from 1550.012nm to 1554.812nm, and the cycle period is 5 seconds, which is convenient to F-P etalon for comparison.

F-P etalon is based on the multi-beam interference principle, and which has the function of chosen wavelength. Different wavelengths have different transmittance; the transmitted ray forms spectral peaks and troughs, and the interval between the peaks is equal. Because of DFB synchronous scanning, the laser beam in FBG and F-P is the same [5]. So they have the same relationship between wavelength and time. Find the time of the peak in the FBG reflection spectrum, and then find the corresponding wavelength of the transmission spectrum of the F-P etalon. And this is the center wavelength of FBG. After finishing these works, the optical signal must be detected, gathered and processed also. We use a photodiode for photoelectric conversion, which is based on the photovoltaic effect of PN junction [5]. An amplifier circuit is to amplify the electrical signal. Amplifying circuit is composed of a bias voltage circuit, voltage regulator circuit and the three amplifier circuit. At last, we use a data acquisition card to collect the data.

3   WAVELENGTH RECOGNITION

The collected data is processed by C language. If we want to get the time of the peaks of FBG reflection spectrum and the wavelength of the corresponding F-P transmission spectrum, firstly we must find the peak position. Peak image approximates the Gaussian distribution. Make the peak data smoothly by Gaussian fitting. Then use the peak searching algorithm to find the peak and get the time. The data of F-P transmission spectra are fitted by a Gaussian. After this, six spots are found by using the peak searching algorithm in one cycle [6-8]. These spots have a binomial distribution, so the relationship between wavelength and time can be by binomial fitting. If the peak time of the FBG reflection spectrum was taken into the relationship, the center wavelength of FBG can be obtained. Figure 2 shows the block diagram of C language.

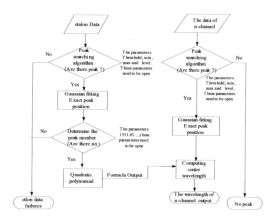

Figure 2.   Block diagram of C language.

4   SOFTWARE EMULATION

LabView is a universal programming system. It is a graphical programming to create the application language application to create with icons instead of lines of text. Not only icons and wires can be used, but also control the object of display graphical by the program. It can load the program written in C language more intuitive and simulate the system. Figure 3 shows the design of simulation diagram.

The function of each position in the figure 3 has been marked. The collected data were called by controls labeled "FBG input data" and "Enter etalon data". The C language program is called by the middle device. Just enter the appropriate parameters, you can read the wavelength values on the display interface. Display interface is divided into three parts, input parameters section, the numerical display section and the waveform display section. FBG center wavelength value and the corresponding point of time and the waveform is displayed on the display section, if the FBG and F-P parameters were entered in the parameter input section.

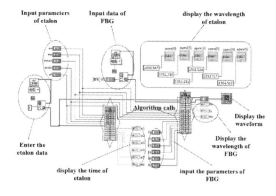

Figure 3.   Design of simulation diagram.

5 TEMPERATURE SENSING EXPERIMENT

For temperature sensing experiments, Figure 4 shows the experimental system. Two FBGs have the center wavelength of 1551.053nm and 1553.612nm respectively. And then put them in the same temperature control box. FBG1 is demodulated by the demodulation system. FBG2 is demodulated by the standard FBG demodulator with the resolution of 2pm. Adjust the temperature control box, temperature range of 0 °C ~ 90 °C. The data of two demodulation system obtained are recorded, and Figures 5 and 6 show the experimental curves.

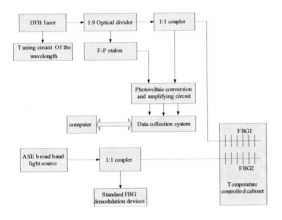

Figure 4.　The diagram of experimental system.

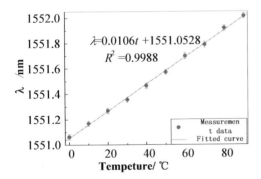

Figure 5.　Relationship between wavelength and temperature Of FBG1.

6 CONCLUSION

We can clearly see that it has a good linear relationship between temperature and wavelength. The temperature sensitivities of FBGs are 10.6pm/°C and

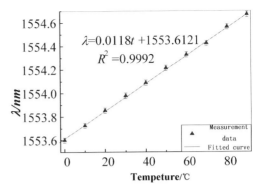

Figure 6.　Relationship between wavelength and temperature Of FBG2.

11.8pm/°C, respectively. From the fitting results, we can conclude that the experimental results and theoretical are consistent, so it is feasible to use this demodulation system for temperature sensing.

ACKNOWLEDGMENT

This work was financially supported by the National Undergraduate Programs for Innovation and Entrepreneurship (20131024019).

REFERENCES

Song, M., S.Yin & P.B. Ruffin, (2000). Fiber Bragg Grating Strain Sensor Demodulation with Quadrature Sampling of a Mach-Zehnder Interferometer. *Applied Optics 39*, 1106–1111.

Nunes L.C.S.. L.C.G. Valente & A.M.B. Braga (2004). Analysis of a demodulation system for fiber Bragg grating sensors using two fixed filters. *Optics and Lasers in Engineering* 42, 529–542.

Jiang B.Q., J.L. Zhao, & C. A. Qin (2011). An optimized strain demodulation method based on dynamic double matched fiber Bragg grating filtering. *Optics and Lasers in engineering* 49: 415–418.

Niu S.L., Y. Liao & Q. Yao (2012). Resolution and sensitivity enhancements in strong grating based fiber Fabry-Perot interferometric sensor system utilizing multiple reflection beams". *Optics Communications 285*, 2826–2831.

Electronic Engineering and Information Science – Wang (Ed.)
© 2015 Taylor & Francis Group, London, ISBN: 978-1-138-02772-5

Study of reverse supply chain based on products' life cycle

L.F. Zhu & P.L. Qiao
Computer science and technology, Harbin University of Science and Technology, Harbin, P.R.China

ABSTRACT: For the service of spare parts supply chain based on the product's life cycle, we establish a reverse supply chain model, the objective function is the optimal location costs, transportation costs and inventory costs, making the total expected cost is minimized. According to mixed-integer nonlinear programming model of the location. We propose to use a method that based on Lagrangian relaxation and heuristic algorithm to solve this problem, and through simulation of the examples to show that the algorithm is accurate and efficient for solving the problem.

KEYWORDS: product's life cycle; reverse supply chain; Lagrangian relaxation.

1 GENERAL INSTRUCTIONS

The introduction of the reverse supply chain, provides a theoretical basis for the reuse of resources, environmental protection and sustainable development. Business scope is a series of activities that mainly involved the disposals of damage products, unsold products and out of the products, as well as integration of remanufacturing and discarded goods processing and so on.

In this article, we show the inventory location model with no capacity limits, integrate the inventory and the location problem in the supply chain. We considered to manage the reverse flow of products, retailers are responsible for collecting the return products and backing to Remanufacturing Center (RC). RC is responsible for remanufacturing the return products to spare parts for retailers. Remanufacturing center needs to maintain a certain amount of semi-finished products for safety stock. We also assume that each retailer returns it back to a corresponding remanufacturing center. In addition, we assume the remanufacturing center must be built in the vicinity of the distribution center to minimize transportation costs. Compared with the forward logistics, supply chain, the reverse logistics system has a higher uncertainty. In this paper, the Lagrangian relaxation combined with heuristic algorithm to solve multiple decision variables of reverse flow product problems to get the optimal allocation strategy under the low cost (Guo, Wang, Wang and Kang. 2010, Abdallah, Diabat and Simchi-Levi (2012)).

Compared with the forward logistics, supply chain, the reverse logistics system has a higher uncertainty. In this paper, the Lagrangian relaxation combined with heuristic algorithm to solve multiple decision variables of reverse flow product problems in the supply chain, in order to get the optimal allocation strategy under the low cost.

2 THE REVERSE SUPPLY CHAIN MODEL

2.1 Problem description

We used R_{il} to represent the amount of returns that customers returned to the retailer i in the number j day, and assumed that the amount of everyday's return is independent and obey the standardized normal distribution $R_{il} \sim N(\mu_i, \sigma_i^2)$. We consider that the inventory strategy of Remanufacturing Center is the economic order quantity (EOQ) model in order that inventory costs can be integrated into the location model. Inventory levels as for the RC, we assume that a product is composed of M+1 components, including a major component and M semi-finished components what are made by factory before. In order to simplify the model, we use an aggregation component instead of the semi-finished components. So we use q to denote inventory cost of the aggregation components, it is proportional to the total inventory cost h, q=vh (v∈(0, 1)) v mainly depends on the proportion of the aggregation components in the whole product value(Fu and Fu. 2010). Now we need to determine the RC's location, the retailer belongs to RC and the best stocks. Model parameters are defined as follows:

2.2 The site selection, inventory model of reverse supply chain

Model parameters
I retailers
K remanufacturing center

l_k annual cost of locating a regional remanufacturing center at site k

d_{ik} the cost of per unit shipped from retailer i to remanufacturing center k

χ the number of days per year

θ inventory cost weight

β transportation cost weight

q subassembly holding cost

F_k fixed cost of an order from RC to retailers

z_α standard normal deviate with p

g_j fixed shipping cost

a_j unit shipping costs

$\overline{L}$ average lead time from RC to retailers

$W_k = \begin{cases} 1 \\ 0 \end{cases}$ if site k is selected as a RC ,the value is 1, otherwise 0

$Z_{ik} = \begin{cases} 1 \\ 0 \end{cases}$ if returns from retailer I are collected by RC k ,the value is 1, otherwise 0

2.3 Model of reverse supply chain cost

$$F_{min} = \sum_{k \in K} l_k W_k + \left(\beta \sum_{k \in K} \sum_{i \in I} \chi \gamma_i d_{ik} Z_{ik} \right) +$$
$$\left(\sum_{k \in K} \sqrt{2\theta vh(F_k + \beta g_k)} \sum_{i \in I} \sum_{i \in I} \chi \gamma_i Z_{ik} + \beta \sum_{k \in K} a_k \sum_{i \in I} \chi \gamma_i Z_{ik} \right) \quad (1)$$
$$+ \left(\theta vh z_\alpha \sum_{k \in K} \sqrt{\overline{L} \sum_{i \in I} \rho_i^2 Z_{ik}} \right)$$

Subject to

$$\sum_{k \in K} Z_{ik} = 1 \quad \forall i \in I \quad (2)$$

$$Z_{ik} - W_k \le 0 \quad \forall i \in I \quad \forall k \in K \quad (3)$$

$$W_k \in \{0,1\} \quad \forall k \in K \quad (4)$$

$$Z_{ik} \in \{0,1\} \quad \forall i \in I \quad \forall k \in K \quad (5)$$

The above system of equations, the objective function (1) is used to represent the total cost of reverse logistics, mainly from the following several parts: 1. Fixed costs of the location of the RCs 2. shipping cost for retailers to RC 3. Inventory cost of the components that RC needed to do the job 4. constraints of the safety, inventory cost of the uncertain risk (2) to make sure every retailer corresponds to a single RC (3) ensure that the RC hasn't corresponding retailers when it hasn't established, (4) (5) constraints 0 and 1 for each decision variable.

3 SOLUTION METHOD

3.1 Lagrangian relaxation

Multiple constraints make problem increasing complexity, in order to reduce the difficulty of solving the model, we are first to simplify the model, we define that:

$$\hat{d}_{ik} = \beta \chi \gamma_i (d_{ik} + a_k) \quad (6)$$

$$K_k = \sqrt{2\theta \chi h (F_k + \beta g_k)} \quad (7)$$

$$\hat{\theta} = \theta h z_\alpha \quad (8)$$

$$\hat{\rho}_i^2 = \overline{L} \rho_i^2 \quad (9)$$

$$\hat{K}_k = \sqrt{v} K_k + v\hat{\theta}\sqrt{\overline{L}n} \quad (10)$$

$$F_{min} = \sum_{k \in K} \left(\begin{array}{c} \sum_{k \in K} l_k W_k + \sum_{i \in I} \hat{d}_{ik} Z_{ik} + \\ \sqrt{v} K_k \sqrt{\sum_{i \in I} \gamma_i Z_{ik}} + v\hat{\theta}\sqrt{\sum_{i \in I} \hat{\rho}_i^2 Z_{ik}} \end{array} \right) \quad (11)$$
$$= \sum_{k \in K} \left(l_k W_k + \sum_{i \in I} \hat{d}_{ik} Z_{ik} + \hat{K}_k \sqrt{\sum_{i \in I} \gamma_i Z_{ik}} \right)$$

Subject to

$$\sum_{k \in K} Z_{ik} = 1 \quad \forall i \in I \quad (12)$$

$$Z_{ik} - W_k \le 0 \quad \forall i \in I \quad \forall k \in K \quad (13)$$

$$W_k \in \{0,1\} \quad \forall k \in K \quad (14)$$

Relax constraints(12) by Lagrangian relaxation:

$$\max_\lambda F_{min} = \sum_{k \in K} \left(\begin{array}{c} l_k W_k + \sum_{i \in I} (\hat{d}_{ik} - \lambda_i) Z_{ik} \\ + \hat{K}_k \sqrt{\sum_{i \in I} \gamma_i Z_{ik}} \end{array} \right) + \sum_{i \in I} \lambda_i \quad (15)$$

Constraints the same as (13)(14).

3.2 Solution steps

Step 1: Finding a lower bound

According to a fix Lagrangian multiplier, location variable W_k and assignment variable Z_{ik} to determine the minimum value of the equation (15).According to method what is introduced by Daskin(Daskin,Coullard and Shen 2002), the problem is deeply decomposed into:

$$V_k = \min \sum_{i \in I} b_i Z_i + \sqrt{\sum_{i \in I} c_i Z_i} \qquad (16)$$

$$b_i = \hat{d}_{ik} - \lambda_i \qquad (17)$$

$$c_i = \hat{K}_k^2 \gamma_i \qquad (18)$$

$$W_k = \left\{ {1 \atop 0} \right. l_k + \tilde{V}_k < 0 \ the \ value \ is \ 1 \qquad (19)$$

$$Z_{ik} = \left\{ {1 \atop 0} \right. W_k = 1, Z_i = 1, the \ value \ is \ 1 \qquad (20)$$

This algorithm is applied to find the lower bound on the objective function, and find the optimal solution of the given Lagrangian multiplier, assume that $(W_k{}^t, Z_{ik}{}^t)$ and an optimal solution is a solution of the relaxation model, the optimal solution can be considered to be the lower bound of the model. If this is a solution of the model, and then, updated Lagrangian multiplier with a standard subgradient optimization.

Step2: Finding an upper bound for the subproblems

Because W_k without any constraints, a lower bound of location variable W_k can be first determined. Only when the $k \in K$, $\sum Z_{ik} > 1$ or $\sum Z_{ik} = 0$, the lower bound can't conform to the constraints. The first case, we need to calculate the additional cost that retailers corresponding RCs. The second case, we need to calculate the additional cost that all the RCs corresponds to the retailers. We could get a feasible solution of the lowest cost in this way.

Step 3: Retailer reassignment

If the current upper bound is better than the best known upper bound, at this time for each retailer reassign alone. For each of the retailers who have been allocated, calculated the incremental cost for another open RC, each of the retailers uses the least incremental cost assign to RCs. If the reassignment result lead to an RC without any assigned retailer, the fixed cost is subtracted from the incremental cost, to make a better solution for this assignment. Note that each reassignment may cause additional reassignment. First of all, the best reassignment may be found and execution stop, and then passed to the next retailers. To iterate over all the retailers, until there is no improvement. If all RCs without assigning returns are closed, the retailer assignment algorithm terminates.

Step 4: RC selection algorithm

Based on the exchange algorithm is applied to the retailers reassignment algorithm, for each open RC, find the best alternative from the closed RC collection. The algorithm executes from the open RC that returned products is minimized. For this RC, the algorithm looks for the retailer with the highest return whose RC at the same position is closed, close the old RC and open a new RC. Then, retailers that correspond to the closed RC assign to open RCs by IC function.If the upper bound is better by exchange of RC, the exchange effect. If not, the algorithm loops to the next highest demand of the retailers of closed RCs. If all open RC check finished, then the algorithm move to the next RC with the lowest reassignment value, and continue. If a loop through all open RCs but not improve, the algorithm terminates.

Step 5: Variable fixing procedure

After the above heuristic algorithm terminates, then we need to fix the variable. node exclusion rule: For the RC that is not opened: if $LB + l_k + V_k > UB$, this RC can't be the optimal solution, LB and UB are respectively as the upper and lower bounds of the corresponding ∂i .Node inclusion rule: For the RC which is opened, $l_k + V_k$ exist a negative value. Thus, if an open RC and $LB - (l_k + V_k) > UB$, this RC can be a component of the optimal solution.

4 INSTANCE AND ANALYSIS

Experimental data are provided by the literature (Tang, Yang and Yang.2008), the database instance is in the Research on a Multistage Stochastic Location-inventory Model. Assuming that the candidate RCs set equal to retailers set, uniform distribution in a coordinate plane, the retailer's return expectation is the ratio of the demand amount of 0.2,the unit fixed cost from retailers to RC is 1000, the unit fixed transport cost from retailers is 1000, the unit fixed transport cost from factory to RC is 500, the service level is 1.96, the delivery lead time is 1, 300 days per year, the number of components need by RC is 1.All related parameters of retailers and RC in the model are given in table 1. Each retailer's average demand and the RC's location costs are random coming out. Then we emulate the algorithm above by matlab, maximum loop 1000 times, the minimum difference between upper bound and lower bound is of 5%. In order to guarantee the universality of the method, we select related parameters that are given random, calculate the optimal cost at this time, and compare the result of the calculation, the calculation time and the number of iterations. Random 10 examples are given in table 2, we can be found that the deviation of the optimal upper bound and lower bound is not more than 5% from the result set, and less than 1% in most cases. The parameters and calculation results are shown in Table 1 and Table 2.

Table 1. Parameters of retailers and remanufacturing centers.

The market	Ordinate	Abscissa	Returns	Fixed cost of construction
1	32	79	431	373360
2	4	13	415	823500
3	58	10	974	619560
4	5	87	103	883320
5	10	92	664	827620
6	12	56	983	633200
7	55	39	509	203240
8	75	71	805	813040
9	75	46	545	397840
10	87	86	714	732380
11	6	45	719	244920
12	63	70	1037	339100
13	15	62	228	375540
14	52	37	464	401500
15	82	10	785	613920
16	1	79	438	298040
17	36	99	799	241080
18	29	7	613	704700
19	27	15	844	685940
20	96	76	233	261700

Table 2. The calculation result.

Diffset	The best upper bound/x10^7	The best lower bound/x10^7	Absolute error/%	Relative error/%	Operation time/s	Loop/times
1	1.1275	1.1273	2056.1	0.0182	756.92	387
2	1.1132	1.1128	3918.4	0.0352	835.65	406
3	1.1088	1.1085	3098.2	0.0279	798.97	368
4	1.1116	1.1111	5302.9	0.0477	431.78	256
5	1.1358	1.1351	6887.5	0.0606	686.90	322
6	1.3210	1.3207	2697.6	0.0204	398.12	278
7	1.1097	1.1096	1011.6	0.0099	1096.88	505
8	1.1266	1.1263	3356.8	0.0298	606.72	287
9	1.2365	1.2363	2097.4	0.0169	608.27	279
10	1.1178	1.1177	966.2	0.0086	467.78	178

5 CONCLUSIONS

This thesis is based on the service of spare parts supply chain for the product's life cycle, part of the reverse supply chain as the research object, mainly including site selection and inventory routing problem that the return products are remanufactured, then as spare parts back into the market.Here we consider a chain that the retail, collect the returns what are set by oneself, and then returns to the remanufacturing center, they will be on the market as spare parts again after processing.

Based on this problem, this thesis established a mixed-integer nonlinear programming model of the location. Therefore, we can use the analytic method that is given by the literature. Finally, we use the heuristic algorithm based on Lagrangian relaxation to solve the model. This thesis also has many places are worth further studying, first of all, the algorithm that used in this thesis belongs to heuristic algorithm, it can't get the determined optimal solution. Secondly, in the process of practical application, the change of parameters will lead to the change of location decisions and produce some deviations.

ACKNOWLEDGEMENTS

This research is supported by research project by Harbin city (2011AA1CG063).

REFERENCES

Guo, L.H., K. Wang,Y. Wang & R. Kang. (2010). Multi-indenture multi-echelon spare part supply chain requirement modeling and simulation. *Computer Integrated Manufacturing Systems.* 10(16), 2038–2043. In Chinese.

Abdallah T, A.Diabat & D.Simchi-Levi (2012). Sustainable supply chain design: a closed loop formulation and sensitivity analysis. *Production Planning & Control.* 23(2–3), 120–133.

Fu, C.H. & Z. Fu. (2010). Inventory routing problem and its recent development: Review. *Journal of Computer Applications.* 30(2), 453–457. In Chinese.

Daskin, M., C.Coullard & Z.Shen (2002). An inventory-location model: formulation, solution algorithm and computational results. *Annals of Operations Research.* 110(1), 83–106.

Tang, K., C. Yang & J. Yang. (2008). Research on a Multistage Stochastic Location-inventory Model. *Journal Of WUT.* 30(5), 795–799. In Chinese.

Electronic Engineering and Information Science – Wang (Ed.)
© 2015 Taylor & Francis Group, London, ISBN: 978-1-138-02772-5

Reinforcement learning of fuzzy joint replenishment problem in supply chain

C.Y. Li, S.H. Zhao, T.W. Zhang & X.T. Wang
School of Computer Science and Technology, Harbin University of Science and Technology, Harbin, China

ABSTRACT: The solution to the Joint Replenishment Problem (JRP) with fuzzy constraint resource was studied by learning algorithm. The fuzzy programming model was constructed to minimize the sum of inventory holding cost, using trapezium fuzzy numbers to represent the fuzzy resource constraints and all of the parameter was given to get the total relevant costs (Tsao Y C. 2014). We solve JRP by unsupervised learning through reinforcement learning, each of the goods' basic cycle was the initial state. Then, we develop a reinforcement learning algorithm for the JRP with resource restriction. A computational experiment is performed to test the performances of the algorithm.

KEYWORDS: Supply chain management; Joint replenishment problem; Reinforcement learning algorithm; Markov decision processes.

1 INTRODUCTION

Now researchers are quite concerned about inventory issues, which inventory management is an important part of supply chain management. The joint replenishment problem is a research problem under the environment in a wide range of items for inventory controlled. It refers to the order under joint decision-making from the supply point ordered a variety of items, replenish inventory, using the appropriate decision makers to judge, and thus meet the needs of the premise to pay the minimum total cost per unit time.

In this article, we use simulation ways to address the joint supplementary model which's needs are fuzzy variables. We use reinforcement learning strategy to adjust varieties of goods' period to sync, to reduce cycle time of check inventory levels and bring the considerable cost savings. First, we make the JRP model to decision-making problems. Then, we solve it by Markov decision and reinforcement learning. Second, the article gives initial space, state space, transfer function setting, etc. The Q was calculated with the system behavior and transition probabilities through semi-Markov decision chain. Then, the article shows a program to solve JRP in detail, that using a reinforcement learning algorithm to solve the constrained joint supplementary question so specific issues. Finally, given numerical simulation examples illustrate the effectiveness.

2 PROPOSED REINFORCEMENT LEARNING ALGORITHM

2.1 The fuzzy joint replenishment problem

We discuss the fuzzy joint replenishment problem, all notations are following:

S major ordering cost
i index of product, i=1,2,…,n
T basic replenishment cycle
k_i integer number that is an integer multiple of T
s_i a per cost of product i in a unit
D_i the demand of product i
b_i unit cost of product i
B most capital that can be invested
γ discount factor $0<\gamma<1$
t the number of stages in discrete time
a decision variables(positive indicates increased, zero indicates unchanged, negative indicates dcrease)
c_o the ordering expect costs of product i.
c_h the expected cost of inventory holding cost of product i.

The fuzzy joint replenishment problem can be described: find a best value of T to get the min of TC.

Set $T_1 = \sqrt{C_1 / C_2}$, $T_2 = B / C_3$, for a collection of k_i, we can get the best value of a basic replenishment cycle:

$$T^* = \min(T_1, T_2) \tag{1}$$

Then $C_1 = S + \sum_{i=1}^{n} s_i / k_i$, $C_2 = \sum_{i=1}^{n} D_i k_i h_i / 2$,

$C_3 = \sum_{i=1}^{n} D_i k_i b_i$, the total cost is expressed as:

$$TC(T, k_i's) = C_1 / T^* + C_2 T^* \qquad (2)$$

We can calculate a set of values for the set of k, then from formula (2), we can calculate T and TC. This paper is: get a set of best values of k_i, then get the min of TC. k_i is the best basic replenishment cycle.

In the fuzzy demand conditions, the actual demand of each product is x_i. $x_i \in [m-q, n+q]$, x_i range from m to n, the value of m and n is changing from x_i, q denotes the uncertain volatility of x_i.

According to fuzzy math, x_i is trapezoidal fuzzy number. Set $f(x) = x$, $x \in C$. C is fuzzy sets. The maximum and minimum numbers of Function $f(x)$ are $f(x)_{max} = m+q$ and $f(x)_{min} = n-q$. Membership function m_f is following:

$$m_f(x) = \frac{f(x) - f(x)_{min}}{f(x)_{max} - f(x)_{min}} = \frac{x - m + q}{n - m + 2q} \quad x \in C \qquad (3)$$

Get $x^* \in C$ and $C_f(x^*) = \max\{C_f(x) | x \in C\}$, x^* is what we will get out. $f(x^*)$ is best value. Show in figure 1.

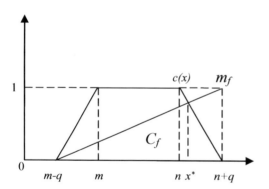

Figure 1.   Condition of maximum point.

Set $m_f(x) = c(x)$, then get the best basic replenishment cycle is:

$$T^* = \min(\sqrt{(S + \sum_{i=1}^{n}(s_i / k_i)) / \sum_{i=1}^{n} x_i^* k_i h_i / 2},$$
$$B / \sum_{i=1}^{n} x_i^* k_i b_i) \qquad (4)$$

The total relevant costs model per unit time is following:

$$TC(T, k_i's) = S / T + \sum_{i=1}^{n} s_i / k_i T + \sum_{i=1}^{n} D_i k_i h_i \, T/2 \qquad (5)$$

Subject to:

$$\sum_{i=1}^{n} D_i k_i T b_i \leq B \qquad (6)$$

To solve this problem, from [6] get:

$$TC(T, k_i's) = S / T +$$
$$\sum_{i=1}^{n} s_i / k_i T + \sum_{i=1}^{n} D_i k_i h_i \, T/2 \qquad (9)$$

x_i means the actual demand rate of i under constrained demand condition. It is a fuzzy number. We use reinforcement learning algorithm to making the value of k_i an array: $s = (k1, k2, ..., k_i, ..., k_n)$, then, this array is the initial value of the original system. Now we use formula (3) to reinforcement learning for the system. The reinforcement learning system model is:

$$Q(s,a) = (1-\alpha)R(s,s',a) +$$
$$\alpha \left\{ r + \gamma \sum_{s' \in S} P(s' | s,a) \max_{a \in A} Q(s',a) \right\} \qquad (8)$$

Using this idea, we modify the reinforcement learning algorithm and develop an new algorithm for solving the constrained JRP. s represent the current time and the next status. $R(s,s',a)$ is a feedback signal which made by a transfer from state s to state s' through action a. $P(s'/s,a)$ is probability that state s take action a to become state s'. α is learning rate of reinforcement learning. After iterative calculation, we can get an array s', this is the best value of ki. The quality of learning outcomes judgment standard is: after each learning, the array $s = (k1, k2, ..., ki, ..., kn)^T$ bring into the equation (7) calculate the basic replenishment cycle T^*. Then, bringing T^* into equation(1) calculate the TC. The lower value of TC, the less cost of the company will pay.

2.2   Solution method

Now we decompose the math model into n single-item inventory problem. Second, we structure into n single-item inventory control problems. Reinforcement learning algorithm contains the Markov process of discrete state and action space. The following will give the supply chain agent actions one by one.

The result of learning is to make sure basic replenishment cycle T and supplementary frequency k_i of item i, k_i is an integer. When k_i is making out, we can find the other decision variable T by equation (3). We define: in the N-dimensional solution space, each dimension on behalf of each arrays k_i in a basic replenishment cycle of each production, the replenishment cycle is $(k_1, k_2, ..., k_i, ..., k_n)^T$. We can find the lower bounds of k_i from the following equation:

$$k_i^{max}(k_i^{max}-1) \leq \frac{2s_i}{D_i h_i T_{min}^2} \leq k_i^{max}(k_i^{max}+1) \qquad (9)$$

Each product replenishment cycle T∈[1,k_i^{max}).

The semi-Markov model has a two-dimensioal state space (s,s'). s defines the inventory position right after an arbitrary demand has occurred, and s' is the changed demand after taking an action. Every transformation takes place in zero time nomater the transformation is 0 to Q or 0 to 1. When take the action a, the transition probability from s to s' are:

$$P(a) = e^{Q(s,a)/C} / \sum_{a' \in A} e^{Q(s,a)/C} \qquad (10)$$

Equation (10) is lattice Boltzmann method(Dawei Tang &Hongwei Wang. 2005).By this equation, we can get the probability for the next step. C is temperature factor, at first, it is a large number, with operation of the system it will gradually decay becomes smaller. Attenuation mechanism as follows:

$$C^{t+1} = \beta C^t \qquad (11)$$

β is range from 0.8 to 0.99.

We choose Semi-Markov average reward technique to solve this question.

For a given period of time 0≤t≤T, the state is sit, take action a, then, the next state is sit+1=sit+a, and kit+1=kit+a, bring into equation (3) for tmin, we can get:

$$V_\gamma^t(k_i) = min \left\{ \begin{array}{c} S/T + \sum_{i=1}^{n}(s_i/a_i)/T + \\ \sum_{i=1}^{n} Ta_i D_i h_i / 2 + \gamma \sum_{i=1}^{n}(c_o + c_h) \end{array} \right\} \qquad (12)$$

The reinforcement learning algorithm:

Step 1: system in the learning phase, Iteration Times is initialized to 1000. State space collection State, action collection {S},set $Q_1(s,a)$=0, set TC=0, set stepnum=0.

Step 2: based on feedback from the system, we can get the current system status s_0.

$$s_0 = C_1 / T^* + C_2 T^* \qquad (13)$$

Step 3: agent randomly selected action a from the action collection {S}, take action a to get the next status s', the reward of the action is following:

$$r(s,s',a) = \frac{T}{2}\sum_{i=1}^{n} a_i D_i h_i + (S + \sum_{i=1}^{n} s_i/a_i)/T \qquad (14)$$

at the same time, set s=s', stepnum++.

Step 4: determine whether the state at this time is terminated, if not, go to step 2, if yes, the process of learning is over. At this time, decision-making system is not dependent on a single agent. The agent will choose an optimal array of each goods replenishment cycle.

2.3 Instance and analysis

Experimental data are provided by the article of Feng Jun, Chenyan Li(Chenyan Wang &Xiaofei Xu & Zhande Chen 2008). There are six products in the joint supplement product management, data of various products are shown in table 1. Major ordering cost S=$200, limit on capital that can be invested B=$25000, the final calculation target is: determine the system's basic replenishment cycle and the min cost of TC.

Table 1. Data of example.

Product i	h_i	D_i	s_i	bi
1	1	1000	45	6.25
2	1	5000	46	6.25
3	1	3000	47	6.25
4	1	1000	44	6.25
5	1	600	45	6.25
6	1	200	47	6.25

γ determines the importance of future rewards for now, by simulation study (Manju S &Punithavalli M. 2011), set γ=0.9. α is learning rate of reinforcement learning, with the program of learning, it decays to 0 from 1. It decays as following:

$$\alpha_{k+1} = \frac{1}{stepnum^{0.51}} \alpha_k \qquad (15)$$

Others: n=6; β=0.99.

All data based on the above, we can calculate TCmin=4251.3570. Other data are shown in Table 2.

Table 2. Data of example.

Product i	1 2 3 4 5 6 7
Optimal cycle	1 1 1 2 2 2 4
Basic replenishment cycle	0.1758
Q(TC)	$4251.3579

With reinforcement learning algorithm further consider the impact of cargo demand, inventory costs and other elements of the decision-making process, we can know the relations inventory relationships and external factors from the model. The binding

781

decision problem itself and the MDP theory, we make the stock model into a Markov decision process to get the real-time feedback of inventory cost. From the reinforcement learning algorithm we can get, Q-learning is different from the solutions which had solved ERP before. It is a search algorithm unrelated to the issue.

3 CONCLUSION

We use a reinforcement learning algorithm, consider the demand for goods, inventory cost and other factors. Then create a math model to perform the interactive relationship between JRP and external factors. Through the reinforcement learning algorithm, we use SMART to solve this math model, in order to get the real-time feedback of inventory cost.

This paper chooses the low dimensional data, and we will study for Markov decision problems with continuous, high-dimensional face dimension disaster. Q learning can learn from the MDP model; estimation online, but it must be to consume the huge computation and storage of cost. Therefore, the convergence rate of Q learning needs further research and improvement.

ACKNOWLEDGMENT

This work was supported by grant No.12541142 from the Research Program of the Education Department of Heilongjiang Province, China.

REFERENCES

Wang, C.Y. ,X.F. Xu &Z.D. chen. (2008).Joint replenishment problem with fuzzy resource constrain, *Computer integrated manufacturing system, 13(2):14–1*. In Chinese.
Tang, D.W. &H.W. Wang. (2005). Application of Intensive Learning Algorithm to Inventory Control in supply Chains. *Chinses Journal of Management,2(3):358–361*. In Chinese.
Jun, F. &C.Y. Li. (2012). PSO-optimization method of joint replenishment problem with fuzzy requirement.*Computer Engineering and Appliactions, 48(35):238–242*. In Chinese.
Tsao Y C. (2014).Joint location, inventory, and preservation decisions for non-instantaneous deterioration items under delay in payments, *International Journal of System Science, (ahead-of-print):1–14*.
Manju S &Punithavalli M. (2011).An Analysis of Q-learning Algorithms with Strategies of Reward Function. *International Journal on Computer Science and Engineering,3(2):814–820*.

Electronic Engineering and Information Science – Wang (Ed.)
© 2015 Taylor & Francis Group, London, ISBN: 978-1-138-02772-5

Joint replenishment and freight problem with fuzzy demand

C.Y. Li, X.T. Wang, T.W. Zhang & C. Wang
Computer science and technology, Harbin University of Science and Technology, Harbin, P.R.China

ABSTRACT: We considered the multiple products Joint Replenishment Problem (JRP) and freight problem with fuzzy demand. The fuzzy programming model was constructed to minimize the total cost, including order cost, inventory holding costs, customer waiting cost and transportation. Using RAND algorithms to solve and taking advantage of a trapezoidal fuzzy number to represent the fuzzy resource constraints. Finally, the simulation, numerical example was given, and comparisons were made between fuzzy resource constraint model and deterministic resource constraint model for 1600 stochastic problems.

KEYWORDS: joint replenishment problem; fuzzy demand; trapezoidal fuzzy number; RAND algorithm.

1 INTRODUCTION

In recent years, the change of market development made a spurt of progress. Reducing inventory cost can bring considerable benefits to the enterprise, so the inventory occupies an important position in manufacturing enterprises or dealer. JRP (Joint Replenishment Problem) is determined for each product order batch size and order cycle in multiple item inventory replenishment process, in order to provide to minimize the total cost per unit time under the premise of internal demand.

Kaspim (Kaspim. 1991) presented RAND heuristic algorithm to solve the Joint Replenishment Problem, through the calculation of the upper and lower bounds of the interval in between m equal sections to get basic replenishment cycle, the RAND algorithm is more than all previously proposed heuristic algorithm with performance. The study of JRP is extended to be with resource constraint by Moon (Moon and Cha 2006). The calculation results show that the performance of genetic algorithm has good expansibility. Khoujam (Khoujam, Michaliwiczz, Satosars. 2000) extended the study to JRP with constraints, and solves the JRP by improving RAND heuristic algorithm and genetic algorithm with determinate resource constraints. In past research, researching on JRP under the quasi-stationary policy had little consideration to the transport of goods, but in fact, the freight problem is an important factor influencing the total cost of the supply chain. In this article, we want to deal with the multiple products Joint Replenishment Problem and the freight problem with fuzzy demand. First, we use trapezoidal fuzzy numbers to represent the fuzzy resource constraints and set the fuzzy programming model. Then, we use a RAND algorithm to solve it.

Finally, we verify the correctness of the model and algorithm through a numerical example.

2 THE FUZZY JOINT REPLENISHMENT PROBLEM

2.1 *The fuzzy joint replenishment problem*

To discuss the fuzzy joint replenishment problem, we introduce the following notation:

i item index $i=1,2,...,n$
S^w major ordering cost
T basic cycle time
S_i^w minor ordering cost
k_i integer number that determines the replenishment schedule of item i
D_i demand rate of item i
S_i^c outbound transportation cost
f_i integer number that determines the outbound delivery schedule of item i
W_i^C customer waiting cost for item i
h_i^w inventory holding cost of item i
K $n \times 1$ vector that consists of k_i, $i=1,...,n$
B limit on capital that can be invested
F $n \times 1$ vector that consists of f_i, $i=1,...,n$

Combined with the actual situation, the problem of hypothesis is as follows: the demand rate of each kind of product is fuzzy quantity, the capital is a determined value, the inventory replenishment time is a basic replenishment cycle, the replenishment cycle for each product is an integer multiple of the basic replenishment cycle, the delivery cycle for each product is an integer multiple of basic delivery cycle. The total cost is the sum of ordering cost (major ordering cost and minor ordering cost), holding cost, customer waiting cost and outbound transportation cost. This paper mainly uses the method of linear programming in Fuzzy Mathematics and takes the demand rate of each product as a fuzzy value x_i. In fuzzy environment, we can make use of cross-docking: the supplier sends R_iD_i units directly to the retailers at the end of the replenishment cycle. Let R denote an $n\times1$ vector consisting of R_i, $i=1,...,n$; $R=(R_1,...,R_n)$. The total cost is shown as equation (1).

$$TC(T,K,F,R) = \frac{S^W + \sum_{i=1}^{n}(s_i^w/k_i)}{T}$$

$$+ \sum_{\{i:f_i\neq1\}} \frac{\left[(k_iT-R_i)^2+(f_i-1)R_i^2\right]D_iw_i^C}{2(f_i-1)k_iT}$$

$$+ \sum_{i=1}^{n}\frac{f_is_i^C}{k_iT} + \sum_{\{i:f_i\neq1\}}\frac{k_iTD_iw_i^C}{2}$$

$$+ \sum_{\{i:f_i\neq1\}}\frac{f_i(k_iT-R_i)^2D_ih_i^w}{2(f_i-1)k_iT}$$

(1)

Visible, as long as a set of values (K, F, R) is given, the fundamental period and the total cost can be through the formula above. Also is to say, the task of this paper is: get a set of values like this. To make the total cost as the minimum value. At the moment, the corresponding basic cycle is the optimal basic cycle which we want. The general situation of the future uncertain resource constraint in the actual plan of policy decision is: The possible range of resource constraints is [a–d,a+d]. Among that, a is the most probable value of the available resource, d is an elastic factor which said the fluctuation range. We can describe the uncertain resource constraints by trapezoidal fuzzy number. The equation (2) is the fuzzy membership function of resource constraint: $f(x)$ shows that the membership degree of belonging to the set, R is the set of real numbers. The graph of the function is shown in Fig.1.

$$f(x)=\begin{cases} \dfrac{x-a+d}{d} & a-d<x\leq a \\ 1 & a<x\leq b \\ \dfrac{-x+b+d}{d} & b<x\leq b+d \\ 0 & others \end{cases}$$

(2)

Under the condition of fuzzy product demand, we assume that the demand rate of each product is x_i, and the possible range of x_i is $[a–d,b+d]$. The objective function is $f(x)=x$, and $x\in C.C$ is a fuzzy collection, and is expressed by trapezoidal fuzzy number. The membership function is given by

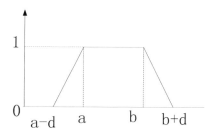

Figure 1. The graph of the function.

$$m_f(x) = \frac{f(x)-\min f(x)}{\max f(x)-\min f(x)}$$

$$= \frac{x-a+d}{b-a+2d}, x\in C$$

Calculate $x^*$by $C_f(x^*)$ = max {$C_f(x)/x\in$ C} $(x^*\in C).x^*$ is the point conditional maximum value, and $f(x^*)$ is the conditional maximum value of $f(x)$. As shown in Fig.2.

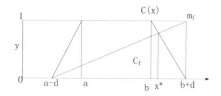

Figure 2. The point of conditional maximum value.

Set $m_f(x)=c(x)$, get $x^*=b(b–a+3d)+d^2/b–a+3d$. Optimal value of the basic cycle is:

$$T^* = \left[\frac{S^W + \sum_{i=1}^{n}\dfrac{s_i^w+f_is_i^C}{k_i} + \sum_{\{i:f_i\neq1\}}\dfrac{f_iR_i^2x_i^*\left(h_i^w+w_i^C\right)}{2(f_i-1)k_i}}{\sum_{\{i:f_i\neq1\}}\dfrac{k_ix_i^*\left(f_ih_i^w+w_i^C\right)}{2(f_i-1)} + \sum_{\{i:f_i\neq1\}}\dfrac{k_ix_i^*w_i^C}{2}}\right]^{1/2}$$

(3)

We solve the JRP model with fuzzy resource constraint by RAND algorithm under the uncertain condition in this paper. The decision variables are T, k_i, f_i, which need to determine.

2.2 Solution method

In this paper, we solve the JRP model with fuzzy resource constraint by RAND algorithm under the uncertain condition. The decision variables are T, k_i, f_i, which need to determine.

The RAND algorithm:

Step1: Compute the bounds for T by equation (4) and equation (5).

$$T_{max} = \left[2(S + \sum_{i=1}^{n} s_i) / \sum_{i=1}^{n} D_i h_i \right]^{1/2} \quad (4)$$

$$T_{min} = \min(2s_i / D_i h_i)^{1/2} \quad (5)$$

Step 2: For given m, divide $[T_{min}, T_{max}]$ into m aliquots of T.

Step 3: Set $r = 0$, set $j = j+1$, set $T_j(r) = T_j$, set $F(r) = (f_1(r), f_2(r), ..., f_n(r)) = (1,1,...,1)$, set $R(r) = (R_1(r), R_2(r), ..., R_n(r)) = (T_1(r), T_2(r), ..., T_n(r))$, set $r = r+1$.

Step 4: If $f_i(r-1)=1$, then calculate the optimal value of k_i by equation (6). Otherwise, calculate the optimal value of k_i by equation (7).

$$k_i(k_i - 1) \le \frac{2(s_i^w + k_i s_i^C)}{T^2 D_i w_i^C} \le k_i(k_i + 1) \quad (6)$$

$$k_i(k_i - 1) \le$$
$$\frac{2(f_i - 1)(s_i^w + f_i s_i^C) + f_i R_i^2 D_i (h_i^w + w_i^C)}{T^2 D_i (f_i h_i^w + w_i^C)} \quad (7)$$
$$\le k_i(k_i + 1)$$

Step 5: Calculate the optimal f_i and R_i by the program as follows, $i=1,...,n$.

Step 5-1: Set f_i=2.

Step 5-2: Set old_f_i=f_i. Calculate R_i by using equation (8).

$$R_i = \frac{(f_i h_i^w + w_i^C) k_i T}{f_i (h_i^w + w_i^C)} \quad (8)$$

Step 5-3: Calculate f_i by using equation (9). If $old\_f_i \ne f_i$, return to Step 5-2. Otherwise, calculate the equation (10). If equation (10)<0, set $f_i(r)=f_i$ and $R_i(r)=R_i$. Otherwise, set $f_i(r)=1$ and $R_i(r)=k_i(r)T_j(r-1)$. Exit.

$$(f_1 - 1)(f_2 - 2) \le \frac{(k_i T - R_i)^2 D_i (h_i^w + w_i^C)}{2s_i^c} \quad (9)$$
$$\le (f_1 - 1) f_i$$

$$TC(f_1, R_i) = TC(f_i = 1, R_i = k_i T)$$
$$= \frac{(k_i T - R_i)^2 D_i (f_i h_i^w + w_i^C)}{2(f_1 - 1) k_i T} + \frac{(f_1 - 1) s_i^C}{k_i T} \quad (10)$$
$$+ \frac{\{R_i^2 - (k_i T)^2\} D_i w_i^C}{2k_i T}$$

2.3 Instance and analysis

Experimental data are provided by Chengyan Li. There are six products in the joint supplement product managements, data of various products are shown

in table 1. Major ordering cost S^W=\$200, limit on capital that can be invested B=\$25000, the final calculation target is: determine the basic replenishment cycle and the min cost of TC. In order to prove the effect of the total cost that is generated under fuzzy demand. We make the comparison according to the above example. As shown in Table 2.

Table 1. Date of example.

i	1	2	3	4	5	6
D_i	10,000	5,000	3,000	1,000	600	200
S_i^W	45	46	47	44	45	47
h_i^W	1	1	1	1	1	1
S_i^C	5	5	5	5	5	5
W_i^C	1.5	1.5	1.5	1.5	1.5	1.5

Table 2. Comparison of JRP under the fuzzy environment and the certain environment.

Product i	Basic cycle T^*	Fuzzy target value TC
JRP under fuzzy demand	0.1881	4115.81
JRP under certain demand	0.2414	4828.89

In order to further comparison of JRP with fuzzy demand and with determine demand. We use the classic examples in literature (Li,Xu and Zhan 2008)). It results in 16 associations of n and S^W, 100 problems with random parameter values are generated for each association, and solved by using the RAND algorithms. The result of the comparison is shown in table 3.

Table 3. The result of the comparison.

n	S^W	Row1	Row2	Row3
10	100	6	40	54
	200	2	36	62
	300	3	38	59
	400	2	20	78
20	100	13	26	61
	200	11	29	60
	300	16	17	67
	400	14	19	67
30	100	9	21	70
	200	23	16	61
	300	28	15	57
	400	27	20	53
50	100	1	33	66
	200	0	27	67
	300	1	34	65
	400	3	41	46

The column Row1, Row2 and Row3 respectively represent the number of combinations that fuzzy constraint target value with fuzzy demand is less than, equal to and greater than it with certain demand. It indicates that the existence of fuzzy factors has greatly influenced on the total cost. Therefore, we must consider it while formulating the plan. To avoid the additional losses because of the difference between plans and the actual is too large.

3 CONCLUSION

Due to the random factors in many aspects to fuzzy resource constraints, we showed that incurs evidently fewer total costs than the determinate demand by getting further scales of transportation economy without evidently affecting the service for customer. So the model of this paper is more suitable for practical application.

ACKNOWLEDGMENT

This work was supported by grant No.12541142 from the Research Program of the Education Department of Heilongjiang Province, China.

REFERENCES

Kaspim. (1991)On the economic ordering quantity for jointly replenished items. *International Journal of Production Research*, 29(1): 107–114.

Moon, I. K.,&B. C. Cha(2006). The joint replenishment problem with resource restriction.*European Journal of Operational Research, ,* 173(1): 190–198.

Khoujam, Michaliwiczz, Satosars. (2000) A comparison between genetic algorithms and the RAND method for solving the joint replenishment problem. *Production Planning & Control*, 11(6): 556–564.

Li, C.Y., X.F. Xu,& D.C. Zhan(2008). Joint replenishment problem with fuzzy resource constraint. *Computer Integrated Manufacturing Systems*, 13(2): 14–1. In Chinese.

Moon. I. K. (2008) The joint replenishment and freight consolidation of a warehouse in a supply chain. *Department of Industrial Engineering, Pusan National University, Korea.*

Author index

788

Electronic Engineering and Information Science 2015 is a collection of contributions from the International Conference of Electronic Engineering and Information Science 2015 (ICEEIS 2015) held in Haikou, China on January 17-18 2015.

The papers in this proceedings volume consist of topics as:
- Electronic Engineering
- Information Science and Information Technologies
- Computational Mathematics and Data Mining
- Image Processing and Computer Vision
- Communication and Signal Processing
- Control and Automation of Mechatronics
- Methods, Devices and Systems for Measurement and Monitoring
- Engineering of Weapon Systems
- Mechanical Engineering and Material Science
- Technologies of Processing.

The content of this proceedings volume would be of interest to professionals in the fields of Electronic Engineering, Computer Science and Mechanical Engineering.

CRC Press
Taylor & Francis Group
an **informa** business

www.crcpress.com

ISBN 978-0-367-73824-2

9 780367 738242